Fire Investigator

Principles and Practice to NFPA 921 and 1033

FOURTH EDITION

Jones & Bartlett Learning
World Headquarters
5 Wall Street
Burlington, MA 01803
978-443-5000
info@jblearning.com
www.jblearning.com

National Fire Protection Association
1 Batterymarch Park
Quincy, MA 02169-7471
www.NFPA.org

International Association of Fire Chiefs
4025 Fair Ridge Drive
Fairfax, VA 22033
www.IAFC.org

International Association of Arson Investigators
2111 Baldwin Avenue, Suite 203
Crofton, MD 21114
firearson.com

Jones & Bartlett Learning books and products are available through most bookstores and online booksellers. To contact Jones & Bartlett Learning directly, call 800-832-0034, fax 978-443-8000, or visit our website, www.jblearning.com.

Substantial discounts on bulk quantities of Jones & Bartlett Learning publications are available to corporations, professional associations, and other qualified organizations. For details and specific discount information, contact the special sales department at Jones & Bartlett Learning via the above contact information or send an email to specialsales@jblearning.com.

Copyright © 2016 by Jones & Bartlett Learning, LLC, an Ascend Learning Company and the National Fire Protection Association®.

All rights reserved. No part of the material protected by this copyright may be reproduced or utilized in any form, electronic or mechanical, including photocopying, recording, or by any information storage and retrieval system, without written permission from the copyright owner.

The content, statements, views, and opinions herein are the sole expression of the respective authors and not that of Jones & Bartlett Learning, LLC. Reference herein to any specific commercial product, process, or service by trade name, trademark, manufacturer, or otherwise does not constitute or imply its endorsement or recommendation by Jones & Bartlett Learning, LLC and such reference shall not be used for advertising or product endorsement purposes. All trademarks displayed are the trademarks of the parties noted herein. *Fire Investigator, Principles and Practice to NFPA 921 and 1033, Fourth Edition* is an independent publication and has not been authorized, sponsored, or otherwise approved by the owners of the trademarks or service marks referenced in this product.

There may be images in this book that feature models; these models do not necessarily endorse, represent, or participate in the activities represented in the images. Any screenshots in this product are for educational and instructive purposes only. Any individuals and scenarios featured in the case studies throughout this product may be real or fictitious, but are used for instructional purposes only.

The procedures and protocols in this book are based on the most current recommendations of responsible sources. The National Fire Protection Association (NFPA), International Association of Fire Chiefs (IAFC), International Association of Arson Investigators (IAAI), and the publisher, however, make no guarantee as to, and assume no responsibility for, the correctness, sufficiency, or completeness of such information or recommendations. Other or additional safety measures may be required under particular circumstances.

05714-0

Production Credits
Chief Executive Officer: Ty Field
President: James Homer
Chief Product Officer: Eduardo Moura
Vice President, Publisher: Kimberly Brophy
Vice President of Sales, Public Safety Group: Matthew Maniscalco
Director of Sales, Public Safety Group: Patricia Einstein
Executive Editor: William Larkin
Associate Managing Editor: Amanda Brandt
Production Editor: Cindie Bryan
Senior Marketing Manager: Brian Rooney
Art Development Editor: Joanna Lundeen
Art Development Assistant: Shannon Sheehan
VP, Manufacturing and Inventory Control: Therese Connell
Composition: Cenveo Publisher Services
Cover Design: Kristin E. Parker
Manager of Photo Research, Rights & Permissions: Lauren Miller
Cover Image: Courtesy of International Association of Arson Investigators
Printing and Binding: Courier Companies
Cover Printing: Courier Companies

Library of Congress Cataloging-in-Publication Data
Fire investigator : principles and practice to NFPA 921 & 1033 / International Association of Fire Chiefs.—Fourth edition.
pages cm
Includes bibliographical references and index.
ISBN 978-1-284-05714-0 (pbk.)—ISBN 1-284-05714-3 (pbk.)
1. Fire investigation—Standards. 2. Fire investigation—Examinations—Study guides. I. International Association of Fire Chiefs.
TH9180.F487 2014
363.37'65–dc23

2014033496

6048
Printed in the United States of America
18 17 16 15 14   10 9 8 7 6 5 4 3 2 1

Brief Contents

Contents

© Photos.com

© Photos.com

© Photos.com

© Photos.com

Resource Preview

Instructor's ToolKit

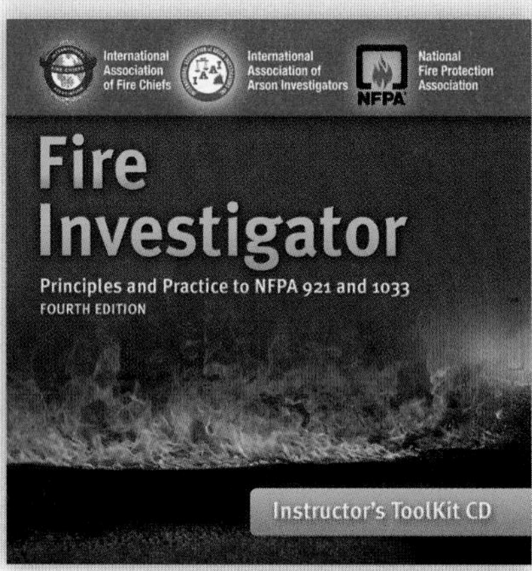

Preparing for class is easy with the resources on this CD. The CD includes the following resources:

- Adaptable PowerPoint Presentations: Provide instructors with a powerful way to create presentations that are educational and engaging to their students. These slides can be modified and edited to meet instructors' specific needs.
- Detailed Lesson Plans: Keyed to the PowerPoint presentations, these complete, ready-to-use lecture-outlines include all of the topics covered in the text. The lesson plans can be modified and customized to fit any course.
- Image and Table Bank: Offers a selection of the most important images and tables found in the text. Instructors can use these graphics to incorporate more images into the PowerPoint presentations, make handouts, or enlarge a specific image for further discussion.
- Electronic Test Bank: Contains multiple-choice questions and allows instructors to create tailor-made classroom tests and quizzes quickly and easily by selecting, editing, organizing, and printing a test along with an answer key that includes page references to the text.

■ Navigate TestPrep: Fire Investigator

Navigate TestPrep: Fire Investigator is a dynamic program designed to prepare students to sit for Fire Investigator certification examinations by including the same type of questions they will likely see on the actual examination.

It provides a series of self-study modules, organized by chapter, offering practice examinations and simulated certification examinations using multiple-choice questions. All questions are page referenced to *Fire Investigator: Principles and Practice to NFPA 921 and 1033, Fourth Edition* for remediation to help students hone their knowledge of the subject matter.

Students can begin the task of studying for Fire Investigator certification examination by concentrating on those subject areas where they need the most help. Upon completion, students will feel confident and prepared to complete the final step in the certification process—passing the examination.

Acknowledgments

© Photos.com

Jones & Bartlett Learning, the National Fire Protection Association, the International Association of Arson Investigators, and the International Association of Fire Chiefs would like to thank the editors, authors, contributors, and reviewers of ***Fire Investigator: Principles and Practice to NFPA 921 and 1033, Fourth Edition***.

Editorial Board

George Codding, IAAI-CFI
Fire Investigator
Louisville Fire Department
Louisville, Colorado

Richard Jones, IAAI-CFI, IAAI-CI, IAAI-ECT, MIFireE
Owner/Chief Investigator
Forensic Investigations Group
Covington, Louisiana

Shawn Kelley
International Association of Fire Chiefs
Fairfax, Virginia

Steven F. Sawyer
Senior Fire Service Specialist
National Fire Protection Association
Quincy, Massachusetts

Authors and Contributors

Michael E. Carlson, Jr. CFI, CFEI, CVFI, CATI, CPP
President
Probe, Inc.
Skokie, Illinois

Steven Carman, IAAI-CFI, ATF CFI (Ret.)
Owner and Principal Investigator
Carman & Associates Fire Investigation
Dunsmuir, California

Dan Choudek
Founder and Owner
OnSite Engineering
Prior Lake, Minnesota

Robert A. Corry IAAI-CFI, IASIU-CIFI
Forensic Investigation & Technical Services, LLC
Brimfield, Massachusetts

Mike Dalton
Detective/Fire Investigator
Knoxville, Tennessee

Michael R. Denney
Idaho Department of Lands
Coeur d'Alene, Idaho

Jay Goetz, PE
3M Product Designer and Forensic Engineer
Deephaven, Minnesota

Daniel Hebert, IAAI-CFI, IAAI-CI, IAAI-ECT
Fire Investigator
Forensic Investigations Group
Covington, Louisiana

Fred Herrera, CFI
Owner/Investigator
Fred Herrera Fire Investigations
San Diego, California

Terry-Dawn Hewitt, BA (Spec.), LLB, CIP, LLM
Attorney-at-Law, Barrister & Solicitor
McKenna Hewitt
Denver, Colorado

Gary Hodson, IAAI-CFI, IAAI-ECT
Senior Investigator
Unified Investigations & Sciences
Provo, Utah

Todd M. Iaeger, CFI-IAAI, CFEI/CVFI-NAFI
Fire Marshal
Reading, Pennsylvania

A. Maurice Jones, Jr.
Supervisor, Fire Protection Systems
Alexandria Fire Department
Alexandria, Virginia

Casandra L. Jones
Business Administrator
Forensic Investigations Group
Covington, Louisiana

David S. Komm, PE, PEng, CFEI
President
Augspurger Komm Engineering, Inc.
Phoenix, Arizona

Jim Kuticka, IAAI-CFI, MO-CFI
Owner/Operator
Kuticka Fire Investigations
Nixa, Missouri

John J. Lentini, CFI, D-ABC
Scientific Fire Analysis, LLC
Big Pine Key, Florida

Hal C. Lyson
Director of Fire Investigations
Fire Cause Analysis
Bismarck, North Dakota

James Mazerat, IAAI-CFI
Consultant
Forensic Investigations Group
Covington, Louisiana

Wayne J. McKenna, BA (Spec), LLB
Attorney-at-Law, Barrister & Solicitor
McKenna Hewitt
Denver, Colorado

William Moylin, IAAI-CFI
Senior Fire Investigator
SEA Limited
Tampa, Florida

Bryce Peck, IAAI-CFI, IAAI-ECT
Fire Investigator
Forensic Investigations Group
Covington, Louisiana

Rodney Pevytoe, IAAI-CFI
Investigator
Kubitz and Associates
Clintonville, Wisconsin

Gene Pietzak, IAAI-CFI
Investigator
Nassau County Fire Marshals
Uniondale, New York

John J. Powell, CFI
Principal/Investigator
C&O Services, Inc.
Portland, Oregon

Ed Rostalski, SCO, IAAI-FIT, IAAI-ECT
Owner/Fire Investigator
Accurate Fire Investigative Services, Inc.
Edmonton, Alberta, Canada

Rob Rush, IAAI-CFI
Captain, Bomb and Arson Squad
Palm Beach County Fire Rescue
West Palm Beach, Florida

Robert Schaal, IAAI-CFI, IAAI-CI
Owner/Consultant/Forensic Investigator
Gulf Coast Fire Investigation, Research, and Education
Mandeville, Louisiana

Joe Sesniak, IAAI-CFI
Forensic Fire Consultants, Ltd.
Chandler, Arizona

Jeff Spaulding, CFI, CFEI
President/Investigator
Fire and Explosion Consultants
Middletown, Ohio

Andrew Thoresen, PE
Oracle Forensics, Inc.
Phoenix, Arizona

Reviewers

James Abner, CFI, CFEI
Fire Marshal
Artesia Fire Department
Artesia, New Mexico

Dr. Norman Arendt
International Society of Fire Service Instructors
Middleton Fire District
Middleton, Wisconsin
Dane County Arson Response Initiative
Madison, Wisconsin

Tim Baker
Program Director, Public Services Careers
Lansing Community College
Lansing, Michigan

Brian K. Blomstrom, MPA, CFO, FIT
Sergeant
Greenville Department of Public Safety
Greenville, Michigan

Justin E. Bouler
Jackson County Fire
Pascagoula, Mississippi

Chip Brantley, BS
Fire Marshal
Alabaster Fire Department
Alabaster, Alabama

Jerry V. Bridges, Sr.
Firefighter and Fire Service Training Instructor
Lincoln Township Fire Department
Lincoln, Michigan

David Bunn
Fire Marshal, Auburn Fire Department
Auburn, Indiana
Ivy Tech College
Fort Wayne, Indiana

Joe Bunn
Deputy Chief of Training (Ret.)
Encinitas, Solana Beach, and Del Mar, California
Fire Service Training Specialist III, CAL FIRE
Sacramento, California

Andrew Byrnes, EFO, MEd
Associate Professor
Utah Valley University
Provo, Utah

Glenn C. Clapp, CHMM, CFPS
High Point Fire Department
High Point, California

Larry Conoyer
Captain (Ret.)
Central County Fire & Rescue
St. Peters, Missouri

Tim Cortez
Division Chief
Casper Fire-EMS
Casper, Wyoming

Todd Eckes
Captain of Operations
Wisconsin Rapids Fire Department
Wisconsin Rapids, Wisconsin

Gary W. Edwards
Fire Chief (Ret.)
Director, Fire Science Program
Montana State University – Billings, City College
Billings, Montana

Brent S. Elliott
Division Chief
Green Bay Metro Fire Department
Green Bay, Wisconsin

Stan Fernandez, IAAI CFI, CI, ECT
Bureau of Fire Prevention & Investigations
Berkeley Fire Department
Berkeley, California

David R. Fischer
Safety Training Officer
NE Lincoln County Fire District
Stillwater, Oklahoma

Stephen R. Gallagher
Program Coordinator, New Hampshire Fire Standards, Training, and EMS
New Hampshire Fire Academy
Concord, New Hampshire

Rob Gaylor
Deputy Chief of Operations
Westfield Fire Department
Westfield, Indiana

Rodney Geilenfeldt II, BS
Fire Captain and Paramedic
Heartland Fire & Rescue
El Cajon, California

Anthony Gianantonio
District Chief
Palm Bay Fire Rescue
Palm Bay, Florida

Bobby Gonder
Fire Investigator
Coeur d'Alene Fire Department
Coeur d'Alene, Idaho

Andrew J. Gonzales
Captain Fire Investigator
City of Hobbs Fire Prevention Bureau
Hobbs, New Mexico

Kevin M. Goodwin
Captain
Central Kitsap Fire & Rescue
Silverdale, Washington

Chris Harner, CFEI
City of Pueblo Fire Department
Pueblo, Colorado

Gary S. Hodson, IAAI-CFI, IAAI-ECT
Senior Investigator
Unified Investigations & Sciences
Provo, Utah

Lonnie Inzer
Fire Science Professor Emeritus, Pikes Peak Community College
Colorado Community College System
Colorado Springs
Colorado

Bradley Iverson
Firefighter, Las Vegas Fire & Rescue
Las Vegas, Nevada
Adjunct Faculty, College of Southern Nevada
Henderson, Nevada

Joseph Knitter
Fire Chief
South Milwaukee Fire Department
South Milwaukee, Wisconsin

Rob Kunze
Lieutenant
Fitchburg Fire Department
Fitchburg, Wisconsin

Keith T. Lenderman, CFI
Assistant Fire Chief (Ret.)
Oakland Community College
Auburn Hills, Michigan

Michael Mentink
Fire Marshal (Ret.)
Industrial Emergency Council
Menlo Park, California

Ed Mezulis
Battalion Chief
Sedona Fire District
Clarkdale, Arizona

Dustin L. Paddack
Fire Marshal
Liberty Fire Department
Liberty, Missouri

Sean F. Peck, MEd, EMT-P
Associate Faculty—Fire Science
Arizona Western College
Yuma, Arizona

Michael D. Penders, MPA
Firefighter and Paramedic
Guilford Fire Department
Guilford, Connecticut

Lee Richardson
Lieutenant
City of Altoona Fire Department
Altoona, Iowa

Tracy E. Rickman
Rio Hondo College Fire Academy
Santa Fe Springs, California

Roger M. Rybicki
Adjunct Professor, Los Medanos College
Pittsburgh, California
Fire Captain, Contra Costa County Fire Protection District (Ret.)
Pleasant Hill, California

Jesse Silva
Captain
Charlevoix County Fire Academy
Charlevoix, Michigan

Steven A. Sommers
Chief
Matoaka Volunteer Fire Department
Matoaka, West Virginia

Bryan K. Thomson
Captain and Fire Investigator
Sacramento Metropolitan Fire District
Mather, California

Scott Ventura
Fire Technology Coordinator
College of the Desert
Palm Desert, California

Craig Yarborough
Training Chairman
International Association of Arson Investigators
Burlington, North Carolina

Gray Young
Coordinator
Louisiana State University Fire and Emergency Training Institute
Baton Rouge, Louisiana

Administration

© Photos.com

© Justin Sullivan/Getty Images

Knowledge Objectives

After studying this chapter, you should be able to:

- Discuss the purpose and development of NFPA 921, *Guide for Fire and Explosion Investigations*. (pp 4–5)
- Discuss the purpose and development of NFPA 1033, *Standard for Professional Qualifications for Fire Investigator* NFPA 1.1 NFPA 1.2 NFPA 1.2.1 NFPA 1.2.2 NFPA 1.3.1 NFPA 1.3.2 NFPA 1.3.3 NFPA 1.3.4 NFPA 1.3.5 NFPA 1.3.6 NFPA 1.3.7 NFPA 1.3.8. (pp 5–6)
- Describe how NFPA documents are developed. (pp 6–9)
- Discuss the value of NFPA 921 definitions. (p 9)

Skills Objectives

There are no skills objectives for this chapter.

Additional NFPA References

NFPA 921, *Guide for Fire and Explosion Investigations*

NFPA 1031, *Standard for Professional Qualifications for Fire Inspector and Plan Examiner*

NFPA 1037, *Standard on Professional Qualifications for Fire Marshal*

CHAPTER 1

FESHE Course Outcomes

Fire Investigation I

4. Define the common terms used in fire investigations. (p 9)

Fire Investigation II

There are no Fire Investigation II (FESHE) course outcomes for this chapter.

You Are the Fire Investigator

© Jones and Bartlett Publishers. Photographed by Glen E. Ellman

While conducting an origin and cause investigation, you identify at least two independent points of origin. Utilizing NFPA 921 as a guide to the investigation, you systematically examine and document each point of origin with photographs. You then collect an ignitible liquid sample from each location, but you do not create a scene diagram that identifies the significant items of evidence, room dimensions, or other potential sources of ignition.

During a departmental review by your supervisor, you are asked why you did not produce a scene diagram as indicated by NFPA 1033. You reply that you usually produce a diagram on "significant" fires only and that your photographs are a better depiction of the fire scene. After providing you with a copy of NFPA 1033, the supervisor asks you to review the job performance requirements (JPRs) in regards to scene processing and documentation.

1. How do NFPA 921 and 1033 differ from each other?
2. What other rules or regulations may affect how an investigator conducts his or her investigation?
3. What is thermometry?

Introduction

NFPA 921, *Guide for Fire and Explosion Investigations*, developed by the National Fire Protection Association (NFPA), can be used by anyone who is charged with the responsibility of investigating and analyzing fire and explosion incidents. People involved in fire investigation or fire analysis and in rendering opinions as to the origin, cause, responsibility, or prevention of such incidents include on-scene investigators, fire analysts, and other experts with special areas of interest, as well as technicians who analyze the evidence found in the course of the fire investigation. (The term *fire investigation* is used frequently in NFPA 921 when the context indicates that the relevant text refers to the investigation of either fires or explosions.) These professionals are found both in the public sector (including fire and police departments and other governmental agencies) and in the private sector (including private investigation firms, engineering firms, and insurance companies).

NFPA 1033, *Standard for Professional Qualifications for Fire Investigator*, was developed by a technical committee of the NFPA to use as the minimum job performance requirements (JPRs) for the fire investigator. At the beginning of each chapter of this text, the applicable JPRs of NFPA 1033 will be cited. It is important for the reader to understand how the information of this text is related to the JPRs of fire investigation.

Fire Investigator Tip

The metric system is employed in fire investigation in the United States and internationally, and is therefore included in values mentioned in this text. Data measurements may vary, however, and the fire investigator must evaluate data with this in mind.

NFPA 921

NFPA 921 was developed by the NFPA and approved by the American National Standards Institute (ANSI). NFPA defines NFPA 921 as a guide, as distinct from a standard, such as NFPA 1033. The factors causing NFPA 921 to be treated in court as a standard is a complex issue and is beyond the scope of this textbook.

Although NFPA classifies 921 as a guide and not a standard, many courts and professional fire investigation organizations recognize NFPA 921 as the "standard of care" in the profession. The purpose of NFPA 921 is to establish guidelines and recommendations for the safe and systematic investigation or analysis of fire and explosion incidents. It also serves as a model for the advancement and practice of fire and explosion investigation, fire science, technology, and methodology. The content of NFPA 921 is not intended to be a comprehensive scientific or engineering text. These are important aspects of fire investigation but are presented at an elementary level only, and the investigator may have to utilize additional resources in the course of an investigation.

The reason for conducting any fire investigation, no matter who is conducting it, is to arrive at the truth in an attempt to prevent an incident's recurrence. Determining what caused the fire, who may be responsible for it, what factors may have contributed to the fire, and the fire's subsequent impact are the objectives. The accurate determination of a fire or explosion cause is fundamental to the protection of lives and property and the prevention of future incidents.

NFPA 921 is designed to produce a systematic, working framework to ensure an effective fire/explosion investigation and origin/cause analysis. Deviations from these procedures are not necessarily wrong or inferior, but do need justification.

Because every fire and explosion incident is in some way different and unique, NFPA 921 is not designed to encompass all of the necessary components of a complete investigation or analysis of any individual case. It is up to investigators (depending on their responsibility, as well as the purpose and scope of their investigation) to apply the appropriate recommended procedures in the guide to a particular incident. The use of the scientific method should be applied in every instance. The scientific method allows for the systematic defining, information gathering, analysis, and testing of a problem posed. (The scientific method is discussed in more detail in the "Basic Methodology" chapter.)

Although NFPA 921 was never originally intended as a document to define legal issues involving civil or criminal litigation of fire or explosion incidents, it has become such in the minds and practice of many fire investigation professionals, including attorneys and trial courts. Many court decisions have cited NFPA 921 or the principles of methodology utilized therein. It is clear with the court's ruling in cases such as *Michigan Millers Mut. Ins. Corp. v. Benfield* and *Daubert v. Merrell Dow Pharmaceuticals* that expert opinion is supported by established and tested principles and not by assumptions, speculation, or years of experience.

Relevance of NFPA 921 to Fire Investigators

One of the common issues raised is that NFPA 921 is not an NFPA standard to which fire investigators must adhere when conducting fire investigations. NFPA 921 is a guide and, as such, does not include mandatory language, such as *shall* or *must*. The distinction between *standard* and *guide* is important because different jurisdictions have varying interpretations on the importance of NFPA 921. Investigators should be aware of rulings in their local jurisdiction.

There is another concern that the fire investigator must consider, however: the concern that NFPA 921 may be referred to as a standard of care comparable to that in the medical field. It must be stressed that NFPA 921 is not an NFPA standard by NFPA definitions. It might be considered by some as a de facto standard, in that it is a widely used and recognized document that provides guidance to the fire investigator. There are other standards-making bodies that define the word *standard* differently.

Investigators do not need to use NFPA 921 all of the time at every investigation. Simply because material is covered in this guide does not necessarily mean that its procedures must be carried out in every case. The experienced investigator must use his or her judgment to determine what actions are appropriate at each investigation to ensure that the proper steps are taken to arrive at the truth. If an investigator chooses not to use NFPA 921, however, he or she must be able to explain the method used and the reasons for using it.

Since the first edition of NFPA 921 was introduced in 1992, there have been significant advances in fire science as applied to fire investigations. Although every effort is made for NFPA 921 to be as complete as possible, there are certainly areas that are not covered in each edition. With an infinite possibility of items that could be involved in a fire, such as different types of fuels, objects burned, or individual sources of ignition, the text cannot address every potential fire scenario; however, through the NFPA process, the guide is a living document, updated regularly to include new material, as described in the Revision Cycles section later in this chapter.

> **Fire Investigator Tip**
>
> NFPA 1033 is a standard that is designed to establish the minimum JPRs for service as a fire investigator, and NFPA 921 is classified as a guide that establishes guidelines and recommendations.

> **Fire Investigator Tip**
>
> Fire investigators should be aware that new editions of NFPA 921 are published periodically, and that the edition that was in effect when the investigator performed a particular task (such as fire scene investigation or a case review) may have been superseded by the time a case comes to deposition or court. This could have an impact if the fire investigator is asked questions regarding why he or she performed a specific task or how his or her investigation progressed.

NFPA 1033

In 1972, the Joint Council of National Fire Service Organizations (JCNFSO) created the National Professional Qualifications Board (NPQB) to facilitate the development of nationally applicable performance standards for uniformed fire service personnel. As a result of that process, the NFPA adopted, in May 1977, NFPA 1031, *Standard for Professional Qualifications for Fire Inspector and Plan Examiner*. In the years that followed, it was decided to create an independent standard to apply to fire investigators only. In June 1987, NFPA adopted NFPA 1033. Since the release of this document, it has been updated on a regular basis.

Relevance of NFPA 1033 to Fire Investigators

Unlike NFPA 921, NFPA 1033 is a standard that is designed to establish the minimum JPRs for service as a fire investigator. NFPA 1033 specifies the minimum knowledge and skills required for a fire investigator to evaluate a fire scene and specifies how to conduct the investigation and document the scene safely. NFPA 1033 also addresses skills related to evidence collection, interviewing of witnesses, report writing, and final presentation of investigative findings. The intent of the standard set forth in NFPA 1033 is applicable to all fire investigations, including wildland, vehicle, and structural fires. Jurisdictions may choose to enforce requirements exceeding those set forth in NFPA 1033.

> **Fire Investigator Tip**
>
> NFPA 1037, *Standard for Professional Qualifications for Fire Marshal*, specifies the minimum JPRs for service as a fire marshal, which includes NFPA 1033.

© Jones & Bartlett Learning. Photographed by Glen E. Ellman

FIGURE 1-1 Fire investigators must be evaluated periodically to ensure that they are meeting all requirements established by NFPA 1033 and the laws in their jurisdiction.

Qualifications for Fire Investigators

NFPA 1033 requires a fire investigator to be at least 18 years of age and possess a high school diploma or equivalent level of education. Because the fire investigator will be in a position to examine the personal property of others, a complete and thorough background and character investigation must be completed by the authority having jurisdiction (AHJ).

As with any profession, fire investigators must meet a basic level of education and training. The fire investigator must meet the JPRs established in NFPA 1033 and by his or her employer, as well as any requirements established by law, to carry out his or her duties successfully. This requires that the fire investigator be evaluated periodically by qualified personnel to ensure that the investigator is meeting those requirements FIGURE 1-1. Training programs must be designed to maximize the instructional process to prepare the investigator to meet his or her roles and responsibilities. The curriculum should be prioritized to allow the fire investigator to obtain mastery-level learning, although it is not necessary for the topics to be instructed in any particular order. The fire investigator's duties require both proficiency in the JPRs and additional knowledge.

Roles and Responsibilities of Fire Investigators

After the investigator's initial training, he or she must remain current with new investigative methodologies, fire protection technologies, and code requirements to determine the fire's origin and cause accurately and to compile meaningful fire statistics that will form the basis of fire prevention codes, standards, and training. This is achieved by attending formal education courses, workshops, and seminars and/or through professional publications and journals. An up-to-date knowledge of the following topics should be maintained as a minimum for the fire investigator:

- Fire science
- Fire chemistry
- Thermodynamics
- Thermometry
- Fire dynamics
- Explosion dynamics
- Computer fire modeling
- Fire investigation
- Fire analysis
- Fire investigation methodology
- Fire investigation technology
- Hazardous materials
- Failure analysis and analytical tools
- Fire protection systems
- Evidence documentation, collection, and preservation
- Electricity and electrical systems

The fire investigator should also strongly consider joining organizations relating to the field, such as the International Association of Arson Investigators (IAAI), the National Association of Fire Investigators (NAFI), the NFPA, and others. The IAAI has state- and country-based chapters worldwide and offers professional credentials such as Certified Fire Investigator (IAAI-CFI®) and Fire Investigation Technician (IAAI-FIT®). NAFI offers the Certified Fire and Explosion Investigator (NAFI-CFEI) credential. Additionally, many states in the United States have certifications and designations designed to assess and document the investigator's professional qualifications and knowledge and adherence to relevant NFPA documents.

Development of NFPA Documents

NFPA is the publisher of almost 300 codes, standards, recommended practices, and guides that are related to fire and life safety. All NFPA documents are voluntary documents, which means that they do not have the power of law unless an AHJ adopts them. A number of NFPA's codes and standards have been adopted into law in various jurisdictions around the world, including the following:

- NFPA 1, *Fire Code*
- NFPA 54, *National Fuel Gas Code*
- NFPA 70, *National Electrical Code*
- NFPA 101, *Life Safety Code*

One of the distinctive features of all NFPA documents is that they are developed through a consensus process, which brings together technical committee volunteers representing varied viewpoints and interests to achieve consensus on the document content. There are more than 200 of these technical committees, and they are governed by a 13-member Standards Council, which is also composed of volunteers.

It is the technical committees—not NFPA staff—who are responsible for the document content. In contrast to a common misconception, the staff liaisons do not write the material that is contained in the document. Their role is to coordinate the development process, and the NFPA's function is to serve as a mechanism for publishing the material that is written or approved by the technical committee volunteers.

Fire Investigator Tip

The variety of expertise of technical committee members helps ensure that NFPA standards and other documents are well-rounded and accurate. In addition to the expertise of the technical committee members, however, the document development process also allows for public input.

Technical Committees

More than 6000 volunteers serve on NFPA's technical committees, with selection of these volunteers based on their background and expertise. The committees are balanced to ensure that no single interest has an overriding representation that may unfairly influence the outcome of the development process.

NFPA uses the following membership categories to fill and balance a committee:

- Manufacturer
- User
- Installer/maintainer
- Labor representative
- Enforcing authority
- Insurance representative
- Special expert
- Consumer
- Applied research/testing laboratory

Technical committees review and respond to all proposed changes to existing documents. Consensus occurs when a majority of a committee accepts a proposed change to the document it oversees. Acceptance by a committee—and its subsequent recommendation for change—requires at least a two-thirds majority vote by written ballot. Ultimately, committee recommendations are voted on at the NFPA's annual meeting, at which the entire association membership has the opportunity to approve or reject the committee recommendations for change to the documents.

Oversight of Technical Committees

The work of the committees is coordinated by NFPA staff liaisons. NFPA staff members appointed to serve as staff liaisons to the technical committees are responsible for ensuring that the committees follow the rules governing committee activities and for coordinating their meetings. The staff liaisons also record the meeting actions and coordinate the publication of committee reports as well as the final document.

Public Input to NFPA Documents

All of the committee meetings and the NFPA annual meeting are open to the public, and anyone, whether an NFPA member or not, can submit public inputs (proposals) to change a document or make public comments on actions taken by the committee. The only step in the process that is limited to NFPA members is the voting that takes place at the annual technical session.

The schedule for public input on NFPA documents varies according to the document. Public input deadlines are posted on the NFPA website, and public input is submitted through a web-based electronic submission system.

Tools and techniques are constantly being updated and modified between editions of NFPA 921. Consequently, newer methods that are not covered in the current edition may be used at a fire scene. As new principles and technologies are evaluated by the NFPA technical committee, they may or may not be included in the regular cycle revisions of the document. For this reason, it is important that fire investigation professionals be diligent in making proposals and comments to the document during its revision cycles.

Revision Cycles

All NFPA documents are revised on a staggered basis in two annual revision cycles—an annual and fall revision cycle. Documents are updated on regular cycles of 3 to 5 years, with anywhere from 20 to 45 documents reporting in a given revision cycle. NFPA 921 has been on a 3-year cycle, and NFPA 1033 has been on a 5-year cycle.

Sequence of Events in Standards Development

TABLE 1-1 describes the current sequence of events for the standards development process (in this context, the term *standard* includes guides such as NFPA 921), which includes the following steps:

1. Input stage
2. Comment stage
3. Association technical meeting
4. Council appeals and issuance of standard.

As soon as the current edition is published, a standard is open for public input.

Table 1-1 Standards Development Process

Step 1: Input Stage	■ Input accepted from the public or other committees for consideration to develop the First Draft ■ Committee holds First Draft Meeting to revise Standard (23 weeks) Committee(s) with Correlating Committee (10 weeks) ■ Committee ballots on First Draft (12 weeks) Committee(s) with Correlating Committee (11 weeks) ■ Correlating Committee First Draft Meeting (9 weeks) ■ Correlating Committee ballots on First Draft (5 weeks) ■ First Draft Report posted
Step 2: Comment Stage	■ Public Comments accepted on First Draft (10 weeks) ■ If Standard does not receive Public Comments and the Committee does not wish to further revise the Standard, the Standard becomes a Consent Standard and is sent directly to the Standards Council for issuance ■ Committee holds Second Draft Meeting (21 weeks) Committee(s) with Correlating Committee (7 weeks) ■ Committee ballots on Second Draft (11 weeks) Committee(s) with Correlating Committee (10 weeks) ■ Correlating Committee First Draft Meeting (9 weeks) ■ Second Draft Report posted
Step 3: Association Technical Meeting	■ Notice of Intent to Make a Motion (NITMAM) accepted (5 weeks) ■ NITMAMs are reviewed and valid motions are certified for presentation at the Association Technical Meeting ■ Consent Standard bypasses Association Technical Meeting and proceeds directly to the Standards Council for issuance ■ NFPA membership meets each June at the Association Technical Meeting and acts on Standards with "Certified Amending Motions" (certified NITMAMs) ■ Committee(s) and Panel(s) vote on any successful amendments to the Technical Committee Reports made by the NFPA membership at the Association Technical Meeting
Step 4: Council Appeals and Issuance of Standard	■ Notification of intent to file an appeal to the Standards Council on Association action must be filed within 20 days of the Association Technical Meeting ■ Standards Council decides, based on all evidence, whether or not to issue the Standards or to take other action

Reproduced with permission from NFPA 921-2014, Guide for Fire and Explosion Investigations, Copyright © 2014, National Fire Protection Association. This reprinted material is not the complete and official position of the NFPA on the referenced subject, which is represented only by the standard in its entirety.

Documents for which there are no properly certified and timely Notices of Intent to Make a Motion are forwarded directly to the Standards Council for action on issuance.

Corrections to Documents

Between revisions of the document, errata and Temporary Interim Amendments (TIAs) may be issued. For example, if a publishing error is found in some part of the document, the NFPA can issue an erratum correcting the error. In the next revision of the document, the error is corrected.

If there are proposed emergency changes that must occur between revisions, a TIA is placed before the committee to address the issue. The committee must conduct a letter ballot on proposed TIAs and a notice for public comments is sent out.

A recommendation for approval is established if three-fourths of the voting members have voted in favor of the TIA, and with this approval and the review of public comments, the Standards Council may issue the TIA. The TIA then becomes a public input during the next cycle of the document. Several errata and TIAs have been issued for NFPA 921 since its inception.

Formal Interpretation of Codes and Standards

A *Formal Interpretation* (FI) is a mechanism for explaining the meaning or intent of any specific provision that is included in an issued NFPA code or standard. FIs are processed through the technical committee that is responsible for the document and must be clearly worded to solicit a "Yes" or "No" answer from the committee.

Confirmation from the technical committee for an FI is achieved through a letter ballot. If a three-fourths majority is not achieved, the FI fails, and the item is placed on the committee's next meeting agenda. FIs of NFPA guides such as NFPA 921, though not unheard of, are unusual because NFPA guides do not contain mandatory provisions.

NFPA 921 Definitions

Definitions of terms within a profession or discipline are necessary for a full understanding of the language by which various practitioners communicate. The definitions contained within NFPA 921 are critically important in that they provide a consistent language for all fire investigators to use when conducting a fire investigation and preparing a case. Without consistent language, ambiguities, misunderstandings, and other related problems may arise. These problems can detract from the purpose of the investigation (to find the truth) by creating the appearance of unprofessionalism.

For example, confusion (and hence misuse of the terms) often arises between the concepts of *backdraft* and *flashover*. To avoid any potential communication problems, the investigator should be familiar not only with the definitions as listed in NFPA 921, but also with the reasoning behind those definitions.

For the purpose of this textbook, we use the definitions that are provided in NFPA 921. You may refer to other documents for clarification.

Wrap-Up

Ready for Review

- NFPA 921 can be used by anyone who is responsible for investigating and analyzing fire and explosion incidents.
- NFPA 921 establishes guidelines and recommendations for the safe and systematic investigation or analysis of fire and explosion incidents.
- The distinction between *standard* and *guide* is important because different jurisdictions have varying interpretations on the importance of NFPA 921.
- In 1972, the Joint Council of National Fire Service Organizations (JCNFSO) created the National Professional Qualifications Board (NPQB) to facilitate the development of nationally applicable performance standards for uniformed fire service personnel.
- NFPA 1033 is a standard that is designed to establish the minimum JPRs for service as a fire investigator. It establishes the minimum knowledge and skills required for a fire investigator to evaluate a fire scene and specifies how to conduct and document the scene safely.
- Because the fire investigator will be in a position to examine the personal property of others, a complete and thorough background and character investigation must be completed by the authority having jurisdiction (AHJ).
- After the investigator's initial training, he or she must remain current with new investigative methodologies, fire protection technologies, and code requirements to determine accurately the fire's origin and cause and to compile meaningful fire statistics that will form the basis of fire prevention codes, standards, and training.
- All NFPA documents are voluntary documents, which means that they do not have the power of law unless an AHJ adopts them.
- More than 6000 volunteers serve on NFPA's technical committees, with selection of these volunteers based on their background and expertise.
- NFPA staff members appointed to serve as staff liaisons to the technical committees are responsible for ensuring that the committees follow the rules governing committee activities and for coordinating their meetings.
- NFPA documents are revised on a staggered basis along revision cycles.
- The following are the criteria used in the latest revision of NFPA 1033:
  - Call for proposals
  - Report on proposals
  - Report on comments
  - Technical committee report session
  - Standards council issuance
- Between revisions of the document, errata and Temporary Interim Amendments (TIAs) may be issued.
- A *Formal Interpretation* (FI) is a mechanism for providing an explanation of the meaning or intent of any specific provision that is included in an issued NFPA code or standard.
- The definitions contained within NFPA 921 are critically important in that they provide a consistent language for all fire investigators to use when conducting a fire investigation and preparing a case.

Hot Terms

Fire investigation The process of determining the origin, cause, and development of a fire or explosion.

Guide A document that is advisory or informative in nature and that contains only nonmandatory provisions. A guide may contain mandatory statements, such as when a guide can be used, but the document as a whole is not suitable for adoption into law.

NFPA 921, *Guide for Fire and Explosion Investigations* A guide that establishes guidelines and recommendations for the safe and systematic investigation or analysis of fire and explosion incidents.

NFPA 1033, *Professional Qualifications for Fire Investigator* A standard that is designed to establish the minimum job performance requirements (JPRs) for service as a fire investigator.

Standard A document, the main text of which contains only mandatory provisions using the word "shall" to indicate requirements and that is in a form generally suitable for mandatory reference by another standard or code or for adoption into law. Nonmandatory provisions shall be located in an appendix or annex, footnote, or fine print note and are not to be considered a part of the requirements of a standard.

© Greg Henry/ShutterStock, Inc.

FIRE INVESTIGATOR *in action*

During a training conference, you and several other fire investigators are discussing the previous lecture regarding the use of new technologies during origin and cause investigations. Most of the investigators are intrigued with applying these technologies to assist them during their investigations but understand that these technologies cannot replace the proper application of NFPA 921 and meeting the objectives defined in NFPA 1033.

The discussion continues with the recent challenges investigators have encountered conducting investigations over the past several years as the "bar" has been raised for fire investigators to conduct a professional, objective, and scientifically based investigation. Some of the investigators describe investigations they had conducted in the past for which they now believe their conclusions as to the origin and cause were likely wrong.

1. NFPA 1033 is a standard that:
- **A.** establishes minimum education requirements for fire investigators.
- **B.** is applicable to all fire investigations.
- **C.** identifies specific knowledge and skills required of fire investigators.
- **D.** All of the above

2. When utilizing NFPA 921 to conduct a fire investigation, the investigator:
- **A.** may apply only certain sections of the guide that he or she finds applicable.
- **B.** will be able to determine the cause of every fire.
- **C.** may not use other sources of information for determining the cause.
- **D.** should never reference the document during the scene examination.

3. NFPA standards and guides are updated every:
- **A.** 2 years.
- **B.** 3 years.
- **C.** 5 years.
- **D.** 3 or 5 years, depending on the document.

4. Which of the following individuals may be a volunteer for an NFPA technical committee?
- **A.** Labor representative
- **B.** Enforcing authority
- **C.** Special expert
- **D.** All of the above

Basic Fire Methodology

© Photos.com

© John Blair/Alamy

Knowledge Objectives

After studying this chapter, you should be able to:

- Describe the scientific method. (p 14)
- Discuss the scientific method as it relates to fire investigation NFPA 4.1.2 NFPA 4.6.5. (pp 14–17)
- Identify the basic steps of the scientific method in the context of a fire investigation NFPA 4.6.5. (pp 17–18)

Skills Objectives

After studying this chapter, you should be able to:

- Conduct a fire investigation following the steps of the scientific method NFPA 4.1.2 NFPA 4.6.5. (p 17)

Additional NFPA Reference

NFPA 921, *Guide for Fire and Explosion Investigations*

CHAPTER 2

FESHE Course Outcomes

Fire Investigation I

10. Describe the process of conducting investigations using the scientific method. (pp 14–18)

Fire Investigation II

There are no Fire Investigation II (FESHE) course outcomes for this chapter.

You Are the Fire Investigator

© Jones and Bartlett Publishers. Photographed by Glen E. Ellman.

While investigating a fire in a child's bedroom, you locate the remains of what you suspect is a small disposable lighter near the foot of the bed. Both of the child's parents state they are smokers and deny having any extra lighters or matches within the home.

1. What investigative steps will you utilize to determine the cause of this fire?
2. What data might you collect and evaluate during your investigation?
3. What hypotheses might you consider?

Introduction

The investigation of a fire or explosion most often involves the identification, collection, and analysis of data that result from the destruction of materials. The ability to recognize these data, or facts, and then to analyze them properly and objectively, is paramount in fire investigation. It is essential that each investigation be carried out in a consistent manner to ensure that all aspects of any given fire scene are addressed. A systematic approach, such as one based on the scientific method, should be employed to ensure thorough physical evaluation of the scene, careful collection of evidence, and complete documentation of the scene and physical evidence, all of which are required for proper analysis and correct determination of results. (Chapter 4 of NFPA 921 covers basic methodology.)

Scientific Method

The scientific method, as defined by NFPA 921, is the systematic pursuit of knowledge involving the recognition and formulation of a problem; the collection of data through observation and experiment; analysis of the data; the formulation, evaluation, and testing of a hypothesis (theory supported by data); and, when possible, the selection of a final hypothesis FIGURE 2-1. This should not be confused with the systematic method for which you gather the data, such as working from the least burn to the most burn or from the outside in toward the origin.

Fire Investigator Tip

The scientific method is a means of analysis that has been used and refined by researchers in the sciences for many years. This method is not exclusive to the fire investigation community but is used by many as a method for problem solving.

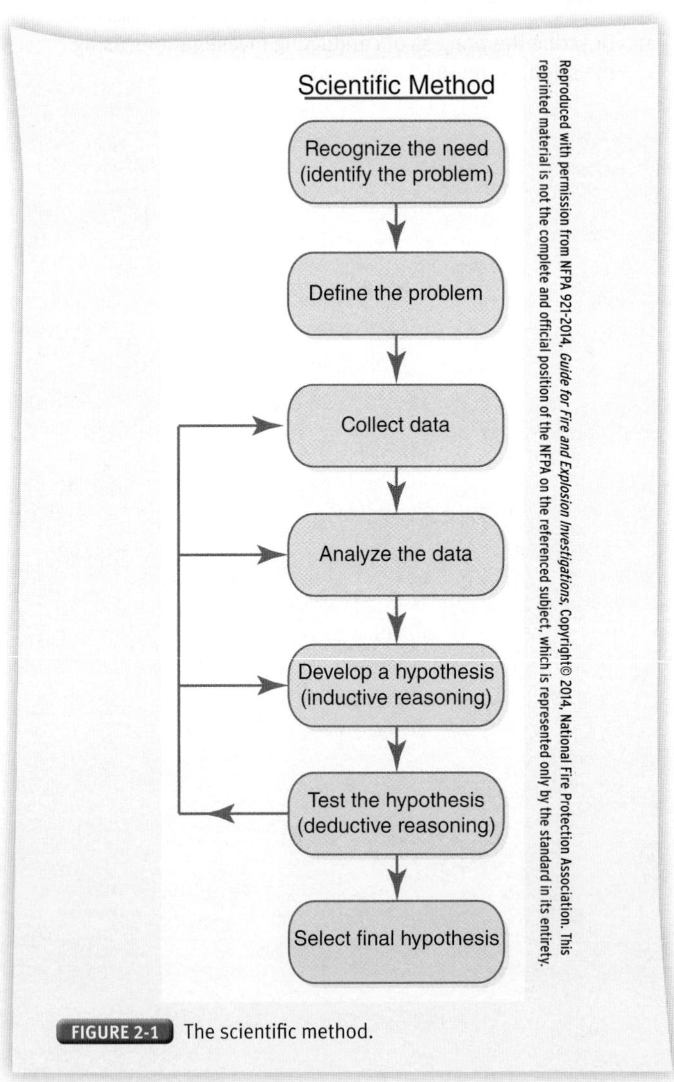

Reproduced with permission from NFPA 921-2014, *Guide for Fire and Explosion Investigations*, Copyright© 2014, National Fire Protection Association. This reprinted material is not the complete and official position of the NFPA on the referenced subject, which is represented only by the standard in its entirety.

FIGURE 2-1 The scientific method.

Fire Investigation and the Scientific Method

Fire investigators have often followed the scientific method but have not defined it by name. When questioned on the method used to conduct an investigation, the investigators often say that they document the fire scene from the outside to the inside or from the least damaged to the most damaged areas. This is one example of only one part of the scientific method—collecting data. The following are steps used to apply the scientific method to fire investigations.

Step 1: Recognize the Need to Use the Scientific Method

The first step is to realize that there is a problem to be resolved—something that is often self-evident because the investigator is notified of an incident and asked to determine its origin and cause. The exact nature of the problem that each investigator needs to address depends on the role and responsibility that the investigator fulfills during the investigation.

An example of this is receiving a call for a house fire in a single-family dwelling on Green Street.

Step 2: Define the Problem

When the investigator arrives on the scene or when the assignment is given, the investigator will have to define the problem to be resolved. This may depend on the role or employment of the investigator. Defining the problem may include determining the origin and cause of the fire, determining what role the furnishings may have played in the spread of the fire, or determining the cause of death of a victim involved. It is important to define the problem so that the investigator can apply his or her resources in the most effective and efficient manner.

An example of defining the problem is if you are tasked to conduct a comprehensive examination of the scene to determine the area and point of origin of the fire and find that the fire.

Step 3: Collect Data

The investigator will begin with a systematic evaluation of the scene and gather all of the data needed to find the answer to the problem that has been defined. Several chapters in NFPA 921 provide information on data collection. These data can include (but certainly are not limited to) the following:

- Recognition of physical evidence, such as fire patterns, fuel loads, ventilation, building construction, and other facts
- Collection of materials, such as debris samples, for laboratory evaluation
- Results of laboratory examinations
- Documentation of personal observations, such as witness statements
- Documentation of the fire scene through photographs, sketches, and notes
- Official reports, such as those of fire and police departments
- The documentation or results of prior scene investigations

An example of data collection is as follows: You note that there is a smoldering mattress, without bed linens, on the front lawn of the dwelling on Green Street. The incident commander on scene has confirmed that everyone is out of the structure. You locate the homeowner and begin to take notes on his observations and the events leading up to the fire. You also note the condition of the homeowner, who appears to be intoxicated. You gather a few other witnesses on the scene and take notes on their observations. You examine the mattress on the lawn more closely and begin to collect and document any additional evidence found at the home. As part of the reconstruction of the scene, the mattress is replaced into the room of origin for comparison to fire patterns found in the room and on the mattress.

The data collected using the scientific method that will be later analyzed are empirical data. Empirical data are factual data that are based on actual measurement, observation, or direct sensory experience rather than on theory.

It is important that investigators properly document and collect the data that will ultimately be used to test and verify their hypothesis. It is understood that memories fade over time, so proper documentation of interviews must occur. Also, once the fire scene has been demolished or repaired, the opportunity to collect additional data from the scene may be lost.

Collecting data also includes literature review, pattern analysis, scene documentation, photography and diagramming, evidence recognition and preservation, and review and analysis of others' investigations. Documentation of the fire scene is a topic that is addressed later in this text.

Step 4: Analyze the Data

At this point, the evidence is examined as objectively as possible to evaluate its meaning and to determine a possible hypothesis for the events that are being investigated. Data that are speculative or not related to the fire scene should not be considered. The analysis is based on the expertise, knowledge, experience, and training of the person doing the analysis.

To continue with the example, during the reconstruction of the scene, you begin to analyze the data found to include the burn patterns around the mattress. To understand the burning properties of the mattress better, you review manufacturing labels and research listed standards.

Step 5: Develop a Hypothesis

Through the analysis of the data, the investigator will develop a hypothesis (or several hypotheses) based on the empirical data to explain the events of the incident. This is referred to as inductive reasoning. The hypothesis is an attempt to answer the "define the problem" section of the scientific method. This task may require the investigator to identify the origin, cause, fire spread, responsibility, or some combination of these as directed by the investigator's assignment.

When you analyze the data, you should use only data that are known to be true (empirical data). Data that cannot be confirmed or are not known to be factual should not be used in the analysis and testing of the data. Analyzing the data will likely provide one or more hypotheses that the investigator can then take to the next step: "testing the hypothesis." If a single hypothesis

Fire Investigator Tip

Although it is physically impossible to collect everything from a fire scene, proper documentation will validate analysis of the data that are collected.

cannot be developed using the scientific method, the investigator may be required to identify another potential hypothesis and list the hypothesis as "unknown" or "more probable."

For example, based on the physical evidence found on the Green Street scene and interviews conducted, the remains of a candle were recovered from the mattress, and interviews confirm that a candle was accidently knocked over onto the mattress by the occupant.

Step 6: Test the Hypothesis

Testing the hypothesis may provide the investigator with the final hypothesis or conclusions. Testing the hypothesis can be done through different forms, but in the fire investigative field, it is most often done by testing and deductive reasoning. Through deductive reasoning, the investigator uses his or her knowledge and skills to challenge or test the hypothesis analytically. Full-scale laboratory or physical testing of the entire fire scene and fire dynamics may not be feasible because the conditions and the time of the fire cannot be accurately reproduced, such as the sequence of ignition of fuels, ventilation influences, and human involvement. Although fire modeling can assist in this type of testing, laboratory testing may be helpful on a smaller scale, such as to evaluate the ignition scenario or the energy of an available heat source.

Through deductive reasoning, the ultimate conclusion is supported, unsupported, or refuted by the complete body of evidence and data. Deductive reasoning is when all known facts are credibly and seriously used to challenge the final hypothesis. This includes evaluating other potential origins or causes against the final hypothesis.

The hypothesis must be supported by the facts. If it is not supported by the facts, the investigator may have to return to one of the earlier steps in the scientific method and repeat the process from that point to ensure that the hypothesis can be supported when it is subjected to challenge.

It is important to understand that *testing the hypothesis* may refer not only to experimental or physical testing, such as in a laboratory. Testing the hypothesis can also be cognitive or analytical. For example, during the testing and analysis of a hypothesis, the investigator will cognitively test the hypothesis on the basis of his or her knowledge and experience. Cognitive testing is the use of a person's thinking skills and judgment to evaluate the empirical data and challenge the conclusions of the final hypothesis. Other investigators may also provide a forum to test or challenge the investigator's hypothesis.

Many investigators do not have their opinions questioned or challenged; however, the critical step is used to determine whether a hypothesis is valid—that is, whether it can withstand reasonable challenge. Testing the hypothesis not only involves the critical examination and expected challenge of others, but also requires critical evaluation by the investigator. This self-challenge requires the investigator to ask questions regarding the hypothesis. In addition to the self-challenge, the investigator should also seek to determine the level of certainty in his or her opinion regarding the hypothesis. An important step in this process is administrative or technical review of the investigator's work. Having one or more knowledgeable persons review your investigative process and conclusions and, when appropriate, pose challenges to conclusions offered helps in the testing of the hypothesis FIGURE 2-2.

© Jones & Bartlett Learning. Photographed by Glen E. Ellman

FIGURE 2-2 Enlist the assistance of a knowledgeable administrator or investigator to review your process and conclusions.

A key question to determine whether a hypothesis is valid is this: What other hypotheses could be supported by the same set of facts? If there are alternative hypotheses that are valid using the facts, it is likely that the investigator has not gathered enough data. In this case, the investigator should obtain more information, reapply the steps in the scientific method, and attempt to disprove the hypothesis. Occasionally, an investigator cannot develop a hypothesis that can be considered valid. Ultimately, the investigator's goal is to find only one hypothesis that is valid to a probability (more likely than not). In situations in which the hypothesis is not supportable to a probability and is only *suspected* or *possible*, the results should be reflected as such—for example, by stating that the results are *inconclusive* or *undetermined*. A hypothesis that is possible is one that is true but where other hypotheses are also true and none are dominant (or more likely to be true than the others). A hypothesis that is suspected is true, but other hypotheses are also true. TABLE 2-1 shows the levels of hypothesis certainty. These levels of certainty relate to how strongly an investigator holds his or her conclusion or final hypothesis. It is based on the investigator's confidence in the collected, analyzed, and tested data. This level of certainty may impact how the findings are used in court or other proceedings. It is recommended that only hypotheses that raise to the level of "probable" be expressed as expert opinions with the certainty required in the legal system.

Other types of questions the investigator should use in testing the hypothesis include the following:

- Is there another way to interpret the facts (data)?
- If yes, why is your interpretation more likely true?
- What are the weaknesses in the hypothesis (analysis of the data)?
- What arguments will someone else (an opposing expert) use to refute the hypothesis?
- Are there facts that contradict the hypothesis?
- Is there research that supports the hypothesis?
- Can the hypothesis be proven to someone else?
- Does the hypothesis make sense?

In the mattress fire scenario, the hypothesis is compared with known and proven data: The fire originated on the mattress. Manufacturing data and lab tests show that the mattress

Table 2-1 Hypothesis Certainty

Level	Explanation
Probable	More likely true than not (more than 50 percent likely to be true). This level of certainty is usually associated with opinions involving the formation of a final hypothesis that cannot be eliminated during analysis and testing and when all other reasonable competing hypotheses have been evaluated and eliminated.
Possible	Feasible (often used when two hypotheses have the same level of certainty). This can be the result of the formation of two rival or competing hypotheses formed from a critical review of available data, but neither hypothesis can be eliminated during the analysis and testing process. A common example of this is open flame ignition of available combustibles versus smoldering ignition caused by carelessly discarded smoking materials. The evidence and data support the formation of both hypotheses, but neither can be eliminated without additional information or testing.
Suspected	Not enough certainty to be considered an expert opinion. This can be the result of an inability to develop enough data to formulate a particular hypothesis or an inability to analyze or test a hypothesis properly.

will ignite only with the application of an open flame. The evaluation of the scene included consideration of other possible ignition sources that were eliminated during the testing of the hypothesis process. Physical evidence was discovered of a candle on the mattress. An interview with the owner revealed that a candle was burning and was knocked onto the mattress, causing it to ignite.

Refer to the "Failure Analysis and Analytical Tools" chapter for additional methods in testing a hypothesis.

The investigator has to approach every incident with an open mind and without presumptions or expectation bias. There should be no preconceived determination or premature conclusions as to what the cause of the fire was, for example. Any preconceived ideas will, consciously or unconsciously, influence the investigator's efforts and should be avoided at all costs to maintain a proper level of objectivity during the investigation. The investigator should also avoid confirmation bias and work to disprove a hypothesis as opposed to prove a hypothesis. Confirmation bias can lead to a failure to properly formulate and test alternative hypotheses or to prematurely discount hypotheses that seem to contradict data without proper assessment. The data from the fire or explosion scene should stand on their own and be appropriately evaluated and tested in relation to the fire or explosion scene. If the hypothesis cannot withstand a reasonable challenge or test, the investigator will need to return to earlier steps in the scientific method, such as collecting additional data, analyzing that data, or redeveloping the hypothesis. If the hypothesis cannot withstand such a challenge, it must be listed as inconclusive. The legal community has stated in most jurisdictions that the investigator's opinion must meet some standard, such as degrees of certainty—often stated as "within a reasonable degree of scientific certainty within the fire investigation profession." Because every fire is unique and destroys data and evidence of its own existence, the certainty may not be an "absolute" or "without any doubts."

Step 7: Select the Final Hypothesis

The final step in using the scientific method is to select the final hypothesis. This step occurs only after all the previous steps have been performed fully and carefully. If all other reasonable hypotheses have been reliably tested and disproven, the investigator may be able to conclude that the remaining hypothesis is the one to adopt as the appropriate final hypothesis. However, the remaining hypothesis must be reliably and fully tested as well and must not be selected simply because it remains after others are disproven. The selected final hypothesis must be independently supported by evidence and consistent with the facts, and it must be challenged as rigorously as all of the others. Shortcuts must never be taken in this process.

There will be instances in which more than one hypothesis remains consistent with all the known facts and none can be excluded so as to allow one to be more probable. There may be other instances where the investigator's hypothesis testing may result in no hypotheses remaining that are supported and consistent with the facts. In either event, further facts, information, and testing may be needed prior to a final hypothesis being reached. In these situations, the investigator may be left with no option except to report the issue being analyzed as undetermined.

The Basic Methods of Fire Investigation

The scientific method is the analytical framework that the fire investigator must apply to assure that his or her conclusions are valid. As a practical matter, the investigator should follow these general steps to assure that the scientific method is applied properly in each case.

1. Receive the assignment: The investigator is notified of the fire loss and goal of the requested investigation. This step can be equated to "recognizing the need" or "identifying the problem" within the framework of the scientific method.
2. Prepare for the investigation: For the given assignment and the nature of the fire scene, the investigator should determine the tools required, safety concerns to be addressed, and personnel needs to complete the task. Additionally, the investigator should evaluate and ultimately document what legal authority will give him or her the right to conduct the investigation. (For additional information about legal authority, see the "Legal Considerations" chapter of this textbook.) This step can be equated to "defining the problem" within the framework of the scientific method.
3. Conduct the investigation: Begin a systematic process of information gathering to include examination and documentation of the scene, interviews, and information research. This is the "collecting data" step within the framework of the scientific method.
4. Collect and preserve evidence: Identify possible evidence and document and preserve it for future testing or legal presentation. This is also the "collecting data" step within the framework of the scientific method.
5. Analyze the incident: Utilizing the scientific method, the collected and available data are analyzed. A hypothesis

is developed and tested. This is the "analyze the data," "develop a hypothesis," and "test the hypothesis" steps within the framework of the scientific method.

6. Draw conclusions: A final tested hypothesis is determined. This is the "select a final hypothesis" step of the scientific method.
7. Report: As determined by the responsibility of the investigator, the process and conclusions are reported in written or oral forms.

Reviewing the Procedure

The fire investigator's work products, such as reports, are often reviewed by others, including the investigator's supervisor. These reviews are generally categorized into different types: administrative, technical, and peer reviews. This is not the same type of review as a review by opposing parties during litigation.

The administrative review is usually carried out within the investigator's organization to ensure the product meets the organization's procedures, such as for content as outlined in the organization's procedure manual, grammar, and spelling. This type of review is not a true critique of the investigation, the conclusions, or the testing of the hypotheses. The reviewer may not have the necessary skills to perform a technical review and may not review all of the data that the investigator has developed or relied on.

The technical review is a critique of the investigator's work and findings. As such, the reviewer must have the skills, knowledge, and background to perform such an evaluation. This type of evaluation requires the reviewer to have access to all of the data on which the investigator relied. The reviewer will challenge and critique the work product and investigative findings. This will include a challenge of testing the hypotheses of the findings. This type of review is beneficial if it adds a second serious challenge to the investigation, but bias may be present because of working relationships and reviewers may be perceived to have an interest in the outcome of the investigation.

There is also a peer review, which is a formal procedure that is generally used on scientific or technical papers. The peer review is considered independent and objective because the reviewers do not have interest in the outcome of the report. The author does not select the reviewers and the reviewers should not be coworkers or part of the same organization as the author. Peer reviewers have the knowledge to evaluate flaws in the process or procedures employed, but likely do not have the resources to question the investigator's data. Because of this, the peer reviewers may not be able to detect the validity of the investigator's conclusions.

A proper and unbiased technical review usually provides the best means to assess the validity of the investigator's conclusion. During litigation, there may be others who challenge the investigator's methods and findings.

Wrap-Up

© Greg Henry/ShutterStock, Inc.

Ready for Review

- The scientific method is a systematic method of problem solving.
- The scientific method includes recognizing and identifying the problem, defining the problem, collecting information, analyzing the information, developing hypotheses, and testing the hypotheses to determine whether the results are reliable and valid in order to reach a final hypothesis.
- Proper documentation will validate analysis of the data that are collected.
- After all evidence and data are collected, the whole body of evidence and data is reviewed using inductive reasoning; evidence is looked at objectively in order to evaluate its meaning and determine a possible hypothesis.
- The hypothesis is challenged or tested through deductive reasoning.
- If a hypothesis cannot withstand a challenge or test, the investigator may need to return to earlier steps of the scientific method.

Hot Terms

Cognitive testing The use of a person's thinking skills and judgment in order to evaluate the empirical data and challenge the conclusions of the final hypothesis.

Confirmation bias The attempt to prove a hypothesis instead of disprove a hypothesis, resulting in a failure to consider alternative hypotheses or the premature discounting of an alternative hypothesis.

Deductive reasoning The process by which conclusions are drawn by logical inference from given premises.

Empirical data Factual data that are based on actual measurements, observation, or direct sensory experience, rather than on theory.

Expectation bias Preconceived determination or premature conclusions as to what the cause of the fire was without having examined or considered all of the relevant data.

Hypothesis Theory supported by the empirical data that the investigator has collected through observation and then developed into explanations for the event, which are based on the investigator's knowledge, training, experience, and expertise.

Inductive reasoning The process by which a person starts from a particular experience and proceeds to generalizations. The process by which hypotheses are developed based upon observable or known facts and the training, experience, knowledge, and expertise of the observer.

Scientific method The systematic pursuit of knowledge involving the recognition and definition of a problem; the collection of data through observation and experimentation; analysis of the data; the formulation, evaluation, and testing of a hypothesis; and, when possible, selection of a final hypothesis.

FIRE INVESTIGATOR *in action*

You are requested to respond to a fire in a commercial laundry. When you arrive at the scene, the incident commander (IC) advises you that he noted "heavy" smoke conditions from the west end of the building, with flames visible from the south exterior windows. The IC reports that fire personnel believe the fire originated within one of the clothes dryers that was in use at the time the fire was discovered.

Several patrons of the laundry are still present at the scene and report they began to notice an odor of smoke within the building approximately 10 minutes prior to seeing flames within the last dryer at the west end of the building. A male patron states that he had been using this dryer to dry his work clothes he had just washed.

When you enter the building, fire patterns indicate the fire originated near the west end of the building where the dryers are located. Five clothes dryers are present in this area, with severe damage noted to the western-most dryer, which the male patron stated he was using. The door to the dryer has failed, and the contents within the dryer are severely charred.

While examining the dryer, you notice a small utility access room located behind the dryers and examine it. You observe minimal damage within the utility space and note various manufacturer stickers and labels on the rear of each dryer. Additionally, you determine each dryer is gas-fired, and a significant accumulation of lint and debris is present in the utility space.

1. What data are present that you may collect?
   A. Obtaining witness statements from the patrons
   B. Documenting the scene with photographs
   C. Recording the manufacturer labels and markings
   D. All of the above
2. Which of the following hypotheses would not be considered probable?
   A. A failure of the clothes dryer
   B. Ignition of built-up lint in the dryer
   C. Ignition of the clothing within the dryer
   D. A failure of a light fixture in the utility space
3. A critique of an investigator's work and findings is considered a(n):
   A. audit review.
   B. administrative review.
   C. technical review.
   D. investigative review.
4. If you suspect the clothes dryer caused the fire, how would you test this hypothesis?
   A. Laboratory testing of the dryer
   B. Cognitive testing of the ignition sequence
   C. Peer review of your findings
   D. All of the above

© Greg Henry/ShutterStock, Inc.

Basic Fire Science

© Photos.com

© Dale A Stork/Shutterstock, Inc.

Knowledge Objectives

After studying this chapter, you should be able to:

- Explain the chemistry of fire NFPA 1.3.7. (pp 22–23)
- Describe phase changes and thermal decomposition NFPA 1.3.7. (p 23)
- Identify and describe products of combustion NFPA 1.3.7. (pp 23–24)
- Discuss the components of fire dynamics NFPA 1.3.7. (pp 24–26)
- Discuss the roles of fuel items and fuel packages NFPA 1.3.7. (pp 26–28)
- Discuss the factors that lead to ignition NFPA 1.3.7. (pp 28–32)
- Describe the process of flame spread NFPA 1.3.7. (p 32)
- Describe the process of compartment fire spread NFPA 1.3.7. (pp 32–35)

Skills Objectives

There are no skills objectives for this chapter.

Additional NFPA Reference

NFPA 921, *Guide for Fire and Explosion Investigations*

CHAPTER 3

FESHE Course Outcomes

Fire Investigation I

5. Explain the basic elements of fire dynamics and how they affect cause determination. (pp 24–26)

Fire Investigation II

3. Describe the chemistry of combustion. (pp 22–23)
4. Explain the nature and behavior of fire. (pp 24–26, 33–35)
5. Identify the combustion properties of liquid, gaseous, and solid fuels. (pp 28, 30–32)

You Are the Fire Investigator

© Jones and Bartlett Publishers. Photographed by Glen E. Ellman

A large fire has occurred in a small house and has spread and ignited a neighboring home. You are advised by fire fighters on the scene that dark, black smoke was emanating from all of the windows and front door when they arrived, and that flames were visible soon after. Fire personnel state that the fire quickly grew to approximately 30 feet (10 m) above the house before it could be extinguished. The damage to the interior of the original house is severe, with most of the contents being consumed during the fire.

1. What are the potential fuel sources for this fire?
2. How does the investigator interpret the color of the flames and smoke?
3. How did the fire extend from the first house to the neighboring home?

Introduction

Fire has been used by humans for thousands of years for many purposes, including heating homes, cooking food, and waging war. In each period of human existence, tragic events have occurred as a result of the destructive power of fire. Every fire investigator must fully understand the fundamental properties of fire and fire development to determine accurately the origin and cause of any fire or explosion event he or she may investigate.

Fire Chemistry

All matter is composed of elements; combinations of elements are known as compounds. Elements are substances that will not break down into simpler substances. There are 92 natural elements, such as carbon, sulfur, and hydrogen. All elements are composed of smaller units, called atoms. Atoms are the smallest unit of an element that can take part in a chemical reaction. Matter may be present in one of three states: solid, liquid, or gas. As it relates to fire, fuels must generally be in the gas or vapor state to be involved in the combustion process, so a phase change or decomposition may be required. The application of heat to matter can effectively result in a phase change—from a solid to a liquid, for example—where no atoms change, or may result in thermal decomposition, as in pyrolysis where atoms are broken down and rearranged to form different matter.

In a fire, there are many chemical reactions, but overall it is an oxidative reaction. Atoms from a fuel are joining with the oxygen in the air, or being oxidized. As the atoms in a material are broken down and rearranged, heat energy is produced and released into the atmosphere. This release of heat energy is referred to as an exothermic reaction. In the fire environment, this heat that is generated serves to help continue the reaction until it is interrupted (uninhibited chain reaction). The molecular rearrangement and breaking down of atoms results in the formation of new substances and the presence of free radicals, which are atoms that have an open bond and are highly reactive. Free radicals are not as stable as regular atoms and easily react with other substances. This continued formation and reaction of free radicals is part of the combustion process. Other reactions, such as some phase changes where the matter is not altered, may be an endothermic reaction where the heat is absorbed to continue the reaction. Although there are many compounds, the most important class for fire investigators is the carbon-based or organic compounds. Of the organic compounds, some of the best fuels are formed of hydrocarbons, compounds that are composed solely of carbon and hydrogen.

Fire is a rapid oxidation process, a chemical reaction resulting in the evolution of light and heat in varying intensities. In general, fire occurs only in the gas phase. Solids, for example, must be heated, causing the material to decay and produce fire gases. This process of decomposition of a material into simpler molecular compounds by the effects of heat alone is known as pyrolysis and is seen when wood, heated sufficiently, breaks down into vapors and char.

For liquids to burn, on the other hand, they must be heated to produce ignitible mixtures in air through a process known as vaporization. For a fire or combustion to occur, four components must be present: fuel, oxidizing agent, heat, and an uninhibited chemical chain reaction. By removing or eliminating one of the four components, the fire can be extinguished. The four components of fire are often displayed as the fire tetrahedron FIGURE 3-1.

Fuel

The first component of the fire tetrahedron, fuel, is the material that will be reduced or consumed during the combustion process. The most common fuels an investigator will encounter are organic fuels, which contain carbon. These fuels include wood, plastics, and petroleum products. Inorganic fuels, such as combustible metals, do not contain carbon. When a solid fuel is heated, the compounds and elements that make up the material begin to break down. This process of pyrolysis produces ignitible gases that will burn on the surface of the material. Similarly, vaporization of liquids causes ignitible

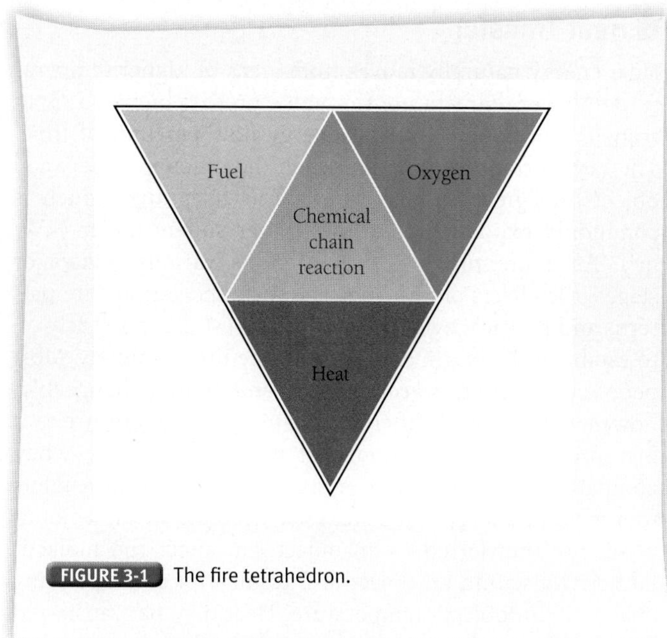

FIGURE 3-1 The fire tetrahedron.

vapors and gases to be produced, which will also burn on the surface of the liquid.

Gaseous fuels are probably the most dangerous of the three physical states because they need only an ignition source for combustion to occur.

Oxidizing Agent

Most fires require an oxidizing agent to support the combustion process. The oxygen in air is the most readily available source of oxygen; however, certain other materials can also oxidize a fire.

Heat

Heat provides the energy needed to create and ignite vapors produced from the fuel source. Heat is a form of energy, which is measured in joules. The rate at which heat is released, known as the heat release rate (HRR), is measured in joules per second or watts.

Uninhibited Chemical Chain Reaction

The last component, an uninhibited chemical chain reaction, provides a self-sustaining event that continues to develop fuel vapors and sustain flames even after the removal of the ignition source. As a fire continues to burn, this exothermic reaction radiates heat back to the surface of the fuel, producing more vapors and continuing the combustion process.

Phase Changes and Thermal Decomposition

The response of materials to heat is quite varied. During a fire, materials may change their physical state as a result of being heated. These physical changes or phase changes include melting (in which solids turn to liquids) and vaporization (in which liquids turn to gases). Both melting and vaporization are reversible upon cooling because the chemical structure of the material does not change. When an irreversible change occurs in either a solid or a liquid, it is referred to as thermal decomposition.

Combustion: Premixed and Diffusion Flames

Flames produced during the combustion process can be categorized as either premixed or diffused. Premixed burning (flame) can occur when fuel vapors such as natural gas are dispersed and mixed with air in the absence of an ignition source. These fuels can produce both fires and explosion events. The fuel and air are already combined and need only a competent ignition source to start the chemical reaction. When the ratio of fuel to air is balanced, the most efficient means of combustion exists. The "perfect" combustion condition is referred to as the stoichiometric ratio. The stoichiometric ratio is a concentration that exists above the lower explosive limit (LEL) and below the upper explosive limit (UEL). Below the LEL, there is not enough fuel to sustain combustion; above the LEL, there is not enough air to sustain combustion.

Most fuels, however, are consumed by diffusion flame burning in which the fuel is pyrolyzed. This process produces vapors that are oxidized, usually above the surface of the material. The best example of the diffusion flame is a candle. With a candle flame, the fuel and the air meet and mix in the luminous zone. Like a premixed flame, the diffusion flame can exist only with sufficient concentrations of oxygen present.

Flames may transition from a premixed to a diffusion flame. An example of this is when the premixed vapors above a gasoline pan are ignited and then are consumed. The flame then becomes a diffusion flame with the burning of the vapors from the pan.

Products of Combustion

When a fuel burns, it can produce chemical compounds that are released as both visible and invisible products of combustion. Hundreds of different compounds can be produced during the combustion process, including carbon monoxide, carbon dioxide, and water vapor. If a fire has a limited amount of air for combustion, an increase in the amount of visible products of combustion (such as soot and smoke) and carbon monoxide will occur.

Smoke is created by the combination of the various products of combustion, dependent on the particular fuel being consumed. These products include solids such as soot or ash, small liquid droplets, and various other fire gases. Some fuels, such as alcohols and propane, will burn without the production of smoke, whereas plastics and other hydrocarbons tend to produce large amounts of dense black smoke. As these products of combustion begin to migrate away from the fire and cool, they accumulate on both horizontal and vertical surfaces. Surfaces that are normally cooler, such as ceramic tile, tend to collect more of these combustion products than warmer surfaces do.

Fire Investigator Tip

Smoke and flame color are not valid indicators of the burning material. Many factors, including suppression activities, can alter the appearance of smoke and flames.

Smoke and Flame Color

The color of smoke should not be relied on as an indicator of the material burning. Although certain fuels may produce a particular color and density of smoke, other factors (including decreasing oxygen levels and ventilation-controlled fires) can produce heavy, dark smoke commonly observed by oil and other hydrocarbons burning. As a fire goes through its various phases, smoke production will generally increase. Firefighting operations can also change the color of smoke by mixing condensing vapors with black smoke, producing a white to gray color. This can cause witnesses and fire fighters alike to believe that other fuels may be burning.

Although fuels may burn at a particular temperature and produce a certain color flame, neither temperature nor flame should be relied on as an indicator of what is burning.

Fire Dynamics

Fire dynamics is the study of all of the behaviors of fire, but the fire investigator will be mainly interested in the interaction of the fire with its surroundings. To understand the behavior of fire, the investigator should have some understanding of fluid flows, heat transfer, ignition and flame spread, fuel packages, heat flux, and the distinction between fuel-controlled fires and ventilation-controlled fires.

Fluid Flows

Fluids include both liquids and gases, but in fires, we are generally focused on gas flows. These flows can be generated by mechanical forces such as fans, but in most cases, the fire creates its own buoyant flow. A buoyant flow is the result of hot gases being less dense than cool gases, so they rise. The primary engine that creates flows in fires is the fire itself. The hot gases created by the fire rise above the fire in a plume. As these gases rise, they entrain or draw in cool air, which causes the velocity of the gas flow to increase. At the same time, the temperature of the plume decreases. The entrainment of air also causes the diameter of the plume to increase, resulting in a cone shape.

When the plume reaches the ceiling of a room, the movement of the fire gases parallel to the ceiling is known as a ceiling jet. The ceiling jet flows along the ceiling until it encounters a vertical obstruction, such as a wall, which impedes its flow.

In compartment fires, the buoyant gases flow in and out through multiple vents. If there is only one vent in the compartment, the gases flow out the upper portions of the opening and the fresh air flows into the compartment to feed the combustion through the lower portions of the vent opening.

Heat Transfer

Heat energy naturally moves from areas of higher temperature to lower temperature through a process known as heat transfer and is measured as energy flow per unit of time. The rate of transfer increases when the differences between objects are greater. This is known as heat flux, which is commonly expressed as kilowatts per square meter (kW/m^2). As a fire progresses through its various phases or stages, the effects of heat transfer create the various fire patterns and physical evidence that the investigator will rely on to establish the origin and cause of the fire accurately. Most people use the terms *heat* and *temperature* interchangeably; however, they have different meanings. Temperature is a measurement of the amount of molecular activity when compared with a reference or standard, whereas *heat* refers to a form of energy. An increase or decrease in the amount of energy transferred to an object can affect the molecular activity within an object and result in a corresponding change in an object's temperature. Heat may be transferred by conduction, convection, or radiation. Fire investigators must be aware of how each mode of transfer can influence the development and spread of a fire.

Conduction

Conduction occurs when solid objects are heated and energy is transferred from hotter to cooler areas through direct contact. An example of this is when a spoon is left in a hot liquid and the end of the handle begins to get warmer. The rate of this transfer is influenced by the difference in temperature and the thermal conductivity of an object. Thermal conductivity (k), the measure of heat that could travel across an area with a temperature gradient, is expressed as one degree per unit of length (W/(m K)). In contrast, heat capacity is a measure of the energy required to raise the temperature of an object one degree of a unit mass (J/kg-K). Simply stated, the more conductive a material is, the faster heat will transfer, and although denser objects tend to be better conductors than lower-density objects, materials with higher heat capacities will need more energy to raise their temperature versus lower heat capacity items. Thermal conductivity can be an important property when assessing how easily a fuel package can be ignited. Thermal conductivity plays an important role in determining thermal inertia, which is defined as those properties of the material that characterize its rate of surface temperature rise when exposed to heat. The thermal inertia is the product of the thermal conductivity (κ), the density (ρ), and heat capacity (c). This mathematical expression is pronounced *kay-row-see*.

Plastic foams have a low density and a low thermal conductivity, so when exposed to heat, the surface temperature increases quickly. Metals, on the other hand, have a high thermal conductivity and a high density, so the surface temperature of a metal object increases more slowly when exposed to the same heat source.

As materials are heated, the surfaces of an object begin to absorb heat that is then transferred through the remainder of the object via conduction. The speed at which an object's surface absorbs heat is determined by its thermal inertia and is most influential during the initial heating of the object. For instance, objects that are thin tend to heat and ignite faster than objects of like materials that are thicker. This effect has a direct impact on ignitibility and flame spread.

Convection

When heat is absorbed from heated gases or liquids by cooler objects or materials, we refer to this as convection. During a fire, heat and flames are initially spread via convection. Convection may occur by one of two mechanisms: naturally or forced. With natural ventilation, the buoyant effect of the heated gases causes the heat to be moved away from the source, typically by rising then spreading laterally through a space. Forced convection is caused by the influence of external sources, including the use of fire streams and ventilation fans.

Radiation

The transfer of heat via electromagnetic waves is defined as radiation. Heat from the fire is transferred in a direct line away from the source object and absorbed by cooler objects, including liquids and gases. Most exposure fires are the result of radiant heat transfer.

The rate of heat transfer from a radiating material is proportional to that material's absolute temperature (the temperature of a material measured in degrees Kelvin or according to the Rankine scale) raised to the fourth power. As an example, a radiating material that doubles its absolute temperature will have a 16-fold increase in radiation.

The rate of heat transfer between two objects is affected mainly by the distance between the two objects, but other factors come into play as well. The relative temperatures of the two objects and the view (i.e., the angle between the radiator and the target) also affect the heat transfer rate. Radiation is transferred more readily between two parallel surfaces than between two perpendicular surfaces. When the target surface is not parallel to the radiator, the energy is spread out over a broader area, reducing the number of watts per unit area.

Once a fire reaches full room involvement, radiation becomes the dominant means of heat transfer. Because of this, an understanding of radiation is critical to a fire investigator's ability to comprehend fire growth and spread. It is important to be able to relate the concept of heat flux to everyday experience. The data in TABLE 3-1 can provide some assistance in this regard.

Fire Investigator Tip

In heat transfer, heat energy moves from areas of higher temperature to areas of lower temperature. This heat transfer impacts the fire patterns and other physical evidence at a scene. The primary types of heat transfer are as follows:

- Conduction: Heat transfer through direct contact of solid objects
- Convection: Heat transfer from heated gases or liquids to cooler objects or materials
- Radiation: Heat transfer through electromagnetic waves

Table 3-1 Effect of Radiant Heat Flux (NFPA 921 Table 5.5.4.2)

Approximate Radiant Heat Flux (kW/m²)	Comment or Observed Effect
170	Maximum heat flux as currently measured in a postflashover fire compartment.
80	Heat flux for protective clothing Thermal Protective Performance (TPP) Test.[a]
52	Fiberboard ignites spontaneously after 5 seconds.[b]
29	Wood ignites spontaneously after prolonged exposure.[b]
20	Heat flux on a residential family room floor at the beginning of flashover.[c]
20	Human skin experiences pain with a 2-second exposure and blisters in 4 seconds with second-degree burn injury.[d]
15	Human skin experiences pain with a 3-second exposure and blisters in 6 seconds with second-degree burn injury.[d]
12.5	Wood volatiles ignite with extended exposure[e] and piloted ignition.
10	Human skin experiences pain with a 5-second exposure and blisters in 10 seconds with second-degree burn injury.[d]
5	Human skin experiences pain with a 13-second exposure and blisters in 29 seconds with second-degree burn injury.[d]
2.5	Human skin experiences pain with a 33-second exposure and blisters in 79 seconds with second-degree burn injury.[d]
2.5	Common thermal radiation exposure while fire fighting.[f] This energy level may cause burn injuries with prolonged exposure.
1.0	Nominal solar constant on a clear summer day.[g]

Note: The unit kW/m² defines the amount of heat energy or flux that strikes a known surface area of an object. The unit kW represents 1000 watts of energy, and the unit m² represents the surface area of a square measuring 1 m long and 1 m wide. For example, 1.4 kW/m² represents 1.4 multiplied by 1000 and equals 1400 watts of energy. This surface area may be that of the human skin or any other material.
[a]From NFPA 1971, *Standard on Protective Ensembles for Structural Fire Fighting and Proximity Fire Fighting.*
[b]From Lawson, "Fire and Atomic Bomb."
[c]From Fang and Breese, "Fire Development in Residential Basement Rooms."
[d]From the Society of Fire Protection Engineering Guide, "Predicting 1st and 2nd Degree Skin Burns from Thermal Radiation," March 2000.
[e]From Lawson and Simms, "The Ignition of Wood by Radiation," pp. 288–292.
[f]From U.S. Fire Administration, "Minimum Standards on Structural Fire Fighting Protective Clothing and Equipment," 1997.
[g]From the *SFPE Handbook of Fire Protection Engineering*, 2nd edition. NFPA, Quincy, MA.
Reproduced with permission from NFPA 921-2014, *Guide for Fire and Explosion Investigations*, Copyright© 2014, National Fire Protection Association. This reprinted material is not the complete and official position of the NFPA on the referenced subject, which is represented only by the standard in its entirety.

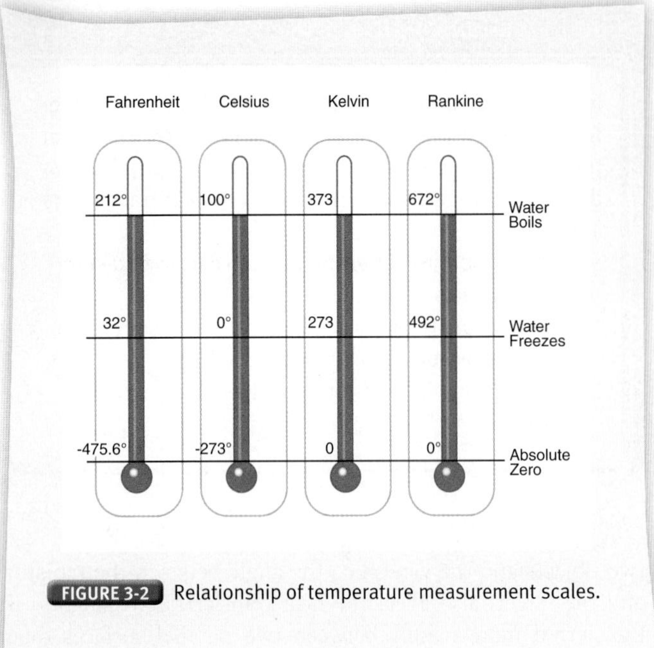

FIGURE 3-2 Relationship of temperature measurement scales.

■ Thermometry

Thermometry is the study of the science, methodology, and practice of temperature measurement. Although thermometry is seldom used directly at the fire scene, it can play an important role in the investigation and analysis of a fire event. During the investigation of suspected spontaneous combustion fires, as experienced in decomposing haystacks, core temperature measurements may be taken at the scene as part of the hypothesis testing process to determine if other sections of material are at critical temperatures. Additionally, an understanding of the appropriate application of thermometry is necessary during the use of fire dynamics equations and formulae. The study is frequently used in fire safety or code compliance cases as well. There are several systems for measuring degrees of temperature, which can be divided into empirical temperature scales and thermodynamic temperature scales.

The Fahrenheit and Celsius (Centigrade) scales are the most common empirical temperature scales and are based on the relative temperatures at which water boils and freezes, along with other empirical comparisons.

The thermodynamic temperature scales are based on the lowest possible temperature of absolute zero, and therefore are called the absolute temperature scales. The scales are based on the fundamental laws of thermodynamics or statistical mechanics, where temperature is measured in Kelvins (K) or using the Rankine (R) scale FIGURE 3-2.

■ Fuel Load

The amount of fuel present is referred to as fuel load. In any fire scenario it is important to consider the fuel load of a given scene. The total fuel load of a given fire will be a reflection of the potential energy; it does not determine how fast the fire develops once ignition occurs.

Fuel Items and Fuel Packages

The materials consumed during a fire are referred to as the fuel items. When fuel items are assembled and placed close to each other, a fuel package is created. If only one specific fuel item exists, it may also be considered a fuel package. If other fuel packages are nearby, ignition via radiant heat can occur. Commonly found fuel packages include the following:

- Furniture and other contents in a dorm or bedroom
- Personal items in a commercial storage space
- Combustible raw materials used in a paper manufacturing facility

The specific fuel items within a fuel package are factors that influence the speed and intensity of a particular fire, as well as the best method of fire extinguishment.

The location and position of a fuel package are also important for the fire investigator to consider. For example, fuel packages that are located next to or near a wall or corner will restrict the amount of air entering the plume and create an imbalance. This imbalance will bend the flame and thermal plumes toward the restricting surface or object. Once the flame and thermal plumes begin to bend toward the wall, they may attach themselves to the wall, restricting air entrainment into the base of the fire. The geometry of the fuel package and its proximity to the wall will determine the extent to which the flame and thermal plume will bend. This decrease in air entrainment will also cause an increase in plume and upper layer temperatures because of the reduction of cooler air entraining into the fire, which dilutes the higher temperatures of the plume. As the plume temperature increases and releases energy to the upper layer, the temperature of the upper layer will naturally increase.

When entrainment is decreased, less air is available to mix with fuel vapor, and the vapor must be transported over a longer distance to mix completely with this reduced quantity of air. This can result in increased flame heights. A fuel package that is located close enough to a wall to restrict the amount of air entrainment will have an increase in absolute temperature in the upper layer compared to the fuel package located away from the wall. Whether the flame height will also increase depends on numerous variables, particularly the geometry of the fuel package. Increased flame heights have been reported, although NFPA 921 states that experiments have shown, there will be no significant increase in flame height. When the fuel package is placed in a corner, air entrainment is further reduced, causing an increase in absolute temperature of the upper layer and a significant increase in flame height. The position of the fire in relation to walls and corners must be considered by the fire investigator when interpreting damage patterns. These effects may also be observed for outdoor fires when similar conditions are present.

Heat Release Rate

The power of a fire is determined by calculating the energy being released by the individual fuels being consumed. This is called the heat release rate (HRR) and is measured in either watts or kilowatts. The HRR of a fuel is often illustrated on a generalized curve to indicate the energy being released during the incipient, free-burning, flashover, and smoldering/decay stages of a fire. FIGURE 3-3 displays an example of an ideal HRR curve.

When a fuel item is burned, the highest value measured is termed the *peak HRR*. The peak HRR is a property of a fuel calculated by multiplying the mass burning rate by the heat of combustion of the fuel. This is also dependent on the total area involved in the combustion process. The mass burning rate or mass loss rate typically relates to how much material is consumed during combustion, measured in grams of material per second per area burning. The heat of combustion relates to how much energy is being released as a result of this reaction and is typically expressed in kilojoules per gram of material. TABLE 3-2 provides general values for fuel packages and may assist the investigator in explaining both fire development and spread issues. Although these values may be helpful to the investigator, they should not be considered absolute because fuel items that may have a similar use can produce varying HRR values. Testing and experimentation is the most accurate method of determining a particular fuel item's peak HRR, but such testing data are usually not available. When using HRR estimates to test fire-spread hypotheses, it is the best practice to estimate at the high end of the ranges shown in Table 3-2.

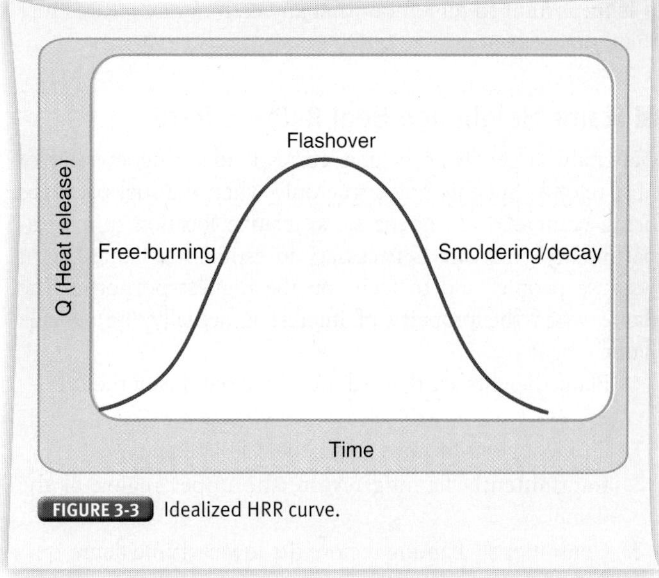

FIGURE 3-3 Idealized HRR curve.

When a fire is unconfined and allowed to grow, the HRR will generally increase as the fuel item becomes more involved. Once flames have spread across the entire surface of the fuel item, the peak HRR will be reached, and the rate will begin to decrease as the fire moves toward the decay phase. In the instance of a compartment fire, the HRRs of all fuel items and packages that are involved are added together to produce a HRR for the entire compartment. The HRR may also increase in the compartment fire as heat is contained and radiated back on the burning items.

Table 3-2 Representative Peak HRRs (Unconfined Burning) (NFPA 921 Table 5.6.3.1)

	Weight		Peak HRR (kW)
Fuel	**kg**	**lb**	
Wastebasket, small	0.7–1.4	1.5–3	4–50
Trash bags, 42 L (11 gal) with mixed plastic and paper trash	2.5	7.5	140–350
Cotton mattress	12–13	26–29	40–970
TV sets	31–33	69–72	120 to over 1500
Plastic trash bags/paper trash	1.2–14	2.6–31	120–350
PVC waiting room chair, metal frame	15	34	270
Cotton easy chair	18–32	39–70	290–370
Gasoline/kerosene in 0.2 m^2 (2 ft^2) pool	19	42	400
Christmas trees, dry	6–20	13–44	3000–5000
Polyurethane mattress	3–14	7–31	810–2630
Polyurethane easy chair	12–28	27–61	1350–1990
Polyurethane sofa	51	113	3120
Wardrobe, wood construction	70–121	154–267	1900–6400

Sources: Values are from the following publications:
Babrauskas, V., and Krasny, J. (1985). *Fire Behavior of Upholstered Furniture.*, NBS Monograph 173 Fire Behavior of Upholstered Furniture, National Bureau of Standards, Gaithersburg, MD.
Babrauskas, V. (2002). "Heat Release Rates," in *SFPE Handbook of Fire Protection Engineering*, 3rd ed., National Fire Protection Assn., Quincy, MA.
Lee, B. T. (1985). *Heat Release Rate Characteristics of Some Combustible Fuel Sources in Nuclear Power Plants,* NBSIR 85-3195, National Bureau of Standards, Gaithersburg, MD.
NFPA 72, *National Fire Alarm Code*, 1999 ed., Annex B.
Reproduced with permission from NFPA 921-2014, *Guide for Fire and Explosion Investigations*, Copyright© 2014, National Fire Protection Association. This reprinted material is not the complete and official position of the NFPA on the referenced subject, which is represented only by the standard in its entirety.

It is important to remember that any estimate of HRR is just that—an estimate.

Flame Height and Heat Release Rate

Generally, larger flame heights correspond to higher rates of heat release, but this holds true only when the fuel packages being compared are in the same relative location in a compartment. Some care is necessary in estimating flame height because people tend to focus on the highest portion of the flame when the property of interest is actually the average flame height.

Flame heights are defined into three regions of the fire:

1. Plume region: the area above the visible flame
2. Intermittently flaming region: the upper region of the flame
3. Continuously flaming region: the lower visible flame

Correlations of flame height versus HRR were determined experimentally using liquid pools, so they may not apply in all cases of burning solid fuels. Further, the diameter of the flame is an important consideration.

Of more importance to the fire investigator is the location of the fuel package within a room. A fuel package burning in the center of the room will have a plume that is able to entrain air from all sides. If that package is moved against the wall, less air is available for entrainment, and, as mentioned above, the flame height may increase. If that package is located in the corner of the room, significantly less air is available, resulting in a much taller flame. This effect is known as the wall effect or the corner effect and can lead to misinterpretation of the origin of the fire because heavier fire damage will be located above fuel packages located against walls or in corners. The effect is illustrated in FIGURE 3-4, which shows three identical sofa cushions 17 seconds after ignition. Also, keep in mind that ceiling height and distance from a plume greatly affect how quickly fire protection devices react to the fire.

Ignition

With any fire investigation, the source of ignition is critical in determining the cause of a fire. An ignition source can be defined as either smoldering (as is the case with hot coals and glowing embers) or flaming (ignition when a match or lighter is the source). A source of ignition may also be characterized as either piloted or autoignition. Sparks, arcs, and open flames are common sources of piloted ignition in which an external ignition source ignites the fuel item. Autoignition, however, occurs when the fuel item is heated to a temperature at which fuel gases being released from the object ignite without a piloted ignition source.

To heat a fuel sufficiently to generate ignitible vapors, there are a number of interrelated factors.

- Form of the fuel: Is it a solid, liquid, or gas? Does it have a lot of surface area that can absorb heat and a relatively small amount of mass?
- Amount or mass of the fuel: How much fuel is present that needs to be heated? The greater the mass, the more heat is needed.
- Proximity of the fuel to the heat source: Generally, the closer the fuel is to the heat source, the faster the fuel's temperature is raised.
- Amount of heat being generated: How much heat is the heat source generating? If it is a relatively small heat source, the fuel needs to be either closer or exposed longer to have its temperature raised.
- Duration of exposure: The longer the duration, the more the fuel's temperature is going to be raised; however, some fuels may not need long exposures to have their temperature raised sufficiently to generate combustible vapors.

For example, a single match cannot heat a large block of wood sufficiently to ignite it, even though the temperature of the heat source may be above the minimum ignition temperature of the target fuel. There is too much mass in the wood, and the match generates insufficient heat. Similarly, a solid piece of wood, such as a 2 by 4, that is exposed to a brief electrical arc would probably not ignite because the duration of heat exposure is not sufficient.

In the initial stages of a fire, an upholstered chair does not ignite from the radiant heat of a fire in a wastebasket that is several feet away. The fire is too small and the distance too great to raise the temperature of the upholstery sufficiently to generate ignitible vapors; however, if the same wastebasket fire is moved to within a foot of the chair, the chair may easily ignite.

Most fuels need to liberate fuel gases or vapors for ignition to occur. When the gases and vapors are released into the atmosphere, they must mix with air to produce an ignitible mixture. Some fuels, specifically liquids, have flash points that produce ignitible vapors at ambient temperatures, releasing vapors into the atmosphere that will ignite if the vapors are heated to their ignition temperature and an ignition source is present.

To heat a fuel item to its ignition temperature, the rate of heat transfer must overcome the rate of heat loss due to convection, conduction, and radiation, as well as the loss of energy as a result of pyrolysis or vaporization. This means that the source of heat energy needs to be greater than the ignition temperature of the fuel item. An exception to this is instances where spontaneous ignition occurs. A list of selected ignition temperatures is shown in TABLE 3-3.

Some materials such as cigarettes, cellulose insulation, and sawdust have higher surface-to-air ratios and will allow for a smoldering fire or solid-phase combustion. In smoldering fires, flames are absent; however, more products of combustion are produced per unit of mass burned. Although these materials may not produce flames when they burn, they can be a competent heat source to other fuels that will produce flaming combustion.

Ignition of Flammable Gases

For a flammable gas to be ignited, it must be present in a concentration that will allow for a piloted ignition from either a

© Jones & Bartlett Learning. Photographed by Glen E. Ellman

© Jones & Bartlett Learning. Photographed by Glen E. Ellman

© Jones & Bartlett Learning. Photographed by Glen E. Ellman

FIGURE 3-4 The effect of location in a compartment on flame height. **A.** A couch cushion burning in the center of a room. **B.** A couch cushion burning against a wall. **C.** A couch cushion burning in a corner.

Table 3-3 Reported Burning and Sparking Temperatures of Selected Ignition Sources (NFPA 921 Table 5.7.1.1)

	Temperature	
Source	**°C**	**°F**
Flames		
Benzene[a]	920	1690
Gasoline[a]	1026	1879
JP-4[b]	927	1700
Kerosene[a]	990	1814
Methanol[a]	1200	2190
Wood[c]	1027	1880
Embers[d]		
Cigarette (puffing)	830–910	1520–1670
Cigarette (free burn)	500–700	930–1300
Mechanical sparks[e]		
Steel tool	1400	2550
Copper-nickel alloy	300	570

[a]From Drysdale, D. (1999). *An Introduction to Fire Dynamics*. Wiley.
[b]From Hagglund, B., and Persson, L. E. (1976). *Heat Radiation From Petroleum Fires*, National Defence Research Inst., Stockholm, Sweden, FOA Report C20126-D6(A3).
[c]From Hagglund, B., Persson, L. E. (1974). *Experimental Study of the Radiation From Wood Flames*, National Defence Research Inst., Stockholm, Sweden, FOA Report C4589-D6(A3).
[d]From Krasny, J. (1987). *Cigarette Ignition of Soft Furnishings—A Literature Review with Commentary*, NBSIR 87-3509, National Bureau of Standards, Gaithersburg, MD.
[e]From NFPA *Fire Protection Handbook*, 15th ed., Section 4, p. 167.
Reproduced with permission from NFPA 921-2014, *Guide for Fire and Explosion Investigations*, Copyright© 2014, National Fire Protection Association. This reprinted material is not the complete and official position of the NFPA on the referenced subject, which is represented only by the standard in its entirety.

spark or a flame. This is referred to as the *flammable range* of a gas and will have both a lower and an upper limit. Below the lower flammability limit, the mixture is too "lean" to burn (not enough fuel); above the upper flammability limit, the mixture is too "rich" to burn (not enough air). The terms *flammability limit* and *explosive limit* are used interchangeably. Reference may be made to LFL and UFL, or more commonly LEL and UEL.

Flammable gases may also ignite without a piloted ignition source if the gas–air mixture is heated to a specific temperature. This temperature is referred to as the autoignition temperature (AIT). The autoignition temperature is the lowest temperature at which a gas–air mixture will ignite in the absence of an ignition source. Factors such as the size and geometry of gas volume and concentration will influence the autoignition of a flammable gas. When the volume of a gas–air mixture increases, the ignition temperature decreases.

Ignition of Liquids

Ignited liquids are one of the fuel sources frequently encountered in fires. The fire investigator needs to have a knowledge base of the characteristics of these types of fuel packages.

Flash Point

The flash point is the lowest temperature at which a liquid produces a flammable vapor. It is measured by determining the lowest temperature at which a liquid will produce enough vapor to support a small flame for a short period of time if a piloted ignition source is present. To sustain burning after the removal of an ignition source, the liquid must be heated to what is known as its fire point. The fire point is usually only a few degrees higher than the flash point, and in some instances may be the same. In most of these cases, this is a distinction without a difference.

When a large pool of liquid is present, ignition may occur only to the portion of the liquid that is heated to its fire point. Flaming combustion may spread to the remainder of the pool after ignition or produce a wick effect in which the fuel is drawn toward the fire.

When liquids are dispersed in an atomized mist, the surface-to-mass ratio will increase and can allow for piloted ignition below published fire points of a bulk liquid. Liquids such as hydraulic fluids have been shown to be ignitible when in the form of a spray, despite having a very high flash point.

Another potential source of ignition occurs when the liquid itself begins to oxidize, creating an exothermic reaction. This may be the result of certain liquids like linseed oil being suspended in porous material, such as cloth rags.

Ignition of Solids

Solids may be ignited by either a smoldering ignition, piloted flaming ignition, or flaming autoignition.

Smoldering Ignition and Initiation of Solid-Phase Burning

Most cellulosic products, including wood and paper, must be pyrolyzed first to produce char. This char will then allow for smoldering or solid-phase burning to consume the remaining material. Some materials, such as carbon and magnesium, are capable of solid-phase burning without being reduced or converted to char.

Certain materials, such as thermoplastics, do not produce a char when burned and are not capable of solid-phase burning. Similarly, materials that change to a liquid when heated will also not support solid-phase burning.

It is important to note that smoldering is a result of an oxidation process and is not caused by an external heat source impinging on the fuel.

Self-Heating and Self-Ignition Some materials are capable of initiating a chemical chain reaction and increasing in temperature just by coming in contact with air. This is termed *self-heating* and usually involves solid fuels.

Even though some materials are capable of self-heating, a condition of thermal runaway needs to occur for self-ignition to occur. In thermal runaway, the heat generated exceeds the amount of heat loss within the material. This will most often occur in the best-insulated regions of the fuel package, mainly the center of the package.

Self-heating may also be the result of materials, mostly organic and certain metals, reacting with oxygen. Metal powders may rapidly oxidize and self-ignite, producing metal oxides as by-products. Additionally, certain chemical reactions such as polymers, resins, and adhesives are known to produce heat while they are forming into solids during the curing process. Spontaneous combustion is found when organic materials that contain fatty acids begin to react with oxygen, generating heat.

Ignition by self-heating of a fuel typically requires a lower minimum surrounding or exposure temperature than the ignition without self-heating. For example, rags soaked in linseed oil can ignite at a surrounding temperature of 68°F, but the pure liquid form of linseed oil has a flash point of 428°F.

Other materials, such as unsaturated molecules that contain double bonds of carbon to carbon, may also begin to generate heat. Oils and other saturated hydrocarbons that contain carbon-to-carbon single bonds will self-heat or spontaneously combust only when the temperature is elevated or they are contained in very large piles. These materials, such as hydraulic and motor vehicle oils, generally do not produce sufficient heat energy to self-ignite.

Materials that can self-heat to the point of ignition must be porous and permeable and must allow for the material to be oxidized. These materials must also be capable of producing a char without melting.

As with any fire, a self-heating material will require sufficient air (oxygen) to support the combustion process. For example, placing rags soaked in linseed oil into a sealed container limits the air supply and removes the risk of an accidental fire.

Other factors that should be considered when evaluating a material's ability to self-heat include the size and shape of the fuel package and the environmental conditions present. Fuels with smaller surface areas are more likely to self-heat than fuels with larger surface areas. This is due to the ability of a smaller surface area fuel to insulate itself and increase the interior temperature of the fuel. Objects that are round or square will sustain self-heating more efficiently than large, flat surfaces of similar mass. FIGURE 3-5 illustrates the critical conditions that must exist in order for spontaneous ignition to occur. The five numbered points of the "star" in the center of the figure show situations in which just one condition is not met and ignition does not occur.

The typical situations in which an investigator is likely to encounter spontaneous heating fires involve commercial clothes dryers with oily laundry, painting operations, and piles of vegetable materials such as cotton bales or haystacks. Wood, like other cellulosic material, is subject to self-heating when exposed to elevated temperatures that are below its normal ignition temperature of 482°F (250°C). The scientific community has not reached a consensus on the self-ignition of wood. Charcoal briquettes have been suspected of self-heating to ignition; however, laboratory testing has shown that the common size of briquettes has not reached sufficient temperatures to self-ignite.

Certain elements may spontaneously ignite when exposed to air. White phosphorus, sodium, potassium, and some finely divided metals such as zirconium may react in this manner. Such elements are known as pyrophoric materials.

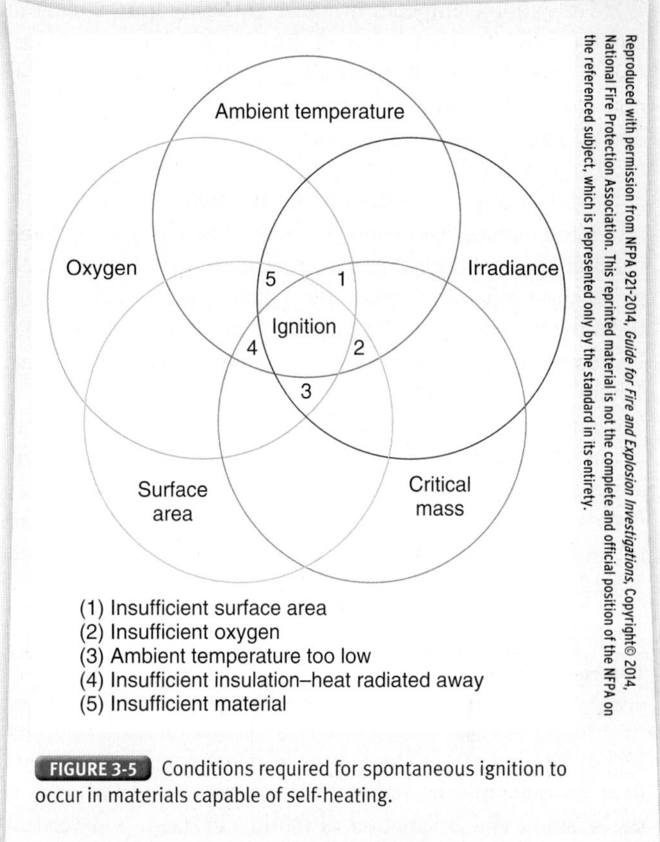

Reproduced with permission from NFPA 921-2014, *Guide for Fire and Explosion Investigations*, Copyright© 2014, National Fire Protection Association. This reprinted material is not the complete and official position of the NFPA on the referenced subject, which is represented only by the standard in its entirety.

FIGURE 3-5 Conditions required for spontaneous ignition to occur in materials capable of self-heating.

Oxidizer Fires A material that is not combustible but can increase the rate of combustion or produce spontaneous combustion when combined with other substances is considered to be an oxidizing agent. Commonly found oxidizers include chlorine tablets for swimming pools and certain fertilizers that may begin to self-heat when combined with organic materials such as hydraulic fluids.

Transition to Flaming Combustion When sufficient heat and vapors are created, smoldering materials may produce flaming combustion when a piloted ignition source is introduced. The amount of time required for this transition is difficult to predict and depends on numerous factors.

When the ignition source for flaming combustion is a smoldering material such as a cigarette, a long period of time may lapse before the first flames are observed. This smoldering heat can also begin heating the fuel and accelerating the initiation of flaming combustion. In other words, a "slow smoldering fire" may burn quite rapidly once the transition occurs. Therefore, it is not valid to eliminate a smoldering ignition source as an option because the fire was perceived to burn rapidly after its discovery.

Piloted Flaming Ignition of Solid Fuels

To create flaming combustion of a solid fuel item, heat must be applied to the fuel to generate flammable vapors through either pyrolysis or vaporization. These vapors are then ignited by a piloted ignition source such as an arc, spark, or flame once their ignition temperature is reached.

The ignition temperature is an engineering approximation, but in general, the temperature of ignition of solids ranges from 518°F to 842°F (270°C to 466°C). Ignition temperatures for non–fire-retarded plastic range from 518°F to 680°F (270°C to 360°C), and wood-based products range from 626°F to 707°F (330°C to 375°C).

Additionally, unlike smoldering fires (which require only a minimum radiant flux of approximately 7 to 8 kW/m^2), piloted ignition requires critical radiant heat flux near 10 to 15 kW/m^2 to sustain flaming combustion. Experiments to determine the minimum heat flux depend on the duration of the testing. Piloted ignition has been observed in some cases after about 1 hour of radiant heating.

Finally, the thickness of the material will also play a role in the ignition of the fuel. Items that are thicker are generally more difficult to ignite than thinner ones. Further, thinner fuels such as papers and thin wood pieces or shavings allow for the heating of more than one side, reducing the time required to ignite the fuel item.

Flame Spread

As a fire grows, flames begin to move across the surface of the fuel at varying rates. These rates of flame spread are dependent on not only the individual fuel properties, but also on the position and orientation of the fuel surfaces. On vertical surfaces, concurrent flame spread is upward, and counterflow flame spread is downward.

On liquids, flame spread is dependent on the temperature of the liquid in relation to its flash point. If the liquid is below its flash point, flames spread across the surface through liquid flow. When the liquid is above the flash point, flame spread is the result of the ignition of gases.

For flames to move across the surface of a solid fuel, the material must be heated to produce fuel gases that ignite and facilitate the flame spread similar to that of the ignition process. Rates of flame spread depend on both concurrent and counterflow flame spread as well as the particular properties of the fuel (thickness, thermal inertia). The counterflow flame spread on thin fuels occurs for downward flame spread. The flame attaches from both sides, with the burning region being fairly short. With thick fuels, the counterflow flame spread normally is downward on a wall or horizontal on an upward-facing horizontal surface. Concurrent flame spread on thin fuels occurs from upward flames and often attaches itself to both sides of the fuel. With thick fuels, the concurrent flame spread is normally for upward flame spread on walls and the underside of combustible surfaces.

Role of Melting and Dripping in Flame Spread

Flame spread may also be the result of melting or dripping materials from a fuel package. When a material melts, it flows with gravity, usually collecting on a horizontal surface below the object. This flow can allow the flames to spread away from the burning fuel, igniting other portions of the burning fuel, or spread to other fuel packages not currently on fire. This may limit the ability of the flame front to spread across the surface of the fuel package. Fire spread may also be accelerated when certain fuels begin to melt and pool. For example, when urethane foams burn, such as a cushion on a recliner, pools of burning liquid are created on the floor underneath the recliner and increase the rate of flame spread.

Role of External Heating on Flame Spread

The rate of flame spread of a given fuel package may also be accelerated when radiant heat produced by other burning objects, as well as radiant heat from the upper gas layers of a compartment fire, impact the given fuel package.

Compartment Fire Spread

The way fire interacts with its surroundings in a compartment fire is different than the way fire in the open acts. Understanding what happens in a confined fire is critical to being able to understand its aftermath.

Whereas flame spread involves the movement of fire across the surface of a fuel item, fire spread refers to the ignition of other fuel items and packages that may be present or located nearby. Fire spread may be a result of either direct flame contact or remote ignition. Fire may spread to a nearby fuel package or to combustible wall and ceiling surfaces through direct flame contact.

Remote ignition can occur through the three modes of heat transfer (conduction, convection, and radiation) discussed earlier.

During a compartment fire, super-heated gases and smoke rise and are confined by the ceiling. As the temperature at ceiling level rises, the radiant heat created begins to accelerate the rate of fire spread as well as the rate of heat release of the burning fuel items.

As the flames or fire plume begin to grow and impinge on the ceiling, the plume begins to spread across the ceiling in all directions as a ceiling jet. This jet continues to spread laterally until it is confined by vertical surfaces such as a wall or partition.

A heated layer of gas and smoke is then created that begins to bank downward into the room. This heat then radiates back on the burning fuel package as well as other "target" fuels located within the compartment FIGURE 3-6.

As the fire continues to burn, hot gases and energy continue to be added to the gas layer, which begins to descend until it either exits the room through an opening or reaches the base of the fire.

The gas layer stops descending when an opening such as a window or door allows for the flow of gases out of the compartment at a rate equal to the hot gases being produced. Unless the fire is contained or extinguished, the hot gas layer continues to increase in temperature and to radiate heat back to other target fuels, ultimately igniting them. A distinct pattern indicating the flow of the hot gases and smoke is created by the gases exiting the room at the top of the doorway, with cooler air entraining to the base of the fire below.

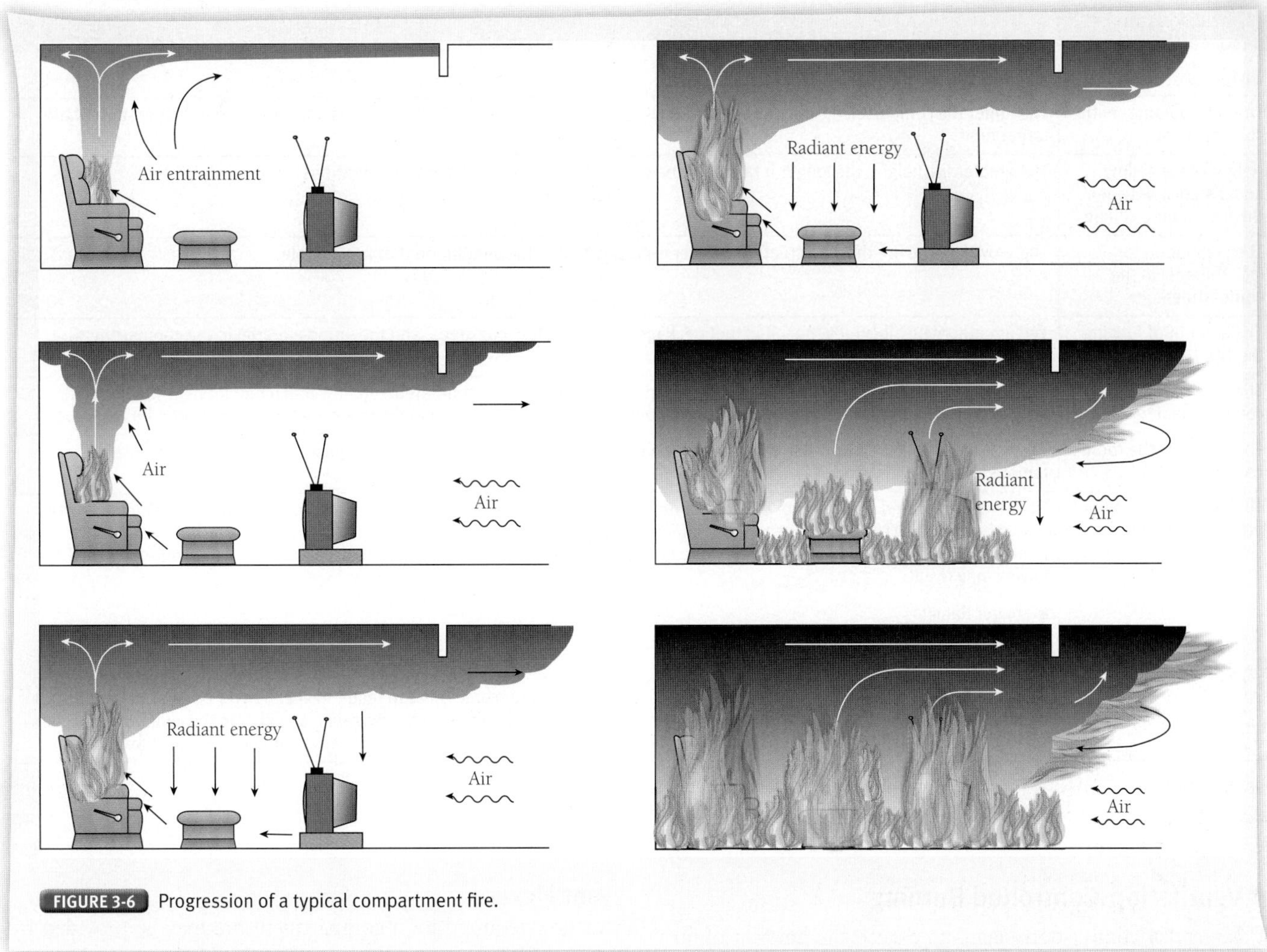

FIGURE 3-6 Progression of a typical compartment fire.

Fuel-Controlled Fire

During the ignition and growth phases of a compartment fire, there is a sufficient amount of air available to allow the fire to continue to burn. When the size of the fire is controlled by how much fuel is burning, this is referred to as fuel-controlled fire.

Flashover

Once floor-to-ceiling flames are produced, the fire begins to grow rapidly. The ceiling layer temperature continues to rise, reaching approximately 1100°F (600°C). A heat flux of approximately 20 kW/m^2 on the floor level is generally considered sufficient energy to bring the transition of the fire. At this point, convected and radiated heat energy impinges on the other combustible items within the room, producing fire gases. These items then ignite at nearly the same time, causing full-room involvement. This event is known as a flashover. Flashover is a transition. The fire progresses from being "a fire in a room" to "a room on fire." After flashover, the room is said to be fully involved. The fire also progresses from fuel-controlled burning to ventilation-controlled burning.

Not all compartment fires can evolve to the flashover stage. Rooms that lack sufficient ventilation or very large compartments such as a warehouse may not reach full involvement. The minimum size of a fuel package that can cause flashover in a given room is determined by the size of the ventilation provided. The time to flashover in residential room fires may be as little as 3 to 5 minutes, or even shorter with nonaccelerated fires. These times will be greatly affected by the HRR of the fuel, fuel package placement, and room geometry.

Using the equation

$$HRR_{fo}(kW) = (750A_o)(h_o)^{0.5}$$

Where HRR_{fo} = HRR for flashover (kW)

A_o = area of opening (m^2)

h_o = height of opening (m)

An approximation of the HRR required for flashover for a compartment with a single vent opening can be generated. In a compartment with a single vent opening the size of a standard door passage (3 feet wide by 7 feet high), a minimum fuel package of about 2000 kW would be required to bring the fire to flashover.

Table 3-4 Factors That Influence Flashover

Factor	Effect
The size (volume) of the compartment	The larger the compartment, the more time required to fill the upper layer with hot gases and to raise the temperature of the target fuels.
Height of the ceiling and distance between the fire and the ceiling	The greater the height, the longer it takes for the compartment to reach flashover conditions.
The ventilation (or lack thereof) in the compartment	The growth of the fire that can create flashover is closely tied to the ventilation that is available.
The amount of fuel in the compartment	The amount of fuel includes the initial fuel package, the target fuel packages, and the lining material in the compartment.
The type of fuel in the compartment	The type of fuel determines the rate at which combustible fuels can be generated and how much radiant heat will be required to raise the temperature of the target fuels to a point where sufficient combustible gases are being released to be ignited.
The layout of the room's contents	Target fuels located closer to the initial fire are heated sooner and release combustible gases sooner than those that are remote from the initial fire.
The location of the fire in the compartment	Fires that are located in the center of a compartment have more air entering the plume and cooling it by entrainment, shorter plume heights, and no restrictions to the geometry of the plume. If the same fire is located against a wall, there is 50 percent less entrainment because of the wall barriers. Therefore, the following may result: ■ Longer flames ■ A faster rise in upper-layer temperatures ■ Less time to flashover If the fire is located in a corner, the entrainment is reduced by 75 percent. This can result in even longer flames, a faster rise in upper-layer temperatures, and a reduced time to flashover than if the fire is located against a wall or in the center of the room. The location of the fire in relation to vertical barriers is important in interpreting the fire development and spread.

■ Ventilation-Controlled Burning

If the rate of combustion begins to exceed the amount of air flow into the compartment, the fire transitions from fuel controlled to ventilation controlled. The hot gas layer begins to darken because of the unburned pyrolysis products as well as an increase in the amount of carbon monoxide.

Unless the fire is extinguished or sufficient heat and smoke is removed from the room, the gas layer continues to extend to the floor, resulting in a flashover and full room involvement. It is important to note that the gas layer may not extend completely to the floor prior to full room involvement. Likewise, a flashover may not necessarily proceed to full room involvement.

One of the major determining factors in whether flashover will occur is the presence of a hot gas layer with sufficient energy to radiate downward to involve exposed fuel packages. It is because of this that fire fighters often ventilate the roof of a structure to allow the hot gas layer to escape before flashover occurs. TABLE 3-4 lists several of the variables that determine whether and how fast flashover will occur.

It is important for the investigator to be able to recognize which rooms in a structure have progressed to full room involvement. Full room involvement can result in the production of low patterns, irregular patterns, and holes in the floor, as is discussed in the "Fire Patterns" chapter in this textbook. Although these patterns might appear suspicious in a compartment that did not become fully involved, in a fully involved compartment, such patterns are common.

Vent Flows

Air flow required for a compartment fire may be provided by mechanical means such as an HVAC or more commonly by natural openings. The hot gas layer in the room, being more buoyant, will flow out the upper regions of the vent opening, and air will flow in the lower region of the vent opening. The line where the flow of the two layers changes is the neutral plane FIGURE 3-7. As the hot gases accumulate within the room, the height of the neutral plane moves downward. A fire in a compartment will have only one neutral plane even if there are multiple ventilation openings. In fires that have multiple vents, it is possible for vents to serve only as a vent for the

Fire Investigator Tip

Investigators should be cautious of statements made by witnesses as they relate to the rate of fire growth. Many times, witnesses describe the rate of fire growth from the time they discover the fire, which may be difficult to compare with the time of ignition. Factors such as the compartment size, fuel configuration, ventilation, and fuel load present all influence the rate of growth. For this reason, rapid fire growth in itself is not a reliable indicator of an incendiary fire.

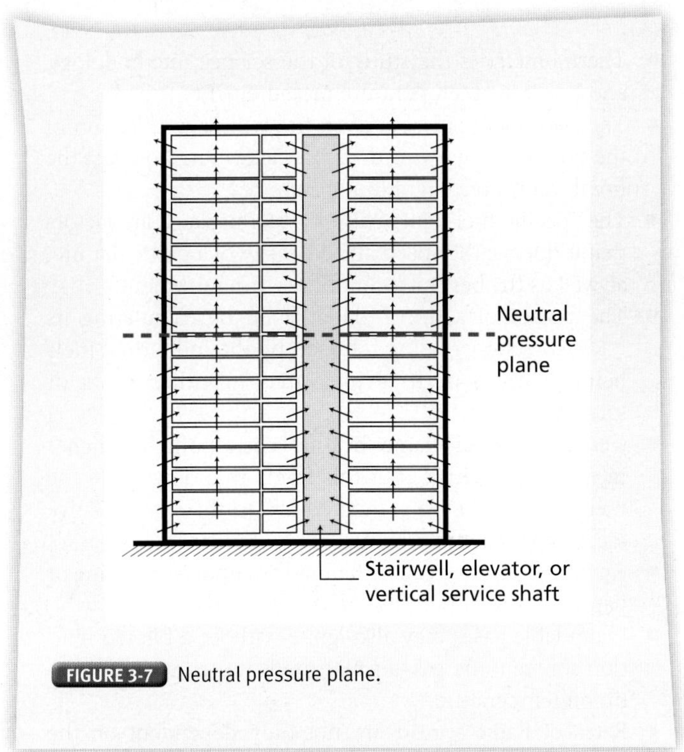

FIGURE 3-7 Neutral pressure plane.

hot gases or as an inflow, depending on the level of the vent to the neutral plane.

With a single ventilation opening, there is a proportional relationship of the inflow of air to the ventilation factor, which is the area of the opening (A_0) multiplied by the square root of the height of the opening (H_0). This ventilation factor is an integral component in evaluating the peak HRR of a fire in a compartment. For example, a standard door opening (3 feet wide by 7 feet high) can support an HRR of about 4000 kW, but if the lower half of the opening is blocked, the opening can support an HRR of only about 1600 kW.

As the fire progresses, the ventilation may change. Windows may fail as the glass cracks out and additional vents are created as the fire may destroy doors or create openings. As these new vents appear, the neutral plane height will change. Openings that were at first venting may transition to inflow.

Wrap-Up

Ready for Review

- Every fire investigator must fully understand the fundamental properties of fire and fire development to determine accurately the origin and cause of any fire or explosion event he or she may investigate.
- For a fire to occur, four components must be present: fuel, oxidizing agent, heat, and an uninhibited chemical chain reaction. This is referred to as the fire tetrahedron.
- The most common fuels an investigator will encounter are organic fuels, which contain carbon (e.g., wood, plastics, and petroleum products).
- Most fires require an oxidizing agent, like oxygen from the air, to support the combustion process.
- The rate at which heat is released, known as the heat release rate (HRR), is measured in joules per second or watts.
- An uninhibited chemical chain reaction provides a self-sustaining event that continues to develop fuel vapors and sustain flames even after the removal of the ignition source.
- During a fire, materials may change their physical state as a result of being heated. These phase changes include melting and vaporization.
- Flames produced during the combustion process can be categorized as either premixed or diffused.
- If a fire has a limited amount of air for combustion, an increase in the amount of visible products of combustion, such as soot, smoke, and carbon monoxide, will occur.
- The color of smoke should not be relied on as an indicator of the material burning.
- Fluid flows, heat transfer, ignition and flame spread, fuel packages, heat flux, and the distinction between fuel-controlled fires and ventilation-controlled fires are all components of fire dynamics.
- Fluids include both liquids and gases, but in fires, we are generally focused on gas flows.
- Heat energy naturally moves from areas of higher temperature to lower temperature through heat transfer, measured as energy flow per unit of time.
- Thermometry is the study of the science, methodology, and practice of temperature measurement.
- The total fuel load of a given fire will be a reflection of the potential energy; it does not determine how fast the fire develops once ignition occurs.
- The specific fuel items within a fuel package are factors that influence the speed and intensity of a particular fire, as well as the best method of fire extinguishment.
- The power of a fire is determined by calculating its HRR, the energy being released by the individual fuels being consumed. It is measured in either watts or kilowatts.
- Generally, larger flame heights correspond to higher rates of heat release, but this holds true only when the fuel packages being compared are in the same relative location in a compartment.
- An ignition source can be defined as either smoldering or flaming, or as either piloted or autoignition.
- Flammable gases may also ignite without a piloted ignition source if the gas–air mixture is heated to its autoignition temperature.
- Rates of flame spread are not only dependent on the individual fuel properties, but also the position and orientation of the fuel surfaces.
- When a material melts, it flows with gravity, usually collecting on a horizontal surface below the object and allowing the flames to spread away from the burning fuel to other fuel packages.
- The rate of flame spread may be accelerated through radiant heat produced by other burning objects and from the upper gas layers of a compartment fire.
- Whereas flame spread involves the movement of fire across the surface of a fuel item, fire spread refers to the ignition of other fuel items and packages that may be present or located nearby.
- A fuel-controlled fire is when the size of the fire is controlled by how much fuel is burning.
- After flashover, the room is said to be fully involved and progresses from fuel-controlled burning to ventilation-controlled burning.
- One of the major determining factors in whether flashover will occur is the presence of a hot gas layer with sufficient energy to radiate downward to involve exposed fuel packages.

© Greg Henry/ShutterStock, Inc.

© Greg Henry/ShutterStock, Inc.

Hot Terms

Air entrainment The process of air or gases being drawn into a fire, plume, or jet.

Autoignition temperature (AIT) The lowest temperature at which a combustible material ignites in air without a spark or flame.

Ceiling jet A relatively thin layer of flowing hot gases that develops under a horizontal surface (e.g., ceiling) as a result of plume impingement and the flowing gas being forced to move horizontally.

Combustion A chemical process of oxidation that occurs at a rate fast enough to produce heat and usually light in the form of either a glow or flames.

Conduction Heat transfer to another body or within a body by direct contact.

Convection Heat transfer by circulation within a medium such as a gas or a liquid.

Diffusion flame A flame in which the fuel and air mix or diffuse together at the region of combustion.

Exothermic reaction Reaction characterized by or formed with evolution of heat.

Fire dynamics The detailed study of how chemistry, fire science, and the engineering disciplines of fluid mechanics and heat transfer interact to influence fire behavior.

Fire point The lowest temperature at which a volatile combustible substance continues to burn in air after its vapors have been ignited (as when heating is continued after the flash point has been determined).

Flash point The lowest temperature of a liquid, as determined by specific laboratory tests, at which the liquid gives off vapors at a sufficient rate to support a momentary flame across its surface.

Flashover A transition phase in the development of a compartment fire in which surfaces exposed to thermal radiation reach ignition temperature more or less simultaneously and fire spreads rapidly throughout the space, resulting in full room involvement or total involvement of the compartment or enclosed space.

Fuel A material that will maintain combustion under specified environmental conditions.

Fuel-controlled fire A fire in which the heat release rate and growth rate are controlled by the characteristics of the fuel, such as quantity and geometry, and in which adequate air for combustion is available.

Fuel items Any articles that are capable of burning.

Fuel load The total quantity of combustible contents of a building, space, or fire area, including interior finish and trim, expressed in heat units or the equivalent weight in wood.

Fuel package A collection or array of fuel items in close proximity with one another such that flames can spread throughout the array.

Heat A form of energy characterized by vibration of molecules that is capable of initiating and supporting chemical changes and changes of state.

Heat capacity The amount of heat necessary to raise the temperature of a unit mass 1 degree, under specified conditions (J/kg-K, Btu/lb-°F).

Heat flux The measurement of the rate of heat transfer to a surface, expressed in kilowatts/m^2, kilojoules/m^2 sec, or Btu/ft^2 sec.

Heat release rate (HRR) The rate at which heat energy is generated by burning.

Heat transfer The transport of heat energy from one point to another caused by a temperature difference between those points.

Neutral plane The line where the flow of the hot gas and cooler air changes.

Oxidizing agent A substance that promotes oxidation during the combustion process.

Phase change The conversion of a material from one state of matter to another that is reversible and does not change the chemical composition of the material.

Plume The column of hot gases, flames, and smoke rising above a fire; also called *convection column*, *thermal updraft*, or *thermal column*.

Premixed burning Burning in which the fuel and oxidizer are mixed prior to combustion, as in a laboratory Bunsen burner or a gas cooking range; propagation of the flame is governed by the interaction between flow rate, transport processes, and chemical reaction.

Pyrolysis Process in which material is decomposed, or broken down, into simpler molecular compounds by the effects of heat alone; pyrolysis often precedes combustion.

Radiation Heat transfer by way of electromagnetic energy.

Stoichiometric ratio The optimum ratio in a fuel air mixture at which point combustion will be most efficient (above the LEL and below the UEL).

Temperature The degree of sensible heat of a body as measured by a thermometer or similar instrument.

Thermal conductivity (k) The measure of the amount of heat that will flow across a unit area with a temperature gradient of 1 degree per unit of length (W/m-K, Btu/hr-ft-°F).

Thermal decomposition An irreversible change in chemical composition as a result of pyrolysis.

Thermal inertia The properties of a material that characterize its rate of surface temperature rise when exposed to heat; related to the product of the material's thermal conductivity (κ), density (ρ), and heat capacity (c).

Thermal runaway Condition in which the heat generated exceeds the amount of heat loss within the material.

Thermometry The study of the science, methodology, and practice of temperature measurement. The study is frequently used in fire safety or code compliance cases.

Uninhibited chemical chain reaction One of the elements of the fire tetrahedron. This element provides for the combination and interaction of the other elements.

Vaporization A phase transition from a liquid or a solid phase to a gas phase.

© Greg Henry/ShutterStock, Inc.

FIRE INVESTIGATOR *in action*

An early morning fire in a restaurant has caused extensive damage to the interior of the building, with significant smoke and soot staining present throughout. Fire patterns inside the building indicate the fire originated in the kitchen area where numerous cooking appliances, stored food products, and equipment were located. The owners of the restaurant state that they had closed the night before noting nothing unusual at that time. When they returned this morning, they observed the interior of the building filled with dark gray smoke, at which time they called the fire department to report the fire.

The investigation determines the fire originated at one of the deep fryers in the kitchen, which was filled with cooking oil. You continue to interview the owners, who state that they turn down the temperature setting of each fryer each night before closing but leave them on so they will be ready to use after only a couple minutes of heating the next day.

1. The lowest temperature at which a liquid produces a flammable vapor is defined as the:
   - **A.** flash point.
   - **B.** autoignition temperature.
   - **C.** fire point.
   - **D.** flame point.
2. Which of the following factors would have influenced this fire's development?
   - **A.** The physical state of the fuel being consumed
   - **B.** The heat release rate of the fuel
   - **C.** Air entrainment to the fire
   - **D.** All of the above
3. You determine that gas-fired burners are used to heat the cooking oil, which is located in a metal tub above each burner. What type of heat transfer is occurring?
   - **A.** Convection
   - **B.** Conduction
   - **C.** Radiation
   - **D.** Direct flame contact
4. Thermometry is:
   - **A.** the study of flame temperatures and heat release rates.
   - **B.** the study of the science, methodology, and practice of temperature measurement.
   - **C.** the study of heat transfer from one fuel to another.
   - **D.** the study of fire dynamics as it relates to fire growth.

Fire Patterns

© Photos.com

© HAYKIRDI/Getty Images

Knowledge Objectives

After studying this chapter, you should be able to:

- Identify and describe fire effects NFPA 4.2.4 NFPA 4.2.5. (pp 42–48)
- Identify and describe fire patterns NFPA 4.2.4 NFPA 4.2.5. (pp 48–51)
- Explain the significance of fire pattern location NFPA 4.2.4 NFPA 4.2.5. (pp 51–52)
- Describe the potential causes of, and misconceptions about, irregular fire patterns NFPA 4.2.4 NFPA 4.2.5. (p 53)
- Describe the geometry of fire patterns NFPA 4.2.4 NFPA 4.2.5. (pp 53–54)
- Describe how to analyze fire patterns to produce a hypothesis NFPA 4.2.4 NFPA 4.2.5. (pp 54–55)

Skills Objectives

After studying this chapter, you should be able to:

- Interpret fire patterns to determine the point of origin NFPA 4.2.4 NFPA 4.2.5. (pp 42–54)
- Analyze fire patterns to produce a hypothesis NFPA 4.2.4 NFPA 4.2.5. (pp 54–55)

Additional NFPA Reference

NFPA 921, *Guide for Fire and Explosion Investigations*

CHAPTER 4

FESHE Course Outcomes

Fire Investigation I

There are no Fire Investigation I (FESHE) course outcomes for this chapter.

Fire Investigation II

2. Interpret a fire scene. (pp 42–55)

You Are the Fire Investigator

© Jones and Bartlett Publishers. Photographed by Glen E. Ellman

As you begin to examine and document the various fire patterns within the room of origin, you note varying levels of damage to structural components and contents of the room. A large portion of the gypsum wallboard calcified and began to fail along the south half of the room, with significant charring noted to the wooden roof trusses near the window in the east wall. The majority of the furnishings and personal items located within the room are constructed of metal, plastic, and wood components and display varying degrees of damage.

1. What is the significance of the calcified wallboard?
2. How could the damaged items within the room assist with identifying the point of origin?
3. Where may you find additional fire patterns and effects within the room?
4. Why is the presence of a window in the south wall significant to you?

Introduction

Chapter 6 of NFPA 921, *Guide for Fire and Explosion Investigations*, provides information relating to the recognition, identification, and analysis of fire effects and fire patterns. The discussion of fire effects and patterns logically follows the discussion of basic fire science in the previous chapter because it is impossible to understand fire patterns without a solid knowledge of fire dynamics.

Fire Effects and Fire Patterns

The collection of fire scene data requires the recognition and identification of fire effects and fire patterns. Fire effects are the observable or measurable changes in or on a material as a result of exposure to the fire. A fire pattern is the visible or measurable physical changes or identifiable shapes formed by a fire effect or group of fire effects.

In the majority of cases, it does not take a fire investigator to determine that a fire has occurred. Even a layperson can easily identify that there has been a fire, just by interpreting visual clues. For instance, people easily identify whether the wood in a fireplace is burned or pristine or whether a discarded cigarette has been smoked. The predominant visual indicator is the presence of burned material.

The role of the fire investigator, however, is more complex than just determining whether a fire has occurred. The fire investigator attempts to recreate the fire development history and then backtracks to identify the origin and the cause of the fire. The primary physical evidence available to the investigator consists of the materials subjected to the fire and the by-products of burning (such as heat, smoke, and soot). It is the understanding of the materials' response to the fire, coupled with knowledge of fire dynamics, that enables the investigator to develop and support a hypothesis of a fire's origin and cause.

The interpretation of fire patterns has traditionally been one of the primary processes used in fire investigation. Knowledge about the meaning of the patterns was typically gained through experience and training. In the evolution of fire investigation, there is now greater emphasis on the verification of the fire dynamics that produced the fire patterns. Many traditional interpretations of the meaning of fire patterns and fire effects have been challenged and found to be incorrect. For example, it was once believed that crazed glass, spalled concrete, and wide-base V patterns were a direct result of the application of an ignitible liquid. These concepts had never been tested under controlled conditions. Crazed glass was found to be created from rapid cooling of heated glass. Other conditions, such as spalled concrete, can be created by conditions such as material expansion or release of moisture from within, in addition to ignitible liquids. The investigator should maintain current knowledge of the fire science literature and the new research pertaining to fire patterns currently being published.

Fire Effects

A fire investigator must have a strong understanding of the effects of fire to identify and accurately interpret fire patterns created. Fire effects include the loss of material, charring, spalling, oxidation, color changes, melting, annealing, alloying, crazing, soot and smoke deposition and removal, calcination, staining, bending, breaking, and distorting.

Melting, distortion, and other fire effects sustained to materials may be very useful to an investigator estimating the temperature at a particular location. Temperatures in most structure fires seldom remain above 1900°F for an extended period. Knowing the melting temperatures of common materials in a fire, the investigator may better understand the minimum temperatures that were present in a given area of the fire. This information may allow the investigator to establish the intensity and duration of combustion at this location. Additionally, the investigator may also be able establish the potential heat release rate (HRR) of the burning fuel items. Keep in mind, however, that many materials have a broad range of responses to increased temperatures and that one type of metal or glass may respond at a considerably different temperature than another type of metal or glass.

It is important to remember that the temperatures developed at a particular location do not necessarily represent the type of fuel that burned there. For example, wood and gasoline burn at essentially the same flame temperature. The turbulent diffusion flame temperatures of all hydrocarbon fuels (plastics and ignitible liquids) and cellulosic fuels are approximately the same, although the fuels release heat at different rates. Accelerated fires burn faster than nonaccelerated fires do, but they do not burn at higher temperatures.

Mass Loss

Although the duration and intensity of a fire can sometimes be determined by the amount of mass loss a fuel item sustains, there are many additional factors—such as the type of fuel being consumed and the effects of ventilation—that must be considered.

As a fuel item is consumed during the combustion process, the mass of the object begins to decrease. Postfire analysis of the damaged areas can be conducted either by using exemplar items or by examining undamaged portions of the object. Additionally, plans, drawings, diagrams, photographs, or even interviews can also be used to determine an object's prefire shape and orientation.

The rate of mass loss also changes during a fire due to various factors, including heat flux, fire growth rate, and the object's heat release rate. Put simply, if a fire continues to intensify and grow, the rate of mass loss also increases.

Char

Char is carbonaceous material that has been burned or pyrolyzed and has a blackened appearance. The most commonly found fuel item that chars is wood, which is present at most structure fires. As the wood is pyrolyzed, it begins to break down, liberating hot gases, water vapor, and other products of combustion. The remaining solid material consists largely of carbon, forming cracks and blisters on the surface of the fuel item.

For many years, fire investigators misinterpreted the appearance of char. It had been suggested in the past that large silver or shiny char blisters (alligator) were a positive indicator of the use of an ignitible liquid, despite the fact that this type of damage can be produced by various types of fires. Numerous test fires and experiments related to this type of charring have found no scientific evidence to support this claim FIGURE 4-1.

The rate at which wood chars varies widely throughout the course of a fire. There is no standard rate of char such as 1 inch every 45 minutes or 3/4 inch per hour. Many factors affect the rate at which wood may char. Such conditions include the following:

- Rate and duration of heating
- Ventilation effects
- Surface-to-mass ratio
- Direction, orientation, and size of wood grain
- Species of wood (e.g., maple vs. yellow pine)
- Wood density
- Moisture content of the wood product
- Any surface coating on the wood

© Vladimir Zanadvorov/ShutterStock, Inc.

FIGURE 4-1 Variability of char blisters on wood boards exposed to the same fire.

- Oxygen density of any hot gases
- Velocity of hot gases
- Edge effects of material

Once the wood has dried, the age of the wood has no bearing on its char and burn rates. It is not possible to make a determination of the duration of a fire based on the depth of char alone; however, depth of char can be used to assess fire movement and the relative rates of heat release to which wood surfaces were exposed. Depth of char to like wood surfaces may also be useful in the determination of a concentrated fuel area, such as one might find when a fuel gas jet is directed on the surface from a leakage point.

Investigators must consider the many variables involved in the fire before using the char to determine which fuels were involved in the fire.

Spalling

Spalling is the chipping or pitting of concrete or masonry surfaces FIGURE 4-2. When concrete, brick, and other masonry items are heated, they can begin to expand. Different areas expand at different rates, causing pitting and cracks on the surface of the material and reducing the

FIGURE 4-2 Spalling on the floor of a garage.

surface tensile strength of the material. Spalled areas may often have a different coloration; the fractured areas may appear lighter in color because of the exposure of the clean subsurface. During a fire, rapid heating can cause moisture in the concrete to heat and expand.

The investigator must be cautious not to interpret either the presence or lack of spalling as an indicator that an ignitible liquid has been applied. The lower boiling points of ignitible liquids tend not to cause spalling directly below the surface of the liquid; however, spalling may be present to adjacent surfaces as the fire rapidly develops and intensifies.

Spalling sometimes occurs on ceilings and walls. Especially with ceilings, care must be taken in interpreting the meaning of the spalling. Many times concrete will spall in areas of high mechanical stresses, so spalling on the ceiling does not necessarily indicate an area of origin.

It is also possible for spalling to be caused by factors other than heat. The investigator should, therefore, make certain that the spalling did not exist prior to the fire.

Oxidation

Oxidation is the basic chemical process associated with combustion. Oxidation of noncombustible materials can produce color and texture changes and fire patterns that could be useful for the fire investigator. A material can display more pronounced evidence of oxidation due to higher temperatures and longer periods of exposure. Postfire oxidation is affected by both the ambient humidity and the length of time the material is exposed to this environment.

Items made of metal also display oxidation as a result of fire exposure. Galvanized steel surfaces may appear dull white from oxidation. Often the fire will affect the corrosion protection, and the surface may rust if exposed to moisture. Steel may appear blue-gray in color. With elevated temperatures the oxidation may appear black, and if the oxidation is thick, it may flake off. Moisture may also bring on the expected rust-colored appearance. Stainless steel may appear dull gray with severe oxidation and display a coloration with mild exposure. Steels that have been heavily oxidized display damage consistent with melting and may require a metallurgic exam to determine whether the object did, in fact, melt. Copper exposed to fire begins to oxidize, creating dark red or black oxide. As the copper object is subjected to higher temperatures and/or longer periods of exposure, the thickness of the oxide increases.

Objects such as rock and soil also change colors as a result of high temperature exposures, turning either yellow or red in color.

Coloration Changes

As an investigator views a fire scene, changes in color of materials will be a source of information. These color changes are not without their limits. All people do not see color with the same perspective. Factors such as lighting conditions, angle of view, and nature of color are all factors to be considered. In addition, coloration changes may occur from nonfire factors such as exposure to sunlight. Fabric dyes may change in appearance related to heat exposure. How a particular dye reacts to heat exposure may vary with the product.

Melting

The melting of a material is a physical change from solid to liquid caused by exposure to heat. Demarcation lines can be produced between the melted and unmelted portions of the material and may be useful fire patterns to the investigator. Each solid material has its own melting temperature. These temperatures can range from just above average room temperature to several thousand degrees. For example, steel has a melting temperature of 2660°F (1460°C), whereas thermoplastics melt at temperatures around 167°F to 750°F (75°C to 399°C). The presence of melted materials can provide insight as to the temperatures that may have been produced during a fire at a particular location FIGURE 4-3. Materials with lower melting points often soften and fail as a result of the heat exposure, whereas metals with higher melting points can provide information as to the flame temperature.

It is sometimes difficult, particularly in the case of steel, to distinguish visually between melting and oxidation. Heavily oxidized steel may assume a bulbous appearance. The only way to demonstrate actual melting of steel is with a metallurgical examination.

The softening of glass items can be useful for the investigator in determining temperatures sustained during a fire. It should be noted, however, that these temperatures can vary.

Alloying of Metals

Alloying, which can look very much like melting, must also be considered by the investigator during postfire examination of damaged metal items. When two metals, one of which is in a liquefied state, come into contact with each other, they may begin to form a new material—an alloy of the two. An

© Stephen Roberts/ShutterStock, Inc.

FIGURE 4-3 Melted materials can provide insight into the temperatures that may have been produced during a fire.

Courtesy of Rodney J. Pevytoe

FIGURE 4-4 Steel I-beams deformed by thermal effects.

example of this is when zinc comes in contact with copper, creating a zinc-copper alloy. The melting temperature of zinc is lower than that of copper; however, the alloying of the two materials now allows the copper to melt at a lower temperature.

Alloying of metals with high melting temperatures does not necessarily indicate the presence of an accelerant or exceptionally high temperatures.

Thermal Expansion and Deformation

When exposed to heat, most materials begin to expand and change shape, either temporarily or permanently. In the case of solid structures, different materials are likely to have different expansion rates. Structural failures can occur as a result of these differing expansion rates.

When steel structures such as columns and beams reach temperatures in excess of approximately 900°F (500°C), they begin to buckle and bend, ultimately failing. As steel is exposed to higher temperatures, the strength and load-carrying capability decrease until deformation of the material occurs. This deformation is the result of the object's inability to support the load placed on it. The load may be the object itself. This should not be confused with melting because the object has not liquefied, merely distorted. The distortion or deformity of an object indicates that the melting temperature was never reached FIGURE 4-4. Bending may also be a result of thermal expansion when, for example, a steel beam is heated but restrained at the ends as when used for the roof supports of a building.

Piping systems, specifically fittings, may deform or change in position. With the expansion and contraction caused by heating followed by cooling, a fitting may appear loose when it was tight in the prefire condition. Sealants used in fitting connections may also be compromised by the fire.

Other materials that may also distort because of thermal expansion are plastered wall and ceiling surfaces. When heated, the plaster may begin to expand and detach from the lath support material. Additionally, drywall spackling and gypsum wallboard may also separate from wall and ceiling surfaces as a result of thermal expansion.

Smoke Deposits

During the combustion process, multiple by-products are released. Smoke contains various particulates, liquid aerosols, and gases. As the hot products of combustion begin to migrate away from the smoke plume, they begin to collect on cooler surfaces in other portions of a structure. These smoke deposits or condensates, especially during smoldering fires, may begin to collect on cooler surfaces such as walls and windows and may be wet and sticky and varying in thickness. The color and texture of smoke deposits are not reliable indicators of the burning or the rate of heat release. A chemical analysis of the smoke deposits is the only method to determine the nature of the fuel. The smoke deposits will not necessarily indicate when the incident occurred and may require additional data.

The nature of soot deposition on certain surfaces of smoke alarms can, in many cases, show whether the smoke alarm sounded during a fire. Enhanced soot deposition, also known as acoustic soot agglomeration, is a phenomenon where soot particulate in smoke forms identifiable patterns on internal and external surfaces of the smoke alarm cover near the edges of the horn and the surfaces of the horn disks themselves.

When a smoke alarm has this or other physical evidence of alarm activation, the investigator should consider more detailed documentation, examination, and collection of the evidence.

Evidence of enhanced soot deposition can be delicate and easily disturbed or wiped away with careless handling or evidence packaging. Care should be taken not to disturb any suspected soot deposits. The evidence may be subtle and difficult to identify, and microscopic examination may be necessary to confirm its presence.

A smoke alarm should be taken into evidence when its performance may be an issue. Prior to collection, it should be photographed in place and should not be altered by applying power, removing or inserting batteries, or pushing the test button.

As a fire continues to grow, it consumes smoke condensates deposited earlier in the fire. Either direct flame contact or radiant heat exposure produces clean areas on noncombustible surfaces that are bordered by areas darkened by the products of combustion and do not exhibit a loss of surface material. This exposure to flame and intense heat begins to oxidize soot deposits, paint, and even the charred paper of wallboard, consuming the carbon present. This is referred to as a clean burn FIGURE 4-5. The demarcation in these areas can guide investigators in determining fire spread, but although clean burns may indicate areas of high heat, they do not necessarily represent the origin of the fire.

Courtesy of Robert Schaal

FIGURE 4-5 Example of a clean burn.

Calcination

Calcination occurs in gypsum wall surfaces when the free and chemically bound water is driven out of the gypsum by the heat of the fire. Gypsum wallboard will react to fire exposure in a predictable manner. The fire will cause a difference in the color and composition of the material, where it may form a line of demarcation. Upon first exposure, the paper surface will darken, char, and then burn off. As the gypsum is exposed to the fire, the organics and other materials will change in color, often becoming whiter in appearance. This condition may, with sufficient exposure, go through the thickness of the material. With exposure of the total thickness of the material, the board becomes crumbly and often will drop off wall and ceiling surfaces, especially when exposed to additional factors during fire suppression. Fire-rated gypsum board will contain additives that strengthen the board even after fire exposure.

The rate of calcination is not a reliable indicator for estimating the time of burning. Likewise, depth of calcination is not a reliable method for determining burn times, although comparing the relative depth of calcinations can be a reliable method for indicating fire spread.

Glass Effects

Glass fragments found to be free of any soot deposits of smoke condensates usually indicate early failure of the glass prior to the accumulation of smoke. This may be the result of rapid heating, damage prior to the fire, or direct flame impingement. Other factors include the location of the heat source relative to the glass and ventilation effects.

Windows coated with thick, oily deposits, including the presence of hydrocarbon products, have in the past been attributed to the use of ignitible liquids. This type of staining may also be the result of incomplete combustion.

Lightbulbs, specifically incandescent, may also be used to determine the direction of heat impingement and fire travel. When sides of lightbulbs 25 W or greater are heated and softened, the gases within the glass envelope push outward toward the heated side. This *pulled* portion of the bulb will "point" toward the direction of the heat source FIGURE 4-6.

In contrast, bulbs of less than 25 W collapse inward on the heated side due to the vacuum placed on the glass envelope. Bulbs that have survived the fire and extinguishment activities may be useful in determining direction of fire travel; however, the investigator must also consider the possibility that the fixture or bulb itself was moved or displaced during or after the fire.

Fractured glass is found in most structure fires. Glass panes that are fractured by an impact object will fracture most often from the point of force in the typical "cobweb" pattern. These fracture lines are straight and numerous. Heat may also fracture glass. Research has shown that a temperature difference of 126°F between the center of a glass pane and the

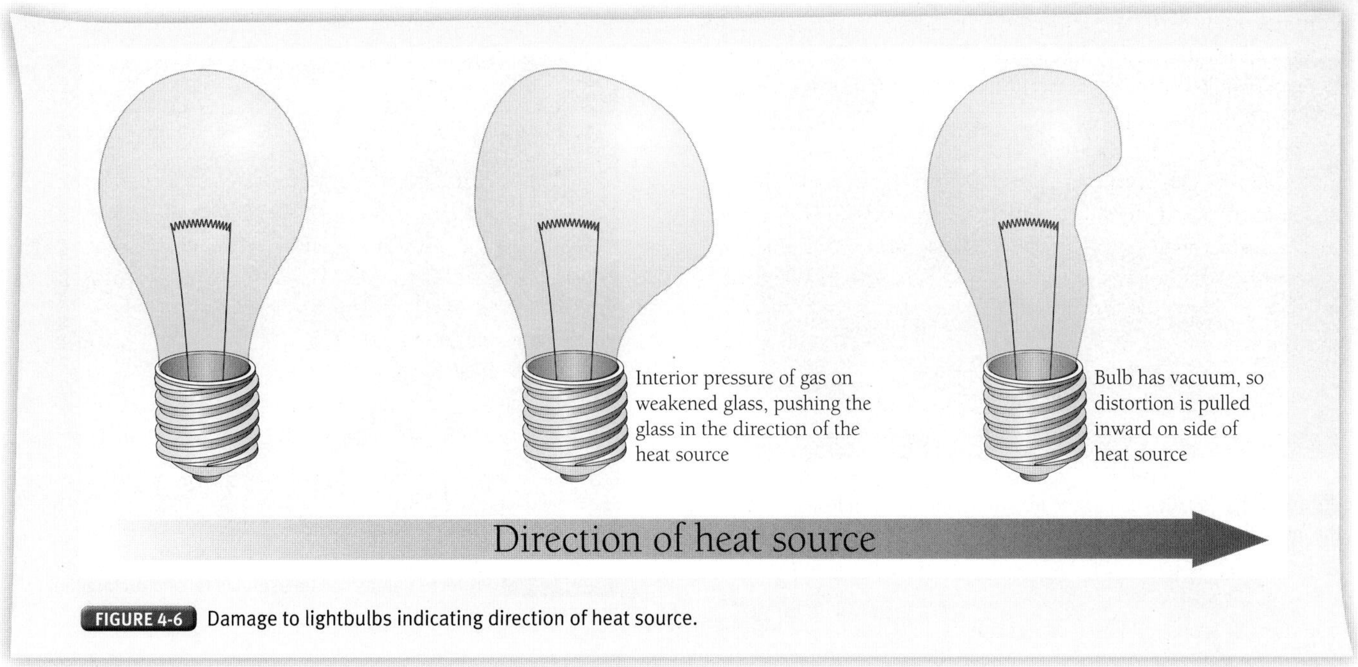

FIGURE 4-6 Damage to lightbulbs indicating direction of heat source.

protected edge of the pane can cause fractures. In glass that has an exposed edge, the glass will break at a higher temperature. The fractures start at the edge of the pane and appear as smooth, undulating lines that may spread and then join together. Glass may also fracture when one side is exposed to flames and the opposite side is cool, causing stress to the glass. Tempered glass, such as is found in commercial buildings and shower and patio doors, will fracture into many small cubes when broken by impact or heat.

With the expansion caused by heat, some windowpanes may pop from their frames. Pressures developed by structure fires are generally insufficient to cause windowpanes to be blown out. When an overpressure occurs from a backdraft, deflagration, or detonation, glass fragments may be many feet away from the building.

When glass is rapidly cooled, such as when water is applied to a heated windowpane, short cracks can be produced on the surface of the glass. These cracks, or crazing, are a complicated pattern of short cracks that can be either straight or crescent shaped and can extend through the entire thickness of the glass. In the past, it was believed that crazing may also be created through rapid heating of the glass; however, research reveals that glass will not craze as a result of rapid heating FIGURE 4-7.

Courtesy of Rodney J. Pevytoe

FIGURE 4-7 Crazed window glass.

Furniture Springs

By examining the damage sustained to furniture springs, the investigator may be provided with clues as to how intense the fire was and its duration and direction of travel FIGURE 4-8. This damage, however, does not provide an indication of the type of fire (such as smoldering ignition) or presence of ignitible liquids as once believed. Annealed springs have been determined through laboratory testing to be caused by the application of heat, causing a loss in tension of the spring. Laboratory tests have revealed that the intensity and time of exposure both play roles in the loss of tensile strength to the springs. Additionally, the presence of a load or weight exposed to heat may also cause a failure of the springs.

By comparing the differences in springs from one area of a mattress or cushion, the investigator can develop a hypothesis regarding the relative exposure to a heat source. If one end of a bed mattress displays significant loss of strength yet the other end is relatively intact, the investigator may hypothesize that the heat source or fire was closer to the damaged end. The investigator must also consider the effects of ventilation,

Courtesy of Robert Schaal

FIGURE 4-8 Relative damage to furniture springs can indicate the intensity, duration, and direction of fire.

© Jones & Bartlett Learning. Photographed by Glen E. Ellman

FIGURE 4-9 Protected areas can be useful in reconstructing the fire scene.

direction of fire travel, and other evidence located in the area of origin to support this hypothesis. Additionally, bedding, pillows, and other objects can protect or shield the springs from exposure or provide fuel items that intensify heat exposure at a particular location on the springs. Another consideration would be when springs have already begun to weaken because of age or extensive use prior to the fire.

Heat Shadowing

Heat shadowing is caused by an object blocking the travel of radiated heat, convected heat, or direct flame to a surface, thus creating a discontinuous pattern that may mask lines of demarcation on that surface due to the interruption of that heat transfer.

Protected Areas

When an object is physically in contact with another material in such a way that the material shields the object from the effects of heat transfer, combustion, or deposition, the object is said to be in a protected area. The investigator will find that protected areas are very useful in reconstructing the fire scene. FIGURE 4-9 shows the protected area beneath an ottoman.

Rainbow Effect

Oily substances that do not mix with water can often be seen floating float on the surface of water, creating an interference pattern that produces a rainbow effect. This effect is present at fire scenes and should not be relied on as an indicator of the presence of an ignitible liquid. Many materials such as asphalt, plastic, and wood products can produce rainbow effects as a result of pyrolysis.

Fire Patterns

Fire patterns are defined as the visible or measurable physical changes, or identifiable shapes, formed by a fire effect or group of fire effects. These physical effects are the primary pieces of information used by an investigator to identify fire patterns. Every fire is different and produces unique fire patterns that must be interpreted by the investigator. Factors such as ventilation, ignition factors, fuel loads, and airflow all influence fire patterns.

Heat, deposition, and consumption are the three basic causes of fire patterns. The investigator should analyze individual fire patterns within the context of the complexity of all the patterns and is advised not to rely solely or too heavily on only one fire pattern. Seldom is one fire pattern definitive. A pattern analyzed in isolation can lead to an incorrect conclusion. A process of analysis—proceeding from the gross patterns to an area of interest and then to an analysis of the discrete smaller patterns—can provide more meaningful information.

The development stage of the fire affects the fire patterns produced in a compartment fire. Patterns produced during the ignition and early growth stages are primarily influenced by the ignition source and fuel. The development of a gas layer changes patterns as radiation from the layer exerts its effect.

Postflashover conditions of full-room involvement can mask original patterns and change the pattern production to a process more influenced by ventilation than by fuel.

Fire pattern analysis should not occur in isolation. Interviews with the owners or occupants can be essential to the investigator's understanding of preexisting conditions (fuel, building construction, occupant activities, etc.), fire suppression activities, and observed fire development. Ignoring relevant information can result in the investigator relying too heavily on fire patterns that may not be relevant to the origin and cause of a particular fire.

When examining a fire scene, the fire patterns that remain can be used by the investigator to determine the sequence of events that occurred during the fire. As the fire continues to grow from the point of origin, other fuel packages can ignite, producing additional fire patterns. When fires increase in size or burn for an extended period of time, the fire patterns located at the point of origin can be more difficult to identify.

Plume-Generated Patterns

Although the patterns left on a single surface are essentially two dimensional, it is important for the investigator to be able to visualize the fire in three dimensions. The pattern that remains on a vertical surface, such as a wall, may be a V-shaped pattern, but the fire plume that created the pattern had a three-dimensional shape. The ability to view a fire pattern and abstractly convert it to the fire dynamic process that produced the pattern is vital to every fire investigation FIGURE 4-10.

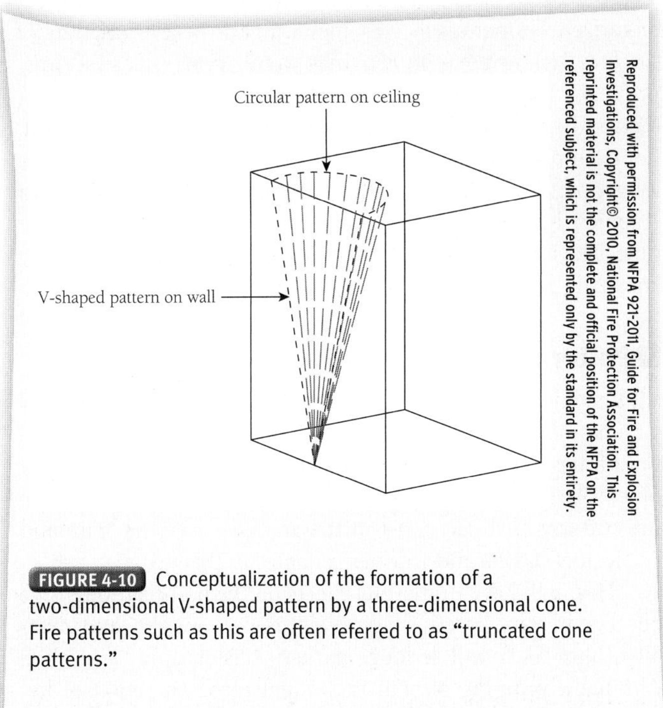

Reproduced with permission from NFPA 921-2011, Guide for Fire and Explosion Investigations, Copyright© 2010, National Fire Protection Association. This reprinted material is not the complete and official position of the NFPA on the referenced subject, which is represented only by the standard in its entirety.

FIGURE 4-10 Conceptualization of the formation of a two-dimensional V-shaped pattern by a three-dimensional cone. Fire patterns such as this are often referred to as "truncated cone patterns."

When a given fuel package is ignited, a column of buoyant hot gases, smoke, and flame rises upward. If the fire is located in a compartment, the plume may reach ceiling height, causing the lateral extension of the plume in a ceiling jet. If there is no ceiling confinement, the plume will continue upward until it cools to the ambient air temperature and diffuses. The creation of a fire pattern will be greatly affected by the heat release rate of the fuel burning. The fire pattern is most pronounced when the surface that displays the pattern was impacted by plume temperatures that were at or near pyrolysis temperatures. The shape of the pattern left on a surface may appear in various ways to include V-shaped, inverted cone, hourglass, U-shaped, circular, or a pointer and arrow pattern.

As the fire plume develops, the size and shape of the pattern will change. Early (incipient) stage fires may produce an inverted cone shape pattern. As the fire progresses, an hourglass pattern may evolve. Eventually, a noticeable pattern may be lost as the full room involvement of a flashover stage fire is reached. Further, early patterns may be partially or fully erased by new fire patterns caused during fire growth and spread, and by heavy burning of large adjacent fuel packages.

Ventilation-Generated Patterns

As pressure is created during the combustion process, hot gases and fire itself may escape through windows, doors, and other restricted openings with an increased velocity and flow over, under, and around combustible items. This increases the rate of damage to materials at those locations. Well-ventilated fires burn with higher heat release rates, which can also cause greater damage. When severe damage is noted at these locations, the investigator must consider these factors as a possible explanation rather than an indication of the point of origin. If embers from a smoldering fire ignite a fuel package that is capable of producing large amounts of heat, the damage may be more severe at this fuel package than the smoldering fire at the point of origin.

Holes burned through floors are an unusual occurrence and often are mistaken to be the result of the application of an ignitible liquid. This damage can also be produced by building materials collapsing and smoldering in the debris, creating holes in the floor. These holes allow the movement of air and increase the burning rate at these locations.

When fire and heat gases are able to escape the compartment, the fire intensifies, changing the shape and magnitude of fire patterns. Localized heavy damage is often found at open windows and doors with combustible items nearby heavily damaged or consumed. Although the damage can be the most severe in the area, the point of origin may not be at this location.

In a compartment fire that is not fully developed, the hot fire gases may escape in the top space between a door and the door frame, causing charring. Cool air enters the compartment under the closed door. In a full-involvement fire, the hot fire

FIGURE 4-11 A portion of a plume pattern generated by the hot gases of the fire.

gases will extend to the floor level and may escape and char a door bottom. Ignitible liquids burning under a door edge will also char the surface.

Hot Gas Layer–Generated Patterns

Prior to a flashover event, the hot gas layer of a fire begins to descend, and radiant heat flux can damage both the upper surfaces of objects and the floor surfaces. If the gas layer descends to floor level, damage may also be present outside the room at floor level due to hot gases escaping under the door itself. The level of descent of the hot gas layer can be determined by examining the line of demarcation present on the walls and vertical surfaces of objects within the room. The damage is usually uniform throughout the room; however, drop-down materials can create isolated areas of damage. Conversely, objects such as dressers and televisions can protect both vertical surfaces and floor surfaces from exposure to the heat gas layer FIGURE 4-11. Protected areas that are generated in this fashion can provide some sequential data to the investigator.

Patterns Generated by Full-Room Involvement

Patterns created by full-room involvement can usually be found on all exposed surfaces throughout the room, all the way down to the floor level. Traditional fire patterns, such as V patterns, may be more difficult to document and analyze if the fire continues to burn for a long time.

When full-room involvement occurs, the damage sustained to the room, including the floor, will be more extensive as radiated and convected heat ignites fuel packages within the room. As the hot gas layer continues to descend to the floor, damage to the objects within the room intensifies (including charring to the underside of objects, intensified damage to corners of the room, ignition of the carpet and flooring, and charring to the underside of doors). Although full-room involvement can produce holes in the flooring, these patterns can be the result of objects or debris burning on the floor or the result of objects creating protected areas and should be considered by the investigator. Damage to the room increases as the fire continues to burn; however, major damage can occur within minutes in extreme conditions due to the effects of ventilation and fuel items available.

Fire Investigator Tip

Holes burned through floors are not necessarily the result of an ignitible liquid. It is possible that building materials collapsed and that smoldering in the debris created holes in the floor.

Suppression-Generated Patterns

Fire suppression actions may create or change fire patterns. Water streams, especially master streams, may change the direction of fire spread, and ventilation actions will affect the fire patterns present. The actions of individuals at the scene and of the fire service should be learned in order to understand whether the actions taken may have caused changes to patterns noted.

Lines of Demarcation

When flame, heat, and smoke impinge on an object or surface, lines or areas of demarcation can be created, producing a border between affected and less affected areas of the object or surface. Numerous factors, including the heat release rate of the fire, fire-suppression activities, ventilation, exposure time, proximity to the fire or heat source, and the material itself, dictate the production of these lines and areas of demarcation. For example, if an object is heated at a lower temperature for a longer period of time, it can display damage identical to that of an object heated at a higher temperature for a shorter period of time. The investigator must always consider these factors when analyzing fire patterns.

Patterns Detected in Fire Victims' Injuries

Although there may be a natural instinct to remove a deceased victim quickly from a fire scene, the body should not be moved until it has been analyzed and documented. As with any other object in the fire, the body could have effects and contain fire patterns that can assist in the analysis. (See the "Fire and Explosion Deaths and Injuries" chapter in this text.)

Heat will have numerous effects on the body, although the effects can vary based on the victim's age, weight, and overall health. Skin will redden, darken, blister, split, and char. Although blistering of skin is thought to be an action of living tissue, it may happen to a limited degree in postmortem exposure. The muscles of the body will dehydrate and begin to contract and shrink. This contraction in the limbs of the fire victim will often lead to the development of the pugilistic pose (boxer's stance). Bone that is exposed will change in color and, with enough exposure, in mass as it is consumed.

Calcination of the bone can occur as the organics of the bone are consumed. The skull can fracture from several actions, including heat and trauma. In general, the body is a poor fuel; body fat can melt and burn but typically requires a porous wick-like material to do so. Bodies found in fire scenes should be treated as evidence and interpreted with respect to the fire patterns present. When possible, the fire investigator should attend autopsies to document and examine the effects of the fire. In fires where individuals survive but receive injuries, the investigator should also strive to document those injuries.

Pattern Location

It is important to remember that the fire is a three-dimensional event, and therefore, fire effects can be three dimensional. The fire investigator needs to examine all of the affected surfaces, including surfaces that are not readily apparent. The scene examination includes looking for large-scale patterns such as aerial views of the roof and small-scale patterns such as the heat effects on the insulation of wiring.

It is important to examine all areas in the building or near the fire to determine whether there are other areas with fire patterns. Fire patterns that are separate from the main body of the fire can be attributable to a number of factors, including fire extension, fire brands, drop down, separate areas of origin, or even previous fires.

Chapter 6 in NFPA 921 offers an extensive discussion of pattern location. As you read that material, it should become apparent why the fire investigator also needs a good understanding of building construction.

Fire patterns will develop on the walls, ceiling, floor, and contents of a compartment. Examination of room surfaces may show a pattern on the surface covering or on the construction material beneath.

Wall patterns most often are in a V, hourglass, or U shape. With ceiling or horizontal surfaces, the investigator may look for areas of greater intensity of damage. Many times this indicates a possible area of fire origin; the investigator is cautioned to remember, however, that a fuel package in a room may create an area of ceiling damage and be located remote from the actual origin area. The investigator should examine areas below these often circular patterns to identify a possible heat source. The transfer of heat by conduction may cause damage to components within walls and ceilings. Heat may conduct through drywall and cause damage to wall studs behind it.

The fire investigator should examine whether fire patterns are present on floors. In compartments with postflashover conditions, the patterns may be obliterated or newly created by the fire condition. Modern carpet (since 1970) sold in the United States has been made to specific fire resistance standards. The dropping of a match or a cigarette to these carpets is generally insufficient to cause horizontal flame spread. Generally, carpet will require the input of greater energy to burn and then may only do so to the point where it reaches the minimum energy needed to support flame spread. In flashover conditions, the radiant heat impact on carpet will cause ignition and burning. Burning between the seams of floor boards is not always proof of the presence of an ignitible liquid. Radiant heat or prefire conditions of the flooring may have in fact been responsible for or contributed to the patterns. Holes discovered in floor areas could be from preexisting conditions, glowing combustion, radiation, or ignitible liquids. Vinyl floor tiles may curl at the edges for several reasons. The burning of ignitible liquids, radiant heat, or natural shrinkage may cause the damage to tile.

In all of these patterns, the investigator should take samples (and comparison samples) for laboratory confirmation when the presence of ignitible liquids is suspected.

Flashover and full-room involvement can produce relatively uniform burning on surfaces in the early stages of full-room involvement. The uniformity quickly dissipates as the duration of full-room involvement increases. Some exposed surfaces may have little or no damage due to ventilation effects and the location of furnishings. Additional care must be taken in evaluating patterns.

Beveling

Beveling is an indicator of fire direction on wood wall studs. The bevel leans toward the direction of travel, as is shown in the left portion of FIGURE 4-12. Note how the beveling changes when the heat source is directed toward the narrow surface of the wall stud, as is shown in the right portion of the figure. Beveling, along with the tendency of wall studs to point toward the heat source when viewed from above, are also known as "pointer and arrow" patterns.

Fire Penetration of a Horizontal Surface

When penetrations are found in horizontal surfaces either from above or below, the investigator should consider several causes, including radiant heat, isolated smoldering objects, and ventilation.

When downward penetrations are located, the investigator must be cautious not to assume that the damage is the result of the application of an ignitible liquid. Downward movement is unusual as heated gases rise; however, when fully developed fires occur, gas may be forced though small openings in the flooring, creating penetrations located after the fire. Likewise, when objects such as furniture pieces burn, they can produce penetrations underneath the objects. Floor penetrations may also be created under collapsed floors or roofs where flaming or smoldering fires may exist.

The direction of fire travel through a horizontal surface is determined by examining the sides of the hole and the slope created by the fire. When the hole is wide and slopes downward from the top surface, the direction of fire travel would be from above. In the case of a fire advancing from below a surface, the sides are wider on the bottom side and slope upward. Investigators should be cautious when determining the direction of fire travel through these penetrations because fire movement may have occurred in both directions, leaving only the last direction of travel visible in the hole FIGURE 4-13.

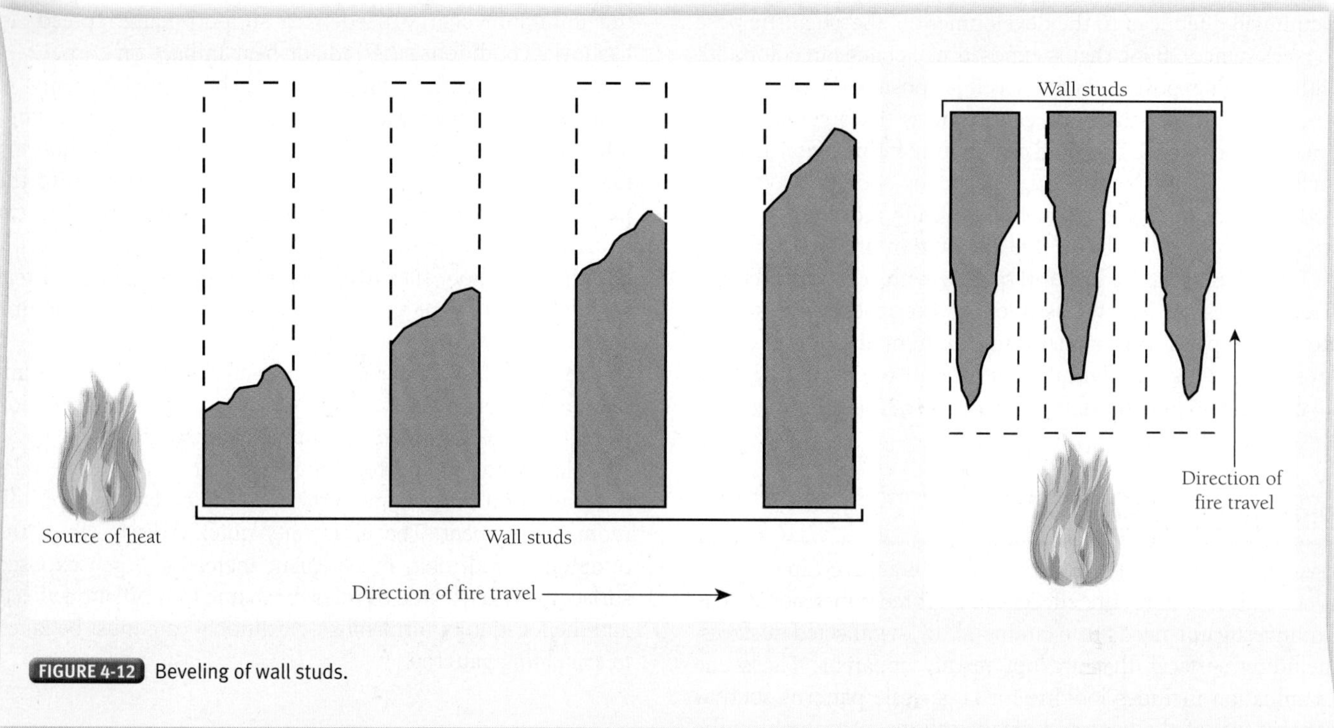

FIGURE 4-12 Beveling of wall studs.

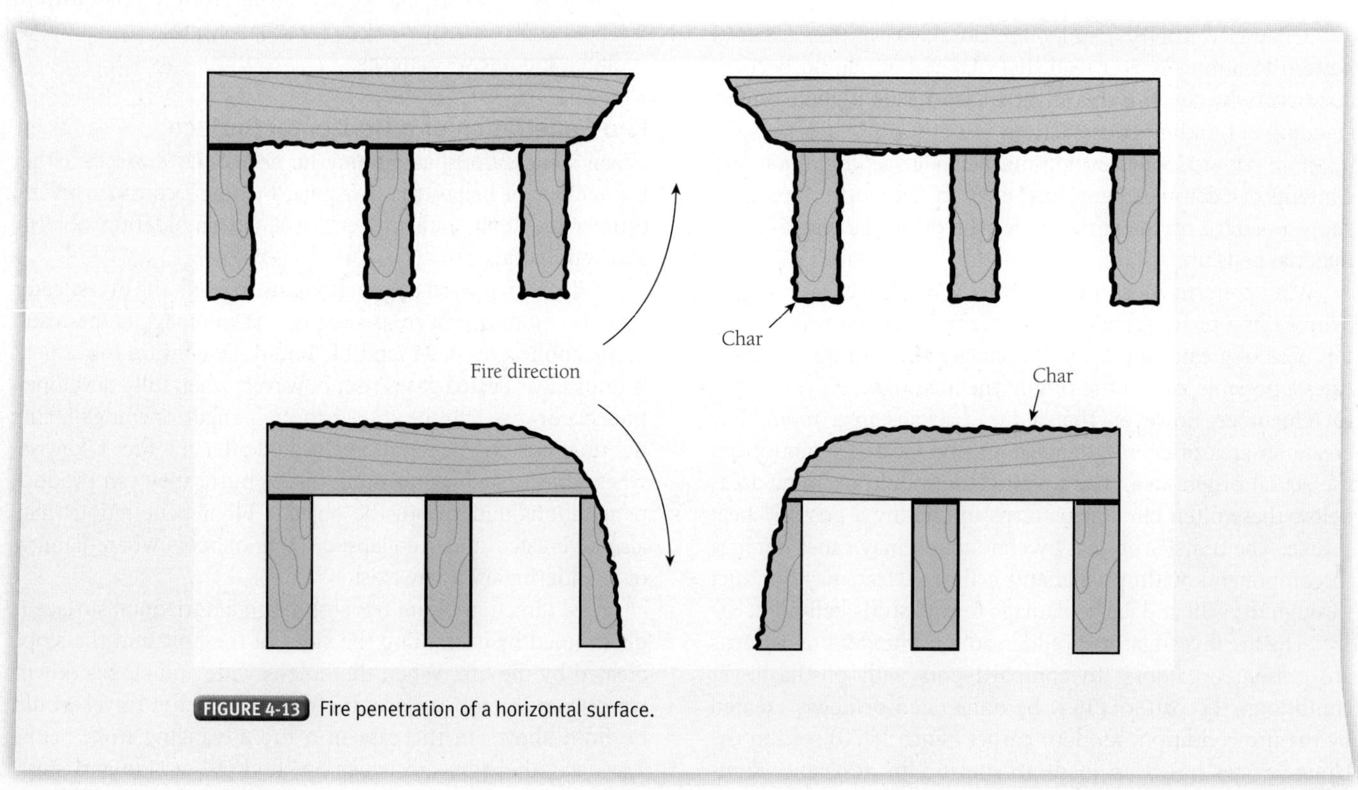

FIGURE 4-13 Fire penetration of a horizontal surface.

Irregular Patterns

NFPA 921 offers more warnings about misinterpreting irregular patterns than it does on any other subject. Simply put, once a fire progresses beyond flashover, and even in some cases of limited fire damage, calling an irregular pattern a "pour pattern" on the basis of its visual appearance alone may lead to error.

The presence of irregularly shaped or "pool" patterns on floors and floor coverings should not be relied on as an indicator of the application of an ignitible liquid. Fires within rooms that have fully developed or are exposed to flashovers, and fires in which there was a long extinguishment time, produce similar patterns despite the lack of an ignitible liquid being present. The lines of demarcation of an irregular-shaped pattern may vary with the properties of the material and the intensity of the heat exposure. A dense material such as hardwood flooring may show sharper lines as opposed to some carpets.

Flaming or smoldering debris, clothing or other irregularly-shaped material, plastic items, and ignitible liquids may all create irregular patterns. The investigator should work to identify the fuel that caused the pattern in question. Should an ignitible liquid be suspected, samples should be submitted for laboratory analysis.

If samples such as carpet or other fire debris are submitted for analysis, comparison samples from undamaged areas should be collected and submitted. Pyrolysis products such as hydrocarbons may be detected even when an ignitible liquid was not present. Therefore, it is always useful to collect a comparison sample, that is, a sample of flooring where the investigator does not believe an accelerant was present.

NFPA 921 contains several photographs of test fires in which the postfire patterns appeared to be identical to patterns produced by ignitible liquids but no ignitible liquid was used. Reading additional reference material on building construction that relates the materials and method of construction to fire pattern development will further your understanding of these patterns.

Pattern Geometry

Fire and smoke will produce a variety of distinctive patterns. These patterns are described based on their geometric shapes. The interpretation of these patterns is not based on scientific research; therefore, alternative interpretations are possible. With the diverse nature of fire patterns, many patterns are possible beyond those described in this text FIGURE 4-14.

Courtesy of Rodney J. Pevytoe

Courtesy of Jamie Novak, Novak Investigations Inc. and the St. Paul Fire Department

FIGURE 4-14 The most common fire patterns. **A.** Partial V pattern created by flames venting from the window opening. **B.** A pattern generated by a fuel package burning on the floor.

> **Fire Investigator Tip**
>
> Once a fire progresses beyond flashover, and even in some cases of limited fire damage, calling an irregular pattern a "pour pattern" on the basis of its visual appearance alone may lead to error.

V-shaped patterns are created by the heat or smoke from the fire plume. The angle of the pattern has several factors that will affect it. These would include the heat release rate of the fuel package, the geometry of the fuel, the effects of ventilation, the combustibility of the surface where the pattern is observed, and horizontal obstructions. The angles of the lines of demarcation are not to be interpreted to indicate the speed of the fire growth or the heat release rate of the fuel alone.

Inverted cone patterns are often created by a vertical flame plume not reaching the ceiling level. It will appear two dimensionally as a triangle shape with its base at the bottom. These fires are often short lived or involve a fuel package of low heat release. Patterns may also be created by leaking natural gas occurring below floor level and rising, burning where the floor and wall intersect.

Hourglass patterns are created with a burning fuel package at its base next to or near a vertical surface. As the fuel burns, the vertical surface will show a pattern lower from the flame zone and also display effects of the upper hot gas zone. The lower zone creates an inverted V connected to a larger V above.

U-shaped patterns display a gently shaped line of demarcation compared with the sharp lines of V-shaped patterns.

Truncated cone patterns, also known as truncated plumes, result from the partial intersection of a cone pattern on both vertical and horizontal surfaces.

Pointer and arrow patterns may be on a series of combustible elements. An example would be a compartment such as an unfinished garage area where a series of wall studs were exposed. These patterns may also occur when the surface sheathing has been destroyed by the fire. By examination of the lines of demarcation on several of the wall studs, a direction of fire travel may be interpreted. The most severe charring to the studs should be expected on the side from which the heat is impacting.

Circular-shaped patterns often are noted on the underside of horizontal surfaces. As the fire plume rises and reaches the surface above, it may leave a circular-shaped pattern. When the heat source is farther from a vertical surface, the pattern created will tend to be more circular in nature.

A doughnut-shaped pattern may be found wherein irregularly shaped areas of burning and damage surround an area of lesser damaged material. This effect is often the result of an ignitible liquid and should be closely examined by the investigator for evidence that an ignitible liquid created the pattern. When these patterns are created by burning liquids, the liquid itself cools the center of the application area while the flames are present along the perimeter.

Saddle burns are characterized as curve-shaped and deeply charred patterns that are often localized on the top of floor joists. These distinct U-shaped patterns may potentially be the result of the application of an ignitible liquid; however, they can also be created by burning materials on floor surfaces, objects that have melted and fallen on the floor, and floor openings that allow for downward ventilation.

Linear patterns are found on horizontal surfaces and may appear as lines of demarcation, extending and elongating across the surfaces.

Long, unusual flame and heat damage at floor level is often caused by the application of an ignitible liquid creating a trailer, which may connect one area to another. Trailers are often used to connect multiple fires or "sets" or used to spread the fire to different floors.

Furniture and other items may shield portions of a room's wall or floor surface, creating a protected area in which materials such as carpeting and flooring sustain little to no damage while the unprotected area is severely damaged. These linear-shaped patterns are sometimes mistaken as trailers; however, they could simply be the result of clothing, boxes, or any other material that may have been present. The investigator should determine what materials or items were present that could have created these patterns.

Linear patterns on flooring may also be seen in areas of the flooring or carpet that were heavily used.

When liquefied petroleum gas and propane burn, they may produce demarcation lines or linear patterns across noncombustible surfaces that direct back to the source. If these fuel gases are dispersed within a room prior to ignition, area patterns may be created and not readily detected by the investigator. This phenomenon can also be created by a rapidly moving fire, such as a flash fire. Flash fires may produce little to no surface damage within a space and may not create an explosion. The areas of greatest damage are likely to be where secondary fuels are ignited. This damage may produce fire movement patterns that are now determined by the secondary fuel, room geometry, and heat release rates. If a flashover occurs, the evidence of a flash fire may be altered or consumed. Flash fires may consume only the fuel gas, causing little damage to the surrounding combustible items. Pocketing of gas may also be present, creating the appearance of unusual movement patterns that make tracing fire spread more difficult. The investigator should obtain accurate witness statements and identify potential fuel and ignition sources when attempting to establish the origin of a flash fire.

As the fire transitions from growth to full development, a flashover event will occur and spread fire rapidly to all exposed combustible surfaces to floor level. Uniform depths of char and calcinations may be present; however, the effects of ventilation and the location of objects such as furniture or fixtures may produce uneven damage to wall and ceiling surfaces.

Fire Pattern Analysis

Fire pattern analysis is the process of identifying and interpreting fire patterns to determine how the patterns were created and their significance. This analysis relies on the investigator having an understanding of fire dynamics and fire development to ensure he or she can properly recognize, identify, and analyze the patterns observed. There are two basic types of fire patterns: movement patterns and intensity patterns. (These types of patterns are defined by the fire dynamics discussed in the "Basic Fire Science" chapter in this textbook.) Systematic appearance of more than one type of fire pattern at a fire scene can be interpreted to lead back to the heat source that produced them. Some patterns may display aspects defining both movement and intensity (heat/fuel). Generally no individual pattern is definitive in and of itself, and each pattern observed should be analyzed in relation to the other patterns at the fire scene. A comparison of the patterns may allow an investigator to determine the path of fire progression, leading back to the area of origin.

Heat (Intensity) Patterns

As a fuel item is exposed to heat and flames, patterns are created, producing lines of demarcation and mass loss that may be useful to the investigator in determining the direction of fire travel as well as the characteristics and amounts of fuel materials present. For example, levels of overall damage and smoke deposition will often assist the investigator in an initial determination of the area of greater fire. Further, when one area or side of an object has lost more mass than other areas, this can often indicate a greater heat intensity coming from the area of greater loss. Often the investigator can compare like materials (such as wall framing studs if they were configured identically) to gain information about where the heat was greater. V- and U-patterns show the presence of heat, although they may simply be indicating a source of heavy fuel rather than the origin of the fire. These lines of demarcation can be examined and may differentiate an area that was exposed to more heat or exposed for a longer duration, resulting in greater consumption of an item in relation to a similar item or greater consumption of one area of an item in relation to another area. This information can be analyzed and compared to the damage and lines of demarcation on other items, allowing the investigator to determine the areas of greatest heat and fire intensity, which will assist in determination of the area of origin.

Fire Spread (Movement) Patterns

As a fire grows and spreads, products of combustion (including smoke and hot gases) flow away from the initial heat source. Fire often leaves distinctive patterns as it travels through interior openings and doorways. Pointer and arrow patterns can be analyzed for information about fire movement within an area. As these products move throughout a compartment, they impact boundary surfaces and contents in varying degrees. This impact and the resulting patterns can be analyzed in relation to each other, allowing an investigator to trace the patterns back to the heat source that produced them. This information is helpful in analyzing fire growth and development, identifying the area of origin, and determining fire causation. The investigator must consider all patterns and other data found at a scene and must consider patterns in the context in which they appear. For example, clean burn from a small plastic object on the floor may suggest a fire originating from the object; however, if it is obvious that the object fell to the floor from a burning area above, the patterns associated with the object have a much different meaning.

Wrap-Up

Ready for Review

- It is impossible to understand fire patterns without a solid knowledge of fire dynamics.
- The primary physical evidence available to the investigator consists of the materials subjected to the fire and the by-products of burning (such as heat, smoke, and soot).
- A fire investigator must have a strong understanding of the effects of fire to identify and accurately interpret the fire patterns created.
- The rate of mass loss changes during a fire due to various factors, including heat flux, fire growth rate, and the object's heat release rate.
- The rate at which wood chars varies widely throughout the course of a fire.
- Different areas of concrete and masonry expand at different rates, causing pitting and cracks on the surface of the material and reducing the surface tensile strength of the material.
- Oxidation of noncombustible materials can produce color and texture changes and fire patterns that could be useful for the fire investigator.

- Demarcation lines can be produced between melted and unmelted portions of a material and may be useful fire patterns to the investigator.
- Alloying, which can look very much like melting, must also be considered by the investigator during postfire examination of damaged metal items.
- When exposed to heat, most materials begin to expand and change shape, and different materials are likely to have different expansion rates. Structural failures can occur as a result of these differing expansion rates.
- The color and texture of smoke deposits are not reliable indicators of the burning or the rate of heat release.
- Calcination occurs in plaster or gypsum wall surfaces when the free and chemically bound water is driven out of the gypsum by the heat of the fire.
- Glass fragments found to be free of any soot deposits of smoke condensates usually indicate early failure of the glass prior to the accumulation of smoke.
- By comparing the differences in springs from one area of a mattress or cushion, the investigator can develop a hypothesis regarding the relative exposure to a heat source.
- Heat shadowing is caused by an object blocking the travel of radiated heat, convected heat, or direct flame to a surface, thus creating a discontinuous pattern.
- The investigator will find that protected areas are very useful in reconstructing the fire scene.
- Oily substances that do not mix with water can often be seen floating on the surface of the water, creating an interference pattern that produces a rainbow effect.
- Factors such as ventilation, ignition factors, fuel loads, and airflow all influence fire patterns.
- Although the patterns left on a single surface are essentially two dimensional, it is important for the investigator to be able to visualize the fire in three dimensions.
- If embers from a smoldering fire ignite a fuel package that is capable of producing large amounts of heat, the damage may be more severe at this fuel package than the smoldering fire at the point of origin.
- Prior to a flashover event, the hot gas layer of a fire begins to descend, and radiant heat flux can damage both the upper surfaces of objects and the floor surfaces.
- Patterns created by full-room involvement can usually be found on all exposed surfaces throughout the room, all the way down to the floor level.
- Fire suppression actions may create or change fire patterns present.
- When flame, heat, and smoke impinge on an object or surface, lines or areas of demarcation can be created, producing a border between affected and less affected areas of the object or surface.
- As with any other object in the fire, a body could have effects and contain fire patterns that can assist in the analysis.
- The scene examination includes looking for large-scale patterns such as aerial views of the roof and small-scale patterns such as the heat effects on the insulation of wiring.
- Beveling is an indicator of fire direction on wood wall studs. The bevel leans toward the direction of travel.
- Calling an irregular pattern a "pour pattern" on the basis of its visual appearance alone may lead to error.
- Fire and smoke will produce a variety of distinctive patterns, described based on their geometric shapes.
- There are two basic types of fire patterns: movement patterns and intensity patterns.

Hot Terms

Beveling A fire pattern that indicates fire direction on wood wall studs. The bevel leans toward the direction of travel.

Calcination A fire effect realized in gypsum products, including wallboard, as a result of exposure to heat that drives off free and chemically bound water.

Char Carbonaceous material that has been burned or pyrolyzed and has a blackened appearance.

Clean burn A fire effect that appears on noncombustible surfaces after any combustible layers (such as soot, paint, and paper) have been burned away. The effect may also appear where soot was not deposited due to high surface temperatures.

Crazing A complicated pattern of short cracks in glass that can be either straight or crescent shaped and can extend through the entire thickness of the glass.

Fire effects The observable or measurable changes in or on a material as a result of exposure to the fire.

Fire pattern The visible or measurable physical changes or identifiable shapes formed by a fire effect or group of fire effects.

Fire pattern analysis The process of identifying and interpreting fire patterns to determine how the patterns were created and their significance.

Heat shadowing A pattern that results from an object blocking the travel of radiant heat from its source to a target material on which the pattern is produced.

Melting A physical change caused by exposure to heat.

Oxidation The basic chemical process associated with combustion.

Rainbow effect A diffraction pattern formed when hydrocarbons float on a surface.

Smoke deposits Hot products of combustion that may adhere upon collision with a surface.

Spalling The chipping or pitting of concrete or masonry surfaces.

© Greg Henry/ShutterStock, Inc.

FIRE INVESTIGATOR *in action*

You are conducting an investigation at a four-unit apartment building that was under renovation at the time the fire occurred. Your exterior examination noted only smoke staining from the windows and doorway of the east unit. Once inside, you observe smoke and soot staining throughout the interior, creating a demarcation line halfway down the walls of each room with a flame pattern present in a hallway leading to two bedrooms and a bathroom. The top half of the bedroom doors display fire patterns indicating flame and heat impingement from the bathroom.

The bathroom door has been consumed with only the door handle being located in the debris. The gypsum wallboard on the left wall is no longer intact, with severe charring present to the wooden wall studs. A large mass of plastic is present along the left wall that you determine to be a bathtub that was to be installed after plumbers finished connecting copper water lines. You note the wall studs and bottom wall plate are almost completely consumed where the water lines come up through the floor below.

1. When you examine the wall studs by the remains of the bathtub, you note they display severe damage and are leaning toward the water lines. This is known as:
   - **A.** charring.
   - **B.** stepping.
   - **C.** beveling.
   - **D.** None of the above
2. As you remove the remains of the bathtub, you note the flooring beneath is undamaged and creating a pattern consistent with the remains of the bathtub. This is known as:
   - **A.** heat shadowing.
   - **B.** a protected area.
   - **C.** an outline pattern.
   - **D.** spalling.
3. After removing the debris from the floor, you note a fire pattern that has created a "U" shape on the top of a floor joist. This is called a(n):
   - **A.** cup burn.
   - **B.** saddle burn.
   - **C.** arrow pattern.
   - **D.** inverted cone pattern.
4. Which of the following would be considered a fire effect?
   - **A.** The damage to the top of the bedroom doors
   - **B.** The remains of the doorknob
   - **C.** The remains of the bathtub
   - **D.** Both A and C
5. The wall studs display silver-colored blisters approximately ¾ inch in depth. This indicates:
   - **A.** the use of an ignitible liquid at that location.
   - **B.** a burn time of 45 minutes to 1 hour.
   - **C.** a slow-burning fire with limited ventilation.
   - **D.** No specific determination can be made from depth of char alone.

Building Systems

© Photos.com

© romakoma/ShutterStock, Inc.

Knowledge Objectives

After studying this chapter, you should be able to:

- Explain the principles of compartmentation and its effect on fire confinement NFPA 4.2.8. (pp 60–61)
- Identify design, construction, and structural elements of buildings and describe their effect on fire development, spread, and control NFPA 4.2.2 NFPA 4.2.3 NFPA 4.2.8. (pp 61–64)
- Identify and describe types of building construction NFPA 4.2.2 NFPA 4.2.3. (pp 65–69)
- Discuss the structural integrity of construction assemblies during a fire NFPA 4.2.3. (pp 70–71)
- Discuss the effects of weather on building systems. (p 71)
- Identify construction materials and describe their effects on fire confinement NFPA 4.2.2 NFPA 4.2.3. (pp 71–72)
- Explain the impact of passive fire protection systems on fire investigation NFPA 4.2.3. (pp 72–73)

Skills Objectives

There are no skills objectives for this chapter.

Additional NFPA References

NFPA 13, *Standard for the Installation of Sprinkler Systems*

NFPA 80, *Standard for Fire Doors and Other Opening Protectives*

NFPA 220, *Standard on Types of Building Construction*

NFPA 921, *Guide for Fire and Explosion Investigations*

CHAPTER 5

FESHE Course Outcomes

Fire Investigation I

6. Compare the types of building construction on fire progression. (pp 65–69)
7. Describe how fire progression is affected by fire protection systems and building design. (pp 70–73)

Fire Investigation II

There are no Fire Investigation II (FESHE) course outcomes for this chapter.

You Are the Fire Investigator

© Jones and Bartlett Publishers. Photographed by Glen E. Ellman.

You have been requested to investigate a fire in a large, two-story home in one of your historical districts. Fire personnel have informed you that when they arrived, flames were visible from the basement and second floor windows in the south exterior walls. They state that the fire was quickly knocked down; however, during overhaul, flaming and smoldering materials were located in various wall spaces along the south half of the first and second floors.

As you examine the interior of the structure, you note severe fire damage within the basement and one bedroom on the second floor; however, the damage to the first floor rooms is minor. The plaster and lath has been largely removed by fire personnel during overhaul operations.

In the south wall of one of the first floor rooms, you note two wall spaces that are severely charred, extending from below the level of the floor to the second floor. While inspecting the wall space you discover that there are no fire stops between the basement and second floor.

1. How does the construction type influence your investigation?
2. Why will interviews with the building owner be important?
3. How do the interior finishes and contents affect fire development?
4. What information could you obtain from the building or renovation plans?

Introduction

The design, materials, and construction type of a structure influence the development and movement of any fire that may occur within it. The influence can be either positive, containing the fire within a compartment, or negative, allowing the fire to spread from the area of origin.

The fire investigator must have an understanding of building construction and building systems, as well as the definitive fire patterns that remain after a fire, to analyze fire growth and movement properly. The application of heat to any product will cause varying degrees of damage and leave various fire patterns depending on the product's composition. Computer models have been developed to help investigators analyze fire incidents. Some of these mathematical, engineering, and graphic models are described in the "Failure Analysis and Analytical Tools" chapter of this textbook.

Building Systems Overview

Some building techniques are a direct result of the analysis of fires, often catastrophic fires, some involving conflagrations that occurred at the turn of the 20th century. Today, more than half of modern code requirements are related to fire protection.

One primary cause of severe fire damage to structures is failure to contain or confine the fire. Basic principles of construction provide that fire should be contained to the room of origin, area of origin (often called *fire area*), or structure of origin.

When a fire does move from the room of origin, it is most often through unprotected or improperly protected openings or construction defects rather than system failures. Failure to close fire doors manually (or impeding their automatic closure), or penetrations of fire resistance-rated assemblies such as walls and ceilings for building or utility services provide ample corridors for fire movement.

However, fires originating in interstitial spaces are not afforded the same level of protection as a fire that develops within a compartment (see Section 7.2.2.5 of NFPA 921, *Guide for Fire and Explosion Investigations*). Interstitial spaces are associated with the spaces between the building frame and interior walls and the exterior façade, and with spaces between ceilings and the bottom face to the floor or deck above (e.g., a cockloft). Fires that originate in these concealed locations may develop undetected and thus move freely and rapidly without barriers to stop their spread.

An investigator needs to have an understanding of active and passive building systems, including manual and automatic fire detection and fire suppression; heating, ventilating, and air conditioning (HVAC) systems; utilities; and building compartmentation, including the use and location of fire resistance-rated assemblies. Knowing where these systems are installed and how they operate will help to determine the system's overall performance and its effect on the fire and will allow identification of any alterations and/or failure indicators. It is important to remember that the system failure may not have occurred through system faults or tampering, but rather, the system simply may have been overcome by the fire development.

An HVAC system operating at the time of a fire can facilitate the spread of smoke and fire through a structure. The investigator needs to examine such systems to determine their status at the time of the fire. Fire investigators should consider using outside resources to help examine systems with which they are not familiar. Mechanical and electrical inspectors or engineers are great assets available to the fire investigator, and they are generally eager to assist in whatever capacity they can.

Safety Tip

The fire investigator may encounter areas where his or her safety may be in question. As an example, the fire may have caused structural collapse or weakening of structural components that can lead to eventual collapse, possible electrical hazards, toxic fumes and dusts, water-filled holes, fire-weakened floors and stairs, falling glass, and other penetration hazards. With all of the dangers associated with the different types of building construction, the investigator should be prepared to enact measures to protect himself or herself before and during the scene examination. Prior to doing any removal or demolition, the scene should be properly documented through photographs, and all interested parties should be contacted to avoid claims of spoliation. Some of the things that an investigator should consider to protect himself or herself are as follows:

- If in doubt, have a structural engineer or building official ensure the structure is structurally sound before entering.
- Shore any areas where the integrity of the structure is in question.
- Remove or demolish walls, floors, or ceilings that may be hazardous.
- Remove glass that poses a threat.
- Pump out any standing water.
- Ensure that all electricity has been discontinued to areas that have fire-damaged electrical components.
- Ensure that there are two means of exit from any area being processed.
- Use monitoring devices to check all areas to be examined, ensuring that the atmospheres are within accepted levels.
- Do not examine any areas containing toxic chemicals without using the proper personal protective equipment respirators/self-contained breathing apparatus and protective clothing.

Following are some examples of the types of building systems that fire investigators should be familiar with:

- Fire dampers, smoke dampers, and combination fire/smoke dampers
- Automatic fire sprinkler systems
- Fire alarm and voice communication components
- Signaling systems for monitoring fire alarms
- Fire doors and windows
- Fire resistance-rated wall, floor, and ceiling assemblies
- Commercial cooking equipment, including hood and mechanical ventilation systems
- Automatic extinguishing systems for commercial cooking
- Standpipe systems
- HVAC and electrical and gas service utilities

Understanding these systems (discussed in more detail in the "Fire Protection Systems" chapter) will help the investigator determine whether the systems functioned properly during the fire event, and if not, why. Possibilities include outright failure and nonmalicious actions, for example, materials blocking a fire door. Malicious actions are another possibility and include purposeful efforts to defeat a system and allow for a larger fire or a delay in notification.

Design, Construction, and Structural Elements

A building's characteristics affect the development, spread, and control of a fire. These characteristics include the type of construction; the integrity and performance of its structural elements under a fire load; and its building systems, including but not limited to active and passive fire protection systems. Everything, from the construction type of the building to the number of doors, windows, and other openings it contains, can affect how the structure responds to a fire or explosion and the level of damage the building sustains (e.g., the amount of structural damage sustained from a low-order explosion is decreased the more windows the building has).

Once a window breaks or a door is opened, the introduction of additional oxygen can greatly increase the fire's size and speed. Exterior winds can rapidly move a fire down unprotected corridors. The investigator must examine and consider the environment and mechanical conditions at the time of the fire that may have impacted the size, speed, and spread of the fire. Ventilation effects can have a tremendous impact on the fire's growth. Additional factors that influence the origin, development, and spread of a fire include the interior layout, interior finish materials, and building services and utilities. FIGURE 5-1 shows the structural components in a typical single-family dwelling.

Building Design

In a building fire, fire spread and development are largely effects of radiant and/or convective heat transfer. (These areas are discussed in the "Basic Fire Science" chapter in this textbook.) In compartment fires, the following factors significantly affect fire spread:

- Room size
- Room shape
- Ceiling height
- Placement and area of doors and windows
- Interior finish and furnishings
- HVAC systems
- Fuel packages and location

Fire Investigator Tip

The fire investigator should be familiar with not only the most common building designs and construction, but also less common buildings or compartments, such as atriums, stadiums, and tunnels, that could have different impacts on fire spread. Special designs may also incorporate special materials.

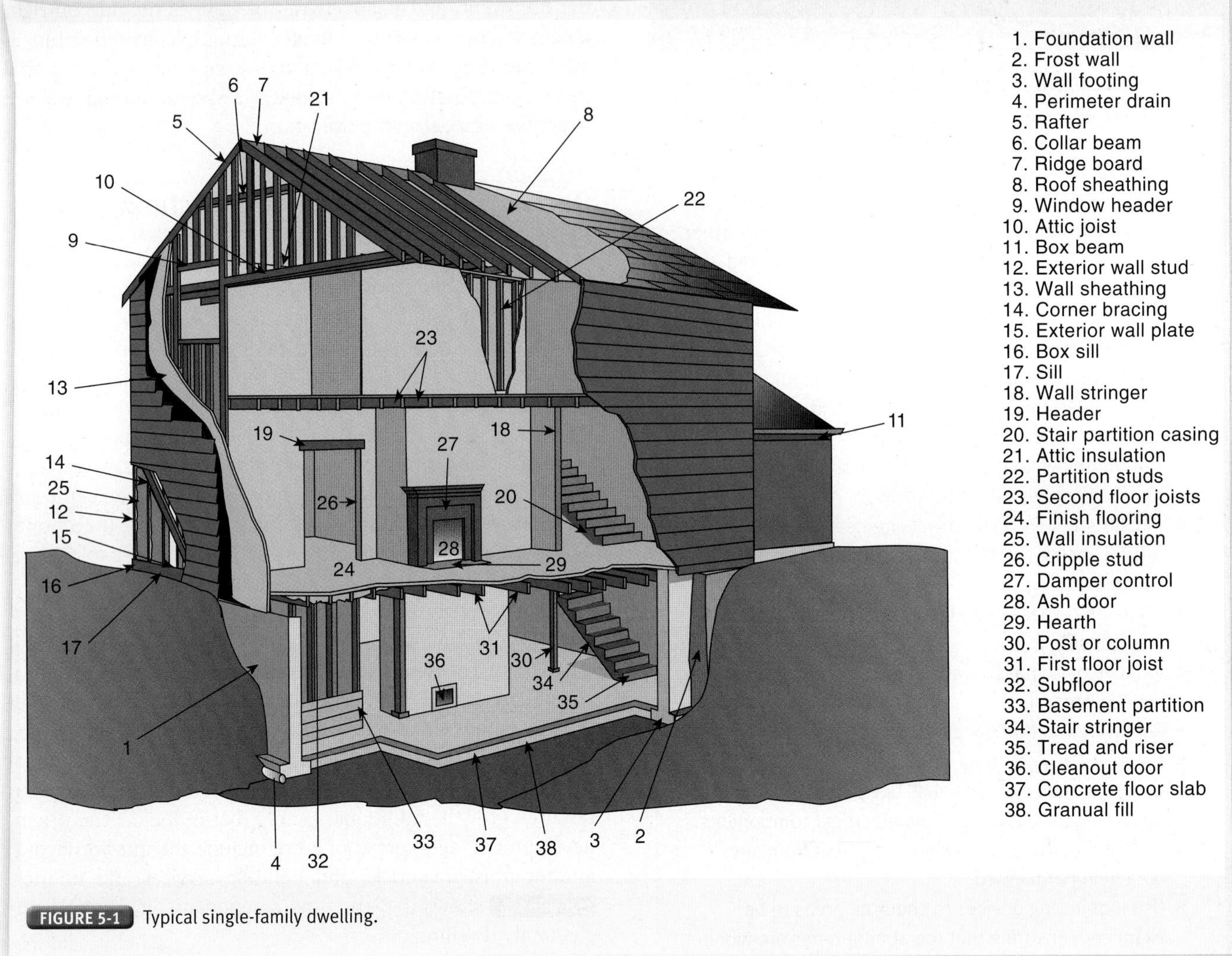

FIGURE 5-1 Typical single-family dwelling.

Limiting the fire spread to a specific area through compartmentation is a primary objective of fire safety design. Investigators should be prepared to encounter a variety of construction techniques and building materials that may have an impact on the fire spread.

Building Loads

Building loads, or forces acting on a structure, are classified as either live loads or dead loads. A live load is a load that can move. A live load is the weight of temporary loads that need to be designed into the weight-carrying capacity of the structure, such as people, wind, furniture, furnishings, equipment, machinery, snow, and rainwater. A dead load is constant and immobile, such as floor slabs, structural columns, and roof coverings. Dead loads are the weight of materials that are part of a building, such as the structural components and mechanical equipment. Building designs and the construction techniques used take into account both of these loads.

There are certain conditions under which a building will no longer be capable of supporting its loads. Loads applied above and beyond the structure's design parameters may create instability and ultimately failure. Examples of such conditions include the following:

- Extreme snow loads
- Extreme wind loads
- Additional contents, such as stock
- Additional mechanical components, such as HVAC and elevator equipment
- Large congregation of people in a limited area
- Water from firefighting operations

Structural Changes

Structural changes can occur either deliberately—during building renovations, for example—or during a fire. As the fire grows and progresses, it may damage structural support elements so that they are no longer capable of supporting their designed loads. An exterior examination of the building should always be performed during an investigation. There are several reasons for doing a survey:

- To determine whether there is any potential structural compromise that may have occurred during the fire that could be a hazard to the investigator if collapse were to take place
- To preserve any evidence noted
- To interpret the fire-damaged areas noted from the exterior

- To identify all potential means of entry and egress, which helps to ensure that the investigator(s) always has a minimum of two ways out of any area being processed
- To identify utilities used in the structure

All of these items are also vital when doing the initial interior survey prior to commencing scene processing.

Room Size

The geometry of the room—its height, width, and specifically the distances of the walls from each other—is an important factor in fire development. Given a similar fuel package and similar ventilation points, a smaller room reaches flashover in less time than a larger room. Flashover is an important mechanism for fire spread beyond the room of origin. Extremely large rooms may never have sufficient heat energy transfer to cause flashover.

Compartmentation

The compartmentation of a structure refers to its subdivision into separate sections or units. A common mechanism of fire spread outside the compartment of origin is horizontal movement through openings such as doors, windows, and unprotected openings and penetrations in walls. Vertical openings such as stairways, utility chases, and shafts also contribute to fire spread outside the compartment of origin.

Investigators need to examine the building's interior walls for several reasons. First and foremost, the walls should be examined to determine which are load-bearing as opposed to the nonbearing walls. Removal of any structural elements along a load-bearing wall could create undue stress on the building, leading to a potential local collapse of the area or the collapse of the entire structure.

Second, the investigator should examine the walls to determine whether they were constructed to resist passage of fire and/or smoke. Fire walls and smoke barriers are constructed of specific materials that have been tested and proven to prevent the passage of smoke or fire. If an interior wall is constructed with masonry construction, it should be capable of stopping the fire from penetrating through to the other side of the wall if designed to do so. If the fire has spread through the wall, the investigator needs to examine the wall for any penetrations that may have allowed for the passage of the fire. It is not uncommon for contractors to install lines or other components and penetrate the compartmented separations without sufficiently closing off the openings in conformance with the local fire/building codes. Smoke and fire dampers in the HVAC system need to be examined to determine whether they functioned properly.

Fire doors should be examined to see whether the automatic closers worked on activation of the fire alarm system and, if they did, whether they closed and latched securely or were propped open with door chocks, etc. If the doors were propped open, determine whether this a normal circumstance for this business, or whether it is an unusual situation. It is also possible that fire fighters propped these doors open during firefighting activities.

Another issue that must be examined is the intensity of the fire. Was the rate of heat release generated from the interior components intense enough to cause a properly designed separation to fail before its intended laboratory-tested rating (keeping in mind whether the fire followed the time-temperature curve)? All of these issues must be determined by the fire investigator to come to a complete and accurate conclusion as to the fire's origin and spread.

Interstitial and Concealed Spaces

Concealed and interstitial spaces exist in most buildings and can provide a mechanism for fire spread beyond the compartment of origin. Interstitial spaces generally lack fire stops. As discussed earlier, these spaces are generally located between the interior and exterior walls and between the ceiling and the underside of the floor above. Interstitial spaces can also contribute to spread of fire in buildings through openings between the building frame and interior walls and the exterior façade and in spaces between ceilings and the underside of the floor or deck above. A lack of fire stops may allow for vertical spread of fire. If fire stops were provided, they should be inspected to ensure that they maintained their integrity to stop or limit the fire spread.

Concealed spaces are another major concern for the fire investigator. Failure to account for concealed spaces can lead to erroneous interpretation of fire patterns. In some construction, concealed spaces are often equipped with fire sprinklers, early detection devices, or fire stops. Building codes sometimes allow concealed spaces if they are constructed of noncombustible material. These concealed spaces become a code-enforcement problem, as well as a fire-spread problem, because they are often used for the general storage of combustible items when they are prohibited from doing so.

Planned Design Versus "As-Built" Conditions

A building's original design may not, and frequently does not, reflect what was finally constructed. Consequently, building plans available to the investigator may or may not reflect the true "as-built" structure. It is important for the investigator to determine the accuracy of any plans in relation to the actual structure because conditions may exist that vary from the original building plans. These changes can be identified through examination of the fire scene, by examination of similar houses built by the same contractor in the same project, or through witness interviews and occupant photographs.

Obtain a copy of the structure's floor plan. If the owner cannot provide the information, often the local fire department has a fire preplan on file. Most departments will provide this information. In some cases, the homeowner/property owner may have a copy of the original drawings. The investigator may also access most building drawings from the respective building department for the building's area or the original architect. These plans are usually kept on file either as hard copies or electronically. Accessing the file from the building department will also determine whether the proper permits were obtained if alterations were made to the interior or exterior portions of the structure. In many cases, walls, wiring, and mechanical systems are changed without the required permits.

■ Building Materials

Building materials can have an impact on a fire's ignition and growth. The types of materials and their chemical composition,

Table 5-1 Building Materials

Characteristics	Influencing Factors
Ignitibility	■ Minimum ignition temperature ■ Minimum ignition energy ■ Time/temperature relationship for ignition
Flammability	■ Heat of combustion ■ Average and peak heat release rate ■ Time to peak heat release rate ■ Mass loss rate ■ Air entrainment
Thermal inertia	■ Reaction to heating ■ Ease of ignition ■ Fuel load
Thermal conductivity	■ Good conduction versus poor conduction
Toxicity	■ Quantity and types of gases produced by a material while burning
Physical state and heat resistance	■ Temperature at which a material changes phase ■ Amount of heat required to ignite the material in its different phases
Orientation, position, and placement	■ Different burning characteristics exhibited by the materials, depending on whether they are vertical or horizontal ■ Flame spread ratings, which can be obtained through the Steiner Tunnel Test (see NFPA 921, Section 7.2.3.7.2)

thermal conductivity, and density all have a definitive impact on the fire's growth and speed. Items commonly used in today's construction, such as plastics and synthetic materials, can greatly elevate heat release rates. Thermoplastics can be transformed from solids to liquids to ignitible gases and can actually ignite into flaming drips and pools when burning, causing drop-down damage and patterns. Thermoset plastics pyrolyze directly to ignitible gases and do not tend to drip and flow. It is important for the investigator to determine the type of plastic material in the area of origin so as not to confuse a drop-down pool of thermoplastic material with a suspected ignitible liquid. TABLE 5-1 includes some of the factors that should be considered.

Orientation, position, and placement of materials make a difference in how the materials react under fire conditions. For instance, carpet is generally placed horizontally on the floor; when carpet is placed vertically on the wall, the flame spread of the carpet is greatly increased.

Orientation, Position, and Placement of Loads

The fuel load is the total quantity of combustible contents of a building, space, or fire area, including interior finish and trim, expressed in heat units or the equivalent of weight in wood. One thing that many investigators overlook is the orientation, position, or placement of fuel loads within a building or area. Where an item is located, or how it is installed, can have an impact on how materials burn and how fast a fire progresses. For example, a trash can that is located 3 feet from a desk that catches fire will have different potentials for fire spread as compared with a trash can located immediately adjacent to the edge of the desk.

Interior finish and composition of furnishings also play a major role in the speed and intensity of a fire. A piece of 2 inch × 4 inch oak wood will burn much slower and create less char than a piece of wood with less density, such as a 2 inch × 4 inch piece of balsa wood.

The Steiner Tunnel Test (ASTM E 84) establishes flame spread indexes commonly used in fire and building codes. The Steiner Tunnel Test burns a sample of the material in a horizontal orientation, suspended on the ceiling of the 24-foot-long Steiner Tunnel. Many of the materials tested in the Steiner Tunnel are not designed or intended to be applied in building designs as wall or ceiling coverings if not installed as designed (e.g., orientation). The actual flame spread of the material as used in construction is often different and might bear no real relationship to its ASTM E 84 flame spread index classification (i.e., Class I, Class II, and Class III) and is therefore not an acceptable interior finish material per the fire and building code. Interior finishes designated in fire and building codes set by these tests are required according to occupancy type and size. An assembly that is not rated does not offer information as to the reaction of the assembly under the standard fire tests.

We have seen the devastating effects when interior finishes do not meet the code requirements. There have been many instances that have resulted in multiple fire deaths when these codes are violated. This could help the investigator explain the speed of a fire or the reason for the loss of life.

■ Occupancy

The investigator should determine whether there were any changes in use or occupancy classification during the life of the building. A change in use or occupancy classification can result in the introduction of fuel loads for which the building's fire protection systems were not designed. A change of occupancy is often accompanied by changes to structure and systems, making the "as-built" even more different than the original plans. In these cases, egress, active and passive fire protection, and other safeguards may not be placed as appropriate.

For example, consider a large, fully sprinklered building that was originally designed to house the paint section of a large home-improvement retail occupancy. The paint section is designed with a sprinkler system that has rack storage protection as well as the standard suppression protection. The area was specifically designed for this high-hazard fuel load. The owners decide to move the paint section to another portion of the store but do not add the additional sprinkler protection. With a paint section housing items such as enamel paints, paint thinners, and strippers, if a fire were to occur in this area, the standard sprinkler protection may be overwhelmed by the fuel load present. (See the "Failure Analysis and Analytical Tools" chapter in this textbook for additional discussion of occupancy and how it can affect fire spread in a building.)

Fire Investigator Tip

It is important to determine whether a building has changed occupancy classification since it was built or renovated. Different occupancies require different fire protection systems depending on load, contents, etc.

Types of Building Construction

There are many construction types employed throughout the world. This text addresses the most common types. The investigator should determine and document the type of construction based on the structural elements of the building. Documentation of structural elements, breaches, structural changes, or other factors that may influence the integrity or fire spread of the structure should be noted. The following discussion of construction types includes a list of features typical of each. (A wood wall stud is commonly called a 2 by 4, reflecting the *nominal* lumber measurement. The actual dimension of the 2 by 4 [2 × 4] is 1¾ by 3½ inches, which is the *dimensional* lumber measurement.)

Ordinary Construction

In ordinary construction, exterior walls are masonry or other noncombustible material FIGURE 5-2. It is referred to as ordinary construction because it is used in a wide variety of buildings. The floor, roof, and partition framing are wood assemblies and use the braced or platform framing methods. NFPA 220, *Standard on Types of Building Construction*, classifies ordinary construction as Type III construction. Open vertical shafts, combustible materials, and multiple ceilings are a few examples of factors that affect fire spread in this type of construction.

Wood Frame Construction

In wood frame construction, exterior walls and load-bearing components are constructed of wood or other combustible materials.

Wood frame construction is generally associated with residential or commercial construction. Lightweight wood construction can be used in any wood construction. Buildings utilizing wood frames are of limited size, with their floor joists and vertical supports generally spaced 16 inches on center. Vertical supports can be 2- by 4- or 2- by 6-inch nominal wood members. These components alone offer little fire resistance; flames and hot gas can penetrate into spaces between joists or studs, allowing the fire to spread. They may be sheathed with a fire-resistive material such as gypsum board, lath and plaster, or mineral tiles. Even non-fire-rated sheathing can provide some fire resistance. It is important to remember that a fire-rated wall may not necessarily stop heat from being conducted through the wall to underlying members or to combustible materials on the other side of the wall. If a building has a masonry veneer exterior over wood framing, it is still considered wood frame construction. Floor joists in lightweight wood frame construction are normally spaced 16 in. (0.4 m) on center but can be spaced up to 36 in. (1 m) on center depending on size and load capacity.

A downside to wood frame construction is the gaps that can be present between the joists or studs, which can allow hot gases to penetrate and fire to spread outside the area of origin. To compensate for this, wood frame construction can be sheathed with a fire-resistive membrane, which can provide up to a 2-hour fire resistance when applied according to ASTM testing standards.

Wood frame construction is classified as Type V construction, as defined in NFPA 220. The characteristics of ordinary construction versus frame construction are shown in TABLE 5-2.

© Ken Hammon/USDA

FIGURE 5-2 Ordinary construction is used in a wide variety of buildings.

Table 5-2 Ordinary Construction Versus Frame Construction

	Ordinary Construction	Frame Construction
Exterior walls	■ Masonry or noncombustible material	■ Wood
Interior walls	■ Wood assemblies ■ Platform or braced frame assembly	■ Wood assemblies ■ Platform or braced frame assembly

Platform Frame Construction

Platform frame construction is used for a majority of modern wood-frame construction FIGURE 5-3. In this method, walls are placed on top of platforms or floors, and thus, the platforms act as a fire stop for floor-to-floor vertical fire spread; however, barriers that are combustible can be overcome in time. Areas of concern with this type of construction are concealed spaces in the soffits (the horizontal undersides of the eaves or cornice) and vertical openings for utilities.

Safety Tip

Checking the floor or roof to see whether it is spongy does not ensure safety with modern lightweight construction.

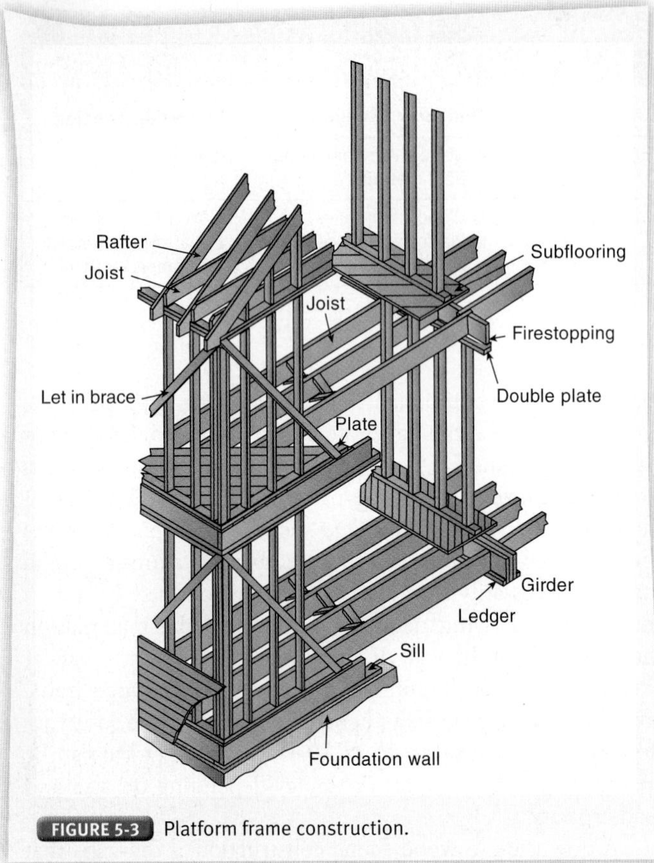

FIGURE 5-3 Platform frame construction.

FIGURE 5-4 Balloon frame construction.

Balloon Frame Construction

In balloon frame construction, the exterior wall studs extend from the foundation wall to the roofline FIGURE 5-4. Because of this feature, building codes have for many years required the installation of firestopping in all vertical channels created by the studs. Firestopping can be in the form of wood boards or by filling the void space with noncombustible materials, such as insulation. If fire stops are properly installed, the fire performance of the building can be similar to that of platform frame construction; however, lack of firestopping can allow uninhibited vertical fire spread or fire ignition by fall down from the attic area. Some forms of firestopping can be removed during the installation of utilities such as building wiring, HVAC, or other services. Balloon-frame construction can cause rapid fire extension through unprotected vertical channels, and horizontal extension is probable due to the open connections of the floor joists to the vertical channels. Open channels also allow for fall down of burning debris and combustion gases to convect downward. There can be more extensive burning at the upper level than where the fire originated. The investigator must identify all avenues of fire travel in this type of construction. Fire can break out in an area remote from the point of origin or bypass one floor to another because of the ability to spread through vertical channels.

Plank-and-Beam Construction

In plank-and-beam construction, larger beams such as 4 by 10 or 5 by 12 inches are used FIGURE 5-5. The beams are more widely spaced at 4 or 6 feet on center, and the beams are

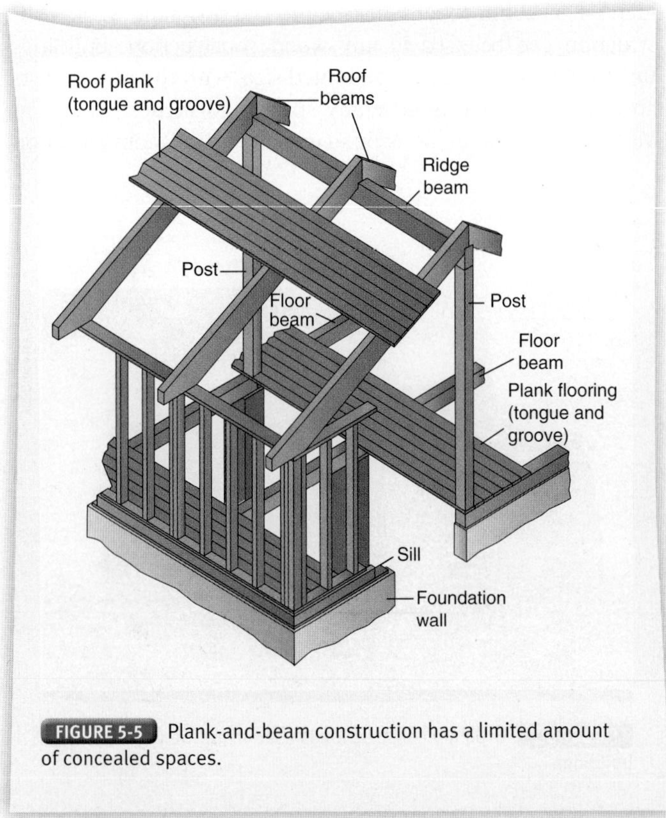

FIGURE 5-5 Plank-and-beam construction has a limited amount of concealed spaces.

Safety Tip

Fires that have burned in balloon frame walls destroy the structural integrity of the building. Collapse is a serious threat.

supported by posts. The floor decking has a minimal thickness of 2 inches (5 cm) and is generally of the tongue and groove type, which helps limit fire spread. This construction type has a limited amount of concealed spaces, and the exterior finish has no structural value. Plank-and-beam construction allows for large spans of unsupported finish material, which may result in failure of structural sections with large frame members still standing. Large areas on the interior often have exposed combustibles that may allow for flame spread.

Post-and-Frame Construction

Post-and-frame construction is similar to plank-and-beam construction. The structure uses larger elements with an identifiable frame or skeleton of timber that is fitted together FIGURE 5-6. An example is a typical barn construction where the posts provide a majority of the support and the frame provides a place for the exterior finish to be applied.

Courtesy of Glenn Corbett.

FIGURE 5-6 Post-and-frame construction.

Courtesy of APA - The Engineered Wood Association

FIGURE 5-7 Heavy timber construction has exterior walls that consist of masonry construction and interior walls, columns, beams, floor assemblies, and roof structures that are made of wood.

Heavy Timber Construction

Heavy timber construction employs structural members that are of unprotected wood, the smallest dimension being 6 or 8 inches (15–20 cm) FIGURE 5-7. Floor assemblies are constructed of 2-inch (5 cm)-thick tongue-and-groove end-matched lumber, and no concealed spaces are permitted. When heavy timber construction is used, building codes require the bearing walls to have a 2-hour rating. Fire spread may be rapid due to the wood frame components, open spaces, and large areas of interior combustibles.

Contemporary log homes use specially milled logs for the exterior walls and for many of the structural elements. The interior construction is generally 2 by 4 wood frame, but the interior finishes are typically made to look like the milled log exterior walls. This inconsistency between construction and materials may cause confusion in cases of rapid fire spread and failure. Another hazard with contemporary log homes is that they often contain large, open spaces, which facilitate fire spread.

Mill construction is an early form of heavy timber construction. Mill construction was influenced and developed largely through the Boston Mutual Fire Insurance Company and other insurance companies when they recognized a need to reduce large fire losses occurring in factories. The exterior walls are of masonry construction, with the columns and beams constructed of heavy timber. Walls were generally assigned a fire rating of at least 2 hours, with interior walls having rated doors between compartmentized areas. The lack of concealed spaces in this type of construction enhances the fire-resistive nature of the building. Protection of vertical openings plays a critical role in limiting the passage of fire. Fire sprinkler systems are provided in this type of construction. Scuppers are provided in the walls for drainage to reduce water damage.

Alternative Residential Construction

In addition to wood frame site-built construction, other forms and materials are being used for residential construction, such as manufactured housing and steel-framed construction.

Manufactured housing employs a construction technique whereby the structure is built in one or more sections. In traveling mode, the sections are 8 feet (2.4 m) or more in width, 40 feet (12.2 m) or more in length. These sections are then transported to the building site and assembled there, where they can be 320 ft^2 (29.7 m^2) or larger. Another type of manufactured housing builds the structure on a steel frame equipped with wheels that allow it to be transported easily to the site. This type of manufactured housing is often referred to as a mobile home.

Manufactured homes are the least expensive to build. Manufactured homes are starting to incorporate residential-type features like steeper pitch roofs, residential-size windows, and covered decks. Manufactured homes consist of four major components or subassemblies: chassis, floor system, wall system, and roof system FIGURE 5-8. The chassis generally consists of two longitudinal steel beams, braced by steel cross members, that receive all of the vertical loads from the wall, roof, and floor. When placed onsite, these loads are transferred to stability devices, which may be piers or footings, or to a foundation. Steel outriggers cantilevered from the outside of the main beams bring the width of the chassis to the approximate width of the overall structure.

Since the mid-1970s, Department of Housing and Urban Development (HUD) Standard homes offer more substantial construction. Construction of HUD-approved structures includes wood studs with interior finishes of, most often, gypsum wallboard. Exterior sidings can be metal, vinyl, or wood. Roof systems in HUD-approved structures are stronger and more solidly built, many times with gable or hip-style roof lines. Roof construction can be similar to that of a site-built home with asphalt paper/shingle roof covering over oriented strand board (OSB) or plywood attached to roof rafters or wood trusses. The wall system is bonded into a complete unit between the roof and floor system by steel ties.

Older units (pre-HUD) are usually constructed of 2 by 2 wood exterior studs covered with asphalt paper as a vapor barrier. The paper is then covered with aluminum siding skin attached directly to the studs. The interior walls are generally wood paneling with no insulation in the wall voids. Roof construction is most often a minimal flat wood frame with a steel exterior skin. This type of construction allows for rapid spread of fire with greater intensities due to the minimalist construction and the lack of interior finishes that resist fire. Newer manufactured homes use gypsum wallboard on walls and ceilings and incorporate code-required smoke alarms. Comparatively, fires in these structures are similar to those found in site-built homes. The fire spread is not as rapid as the pre-HUD homes. The use of early detection devices wired to the house power reduces the chance of a large fire causing higher monetary loss and lowers the risk for death and injury from fire.

A modular home is sometimes referred to as a system-built home or prefabricated home. Modular homes are similar to manufactured homes in that they are built in a controlled environment, utilizing technology similar to that found in newer site-built homes. Modular homes are prefabricated houses that consist of multiple sections manufactured in an offsite facility and then delivered to a location of choice. The modules are assembled into a single residential building using either a crane or trucks. Modular homes do not have axles or a frame, but rather are built on a contractor-provided foundation. Modular homes can be multilevel or single floor in design and can be customized to suit the buyer's preferences. Because of the type of construction, fires in modular homes will have burn tendencies that are similar to site-built wood frame homes.

Steel-framed residential construction is becoming more common. Site-built, panelized, and pre-engineered systems utilize steel framing methods. Steel-framed construction has characteristics similar to those of wood frame construction but is noncombustible; however, steel framing can lose its structural capacity during exposure to extreme heat FIGURE 5-9.

© Leif/ShutterStock, Inc.

FIGURE 5-8 Manufactured home.

© Harold Koch/AP Photos

FIGURE 5-9 Steel will begin to lose its strength at temperatures of 1000°F (537°C).

Manufactured Wood and Laminated Beams

Laminated beams are structural elements that have characteristics similar to solid wood beams. They are composed of many wood planks that are glued or *laminated* together to form one solid beam and are generally for interior use only. They are commonly referred to as *glulam beams*. They will behave like heavy timber until failure. Effects of weathering decrease their load-bearing ability. The investigator should document the size of individual members as well as the overall size of the beam.

Wood I-beams have smaller dimensions than floor joists and therefore can burn through and fail sooner than dimensional lumber. Openings in the web for utilities may reduce the structural integrity of the web. I-beams must be protected by gypsum board to help delay collapse under fire conditions.

Lightweight wood trusses are similar to other trusses in design. Individual members are fastened using nails, staples, glue, or metal gusset plates (gang nail plates) or wooden gusset plates. Truss failure can occur from gusset plates failing even before wood members are burned through. When one member of the truss fails, the other members take on additional loads and may become stressed, which may cause the entire truss to fail.

Rating of Wood Frame Assemblies

A fire rating is accomplished by covering the wall with a noncombustible finish, commonly gypsum board, though remember that a fire-rated assembly may conduct heat through to the underlying combustible members. Field installations can vary from those used in testing and evaluating the fire integrity. Therefore, a fire-rated wall in the field may fail sooner than expected depending on the quality of installation.

Noncombustible Construction

Noncombustible construction is used primarily in commercial, industrial, and high-rise construction. The building materials used in this type of construction do not add to the fuel load. Columns and walls may be built of materials that are strong in compression but weak in tension. Examples of these materials are brick, stone, cast iron, or non-reinforced concrete.

Ductile materials such as steel will deform before failure during fire conditions, and this deformation can be elastic or plastic. In the elastic range, a material deforms and then assumes its original shape with no loss of strength after the load is removed. In the plastic range, a material is permanently deformed but may continue to bear the load. In either event, elongation or deformation of structural materials can produce building collapse or damage.

Metal Construction

Metal is commonly found exposed in unfinished spaces and may fail in a period as short as 3 minutes during flashover. Fire-rating tests are not necessarily indicative of how the member will perform under fire conditions. These tests are performed in a strict, regimented manner under calculated conditions. Hostile fire scenes are unpredictable and can generate conditions that differ from the laboratory tests. Investigators need to familiarize themselves with the conditions of the building and with its interior finishes and ventilation effects and determine whether there may have been exposed metal or steel that could have contributed to failure sooner than one might expect. TABLE 5-3 describes some of the features of metal construction.

Table 5-3 Steel Versus Masonry/Concrete

Material	Factors
Steel	Can conduct electricity. Good conductor of heat. Loses its ability to carry a load well below temperatures encountered in a fire. Can distort, buckle, or collapse as a result of fire exposure. Amount of distortion depends on factors such as heat of the fire, duration of exposure, physical configuration, and composition of the steel.
Masonry and concrete	Due to mass will generally absorb more heat than steel. Good thermal insulators. Do not heat up quickly and do not transfer heat through them as easily as steel. Masonry and concrete will carry a load much longer at equal temperatures when compared with steel.

Concrete and Masonry Construction

Concrete and masonry constructions have inherent fire resistance because of their mass, high density, and low thermal conductivity FIGURE 5-10. They are strong under compression but weak under tension. Structural failure often occurs at connection points. Failures in these types of materials relate to reinforcement failure as a result of heat transfer through the concrete or masonry or surface spacing and exposure of the reinforcement to fire temperatures. Failure in the steel connections between components results at temperatures well within the range found in structure fires.

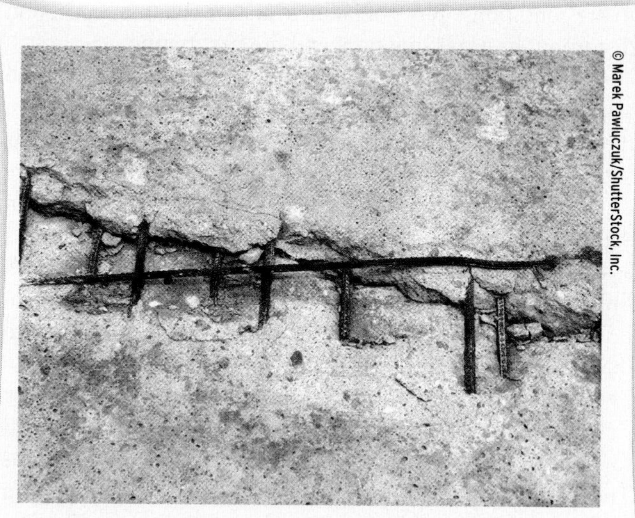

© Marek Pawluczuk/ShutterStock, Inc.

FIGURE 5-10 Concrete is noncombustible and provides thermal protection around steel reinforcing rods. In this case, the thermal protection has been compromised.

Construction Assemblies

Assemblies are defined as manufactured parts put together to make a completed product and may or may not be fire-resistance rated. Assemblies are a collection of components, such as structural elements, that form a wall, floor, or ceiling. Doors form a part of a larger unit and can be considered a component. Most nonrated assemblies do provide some resistance to fire or smoke. Failure of one assembly may lead to failure of another. The way that assemblies react in a fire often influences how a fire grows and spreads. Because assemblies are designed as a complete unit, the integrity of the unit and its ability to perform during a fire are dependent on the unit being manufactured, installed, and maintained in the form for which it was intended. Variables can compromise the assembly and cause it not to function according to its design and testing. Examples are fire doors blocked open between one compartmentalized fire area to another or a fire-rated wall that separates compartmentalized fire areas where the rated wall is violated above the ceiling tiles by construction crews who knock holes in the wall to pass utilities through from one side to the other without properly protecting the penetrations. Holes left in the walls, ceiling tiles missing or not clipped down, areas through which ductwork is passed without proper fire and/or smoke dampers, and so forth can allow for passage of fire.

Fire-rated assemblies are rated for a specific fire test criterion and under test conditions. The actual fire conditions found in the structure may be more severe and may cause the assembly to fail at a time that is actually less than the hourly rating assigned as a result of the fire test. Fire investigators need to recognize rated assemblies and examine them after a fire to identify any flaws in a component in the overall assembly.

Floor, Ceiling, and Roof Assemblies

Floor, ceiling, and roof assemblies are of particular concern to the fire fighter as well as the fire investigator. These assemblies are among the first to fail when structural elements are exposed to fire conditions. Types of failures include collapse, deflection, distortion, heat transmission, and fire penetration.

Factors that can affect the failure of floor, ceiling, and roof assemblies include type of structural element, protection from the elements, span, load, and beam spacing. Added loads, such as water during firefighting operations, can contribute to failure. Penetrations in assemblies for utilities are common. Although penetrations are required to be sealed in a fire-rated assembly, often they are not. Floor assemblies are tested for fire spread from below, not from above. Roof stability can be a critical factor during firefighting operations and for fire dynamics.

Potential collapse should be a vital concern to the fire investigator on a fire scene. Structural elements are not intended to maintain their strength when subjected to fire conditions. The protection afforded to structural elements, such as gypsum wallboard, is intended to prevent the immediate involvement of the structural elements; however, once this protection fails and the structural elements are affected, their ability to maintain their load-bearing capabilities is diminished.

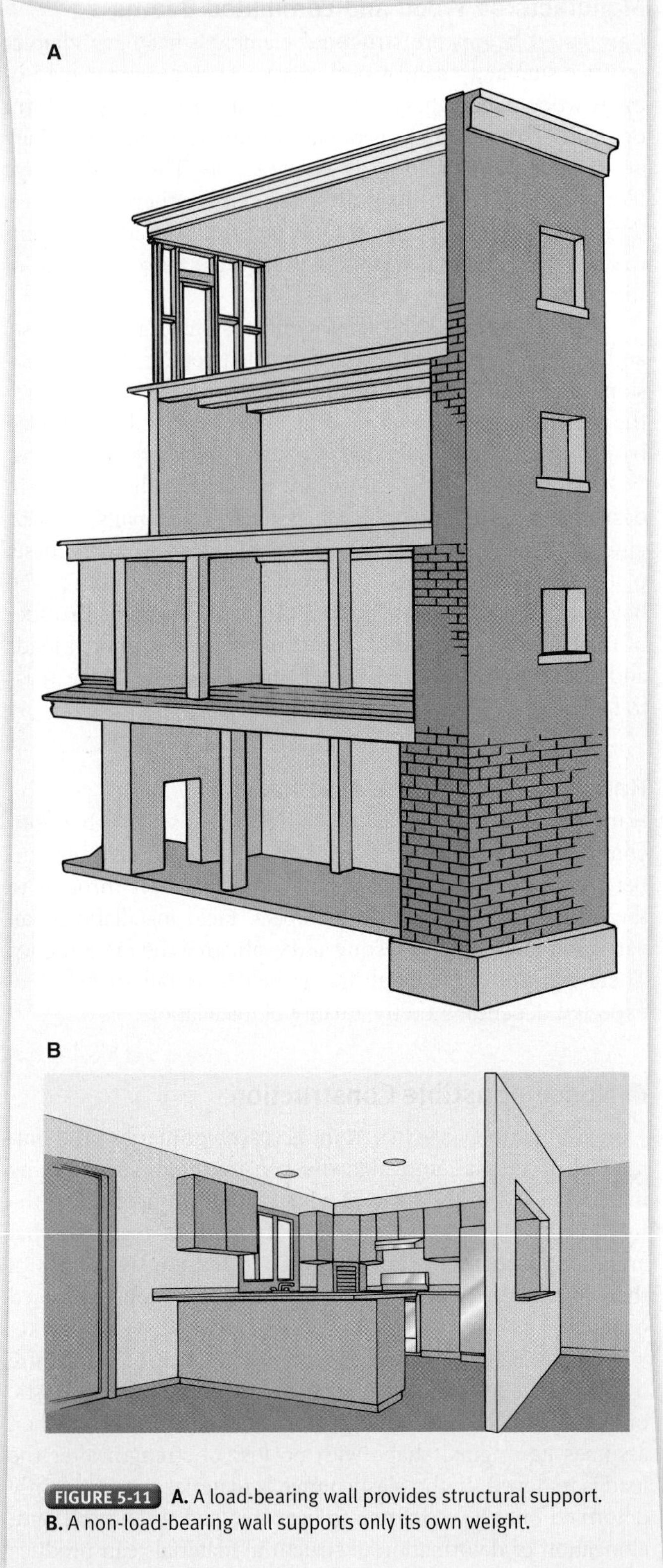

FIGURE 5-11 **A.** A load-bearing wall provides structural support. **B.** A non-load-bearing wall supports only its own weight.

Walls

Walls serve as barriers to fire and smoke spread and can be made to a wide variety of standards and in a wide variety of types. They may or may not be fire-resistance rated or load-bearing **FIGURE 5-11**.

Penetrations in these assemblies for utilities are common and are required to be sealed in a fire-rated assembly. A fire wall separates buildings or compartmentalizes interior areas of large buildings to prevent the spread of fire, while creating a fire-resistance rating and structural stability. A fire barrier also resists the passage of fire and smoke. Fire walls and fire barriers normally do not need to meet the same requirements as smoke barriers, but may have to should they also be constructed as a smoke barrier. A smoke barrier is a continuous membrane, either vertical or horizontal—such as a wall, floor, or ceiling assembly—designed and constructed to restrict movement of smoke. Fire barrier walls or fire walls using gypsum board will use a type X gypsum wallboard. Non-fire-rated walls can provide varying levels of fire resistance. An assembly without a rating means that no test was done for that type of wall, ceiling, or floor component. As such there is no recognized time frame to determine when this component may fail, although they may provide some resistance to the spread of fire within the building. Assemblies with smoke damper systems are designed to restrict the passage of smoke. Such barriers may or may not have protected openings to protect from the passage of fire.

■ Doors

Doors can serve as a critical factor to limiting fire spread throughout the structure. Doors may be made of a variety of materials and may be fire rated or non–fire rated. Any opening in a fire-rated wall or partition is required to have a fire-rated door and associated door assembly installed. Fire-rated assemblies include rated frames, hinges, closures, latching devices, and, if provided (and allowed), glazing. Fire doors are built from a variety of methods and materials and have a fire protection rating. Fire doors may be constructed of solid wood, steel, or steel with an insulated core of wood or mineral material. The insulating value of fire doors aids egress, particularly in multistory buildings, and provides some protection against autoignition of combustibles near the opening's unexposed side. They must be closed to provide an effective barrier to smoke, heat, and fire. Any door installed in a fire-resistance-rated wall assembly should be installed as a fire-protection-rated door assembly FIGURE 5-12.

The hourly rating of the door will be dependent on the rating of the fire wall and will usually be less than the wall system rating. A fire-rated door assembly must have the following properly operating components rated as part of that system:

- Hinges
- Closures
- Latching devices
- Glazing

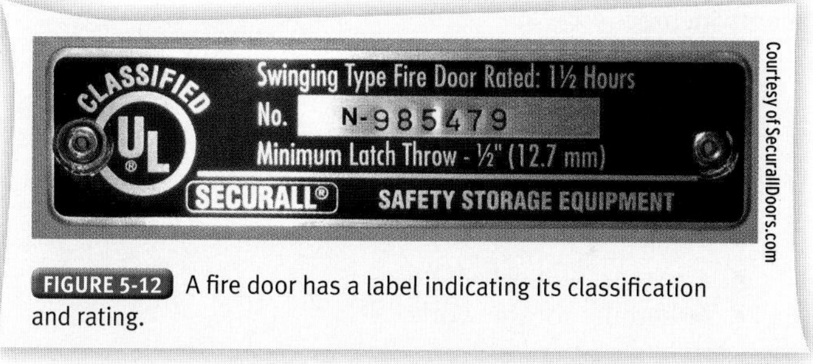

Courtesy of SecurallDoors.com

FIGURE 5-12 A fire door has a label indicating its classification and rating.

See NFPA 80, *Standard for Fire Doors and Other Opening Protectives*, for more information on fire doors.

■ Concealed Spaces

Concealed spaces commonly have areas where penetrations are used to provide access for HVAC systems, plumbing, electrical, computer and/or telephone lines, and other functions. These penetrations are required to be sealed to meet the rating of the wall or floor through which it passes. Many times contractors or others come along after construction and create openings in these rated separations without properly sealing them. Investigators should examine for these possibilities, which may have allowed passage of fire or smoke from one protected area to another.

Effects of Weather on Building Systems

Severity of weather may require an investigation to be delayed. Weather may also necessitate that special clothing and equipment be used to perform the duties safely and effectively. Extreme weather such as heavy snowfalls or rains collecting on flat roofs that do not effectively drain can create unanticipated loads on structural components, creating collapse potential. This is especially likely after a fire has occurred that may further weaken the structure.

During the fire, wind direction and speed, humidity levels, and ambient temperatures can have an effect on fire development. Higher humidity levels can raise fuel moistures, while higher temperatures can speed the evaporation of the fuel moistures. Wind direction can have an impact on ventilation effects of the fire. Wind speed, such as that found in hurricanes or tornadoes, can exert a total wind force of several tons against a wall. Extreme forces or loads may require additional bracing to the building before entry to provide adequate safety for the investigators performing the origin and cause examination.

Construction Materials

Unprotected structural steel loses strength at high temperatures and should be protected from exposure to heat produced by building fires. There are a variety of methods that can provide the appropriate insulation needed to protect the steel. One method to protect the structural steel is encasement with materials such as poured concrete or board systems that provide a barrier. Examples of board systems are calcium silicate and gypsum. Surface treatments such as spray-on fireproofing or intumescent coatings may also be used.

A less common form of protection is filling hollow members, which acts as a heat sink to reduce the temperature increase in the steel element. Reinforced concrete can also be used as a protective coating but can affect the fire resistance of the concrete, including the concrete's density, aggregates, and moisture content. Wood encasement may or may not provide reasonable structural integrity during a fire. Size and moisture content of the wood are factors to be considered, as is the presence of fire retardant in the wood to delay ignition and reduce combustion.

Impact of Passive Fire Protection Systems on Investigation

Part of the investigator's job is to assess whether a building is provided with a passive fire protection system. Passive fire protection can best be defined as fire-resistance-rated wall and floor assemblies used to form fire-rated compartmentalized areas meant to control the spread of fire. These can be in the form of occupancy separations, fire partitions, or firewalls used to keep fires, high temperatures, and flue gases within the fire compartment of origin, to aid firefighting and evacuation processes. The investigator must determine whether any of these provided systems failed, and if so, how and why. There should also be attention paid to how damage to the systems may have aided in the development of the fire's growth. Problems associated with failure may include:

- Improper design
- Inadequate installation
- Change in occupancy and associated hazards
- Breaches in compartment walls or damage to applied coatings after initial system installation

Design and Installation Parameters of the System

When a passive fire protection system is identified, the investigator must evaluate whether each system and/or component had a role in the evolution of the fire. If it has, an analysis of construction quality, the materials used, and thickness of each component has to be evaluated and documented through testing, listings approvals, and/or certifications from recognized testing laboratories, in conjunction with a comparison of all applicable codes. The investigator should take into account the date of the construction and which codes, standards, guides, and manufacturers' instructions were in effect at that time, or whether an update to the most current code would have been required. There should also be an analysis of the protection of openings and doors, including the fire rating of the door, door frame, door hardware, and construction surrounding the door. Windows in the protected openings must also be evaluated, including window glass (type, thickness, and number of panes), opening mechanisms/hardware, and frame and construction surrounding the windows.

Ductwork penetrations in a fire-rated wall/partition/floor and ceiling must be examined to determine any impact on the fire evolution. If smoke and/or fire dampers were required, did they operate as required? All penetrations must be examined to determine if they complied with applicable codes, standards, and manufacturers' instructions.

Documentation and Data Collection

When a passive fire protection system is present at a fire scene, it is important to ensure that proper documentation is included in your investigative process. Design plans, design specifications, as-built drawings, equivalencies or alternate levels of protection approved by the authority having jurisdiction, and other documents relating to the assembly should be noted. Copies of building and fire permits; all available invoices of work to the system; measurements; diagrams and/or photographs; and maintenance, inspection, and testing records should be obtained to properly document the fire scene. The investigator should also document any occupancy changes that may have occurred since the original installation to ensure that the passive fire protection system meets the current occupancy and fuel load.

Care should be taken by the investigator to ensure that no spoliation of evidence occurs and that the passive fire protection system is protected to avoid spoliation issues.

Code Analysis

As a fire investigator, it is necessary to be familiar with codes and where to find information on the installation of passive fire protection systems. Some of the codes you may need to be familiar with are listed here; you can also consult with the local building official and fire prevention staff to assist you:

- Building codes
- Fire codes
- Property maintenance codes
- NFPA 101, *Life Safety Code* (if adopted by the authority having jurisdiction)
- Localized code amendments

Permit applications and/or building permits can provide dates as to what codes may be in effect at the time of the installation of the passive fire protection system. Both the name of each code and the specific edition are important in determining whether the installation met the code at that time. Building codes will include the following requirements:

- Type of construction
- Fire-resistance-rated construction
- Requirements for glass and glazing, gypsum board and plaster, etc.
- Specific examples for ASTM and Underwriters Laboratory

Design Analysis

Items that may be found in the approved construction documents include:

- Drawings
- Specifications
- Design calculations
- Floor plans
- Details
- Cross-section elevations
- Schedules for the work

Installation Analysis

Just because a system was constructed in accordance with the approved construction documents does not necessarily mean that it complies with the applicable codes. There should always be an examination of the codes in reference to the approved local documents. There should also be an examination of the building permits and the inspection sign-off process.

Computer Modeling

If computer modeling or calculations are used to assess the performance of the passive fire protection system, the thermal properties of the walls, ceilings, and floors will likely be needed for the evaluation.

Testing and Maintenance

The fire investigator should assess whether there were any requirements for periodic testing or examination of the passive fire protection system, including fire prevention inspections; however, the building owner may be required to maintain the systems by ensuring that they are not damaged or breached. During the origin and cause determination, the fire investigator should examine the components in TABLE 5-4.

Table 5-4 Structural Damage Considerations for Passive Fire Protection Systems

Component	Considerations
Penetrations	■ Were they properly sealed?
Joint systems	■ Did the system fail by not resisting the passage of fire as expected by its fire rating? ■ Was it securely in or near the joint for the entire length? ■ Did the building code contain exceptions for requiring fire-rated joint systems?
Fire doors	■ Were all the necessary components included in the assembly? ■ Did the closing device perform properly? ■ An examination of the door label should be recorded if available.
Fire windows	■ Were the glass and glazing the proper thickness? ■ Was wire glass required or not? ■ Was the framing appropriate for the window rating?
Duct and transfer openings	■ Were any fire dampers, smoke dampers, fire/smoke dampers, and/or ceiling radiation dampers required? If so, did they operate as required when activated by fire, smoke, or automatic activation? ■ Examine the actuating device/method to ensure that it operated as needed.
Fireblocking and draftstopping	■ Did the structure require fireblocking or draftstopping in combustible concealed spaces? ■ Was the fireblocking or draftstopping installed in the correct locations? ■ Was the material used appropriate for the rating and was it the proper thickness?

Wrap-Up

© Greg Henry/ShutterStock, Inc.

Ready for Review

- The design, materials, and construction type of a structure influence the development and movement of any fire that may occur within it.
- Modern design considerations and construction features are a direct result of the analysis of (often catastrophic) fires.
- A building's characteristics affect the development, spread, and control of a fire. These characteristics include the type of construction; the integrity and performance of its structural elements under a fire load; and its building systems, including but not limited to active and passive fire protection systems.
- The investigator should determine and document the type of construction based on the structural elements of the building. Documentation of structural elements, breaches, structural changes, or other factors that may influence the integrity or fire spread of the structure should be noted.
- The way that assemblies react in a fire often influences how a fire grows and spreads. Because assemblies are designed as a complete unit, the integrity of the unit and its ability to perform during a fire are dependent on the unit being manufactured, installed, and maintained in the form for which it was intended.
- Weather can impact systems through wind spreading the fire, rain and snow increasing the load, increased humidity raising the fuel moisture, increased temperature speeding evaporation, and more.
- Fire spread is impacted by materials used in construction, such as steel, reinforced concrete, and wood.
- Passive fire protection can best be defined as fire-resistance-rated wall and floor assemblies used to form fire-rated compartmentalized areas meant to control the spread of fire. These can be in the form of occupancy separations, fire partitions, or fire walls used to keep fires, high temperatures, and flue gases within the fire compartment of origin, and to aid firefighting and evacuation processes.

Hot Terms

Assemblies Manufactured parts put together to make a completed product.

Balloon frame construction A construction type in which the exterior wall studs go from the foundation wall to the roofline. The floor joists are attached to the walls by the use of a ribbon board, which creates an open stud channel between floors, including the basement and attic.

Compartmentation A concept in which fire is kept confined in its room of origin, minimizing smoke movement to other areas of a building.

Dead load The weight of materials that are part of a building, such as the structural components, roof coverings, and mechanical equipment.

Fire barrier A wall, other than a fire wall, having a fire-resistance rating. Fire walls and fire barrier walls do not need to meet the same requirements as smoke barriers.

Fire wall A wall separating buildings or subdividing a building to prevent the spread of fire and having a fire resistance rating and structural stability.

Heavy timber construction A construction type in which structural members (i.e., columns, beams, arches, floors, and roofs) are made of unprotected, solid, or laminated wood, with large cross-sectional areas (8 or 6 inches [200 or 150 mm] in the smallest dimension, depending on reference).

Interstitial spaces The space between the building frame and interior walls and the exterior facade and with spaces between ceilings and the bottom face to the floor or deck above.

Laminated beams Structural elements that have the same characteristics as solid wood beams. They are composed of many wood planks that are glued or *laminated* together to form one solid beam. Designed for interior use; the effects of weathering decrease their load-bearing ability.

Lightweight wood trusses Similar to other trusses in design. Individual members are fastened using nails, staples, glue, metal gusset plates, or wooden gusset plates.

Live load The weight of temporary loads that need to be factored into the weight-carrying capacity of the structure, such as furniture, furnishings, equipment, machinery, snow, and rainwater.

Manufactured housing A construction technique whereby the structure is built in one or more sections. These sections are then transported to the building site and assembled on the site.

Mill construction An early form of heavy timber construction influenced and developed largely through insurance companies that recognized a need to reduce large fire losses in factories.

Modular home A dwelling constructed in a factory and placed on a site-built foundation, all or in part, in accordance with a standard adopted, administered, and enforced by the regulatory agency, or under reciprocal agreement with the regulatory agency, for conventional site-built dwellings.

Noncombustible construction A construction type used primarily in commercial and industrial storage and in high-rise construction. The major structural components are noncombustible; the structure itself will not add fuel to the fire. Examples of these materials are brick, stone, steel, masonry block, cast iron, or non-reinforced concrete.

Ordinary construction A construction type in which exterior walls are masonry or other noncombustible material and the roof, floor, and wall assemblies are wood.

© Greg Henry/ShutterStock, Inc.

Plank-and-beam construction A construction type in which a few large members replace the many small wood members used in typical wood framing; that is, large dimension beams, more widely spaced, replace the standard floor and/or roof framing of smaller dimensioned members.

Platform frame construction The most common construction method currently used for residential and lightweight commercial construction. In this method of construction, separate platforms or floors are developed as the structure is built.

Post-and-frame construction Similar to plank-and-beam construction, the structure uses larger elements. An example is a typical barn construction where the posts provide a majority of the support and the frame provides a place for the exterior finish to be applied.

Smoke barrier Continuous membrane, either vertical or horizontal, designed and constructed to restrict movement of smoke.

Soffits The horizontal undersides of the eaves or cornice.

Steel-framed residential construction A construction type with characteristics similar to those of wood frame construction but is noncombustible. Steel framing can lose its structural capacity during extreme exposure to heat.

Wood I-beams Constructed with small dimension or engineered lumber, as the top and bottom chord, with oriented strand board or plywood as the web of the beam.

Wood frame construction A construction type in which exterior walls and load-bearing components are wood. This type of construction is often associated with residential construction and contemporary lightweight commercial construction.

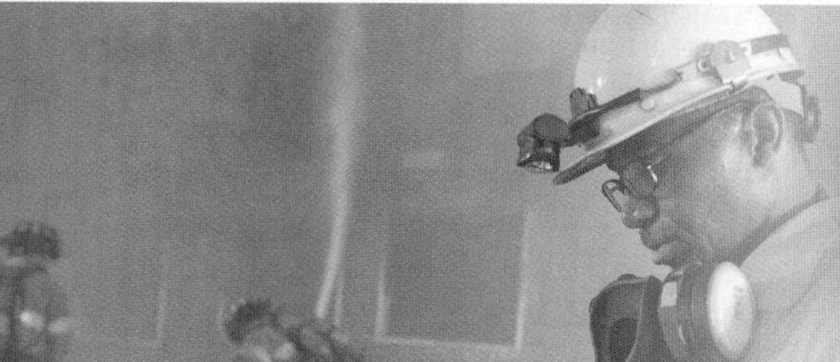

FIRE INVESTIGATOR *in action*

You are investigating a fire that occurred in a three-story brick warehouse built in the 1880s that had been remodeled to accommodate a woodworking shop on the first floor with individual storage spaces on the second and third floors. The structure is severely damaged with portions of the second and third floors having collapsed into the first floor. Due to the damage sustained to the structure, the building department has deemed the structure unsafe to enter.

After removing portions of the collapsed floors, constructed of large wooden beams and decking, you are able to enter the structure. You note severe flame and heat damage within the woodworking business with fire movement patterns indicating the fire originated within a dust collection unit that had been installed in an old freight elevator shaft.

Once you are able to access the remaining portion of the second floor, you locate a large number of cardboard file boxes containing paper records. The boxes are severely charred; however, legible documents that remain indicate that they are associated with a plumbing company that is renting the space.

1. Which of the following factors contributed to the vertical fire spread?
   A. Construction materials
   B. Doors and windows
   C. The elevator shaft
   D. The stored file boxes
2. What is the most likely construction type for this structure?
   A. Wood frame
   B. Mill construction
   C. Ordinary
   D. Fire resistive
3. The stored boxes located on the second floor are considered what type of load?
   A. Dead load
   B. Live load
   C. Fixed load
   D. Static load
4. Which concept in building construction is intended to keep the fire within the room of origin and minimize smoke movement to other areas of a building?
   A. Fire blocking
   B. Draft stopping
   C. Compartmentation
   D. All of the above

© Greg Henry/ShutterStock, Inc.

Fire Protection Systems

© Photos.com

© Brendan Byrne/age fotostock

Knowledge Objectives

After studying this chapter, you should be able to:

- Identify and discuss the operational characteristics of common fire alarm and detection systems NFPA 4.2.8. (pp 78–89)
- Identify and discuss the operational characteristics of water-based fire suppression systems NFPA 4.2.8. (pp 89–97)
- Identify and discuss the operational characteristics of non-water-based fire suppression systems NFPA 4.2.8. (pp 97–100)
- Discuss the components of fire suppression system analysis NFPA 4.2.8. (pp 100–105)
- Discuss how to document fire protection systems NFPA 4.2.8. (pp 105–107)

Skills Objectives

After studying this chapter, you should be able to:

- Document fire protection systems during a fire investigation NFPA 4.2.8. (pp 105–107)

Additional NFPA References

NFPA 13, *Standard for the Installation of Sprinkler Systems*

NFPA 13D, *Standard for the Installation of Sprinkler Systems in One- and Two-Family Dwellings and Manufactured Homes*

NFPA 13R, *Standard for the Installation of Sprinkler Systems in Low-Rise Residential Occupancies*

NFPA 25, *Standard for the Inspection, Testing, and Maintenance of Water-Based Fire Protection Systems*

NFPA 72, *National Fire Alarm and Signaling Code*

NFPA 921, *Guide for Fire and Explosion Investigations*

NFPA 2001, *Standard on Clean Agent Fire Extinguishing Systems*

CHAPTER 6

FESHE Course Outcomes

Fire Investigation I

7. Describe how fire progression is affected by fire protection systems and building design. (pp 78–107)

Fire Investigation II

There are no Fire Investigation II (FESHE) course outcomes for this chapter.

You Are the Fire Investigator

© Jones and Bartlett Publishers. Photographed by Glen E. Ellman

You are investigating a fire that occurred overnight at a local restaurant. The incident commander has informed you that several employees were still in the building preparing to close for the night when they discovered a fire within one of the deep fryers in the kitchen. He states that the employees attempted to extinguish the fire with a fire extinguisher but were unsuccessful. When fire personnel arrived, flames were visible from the roof directly above the kitchen area.

Upon examining the damage within the kitchen area, you note several dry chemical suppression system nozzles suspended above the various cooking appliances. A visual inspection of the system notes that several fusible links have melted, which should have caused the suppression and fire alarm system to activate. Additionally, you note the remains of a wall-mounted fire extinguisher located near the doorway to the kitchen area.

During your interview with the employees who discovered the fire, you learn they had left the deep fryers on while they continued to clean the dining area. They state that when they returned to the kitchen, the deep fryer was already on fire with flames spreading across the back wall. They state that they went back into the dining room and obtained a small extinguisher located near the front door; however, they were unable to reenter the kitchen as the entire room was full of smoke at that time.

1. What information will you need to obtain about the dry chemical suppression systems?
2. What information will you need to obtain about the appliances and cooking medium?
3. What is the significance of human behavior as it relates to your investigation?
4. How will a review of previous inspection reports assist you with your investigation?

Introduction

Buildings equipped with active fire protection systems offer investigators a resource that can be a valuable analysis tool. The type of system and the code and standard in force when the system was installed are factors that will determine how much information the system is designed to provide. However, the information even the most basic systems provide the investigator could help to determine when and where a fire started and how the activation and operation of the system affected the fire.

Fully utilizing this resource requires the investigator to acquire and develop basic knowledge of the various types of active fire protection systems; the associated devices and components; the design, installation, and operational characteristics for each type of system; and requirements for recurring inspection, testing, and maintenance of these systems. This includes familiarization with and analysis of design documents; manufacturers' documentation and data sheets; and inspection, testing, and maintenance reports and records. Also, and perhaps most important, the investigator should understand the importance of reviewing, analyzing, and utilizing the system activation history available from public emergency communication and dispatch centers, proprietary monitoring facilities, third party monitoring service providers, and the event data recorded and maintained in microprocessor-based fire alarm control units.

Active fire protection systems are categorized into three general areas: fire alarm and detection systems, fire suppression systems, and control and management systems. (The roles of control and management systems are briefly discussed in conjunction with the other types of fire protection systems). Each is further subdivided based on its operational characteristics, the suppression agent used, or the specific task performed when activated. When a building or structure is equipped with a number of different active fire protection systems, usually all of these systems interface, communicate, and interact in some manner.

A fire alarm system is designed to cause a response by building occupants through an assemblage of interconnected components and devices that initiate an alarm and notify by audio and visual means. Fire suppression systems use water and other additives, a chemical agent, or a gaseous agent to control, and in many instances extinguish, a fire. Water-based suppression systems are generally considered fire sprinkler systems, and suppression systems that use a chemical agent or gaseous agent are generally considered special hazard or non-water-based suppression systems. Control and management systems typically interface with other building systems to isolate, contain, or remove smoke and gases by starting fans, closing doors and fire dampers, and shutting down equipment. In addition, these systems can take control of elevators, placing them at predetermined locations, and unlock exit doors to permit emergency responders access into and throughout a building.

This chapter provides the investigator with basic information about the different system types, categories, components, design, installation, and operational characteristics to help develop the necessary skills and knowledge to interact with active fire protection systems. However, there are numerous resources available that provide more information about these systems and their components that would be of great benefit to the investigator or other interested party. Forming a solid knowledge base will help the investigator develop methods and strategies to probe, analyze, document, and draw conclusions concerning the role the fire protection system played in the detection, notification, containment, and suppression of a fire.

Fire Alarm Systems

General Information

Purpose of Systems

Surviving a structure fire depends on the amount of time between the fire starting, the occupants realizing there is a fire, and the occupants getting out of the building. A properly designed, installed, tested, and maintained fire alarm system increases building occupants' chances of survival by alerting them to take action. In addition, a fire alarm system can provide direct notification to emergency communication dispatch centers and to proprietary and third-party service providers who initiate the appropriate response without the need for an individual to call in the fire emergency. This alert will expedite the emergency response that in many instances will help limit property damage, and in some instances will save a life.

System Components

A fire alarm system is an assortment of interconnected components and devices that, once activated, outputs audible and visual signals to alert building occupants of a fire emergency to initiate a response. Three basic component groups make up a fire alarm system: the control unit, the initiating devices, and the notification appliances. Within these groups are a number of different devices and components that perform a variety of functions. Associated with the control unit are primary and secondary power supplies that power the unit, the initiating device circuits, and the notification device circuits. Initiating devices such as smoke and heat detectors may automatically respond to a fire condition and initiate an alarm, while a manual fire alarm box requires a person to operate the device. Notification appliances such as bells, horns, and strobe lights provide the audible and visual warning for the occupants to react.

General System Operation

The operation of a fire alarm system starts when an initiating device automatically detects or reacts to a fire condition or when a building occupant discovers a fire and manually initiates the alarm. Once the initiating device operates, a signal transmits to the fire control unit that activates the notification appliances. In addition to notifying the occupants, many systems send a signal to an on- or offsite monitoring location, where operators receive the alarm signal information and take the appropriate action. Most modern fire alarm systems interface with other building control and management systems and fire protection systems to initiate additional life safety operations. These operations might include starting fans to pressurize stairways, elevator shafts, and vestibules; closing smoke dampers and doors to compartmentalize a building; capturing elevators and moving them to predetermined floors; and unlocking certain doors to ensure safe ingress and egress.

Although some system technology dates back 50 years or more, even the oldest and most basic systems can provide information that is useful during an investigation. When an alarm activates, the fire alarm unit can provide information that might be as simple as a light marked with a location identifier, such as the floor number, and the device that activated. However, modern fire alarm units can provide very comprehensive information, including graphic and touch screen units that display the building footprint and/or alphanumeric information readouts that show the device type, type of alarm signal, floor, and the exact location of the device on the floor. A basic understanding of system operation will greatly assist the investigator and provide useful and realistic information.

Key Components of Systems

Fire Alarm Control Unit

There are two types of fire alarm control units (FACUs) still in use today: the conventional technology and the addressable technology units. The conventional technology unit has been around for more than 50 years, but even with technological advances provides only basic information, performance, and monitoring capabilities. There are still limited uses for this type of unit, especially in small buildings where there are few alarm system devices and it does not take much time to move through the building and locate the event. Addressable technology units use state-of-the-art computer and circuit technology to provide first responders extensive information concerning the fire event to monitor the system integrity for conditions that could inhibit the system and components from operating properly, and to provide an interface to other building fire protection and life safety systems.

Fire alarm control units generate three different types of output signals, and each has a specific purpose. The alarm signal alerts building occupants to a fire emergency with the expectation that the occupants will leave the building, call the fire department, or take any other necessary action to ensure life safety. The trouble alarm signal is to alert the responsible party (typically the building engineer or maintenance personnel) that there is a problem with the fire alarm system integrity. System integrity problems include ground faults, breaks in system wiring, power or component failure, communications problems, and device removal. The supervisory alarm signal is to alert the responsible party (typically the building engineer or maintenance personnel) that the normal (ready) status of the system has changed. Changes in normal status include closing electronically supervised fire suppression system control valves, high or low air pressure levels for dry pipe sprinkler systems, and air and water temperatures that are outside of the appropriate range. Additional changes include, water tanks with a low water level, water pressures that are too low or high, and any operation of associated fire protection equipment during non-fire conditions, such as starting a fire pump to exercise the pump and associated components.

To clear any of the three signals—and in most cases reset the fire alarm control unit—the condition that caused the signal must be investigated and resolved.

Power Supplies

The power supply is an integral part of the fire alarm unit and supplies the electrical power to the unit, system components, and system devices. Two power sources are required:

Fire Investigator Tip

The fire investigator should analyze all alarm, trouble, or supervisory signal information that is provided by the fire alarm control unit or any offsite monitoring facility to determine the status of the system and the sequence of events prior to, during, and after the fire.

Fire Investigator Tip

It is common during building renovations to upgrade or add to the fire alarm system. Many times, the power supply for the fire alarm unit does not have the capacity to support the additions. When adding any circuits, components, and devices, it is important to consider the effect on the power supply to support these additions, because a failure to evaluate and upgrade could lead to a system malfunction. To address this problem, installation of power expander panels provides the additional capacity to supplement the main power supply and support the system expansion. Power expander panels provide additional power, are typically located in utility closets in the area of the building where needed, and require a secondary/backup power source just like the main fire alarm control unit.

the primary power source is usually from the local commercial power utility, and the secondary or backup power source is usually rechargeable batteries. In some installations associated with factories or industry, natural gas, diesel fuel, or steam engine-driven generators act as both the primary and secondary power sources. The design capacity of both the primary and secondary power supplies takes into account the current draw from all the system circuits, components, and devices, plus some additional capacity for a margin of safety. This is especially important for the secondary source because code requirements establish the minimum time that a system must be able to sustain and operate on secondary power without failure.

Initiating Devices

Initiating devices interact with the fire alarm control unit through automatic or manual means to activate the fire alarm system. Automatic initiating devices activate the system when detecting changes in the monitored environment and include various types of fire detectors, water flow detectors, and component monitoring devices. Fire detection devices can be spot-type or line-type or use video images. The spot-type detector is a single element device, such as a smoke or heat detector, that is usually installed on the ceiling and is designed to protect a certain square footage area as determined by the manufacturer. The line-type detector runs along a continuous route throughout an area of a building to cover a hazard by using some type of electrically or thermally sensitive cable or light beam. The video detector uses high-definition video cameras to analyze digital images for changes in the pixels due to smoke and flame generation. Water flow and monitoring initiating devices are discussed later in this chapter.

Manual initiating devices signal the fire alarm system when a person operates the device. Manual initiating devices are limited to manual fire alarm boxes and electronic valve supervisory devices. Manual fire alarm boxes, also known as manual pull stations, all operate in the same basic way, but there are some slight differences related to the type of signal that is generated and how some boxes must be operated. The differences include coded and noncoded signaling, discussed later in the section, and the need for one action (single action) or two actions (double action) to operate the alarm box. Operation of an electronic valve supervisory device, also known as a tamper switch, involves turning a wheel or handle that, once moved a certain distance, initiates a supervisory alarm.

■ Smoke Detection

Smoke detectors rely on the smoke particles generated from a fire, not the heat, to initiate an alarm signal. Because sufficient smoke develops before other detection systems operate, smoke detectors are considered life safety devices. Smoke detectors detect smoke by one of two operating principles: by using a radioactive element or by obscuring or scattering light. Both types of detectors are reliable, but depending on the type of fire and conditions, one type of detector might respond faster than the other. The choice of a detector is based on a number of factors, including the environment, the construction and design characteristics of the building, building contents and the burning characteristics of the contents, the shape of the room and the ceiling height, and airflow and ventilation conditions. These factors determine which operating principle and device would be best for the conditions.

Smoke Detection Versus Smoke Alarms

To many people, smoke detectors and smoke alarms are the same device, but one is installed primarily in commercial occupancies and the other primarily in residential occupancies. A smoke detector is a component of a fire alarm system that is usually installed in commercial occupancies. It initiates an alarm through the fire alarm control unit and relies on the fire alarm control unit to supply its power. A smoke alarm is a detector that receives power from a house circuit, has a battery for backup power, has an integrated sounding device, and requires interconnection with the other installed smoke alarms. This arrangement is called multi-station smoke alarms, with the idea that if one sounds an alarm, they all sound an alarm. Some smoke alarms, called single-station smoke alarms, can be standalone and powered only by a battery. In some situations, the smoke alarm is part of a combination system where security, medical, gas detection, and fire signals all report to a monitoring panel that notifies a third-party company to receive

Fire Investigator Tip

Smoke detectors are still one of the best early warning life safety devices, but they do have limitations. For example, spot-type smoke detectors cover only a certain area reliably and, over time, may become overly sensitive or not sensitive enough. This could be problematic because the level of sensitivity must be appropriate for the environment in which detectors are installed. If too sensitive, false alarms will happen; if not sensitive enough, detectors may not react rapidly enough to provide an appropriate warning.

the alarm and take the appropriate action. In other situations, the smoke alarm is a part of a complete fire alarm system that operates in a similar manner as a commercial system.

Ionization

Ionization smoke detectors use a small amount of radioactive material and two electrically charged plates to detect a fire. The radioactive material charges the air particles within the detector's sensing chamber to make the air conductive. Because the air is conductive, a small but measurable current is created by the ionized particles that attach to the oppositely charged plate. When entering the sensing chamber, the smoke particles attach to the charged air particles and reduce the rate of current flow between the plates. The detector activates once the current flow reduces to a predetermined level FIGURE 6-1. Ionization detectors are typically spot detectors best suited for fires that rapidly develop more flame than smoke, such as a burning piece of paper.

Photoelectric

Photoelectric detectors use a light source and photoelectric cell receiver to initiate an alarm. Within the sensing chamber, there are a light source and receiver installed at an angle to each other. When smoke or other particulate matter enters the sensing chamber, the light source refracts off the particles into the receiver and initiates an alarm FIGURE 6-2. Photoelectric smoke detectors are typically spot-type detectors that are able to respond more quickly to smoldering fires because the smoke particles tend to be larger than are produced by rapidly developing flaming fires. Flaming fires do not refract as much light because the particles tend to be smaller, resulting in delayed alarm initiation.

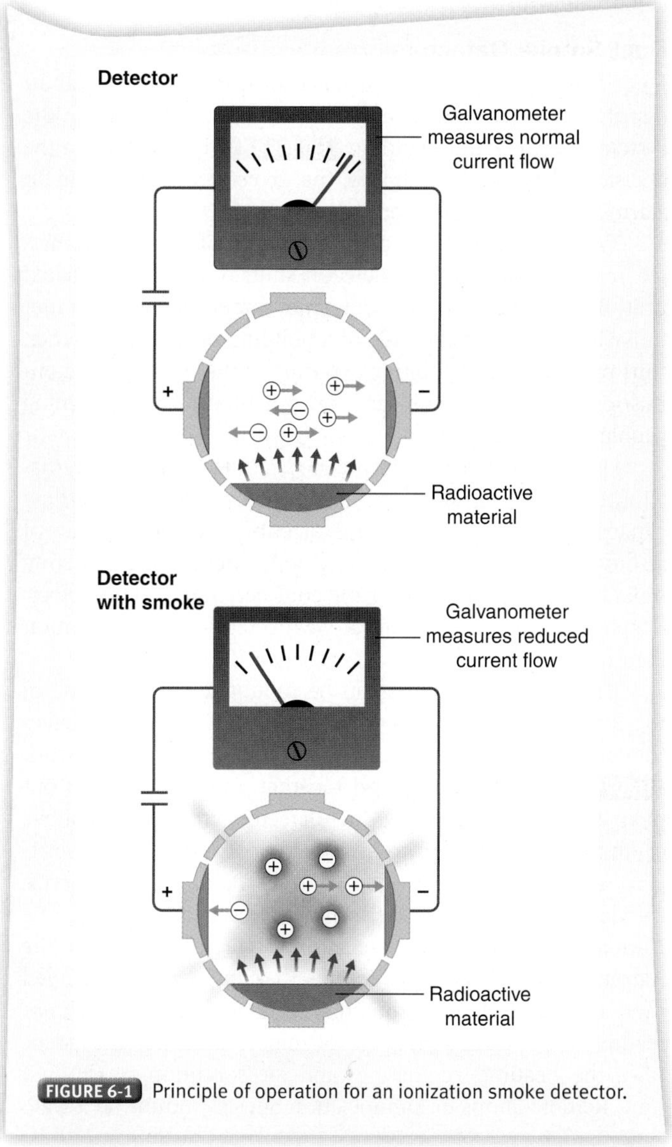

FIGURE 6-1 Principle of operation for an ionization smoke detector.

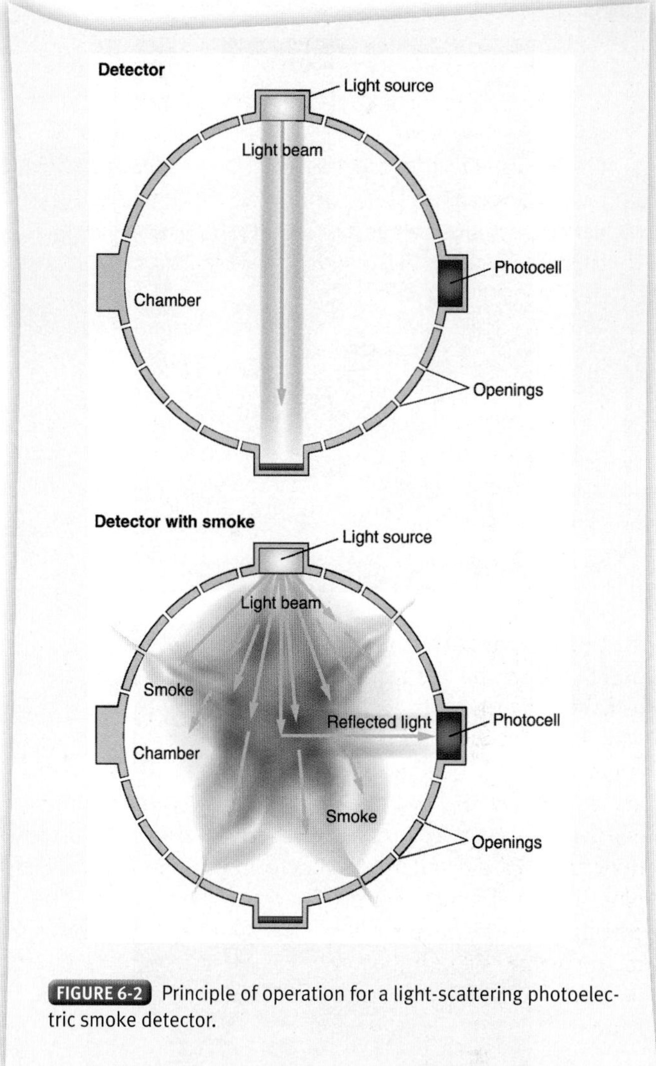

FIGURE 6-2 Principle of operation for a light-scattering photoelectric smoke detector.

Air Sampling (Aspirating)

Air sampling smoke detection, also known as aspirating detection, is one of the more sophisticated types of smoke detection in use. Air sampling smoke detection systems constantly draw and analyze air into a detection chamber through piping with sampling holes that is located in the area of the building being protected by the system FIGURE 6-3. By constantly drawing air for analysis, the system identifies undetectable by-products that form when a material breaks down during the pre-combustion stage of a fire and does so long before an occupant would notice.

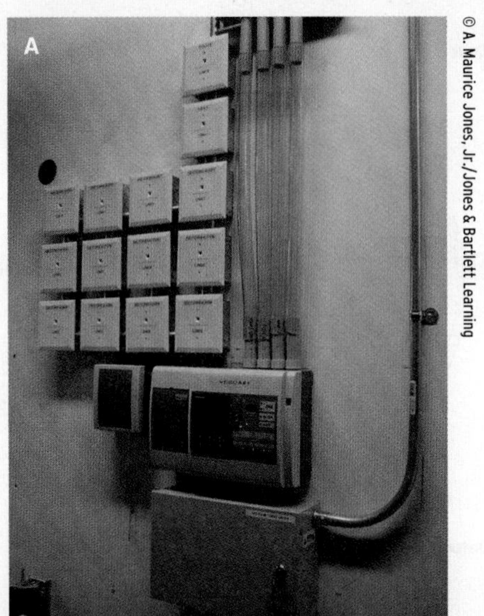

© A. Maurice Jones, Jr./Jones & Bartlett Learning

© A. Maurice Jones, Jr./Jones & Bartlett Learning

FIGURE 6-3 **A.** Air sampling smoke detection system control panel/analyzer. **B.** The ends of these tubes have small holes to capture air samples that are analyzed by a control panel/analyzer.

There are different detection methods and operating principles employed to detect the products of combustion at the very earliest stages. Methods include use of a light source and receptors to measure how much light is reflected by the particles in the air; creation of an environment within the detector chamber where condensation forms on any smoke particles that may be present; and utilization of spot-type filtered photoelectric detection with air sampling technology in challenging environments. One of the big advantages of air sampling smoke detection is that once the detector analyzes the particles and determines there may be a problem, the system can activate different levels of alarm. The first level alarm indicates the system detects an abnormal air characteristic, the second level indicates there is potentially a fire, and the final level indicates that there is a fire in progress.

Air sampling detection systems can be standalone or a part of an overall fire protection strategy. Installations of air sampling smoke detection systems are typically in high-value operations where concealed detection is important, maintenance access is limited, or critical processes cannot be disrupted. Historic buildings, telecommunications facilities, computer/server operations centers, production facilities, electrical substations, semiconductor manufacturing facility clean rooms, warehouses, aircraft hangars, textile mills, and atriums are some of the occupancies where this type of detection would be appropriate.

Projected Beam

Projected beam detectors send a light source to a receiver that spans a protected area. When the receiver senses a reduction in light intensity due to smoke or some other obscuration of the light source, an alarm initiates FIGURE 6-4. The projected range can be a little over 300 feet between the transmitter and receiver, covering a width of about 60 feet. This type of smoke detector is suitable for locations where spot detection would be impractical or not effective. Very high ceilings found in atriums, concert halls, warehouses, gymnasiums, and factories are suitable for this type of detection.

Duct Smoke Detectors

Duct smoke detectors are installed on many commercial air distribution systems to sample the air as it moves through the system ductwork in a building FIGURE 6-5. Depending on the decision of the code authority, the detector could initiate an alarm signal or a supervisory signal.

When the detector senses some type of particulate matter in the airflow, the duct detector shuts down the associated unit. By shutting down the unit, smoke and toxic gases cannot move as easily from one side of a building or floor to another, thus preventing or inhibiting exposure to the products of combustion that create significant and potentially life-threatening problems for occupants.

Duct smoke detectors are not required on all air distribution systems. Required installation depends on airflow capacity based on the amount of cubic feet per minute of airflow the unit delivers. In addition, duct smoke detectors must be able to operate over the complete range of air velocities, temperatures, and humidity expected for the installation conditions.

Duct smoke detectors can be mounted on the inside of the ductwork or on the outside of the unit where the detector connects to air sampling tubes located inside the ductwork FIGURE 6-6. Some are ganged together and installed in front of an air duct intake. Many duct detectors are mounted above ceilings or are too high to see, making it difficult to investigate detector activation; inspect, test, maintain, and service the detectors; and, in some instances, reset the detector. The National Fire Protection Association's NFPA 72, *National Fire Alarm and Signaling Code*, requires that any detector installed more than 10 feet above the finished floor or in a location not readily visible have some type of remote indicator that identifies the heating, ventilating, and air conditioning (HVAC) unit. Remote lamps or annunciation devices mount on a wall,

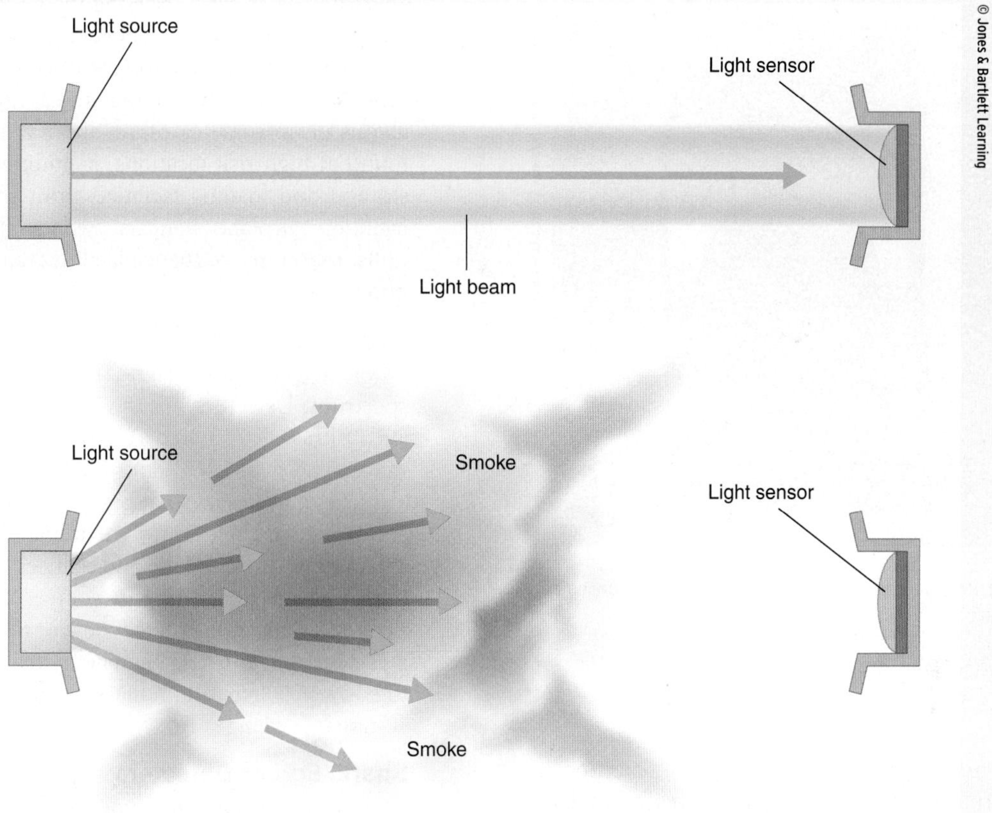

Light source
Light sensor
Light beam
Light source
Smoke
Light sensor
Smoke
© Jones & Bartlett Learning

FIGURE 6-4 Principle of operation for a light-obscuration photoelectric smoke detector.

ceiling, or as close as possible to the detector so the detector can be quickly located by first responders or building maintenance personnel FIGURE 6-7.

Heat Detection

Heat detectors are some of the most reliable and oldest detection devices in use. This type of detector is suited for challenging environments, where temperatures fluctuate, where the fire

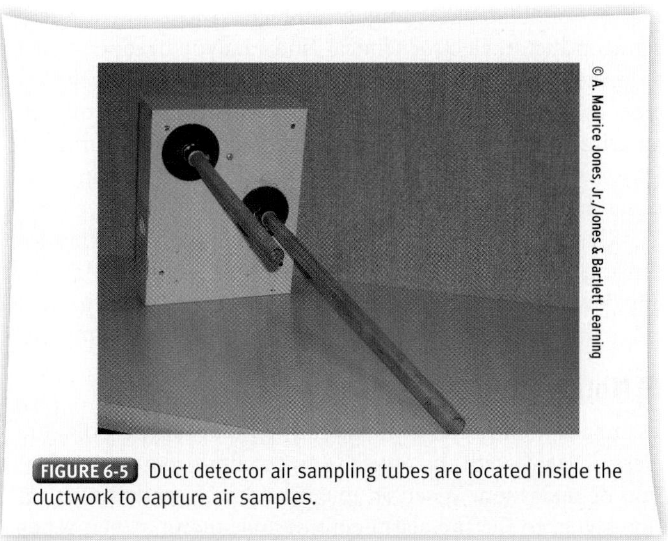

© A. Maurice Jones, Jr./Jones & Bartlett Learning

FIGURE 6-5 Duct detector air sampling tubes are located inside the ductwork to capture air samples.

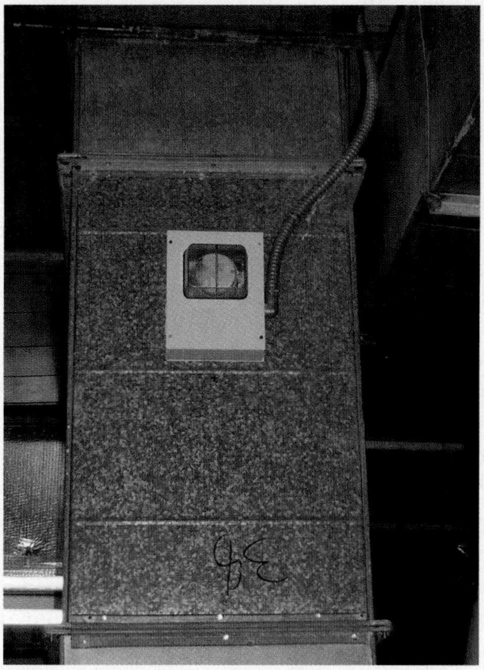

© A. Maurice Jones, Jr./Jones & Bartlett Learning

FIGURE 6-6 Duct detector mounted on an air duct line.

© A. Maurice Jones, Jr./Jones & Bartlett Learning

FIGURE 6-7 Remote indicator for difficult to locate or concealed duct detector.

will have a high heat output, or where fire detection speed is not a concern. Heat detectors detect changes in heat by sensing predetermined fixed temperatures or sensing specified rates of temperature change. There are three operating principles for heat detectors where the fuel, hazard, and associated rate of fire development determine which type of detector—fixed temperature, rate of rise, or rate compensation—should be installed. Fixed temperature detectors operate when the sensing element temperature rating is reached, rate of rise detectors measure temperature changes over a fixed amount of time, and rate compensation detectors react to predetermined air temperatures. Most heat detectors are spot-type, but line-type detectors could be more appropriate for certain hazards and environments.

Because heat detectors rely on the heat generated by the fire, not on the smoke that forms, heat detectors are typically slower to react to fires than other types of detectors, especially smoke detectors. For this reason, heat detectors are not considered life safety devices and should not replace smoke detectors unless approved by the reviewing authority. When a heat detector substitutes for smoke detectors, it is usually due to ambient conditions and environments that could make a smoke detector malfunction.

Radiant Energy-Sensing Fire Detector

Radiant energy-sensing detectors do not depend on smoke and heat plumes to operate, but instead detect specific portions of the visible and invisible light spectrum produced by flames, sparks, or embers. These detectors are suited for environments where exposure to non-fire-related sources of radiant energy is generally greater, such as in manufacturing and industrial settings. Radiant energy-sensing detectors are typically installed to protect high-risk and high-value facilities where rapid and sensitive detection is critical and where remote detection of small fires is crucial. Selecting the appropriate type of radiant energy-sensing detector depends on careful evaluation of the environment and fuel type present, with attention directed to other sources of non-fire radiant energy that could cause false alarms or, in some cases, no alarm at all. These detectors are found at high-risk locations, including petroleum processing, storage, and loading facilities; print shops; paint application facilities; airplane hangars; woodworking facilities; textile mills; museums; ordnance facilities; and, in some cases, computer rooms.

Flame Detectors

Flame detectors are designed to detect a certain part of the light spectrum. Detectors operate in the ultraviolet (UV), infrared (IR), ultraviolet/infrared (UV/IR), or infrared/infrared (IR/IR) range. The dual UV/IR and IR/IR detectors have two sensing elements and require that both elements detect a specific light spectrum band before an alarm initiates. Dual sensing improves detection of an actual fire and reduces the possibility of detecting a non-fire radiant energy source. Radiant energy detectors must have an unobstructed view of the protected area and be close enough to detect the energy to initiate an alarm. When this is a problem, additional detectors are required to ensure complete coverage of the hazard area.

Spark/Ember Detectors

Spark/ember detectors are not fuel-specific and look for sparks or embers that may be present during a manufacturing process. Typically, spark and ember detectors are installed on, in, or near conveyor belts or duct lines where solid materials move through a manufacturing process and the possibility of a spark and ember starting a fire is high. Spark/ember detectors operate in the IR range and must be installed in a closed, dark environment. Once a spark or ember is detected, some type of fire suppression—typically a water spray nozzle connected to an ultra-high-speed water spray system—activates.

Other Types of Detectors

Gas-sensing fire detection is designed to detect a specific gas, any type of toxic gas, or a variety of gases and vapors associated with processes utilizing hydrocarbons. There are a number of different sensing technologies, including infrared, semiconductor, electrochemical, and catalytic bead, installed to protect commercial, industrial, and residential occupancies, including high-hazard environments; off-shore oil and gas rigs; oil, gas, and petrochemical facilities; gas turbines; HVAC air intakes; oil and gas wells; and enclosed air handling ductwork.

Gas-sensing fire detectors can be manufactured only for sensing gas or as a combination gas/smoke detector. As with any device, inspection, testing, and maintenance in the form of calibration is necessary to ensure the device operates properly.

Notification Appliances

Occupant notification is the most important function of a fire alarm system, with the objective to prompt occupant evacuation or relocation. When an initiating device sends an activation signal to the fire alarm control unit, the unit then sends a signal to the notification appliances to produce audible and visual building alarms. There are different types of notification appliances that provide the audible and visual alerts to the occupants, including horns, bells, speakers, chimes, sirens, buzzers, various types of lamps, strobe lights, and any number of these devices in combination.

The type of notification appliance, number, and location will depend on many factors, including the code requirements; use and occupancy condition of the building; the building layout; room sizes; environment; and how much, what type, and to whom the audible and visual information needs to be communicated. Once all the factors are considered, the notification signals may propagate throughout the building, known as public mode notification, or only to a certain monitored location of a building, known as private mode notification. Public mode is appropriate for occupancies where people are capable of taking action upon alarm activation, including office buildings, stores, and movie theaters. Private mode requires that the alarm transmit to a location where trained individuals receive the alarm, interrupt the alarm, and then take the appropriate action. Private mode is appropriate for occupancies where movement is restricted or where people are mentally or physically unable to respond, such as hospitals, nursing homes, and correctional facilities.

Once operating, notification appliances may propagate a coded signal, a noncoded signal, or a textural signal. Coded signals transmit a predetermined number of audible or visual patterns for a specific interval that in some cases helps to identify the location of the initiating device. Noncoded signals transmit a constant audible or visual signal that propagates until the initiating device and system reset. Textural signals can be a prerecorded voice message heard from speakers, alphanumeric information on a fire alarm or annunciation panel, or a symbol or picture that flashes on a display screen at a constantly attended location where trained individuals monitor incoming signals.

An annunciation panel is a type of notification or indicating panel that provides critical information to emergency responders by light panel, LCD/alphanumeric text, graphic display, or touch-screen video monitor. This panel is usually located at the building entrance and interfaces with the fire alarm control unit to provide critical information. Information can include the type of signal initiated, the system or device that activated, and the location of the system or device, including floor level and zone. An annunciation panel is extremely valuable to the investigator because the information provided by the panel can help isolate the incident to specific systems, devices, and areas of a building and the types and numbers of signals generated during the incident.

Operation and Installation Parameters of the System

FACU Features

The FACU is one of the most important parts of the fire alarm system. Frequently described as the "brains" of the system, the unit is where the fire alarm system components and devices interface and interact. Its function is to receive component input signals; to interpret and manage the component input signals; and then to deliver the appropriate alarm, trouble, or supervisory output signal. This includes providing power to support all system devices, processing automatic and manual input signals from initiating devices, monitoring the integrity of the system circuits and power supplies, powering and activating the notification appliances, and interfacing with and activating other building and control systems during a fire emergency.

The type of system and when the fire alarm system was manufactured and installed will determine the operational capabilities of the fire alarm system. Some systems perform only basic notification functions, whereas others interact and control different fire protection and building systems. Modern FACUs are capable of storing information in the panel memory circuits, sending information to a printer, or sending information concerning system status and the signal input and output events to a proprietary or third-party monitoring company, acknowledging the alarm condition, silencing the alarm condition, and resetting the system.

Fire Investigator Tip

Valuable information can be gained by interviewing the first on-scene responders who viewed the information communicated by the annunciation panel because that information establishes the status of the fire protection systems when they arrived.

Location and Spacing of Devices

NFPA 72, manufacturers' guidelines, and, in some cases, local codes govern the installation of initiation devices and notification appliances. Design professionals, contractors, code officials, and inspectors use NFPA 72 to establish the minimum requirements for performance, reliability, and quality of fire alarm and detection systems. In addition, there is significant information regarding the application, design, installation, inspection, testing, and maintenance requirements for these systems. NFPA 72 also provides information concerning components, hardware, system types, calculations, power requirements, and supplemental integration and interface guidance. However, clearly stated in NFPA 72 is the disclaimer that it is not an installation specification or approval guide, nor does it provide the methods necessary to achieve the requirements. That is one reason why it is critical during the evaluation of a building to consider the environment, design, features, and construction characteristics, including room layout and ceiling height, for conditions that could create operational problems for the fire alarm and detection system.

Internal System Communication

Most fire alarm and detection system components are tied together with wire or cable designed specifically for the purpose and installation parameters unique to fire alarm systems. The wiring is the transmission medium to communicate,

Fire Investigator Tip

Code requirements, installation conditions, and manufacturers' guidelines determine the type of wire and cable that must be installed and where it can be installed so it is protected from damage and does not contribute to a fire.

initiate, activate, and power the system components, circuits, and devices. The use of wire to connect the system components has been the standard since the first fire alarm systems were designed and installed. Now, however, wireless systems and devices are meeting the same standards and requirements as hard-wired systems, leading to increased approval and acceptance throughout the fire protection and building industries.

Means of Transmission

There are many approved systems and means to communicate between a protected property fire alarm system and third-party monitoring center. To ensure the signal is received at the monitoring facility, two methods of transmission are required; this dual transmission method helps to ensure that if the primary transmitting link fails, the system will automatically switch to the backup. The most frequently used and approved primary and backup methods include hard-wire and fiber optic communication, one-way radio transmission, two-way radio frequency multiplex, digital cellular technology, or Internet protocol. Both the primary and backup communication links require supervision and must report any failure for investigation. At no more than 24-hour intervals, both links require automatic testing to ensure integrity and continuity of the link.

Systems Monitored and Controlled

Central Station

Central station service establishes a high level of timely, efficient, and appropriate handling of any fire protection system-related issues at a protected property it serves. A central station service provider or local service company that is under contract is responsible to provide services at a protected property and at an offsite monitoring facility. Their responsibilities include installation, testing, and providing response to the site service; signal monitoring; retransmission of the signals; and record keeping. Central station service providers must meet strict performance requirements and be audited by certifying agencies to verify that standards and requirements are met.

Once the central station receives and recognizes a signal, the station personnel must take the appropriate action. If the signal is an alarm, the central station notifies the fire department and then dispatches a responder who must arrive within 2 hours of the alarm to silence and/or reset any systems. If the signal is supervisory, a responder must arrive within 2 hours; however, for trouble signals, arrival is required within 4 hours. Supervisory or trouble signals require the onsite responder to investigate within 30 minutes of arrival.

Proprietary Station

Proprietary supervising station fire alarm systems establish building and property owners as the responsible parties to monitor the fire alarm system signals at their facilities. Typically, the proprietary system consists of a number of fire alarm systems on a property or spread throughout an area or region in buildings that are under the control of a single owner. All signals are transmitted to a central monitoring location on the property, to a regional monitoring location, or to a central monitoring location somewhere in the county. Staffing of the monitoring facility is provided 24 hours a day by trained personnel who monitor and transmit fire alarm signals to the fire department and, if established, the facility's emergency response personnel. Monitoring facility personnel also notify the emergency response personnel or service personnel of any trouble or supervisory signals, regardless of whether they are located at the site or at a remote location.

Proprietary systems also can be set up to handle all types of building control functions, including elevator control, smoke control, stair tower pressurization, stair tower and exit door lock release, and shutdown of manufacturing or other processes. The building or room housing the monitoring equipment must be constructed of fire-resistant materials and be separated from any hazardous processes to ensure survival if there is a fire or other facility emergency.

Remote Station

The remote supervising station fire alarm system reports alarm, trouble, and supervisory signals generated by a protected property fire alarm system that signals an offsite location that is distant from the protected property. Some fire departments permit signals to go directly to a fire or public safety communications center via approved fire department radio frequencies, a dedicated one-way (outgoing only) telephone line, or dedicated communication circuits. When the signals do not go to a fire or public safety communications center, they must go to an approved location where trained individuals monitor the signals at all times. Fire alarm signals require that personnel monitoring the system immediately contact the fire department, whereas a trouble or supervisory signal is reported to the owner or owner's representative for immediate investigation and resolution of the problem.

Analysis

If the investigator discovers the building is equipped with any fire protection system during the course of an investigation, these systems must also be thoroughly documented and analyzed to determine the role played, if any, during the fire. Fire investigators must take great care to utilize proper, accepted, and relevant resources, techniques, strategies, and methods when conducting an investigation. Investigators should be knowledgeable and fluent with the content and application of relevant industry and court-accepted documents such as NFPA 1033, *Standard for Professional Qualifications for Fire Investigator*, and NFPA 921, *Guide for Fire and Explosion Investigations*. These and other documents provide qualifications and systematic guidance to conduct a fire investigation.

As outlined in NFPA 921, fire investigations should use an organized, systematic approach that follows an analytical process to determine the cause of the fire. The recommended systematic approach noted in NFPA 921 is the scientific method.

The scientific method consists of several steps to assist the investigator in establishing reliable conclusions regarding a fire's origin, cause, and responsibility. Two critical components of the scientific method outline the need to collect and analyze data. When a fire occurs within a building, the building, building systems (including any fire protection systems), and building contents must be fully documented and analyzed. Additionally, all witnesses and first responders should be interviewed to gather pertinent information. At a minimum, the investigator should have the resources available to derive appropriate and relevant information and analysis related to the role of any fire protection system found within a burned structure.

Installation Considerations

When performing postfire analysis, the investigator should attempt to gather all related fire alarm system documentation. Documentation should include the manufacturer's data sheets; design drawings; installation, inspection, test, service, and maintenance records; and any dispatch and response records to the location. In addition, the investigator should determine which building and fire codes and referenced design standards were in force at the time of installation. Review and analysis of this information will provide the investigator with a basis to compare the original system with the installation at the time of fire. The importance of this review and analysis relates directly to determining whether there were changes in the use and occupancy, renovations resulting in changes to the building construction and building features, and if any possible change affected the performance of the initiating devices and notification appliances.

As part of the review and analysis, the investigator should consider interviewing personnel from the authority responsible for review and approval of the system design and installation, including acceptance testing, and any design professionals or contractors associated with the design and installation, whether responsible for the initial installation approval or periodic inspection, testing, service, and maintenance approval.

Operability

A very important part of any investigation is verifying the fire alarm system status, operability, and functionality at the time of the fire. Establishing whether the system was in normal ready service or showing trouble or supervisory alarms could play a considerable role in determining how the system responded to the fire event. A trouble or supervisory alarm could render some or all of the system incapable of performing properly. In addition, investigators should determine if the system was powered by the primary power source or the secondary power source at the time of the incident. This is important because when the system relies on a secondary power source such as batteries, it could experience inadequate operability due to limited power reserves. Fire alarm systems are required to be operational in normal ready condition for a certain amount of time after loss of primary power, but they can operate only for a limited amount of time when on standby power in full alarm mode, which means limited notification.

Another very important part of the investigation is determining whether the initiating devices functioned as required by activating in a timely manner and within the design parameters for the conditions to which they were exposed.

Analysis of Smoke Alarm Response

It is crucial for the investigator to ascertain and understand the incident timeline, including how and when the fire was detected or discovered and how the fire department was notified. This is especially important if the fire alarm system or smoke alarm activation was the initial source of incident alarm because this information can help determine area of origin and whether notification appliances or smoke alarms sounded. Determination that a particular initiation device or smoke alarm activated and sounded can be gathered from first-arriving fire fighters, building occupants, the FACU, emergency 911 dispatch centers, and proprietary or third party monitoring centers.

Investigators should interview occupants and fire fighters to document fire growth and spread and should attempt to correlate those observations with sounds and/or sights associated with fire alarm system notification devices. Sounds and/or sights can vary from loud tones or ringing bells to voice directions and flashing strobe devices that may confirm system activation. The investigator should examine fire alarm system devices for signs of contamination by smoke or soot or thermal impact, which may indicate which device(s) operated earliest or if they failed to operate.

Analysis of Smoke Deposition

Fire investigators should examine, analyze, and document smoke detector, smoke alarm, and other system components for soot accumulation, soot patterns, and thermal impact in much the same manner as other fire patterns found on various surfaces. Excessive buildup of soot, smoke staining, and the coloring of soot deposits may assist the investigator in determining fire origin, spread, growth, and behavior. In addition, the investigator should account for localized or specific fuel packages and ventilation effects that may be in proximity to affected or exposed system devices. Analysis of the conditions may assist the investigator in determining whether a smoke detector and smoke alarm device activated and operated during the event.

Alarm Response Time

Understanding and verifying fire alarm system initiating device response time is an important duty for the investigator. It involves a thorough scene examination to collect, document, and analyze the findings and data from the fire scene. The collected data can be utilized to develop a computer model of the fire scene and fire alarm system device response based on examination and data inputs. These models can be of assistance in confirming or disproving fire alarm system activation; however, caution must be exercised when using computer models because their correct use may require a significant amount of knowledge, understanding, and experience.

Estimation of Fire Size

Understanding the effects of fire size and propagation in relation to system response, especially regarding initiation devices such as smoke detectors, can assist the investigator in determining fire size, origin, growth, and spread. As noted earlier, it is important for the investigator to account for fire alarm initiation device activation and equally important to account for nonactivation of initiation devices. By verifying these actions, the investigator may be able to use the fire alarm system as a tool to determine the size of the fire at a particular point in the fire event. Numerous sources of information are available to help the investigator explore this option. Manufacturers' specifications and data sheets, design standards, and calculations can be utilized to estimate system characteristics and responsiveness regarding fire size. If a system can be shown to have activated and operated appropriately, the fire at the time of the alarm could be established as the minimum fire size. In contrast, if the system did not activate, assuming proper and correct installation, the fire could be established as the maximum fire size because fire growth did not reach a level to activate the system.

Development of Timeline

An active fire alarm system can provide the fire investigator with multiple data points. The fire alarm system control panel and annunciation panels are good sources of system response and activity information if not damaged or destroyed by fire. No matter the severity of the incident, the amount of damage to the fire alarm system and components, and whether the primary or secondary power has been disconnected or compromised, investigators should photograph the fire control panel and annunciation panel to capture the indications and events that took place FIGURE 6-8. The reason is that if there is a loss of primary power, some secondary power sources will sustain the system for only a limited amount of time, and once the secondary source expires, system data could be lost.

The amount of information available will vary based on the age and type of fire alarm system. Some systems and panels will provide only general and basic information concerning the device that activated and the general location, whereas others will indicate system status and the type, time, and specific location of an initiation device activation. The information may be presented as indicating lights associated with a device or zone, alphanumeric text, or a combination of both. Additionally, verification of system monitoring provided by remote, proprietary, and third-party offsite companies can supply the investigator useful information concerning when and how the fire alarm system responded.

Investigators are cautioned not to overstep their abilities regarding the use or operation of fire alarm control and annunciation panels. If the investigator does not possess knowledge or competency on the use of such devices or systems, or the system has been damaged, there is a risk of corrupting, altering, or losing data. Therefore, trained and competent individuals should determine the operational status and capabilities of the system so that the data can be safely accessed. Understanding these responsibilities can allow for the appropriate application of a wide range of legal tools available to the investigator to examine fire alarm system data.

It is very important that investigators interview occupants, witnesses, and fire responders to document observations related to fire growth and spread, and to correlate those observations to the data gathered from the fire alarm system panels. This includes any information concerning sights and sounds associated with fire alarm system occupant notification device activation, location of the interviewees, and whether the interviewees took any action to initiate an alarm, such as activation of a manual fire alarm box. The use of fire alarm system activation data and information obtained through interviews can be helpful in the creation and verification of a timeline for the incident to determine fire origin, growth, and spread.

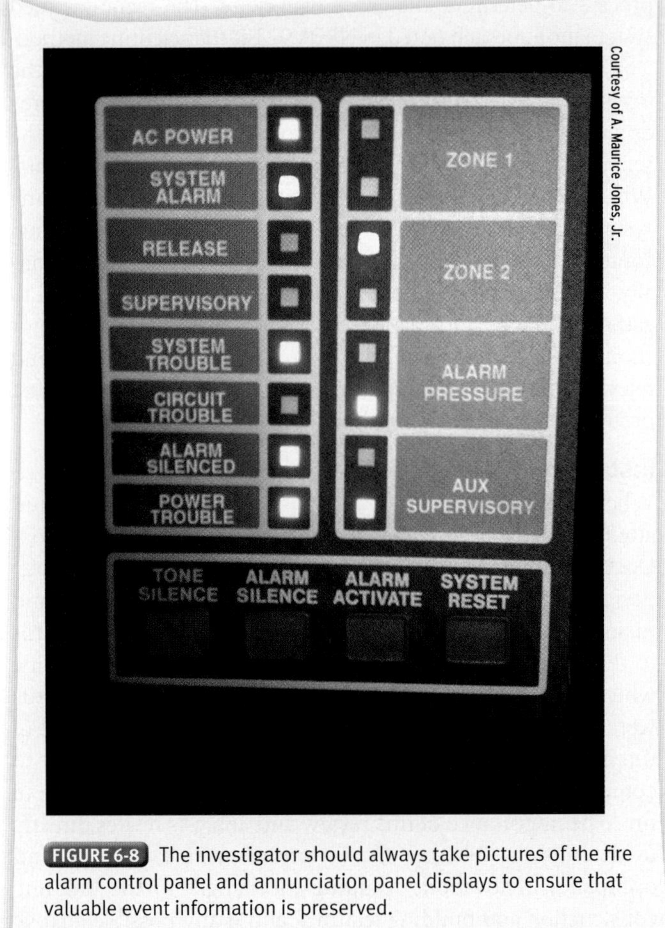

FIGURE 6-8 The investigator should always take pictures of the fire alarm control panel and annunciation panel displays to ensure that valuable event information is preserved.

Thermal Damage

All fire alarm system components have a temperature range established by the manufacturer in which they can properly operate. The investigator must evaluate the entire system for signs of damage from smoke, heat, gases, and flames to determine whether there was any effect on the operation of the system. For example, thermal damage to fire alarm system notification appliances, including smoke alarms, could affect the ability of the device to produce an audible alarm signal or any alarm signal. Once any device is exposed to an environment that forces the device to operate outside of its design parameters (including temperatures outside of the established

> **Safety Tip**
>
> During the course of any investigation, the investigator should not attempt to access or tamper with the internal components of the fire alarm system without possessing the technical expertise or working under the supervision of a trained individual to avoid injury from electrical shock and potential damage to the system or components where valuable data could be lost. This is especially important when trouble or supervisory signals are present related to systems' power supply issues.

operational range), they are susceptible to decreased performance or failure.

Fire Alarm Effectiveness

Fire alarm systems, when properly designed, installed, inspected, tested, serviced, and maintained, have been proven to be highly effective in early occupant notification when fire occurs. However, even systems that are in proper working order might require an investigator to explore situations where occupants did not respond to system activation and the subsequent alarm signal. Many factors can influence occupant reaction, ranging from a person's response to sounds when sleeping, their ability to hear, whether they are impaired, and whether they have a disability. It is imperative that the investigator decipher reasons associated with inaction by occupants to verify system effectiveness. (See also the "Fire-Related Human Behavior" chapter.)

Impact on Human Behavior

The presence of a fire protection system could have a substantial impact on human behavior, both good and bad. Some individuals will always respond to a fire alarm signal, while the presence of these systems might give others some sense of false safety, resulting in delayed or no response. In addition, some individuals might respond better to a certain type of audible notification, be it a horn, bell, or voice message. There are also times when building occupants fail to respond to fire alarm activations due to repeated past alarms that were found to be false or if the occupant did not receive clear direction regarding what, where, and when to evacuate. Human behavior prior to and during a fire is important for the investigator to understand. Crucial information must be gathered and analyzed regarding an occupant's behavior to fire alarm activation. It is important for the investigator to establish occupant awareness and response prior to and during fire alarm system activation.

Water-Based Fire Suppression Systems

Purpose of Systems

The purpose of a water-based fire suppression system is to provide property protection and life safety by delivering sufficient amounts of water over the fire area at the appropriate rate in an effort to control, and in some cases extinguish, the fire. This is accomplished by the automatic activation of a device that reacts to changes in temperature, activation of a fire alarm and detection system automatic initiating device, and, for some types of systems, activation by manual fire alarm box.

General System Operation

The main extinguishing mechanism for water-based fire suppression systems is to reduce the heat by applying sufficient amounts of water to cool the fire area and prevent the fire from spreading. A sufficient amount of water depends on a number of factors because different materials release different amounts of energy. Therefore, to control a fire, there needs to be an adequate and reliable water source that can deliver water at a sufficient rate over a certain area.

Another extinguishing mechanism involves mixing a foam concentrate with water to create a foam solution blanket that not only cools, but also isolates the fuel from the oxygen and the fire. In this case, the amount of water must also be adequate for the hazard, but the foam and water must properly mix because too much water may result in a solution that cannot control or extinguish a fire, or too much foam concentrate may result in running out of foam before the fire is controlled or extinguished.

Types of Water-Based Systems

Water-based sprinkler systems can be categorized into four system types: wet, dry, preaction, and deluge TABLE 6-1. These are the most commonly installed systems in commercial and industrial settings, and most of the other water-based fire sprinkler systems are some form of these systems.

All of these systems use NFPA 13, *Standard for the Installation of Sprinkler Systems* as the design and installation standard. The installation of a particular type of system mainly depends on two factors: the hazard it must protect and the operating environment. In many instances, the four types of system will be appropriate for the hazard and conditions, but other NFPA standards may need to be referenced to deal with special hazard situations.

Specialized Sprinkler Systems

There are some situations where the use of a standard automatic fire sprinkler system is not appropriate for the applications and conditions of a particular building, process, or hazard. In these instances, specialized sprinkler systems, including low-, medium-, and high-expansion foam systems; compressed air foam systems; foam–water sprinkler/foam–water spray systems; and water spray fixed systems are installed to protect the hazard TABLE 6-2. These systems are used where the conditions and applications of standard automatic fire sprinkler systems are not appropriate, where a supplemental suppression agent is needed, or where specialized equipment and components are required for application.

There are some similarities among the four most prominent types of automatic fire sprinkler systems and these types of specialized systems: water is the primary suppression agent,

Table 6-1 Types of Water-Based Systems

Type	Description	Application
Wet pipe sprinkler system	■ Has water in the piping at all times ■ Preferred because upon activation of a sprinkler head exposed to sufficient heat, water immediately flows onto the fire	■ Environment must be able to maintain 40°F or higher so water does not freeze and damage or break the fittings, piping, and components ■ Commonly installed in stores, offices, and homes
Dry pipe sprinkler system	■ Has pressurized air or, in some rare cases, a gas such as nitrogen in the system pipes ■ The air or gas holds pressure on the dry pipe valve clapper so it does not release the water into the system until necessary. ■ A fire sprinkler head must activate to release the pressurized air or gas in the piping. ■ Once sufficient pressure is lost, the valve opens and floods the piping with water. ■ Water must reach open sprinkler head(s) within a certain timeframe (typically 60 seconds) to be effective.	■ Mostly used where environment is not able to maintain at least 40°F. ■ Commonly installed in unheated warehouses, loading docks, attics, and parking garages
Preaction sprinkler system	■ Interfaces with a dedicated fire alarm control unit and initiating devices to release a preaction valve ■ Initiating devices are located in protected area ■ Three operational configurations are available, each requiring one or more activation events to take place before water enters the piping (e.g., activation of a fire sprinkler head, activation of a smoke or heat detector, or operation by manual means). The need for multiple activation events to take place provides a higher level of certainty that there is an actual fire.	■ Protect high-value occupancies including telecommunication centers, server farms, computer operation centers, freezers, document storage centers, museums, and libraries
Deluge sprinkler system	■ All sprinkler heads or nozzles are open to the atmosphere. ■ Interfaces with electric, hydraulic, or pneumatic initiating devices to release the deluge valve and flow water from all the open sprinkler heads or nozzles onto the hazard ■ Provides large amounts of water over the hazard to prevent a rapidly growing fire from spreading ■ Requires an adequate water supply, and frequently a fire pump is necessary to deliver the required water volume and pressure to meet the system design demand ■ Commonly mixes a foam concentrate with water to protect certain hazards	■ Protect high-hazard occupancies where it is necessary to flood an entire area to prevent the fire from spreading (e.g., aircraft hangars, power plants, woodworking facilities, cooling towers, refineries, explosive and ordnance factories, and chemical storage or processing facilities.

and it must be from an automatic supply; the components are the same or similar; and all components must automatically activate and use the same types of system valves. However, specialized systems have some key differences, including specialized spray nozzles instead of fire sprinkler heads, the use of foam agents with water in certain types of systems, the need for specialized components to mix the foam solution and water appropriately, and the design and application methods.

Residential Sprinkler System

The benefits of a residential fire sprinkler system derive from a change in the overall goal from property protection, established for commercial fire sprinkler systems, to life safety by reducing the system water supply requirements, creating a special type of sprinkler head, and placing the fire sprinkler heads in the locations where a fire is most likely to start. Each contributes to achieving the design goals for residential fire sprinkler systems, which are to prevent flashover in rooms with sprinklers where the fire started and to provide a better chance for the occupants to escape or be rescued.

There are many similarities between commercial fire sprinkler systems and residential systems, but there are many differences. First, there are two NFPA standards that deal specifically with residential occupancies: NFPA 13D, *Standard for the installation of Sprinkler Systems in One-and Two-Family Dwellings and Manufactured Homes* and NFPA 13R, *Standard for the Installation of Sprinkler Systems in Low-Rise Residential Occupancies*. These standards establish requirements that are different from NFPA 13 for water supply, design, installation, components, and testing. The majority of residential sprinkler systems are wet pipe systems and use similar types of components as commercial systems. However, the components tend to be smaller because less water is needed based on the design requirements to supply the system and make it work properly.

Some NFPA 13D and some NFPA 13R systems share the water supplied by an appropriately sized water line fed from a reliable community water system, but most water purveyors require separate domestic and fire service water lines. There are instances where the water supply is from an adequately sized stored water source that has an automatically operated pump. This is common in communities where there is no water system, no reliable automatic water source, or well water as the main source of water. As part of the fire sprinkler system, a storage tank of adequate capacity with an automatic

Table 6-2 Specialized Water-Based Systems

Type	Description	Application	NFPA Standard
Water mist	■ Specialized sprinkler heads and spray nozzles discharge a very fine spray mist of water droplets where 99 percent of the droplets discharged from these systems must be 1000 microns (1 mm) or less in size. ■ Extinguish by cooling, displacing oxygen, blocking the radiant heat. ■ Systems classified based on the working pressure of the system (low, intermediate, or high) and whether there is one pipe (single fluid water only) or two pipes (twin fluid water and gas) supplying the head or nozzle. ■ Advantages include discharging less water than a conventional sprinkler system. Some systems are self-contained and are designed for complete fire extinguishment. ■ Disadvantages include higher design and installation costs compared to other sprinkler or suppression systems. The self-contained tank systems only having a finite amount of water, the need for reserve tanks, the need for each specific application to be tested and approved, and the possibility of nozzles clogging when water quality is not exceptional.	■ Land, sea, and air applications including computer rooms, data centers, laboratories, archive storage, museums, historic buildings, electrical switchgear rooms, telecommunication facilities, tunnels, underground mass transit facilities, boats, yachts, navel ships, passenger ships, offshore operations, aircraft cargo spaces, and aircraft hangers	NFPA 750, *Standard on Water Mist Fire Protection Systems*
Low-, medium-, and high-expansion foam	■ Protects hazards that involve flammable and combustible liquids. ■ Foam is less dense than fuel or water and is able to flow freely over the liquid surface. ■ Foam produced from foam concentrate that is mixed with water to produce a foam solution. ■ Foam concentrate is available in 1 percent, 2 percent, 3 percent, and 6 percent mix ratios. ■ Each percentage represents the amount of concentrate to mix with water where the resulting mixture equals 100 percent. ■ The foam concentrate and water mix to form the foam solution, but air must be introduced by mechanical means to form the foam bubbles that are discharged to protect the hazard. ■ Foam products fall into three categories that are defined in the expansion ratio of the foam: low, medium, and high expansion. ■ Expansion is a function of properly mixing the manufacturer's defined percentage of foam concentrate and water. ■ Low-expansion foam increases from a 2:1 ratio of foam to water up to a 20:1 ratio. Medium-expansion foam increases from a 20:1 ratio up to a 200:1 ratio. High-expansion foam increases from a 200:1 ratio up to a 1000:1 ratio. ■ There are different types of foam concentrates, including aqueous film-forming foam (AFFF), alcohol-resistant aqueous film-forming foam (AR-AFFF), film-forming fluoroprotein foam (FFFP), protein foam, and fluoroprotein foam. ■ Some foam is designed for specific applications, whereas other foams have characteristics that make them suitable for use with many different hazards.	■ Flammable and combustible liquid outdoor storage tanks, containment dikes, interior flammable and combustible liquid storage, fuel loading racks	NFPA 11, *Standard for Low-, Medium-, and High-Expansion Foam*
Compressed air foam systems	■ Combines air or nitrogen, water, and foam concentrate to create high-momentum, greater-expansion-ratio foam. ■ Foam concentrate is normally stored in non-pressurized tanks that range from 5 to 500 gallons (19 to 1893 liters). ■ High-pressure air cylinders supply the air or nitrogen coupled with a high-pressure manifold and pressure regulation devices. ■ Compressed air foam systems use mixing chambers to combine the solution that feeds the piping and flows to the specialized nozzles that deliver the foam.	Flammable and combustible liquids, liquid spill fires, three-dimensional fires, ships, wildland fires, and some structural firefighting	NFPA 11, *Standard for Low-, Medium-, and High-Expansion Foam*
Foam-water sprinkler/ foam-water spray systems	■ Uses foam and water as companion suppression agents; the foam-water solution is used as the primary agent and the water used as the secondary agent ■ When the foam supply from these systems is depleted, the water supply continues until it is shut off ■ Primary design goal is to extinguish fire, but also appropriate for prevention, control, and exposure protection. ■ Foam-water sprinkler system protects areas of a building or structure and equipment ■ Foam-water spray system uses directional foam application to protect a specific hazard or piece of equipment within a building or structure.	Class B flammable and combustible liquid hazards including aircraft hangars and petroleum dispensing and storage facilities; acceptable for use with certain Class A hazards	NFPA 16, *Standard for the Installation of Foam-Water Sprinkler and Foam-Water Spray Systems*
Water spray fixed systems	■ Designed to deliver a concentrated and directed water spray pattern onto the surface of the hazard, into an area within, around the hazard for the purpose of fire control, prevention, extinguishment, or exposure protection. ■ Requires one adequate automatic water supply released by automatic or manual activation. ■ Nozzles are either automatic or open configurations. ■ Automatic nozzles use the same basic technology as automatic sprinkler heads (heat sensitive glass bulb/fusible link) and operate when a predetermined design temperature rating is achieved. ■ The open nozzle is open to the atmosphere.	Equipment or structural members surrounding or supporting the equipment, electrical transformers, oil switches, motors, cable trays, paper, wood, textiles, flammable liquid and gas materials, and certain hazardous solids	NFPA 15 *Standard for Water Spray Fixed Systems for Fire Protection*

fill line holds the water until needed. This arrangement is predominantly associated with NFPA 13D systems, not NFPA 13R, due to the size and capacity of the storage tank that would be needed for an NFPA 13R system. Both NFPA 13D and NFPA 13R standards outline a number of approved water sources, but no matter the source, it must be automatic and supply an adequate amount of water to the system. Two other important distinctions between NFPA 13 and NFPA 13D/NFPA 13R systems are the use of a residential sprinkler head and the locations of residential sprinkler heads within a residence or dwelling unit. The first distinction is that residential sprinkler heads must meet a different testing criterion than a standard commercial fire sprinkler head. The residential head must prevent the harmful fire by-products and the associated temperatures from reaching levels that could inhibit safe passage or rescue of the occupants. The second distinction is that NFPA 13D and 13R systems exempt head installation in certain locations within a residence. These locations are deemed to be less likely to be the room of fire origin and, statistically, the likelihood of a fire in these spaces is low. Both standards list a number of locations that do not require fire sprinkler heads, including bathrooms that are less than 55 ft^2, closets and pantries that are no greater than 24 ft^2 and where the lowest dimension does not exceed 3 ft, and the ceiling construction is of limited or noncombustible materials. Also included are garages, open attached porches, carports, attics, equipment rooms, and concealed spaces.

Key Components of Water-Based Systems

Although there are specialized water-based systems that employ components specific to the type of system, all water-based fire protection uses the same basic components, including automatic sprinkler heads or nozzles, piping, pipe support and stabilization assemblies, fittings, gauges, and various types of valves. These components form the building blocks of the various systems, and in most cases, the major difference between systems is the type of system valve, whether an alarm valve is for a wet system, dry valve, preaction valve, or deluge valve.

Of great importance is the requirement that all components be listed, approved, or labeled by a nationally recognized evaluation agency to ensure the components perform as required when tested under the conditions in which the components must operate. Without this oversight, components could fail, leaving the building without a fire sprinkler system.

Sprinklers and Spray Nozzles

Automatic sprinkler heads and nozzles are spray devices that distribute water over a limited area at a designated flow rate. Their purpose is to distribute water over an area to reduce the heat from a fire by controlling and limiting fire growth beyond the early stage of fire development.

Sprinkler heads are heat-activated devices that operate at predetermined temperatures. The heat generated by a fire rises to the ceiling, and everything, including the sprinkler head, will be subject to the higher temperatures. When the sprinkler head reaches the predetermined activation temperature of the heat-sensitive element on the head, it releases so that the disk holding back the water falls away from the head and allows the water to flow from the head and spray onto the fire. Each head operates independently; therefore, only the heads that reach the predetermined activation temperature will operate and flow water.

An important characteristic of any sprinkler head is the listing or approval obtained by the manufacturer through a nationally recognized testing and certification agency. The listing and approval are given once the evaluation process certifies the head will operate and perform as intended. In addition, the listing or approval specifies the application, design, and installation criteria for the head. Fire sprinkler heads installed outside of their listing or approval specifications will likely result in a failure of the head to perform properly and the subsequent failure of the entire fire sprinkler system to control the fire.

Sprinkler heads are defined by characteristics that include temperature rating, orientation, and K-Factor and whether there is a coating on the head so it can be used in a particular environment. Temperature ratings range from 135°F to 650°F. Sprinkler head temperature ratings are determined by either the color of the liquid in the bulb, factory applied paint or coating on the frame or deflector, or engraved on the frame or deflector TABLE 6-3. There are many types of sprinkler heads that fall into different categories and subcategories, but, generally, all heads are oriented in the pendent, upright, or sidewall positions. When performing calculations, the K-Factor is a value assigned by the manufacturer to the head that represents the orifice size. The larger the K-factor number, the larger the orifice, meaning more water will flow from the head.

Table 6-3 Temperature Ratings, Classifications, and Color Codings (NFPA 13, Table 6.2.5.1)

Maximum Ceiling Temperature		Temperature Rating		Temperature Classification	Color Coded	Glass Bulb Colors
°F	°C	°F	°C			
100	38	135–170	57–77	Ordinary	Uncolored or black	Orange or red
150	66	175–225	79–107	Intermediate	White	Yellow or green
225	107	250–300	121–149	High	Blue	Blue
300	149	325–375	163–191	Extra high	Red	Purple
375	191	400–475	204–246	Very extra high	Green	Black
475	246	500–575	260–302	Ultra high	Orange	Black
625	329	650	343	Ultra high	Orange	Black

Reproduced with permission from NFPA 921-2013, Guide for Fire and Explosion Investigations, Copyright© 2013, National Fire Protection Association. This reprinted material is not the complete and official position of the NFPA on the referenced subject, which is represented only by the standard in its entirety.

© A. Maurice Jones, Jr./Jones & Bartlett Learning

FIGURE 6-9 With close observation, a person can determine a number of sprinkler head characteristics, including the manufacturer, the model, the temperature rating, the K-factor, the year of manufacture, the sprinkler identification number, and the listing or approval by a third-party testing agency.

Each sprinkler head has specific characteristics and considerable information applied to the head to inform the end user of the type of head, temperature rating, listing or approval, model number, K-factor, year manufactured, and sprinkler identification number FIGURE 6-9. Nozzles also must carry a listing or approval by a nationally recognized testing and certification agency. Nozzles are generally used for special hazard applications and are designed to apply a special water spray pattern, but most do not have a heat-sensitive element. Depending on what is being protected, nozzles surround the hazard or equipment and apply large amounts of water directly onto the hazard or equipment.

Piping and Fittings

Joined pipe and fittings form a conduit to move the water from the source to and out of the fire sprinkler heads. Pipe and fittings are required to meet or exceed certain standards from a number of nationally recognized organizations to be acceptable for use as part of a fire sprinkler system, including the American Society of Testing and Materials (ASTM), American Society of Mechanical Engineers (ASME), American National Standards Institute (ANSI), and American Welding Society (AWS). In addition, the pipe and fittings must be able to handle the maximum permitted system working pressure of 175 psi or carry an appropriate rating for the anticipated pressure above 175 psi. There are many different materials approved for pipe, including ferrous, black steel, galvanized steel, copper, certain alloy materials, and CPVC.

Fittings connect piping, valves, sprinkler heads, and other components by use of elbows, tees, flanges, crosses, unions, couplings, bushings, plugs, and reducers that are made of cast and malleable iron, steel and wrought carbon steel, alloy steel, copper, and CPVC.

Joining methods include screw thread, grooved, flanged, welded, soldered, heat fusion, brazed, pressure, or glue.

Fire Investigator Tip

Installation of a sprinkler head with a temperature rating that is too low relative to the normal high ambient temperature could result in unwanted sprinkler head activation; however, if the temperature rating of the head is too high, delayed activation could allow the fire to spread to the point where it would be difficult to control.

System Valves

At least one water control valve is required to be installed on all automatic fire sprinkler systems. The purpose of the valve is to permit or prevent water flow into and through the system. Under normal conditions, the control valve or valves should be fully open and remain open to permit water to flow freely into and through the system FIGURE 6-10. However, in the event the system needs modification, maintenance, service, or repair, the system can be shut down. Once the work has been completed, any valve that was closed must be reopened to place the system back in service.

In order to know if the water control valve is open or closed, all automatic fire sprinkler system water control valves are required to be indicating-type valves. Indicating-type valves allow a person to look at the valve and determine if the valve is open, partially open, or shut. In addition, all indicating valves must be identified with a permanent sign and indicate the area of the facility the valve is controlling. Fire sprinkler system indicating valves are usually 2 inches or larger and include the outside screw and yoke type (OS & Y), butterfly indicator type, wall post indicator type (WPIV), and post indicator type (PIV).

Courtesy of A. Maurice Jones, Jr.

FIGURE 6-10 Many wet-based fire suppression system performance failures are due to a closed control valve. This valve was closed and the electronic monitoring device was disabled so that no supervisory alarm would sound.

Safety Tip

Water under pressure can be very dangerous; therefore, the investigator should not attempt to operate any fire suppression system water control valve before analyzing the location and understanding the purpose of the valve's position relative to the event being inspected or investigated. By opening a valve previously closed, there could be considerable water damage if the valve is closed due to a broken sprinkler line or open/activated sprinkler head. By closing a previously open valve, part or all of the system would be shut down, thus preventing water from flowing if the system activates. If the investigator does not possess the knowledge to address the situation, he or she should contact a technical expert or be under the supervision of a trained individual to avoid flooding, damage, injury, or a fire that grows out of control due to no water supply.

Proportioners

Expansion foam systems require a mechanical device called a proportioner to mix the foam concentrate with the water to achieve the proper ratio. There are different proportioning devices, including the venturi, pressure, and balanced pressure proportioners. The venturi proportioner flows water over an open orifice within the proportioning device, which creates a lower pressure that draws the foam concentrate out of the storage arrangement and into the water stream. The pressure proportioner directs the incoming water supply into a storage tank to exert pressure on a collapsible bladder inside the tank holding the foam concentrate, pushing the concentrate out of the tank to the proportioner for mixing. The balanced pressure proportioner uses a pump and an atmospheric tank to flow the foam concentrate into the water supply. The proportioner achieves the correct ratio by balancing the foam and water pressures.

Water Supply

One of the most important elements of any water-based fire protection system is access to at least one automatic water supply. Automatic water supplies do not require a person to interact with any part of the system for water to flow. The water can come from one or a number of different sources including wells, rivers, lakes, municipal water systems, storage tanks, reservoirs, and cisterns, but the automatic water source must support the system design with adequate pressure and flow for the duration necessary to ensure successful control or extinguishment of a fire. If any of these elements is deficient, the system will not perform as required and will fail to protect the intended property or hazard. Water supplies should be reevaluated periodically to ensure no changes affected the pressure, flow, and duration.

Operation and Installation Parameters of the System

Essentially, all water-based fire suppression systems are some type of fire sprinkler system. The basis for system design depends on many factors, including the hazard level, occupancy and use of the building or structure, storage arrangement, the amount and type of fuel load, the potential rate of heat release, processes, and construction materials. These factors play a major role in selecting the design approach, type of fire sprinkler system, and the types of heads that will be installed to protect the hazard.

Location and Spacing of Sprinklers

Automatic fire sprinkler heads are not randomly placed throughout buildings. They are located and spaced so that upon activation, they discharge a sufficient amount of water over a defined area to control a fire within minutes of the fire starting. However, before any head is installed, an evaluation of the hazard must be undertaken by the design professional. The evaluation for most designs starts with classifying the occupancy, commodities, or storage arrangement that the system must protect. Any one or a combination of occupancy, commodity, and storage arrangements produces certain combustible characteristics that will output energy through fire development. Based on the level of energy output, the system design must deliver enough water to absorb the energy from the fire so it does not grow any larger. Evaluating the combustibility, amount of combustibles, and rate of heat release of a commodity, material, or product are critical to determining the hazard classification and design approach.

Head location and spacing are functions of the physical environment and the requirements outlined in the adopted codes and standards that describe how much area a head must cover to protect the hazard. When the hazard is determined to be high, the sprinkler heads are placed closer together. When the hazard is determined to be low, the sprinkler heads can be placed farther apart. Spacing can range anywhere between 90 square feet per head for high-hazard environments to 225 square feet per head for low-hazard environments. These spacing requirements are for fully sprinklered commercial or industrial occupancies. There are a number of exceptions in the adopted codes and standards for commercial and industrial occupancies, but as previously discussed, certain rooms and areas in residential occupancies and dwelling units can automatically omit sprinkler heads.

Another factor in determining sprinkler head location and spacing is the choice of the sprinkler head that will protect the hazard. Sprinkler heads must undergo the same evaluation and approval process as any other piece of fire protection system equipment and, when specified for use, the use and installation must follow the manufacturers' specifications and the requirements of the adopted codes and standards. Because there is a large variety of sprinkler head types available, certain conditions and situations may allow for the use of a head that can exceed 225 square feet spacing as long as it meets the manufacturers' specifications and is within the requirements of the adopted codes and standards.

As important as proper head spacing is to protecting the hazard, the location of the head relative to any component of the structure is critical to ensure successful activation and water discharge. There are installation requirements in the codes and standards that determine how far away from walls, ceilings, obstructions, and equipment heads should be to work properly, but there are exceptions that provide relief for specific conditions. Generally, sprinkler head deflectors can be located between 1 and 12 inches from any part of the structure, but some special storage situations require between 18 and 36 inches.

Pipe Size and Arrangement

The pipe size and arrangement of a fire sprinkler system are determined by evaluation of the inherent hazard level of the occupancy as outlined by the adopted design standard. Once the level of hazard is determined, a decision must be made whether to use a pipe schedule for the design or to use the hydraulic design method.

Pipe schedules are a list of pipe sizes and the number of fire sprinkler heads that the pipe can support based on the hazard classification. Specifically, a pipe schedule spells out that a certain size pipe can have a certain number of total heads attached. Large pipes can have more heads than small pipes, but the total number of heads attached to a system is fixed by the largest pipe size and hazard classification. As the pipe size decreases, the number of heads that can be on the system and a specific sized pipe decreases. These systems were common until the 1970s, but with the advent of hydraulic design, new pipe schedule systems are rarely installed because of the size limitations imposed by the sprinkler standards and the limits to expanding existing pipe schedule systems.

Now, the vast majority of systems are calculated based on the hydraulic design method. Unlike the pipe schedule, where the number of heads on a pipe is based on pipe size, this method establishes pipe size throughout a system by calculating the friction loss at certain points in the system related to the available water pressure and flow at those points. Based on these calculations, it is common for pipes throughout a system to be of similar size. If the calculations determine that smaller pipes are able to support the pressure and flow demand to meet the system water requirements, then smaller pipes will be used because it reduces the cost of installation.

Sprinkler Coverage and Distribution

There are many different types of sprinkler heads manufactured to handle almost any design challenge, hazard, or environmental condition. Heads are installed in one of three basic installation orientations: pendent, upright, and sidewall. Pendent heads discharge water downward, upright heads discharge water upward, and sidewall heads discharge water horizontally. Sprinkler heads are designed and manufactured to provide unique water discharge patterns and water droplets sized for the specific environmental and installation conditions, including heads made for residential occupancies, rack storage, attics, and fur vaults. The variety provides the design professional with many options to meet the protection goal because the head that is chosen must be able to provide sufficient flow and cover a certain area to control a fire.

When a sprinkler head activates, it develops a water discharge pattern over a certain area and at a specific flow rate to meet the design objective. Meeting the design objective requires that an adequate water supply be available. If the water supply is insufficient, the fire is unlikely to be controlled. The discharge pattern, coverage, and flow rate are a function of the sprinkler head design. Two sprinkler head components—the orifice and the deflector—are critical to developing the discharge pattern and flow rate. The size of the orifice determines how much water flows out of the head, and the deflector determines the direction of water flow, water reach, and the size of the droplets. When the hazard level is high, fire development is rapid and intense. Therefore, specifying a head that can discharge large water droplets will provide the necessary protection to penetrate the fire plume and control the fire. Critical to the water delivery is sufficient pressure, which must be at least 7 psi at the most remote head in order for proper flow and discharge to take place.

Water Flow Rate and Pressure

When designing a fire sprinkler system, one of the first things that must be determined is the hazard level of the occupancy. Based on this determination, the minimum amount of water flow and required pressure that are needed to reach the remote area can be established. If the water supplied to this area is adequate, it should be better than adequate in other parts of the system because they are closer to the water source. Pressure is critical because sprinkler heads require a certain amount of pressure to operate properly as defined by the manufacturers' specifications. If the operating pressure is too low, the head will not flow enough water to control the fire.

The amount of water required to flow from the head within the remote area is called the design density. The design density value is based on years of testing, where fires were set and water was applied to determine how much water it would take to control a certain size and type of fire. These values are represented on density/area curves published in NFPA 13. These curves provide a graphic representation of the minimum amount of water that needs to flow over a specific area that the fire sprinkler head must provide within the remote area. There are curves for light, ordinary, high-hazard and other challenging situations FIGURE 6-11. There are a number of points on each curve where the design professional can select a density measured in gallons of water per minute per square foot (gpm/ft^2) over a given area. These curves are designed to balance the water flow and area covered by establishing minimum and maximum density and area parameters. Most design professionals will flow as much water as the curves will permit, and properly performed hydraulic calculations will provide a more accurate indication of pressure and flow throughout the system, allowing for more design flexibility.

Activation Mechanisms and Criteria

There are different ways fire sprinkler systems activate to flow water onto a fire. Many fire sprinkler systems, such as wet and dry sprinkler systems, use heat-activated sprinkler heads. Once

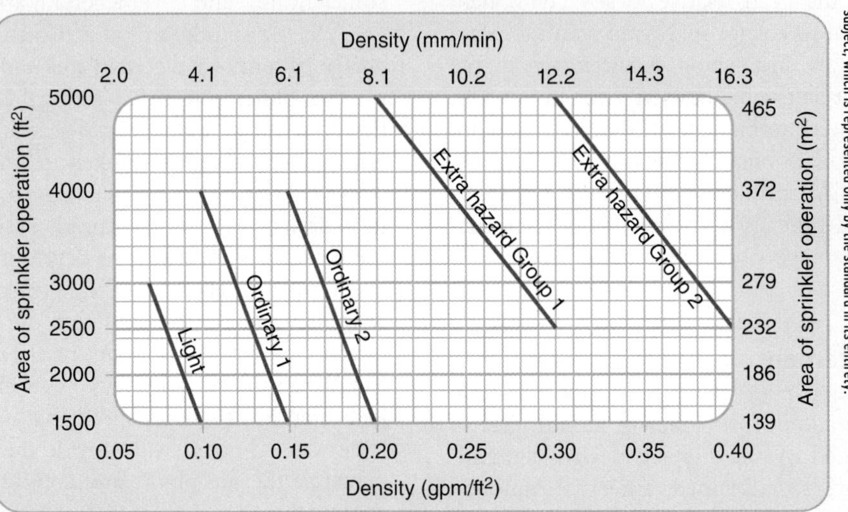

Reproduced with permission from NFPA 921-2013, Guide for Fire and Explosion Investigations, Copyright© 2013, National Fire Protection Association. This reprinted material is not the complete and official position of the NFPA on the referenced subject, which is represented only by the standard in its entirety.

FIGURE 6-11 NFPA 13 density/area curve.

the predetermined temperature is reached in these heads, the heat-activated element releases to allow the disk covering the head orifice to fall away and flow water. Typically, the heat-sensitive element is a glass or frangible bulb, a fusible solder link, or a chemical pellet.

The choice of a sprinkler head with a heat-sensitive element is based on a number of factors and it is essential to use the appropriate head for the situation. Factors include the occupancy condition, the environment, the water supply, and the protection goal. These factors could have an impact on how rapidly a head activates; therefore, it is important to choose a head that is suitable for the situation. The amount of time a head takes to activate once it is exposed to temperatures above the predetermined activation temperature is called the response time index (RTI). A lower RTI indicates that the head should operate more rapidly. A head with a low RTI number is a must for certain occupancy conditions because the system is designed to protect the occupants (e.g., in residential occupancies), or where the sprinkler system relies on a certain type of head to protect the property (e.g., high-piled storage warehouses).

Other systems, such as deluge systems, use open sprinkler heads that are dependent on the actions of a human to release the water or the interaction with a heat-activated device such as a heat detector to initiate water flow. Still other systems, such as the preaction systems, utilize closed sprinkler heads and fire detectors to release the water where the interaction between the sprinkler heads and fire detectors controls the release of the water.

Systems Monitored and Controlled

Fire sprinkler systems are monitored and controlled by various devices and systems to notify those responsible that the normal status of the system has changed. Status changes could be for a water flow condition, a change in air pressure, a valve closure, a temperature change, or a water level change. Typically, the monitoring is performed in conjunction with a fire alarm system. These notifications can be local or sent to an offsite organization that will notify the responsible parties.

The water flow switch is found on wet pipe sprinkler systems. A sprinkler system may have one water flow switch or multiple switches, depending on the size of the building and the number of systems within the building. It utilizes a flexible round plastic paddle or vane that inserts into the piping and is slightly smaller than the inside diameter of the pipe. The paddle moves with sustained water flow to activate the switch and, once activated, a signal is sent to the fire alarm panel to sound a fire alarm. The switch returns to the normal, ready position once the water flow is stopped, leaving only the fire alarm system to be reset and put back in service.

Because the velocity and force of the water rushing through the pipe could blow the paddle off the switch arm and create a blockage somewhere in the pipe, water flow switches are not installed on dry, preaction, or deluge systems. Flow switches have an adjustable delay mechanism to compensate for fluctuations in water pressure surges between the sprinkler system and the water supply. The delay mechanism device is usually set to between 30 and 60 seconds because no delay will result in multiple false alarms due to water surges.

The alarm pressure switch is used to detect water flow in dry, deluge, preaction, and older wet sprinkler systems that use mechanical bell alarms. Pressure switches operate when the preset threshold either rises above or drops below the pressure setting. There is no time delay mechanism; once the pressure threshold is reached, a signal is sent to the alarm panel.

Additionally, unlike the water flow switch, the pressure switch may require manual reset.

Not only can pressure switches be used to monitor water flow, they can be used to monitor air pressure changes in dry and preaction systems and pressure tanks. When conditions change, a pressure switch can generate a supervisory signal that notifies of low and high air-pressure conditions.

Electronic supervision of water control valves through the fire alarm system is critical because one valve in the wrong position could render an entire fire suppression system out of service. Supervision is through a device known as a tamper switch. The switch is either integrated as part of the valve or externally attached to the valve. With as little as a few turns of the valve handle, a signal is sent to the fire alarm control panel that, in turn, initiates a local supervisory signal and in some cases notifies an offsite party.

Other monitoring devices that are critical to ensure the system will perform as required include air and water temperature sensors and water level sensors. Temperature sensors detect drops or increases in buildings and systems where temperatures out of the norm could render a system useless. Water level sensors monitor the amount of water in a storage tank and, if the levels fall or rise above the normal levels, notifies the responsible party to take the appropriate action.

Non-Water-Based Fire Suppression Systems

General Information

Non-water-based fire suppression provides an alternative to water-based fire suppression when water may not be the best choice for the hazard involved. This type of suppression is also referred to as Special Hazard Fire Suppression because it uses gaseous and chemical-based agents to protect special and specific hazards, processes, and equipment.

Purpose of Systems

When choosing a suppression agent, water is usually the first and obvious choice, but there are conditions and situations where water would have little or no effect, would require an additive to work appropriately, would not be appropriate for the hazard, and in some cases would make the situation worse. Under those circumstances, the use of non-water-based fire suppression provides the means to handle and control challenging fire conditions that would otherwise be more problematic or inappropriate when using water-based suppression. The most common non-water-based systems use wet and dry chemicals, dry powders, and gaseous agents.

System Components

The building blocks of many non-water-based fire suppression systems start with extensive laboratory testing, where engineers study and analyze how the agent and associated systems handle certain fire conditions and scenarios. Based on the data, manufacturers can provide engineered and pre-engineered systems that provide component type, size, proper location, design criteria, and design limitations. Unlike water-based fire suppression, where the majority of the components are interchangeable, non-water-based system components are specific to the type of system and the agent in use and cannot be interchanged.

Suppression Agents

Gaseous suppression agents are divided into three basic categories: halon, halon replacements, and inert gases. Chemical agents are either dry or wet chemical or chemical-based dry powders.

Halon and Halon Replacements

Halogenated hydrocarbons (halons) mix carbon and one or more elements from the halogen series in the periodic table: fluorine, chlorine, bromine, or iodine. The chemical makeup of a halon gas is based on a numbering system that lists the number of atoms for each element that make up the agent. For example, Halon 1211 is a mixture of one part carbon, two parts fluorine, one part chlorine, and one part bromine, while halon 1301 is a mixture of one part carbon, three parts fluorine, zero parts chlorine, and one part bromine. Because neither of these halons uses iodine, no corresponding number attaches to the formula. Halon 1211 and 1301 are the only two halons still in use with fire protection systems or fire extinguishers.

Halon agents are odorless, colorless, noncorrosive, nonconductive, and nontoxic at low concentration levels. However, due to the agent's contribution to the depletion of the ozone layer in the atmosphere, the 1987 Montreal Protocol, an agreement signed by almost every country in the world, banned the manufacturing or importation of new halon products. Reclaimed gas and existing reserves are still permitted to be used, so it may be a number of years before all halon gases are depleted. However, eventually all halon systems will be obsolete because manufacturers have stopped making replacement parts for systems and offer little or no technician support for old halon systems.

To overcome this problem, manufacturers started to develop new extinguishing agents called halocarbons (clean agents). The goal was to address the environmental concerns and potential toxicity by producing gases that were environmentally safe, not electrically conductive, and left no residue upon evaporation. Although successful in addressing the environmental concerns, clean agents were never a direct replacement for halon because it took more clean agent and different components to provide the same level of protection.

Clean agents fall into two categories: halocarbon-based agents or inert gas–based agents. Halocarbon-based agents extinguish fires by eliminating the heat from the flame's reaction zone or by interrupting the uninhibited chain reaction. Halocarbon-based agents group into the five sub-categories FC, FK, FIC, HFC, and HCFC. These designations identify the chemical composition of the mixture that may include carbon, chlorine, bromine, fluorine, iodine, and hydrogen. HFC-227ea (heptafluoropropane, C_3F_7H), is one of the most commonly used clean agents, but it has the trade name FM-200.

Inert agents reduce the oxygen in the protected area to a level that will not sustain combustion. Inert gases are identified as IG and are a mixture of helium, neon, argon, nitrogen, and small amounts of carbon dioxide. IG-541 (N_2ArCO_2) is a

Fire Investigator Tip

Timely investigation is important because some dry chemical agents leave corrosive residue that, when exposed to moisture, will cause damage, especially to electronic equipment and components , potentially contaminating evidence.

Fire Investigator Tip

Manufacturers of dry chemical systems and agents have stopped making fixed dry chemical extinguishing systems and parts for commercial kitchen applications. This is due to dry chemical agents losing approval for failing to meet the UL 300 test standard for control and extinguishment of commercial kitchen fires.

combination of nitrogen, argon, and about 8 percent carbon dioxide but has the trade name Inergen.

Inert Gases

With few exceptions, carbon dioxide (CO_2) gas is a very good extinguishing agent for almost all combustible materials, flammable and combustible liquids, and electrical fires. CO_2 is an odorless, colorless, noncombustible, nonconductive gas that leaves no residue. What makes CO_2 such a good suppression agent is that it displaces and reduces the level of oxygen below the 15 percent of the air content that is necessary to sustain combustion. Because CO_2 is about 50 percent heavier than air, it will separate the air from the fuel, making it especially good when dealing with a deep-seated fire. However, the gas is dangerous because asphyxiation is probable to anyone directly or indirectly exposed to discharging CO_2 in a confined space; therefore, any person working in the area should leave once a predischarge warning activates.

The pressure and temperature of the environment determine if CO_2 exists as a gas, solid, or liquid. When used as a suppression agent, it is stored in tanks as a highly pressurized liquid that expands to a gas and combines with moisture in the air to form a cloud that consumes the hazard.

Dry Chemical

Dry chemical extinguishing agents are small, solid, powdery particles that extinguish the flame by covering and smothering the protected surface area of a fire, cutting off the oxygen to the fuel to prevent reignition and flame spread to adjacent areas. Dry chemical extinguishing agents adhere to, coat, and insulate the surface area of the hazard where other types of agents would run off and be ineffective. In addition, dry chemical agents do not react with flammable liquids and gases and they are not conductive. The particles use pressurized nitrogen, carbon dioxide, or air as the transport medium and are suspended in the gas to facilitate flow and distribution of the agent that makes dry chemicals suitable for use with fire extinguishers and hose lines. Upon activation, the particle/gas mixture flows out of the storage container into the piping network and out of nozzles. Although not considered dangerous or toxic, the discharging agent is under pressure and creates a cloud that limits visibility and can cause respiratory problems. The hazard and protection requirements determine the type of nozzle and placement of the nozzle.

Safety Tip

Before entering any area where a gas-based suppression agent has discharged, especially in an area protected by a CO_2 system, all emergency responders and investigators should wear self-contained breathing apparatus (SCBA) and take air quality readings to determine the amount of oxygen, residual gas-based suppression agent, or other gases present to ensure the area is safe to enter.

Sodium Bicarbonate–Based Dry Chemicals

Sodium bicarbonate–based agents are also known as regular or ordinary dry chemicals. Similar to baking soda, sodium bicarbonate is one of the oldest and most widely used dry chemical agents. Sodium bicarbonate works well on Class B fires and on some grease fires related to cooking. When used to protect cooking operations, the agent chemically reacts with the oils and fats to create a soap-like liquid that covers the surface of the hazard. The reaction between the oils, fats, and dry chemical agent is called saponification. However, the use of dry chemical agents related to cooking has greatly decreased because larger, energy-efficient cooking equipment now heats cooking oils to higher temperatures, and the dry chemical agents cannot provide sufficient cooling and isolation to prevent reignition. Sodium bicarbonate is generally not recommended for use on Class A fires but it does work well on Class C fires. To differentiate from other dry chemical agents, the sodium bicarbonate is colored blue or white.

Dry Chemicals Based on the Salts of Potassium

Potassium bicarbonate, potassium chloride, and urea-based potassium agents are considered more effective than sodium bicarbonate because lesser amounts are needed to provide much higher fire extinguishing capabilities. These agents work well on Class B fires except those associated with cooking operations. The agents are also effective on Class C fires but are not recommended for Class A fires. Potassium bicarbonate is purple in color to differentiate it from other dry chemicals.

Multipurpose Dry Chemical

Ammonium phosphate is a multipurpose dry chemical that works very well on Class A, B, and C fires. However, it does not work well on fires associated with cooking operations, especially involving deep-fat fryers. When heated, the agent decomposes and forms a molten residue that sticks to the heated material to isolate the oxygen from the material. Ammonium phosphate is yellow to differentiate it from other dry chemicals.

Foam-Compatible Dry Chemicals

Most dry chemical agents are compatible with foam agents, but to ensure compatibility, the manufacturer should be consulted. When known to be compatible, dry chemicals are sometimes paired with other suppression agents, such as foam or clean agents, as part of a twin agent suppression system. Many of the twin agent systems are portable and are used in industrial settings.

Wet Chemical

Wet chemical extinguishing agents are proprietary water-based solutions. Water is mixed with potassium acetate, potassium carbonate, potassium citrate, or, in some instances, a mixture of agents and other additives to form an alkaline solution. Just like the ordinary dry chemical agents, wet chemical agents react with the cooking oil or fat to form a soapy foam blanket (saponification) that provides separation between the oxygen and fuel. The soapy foam blanket decreases or eliminates the fuel vapors by separating, smothering, and cooling the fuel to prevent reignition. Whereas industrial kitchens use a number of different suppression agents, commercial kitchens use wet chemical extinguishing agents because they are the most effective agent for Class K fires that could start in appliances such as deep fat fryers; ranges; griddles; grills; woks; or char, chain, and upright broilers.

■ Key System Components

Suppression Agent Supply

The containers that store the suppression agent must be made of materials (usually metals) that will not react with the agent and will be appropriately sized to store the amount of agent needed to protect the hazard. The containers must also be able to handle the pressures that will be encountered; this is especially true for gas-based agents that can be pressurized to high levels in the containers. As with all fire protection components and systems, they must be listed, approved, or labeled by a nationally recognized testing and certifying agency.

Pressure Sources

Depending on the type of system and agent, some containers will store only the agent and there will be a secondary pressure cartridge or pressure container that, upon activation, either pushes the agent out of the container or activates a valve to release the agent from the container. Usually, the pressure cartridge or container is filled with nitrogen or carbon dioxide gas as the propellant. This is common with dry chemical systems and some wet chemical systems.

Distribution Piping

The material used for piping must be noncombustible and compatible with the agent that will flow through the pipe. In addition, the piping must be able to handle the anticipated storage and operating pressures and the environment the piping will be protecting that could be corrosive or noncorrosive. Depending on the type of system and agent, the materials most often used are some type of metal that includes black iron, chrome-plated or stainless steel, galvanized steel, copper, and brass.

Valves, Hoses, and Fittings

Any valve, hose, fitting, or component that is part of a fire suppression system must be listed, approved, or labeled for use with the system from a nationally recognized testing and certification agency. Valves, hoses, fittings, or other components that are installed but not listed, approved, or labeled for use with the system could cause the system not to work properly or to fail completely, resulting in an out-of-control fire.

Distribution Nozzles

A nozzle distributes the extinguishing agent onto the hazard. Each system manufacturer makes different types of nozzles for its specific suppression system because each nozzle is designed for an intended application that requires listing, approval, or labeling for the particular application. Nozzles must be made of noncombustible, corrosion-resistant materials that will not deform or fail if exposed to fire. In addition, the nozzles must be permanently marked for proper identification and installed to ensure the correct nozzle is located to operate without failure at the anticipated discharge pressures and to protect the hazard.

Actuation System

Suppression systems use automatic and manual actuation methods to release the extinguishing agent onto the hazard. Actuation systems include fusible links, heat detectors, smoke detectors, fire sprinkler heads, and manual fire alarm boxes. Usually, these actuation devices tie into a fire alarm and detection system that initiates an alarm signal. Fusible links are common with wet and dry chemical systems, whereas heat and smoke detectors are common with gas-based systems. Heat detectors, smoke detectors, and fire sprinkler heads are usually associated with expansion foam systems.

System Monitoring and Control

Just like fire sprinkler systems, fire suppression systems are monitored and controlled by various devices and systems to notify those responsible that the normal status of the system has changed. Status changes could be for system activation, a change in container pressure, a valve closure, a temperature change, or a system leak. Typically, the monitoring is performed in conjunction with the building fire alarm and detection system, with notification sent locally or to an offsite organization that will notify the responsible parties. Most of the gas-based systems require a secondary fire alarm panel that handles special safety functions such as prewarning alarms to ensure occupants are able to exit the hazard area before the agent discharges.

■ Operation and Installation Parameters of the System

Location and Spacing of Nozzles

Nozzle location and spacing are determined by the manufacturer's installation guidelines and the applicable codes and standards, but the type of nozzle and the hazard will have an impact on the placement. Since many non-water-based fire suppression systems are engineered or pre-engineered systems, it is important to follow precisely the system design and nozzle installation requirements to ensure appropriate coverage of the hazard. Usually, the higher the hazard level, the more nozzles, spaced closer together, are required to protect the hazard.

Pipe Size and Arrangement

Depending on the suppression system, pipe size and arrangement may be pre-engineered based on laboratory testing by the manufacturer or, if not pre-engineered, hydraulically calculated. Pre-engineered systems have the advantage of protecting known hazards that generally do not change, making it easy to follow the manufacturer's guidelines. However, when a hazard changes due to the size of the protected area, the size of the hazard, or the environmental conditions, the use of hydraulic calculations determines the pipe size and location. The determination is based on the flow rate and friction losses associated with the pipe size and length, fittings, and other system devices and components as the agent flows through the pipe and out of the nozzles.

Nozzle Coverage and Distribution

Nozzles are designed to deliver the agent in a manner that provides the appropriate coverage for the hazard based on the nozzle's discharge characteristics and the pressure that is available. There are a number of different nozzles available to protect most hazards that can be used for local application or total flooding systems. Local application systems use nozzles that are directed onto or into equipment, processes, operations, or specific areas to smother, cool, and extinguish the fire. Nozzles used with total flooding systems deliver the agent to an enclosed hazard or area within a structure. The numbers of nozzles and the amount of agent needed to protect a hazard or area depend on the volume of the area, the manufacturer's design and installation requirements, and the type of hazard.

Activation Mechanisms and Criteria

Non-water-based fire suppression systems use automatic and manual actuation methods to release the extinguishing agent onto the hazard. Different types of fire detectors can be used that tie into a fire alarm and detection system; activation also can be through detection devices integrated with the suppression system. Some systems provide a manual means to activate the extinguishing system, but most activations are automatic.

Systems Monitored and Controlled

Fire suppression systems are monitored and controlled by various devices and systems to notify those responsible that the normal status of the system has changed. Status changes could be for system activation, a change in container pressure, a valve closure, a temperature change, or a system leak. Typically, the monitoring is performed in conjunction with the building fire alarm and detection system with notification sent locally or to an offsite organization that will notify the responsible parties. Most of the gas-based systems require a secondary fire alarm panel that handles special safety functions such as pre-warning alarms to ensure occupants are able to exit the hazard area before the agent discharges.

> **Fire Investigator Tip**
>
> The investigator must have familiarization or appropriate resources to accurately identify and analyze the prevailing codes and standards adopted at the time of system installation, any amendments to the adopted codes that changed model requirements, and any approved or allowed deviations from set requirements, whether via code amendments or insurance design requirements.

Analysis

■ Code Analysis

Codes and standards adopted by a municipality become gateway documents that establish the requirements for water-based and non-water-based fire suppression systems for certain hazards. Typically, the codes establish the requirement to install a system and the adopted standard that must be used to design, install, inspect, test, and maintain the system. These codes and standards establish requirements for the built environment from design through construction and for the remainder of the life of the system. Some municipalities have adopted NFPA codes and standards, whereas other municipalities have adopted the International Code Council (ICC) series codes that reference NFPA standards. However, it is very important for the investigator to determine which codes and standards were in effect at the time the building was built and which codes and standards were in effect when the fire occurred. The requirements could be quite different, and this information may directly affect when, what, why, and how both water-based and non-water-based fire suppression system reacted during the fire incident. Generally, if the building has not changed use or occupancy condition from the time it was built, the codes and standards in place stay in place, but performing research to find this information is critical to making the appropriate determinations. Establishing the code requirements also involves determining if any local code amendments, modifications, or insurance-based requirements exist and, if in place, their impact on the system.

> **Fire Investigator Tip**
>
> Of great importance in any investigation is the need to verify system inspection, testing, and maintenance requirements; procedures; and compliance based on code requirements derived from manufacturers' requirements and applicable NFPA standards.

Both water-based and non-water-based fire suppression systems are typically required to meet criteria established by NFPA codes and standards. NFPA standards match the type of system needed to the type of occupancy and the hazard needing protection. It is important for the investigator to know that there are NFPA standards that deal with every type of fire protection system. Use of the appropriate standard for the system installed is critical TABLE 6-4.

The ICC-established model code series provides for a platform of minimum requirements applicable to the built environment. These minimum requirements can be (and many times are) amended to reflect particular municipality emphasis, which varies from locality to locality.

Design Analysis

For the fire investigator to interpret the actions and effects that a fire suppression system had during and against a fire, the investigator must have an understanding of the basis for the system design, functionality, and capabilities.

The design of a system takes a number of factors into consideration relative to the hazard it must protect. Assuming that the installed system was approved and installed as designed, it is up to the investigator to evaluate whether the installed system was still able to protect the hazard. This is a critical part of any investigation because it is common for the hazard level to change but for systems to say the same. If the hazard level increases, the system may not be designed to deliver the amount of suppression agent necessary to control or extinguish a fire. This type of evaluation may be beyond an investigator's training; therefore, consultation with a fire protection engineer, design professional, or design technician may be necessary. Additional information is available through the manufacturer, building codes, and NFPA design and installation standards. At a minimum, the investigator should have the knowledge and ability to access these documents and the appropriate resources for technical assistance as needed.

Table 6-4	List of NFPA Fire Protection System Standards
NFPA 10, Standard for Portable Fire Extinguishers	
NFPA 11, Standard for Low-, Medium-, and High-Expansion Foam	
NFPA 12, Standard on Carbon Dioxide Extinguishing Systems	
NFPA 12A, Standard on Halon 1301 Fire Extinguishing Systems	
NFPA 13, Standard for the Installation of Sprinkler Systems	
NFPA 13D, Standard for the Installation of Sprinkler Systems in One- and Two-Family Dwellings and Manufactured Homes	
NFPA 13R, Standard for the Installation of Sprinkler Systems in Low-Rise Residential Occupancies	
NFPA 14, Standard for the Installation of Standpipe and Hose Systems	
NFPA 15, Standard for Water Spray Fixed Systems for Fire Protection	
NFPA 16, Standard for the Installation of Foam-Water Sprinkler and Foam-Water Spray Systems	
NFPA 17, Standard for Dry Chemical Extinguishing Systems	
NFPA 17A, Standard for Wet Chemical Extinguishing Systems	
NFPA 20, Standard for the Illustration of Stationary Pumps for Fire Protection	
NFPA 22, Standard for Water Tanks for Private Fire Protection	
NFPA 24, Standard for the Installation of Private Fire Service Mains and Their Appurtenances	
NFPA 25, Standard for the Inspection, Testing, and Maintenance of Water-Based Fire Protection Systems	
NFPA 70, National Electrical Code	
NFPA 72, National Fire Alarm and Signaling Code	
NFPA 750, Standard on Water Mist Fire Protection Systems	
NFPA 2001, Standard on Clean Agent Fire Extinguishing Systems	

Hazard Protection

The selection of a water-based or non-water-based fire suppression system is predicated on the inherent hazard(s) throughout a property. In addition, some requirements to install a particular type of system may be from a building code requirement or an insurance company requirement. Once the hazard is determined, the design of the system will consider the appropriate delivery system, how the system detects and activates, how much suppression agent is required, how it will be applied, the rate of application, and the concentration. Hazards change frequently; therefore, it is important for the investigator to evaluate the hazard and conditions and then to examine and analyze the type of system installed relative to the hazard encountered to determine if the system was adequate for the task. Knowledge of the hazard requiring protection can allow for a more thorough and focused examination of the water-based or non-water-based fire suppression system related to its operation and functionality during a fire incident.

To illustrate this point, years of laboratory testing and analyzing fire scenes and fire scene data produced tables that classify the hazard in relation to the amount of fuel load and the potential output of the fuel load. Based on the information, NFPA 13 devised tables that classify the hazard in relation to the amount of fuel load and the potential output of the fuel load. The resulting density/area curves list the amount of water in gallons per square foot over a certain area in square feet that

Fire Investigator Tip

It is imperative for the investigator to acquire and analyze system design data against installed system functionality during postfire examination. This includes verification that the system design characteristics, occupancy classification, and product or commodity protection agree with design standard, approved system drawing, building occupancy, and structural layout.

is needed to control or extinguish a fire for the hazard condition. The hazard condition categories are as follows:

- Light
- Ordinary (Groups 1 and 2)
- Extra Hazard (Groups 1 and 2)
- Special Occupancy

The idea is to evaluate the fuel load for the occupancy condition and to correlate the condition to the hazard classification. For example, a school is designed as a Light Hazard building or occupancy and therefore would not require as much water to discharge from the sprinkler heads and control a fire as a sawmill, classified as Extra Hazard Group 1, would.

One of the interesting issues that arises with water-based fire suppression systems directly relates to changes in the occupancy or use of the building that could affect the ability of the fire suppression system to control a fire. It is very common for a building to be designed and built for a specific use or occupancy condition, but later change to a different use or occupancy condition. The change may not appear to be a problem, but if the use or occupancy condition increases hazard level and the water-based fire suppression system is not changed to handle the hazard, a fire could overwhelm the suppression system and lead to a catastrophic loss. For example, if a building was once a public laundry but was purchased and became a dry cleaner, the water-based fire suppression system would need to be changed to support the increased hazard level because a laundry classifies as an Ordinary Group 1 hazard, but a dry cleaner is Ordinary Group 2. Although it would appear that the use or occupancy condition is the same or very similar, a dry cleaner has a greater fuel load because of the amount of clothes (and in some cases chemicals), whereas a laundry uses water and soaps that pose a lesser hazard.

Placement

As with other types of building systems, water-based and non-water-based fire suppression systems must be properly designed, installed, inspected, tested, and maintained for optimum performance during a fire incident. Water-based and non-water-based fire suppression systems are designed to control or fully suppress a fire. To achieve that design objective, sprinkler head and nozzle placement is critically important. During the course of the fire scene analysis, the investigator should document sprinkler and nozzle location and spacing relative to walls, ceilings, storage arrangements, shelving, light fixtures, ductwork, and any other condition that may have affected the proper operation of the sprinkler head or nozzle. As part of any investigation, the fire investigator must determine whether sprinkler heads or nozzles were obstructed, damaged, or improperly placed, allowing for inadequate control and uninhibited fire spread. Additionally, the investigator should document and examine the hazards, fuel packages, or processes protected by the sprinkler heads or nozzles to determine effectiveness.

The investigator should verify which design standard was utilized for the system and review the approved plans, reference standards, and manufacturer's data sheets for comparison to the installed system. Of special importance is review of the manufacturer's data sheets for the sprinkler heads or nozzles. The investigator must be able to establish whether the installed system was capable of handling the fire, or if any system manipulation occurred prior to the fire that may have enabled fire propagation. Many issues can affect system performance, but sprinkler head and nozzle placement is critical to successful distribution of the suppression agent.

Capacity

Water-based and non-water-based fire suppression systems are designed to deliver predetermined quantities of water or agent, typically through sprinkler heads and nozzles over a given area, and to provide additional water capacity for fire hose streams. Appropriate system design is hazard classification and occupancy-type based, and it is imperative that the investigator understand what capacity the system was designed for and how the installed system reacted during the fire incident. For example, NFPA 13 states that an educational facility is classified Light Hazard and, based on the density/area curve for light hazard, requires 0.10 gallons per square foot over 1500 square feet of area. In other words, every sprinkler head within the designated 1500 square foot area must provide at least 0.10 gallons per square foot. This information is part of the hydraulic data nameplate that states the important design characteristics of the system FIGURE 6-12. As a requirement of system installation, a hydraulic data nameplate is required to

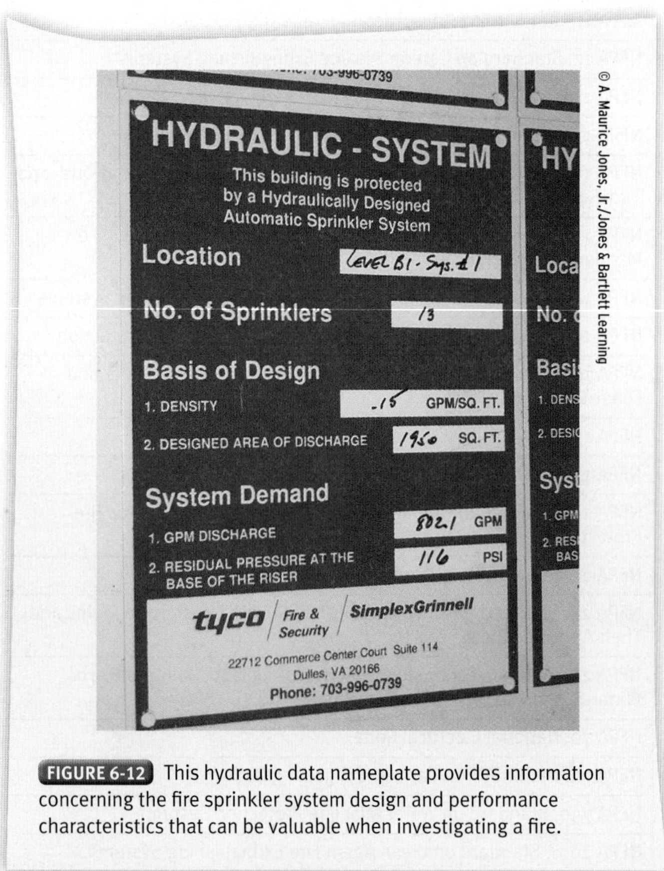

© A. Maurice Jones, Jr./Jones & Bartlett Learning

FIGURE 6-12 This hydraulic data nameplate provides information concerning the fire sprinkler system design and performance characteristics that can be valuable when investigating a fire.

be posted on or near the system valve to which the information applies. This can be very valuable information in determining if the system that was installed was appropriate for the hazard condition at the time of the fire.

During postincident examination, the investigator must document the number of devices that operated during the fire. Depending on the number and location of devices that activated, the investigator must determine if system design was appropriate for the hazard level or if other factors were present that compromised system performance. It is critical for the investigator to determine and document all mitigating items and factors, including appropriate and allowable fuel loads, the possibility of system tampering such as closed valves, water supply reductions, increased water flow due to more than anticipated sprinkler or nozzle activation, structural and nonstructural obstructions to device operation for proper water delivery, and interjection of ignitible liquids or other incendiary compounds to create a larger fire or retard system performance.

Coverage

Scene documentation plays a crucial role for the investigator in determining system coverage levels. The amount of coverage that a fire suppression system provides is dependent on a number of factors, including when the building was built, the codes in force at the time of construction, and the design standards that applied to the type of system. Therefore, in order to analyze and assess flame spread and fire damage, it is important for the investigator to determine the extent of suppression system coverage within a structure. The type and extent of suppression system coverage will have a direct impact on flame spread and fire damage. For example, a fully covered building will minimize flame spread and fire damage, whereas a building with limited or no coverage will probably suffer extensive flame spread and heavy fire damage. Assistance in determining the requirements for coverage can be obtained through the local code office that provided review, approvals, and inspections on the systems.

Total (Complete) Coverage

It can be very helpful for the investigator to understand and retain basic code and design standard terms and definitions. If a system is designed for total or complete coverage, the design requirement is inclusive of every space within a structure, including concealed areas and spaces, unless there is a specific allowable exception. Generally, any such exception will require passive protection such as fire-rated construction in lieu of fire suppression system coverage. Total coverage systems are common in many commercial properties that exceed certain code-defined thresholds, and, again, assistance in determining the requirements for coverage can be through the local code office that provided review, approvals, and inspections on the systems.

Partial or Selective Coverage

Partial or selective coverage is based on use and occupancy conditions in a building that may not require full or any coverage. This level of coverage allows for portions of a building that may have a higher hazard level to be protected by a water-based fire suppression system. The portions protected or covered are designated by specific code and standard requirements.

Local Coverage

There are times when the fire investigator may encounter a fire suppression system that protects only a piece of equipment or machine, not any part of the building around the equipment or machine. Typically, these situations exist due to specific insurance company requirements or owner preference. In these instances, the investigator must approach systems designed for local coverage no differently from code-required coverage; design parameters must be investigated for appropriateness and proper functionality.

Installation

In accordance with NFPA 1033, the fire investigator shall have and maintain an up-to-date basic knowledge of a variety of topics inclusive of fire protection systems. The fire investigator must have, at a minimum, the knowledge, capabilities, or appropriate resources to properly examine, document, and analyze fire protection systems. Based on scene conditions, it may be practicable to engage an engineer or design professional for assistance in completing system and installation analysis.

It can be very helpful for the investigator to be familiar with appropriate technical resources and the system design and installation standards to assist in the examination of fire suppression systems because they may require complicated and detailed engineering analysis. The investigator should interview occupants nearest the system during activation to understand any details regarding the fire and system functionality. Engineers, design professionals, manufacturers' representatives, personnel from the regulatory agency that reviewed and approved the design and inspected the system, and the installing contractor(s) should also be interviewed to ascertain system information.

Examination, analysis, and documentation of a water-based fire suppression system installation should begin with the incoming water supply. For example, a fire sprinkler system examination should start where water is provided to the system, whether it is from the local water purveyor, tank, or other resource. One of the first things to check is whether the main water supply valve is open to confirm that there is an available water supply. The examination then progresses along the system supply line to confirm that system control valves are open. Once the water supply is confirmed, the analysis identifies and examines components such as raisers and main supply piping, control and check valves, smaller branch or sprinkler line piping, and water delivery devices such as sprinkler heads or nozzles for proper type and orientation. The investigator must verify proper component installation and operation and examine and document improper component installation and operation throughout the system. Examples of improperly installed components are incorrect sprinkler head orientation, delayed sprinkler operation due to the use of improperly rated sprinkler heads, check valves installed backwards, and fully or partially closed water control valves.

System Performance

The investigator must analyze and document how the system performed or if it performed at all. It can be very helpful for the investigator to have appropriate technical resources to assist in the examination of suppression systems, as some require complicated and detailed engineering analysis. As with any other aspect of an investigation, the investigator could interview engineers, design professionals, manufacturers' representatives, and relevant personnel to assist with determinations related to cause and effect. In addition, the installing contractor should be interviewed to ascertain system information, conditions during installation, and anomalies present that may have affected system response. Understanding and establishing the relationship of system performance to fire impact can assist the investigator in better establishing timelines, acquiring information, and compiling data.

Evaluation of system performance is similar to installation analysis requiring the investigator to apply a thorough failure analysis process by examination, analysis, and documentation of the suppression system's role in the incident. Improper installation will affect system performance; therefore, the investigator must verify whether the system operated correctly or, if the system failed to operate correctly, why. The investigator is cautioned not to overstep his or her knowledge, skills, and abilities. As noted earlier, the investigator should have appropriate resources at hand to assist in completing proper system and performance analysis.

Testing and Maintenance

One of the main reasons that fire suppression systems fail is improper inspection, testing, and maintenance. Systems must be properly tested and maintained at regular intervals in accordance with manufacturers' recommendations and adopted standards to assure and verify operation and functionality.

As part of any investigation, the investigator should request all test records, maintenance records, and documentation for the system. Water-based systems should be inspected, tested, and maintained based on the requirements in NFPA 25, *Standard for the Inspection, Testing, and Maintenance of Water-Based Fire Protection Systems*, which establishes minimum intervals for the different systems and components. In addition, other NFPA standards that are specific to the type of system provide inspection, testing, and maintenance guidelines. This information is critical in determining if the minimum required inspections, tests, and maintenance were performed so the investigator can qualify when, how, and who was involved with the system. In addition to reviewing records kept by the owner, owner's representative, or the testing and maintenance contractor, the investigator should document whether the required inspection tags were attached to the system. The investigator should compare and contrast records, seeking dates, conditions found, corrections needed and completed, and status of system(s).

Fire investigators should ascertain whether a system was connected to an off-site monitoring company and should acquire data related to system operation, functionality, and notification.

Origin and Cause

The process of fire investigation is rooted in a systematic approach to determine fire spread, origin, cause, and responsibility. The accepted systematic approach requires the investigator to develop and test different hypotheses utilizing inductive reasoning explaining the fire's origin, cause, and spread based on scene observations, findings, and the relative and cogent investigation details that can stand up to challenge either physically or analytically.

Investigators can utilize information derived from system examination to bolster and substantiate, or discard, hypotheses. Conversely, system data can be used to exclude hypotheses. For example, if fire spreads in a structure protected with a fire suppression system, a detailed examination of the system indicating how and why the system failed to operate and suppress or control the fire can assist the investigator in developing a hypothesis or conclusion related to fire origin, cause, spread, and responsibility.

Timelines

To develop a hypothesis relating to fire origin, cause, and spread within a building protected with a fire suppression system, investigators must account for all relevant data associated with system operation. This includes any data gathered through electronic monitoring of system functions available in the activity logs of modern fire alarm system control panels and onsite proprietary and offsite monitoring companies. The investigator must correlate all data sources and scene evidence to formulate and substantiate a hypothesis or hypotheses for the fire incident. These data can assist the investigator in developing a timeline for the fire that can further assist in explaining fire ignition and spread. Beyond the physical examination of the system, investigators must ascertain all data related to electronic monitoring of system functions, if provided.

Most water-based and non-water-based systems are required by code to be electronically monitored at a communications center, proprietary facility, or third-party contractor. Monitoring is used to communicate changes in the normal ready state of a system. Changes to be monitored include detection of agent movement within a suppression system, indicating flow and operation. Other ancillary functions must be monitored as well, including fuel control, power shutoff, control valve positioning and/or movement indicating whether valves are in the open or closed position, and changes in temperature, air, gas, and water pressure, as well as other ancillary functions such as fuel control or power shutoff.

To develop a hypothesis and timeline relating to origin, cause, and fire spread within a building, investigators must account for all relevant data associated with system operation. This includes any data gathered through electronic monitoring of system functions from the data available in the activity logs of modern fire alarm system control panels and onsite proprietary and offsite monitoring companies. The investigator must correlate all data sources and scene evidence to formulate and substantiate a hypothesis or hypotheses for the fire incident.

Estimation of Fire Size

Investigators, through analysis of initial and subsequent system component activation and the system response, can estimate the size, origin, growth, and spread of a fire. In conjunction with information provided by the fire alarm system, the investigator can account for system activation time as well as nonactivation of initiation devices and fire suppression system components. For example, one method to estimate the size of the fire is based on the activation of fire sprinkler heads or nozzles that use frangible bulbs or fusible link elements. The activation point of a head can be estimated by the RTI of the head or nozzle. In conjunction with information provided by the fire alarm system, the investigator can determine system activation time. Of equal importance, however, is determining non-activation of initiation devices, sprinkler heads, or nozzles, and other system components. By verifying these actions, the investigator may be able to use this information to determine the size of the fire at a particular point in the fire event. Sources of information available include manufacturers' specifications and data sheets, design standards, and calculations, all of which can be utilized to estimate system characteristics and responsiveness regarding fire size. If a system can be shown to have activated and operated appropriately, the fire at the time of the alarm could be established as the minimum fire size. In contrast, if the system did not activate, assuming proper and correct installation, it could be established as the maximum fire size.

Fire Modeling

The field of fire investigation is ever-changing and is similar to other analytical fields, particularly regarding the advancements and influences of technology. Fire investigation relies on scientific methodology that must be performed in accordance with applicable science and engineering principles and processes. A technological tool utilizing such principles to better understand and analyze fire phenomena is computer fire modeling.

Computer fire models utilize mathematical data and inputs to predict fire behavior and fire conditions, as well as assist in establishing fire growth and propagation timelines based on expected fire size derived from available fuels. Fire models can provide investigators additional details to explain system performance. Computer fire models can forecast system operation times and behavior and offer an understanding of how the system affected the fire and surrounding area, given the size and characteristics of the model fire.

Computer fire models are additional tools that can assist the investigator in substantiating decisions and hypotheses. However, investigators must use caution not to exceed their abilities or skills regarding the use of computer fire models. Using this investigation tool requires significant training, education, and experience for reliable and valid conclusions that can stand scientific challenges. If the investigator does not possess such a background, outside resources must be utilized to complete the model.

Impact on Human Behavior

All factors regarding human behavior must be examined, including analysis of the interaction with the fire and the fire suppression system. It is very important for the investigator to analyze occupant awareness and response prior to and during system operation.

Buildings protected with any fire suppression system that is properly designed, installed, inspected, tested, and maintained offers the occupant a significantly better chance of survival from fire than a buidling without. When interfaced with a fire alarm system, not only does activation of the system alert occupants to hostile fire conditions, but fire suppression systems can also control and/or suppress the fire, allowing for increased survivability with greater time to respond appropriately to the conditions and circumstances of the fire event. Having adequate time to react and respond is the single most important factor in survivability, but occupants must not take for granted that a system is in place and that they have more time to respond because these systems do not remove smoke and gases that could be in the air prior to system activation.

Exposure to some types of suppression agents may inhibit visibility or breathing, disorient occupants, or impair a person from seeking refuge, escaping, or exiting an area. Although very rare, fire investigators must be aware that death and injury can occur even when systems operate correctly and that all relevant aspects of human behavior must be documented and analyzed regarding interaction with the fire and the system. It is very important for the investigator to analyze occupant awareness and response prior to and during system operation. This information can assist the investigator in determining fire origin, growth, spread, and cause.

Documentation of Fire Protection Systems

Design Documentation

In accordance with NFPA 1033, the fire investigator shall have and maintain an up-to-date basic knowledge of a variety of topics inclusive of fire protection systems. To do so, the fire investigator must have knowledge, capabilities, or appropriate resources to properly examine, document, and analyze fire protection systems. Based on scene conditions, it may be good practice to engage a professional engineer or other topic expert for assistance in completing system and installation analysis.

The initial approach to examine, document, and analyze a fire protection system is to recognize which codes and standards were utilized in laying out and installing the system, and to acquire the design documents for review and comparison. The investigator must verify the edition of the codes and standards in force at the time of design and installation, as well as any additional design information, past or current, that played a role in the system's existence.

Typically, fire protection systems are designed and installed in accordance with NFPA standards. However, other sources for system design can be referenced and used, such as FM Global, which publishes a large number of data sheets, best practices, and design standards for use at their insured facilities. Property insurance companies may also require a specific coverage or design, though this perspective is typically rooted in NFPA or FM Global design criteria. The design information needed by the investigator can be acquired from building owners, system and building designers, system installers, or local regulatory personnel. The investigator must understand his or her respective roles, understand the scope of the investigation (e.g., public or private), and be aware of the tools at his or her disposal to acquire the needed documentation.

■ Permit History

Generally, fire protection systems are required and regulated by fire and building codes that reference design standards specific to the type of system proposed. The permitting process is critical to ensure that there is a review of the proposed system and that the design meets the minimum code and standards. Typically, the fire and building codes require a permit application, plans, calculations, and manufacturers' data sheets be submitted for review and approval by local or state fire or building officials.

It is important for an investigator to understand the permitting process and whether permitting is a requirement for system installation. It is rare that a jurisdiction would not require a permit; therefore, the investigator will need to determine how and where to acquire all known permit information for a fire protection system involved in a fire scenario or other incident. Permit data can assist the investigator in establishing a true understanding of system functionality. Of significant importance for the investigator is the verification that system permitting and approvals, if required, were completed. The permitting files referenced may include initial permits, design drawings, additional requirements, rejections of initial system design, and installation inspection and testing documentation.

■ Invoices and Contracts

As noted, investigators should thoroughly collect and analyze all data available to properly determine fire origin and cause and to identify mitigating circumstances or factors that may have played a role in fire propagation and spread. The investigator should request and examine all records associated with fire protection systems, including, but not limited to, invoices and contracts relating to the installed system.

■ Installation Documentation

Fire protection systems are designed and installed in accordance with the adopted codes and standards that necessitate the creation of design documents for review and approval by the approving authority. These documents are the initial platform for system construction. Upon completion of system installation, it is common to have "as-built" drawings created, documenting changes in the design and installation. However, as-built drawings should be reviewed by the approving authority to ensure that any changes did not violate the code or referenced standard. In any case, the investigator should acquire the as-built drawings for comparison to the installed system to make sure the as-built drawings match the installed system.

■ Inspection and Maintenance Records

Fire protection systems are required to be inspected, tested, and maintained periodically based on the type of system and the adopted codes or standards related to the system. Beyond system design and installation requirements, periodic inspection, testing, and maintenance records are critical to verify system functionality. It is of significant importance for the investigator to acquire inspection, testing, and maintenance records because these records provide a profile of ongoing system compliance and functionality. Proper documentation will provide a good indicator of the likely performance of a system under fire conditions. These documents may be in the possession of several sources, including the building owner; building representative; an insurance company; a professional engineer; the contractor who performed the inspections, tests, and maintenance; or local regulatory agencies such as the Fire Marshal Office.

■ Product Literature

Any time a fire protection system is proposed, designed, and installed, the system must have data and specification information from the manufacturers of all parts and appurtenances utilized. This manufacturer data or product literature provides a variety of information and guidance relating to system equipment, installation, and performance. Product literature can provide valuable insight and understanding for the investigator regarding system components and operation. Moreover, the data found can confirm equipment capabilities and limitations and can explain and verify equipment testing and certification standards completed by certifying agencies such as FM Global, Underwriters Laboratories (UL), or other nationally recognized testing companies. The investigator should acquire and review system literature, both current and referenced at the time of system installation, to gain a better understanding of the requirements when the system was installed versus any significant changes after installation. It could be critical to understand any flaws, failures, or needed corrections relating to system performance that may have been found and noted post system installation.

■ Alarm/Activation History

It is important for investigators to identify and understand the types and capabilities of fire protection systems found within buildings that have experienced hostile fire conditions. Of particular importance are fire alarm systems. Beyond notifying building occupants of fire, these systems are often monitored by an offsite monitoring company, where system data are recorded and maintained. If data are available, the investigator should retrieve the data as soon as possible through appropriate methods.

The investigator should also have a basic understanding of the fire alarm system's capabilities found during scene examination. For example, is the system a zoned system (provides information concerning a general area of the building) or an addressable system (provides information concerning a specific device and location, time of incident, and condition of the device)? Addressable fire alarm panels are microprocessor-based and therefore hold considerable information and history about any events related to the fire protection systems that are monitored by the fire alarm panel. The investigator should determine if the fire alarm panel has any system information relating to the fire that can be analyzed. If fire alarm system examination exceeds the investigator's knowledge, skills, and abilities, the investigator must have appropriate resources available—such as outside engineering experts or systems technicians—to assist with the completion of this task. In addition to verifying activation, the investigator should confirm whether the system was operating on primary power or on a secondary power source, such as battery backup or generator.

Spoliation Issues

Fire scenes yield evidence ideally utilized to determine fire origin, cause, spread, and responsibility. During the course of their duties, fire investigators must be aware of proper procedures, methods, techniques, and guidance to identify, collect, and care for fire scene evidence in accordance with acceptable recommendations and practices. Investigators must consistently be vigilant to avoid spoiling evidence during scene documentation and examination.

Fire scene evidence can be manifested in many ways and items. Fire protection systems within the fire scene can be considered physical evidence and must be properly protected and preserved as such. As previously discussed, modern addressable fire alarm panels contain system event history that is important to preserve but must be obtained by trained individuals who will not lose or compromise the data.

Evidence related to fire protection systems is not limited to the physical system itself; rather, evidence can consist of any documentation associated with a system, such as drawings, reports, contracts, and alarm data. The investigator must apply significant effort not to destroy, alter, manipulate, or misplace documentation evidence. It is important for the investigator to understand fully his or her scope of duty and ensure that all fundamental aspects of evidence documentation are thoroughly executed. Evidence documentation should include the use of photographs, evidence logs, copies, and proper collection materials and methods. Regardless of the evidentiary item, the investigator must properly maintain a chain of custody for the evidence.

Wrap-Up

© Greg Henry/ShutterStock, Inc.

Ready for Review

- Buildings equipped with active fire protection systems offer investigators a resource that can be a valuable analysis tool. The type of system and the code and standard in force when the system was installed are factors that will determine how much information the system is designed to provide.
- A fire alarm system can provide direct notification to emergency communication dispatch centers and to proprietary and third-party service providers who initiate the appropriate response without the need for an individual to call in the fire emergency. This alert will expedite the emergency response that in many instances will help limit property damage, and in some instances will save a life.
- The purpose of a water-based fire suppression system is to provide property protection and life safety by delivering sufficient amounts of water over the fire area at the appropriate rate in an effort to control, and in some cases extinguish, the fire.
- Non-water-based fire suppression provides an alternative to water-based fire suppression when water may not be the best choice for the hazard involved. It is also referred to as Special Hazard Fire Suppression because it uses gaseous and chemical-based agents to protect special and specific hazards, processes, and equipment.
- Analysis is similar for water-based and non-water-based fire protection systems, including code analysis, hazard protection, system placement and installation, testing and maintenance, and other considerations.
- Documentation of fire protection systems includes their design, permit history, invoices and contracts, installation documentation, inspection and maintenance records, product literature, alarm/activation history, and any spoliation issues.

Hot Terms

Active fire protection system Any type of fire protection system that automatically activates upon detection of a fire condition.

Addressable technology Fire alarm and detection systems in which devices and system components are assigned a data address, are capable of two-way communication, and can perform certain activities based on signal input.

Air sampling smoke detection A method of smoke detection designed to draw air from the protected area into a detection chamber for analysis.

Alarm pressure switch A type of initiating device installed on all types of wet-based fire suppression systems; the device initiates a fire alarm signal once a water pressure threshold is met.

Alarm signal A warning signal that alerts occupants of a fire emergency.

Annunciation panel A device that uses indicating lamps, alphanumeric displays, or other means to provide first responders with status, condition, and location information concerning fire alarm system components and devices.

Balanced pressure proportioner A proportioner that uses an atmospheric tank and pump to introduce the foam concentrate into the system with the water supply.

Central station service A facility that receives fire alarm signals from properties that is staffed with trained, qualified, and proficient personnel who take the appropriate action to process, record, respond, and maintain monitored fire alarm systems.

Coded signal A signal that generates a predetermined visual or audible pattern to identify the location of the initiating device that is operating.

Conventional technology The technology used in fire alarm and detection systems that provides limited communication between the fire alarm control panel and system devices and, upon activation of an alarm, provides only general information concerning the device and its location.

Deluge sprinkler system A type of automatic fire sprinkler system equipped with open fire sprinkler heads that simultaneously discharges water from all sprinkler heads.

Density/area curves Graphs that establish the relationship between the required amount of water flow from a sprinkler head (density) and the area that must be covered by the water for different hazard classifications.

Design density The minimum predetermined amount of water that must flow from the fire sprinkler heads in the most hydraulically demanding part of the fire sprinkler system (remote area) to control or extinguish a fire.

Dry chemical extinguishing agents A dry powder suppression agent made from sodium bicarbonate–based, potassium-based, or ammonium phosphate chemicals that covers and smothers the burning material.

Dry pipe sprinkler system A type of automatic fire sprinkler system equipped with automatic sprinkler heads that has pressurized air or nitrogen in the pipes until a sprinkler head activates to reduce the air or nitrogen pressure, permitting the dry pipe valve clapper to open, flooding the system piping with water.

© Greg Henry/ShutterStock, Inc.

Duct smoke detectors Smoke detectors that sample the air moving through the air distribution system ductwork or plenum; upon detecting smoke, the detector sends a signal to shut down the air distribution unit, close any associated smoke dampers, or initiate smoke control system operation.

Electronic valve supervisory device A device or switch that integrates with or attaches to a water control valve that detects movement or changes in the position of the valve and then sends a signal that the normal open position has changed.

Fire alarm control unit (FACU) Equipment that monitors the integrity of the fire alarm system's circuits and devices, processes manual and automatic input signals from the initiating devices, drives the notification appliances, provides an interface to control or activate other fire protection and building systems in a fire emergency, and provides the power to support all of the system devices.

Fire alarm system A group of components assembled to monitor and annunciate the status of automatic and manual initiating devices so building occupants may respond appropriately.

Flame detectors Radiant energy detectors that sense specific portions of the visible and invisible light spectrum.

Foam concentrate A condensed form of the foam product.

Foam solution A solution that results from foam concentrate and water mixing in the correct proportions.

Friction loss As water flows away from the source through a water system, a progressive pressure drop due to changes in the direction of the pipe, length of pipe, type of pipe, size of the pipe, types and sizes of the fittings, and other components in line with the piping.

Gas-sensing fire detection A type of detector designed to sense a specific gas, toxic gases, or a variety of gases and vapors produced when processing hydrocarbons.

Halocarbons (clean agents) A chemical compound made up of carbon and hydrogen, chlorine, fluorine, bromine, or iodine.

Halogenated hydrocarbons (halon) A chemical compound mixture of carbon and one or more elements from the halogen series of elements (fluorine, chlorine, bromine, or iodine).

Heat detectors Initiating devices that operate after detecting a predetermined fixed temperature or sensing a specified rate of temperature change.

Hydraulic data nameplate A permanent and rigid sign posted on or near a fire sprinkler system riser that provides information including the hazard, design density, size of the remote area, required gallons and pressure for the water supply, and location that the system is serving.

Hydraulic design A mathematical method of determining flow and pressure at any point along the sprinkler system piping for the purpose of determining pipe size throughout the system.

Inert gas A gas that could contain a mixture of helium, neon, argon, nitrogen, and small amounts of carbon dioxide and does not contaminate, react, or become flammable.

Initiating device A system of components and devices that provide a manual or automatic means to activate fire alarm and supervisory signals.

Ionization smoke detectors Smoke detectors that use a small amount of a radioactive material that electrically charges air particles to produce a measurable current in the sensing chamber; once smoke enters the sensing chamber, a current drop below a predetermined level initiates an alarm.

K-Factor A number assigned to represent the discharge coefficient for the orifice of a sprinkler head that is used when calculating the water flow or water pressure at a specific location in the fire sprinkler system.

Line-type detector A type of detection device that uses a tube or wires running in different directions as the sensing element to provide coverage over a wide area.

Manual fire alarm box A type of initiating device that requires a person to pull a handle on the device to initiate an alarm signal.

Manual initiating device A type of initiating device that requires a person to make physical contact with the device in order to operated the device.

Noncoded signal A constant visual or audible signal that remains activated until reset.

Non-water-based fire suppression A type of fire suppression system that uses agents that are gas or chemical based.

Notification appliances Constant visual or audible signals that remain activated until reset.

Photoelectric detectors Smoke detectors that use a light source and receiver within the sensing chamber to initiate an alarm when the light source reflects or is obscured by smoke particles.

Pipe schedule A list of pipe sizes and the number of fire sprinkler heads that the pipe can support based on the hazard classification.

Preaction sprinkler system A type of automatic fire sprinkler system equipped with automatic fire sprinkler heads that interfaces with fire detection equipment and requires a fire detector to activate and an automatic fire sprinkler head to operate to flow water.

Pressure proportioner A proportioner that redirects some of the water supply into the foam concentrate tank to either exert pressure on a collapsible bladder or push the concentrate out of the tank to the proportioner for mixing.

Private mode notification An alarm signal that is sent to a location within a facility so that trained individuals can interpret and implement the appropriate response procedures.

Projected beam detectors Fire detectors that project a light beam to a receiver over a hazard; once the beam is obscured or scattered, it activates an alarm.

Proprietary supervising station fire alarm system A group of fire alarm systems at one location or multiple locations that are under the constant supervision and monitoring by the property owner's trained personnel.

Public mode notification An alarm signal that propagates throughout a building to audibly or visually alert the building occupants so they can take appropriate action.

Radiant energy–sensing detectors A type of fire detector that is not dependent on smoke and heat plumes to operate; instead, the detector looks for specific portions of the visible and invisible light spectrum produced by flames, sparks, or embers.

Regular or ordinary dry chemicals Dry chemical agents rated only for Class B and C fires.

Remote area The minimum square footage of the most hydraulically remote pipe in a fire sprinkler system where the minimum design density must be available from all heads in that area.

Remote supervising station fire alarm system A fire alarm system in which alarm, trouble, and supervisory signals transmit to a location that is remote from the protected premises and staffed with individuals trained to take the appropriate action.

Residential fire sprinkler systems A type of automatic fire sprinkler system equipped with fast response automatic sprinkler heads specifically made for low heat release and low water pressures.

Response time index (RTI) The amount of time a head takes to activate once exposed to temperatures above the predetermined activation temperature.

Smoke alarm A single- or multi-station smoke detector with an integrated power source, sensing device, and alarm device.

Smoke detectors Fire detectors that sense visible and invisible particles of combustion.

Spark/ember detectors Radiant energy detectors that sense specific portions of the visible and invisible light spectrum.

Spot-type detector A type of detection device that provides coverage in a specific area where the sensing element is in a fixed location.

Supervisory alarm signal A type of fire alarm system alert signal that indicates when the normal ready status of other fire protection systems or devices connected to or integrated with the fire alarm panel has changed.

Tamper switch A device that detects changes in the normal position of a fire protection system valve and causes a supervisory signal when the valve is off normal.

Textural signal A messaging signal that provides constant and specific information via voice, pictorial, or alphanumeric means.

Trouble alarm signal A type of fire alarm system alert signal that indicates there is a problem with the system's integrity, such as a power or component failure, device removal, communication fault or failure, ground fault, or a break in the system wiring.

Venturi proportioner A proportioner that uses water moving over an open orifice to create a lower pressure at the opening, which draws the foam into the water stream.

Video detector A type of fire detector that uses high-definition video cameras to analyze digital images for changes in the pixels due to smoke and flame generation.

Water control valve A device used to control the flow of water.

Water flow switch A type of initiating device installed on a wet pipe sprinkler system; a paddle inserted into the pipe moves to initiate an alarm signal when there is sustained water flow through the system piping.

Water mist system A fixed fire protection system that uses specialized nozzles to discharge a very fine-spray mist of water droplets that extinguish by cooling, displacing oxygen, or blocking radiant heat.

Wet chemical extinguishing agents Suppression agents that mix water with potassium acetate, potassium carbonate, potassium citrate, and, in some instances, a mixture of these agents and other additives; used primarily to suppress Class K fires.

Wet pipe sprinkler system A type of automatic fire sprinkler system equipped with automatic fire sprinkler heads that has water in the pipes at all times so when a sprinkler head activates, water flow is immediate.

© Greg Henry/ShutterStock, Inc.

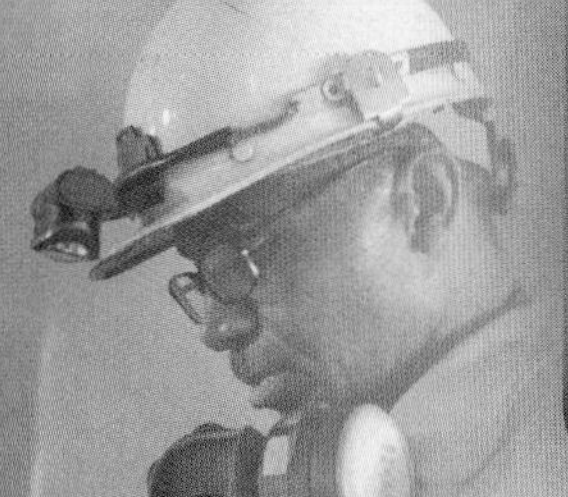

FIRE INVESTIGATOR *in action*

You have been requested to investigate a fire in a large commercial office building. The fire was first reported by an alarm system monitor company that indicated they initially received an alarm from a duct detector located within the ventilation system for the second floor offices. Approximately 2 minutes after the duct detector went into alarm, the alarm company received a water flow alarm from the sprinkler system that protects the second floor.

Fire personnel state that when they arrived, they could hear the alarm system sounding and noted water running through the ceiling near the center of the first floor. After determining the fire was out, fire personnel silenced the alarm system and turned off the water to the sprinkler system via the main control valve.

Your examination of the interior of reveals the fire was contained to a storage room located near the middle of the second floor and had been extinguished by three sprinkler heads. Upon removing the debris and reconstructing the area of origin, you locate the remains of a metal trash can containing charred paper and waste as well as the remains of numerous cigarette butts. As you begin to interview the various employees who were working on the second floor, you notice a young female standing by herself smoking a cigarette.

1. Which of the following is not considered a basic component group of a fire alarm system?
   - **A.** Notification appliances
   - **B.** Control unit
   - **C.** Initiating device
   - **D.** Suppression system
2. What type of signals and information would most likely be indicated on the fire alarm or annunciation panel?
   - **A.** First floor sprinkler and duct detector alarm
   - **B.** First floor sprinkler and duct detector alarm, second floor valve supervisory, first floor system silenced
   - **C.** Second floor sprinkler and duct detector alarm, second floor valve supervisory, first floor system silenced
   - **D.** Second floor sprinkler and duct detector, first floor valve supervisory, first floor system silenced
3. Which of the following sources of information may be useful in determining the effectiveness of suppression and detection systems?
   - **A.** Manufacturer documentation
   - **B.** Inspection records
   - **C.** Permits
   - **D.** All of the above
4. Which type of sprinkler system was most likely installed in this building?
   - **A.** Dry system
   - **B.** Preaction system
   - **C.** Deluge system
   - **D.** Wet system

Electricity and Fire

© Photos.com

© Jones and Bartlett Learning. Photographed by Kimberly Potvin.

Knowledge Objectives

After studying this chapter, you should be able to:

- Explain basic electricity. (pp 114–120)
- Identify and describe the components of a building's electrical system NFPA 4.2.8. (pp 121–127)
- List the conditions that must exist for ignition from an electrical source. (pp 127–130)
- Describe how to interpret damage to electrical systems. (pp 130–135)
- Explain static electricity. (pp 135–136)

Skills Objectives

After studying this chapter, you should be able to:

- Complete calculations based on Ohm's law. (pp 114–115)
- Determine whether a circuit has proper overcurrent protection. (pp 123–125)
- Identify which circuits have overcurrent or are overloaded based on a blown fuse or tripped circuit breaker in a panel. (pp 128–129)
- Examine fire-damaged electrical conductors and determine whether the damage is the result of electrical activity or a result of the fire. (pp 130–135)

Additional NFPA References

NFPA 70, *National Electrical Code (NEC)*

NFPA 70E, *Standard for Electrical Safety in the Workplace*

NFPA 77, *Recommended Practice on Static Electricity*

NFPA 921, *Guide for Fire and Explosion Investigations*

CHAPTER 7

FESHE Course Outcomes

Fire Investigation I

8. Discuss the basic principles of electricity as an ignition source. (pp 127–130)

Fire Investigation II

6. Analyze electrical causes of fires. (pp 130–136)

You Are the Fire Investigator

© Jones and Bartlett Publishers. Photographed by Glen E. Ellman

You are investigating a fire in a single-family residence that originated in the utility room of the basement. Severe flame and heat damage is present throughout the room, which also contains the electric service panel for the structure. The electric service panel is severely damaged with numerous electrical circuits exiting the top of the panel.

As you begin to examine each circuit, you locate a circuit along the south wall of the room that is attached to the remains of an electrical outlet. One of the conductors attached to the remains of the outlet is broken and displays damage you believe to be caused by an electrical arc. The remaining circuits are all intact; however, the protective insulation has been consumed with sporadic parting arcs present along the length of several conductors.

In interviewing the homeowner, he informs you that a sump pump located on the opposite end of the basement had been connected to the outlet using an extension cord. He states that the sump pump had been making more noise than usual and running more often due to melting snow.

1. Why is it important to understand electricity and electrical systems while investigating this fire?
2. How will the size of the electrical conductors influence your investigation?
3. What resources would you utilize to assist in determining what caused the fire?
4. What conditions must exist for this fire to have been caused by a failure of the outlet?

Introduction

This chapter provides an introduction to basic electricity and electrical systems. Determining whether electrical damage was the cause or the effect of a fire is never easy. It requires at least a fundamental understanding of electricity and, more specifically, knowledge of the electrical units. The units of electricity can generally be listed as voltage (volts), current (amps), resistance (ohms), and power (watts [W]). One practical way to establish how the units function is to perform tests of basic circuits, such as a switched light, with an appropriate measurement instrument to gain a feel for important relationships and concepts. These include the proportionality of voltage and current (increasing the voltage increases the current) and the concept of a circuit (open the switch and the lamp goes out).

If the investigator is not qualified to analyze the electrical equipment, a qualified individual (usually an electrical engineer) should be contacted to assist. More than the usual amount of caution is recommended in dealing with electrical systems. These systems should initially be approached and treated as if they are energized, or "live." NFPA 70E, *Standard for Electrical Safety in the Workplace*, is a critical tool for protecting electrical workers from electric shock, arc-flash, and arc-blast hazards while doing electrical work and should be used as a guide prior to approaching circuits of unknown energized status. NFPA 70E also has several tables that help electrical workers select the correct type of personal protective equipment (PPE) to wear, based on proximity and the task that they are performing. There are five different hazard/risk categories (HRCs): 0, 1, 2, 3, and 4.

Before analyzing an electrical circuit or equipment, the HRC of the installation should be assessed, and the investigator should ensure that the power to the system has been deenergized and an electrically safe work condition has been created. For example, workers must wear PPE specified by the tables in NFPA 70E whenever they are within the flash protection boundary (typically 48 inches [1.2 m] for 600-volt equipment), whether or not they are actually touching the live equipment. This includes tasks such as voltage testing to verify whether power has been turned off. Such exposure is considered "live work" that requires workers to wear PPE, including flame-resistant (FR) clothing. This PPE is for arc-flash and arc-blast protection, not protection against electric shock.

Basic Electricity

In an effort to explain the basics of electricity, a comparison to hydraulics is often used. However, please note that hydraulic and electrical systems are not entirely analogous. First, when making the comparison, it is important to envision the hydraulic system as a closed system that circulates water back to the source. A pool filtration system is a good analogy: water flows from the pool (the source), through the filtration (the circuit), then back to the pool (return/neutral). Comparisons to open systems that discharge water, such as a fire hose, are not good examples. NFPA 921, *Guide for Fire and Explosion Investigations*, contains a section on basic electricity, including graphics to help explain simple electrical circuits and electricity. TABLE 7-1 compares a hydraulic system to an electrical system.

Ohm's Law

The basic law of electricity for a resistive circuit (one that does not contain inductance and capacitance) is called Ohm's law. It defines the relationship among voltage, current, and resistance. If two of these three values are known, it is possible to determine the third. Knowledge of the values of these

> **Safety Tip**
>
> PPE is sometimes required for work on electrical systems that are not live.

Table 7-1 Hydraulic System Comparison to Electrical System

Hydraulics	Electricity
A **pump** creates the force that moves the water.	A **generator/battery** creates the electromotive force that moves the electrons.
Pressure is measured in pounds per square inch (psi) and is measured with a pressure gauge.	**Voltage (*E*)** is measured in volts (*V*) and is measured with a voltmeter.
Water moves through the pipes and does the work.	**Electrons** move through the conductors and do the work.
Flow is measured in gallons per minute (gpm) and is measured with a flow meter.	**Current (*I*)** is measured in amperes or amps (*A*) and is measured with an ammeter.
A **valve** controls the flow of water: ■ Open: water flowing ■ Closed: no water flowing	A **switch** controls the flow of electricity: ■ Off: open circuit; no electricity flowing ■ On: closed circuit; electricity flowing
Friction is the resistance of the pipe or hose to the water moving through it and is measured in psi.	**Resistance (*R*)** is the opposition of the conductors to the electrons moving through it and is measured in ohms (Ω).
Friction loss is the amount of pressure lost between two points in a pipe layout.	**Voltage drop** is the amount of voltage drop between two points in a circuit.
Pipe or hose size is measured in inches, inside diameter: ■ Larger pipe = greater flow ■ Smaller pipe = lower flow	**Conductor size** is given in AWG or wire gauge size: ■ Larger wire = greater current ■ Smaller wire = lower current

parameters and their relationship to each other is essential to understanding whether enough heat is present to ignite a fire or contribute to its ignition. The Ohm's law components are shown in TABLE 7-2.

Voltage is often represented in equations by either the letter *E* or *V*. The two are thus interchangeable in equations. When referring to the units of voltage, however, *V* is always used. Current always appears as *I* in equations, with units in amperes abbreviated as *A*.

Ohm's law is:

$$\text{Voltage} = \text{current} \times \text{resistance} \qquad (E = I \times R)$$
$$\text{Volts} = \text{amperes (amps)} \times \text{ohms}$$

By using this law, it is possible to rearrange the values in this equation to determine current and resistance, as follows:

$$\text{Current} = \frac{\text{voltage}}{\text{resistance}} \quad \left(I = \frac{E}{R}\right) \quad \text{Amps} = \frac{\text{volts}}{\text{ohms}}$$

$$\text{Resistance} = \frac{\text{voltage}}{\text{current}} \quad \left(R = \frac{E}{I}\right) \quad \text{Ohms} = \frac{\text{volts}}{\text{amps}}$$

One of the more useful measurements in working with postfire circuits is the measurement of resistance. Usually, once the investigator knows the input resistance, he or she can determine the current if the voltage is fixed and can estimate the overall power generated. The power (discussed in the next section) can then be used to find localized heating or higher respective heat fluxes that could increase the mass of a combustible to corresponding ignition temperatures. The electrical resistance of a *simple* heating appliance is measured with a special meter called a multimeter or VOM (volt/ohm/multimeter), by taking a reading with the probes across the two spades of the male plug of an appliance power cord. The result is the resistance of that appliance's heating element in ohms. As stated earlier, the investigator can then divide the voltage by the ohms to determine the current (in amps) used by that particular appliance.

The investigator can determine the current of all of the appliances (if they are simple resistive devices) on a particular circuit by using this method. The investigator can also determine what the resistance should be for a particular appliance by obtaining the voltage and current and calculating the resistance. These are often printed on the appliance's nameplate, so a measurement may not be required. The investigator divides the voltage by the amperage to determine what the resistance should be, as indicated in Table 7-2 and applying Ohm's law.

Power

Knowledge of voltage, current, and resistance leads to another relationship in defining the flow of energy. Electrical power

Table 7-2 Ohm's Law Relationship

Value	Symbol	Units
Voltage	*E*	Volts
Current	*I*	Amperes (amps)
Resistance	*R*	Ohms
Power	*P*	Watts

Note the difference between symbols and units, as well as the relationship between the unit values when analyzing electrical circuits.

Fire Investigator Tip

Often the heating element cannot be measured directly from the male plug spades because of internal switches, relays, or fire-damaged conductors. Direct connections to the heating element internal to the appliance are needed.

refers to the rate of doing work in an electrical circuit, such as in a hair dryer, electric motor, or lightbulb. Electrical power is measured in watts (or joules per second). Another unit for power that most people are familiar with is horsepower, which is used to express the rate of doing work for some mechanical objects. A third unit for power, used to describe heat sources such as furnaces, is BTU/hour. BTU stands for British Thermal Unit.

The symbol for power is *P*. The formula for determining electrical power is as follows:

Power = voltage × current $(P = E \times I)$

One kilowatt is equal to 3415 BTU/hour and 1.34 horsepower.

Both mechanical and electrical power are typically given on the nameplate of motors. Mechanical (output) power for motors is usually expressed in horsepower. Electrical (input) power is expressed in kilowatts. The rated electrical power is higher than the horsepower (converted to kilowatts) by the amount equal to the losses of the motor (inefficiency), which generates heat. A discussion of this heat source is beyond the scope of this text, but further information, including estimating motor and bearing temperature rise under different operating conditions, can usually be obtained by contacting the manufacturer of the motor and/or the equipment manufacturer.

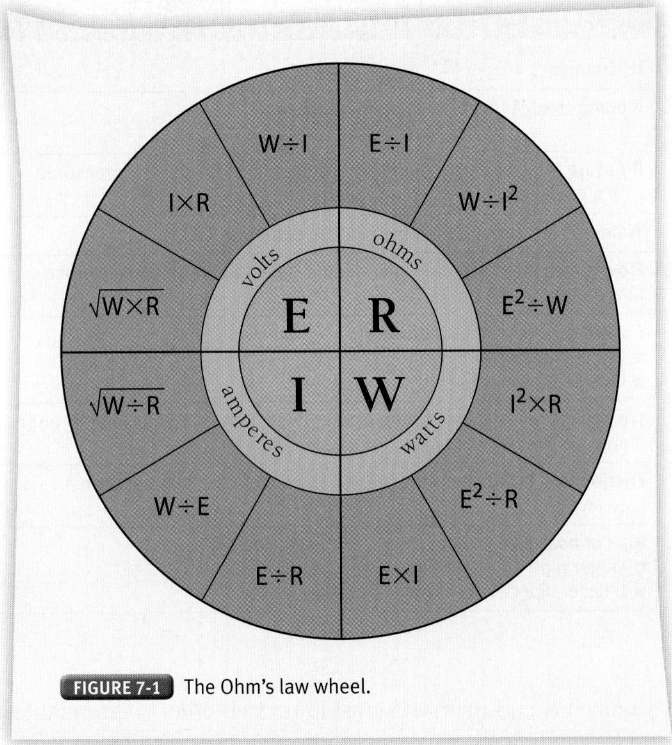

FIGURE 7-1 The Ohm's law wheel.

Relationship Among Voltage, Current, Resistance, and Power

It is important for the investigator to understand the relationship among voltage, current, and resistance and how to calculate power from these. This knowledge will help the investigator understand the potential of electricity to be a fire cause. For example, knowing voltage and resistance, or voltage and power, or power and resistance, will allow the investigator to determine how much current or energy was used by an appliance, whether a circuit was overloaded, or whether the overcurrent protection was properly matched to the power requirements for the circuit. The Ohm's law wheel is useful for determining one parameter in terms of any two others FIGURE 7-1.

Ideally, the investigator should be able to calculate the total power requirements of a circuit with multiple loads on it by inspecting the equipment, determining the power requirements of each piece of equipment, and then totaling all of these values. The nameplate information on the appliances provides some information regarding important electrical parameters. Often, these values are given in amps, watts, or voltage. Keep in mind that these are typical maximum values (the upper end of nominal state voltage) and assume active operation of the equipment during peak power utilization. Obviously, a washing machine is not going to draw full rated current unless it is in the agitation or spin cycle, where the power required to overcome inertia is highest. Because power use is dynamic and can vary over time depending on the action of the appliance, branch circuit overload situations can develop with multiple different loads operating intermittently.

By using Ohm's law, we see that when a fault occurs in a circuit, the circuit resistance decreases and the current increases to an abnormally high value; however, if sufficient resistance exists in the electrical current path, the short circuit current may continue. Thus, even though the resistance of a fault in a circuit might be low, it can still be high enough to dissipate significant energy while drawing lower amperage than is required for the circuit protection device to trip. For example, if a fault develops across a carbon path between two conductors and there is 100 ohms of resistance, the current draw for that particular fault would be 1.2 amps (amps = volts/ohms), and the circuit protection may not trip; however, if a fault occurs where there is very little resistance, such as 0.2 ohms, there will be up to 600 amps flowing through that fault, and the circuit protection should immediately open. In the former case, more than 100 watts is dissipated in the fault path, and this can generate sufficient heat to ignite nearby combustible material.

FIGURE 7-2 illustrates the relationship among voltage (*V*), current (*I*), resistance (*R*), and power (*P*). Note that *R* must be measured with V turned off (0 *V*).

Overload Situations

If the power needs in a circuit exceed the circuit's current carrying capacity, then an overload situation may occur. Overload is the persistent operation of equipment in excess of its normal,

Fire Investigator Tip

Ignition by electrical energy requires an energized circuit, sufficient heat and temperature, and proximity to combustible material.

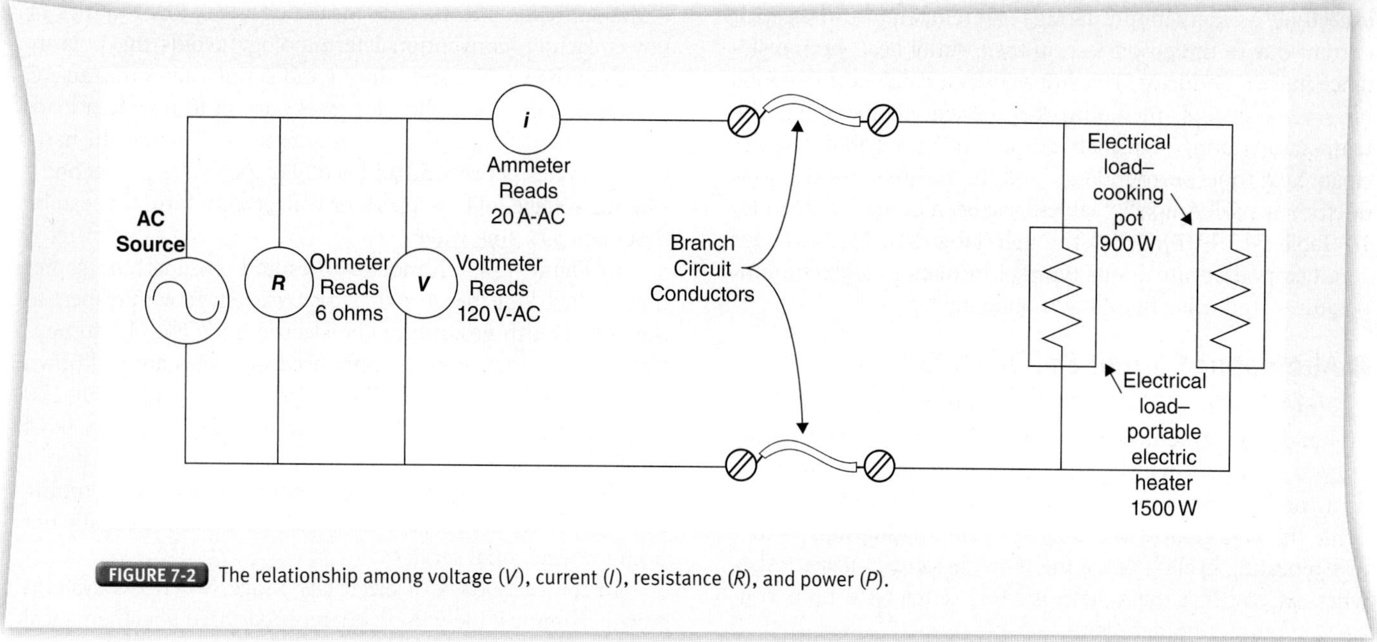

FIGURE 7-2 The relationship among voltage (*V*), current (*I*), resistance (*R*), and power (*P*).

full-power consumption rating or of a conductor in excess of rated ampacity, which (when it persists for a sufficient length of time) would cause damage or dangerous overheating. A direct, short-duration fault, such as a short circuit or ground fault, is not an overload but rather an overcurrent.

An overload can cause the various components in a circuit to overheat. If the overloaded section of conductor/contact surface area or component has sufficient temperature, enough duration, and proximity to a fuel, ignition may occur; however, all three factors (oxygen is assumed to be present) must be considered when determining whether an overloaded circuit started the fire. The investigator must examine the circuit and the appliances on that circuit to determine whether an overload occurred. Overloading of the wiring in a house circuit by only a few amps will not usually open the overcurrent protection device or cause overheating because there is a safety factor built into most electrical circuits. Overloading a circuit significantly, to the point of overheating, usually requires the circuit current to exceed the rating by a greater margin, something that an overcurrent protection device is designed to prevent. It is important to check the overcurrent protection device if an overload is suspected to ensure that the proper ampere rating was chosen.

A properly designed electrical distribution system will take into account the likely maximum amperage for each branch circuit based on occupant usage and will prevent overload situations. The investigator should consider anecdotal information, such as occasional breaker tripping, as a warning sign under normal usage conditions because of the possibility of faulting in the circuit or appliance.

Problems can occur in household wiring that is poorly connected, such as a high-resistance connection at an outlet, electrical switch, or other mechanical connection where there is enough resistance at the connection point to prevent exceeding the circuit protection trip point. This type of connection may cause heating at the connection point and possibly ignite nearby combustible material, despite the use of properly sized circuit protection devices. Such heating usually requires significant amounts of current, usually well beyond the branch circuit rating.

Wire Gauge

The diameter of wire is commonly measured in sizes given by the American Wire Gauge (AWG). The smaller the AWG number, the larger the wire diameter. Common household wiring is 14 and 12 AWG (if copper). There are specific electrical circuits in the home that use larger amounts of electrical currents and are served with larger wires. Appliances such as an electric range, dryer, or water heater are often supplied with 6, 8, or 10 AWG wire.

Wire gauge in an electrical system is similar to pipe diameter for a water system: Larger wire (smaller gauge number) equates to larger diameter pipe. A blockage in a water pipe equates to a nick or cut in an electrical wire. The amount of water flow (electrical current) is reduced for fixed pressure (electrical voltage).

NFPA 70, *The National Electrical Code*, requires that the size, type, and amount of the conductor be labeled on the insulation. For instance, residential wiring will be designated "NM-B 12/2 with ground," which means that it is nonmetallic cable, 12 AWG, and with either two current carrying conductors and a grounding conductor or a hot, a neutral, and a grounding conductor.

Ampacity

Amperage is the amount of current flow measured in amperes (amps or A) and is similar to the flow of water in gallons per minute. Ampacity is the current, in amperes, that a conductor can carry continuously under the conditions of use without exceeding its temperature rating. The temperature rises with current due to the power lost, in the form of heat, from resistance in the conductor. The ampacity of a conductor depends on several factors, including the ambient temperature, the temperature rating of insulation, the conductor's ability to dissipate heat to its surroundings, and the resistivity of the conductor material. Ampacity tables have been generated in NFPA 70 Table 310.15(B)(16)(a) through Table 310.15(B)(21) for conductors used in various types of insulation, ambient temperatures, and cable bundling/routing methods.

Alternating Current and Direct Current

It is important to understand that most electrical current used in buildings, structures, and dwelling units is alternating current (AC), in which current flows in and out from the upstream electrical source in a cycle (usually a transformer) while the voltage is also changing by alternating from – to + in a repeating cycle—hence the term *alternating current*. Also, when we say that the system is "AC," it refers to both voltage and current because with a simple (resistive) load, both of them alternate in a similar fashion.

In the case of a residence, the service panel is considered the local electrical source. With a simple load, the voltage delivered to the transmission lines drives current out from the source and then back in a repeating cycle. One voltage cycle includes both a positive and a negative component, going from peak positive voltage to peak negative voltage during the cycle FIGURE 7-3. Unless peak values or reactive effects are being discussed (examples include insulation breakdown ratings or power factor), conventional terminology avoids the polarity switch of the voltage and current and simply states that an AC voltage of a given amplitude causes current to flow to a load. In the United States, the AC frequency is measured in hertz (Hz) or cycles per second and is 60 Hz (60 cycles per second). The AC voltage follows a path or voltage waveform that can be described as a sine wave.

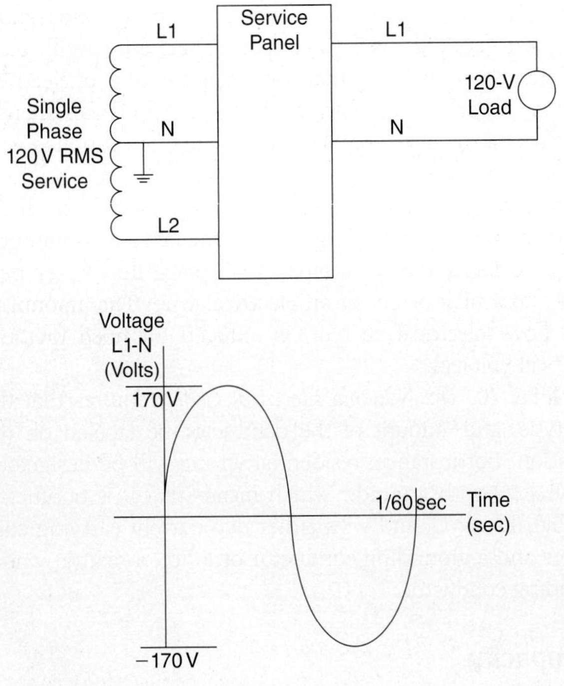

In Figure 7-3, L1 and L2 wires and terminals are sometimes called "hot" because they contain voltage with respect to the power/earth ground for the electric grid. The "L1 to neutral" configuration is commonly used in homes and is known as a 120-volt AC system. It has a peak voltage of about 170 volts. Using an "L1 to L2" configuration, 240 volts (peak of 340 volts) can be derived. Furthermore, by using transformers, different voltages having waveforms with various timing or phase relationships can be derived, but these are usually not used for residential service.

In the early days of electricity usage, when AC systems were becoming widely used, engineers devised a mathematical computation to equate the voltage level of an AC system to that of the more familiar direct current (DC) system. (DC is discussed in more detail following this section.) This computation is called root mean square (RMS).

For standard residential AC voltage source, the RMS value is equal to the peak voltage divided by the square root of 2 (or 0.707), as follows:

$$V\ RMS = \frac{V_{peak}}{\sqrt{2}}$$

The RMS value of AC voltage is typically used and will be assumed throughout the rest of this book unless otherwise specified. Thus, 120 V in residential use is actually 120-V RMS, with peak voltage of 170 V.

The AC voltage thus changes 60 times a second from positive to negative (60 Hz). The AC current can follow the voltage and typically also looks like a sine wave but can vary, depending on what type of a load it is. If the equipment or load is resistive (such as a heater), the current follows a sine wave. If it is a complex device such as a piece of electronic equipment, the current can be quite different because the internal circuits use different amounts of current at different times during the AC cycle. This can cause peak currents to flow that are significantly greater than the sine wave would predict, but they usually occur for very short durations of time and so do not normally stress the electrical system. Controlling (forcing) the current waveform to look more like a sine wave is desirable, and recently manufactured appliances and electronics are required to meet Underwriters Laboratories (UL), the Canadian Standards Agency (CSA), and the European Committee for Standardization (CEN) regulations for load-created noise and harmonics that may require doing this. Advances in power electronics technology have provided cost-effective components to do this.

It is important to understand that modern electronic equipment often contains complex electronics with power, analog, and digital components inside them. There is often

FIGURE 7-3 Single-phase AC sine wave—120 V RMS residential use.

also a microcontroller or application-specific integrated circuit (ASIC) component that acts as the brain for the operation of the equipment. What this means for you, the investigator, is that simple calculations of power are not possible and that you must read the manuals supplied by the manufacturer to learn about power-up and power-down sequencing and about different modes the equipment can run in. A modern washing machine is an example of this kind of equipment. The power sequencing and mode switching are handled by the ASIC, and the power consumption is not steady but intermittent. Thus, overcurrent protection and conductor size are selected based on typical maximum operating power levels.

A direct current (DC) system usually has current flow from the source to the load and back via the circuit return path with one polarity only. This type of system is found in fixed installations and devices requiring stable, controlled voltage levels, including some appliances and control systems such as those used in industrial settings. The voltage is held at a constant value, rather than varying with a sine wave. Mobile or portable equipment such as electric vehicles and wheelchairs also use DC voltage. Some portable equipment can have two power sources—one AC and one DC—depending on how it is configured. An example is a wheelchair or an industrial floor cleaner that uses a battery when in normal operation but uses AC power when the batteries are being charged. The investigator must keep in mind electrical sources and loads that are in play in order to determine or rule out possible ignition sources of a fire. Thus, a battery charger found connected to batteries aboard a wheelchair may not require the chair motors to be energized, so they can be ruled out. The charger, however, is getting power from two sources—the AC line and the DC voltage, which is back-fed from the batteries. Both of these sources should be examined as potential ignition sources if the charger is found connected.

Single-Phase Service

A single-phase AC 120-V system, as it enters the building, requires three conductors: two line (L1/L2) conductors, sometimes called hot legs, and a neutral conductor, which is grounded near the source transformer located outside the building. A grounded distribution system has the neutral grounded at regular intervals if poles are used or at underground distribution cabinets. As previously discussed, the voltage between either the L1/L2 conductor and the neutral or ground is 170 V peak or 120 V RMS. If you measure the voltage across the L1 to neutral and the L2 to neutral conductors, they appear to mirror each other. The voltage between the two L1/L2 conductors is 340 V peak or 240 V RMS. This single-phase system is frequently found in residential buildings (single-family houses) and small commercial buildings.

Single-phase cables can be delivered to the structure either overhead or underground. Wiring coming in from an overhead pole is called a service drop. In a triplex service drop, the hot conductors may be wrapped around the neutral, which is typically not insulated FIGURE 7-4.

Another configuration has the L1/L2 and neutral conductors separated as they come into the structure. Wiring coming in underground is called a service lateral, in which the neutral is always insulated. The voltage between either of the L1/L2 or hot conductors and the neutral or ground is 120 V. The voltage between the two L1/L2 or hot conductors is 240 V. The three conductors in a service entrance are typically multi-stranded aluminum and of a larger gauge than is typically found in branch circuits. The voltage between the L1/L2 conductors is a 240-V sine wave FIGURE 7-5.

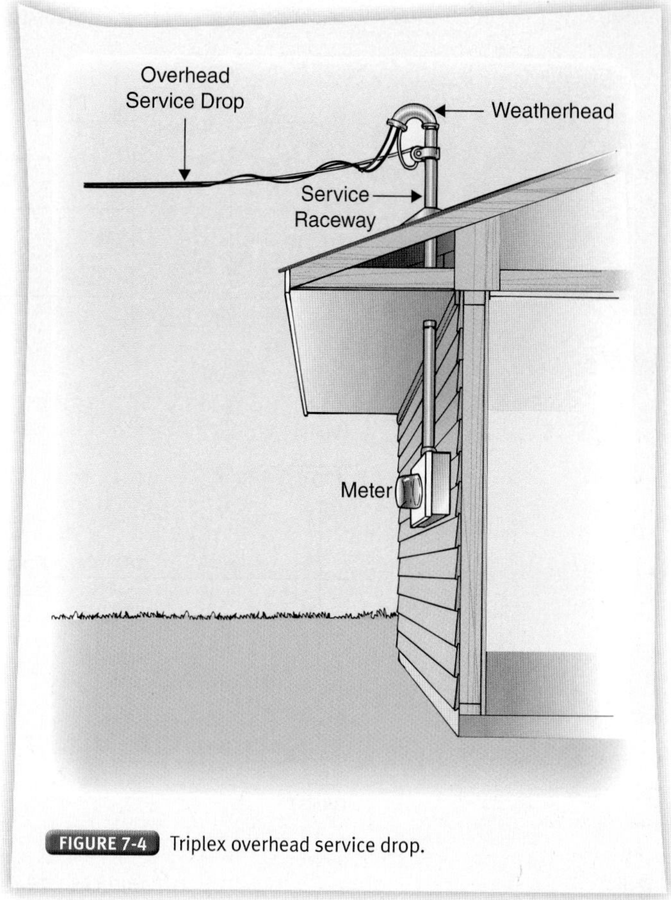

FIGURE 7-4 Triplex overhead service drop.

Three-Phase System

In a three-phase system, electric power is sourced on three conductors with respect to one ground or neutral. The sine wave of the three conductors is out of phase so that when one conductor has voltage at its highest voltage potential, the other

Fire Investigator Tip

Knowledge of the basic principles of electrical energy storage and flow is critical for an understanding of how electrical components and systems can fail catastrophically or be compromised, resulting in a fire.

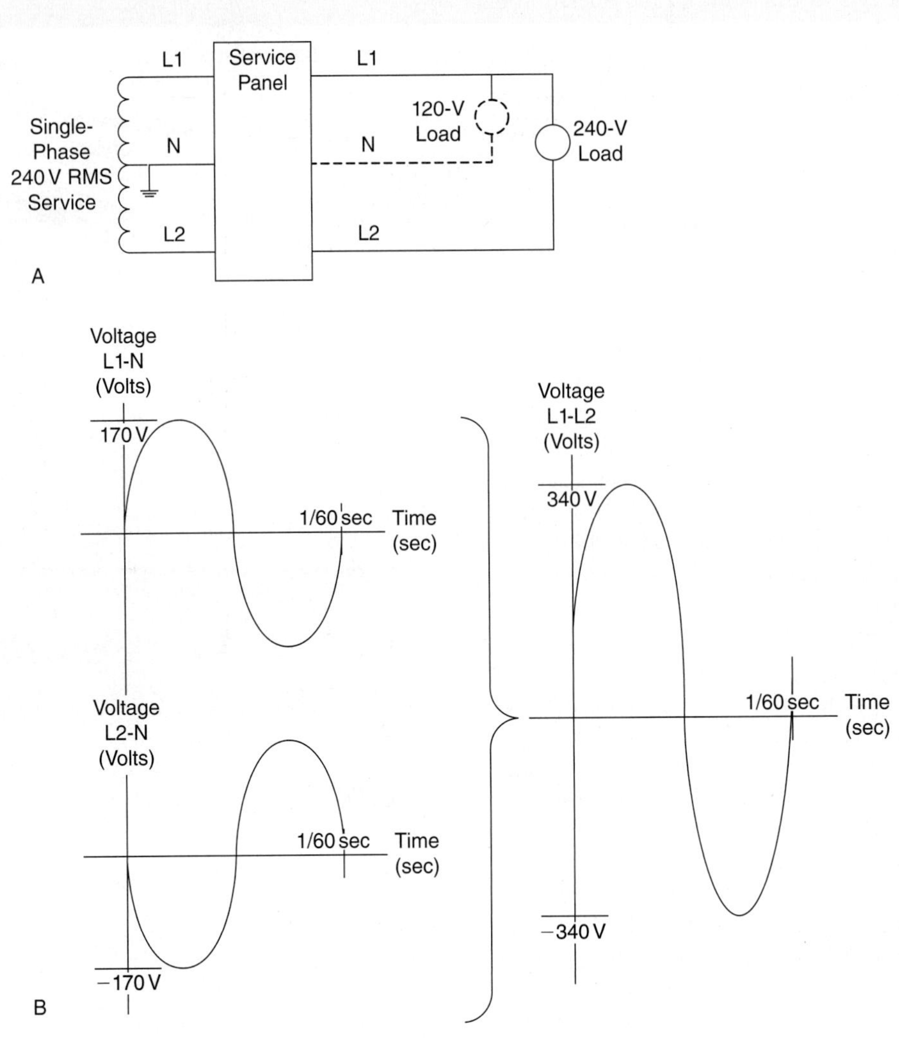

FIGURE 7-5 Single-phase AC sine wave, 240 V RMS, residential use. **A.** Service entrance. **B.** Waveforms.

two conductors are at another position in the sine wave. This phase shift happens to be in 120 electrical degree intervals. Thus, each successive conductor's peak voltage is reached 120 degrees from that of the previous one FIGURE 7-6.

Three-phase systems usually require four conductors: three insulated hot conductors and one isolated conductor that is neutral and is grounded. The three-phase current-carrying conductors and the neutral conductor may also carry current for some loads. Three-phase systems are frequently found in industrial and large commercial buildings, as well as large multifamily dwellings. The AC voltage between any of the L1/L2/L3 or hot conductors can be 480 V, 240 V, or 208 V. The voltage between one of the L1/L2/L3 conductors and the neutral can be 277 V, 240 V, 208 V, or 120 V. Remember that these are RMS values; the values shown in Figure 7-6 are peak values and so are different. A common configuration in large commercial and industrial occupancies is 480/277 V (see Figure 7-6 for service connections and waveforms), where 277 V is used for lighting. The power grid source is shown in Figure 7-6 as a Y transformer, but it could also be a Delta transformer, or center-tapped Delta transformers. Transformer types and characteristics are an advanced topic and are not discussed in detail here, but the investigator should know that transformers are used to step down (or step up) voltages to match the needs of the home or business being fed.

In large buildings, there may be more than one service entrance for electrical power. In some buildings with very high electrical demands, the service entrance may have very high voltage (e.g., 4000 V), and transformers inside the occupancy will reduce, or step down, the voltage.

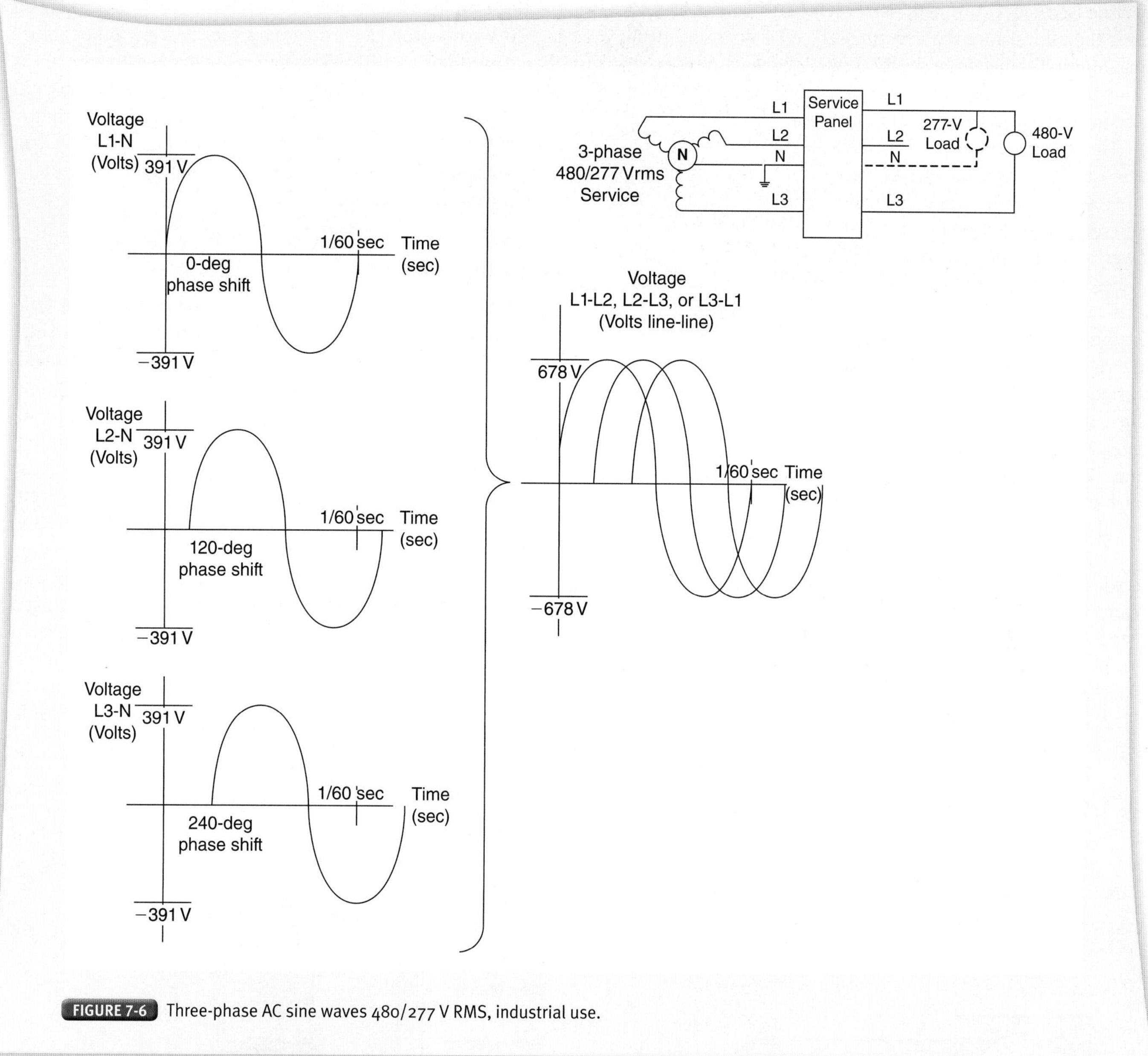

FIGURE 7-6 Three-phase AC sine waves 480/277 V RMS, industrial use.

Building Electrical Systems

This section describes the components of a building's electrical system. The systems described are for residential and small commercial buildings. Larger buildings may have more complex systems.

■ Service Entrance (Meter and Base)

The service entrance is the point where the electrical service enters a building. It is often the transition point between the utility-provided power and the private owner's electrical distribution service. It is important to note that there is usually no overcurrent protection between the utility transformer and the service entrance. Although there is typically a fuse on the primary (high voltage) side of the transformer, its purpose is to limit local transformer damage and to protect the rest of the grid. Without overcurrent protection at the service entrance, any fault that occurs can result in sustained high-energy faulting, possibly resulting in significant damage or even destruction of the service entrance equipment and cabling.

A service entrance consists of the following components:

- Weatherhead: The point where service entrance cables connect to the structure, which is designed to keep water out of the conduit that carries the wire conductors. An underground service entrance does not have a weatherhead because the conductors enter from underground directly into the main panel.
- Meter base: The component where the service cables come in through the weatherhead and go down the conduit to connect to an electric meter that measures the amount of electricity being used.
- Meter: A watt-hour meter that plugs into the meter base to measure the flow of electricity.

The service equipment is most often located close to where the cables enter the structure. The *NEC* does not specify a maximum distance from the location of the main disconnects to the point where the service cables enter the structure, but it is recommended to be as short as possible.

The service equipment includes the main breaker and overcurrent protection devices (fuses and circuit breakers), which are housed in electrical panels near the entrance. The main disconnect provides the mechanism for shutting off power and provides protection using overcurrent protection devices. There may be more than one main disconnect in some electrical services. The main electrical distribution panel also distributes the power to different locations inside the structure via branch circuits FIGURE 7-7. This panel is sometimes referred to by different names, such as the circuit breaker panel, the service panel, the distribution center, or the load center. If the main disconnect is included, it is called the main panel. If a separate panel is used, it is called a subpanel. In larger structures, it is common to find the main panel feeding several smaller panels.

Certain neutral connection problems can occur that create electrical hazards in the structure. For example, in a 120/240-V system, when the neutral conductor is inadvertently not connected at the service panel or out on the transformer or if the neutral line becomes severed or disconnected, a situation called a floating or an open neutral condition can lead to unbalanced voltages at the loads. When this happens, each leg could have from 0 to 240 V across it. There will still be 240 V between the two legs, but instead of the voltages of the two legs being fixed at 120 V to neutral each, they may vary to some other values that may add up to 240 V. This can cause damage to appliances and sensitive circuits in the building and pose a shock hazard to occupants. Symptoms include lamps burning too brightly or dimly or appliances that surge, overheat, or simply fail. This condition does not result from improper grounding of the building electrical service.

Grounding

Grounding is the mechanism for making an electrical connection between the system and the earth (ground). A solid ground connection is needed for safe return pathway if a fault occurs. Without a ground connection, the energy delivered to a circuit or load may flow in an undesirable location or path (such as a human body). When a "hot" or charged conductor contacts a grounded component, such as the metal case of an appliance or water pipes, unimpeded electrical current flows to ground (fault current), and the overcurrent protection devices would open, causing the electricity to cease flowing through the circuit. The investigator should be aware of and take into consideration that electrical systems have changed over the years and that different requirements may have been in place at the time of installation. If the ground was not in place when the charged conductor faulted to the metal case of an appliance or a water pipe, then this metal component could become electrically energized.

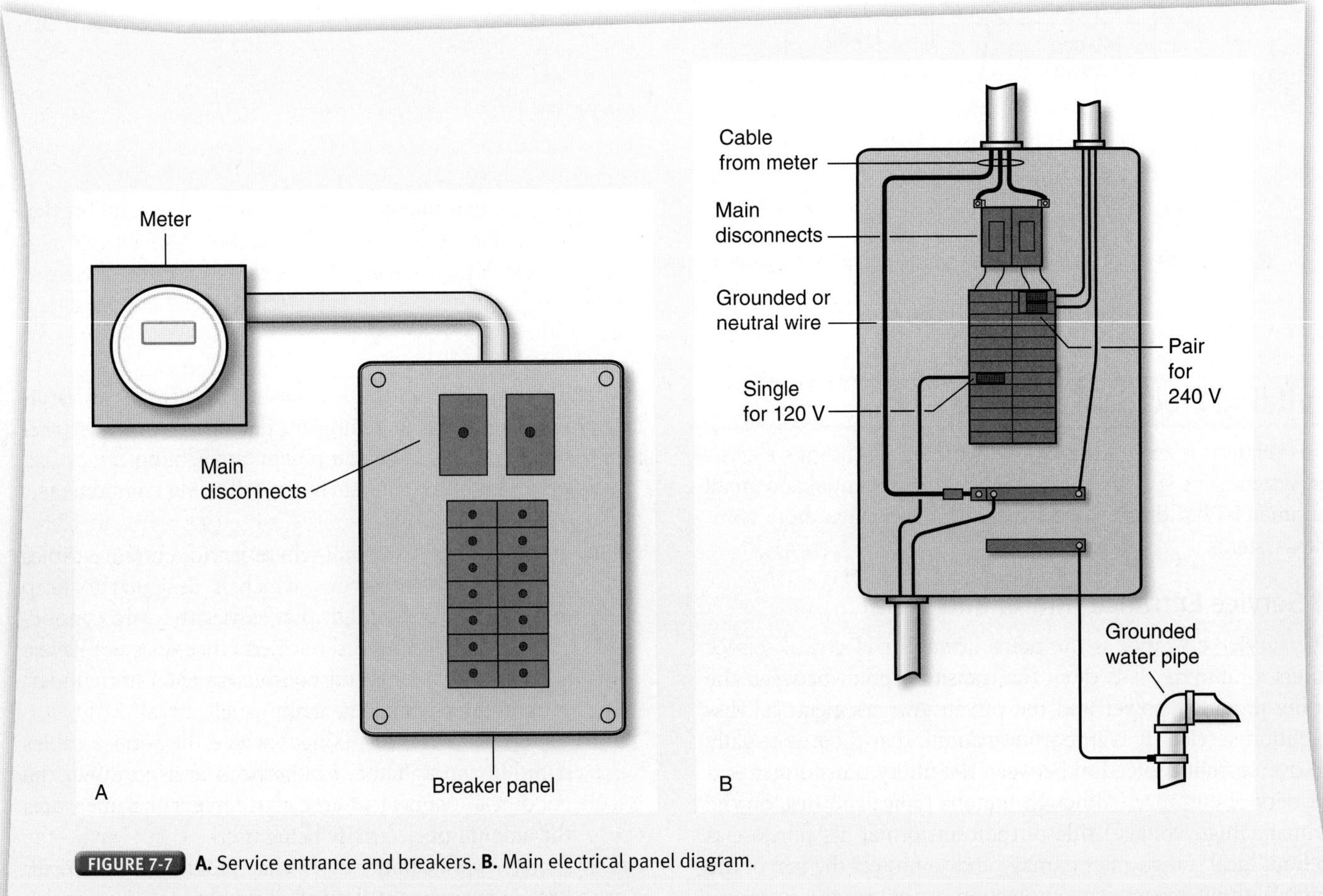

FIGURE 7-7 **A.** Service entrance and breakers. **B.** Main electrical panel diagram.

If a person (electrically conductive and connected to the ground) comes into contact with this ungrounded circuit, he or she may become the ground path to carry the current to ground. This scenario may occur when, for instance, someone touches the conduit that is not grounded and becomes the ground path. Additionally, the ground path could have moved from its original location because of the earth settling or other conditions.

Current *NEC* requirements specify two alternate methods of creating a ground at the service entrance:

1. Connecting the breaker panel/fuse panel/service entrance panel to a bare metal cold water pipe that extends at least 10 ft (3 m) into the soil
2. Connecting the breaker or fuse panel to a grounding electrode that may include a galvanized steel rod or pipe and/or a copper rod that is at least 8 ft (2 m) long and must be driven into the soil

In either case, there must be a secure bond between the panel and the pipe. This is often accomplished by connecting the grounding block in the service panel to the grounding rod or pipe with a copper or aluminum conductor and using proper connecting clamps to connect the conductors to the ground. Also, a minimum of two grounding electrodes must be connected to compose an effective ground. The investigator should consult *NEC* guidelines for allowable electrodes.

Because electricity *typically* flows from high potential (voltage) to ground and takes all available paths that allow that current flow, it is important to have a good grounding system to draw any fault current away from any unintended paths. Although these low currents are usually not sufficient to start a fire, the potential of a personal injury still exists. In addition, all "containment" elements of the distributed electrical service, such as junction, switch, and outlet boxes, must be connected to the grounding conductor distributed with the current-carrying conductors or by means of a conduit. These boxes then act to shield combustible materials from possible arc and fault events at these points of connection. All connections and splices must be made within the box. Occasionally, during the investigation of a fire, wirenuts or terminals appear to be outside a junction box. In reality, however, the connection points likely ended up outside the junction box from the extent of the fire damage and/or collapse of the structure.

Overcurrent Protection

Overcurrent protection is provided by a series-connected, current-interrupting device that prevents damage that can occur to conductors (and possibly to the devices/appliances on the circuit) when excessive current flows. Stated differently, the objective of an overcurrent device is to stop the flow of electricity in the circuit when an abnormally large amount of current flow is detected. The overcurrent protection devices can be either resettable or nonresettable. Most modern overcurrent protection devices are resettable, meaning that they can be re-used to protect the circuit after they have been tripped. Once a device has been tripped, the cause must be indentified, whether it was a ground fault, overload, or short circuit. Nuisance/end-of-life tripping is also possible (e.g., GFCIs routinely trip as they age for reasons other than an actual ground fault). These protective devices have two current ratings: regular current rating and interrupting current rating. Circuit breakers should not be reset until it is determined that they tripped due to an overload versus a short circuit or ground fault; this is done to ensure the conductors have not been damaged.

Regular current rating is the maximum amount of current that the protection device allows the circuit to carry under normal conditions. Above this value, the device opens, stopping the flow of current. Common values found in household use are 15 A, 20 A, 30 A, and 50 A. The time required for the device to open depends on the duration of the current, the amount of overcurrent present, and the ambient temperature. It is possible for an overcurrent protection device to carry a small amount of overcurrent for an extended time. It is also possible, in high ambient temperatures, for an overcurrent device to trip at a slightly lower current threshold (rating) over an extended period of time.

Interrupting current rating is the maximum amount of current the device is capable of interrupting. This can be a large amount of current that seems impossible to reach; however, contacts can weld shut at high currents, and it would be catastrophic if this large current were not interruptible.

There are several types of circuit protection devices. The most common are fuses and circuit breakers. NFPA 921 section 9.6.2.1 defines fuses as "nonmechanical devices with a fusible element in a small enclosure." TABLE 7-3 provides examples of various fuses.

Circuit breakers open when exposed to heat or overcurrent that exceeds their rated trip current. Properly sized circuit breakers do not exceed the ampacity of the conductors FIGURE 7-8. NFPA 921 Section 9.8.5.1 defines a circuit breaker as "a switch that opens either automatically with overcurrent or manually by pushing a handle."

Whereas it may take several minutes for the circuit breaker to trip for a small increase in current flow over its rating, it reacts more quickly to a large current flow over the rating of the breaker. This phenomenon is described by the time current curve of the circuit breaker or fuse, which is available from the manufacturer.

The time current curve (also known as the *characteristic trip curve*) for breakers and fuses defines the amount of time required for a device to interrupt at a specific level of current. The more current there is, the faster it will trip. For example, *The American Electrician's Handbook* states that a typical 100-A breaker will take about 30 minutes to trip at 135 A and 10 seconds to trip at 500 A.

In addition to standard circuit breakers, two other protection devices are used in special locations that include, but are not limited to, bathrooms and bedrooms. These devices are the ground-fault circuit interrupter (GFCI) and the arc-fault circuit interrupter (AFCI). GFCI breakers respond to overcurrent conditions like normal breakers and also look for an imbalance of the current between the energized and neutral conductors. More specifically, if the electrical current leaving the GFCI is greater than the current that is returning to the GFCI, current must be flowing on some other path (a ground fault).

Table 7-3 Types of Circuit Protection Devices

Type	Application	Amps	Notes
Plug (also referred to as Edison)	■ Older application; not used in new installations	30 A or less	■ Possible to overfuse (e.g., put a 30-A fuse in a circuit that would require a 15 A) because fuses with different ratings are all interchangeable ■ Possible to bypass the fuse by using a penny to complete the circuit
Type S	■ New installations		■ Designed to make bypassing more difficult ■ Can be used in older installations (Edison bases) with an adapter
Time-delay fuses	■ Allow short duration overcurrent ■ May be up to six times normal current		
Cartridge fuses	■ Fast action or time-delay ■ Single use or replaceable element ■ Can be used for high-current loads (water heaters, ranges, etc.) ■ Greater than 100 A are found in commercial or industrial occupancies	Circuits greater than 30 A	

Once this condition is detected, the GFCI breaker trips, interrupting the current flow and thus interrupting the circuit. AFCI breakers monitor the circuit for abnormal conditions/waveforms and are now required in new residential dwellings. Local (usually the state electric board) requirements should be consulted to determine compliance with *NEC* regarding the required locations and specified incorporation dates for the use of AFCIs. Arcing at loads or in circuits can cause fires because the electrical arc can reach upward of 35,000°F (19,426°C); AFCIs discriminate between normal circuit current flow and arcing current flow to a load and open when arcing conditions are detected. Normal switch closure can also generate a small amount of arcing at the switch, but the AFCI does not respond (open the circuit) because of that.

TABLE 7-4 provides a list of the types of resettable circuit protection devices.

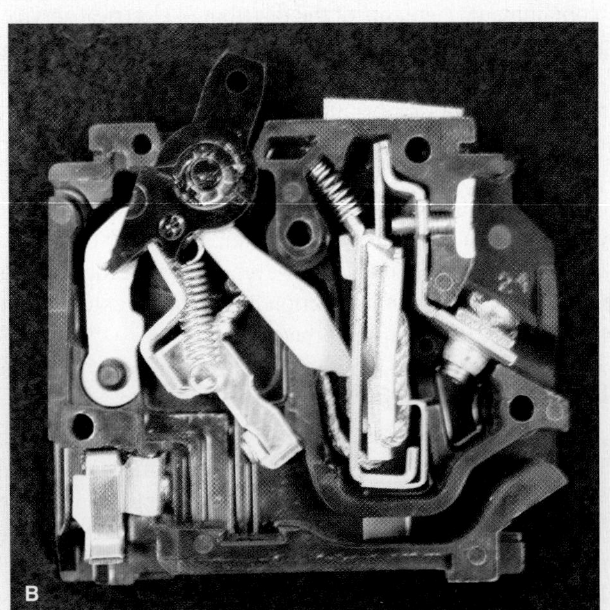

FIGURE 7-8 **A.** A 15-A residential-type circuit breaker in the closed (ON) position. **B.** A 15-A residential-type circuit breaker in the open (OFF) position.

Table 7-4 Types of Renewable Protection Devices

Type	Amps	Notes
Main breakers	100 A to 200 A (residential)	■ Interrupt the circuit at the main disconnect panel ■ Pair of breakers ■ Handles fastened together
Branch circuit breakers	15 A or 20 A (general lighting and receptacle circuits); 30 A, 40 A, or 50 A (ranges, water heaters, etc.)	■ Switch ■ Used on the branch circuits that distribute electricity throughout the structure ■ Rating generally imprinted on the handle ■ Cannot be manually tripped (can be placed in the "off" position manually); can be tripped by physical contact or external heat ■ Will trip even if the handle is locked in the "on" position ■ May trip to the "off" position ■ Have body of phenolic plastic ■ Body does not melt or sustain combustion ■ Can be destroyed by fire impingement ■ Do not rely on visual inspection to determine the status of the breaker
GFCIs		■ Monitor the electrical flow for abnormal conditions ■ Used in locations where a person might become grounded while using appliances, such as in the: • Bathroom • Bedroom • Kitchen • Patio/garage

Circuit Breaker Panels

Heat or electrical faulting can damage internal components of the circuit breaker panel. If made of standard plastic, insulation and molded circuit breaker components can melt and allow energized conductors or bus bars to contact each other or ground. High-temperature thermoset plastics will not melt, but will become brittle at high temperatures. The service entrance cables to the building enter the panel and are distributed to bus bars. Because these components are usually unprotected, significant arcing and damage, including blow holes in the panel wall, can occur if the separations and insulation are not maintained. Separately insulated conductors in a metal conduit, if overheated to the point of insulation breakdown, can also arc and cause a blow hole in the conduit. Analysis of the circuits inside and outside an arc-damaged panel is needed to determine whether arcing inside the panel was the source of ignition or was the result of heat impingement on the panel by the fire.

Branch Circuits

Branch circuits are the circuits that distribute the electricity from the circuit breaker or fuse panel throughout the building. Each circuit should have its own overcurrent protection device (either a fuse or a circuit breaker). These are called branches because they have a single connection point in the service panel with conductors fanning out to more than one load in the building, much like the branches of a tree. TABLE 7-5 lists the conductors found in residential branch circuits.

Conductors

There are many sizes of conductors. The AWG number or gauge refers to the wire diameter. The larger the gauge, the smaller the wire diameter. For instance, 15-A circuits should use no less than 14-AWG wire (12 AWG if aluminum), whereas 20-A circuits should use no less than 12-AWG wire (10 AWG if aluminum). Higher power circuits use 8- and 6-AWG wire. The AWG relates to size, which determines the ampacity of the conductor. The type of insulation, bundling of conductors, method of cable conveyance (e.g., proximity to insulating materials, embedded in walls), and ambient heat also affect required conductor size. Conductors may be larger than required but cannot be smaller than required. Undersized conductors have more resistance to larger current flow and therefore can generate more heat when their ampacity is exceeded. Oversized circuit protectors (sometimes called overfusing) are

Table 7-5 Conductors in a Typical Two-Wire Plus Ground Branch Circuit

Type of Conductor	Function	Grounding Protection	Description	Notes
Ungrounded	Hot	Attached to the protective device	Carries the current to the load	All circuits must have this conductor.
Grounded	Neutral	Attached to the grounding block	Returns the current to the source	All circuits must have this conductor.
Grounding	Ground	Attached to the ground	Provides protection and allows the current to go to ground	This is not required to make a circuit function. Grounding may be provided through a metallic conduit instead of a separate conductor.

a dangerous condition that occurs when the circuit protection (fuse or circuit breaker) rating significantly exceeds the ampacity of the conductor, leading to a condition where increased heat can occur in the conductors.

The three most commonly used conductors are copper, aluminum, and copper-clad aluminum conductors.

Copper Conductors

Pure copper is used in copper conductors. The melting temperature is approximately 1980°F (1082°C); however, surface melting may occur below this because formation of copper oxide on the surface can leave an unmelted core. When fire attacks a copper conductor, the conductor is heated in a larger area, along its length (as opposed to an electrical arcing event that is usually confined to a smaller, localized area on the conductor). When fire attacks a copper conductor from the exterior, the investigator will observe surface melting and formation of copper oxide.

Although it is not always easy to determine whether a conductor has melted from the heat of the fire or melted as a function of electrical arcing, as a general rule, copper conductors form pointed ends, globules, and thinned areas when heated from fire. By contrast, as a general rule, arced conductors display a clear line of demarcation where the conductor will appear normal, then suddenly melt. This abrupt melting is often accompanied by a melted spot (or sometimes divot) on an adjacent wire. It is also important to note that copper conductors can have a variety of different appearances in postfire conditions. Some conductors lose their insulation and oxidize, and the surface is blackened with cupric oxide. Other conductor surfaces may have no oxide or may be covered with reddish cuprous oxide. Other chemicals may also be present that can affect the coloration and condition of the conductors.

A copper conductor's melting temperature may be affected by "alloying," or eutectic melting, which occurs when other metals, such as aluminum or zinc, melt and come into contact with the copper and change its melting temperature, usually lowering it. This frequently happens during a fire when metals drip or splash onto or come into contact with the copper conductors. The investigator can often, but not always, determine that this has occurred by noting a color change on the copper melt area. This change may be brass or silver in color. This is not a cause of the fire but rather a result of the fire.

Aluminum Conductors

Aluminum in its pure form is used in conductors. Primarily used as service-entrance conductors on new installations, aluminum conductors may be used as branch circuits on some older installations. Circuits that supply power to larger appliances (e.g., water heaters, ranges, ovens, air conditioners) are also often made of aluminum. An aluminum conductor's conductivity is lower than copper's and must be two AWG sizes larger than copper for the same ampacity (10 AWG aluminum = 12 AWG copper). Its melting temperature is approximately 1220°F (660°C). Aluminum oxide may form on the surface, but it does not mix with the aluminum; therefore, surface melting does not occur. Aluminum can melt all of the way through the wire and can flow through the skin of aluminum oxide. The heat of fire often melts aluminum conductors entirely, ill-affecting the ability to determine whether the conductor arced.

Fire Investigator Tip

When metal melts onto metal with the same composition (e.g., copper on copper), it is considered extraneous melting. This situation is uncommon, requires microscopic testing to confirm, and may or may not have had an electrical cause.

Copper-Clad Aluminum

A less common type of conductor is copper-clad aluminum. As the name implies, the aluminum conductor is encapsulated in copper. Its melting characteristics are similar to those of aluminum.

Insulation

Insulation on the conductors prevents faulting or leakage via unwanted paths. Insulation includes any material that can be applied readily to conductors, does not conduct electricity, and retains its properties for an extended time at elevated temperatures. Air can be an insulator, even for high voltages, when the conductors are kept separate. An arc can occur when dust, pollution, or products of combustion contaminate air or the insulation. The conductor's applications and insulation ratings are marked onto the insulation to identify its type and list its temperature rating. Additional information of the conductor's construction and applications can be located in Table 310.104 (a) of the *NEC*.

Polyvinyl chloride (PVC), a common insulator, can become brittle with age or heat and may give off hydrogen chloride in a fire. Rubber was a common insulating material until the 1950s. It can become brittle with age, and once it becomes brittle, it can easily be broken off the conductor. Rubber chars and leaves an ash when exposed to fire. Other materials used as conductors include polyethylene and related polyolefins. These materials are commonly used on higher current circuits in residential applications. Nylon jackets may be used to increase thermal stability.

The color of the conductor typically indicates the function it provides: Green typically indicates an equipment grounding conductor; white or light gray indicates a grounded conductor (neutral); and any color except green, white, or light gray indicates a hot or charged conductor. In 120-V circuits, the Line (L1 or hot) conductor is often black. In 240-V circuits with nonmetallic cable, the two hot conductors (L1, L2) are usually black and red. Caution is needed because in certain applications white or gray conductors could be hot conductors when used in three- or four-way switches and therefore could be energized. The investigator is advised to consider all conductors as being energized until they are proven otherwise. For this reason, a voltage sniffer is a recommended tool for the investigator to carry.

Outlets and Special Fixtures/Devices

Circuits will terminate at or connect to switches, receptacles, or appliances. Switches are used to control the flow of current to receptacles and/or appliances and will have either screw terminals, push-in terminals (i.e., back-stabbed), or both. Permanent lighting fixtures are typically attached to electrical boxes in the wall or ceiling with a wall switch or dimmer to control the lighting circuit.

Back-stabbed (push-in terminal) receptacles are those in which the conductor is not attached to the screw terminal but is pushed in the back of the receptacle. Because of their small surface contact area, some push-in styles have created problems, such as resistance heating. As was previously discussed, terminals or connection points are the most likely areas for resistance-type heating to occur because of faulty connections. Switches are located in the black (hot) conductor side of lighting circuits, not in the white (neutral) side. Industrial settings typically use contactors for switching loads. Relays or manual switches are used to energize the contactor coils.

There are two types of 120-V receptacles: duplex receptacles on 15-A and 20-A circuits (two outlets per receptacle) and single receptacles on 30-A or greater circuits. Receptacles must now be polarized, in that the larger neutral blade of the plug can be inserted only into the larger neutral opening on the receptacle. Receptacles also must be grounded in new installations. Receptacles may have screw terminals, push-in terminals, or both. Hot conductors (usually black) should be attached to the brass screw. Neutral conductors (white) should be attached to the colorless screw. The green screw is for ground.

Special fixtures for lighting are usually connected to junction boxes in the wall or ceiling. Sometimes these fixtures contain thermal cutout switches, which help prevent overheating of the device. Often these types of fixtures are designated as IC fixtures, meaning that they are insulation-contact (IC) rated.

GFCI outlets are used in bathrooms, kitchens, or other wet locations. These outlets monitor the amount of current going in and out and trip if the current differential exceeds 5 mA. Ideally, GFCI outlets are intended to interrupt current flow before there is sufficient fault-current that could result in bodily injury. Explosion-proof outlets and fixtures should be used in all installations that are exposed to explosive atmospheres. Part III of Article 501 of the *NEC* discusses the equipment used for hazardous locations.

Ignition by Electrical Energy

The following conditions must exist for ignition from an electrical source:

- Wiring must be electrically energized (power must be on).
- Sufficient heat and temperature must be produced (enough thermal energy is produced to ignite the first fuel; see the "Fire Cause Determination" chapter in this textbook).
- The combustible material must be a material that can be ignited by the electrical energy produced by the electrical failure.
- The heat source and the combustible fuel must be close enough (sufficient thermal energy transfer) for a sufficient period of time to allow the combustible material to generate combustible vapors. Generally speaking, a simple arc does not produce enough energy to ignite most ordinary, higher-density combustibles.

Some methods that may generate sufficient thermal energy include the following:

- Resistance heating: an electrical event that occurs when electrical current flows through something that provides high resistance to current flow, such as a heating element (intentional) or a resistive connection (unintentional).
- Ground fault: an electrical event where equipment failure or poor grounding facilitates excessive current flow on an unintentional path as opposed to the intended circuit path.
- Parting arcs: arcing that occurs when wires or switch contacts are separated and current flow is interrupted. This will create a momentary arc of very short duration, with accompanying sparks.
- Excessive current flow: when abnormally high amounts of electrical current flow through an appliance or conductor, faciliating overheating (e.g., a 14-AWG wire carrying 50 A instead of 8 A).
- Lightbulb over lamping, close proximity of combustibles to heaters, or cooking equipment.
- Heating in a confined space such as a heating element that is normally exposed to ambient or cooling air but that is somehow trapped under blankets or bunched up. Another example of confined heating is a heat-producing object such as a motor or ballast that is covered in dust, insulation, etc.

Electrical current flows from the source to the load and back via any path that allows this flow, even if it is not the intended path. If a direct connection is made to a grounded or grounding element, it is considered a short circuit. When electricity flows through most materials, including wires and conductors, heat is produced because of the inherent resistance of the material it passes through. The heat is dissipated under normal conditions, if used within the safety margins calculated for the size of the conductors (ampacity). Sometimes this heating is desirable, such as in the heating element on a stove. Other times it is not desirable, such as current flowing through an undersized conductor.

Heat-Producing Devices, Excessive Current, and Poor Connections

Some devices are designed to generate heat. When this is the case, the design of the devices also allows for the safe dissipation of the heat; however, in cases where the heat is not allowed to dissipate safely and where there is a combustible fuel nearby, ignition may be possible. For example, if a cloth or piece of clothing is placed over a halogen lamp, the heat that is absorbed by the material may begin to produce combustible vapors that are then ignited if the heat is high enough (autoignition) or by arcing nearby (piloted ignition). An incandescent lamp, even a small wattage one, can reach temperatures that will ignite combustible material. This self-heating is increased by immersing the lamp in cellulose or wrapping it in clothing. In some cases, a heat-producing device may deliberately be used to cause a fire (arson), either by sabotaging its safety features or by placing it in proximity to a combustible fuel load.

In other cases, the device may malfunction, allowing the heating elements to become hotter than they are designed to be, causing a fire. An example could be a deep-fat fryer whose controls fail, allowing the grease to heat up to its ignition temperature.

Poor connections are another heat producer; they can allow heating at the point of the poor connection and also allow the formation of an oxide interface at the poor connection. This oxide increases the resistance of the connection, which contributes to the heat. This type of connection interface causes resistance heating. The circuit is functional, but there is increased resistance at this point. A hot spot can develop that can glow. If there are combustible fuels in proximity to this spot, they can possibly ignite. Structure wiring connections are required to be made within a special electrical box or appliance enclosure, which can reduce the chances of igniting a combustible if these hot spots occur. With currents of 15 to 20 A, for example, power of up to 40 W can be produced at a poor connection with temperatures exceeding 1000°F (538°C). This may not seem significant, but the heat is very focused, rather than spread over a long distance. A 60-foot (20-meter) cord with 14-AWG wire has roughly the same resistance as the poor connection of this example.

Overcurrent

In a properly designed circuit, overcurrent is the temporary excessive flow of current through a conductor. For overcurrent to happen, either a fault must occur (such as a short circuit) or too many loads must be placed on the circuit (e.g., an excessive number of appliances). Circuits can handle these types of momentary overcurrents unless they persist.

As discussed earlier in this chapter, overload is the *persistent* excessive flow of current through a conductor. Factors that influence the potential for fire include the magnitude of the overload and the duration of the overload condition. Overcurrents that are large and that persist can cause the conductor to melt the insulation and can heat the surrounding material.

Undersized cords can produce overheating if used far beyond their imparity. A 1200-W appliance requires a 12-AWG wire connecting it to a 120-V outlet. If an 18-AWG is used instead of 12 AWG, the normal 10-A current draw is much greater than the 5-A rating of the 18-AWG wire, but because the breaker is sized at 20 A, the circuit will continue to function, allowing the undersized cord to increase in temperature. If it is bunched, coiled, or covered, it can cause an overheat condition, leading to ignition FIGURE 7-9. Overloads that cause fires are uncommon in properly sized circuits with operating overcurrent protection.

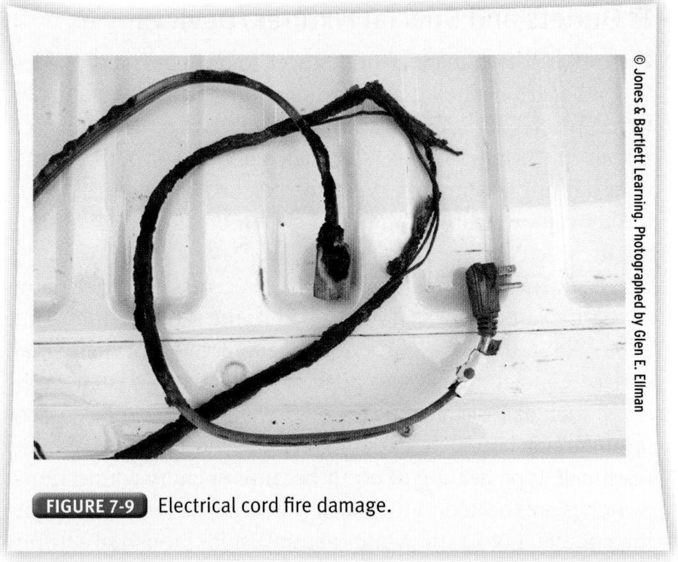

© Jones & Bartlett Learning. Photographed by Glen E. Ellman

FIGURE 7-9 Electrical cord fire damage.

The temperature of a heat-producing device, or generated from an undersized conductor, can degrade the insulation between conductors. This situation is aggravated if the conductors are not in free air—for example, if they are covered with blankets. If the overcurrent is allowed to continue, carbonization of insulation occurs, creating a path for significant current to flow. When heated to about 392°F (200°C), PVC insulation begins to pyrolize, and combustible vapors escape. If an arc then occurs because of insulation breakdown in the high heat environment, a fire can result if suitable combustible material is nearby.

In FIGURE 7-10, the overcurrent protection of an air conditioner fuse disconnect has been bypassed by two copper pipes.

Courtesy of OnSite Engineering & Forensic Services, Inc.

FIGURE 7-10 Overcurrent protection bypassed.

This could allow a dangerous overload condition to develop because of a fault or locked rotor condition in the air conditioner compressor motor.

Contaminated water or corrosion at connection points or near bare conductors can cause or contribute to a carbon tracking process. This weakness in the normal insulation properties of the device or conductors allows leakage current to flow. With time, heat, or vibration, this weakness can develop into a short circuit to a nearby ground connection.

Arcs

An arc is a high-temperature discharge across a gap where the conductor is missing. The temperature of the arc can be several thousand degrees, depending on current, voltage drop, and the metal involved. For the spontaneous initiation of an arc to occur across an air gap, there must be a relatively large voltage potential. In 120/240-V systems, arcs do not occur spontaneously under normal conditions. A special type of arc can occur, however, when the conductors separate while current is flowing, even in 120/240-V systems. This is called a *parting arc*. Because of their typically short duration, arcs may not be competent ignition sources unless ignitible vapors or gases are present, as was discussed with the cord with degraded insulation.

Solid fuels are much less likely to ignite from an arc due to its short duration. Arcing is brief and localized because of the cycling of AC voltage from peak positive to peak negative value, going through zero in between. Adjacent combustible fuel that has a low surface-to-mass ratio and that cannot be heated sufficiently to sustain combustion (e.g., a wood 2 by 4) will likely not ignite. Fuels with a high surface-to-mass ratio, such as a gas or vapor, can sustain combustion, but they must also have an acceptable ratio of air to gas. This ratio must fall within two levels defined as LEL (lower explosive level) and UEL (upper explosive level). Fuels that may ignite from an arc include cotton batting, dust, tissue paper, volatile gases and vapors, and lint. Short-duration arcs often have insufficient energy available to ignite solid fuels under normal conditions. A conductor-to-conductor fault—for example, a metal knife or staple piercing an energized extension cord—will short circuit the branch and open the circuit protection device rather than cause an arc.

High-Voltage Arcs

High voltages exist at the transformer connection near the service entrance of the building and can accidentally be impressed on conductive elements, which enter or are part of the building. Lightning can also cause high-voltage surges at the building's electrical service entrance. These events impress large overvoltage on the wiring, which is designed for a few hundred volts and will cause insulation breakdown with arcing at thousands of volts.

Parting Arcs

Parting arcs are brief discharges created when the electrical path is opened while energized. Examples include opening a switch, pulling a plug, and motors with brushes, which can produce continuous arcing. Pulling a plug while current is flowing in the conductors often causes a visible arc. Once the conductor separates, the heating stops because the current flow has stopped. Contacts with surfaces that open and close are exposed to this kind of arc and often show pitting and roughness. Parting arcs occur in arc welding, which typically uses DC voltage, which allows the arc to continue because the current does not pass through zero.

Parting arcs can also occur within thermal cutoff devices. These devices are often placed inside motor casings on the windings of the motor, on transformer windings, or on heating elements or chambers. Because this device is thermally actuated, it will open the circuit if it gets hot enough, typically 275°F (135°C) or so, and the motor or element can cool down before the device resets itself and recloses the circuit. These devices are designed for many thousands of cycles of operation but can fail at the extremes of operation at elevated temperatures or if the protected component is confined by insulating dust. When this happens, arcing is accompanied by sparking as more material is removed from the contact surfaces. Additionally, the metal contacts can fatigue to the point of breakage. At this point, the cutoff action may not occur, or the device may simply heat up or increase the series resistance of a motor circuit, allowing it to stall or not start. Any of these failure modes can have catastrophic results.

Short-circuit or ground fault may cause metals to melt at the initial point of contact. An arc occurs as the metals part, and the arc is quenched immediately. It may throw sparks. Some power supplies and motor drives with large capacitors will arc when plugged in because of the high inrush current.

Arc Tracking

Arc tracking has been observed in high-voltage systems and has been demonstrated experimentally in 120/240-V systems. It may occur on surfaces of noncombustible materials and lead to the development of a path of electrical current over time. The arc path or arc tracking will start small and may continue to grow in size. The phenomenon begins when the surfaces become contaminated with salts, conductive dusts, and liquids. This contamination causes degradation of the base material. Arc tracking can also occur when a surface is contaminated with water containing material such as dirt, dusts, salts, or mineral deposits. Tap water may contain these types of contaminants. Often, the arcing will create sufficient heat to dry the wet path and stop the flow of current. If the water is replenished, thereby reestablishing the current, deposits of metals or corrosion can form. Arc tracking is more pronounced in DC current flows.

In low-voltage systems with feedback temperature control, circuit board contamination can cause erroneous signals to be read in sensitive analog circuits. These circuits are sometimes coated with moisture-resistant coatings to prevent contamination from reaching printed circuit (PC) board surfaces, but if not, contamination can disrupt control sequences in heating appliances and lead to malfunction, even fires. A thermistor reading in error can drive a heating element to an over-temp situation if contamination occurs in the sensitive portions of the measurement circuit. (Thermistors are sensors used to measure temperature by change of their resistance.) Even though the thermistor itself may be potted or enclosed, the measurement circuit on the PC board it connects to can be contaminated by moisture if not protected.

Sparks

Sparks are metal particles thrown out by arcs. When a high-current ground fault occurs (low resistance), hundreds or thousands of amperes may be flowing. The energy is sufficient to melt the metal and throw out sparks. Protective devices generally open almost immediately, so the event should happen only once. If the metal involved is copper or steel, the particles cool as they fly through the air. If the metal involved is aluminum, the particles may burn as they fly through the air and are more likely to ignite nearby combustibles. In general, however, sparks are not competent ignition sources for ordinary combustibles, although they can ignite fine fuels. The size of the spark particle is important because it determines the total heat content. Arcing in service entrance conductors can produce more and larger arcs than branch circuits.

High-Resistance Faults

High-resistance faults occur when the current flow through the fault is not sufficient to trip the protective devices, at least in the initial stage, and the current keeps flowing. These types of faults develop more heat, depending on how long the fault continues, and may be capable of igniting combustibles. It is difficult to find evidence of a high-resistance fault after a fire because the ensuing fire and heat damage mask the evidence of the original failure. Arcing through charred insulation of conductors is a high-resistance fault that may not open the circuit protection. Multiple events with sustained arcing can occur under these conditions.

Interpreting Damage to Electrical Systems

Because of the high potential energy of electrical circuits, significant observable damage will occur from faulting or arcing elements and components in the circuits. The ability to differentiate this damage from that which is due to the heat of fire is a key skill for investigators to develop and can often make the difference between success and failure in determining whether electricity played a role in causing the fire.

Arc Mapping

Arc mapping is often an important part of a fire investigation for both large- and small-scale events. It can be helpful in defining the general area of origin in the structure, and the investigator can use it to find the origin within a building or an appliance. Arc mapping requires a careful examination of the electrical circuits to identify arcing that occurred within the circuit. Mapping those arcs with each of the circuits can help identify an area that needs additional examination.

Downstream (electrically distal from the source) from the severance point of a conductor, power is stopped; therefore, additional electrical damage cannot occur. Unless damaged by heat, the conductors and insulation will likely remain intact downstream from the severed conductors. Upstream from the first severance point, the circuit may remain energized if the circuit protection does not open, and the circuit may continue to arc upstream. On conductors with multiple arcing, the first arcing most likely occurs farthest from the power source, and additional arcing will occur sequentially as you move toward the power source. The investigator should attempt to find as much of the conductor as possible to identify first arcing. This will identify the first failure point on the conductor. The identification of the first points of failures on conductors will help to identify the following:

- Where the fire first attacked the circuits
- A possible area of origin
- A possible cause of the fire (if caused by an electrical event)

Branch circuits in conduit may also have visible holes, called blow holes, that indicate significant arcing and energy release inside the enclosure/conduit. Electrical panels can also have blow holes occur in the metal walls during a fire.

Short Circuit and Ground Fault Parting Arcs

A short circuit will occur when current passes from an energized conductor to a grounded or oppositely charged conductor or to a metal grounded object. A surge of current results. Melting occurs at the point of contact, creating a small void and a parting arc. A parting arc melts metal only at the point of contact; adjacent or surrounding areas are unmelted. Thus, the investigator can reason that if the adjacent areas are melted, the fire caused it, or possibly masked the initial arc damage. It may be difficult to identify the area of the initial parting arc if subsequent melting occurred. The surface of an arced contact point typically appears to be notched or beaded and is melted under microscopic examination. A companion point of damage confirms that the fault occurred between the two points of arc damage. For an arc to occur on insulated conductors, the insulation had to fail in the first place. A significant fault may cause the wires to separate. This is termed arced and severed and can break the wires into segments, which may drop and become lost. The investigator must determine how the insulation failed or was removed and how the conductors came into contact with each other.

Damage to stranded cords appears less consistent than for solid conductors. Cords used in lamps and appliances may have several strands notched or severed, or all strands may be severed or fused together.

There are many ways to interpret the damage to electrical conductors, including arcing through char, parting arcing, and so forth. Examples of this damage are shown in TABLE 7-6.

Arcing Through Char

Insulation exposed to fire may char. As it chars, it may become more conductive, which will allow sporadic arcing between energized conductors. This arcing may leave surface melting at spots and can melt through the conductor, depending on the duration and repetition of the arcing FIGURE 7-11. If the arcing persists, the conductors may become severed, usually forming beads on the ends, which may also weld together. Because the heat caused by arcing wires is intense in a very small area

Table 7-6 Types of Conductor Damage

Mode of Damage	Effects	Result	Cause of Fire?
Arcing through char	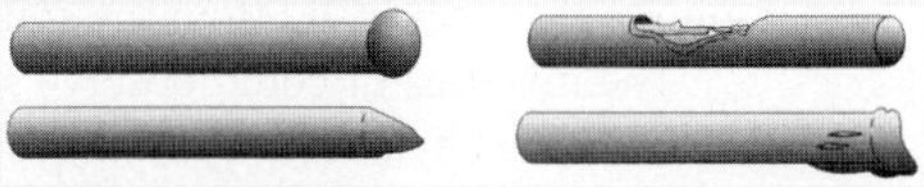	Direct fire heating	No, always a result of fire
Parting arcing		Heating at about 400°F (250°C) but no direct fire	Usually not
Overcurrent	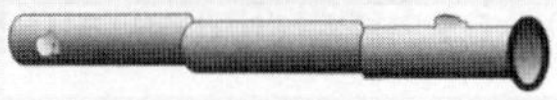	Short circuit or failure in a device plus failure of overcurrent protection	Yes, but also may be a result of fire
Fire	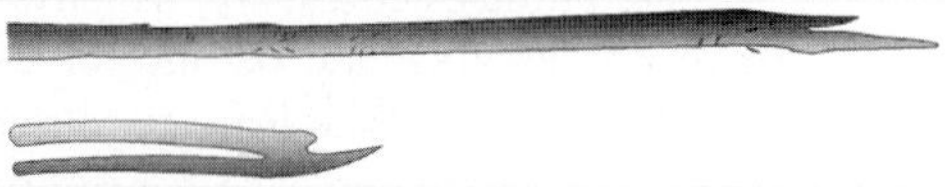	Cable exposed to existing fire	N/A
Heating connection		Connection not tight	Yes
Mechanical	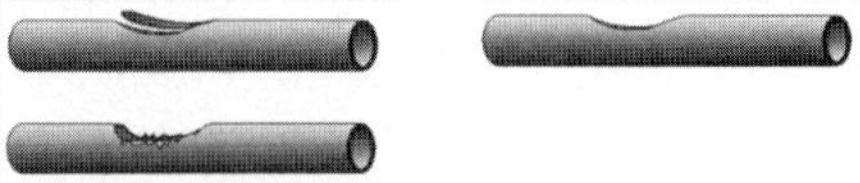	Scraping or gouging by something	No
Alloying		Melted aluminum on the wire	No

around the point of first contact, it is called *localized heating*. Melting caused by nonlocalized heating has a different, more gradual, or globular appearance from beads formed by arcing through char. If the melted conductors are in conduit, holes can be melted in the conduit.

Often, the investigator will notice multiple points of arcing. Several inches of conductor may be destroyed. Ends of conductors may be severed and have beads on the ends; the beads may weld two conductors together. In conduit, holes may be melted in the conduit at one point or along several inches. TABLE 7-7 provides some general indicators to help determine whether the damage to the conductor is from the fire, arcing, or overload. This damage, by itself, does not necessarily indicate whether it was or was not the cause of a fire.

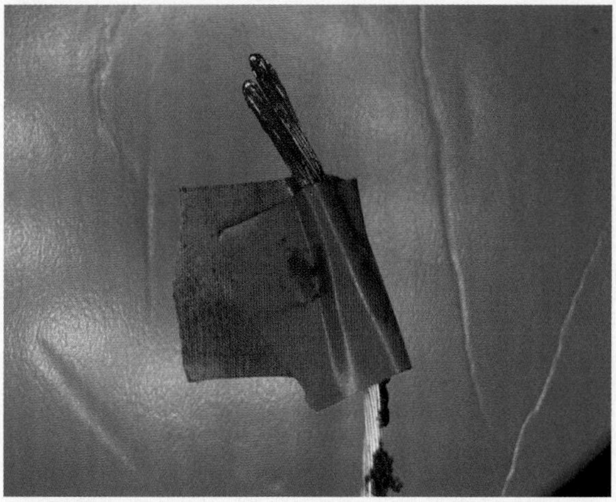

Service-entrance conductors have no overcurrent protection, and several feet of conductor can be partly melted or destroyed if electrical activity occurs. Arcing can continue on service-entrance conductors until the power is cut by severing of wires. The primary side fuse of the feed transformer may blow if the secondary side current is high enough, usually several hundreds of amps. A 2-A fuse at 12,000-V primary on a 35-KVA transformer, for example, will allow a sustained 200-A secondary side current without blowing. Conductors used at the service entrance connection are typically made of aluminum, which has higher resistivity than copper, and, thus, higher gauge wire is used. Faulting at that point often results in temperatures high enough to melt aluminum, and significant melting occurs because of the high current.

Courtesy of Jason Mignano, CFEI

FIGURE 7-11 Arcing through char.

Table 7-7 Potential Indicators of Conductors

Beads	Globules
Localized heating	Nonlocalized heating, such as overload or fire melting
Distinct line of demarcation	

Overheating Connections

Poor connections are likely places for overheating. The probable causes are loose connections and/or the presence of oxides at the point of connection, creating resistance. Poor connections can often be verified by color changes at the point of the connection. At the point of the poor connection, portions of the metal connection may be found to be deformed or destroyed. Poor contact connections can cause heating, oxidation, pitting at points of glowing, arcing, and even destruction of metals FIGURE 7-12.

Overload

Overloads, as explained earlier, are overcurrents that are large enough and persistent enough to cause damage and create danger of a fire. This situation occurs when currents exceed the rated ampacity. The amount of damage depends on the degree and duration of the overcurrent. The most likely place for an overload to occur is on stranded cords such as extension cords for large-draw appliances (e.g., air conditioners, space heaters, and refrigerators). This is unlikely to happen on circuits with proper overcurrent protection. The effects of an overload cause internal heating on the conductor, which occurs along the entire length of the conductor (except where it has been thinned) and may cause sleeving. If the overloading is severe, the conductor can ignite fuels in the vicinity and melt the conductor. Once the conductor melts, the current stops flowing, and heating stops. Overcurrent melting is not necessarily indicative of ignition, however, because the fire itself may cause overcurrent to the circuit. Effects of overcurrent may be seen as sleeving of the insulation on the conductor. This overheating and evidence of sleeving are usually observed along the whole length of the circuit of the cord, as the overload occurs along the whole length of the conductor. There may be areas of limited overloading in appliances and short conductor paths.

Overload can bring the conductor to melting. The first point to sever stops current and heating. Other neatly melted spots freeze as offsets. If the overload is caused by downstream faulting, the farthest arc from the source should be found.

The investigator could observe some effects on wires that are not caused by electricity. These effects, which include conductor surface colors (dark red to black oxidation, green, or blue), are of no value in determining cause but are always present in fires.

Melting by Electrical Arcing

When arcing occurs, very high temperatures are produced at the site of the electrical arc, which will melt the conductors in a very small area, and because of fast overcurrent action, the duration is short. The damage caused by the arc is thus localized, and a sharp line of demarcation occurs between the point of damage and surrounding conductor material. Gloves can be used that snag on the small notches on the conductor caused by the arc. The process involves running the gloved hand over the wires until it catches on the notch, at which point further investigation will identify whether the surface roughness that is felt is caused by an arc or another event. Often, magnification is required to detect this damage. Sometimes arcing produces sparks and dislodges material, which then collects on nearby surfaces. Further metallurgical analysis can be performed on wire segments containing suspicious notches to confirm melting patterns consistent with arcing.

Melting by Fire

Solid copper conductors first become blistered and distorted on the surface FIGURE 7-13. Some copper flows on the surface, and some hanging droplets form. Further melting allows thin areas to form necking and droplets, and the surface of the conductor becomes smooth. Resolidified copper forms globules that are irregular in shape and size and are often tapered and may be pointed. No distinct line of demarcation is present.

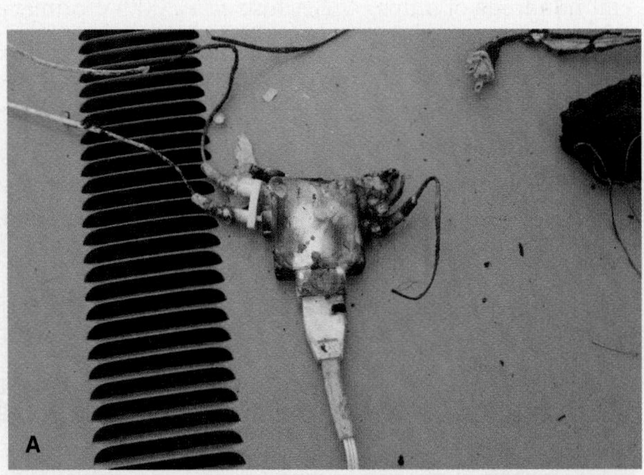

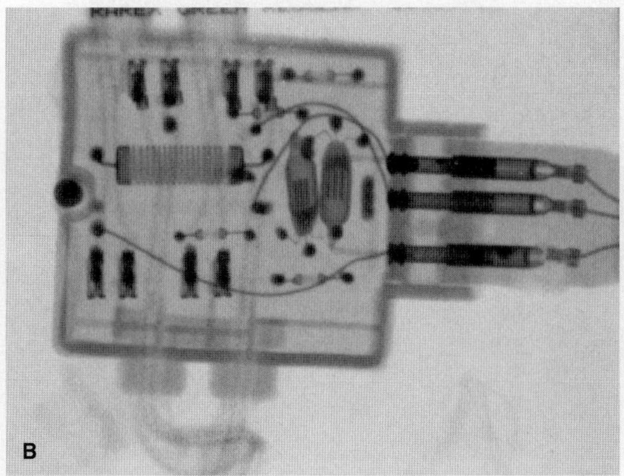

FIGURE 7-12 **A.** Photo of poor contact connections. **B.** X-ray of poor contact connections.

Courtesy of Mike Dalton

FIGURE 7-13 When melted, solid copper conductors first become blistered and distorted on the surface, later forming droplets before the surface becomes smooth.

Electrical damage is usually very localized and well defined because the electrical energy to melt the conductor is localized.

Stranded conductors melted by fire become stiff as they reach melting temperature. Further heating melts the individual strands together, and the surface becomes irregular, showing individual strands. Continued heating creates conditions similar to those in solid conductors. In large-gauge stranded conductors, the individual strands may fuse, separate, and have bead-like globules.

Aluminum conductors have low melting temperatures. They melt in any fire and solidify into irregular shapes. They are generally of little help in determining a fire's cause.

Alloying

Alloying is the combining of metals of different physical properties, including melting temperature. Aluminum–copper alloying can begin with aluminum dripping onto the surface of a copper conductor and sticking lightly to the surface. If heating continues, the melted aluminum can penetrate the copper's surface and form an alloy. The alloy has a melting point lower than that of pure copper or pure aluminum. The investigator observes an alloy spot as either a rough gray area or shiny silvery area. Copper–aluminum alloy is brittle and may break easily. If the alloyed material drips off during the fire, the investigator notes a pit lined with the alloy. Chemical analysis can verify the existence of an alloy. Some other common alloys include brass (copper–zinc) and bronze (copper–tin). Silver is often used on the contact matting surfaces of electromechanical switches such as relays, thermostats, and contactors. Copper and silver on these surfaces can form an alloy at temperatures below their respective melting points.

Mechanical Gouges

Mechanical gouges can be distinguished from arcing marks by microscopic examination. They usually show scratch marks, dents in the insulation, and deformation of the conductors. They do not exhibit fused surfaces caused by electrical energy.

Considerations and Cautions

Some prior beliefs have been disproved by laboratory studies, based on knowledge of chemical, physical, and electrical sciences. Investigators thus need to be aware of the underlying research and studies supporting hypotheses that are used to describe electrical failures. The following discussion provides some caution for some of these hypotheses.

Undersized Conductors

Conductor sizing is an important safety factor in the design of circuit ampacity; however, applying 20 A to a 15-A circuit does not always result in overheating. In fact, significant overcurrent may cause increased heating but not necessarily fire ignition. The investigator must carefully evaluate circuits that have more current flow than the *NEC* allows. The investigator must also take into account the specific circumstances of the fire. Generally speaking, an undersized conductor in and of itself is not indicative of the fire cause. For example, a 20-A current flow in a 14-AWG conductor would not cause the wire to heat sufficiently to cause ignition if the wire were bare and uninsulated. However, fires have been observed in attics where an overcurrent of this magnitude exists (20 A on a 14-AWG wire) when the wiring is underneath cellulose insulation.

Nicked or Stretched Conductors

Nicked or stretched conductors that reduce a conductor's diameter in a localized area usually cannot be said to cause heating at that location under normal load conditions; however, the mechanical strength of the conductor is compromised, and flexing it near the damage could weaken it to the point where the conductor could fracture and open and arcing could occur. It is not easy to stretch copper conductors in good condition sufficiently to reduce the cross-sectional area without breaking the conductor.

Deteriorated or Damaged Insulation

Insulation deteriorates with age and heating. Rubber deteriorates more quickly than thermoplastic. It may become brittle and will crack when bent. Cracks do not allow for conduction of electricity unless a conductive physical object or fluid should enter the cracks; however, the insulating strength of air (in a crack) is less than that of insulation, leading to a greater potential for arcing.

Mechanical damage and vibration also deteriorate insulation. In FIGURE 7-14, the electrical installer had cut through insulation, and a ground fault to the electrical panel occurred.

Overdriven or Misdriven Staple

Properly driven staples do not damage the conductor to the point of faulting immediately; however, over time, if overdriven, a high-resistance contact can occur as the insulation gives way to pressure. This type of high-resistance contact can occur months or years after installation of the wire. Even if evidence of this kind of a fault exists, there is still a question of whether this is an ignition source rather than the effect of

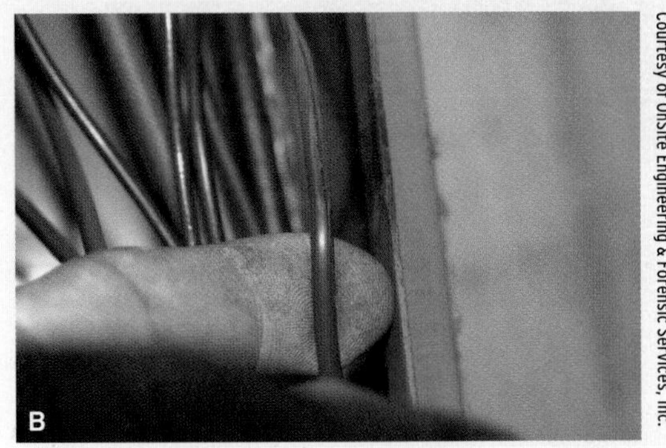

FIGURE 7-14 Ground fault from damaged (cut) insulation.

an advancing fire. Given that the steel staple will heat easily within an attic fire, the insulation on the wiring immediately under the staple can be affected early in the fire, allowing arcing immediately under the staple. For this reason, arcing is commonly observed in the immediate area of staples and commonly misdiagnosed as a fire cause. A real-world example of this phenomenon is sitting in a sauna with a necklace on. The necklace burns your skin as it heats quicker (thermal diffusivity) than your skin.

Improperly positioned staples driven at an angle can cut across several conductors, creating a short circuit. This occurrence should be evident after the fire at the point where the staple was misdriven. Overcurrent protection should stop the flow of electricity and stop any heating. A parting arc from this event normally would not be sufficient to ignite the insulation or supporting wood. If a leg of the staple penetrates the insulation and contacts an energized conductor and a grounded conductor, a short circuit will occur. If the staple severs an energized conductor and a high-resistance fault develops, a heating connection may be formed.

Short Circuit

Short circuits or conductor-to-conductor faults generally cannot generate sufficient heat to ignite combustibles nearby before the overcurrent protection activates. If the overcurrent protection does not function, then an overload may occur, which, along with arcing, may generate sufficient heat to ignite nearby material.

Beaded Conductors

A beaded conductor is not indicative, in and of itself, of the cause of a fire. Beading may be the result of the fire damaging the wire and then creating a failure.

■ Collecting Evidence

Damaged conductors are potential evidence and should be documented at the fire scene before being disturbed. The investigator should be mindful of spoliation of evidence and notification issues whenever handling or removing electrical equipment (see the spoliation discussion in the "Legal Considerations" chapter of this text) and should not disturb such equipment without fully considering the need and advisability of doing so. Included in the documentation should be the location of the damage in the room; the switches, outlets, connections, and branch circuit that the damaged conductor is connected to; and the state of the overcurrent protection for that branch. A photograph and a sketch of the location of the damaged conductor in its found position are useful in later analysis.

The investigator should affix bright flags or colored tape to points near the damage in a room and photograph the affected parts of the circuit from several feet away to add overall perspective to their placement in the room. Close-up macro photos of these sites should be taken if moving the conductor could cause more damage. For lengths of conductor removed, locator markings can be added near the cuts made with corresponding markings made on the remaining segments left in the room. This will allow the investigating team to return to collect more evidence if necessary, and the location markings can be used to create a grid drawing of the conductors in the room.

Creating a sketch of the room while on the scene is an important task that will help in further analysis and in writing the report. Include the sites of damage and the places where the conductors were cut, as well as other connected and nearby conductors, switches, lamps/loads, and outlets on the sketch. The overcurrent protection device and all switch settings associated with each circuit connected to both the damaged and nearby conductors should be included in this documentation as possible proof that the circuit was energized (or not) at the time of the fire. Sometimes removal of part of a circuit, including a switch or outlet connection, is warranted in addition to the damaged portion. This can be kept intact and will help with explaining the circuit's function and active state at the time of the fire or in assessing damage done to the circuit in future analysis.

If the damaged conductor is to be cut away from the circuit, make sure that the cuts you make are far enough away from the damage that they are in the "good" portion without melting, if possible. Avoid cleaning the conductors on-scene because material cleaned from the surface may be needed for future analysis. Individual pieces of evidence should be carefully labeled and packaged separately for transport to be

Fire Investigator Tip

Electrical systems are rarely simple and usually complex. Furthermore, new product advances in electronic controls increase the challenge, usually requiring additional design information such as schematics and theory of operation. By knowing what areas to focus on and the characteristics of electrical energy flow, the investigator can break the system down into workable tasks and, often as a part of a team of investigators, converge on a valid conclusion. New product advances in electronic controls increase the challenge, usually requiring additional design information such as schematics and theory of operation.

archived and studied later. Labels and locations where found should be affixed to the evidence. Long copper conductors can be carefully coiled, if they are not brittle and if the damage points are not bent in the process of arranging them into a coil, and placed into plastic bags or shrink-wrap, which should be used to protect the conductors from further damage caused by abrasion, kinking (tight bends made in the conductor), or nicking. Lengths of conduit and rigid conductors should not be bent if possible. As with other evidence that is damp or wet, each item taken must be dried out before storage. This can be done at the end of transport to the holding facility. Leaving the bags open to "breathe" will help reduce the effects of moisture on degrading the evidence. Place these bags into tubs or bins with tight covers to prevent wind damage during transport.

Static Electricity

Static electricity results from buildup of a stationary charge caused by the physical contact and subsequent rubbing or movement of one object on the other. Examples of situations that can create static electricity include walking across a carpet, conveyor belts moving over rollers, and flowing liquids.

Static discharge can occur when two charged surfaces pass over each other and a sudden recombination of the separated positive and negative charges creates an electric arc. A static charge can build up over time. It is difficult to prevent static electricity absolutely. Conditions must be ideal for it to be a competent ignition source. It cannot be a competent ignition source unless the discharge occurs in an ignitible atmosphere.

Static can be generated by movement of liquids in relation to other objects and can occur in flow through pipes, mixing, pouring, pumping, spraying, filtering, and agitating. Static may accumulate in the liquid, particularly with liquid hydrocarbons. If sufficient static builds up, an arc may occur, and if there is a flammable vapor–air mixture, ignition results. Lower liquid conductivity implies a higher ability to create and hold a charge. Surface charge on the surface of the liquid is of the greatest concern.

Relaxation time is defined as the amount of time for a charge to dissipate. It can range from several seconds to several minutes and is dependent on the conductivity of the liquid, the rate at which the liquid is being introduced into a tank, and the manner in which the liquid is being introduced into the tank. If the electrical potential between the liquid and the metal tank shell becomes sufficient, the air may ionize, and an arc may form between the liquid surface and the tank shell; however, an arc between the surface and the shell is less likely than an arc between the surface and a projection into the tank. Projections are known as spark promoters. No bonding or grounding can remove these static charges. If the tank or container is ungrounded, an arc may occur between the tank and a nearby object.

Switch loading occurs when a liquid is introduced into a tank that held a liquid of different properties. Static discharge can ignite the vapors from the more flammable liquid.

Spraying operations can produce significant static charges on the surfaces being sprayed and on an ungrounded spraying nozzle or gun. If the material being sprayed is flammable, then ignition can occur. High-pressure spraying operations have a greater potential for generating static than low-pressure spraying operations do.

Static can build up when a flowing gas vapor is mixed with metallic oxides, scale particles, dust, and liquid droplets or spray. When such a gas is directed against a conductive, ungrounded object, a static buildup may occur. If the static charge is sufficient and an ignitible atmosphere is present, then an arc may occur to another nearby grounded object, and the gases will ignite.

The minimum electrical charge required for ignition of a dust cloud is 10 to 100 millijoules (mJ). This range of charge energy encompasses and can be below the amount of energy in a typical static arc from the human body, which can reach 20 to 30 mJ. The human body can accumulate charges in atmospheres less than 50 percent relative humidity. The charge can be as high as several thousand volts.

Charges can build up when layers of clothing are separated, when layers of clothing are moved away from the body, or when layers of clothing are removed entirely. This is common when the layers are of dissimilar fabric. Ignition sources can occur in ignitible atmospheres from synthetic garments and from removing outer garments.

To be capable of causing ignition, the energy released in the discharge must be at least equal to the minimum ignition energy (MIE) of the ignitible mixture, which can be as low as 350 V in air. Some of this energy is required to heat the conductive surfaces that will arc, and thus, the minimum amount of energy required for static discharge ignition is realistically 1500 V. Dusts and fibers require 10 to 100 times more energy to ignite than gases and vapors do.

Controlling Accumulations of Static Electricity

Static charges can be removed or dissipated through humidification and by bonding and grounding. Humidification refers to the moisture present in a material. The more moisture there is in a material, the more conductive it will be and the less likely it will accumulate static charge. The moisture content is related to the relative humidity of the surrounding air.

In atmospheres with high relative humidity (humidity of 50 percent or more), materials and air reach equilibrium, contain sufficient moisture to be adequately conductive, and do not have significant static electricity accumulation.

In atmospheres with low relative humidity (humidity of 30 percent or less), materials dry out and become good insulators, and static accumulations are likely. Conductivity of the air itself is not changed by humidity.

Bonding refers to the electrical connection of two or more conductive objects in such a way as to reduce the electrical potential differences between them. Grounding is the process of electrically connecting an object to ground. It reduces the electrical potential differences between objects and the earth. Objects such as pipes and tanks that are embedded in the earth are naturally bonded. The grounding and bonding should be tested to determine their effectiveness.

Conditions Necessary for Static Arc Ignition

Five conditions must be present for static arc ignition:

1. A means of static charge generation
2. A means of accumulating and maintaining the charge
3. A static electric discharge arc with sufficient energy
4. A fuel source with the right air mixture and with a small-enough ignition energy
5. Co-location of the arc and fuel source

According to NFPA 921 Section 9.12.5.2.4, if static arc ignition is suspected as an ignition source, an electrical inspector with proper qualifications (in compliance with NFPA 77, *Recommended Practice on Static Electricity*) should review the circuit grounding paths and related connections.

Investigating Static Electric Ignitions

Investigating static electric ignitions often requires the gathering of circumstantial evidence. The five conditions listed previously here must exist. The investigator must determine how the static electricity was generated by identifying the materials involved and determining the materials' conductivity, motion, contact, and separation. Additionally, the investigator must identify how the charge was able to accumulate through the presence or absence of grounding, bonding, or conductivity of the material. The investigator must determine the relative humidity, identify the potential location of the static arc, determine whether the arc contained sufficient energy to be a competent ignition source, and determine whether the arc and the fuel were located in close proximity. Eyewitness reports and circumstantial evidence should help determine, as exactly as possible, the location of the arc.

Also important are the actual discharge energy and the fuel present for ignition. Although arcing can occur at lower discharge voltages, 1500 V has been shown to be a lower bound for ignition to occur. Dusts and fibers typically require 10 to 100 times the energy of vapors and gases (in ideal air–fuel mixture) to ignite.

Lightning

Lightning bolt characteristics include the following:

- Core energy plasma of 1/2 to 3/4 inch (1.27 to 1.9 centimeters) diameter
- Surrounded by a 4-inch-thick (10.2-centimeter) channel of superheated ionized air
- Average 24,000 A, but can exceed 200,000 A
- Potentials that range up to 15,000,000 V

Lightning often strikes the tallest object and follows it as a path into the ground. Lightning can enter a structure in four ways: striking a metal object on top of the structure, striking the structure, striking a nearby tall structure or even the ground and moving horizontally to the building, or striking overhead conductors and being conducted into the building via the conductors. Lightning strikes carry high electrical potentials (hundreds of thousands of volts) and large currents (thousands of amps), and extremely high heat energy and temperatures are generated. The damage may be displaced or may explode building features. The investigator may observe damage resulting from a lightning strike as damage to the structure or to the electrical system. Investigators must pay particular attention to any point where the building object may be grounded. Both line-voltage (120 V AC) and low-voltage (<50 V DC) systems in buildings are susceptible to lightning strike damage. Some equipment may contain surge protectors and surge monitors. These devices are useful only if they are able to absorb or deflect the energy levels present on the equipment power conductors and signal interfaces. Lightning detection networks and lightning strike reports may assist in determining the exact time and location of the strike.

© Greg Henry/ShutterStock, Inc.

Wrap-Up

Ready for Review

- Determining whether electrical damage was the cause or the effect of a fire requires at least a fundamental understanding of electricity and, more specifically, knowledge of the electrical units: voltage (volts), current (amps), resistance (ohms), and power (watts [W]).
- Understanding the basic principles of electricity, including Ohm's law, calculation of power, and current flow in a circuit, helps to quantify the electrical energy available in a circuit. With the amount of energy known, you can focus on the distribution (conductors) and load part of the circuit.
- Building electrical systems include the service entrance, grounding, overcurrent protection, circuit breaker panels, branch circuits, conductors, outlets, and special fixtures/devices.
- The following conditions must exist for ignition from an electrical source:
  - Wiring must be electrically energized.
  - Sufficient heat and temperature must be produced.
  - The combustible material must be a material that can be ignited by the electrical energy produced by the electrical failure.
  - The heat source and the combustible fuel must be close enough for a sufficient period of time to allow the combustible material to generate combustible vapors.
- Because of the high potential energy of electrical circuits, significant observable damage will occur from faulting or arcing elements and components in the circuits.
- Static electricity results from buildup of a stationary charge caused by the physical contact and subsequent rubbing or movement of one object on the other.

Hot Terms

Alloying The mixing of two or more metals in which one or more of the metals is in a liquefied state.

Alternating current (AC) The voltage varies in time with a sine wave at 60 cycles per second. A sine-wave current results if the load is resistive.

Ampacity The current, in amperes, that a conductor can carry continuously under the conditions of use without exceeding its temperature rating.

Amperage The unit of electric current that is equivalent to a flow of one coulomb per second; one coulomb is defined as 6.24×10^{18} electrons.

Arc A high-temperature luminous electric discharge across a gap where the conductor is missing or through a medium such as charred insulation.

Arced and severed Occurs when wires are arced off on the ends and possible segemented wires result.

Arc-fault circuit interrupter (AFCI) Designed to protect against fires caused by arcing faults in home electrical wiring. The circuitry continuously monitors current flow.

Arc mapping The analysis of the locations where electrical arcing has caused damage and the documentation of the involved electrical circuits.

Blow holes Visible holes in the conduit or metal panels of branch circuits. They are an indication of significant arcing and energy release inside the enclosure or conduit.

Branch circuits The individual circuits that feed lighting, receptacles, and various fixed appliances.

Direct current (DC) The voltage is a steady level that does not vary with time. The current will also be a steady level if the load is resistive.

Electrical power The rate of doing work (using electrical energy) in an electrical circuit, such as in a hair dryer, electric motor, or lightbulb.

Eutectic melting Any combination of metals with a melting point lower than that of any of the individual metals of which it consists.

Ground-fault circuit interrupter (GFCI) A device intended for the protection of personnel that functions to de-energize a circuit or portion thereof within an established period of time when a current to ground (possibly a human body) exceeds the values established for a Class A device.

Grounding A conducting connection, whether intentional or accidental, between an electrical circuit or equipment and earth or to some conducting body that serves in place of the earth.

High-resistance fault An unintended path for electricity that allows sufficient current for heating to occur, but not enough to reset the current protection.

Horsepower A unit of power that is used to express mechanical energy use or production. One kilowatt is equal to 1.34 horsepower.

Hot legs The two insulated conductors of a single-phase system (or three if three-phase power is used), sometimes referred to as L1, L2, or Line voltages.

Interrupting current rating The maximum amount of current the device is capable of interrupting.

Main disconnect Provides the master shutoff mechanism for the overall system power and provides high current level protection of downstream overcurrent protection devices and wiring.

Meter A watt-hour meter that plugs into the meter base to measure the amount of electricity consumed at a site.

Meter base The component where the service cables come in through the weatherhead and go down the conduit to connect to an electric meter that measures the amount of electricity being used.

Ohm's law A basic law of electricity that defines the relationship among voltage, current, and resistance. If two of these three values are known, it is possible to determine the third.

Overcurrent Any current in gross excess of the rated current of equipment or the ampacity of a conductor; it may result from an overload, short circuit, or ground fault.

Overfusing A dangerous condition that occurs when the circuit protection (fuse or circuit breaker) rating significantly exceeds the ampacity of the conductor, leading to a condition in which increased heat can occur in the conductors.

Overload Operation of equipment in excess of normal, full-load rating or of a conductor in excess of rated ampacity that, when it persists for a sufficient length of time, could cause damage or dangerous overheating.

Regular current rating The level of current above which the protective device will open, such as 15 A, 20 A, or 50 A.

Relaxation time The amount of time for a charge to dissipate.

Resistance heating Occurs when current flows through a path that provides high resistance to current flow, such as a heating element (intentional) or a resistive connection (unintentional).

Resistive circuit A circuit that does not contain inductance and capacitance.

Root mean squared (RMS) A mathematical computation to equate the voltage level of an alternating current (AC) system to that of a more familiar direct current (DC) system.

Service drop The overhead service conductors from the last pole or other aerial support to and including the splices, if any, connecting to the service-entrance conductors at the structure.

Service entrance The point where the electrical service enters a building.

Service equipment The necessary equipment, usually consisting of a circuit breaker(s) or switch(s) and fuse(s) and their accessories, connected to the load end of service conductors to a building or other structure, or an otherwise designated area, and intended to constitute the main control and cutoff method of the supply.

Service lateral Wiring coming in underground.

Sine wave The waveform that AC voltage follows. An example is 120-V AC, which has 170-V peak, crosses 0 to −170 V, and repeats this cycle 60 times per second.

Static electricity The electrical charging of materials through physical contact and separation and the various effects that result from the electrical charges formed by this process.

Switch loading A term used to describe a product being loaded into a tank or compartment that previously held a product of different vapor pressure and flash point.

Thermistors Sensors used to measure temperature by change of their resistance.

Time current curve The amount of time required for a current protection device to interrupt at a specific level of current.

Voltage sniffer A noncontact voltage monitor that outputs a beep or turns on a light when voltage is present or nearby (within about an inch or less).

Weatherhead The point where service entrance cables connect to the structure, which is designed to keep water out of the conduit that carries the wires.

© Greg Henry/ShutterStock, Inc.

FIRE INVESTIGATOR *in action*

You are investigating a fire in a small beauty salon. The fire has caused extensive damage throughout the business, which was closed at the time of the fire. Fire personnel are finishing extinguishing the fire when you begin to speak with the business owner.

The business owner states that there had been no problems with the building that she was aware of; however, she did occasionally "trip" a breaker when too many devices were being used at the same time. She states that she had reset the breaker to an outlet at the first style station several times the previous day.

Once inside, you begin to examine the interior of the salon and identify the first station as the area of origin. After systematically removing the debris, you locate the remains of several handheld hair dryers, curling irons, and a wax heater connected to a severely damaged power strip that has melted to the floor. The outlet that the power strip is connected to is also severely damaged.

1. Which of the following events may be a cause of this fire?
   - **A.** An overload
   - **B.** Static discharge
   - **C.** Arc through char
   - **D.** Both A and C
2. Ignition by electrical energy requires all of the following conditions, except:
   - **A.** an energized circuit.
   - **B.** sufficient heat and temperature.
   - **C.** an electrical arc.
   - **D.** proximity to combustible material.
3. Arc mapping can be useful for the investigator in:
   - **A.** identifying the wire gauge of a conductor.
   - **B.** determining the area or appliance of origin.
   - **C.** establishing the amount of current present in an electrical circuit.
   - **D.** All of the above
4. The amount of current flow within an electrical circuit is measured in:
   - **A.** volts.
   - **B.** watts.
   - **C.** ohms.
   - **D.** amperes.
5. If you were going to collect electrical components as evidence, you should:
   - **A.** create a sketch of the area indicating the location of the evidence.
   - **B.** mark and photograph locations where you must cut electrical circuits.
   - **C.** cut the electrical circuits outside the areas of damage.
   - **D.** All of the above

Building Fuel Gas Systems

© Photos.com

© Brian Brainerd/The Denver Post/Getty Images

Knowledge Objectives

After studying this chapter, you should be able to:

- Describe the elements of fuel gas systems and how they can contribute to fire and explosion investigations. (pp 142–143)
- Discuss the characteristics of fuel gases. (pp 143–144)
- Discuss the components of natural gas systems (p 144).
- Discuss the components of LP gas systems. (pp 144–146)
- Identify and describe components common to fuel gas systems. (pp 146–148)
- Discuss common piping for fuel gas systems in buildings. (pp 148–149)
- Discuss common appliance and equipment requirements. (pp 149–150)
- Identify fuel gas utilization equipment. (pp 150–151)
- Describe how to investigate fuel gas systems. (pp 151–154)

Skills Objectives

After studying this chapter, you should be able to:

- Conduct a fire and explosion investigation for fuel gas systems. (pp 151–154)

Additional NFPA References

NFPA 54, *National Fuel Gas Code*

NFPA 58, *Liquefied Petroleum Gas Code*

NFPA 70, *National Electrical Code*

NFPA 921, *Guide for Fire and Explosion Investigations*

CHAPTER 8

FESHE Course Outcomes

Fire Investigation I

There are no Fire Investigation I (FESHE) course outcomes for this chapter.

Fire Investigation II

There are no Fire Investigation II (FESHE) course outcomes for this chapter.

You Are the Fire Investigator

© Jones and Bartlett Publishers. Photographed by Glen E. Ellman

At the scene of a fatal explosion, you begin to document the scene and identify items of potential evidence. You observe a severely damaged excavator near a trench that has been excavated between the remains of a house and the street. As you inspect the trench, you observe the water line and a natural gas line, both of which are intact but have gouges and are bent at several locations. A neighbor who witnessed the explosion has asked to speak with you because he believes he knows what happened.

1. What questions would you ask the witness?
2. Which components of the fuel gas system would be significant to your investigation?
3. Which resources would you obtain to conduct this investigation?
4. Where would you expect the leak to have occurred?

Introduction

During every fire scene examination, the presence and condition of fuel gas systems need to be examined and documented. Fire and explosion incidents are often the result of failures within these systems. When an incident occurs involving a fuel gas system, it frequently leads to dramatic events including explosions and intense fires, often totally destroying the structure and causing serious injuries to those within or nearby. The fire investigator should have knowledge of fuel gases and the systems that utilize them. There are three governing documents considered to be the primary resources for developing an understating of building fuel gas systems:

- NFPA 54, *National Fuel Gas Code*
- 49 CFR Part 192, "Transportation of Natural and Other Gases by Pipeline: Minimum Safety Standards"
- NFPA 58, *Liquefied Petroleum Gas Code*

FIGURE 8-1 Natural gas explosion scene.

Fuel Gas Systems

A building's fuel gas system can be used to control the indoor climate, to heat water, for cooking, or to provide energy for manufacturing processes. The two most common fuel gases used within a structure are natural gas and commercial propane.

Impact of Fuel Gases on Fire and Explosion Investigations

Fuel gas systems to either a building or an appliance can act as the initial fuel source, the initial ignition source, or both the fuel and ignition source. It can also cause the fire to spread rapidly through the structure. When the fuel delivery system fails and the fuel comes into contact with an ignition source, the fuel gases can act as the source for the original ignition sequence by acting as the first fuel FIGURE 8-1. The fuel gases then can act to influence the spread and growth of the fire. If fuel gases become involved in a fire, they are usually a factor in the fire's spread and growth.

Fuel Sources

Fuel involvement most frequently results from compromised fuel delivery or containment systems. The fuel usually escapes from the containers that hold the bulk fuel supply, the piping from the bulk supply to the appliance, or systems in the appliance itself. Fuel that has escaped is often referred to as fugitive gas. Fugitive gas readily mixes with the air and ignites easily, even by ignition sources that have very little energy. The ignition temperature of most fuel gases ranges from 723°F to

Safety Tip

Because of the amount of damage that can occur at an explosion or fire incident, always examine the scene for potential hazards. Holes, voids, and even residual gas may exist, posing a significant risk to those conducting the investigation.

1170°F (384°C to 632°C). The minimum ignition energy for these gases can be as low as 0.2 mJ.

Ignition Sources

Pilot lights and open flames from the appliances served by fugitive gases often function as ignition sources. Ignition sources can also be from static arcs from the appliances, electrical arcs, arcs from switches or contacts in appliances, or other electrical equipment. Often, it is difficult to isolate the source of ignition for fugitive fuel gases because there are so many possibilities.

Fugitive fuel gases may develop around gas appliances or equipment that is served by the fuel gas. Once these appliances become involved, the fire will ignite combustibles in the vicinity of these appliances.

Gas Systems as Ignition and Fuel Sources

Fuel gas systems and apparatus, particularly those with burners and pilot lights, can serve as both the source of the fuel and as the ignition source for the fire or explosion. This could occur when, for example, gas escapes from a damaged pipe or mechanism in a water heater, eventually reaching the pilot light where it is ignited.

Additionally, fuel gases can serve as both the first fuel and the fuel for the fire. In such a circumstance, after providing the initial fuel for an ignition, the gas system continues to emit additional fuel for the fire to grow. The amount of energy provided by the fugitive fuel gases depends on the size of the leak or the amount of fuel that is escaping. Fugitive fuel gases can greatly accelerate the spread and growth of a fire, especially if the leak continues during the fire growth and there are combustibles in the area that will allow for fire growth. Fuel gases that are simply burning into the air with no additional combustibles nearby will not accelerate fire growth.

Additional Fire Spread

Fire can spread to other areas of the building because of this fugitive fuel gas. They may appear as separate fires, but in reality, they are related to the release of the gas. This phenomenon often occurs when the gas is released for a period of time prior to ignition and is allowed to pool in other areas. It can also occur when the gas is released under pressure, spreading the fire to other areas of the building. TABLE 8-1 describes some of the characteristics of natural gas and propane (liquefied petroleum [LP] gas), the fuels that are most common in residential use. (For more information, see NFPA 54 and NFPA 58.)

Table 8-1 Characteristics of Natural Gas and Propane

	Natural Gas	Propane
Composition	Hydrocarbon gas	95% propane and propylene
	Primarily methane	5% other gases
Density	Lighter than air	Heavier than air
	Vapor density of 0.59 to 0.72	Vapor density of 1.5 to 2.0
LEL	3.9 to 4.5 percent	2.15 percent
UEL	14.5 to 15 percent	9.6 percent
Ignition temperature	900°F to 1170°F (483°C to 632°C)	920°F to 1120°F (493°C to 604°C)
BTUs (per cubic foot)	1030	2490
Delivery	Delivered via a distribution system directly to the customer	Stored in tanks on the customer's property and delivered via tank trucks

Characteristics of Fuel Gases

The investigator must be aware of the individual properties of the fuel gas that they suspect may have caused or contributed to a fire or explosion event. An understanding of each fuel gas's properties will allow the investigator to interpret fire and damage patterns accurately.

Natural Gas

Natural gas is a mixture of gases that is often recovered from underground pockets during drilling operations for crude petroleum. Comprised largely of methane (72 to 95 percent), other gases such as nitrogen, ethane, propane, butane, hexane, carbon dioxide, and oxygen can be found in varying percentages.

Pure natural gas is lighter than air and has a specific gravity of 0.59 to 0.72. The flammable range is between 3.9 and 15 percent, with an ignition temperature of 900°F to 1170°F (482°C to 632°C).

Commercial Propane

Propane, unlike natural gas, is recovered during the refining process of petroleum. Because of propane's ability to be liquefied at moderate pressures and normal temperature, liquefied propane (LP gas) can be more easily transported, stored, and used than natural gas, making it more suited to remote locations or portable consumer use. Commercial propane, at a minimum, is 95 percent propane and propylene, with the remaining 5 percent consisting of other gases.

Propane (undiluted) has a specific gravity greater than air (1.5 to 2.0), a flammable range of 2.15 to 9.6 percent, and an ignition temperature of 920°F to 1120°F (493°C to 604°C).

Other Fuel Gases

Other gases may also be of interest to the investigator when conducting commercial or industrial investigations. Commercial butane largely comprises butane and butylene (minimum 95 percent), whereas the remaining 5 percent consists of other gases. Propane HD5 (95 percent propane) is used as a motor fuel and is subjected to more restrictive requirements than is commercial propane.

Specific manufacturing processes may also use or produce other gases such as coke oven gas, hydrogen, and acetylene.

Odorization

By law, odorization is added to LP gas and commercial natural gas. There are cases in which natural gas does have an odor directly from a well head, but LP gas and natural gas frequently have no odor themselves. The supply company adds it as a safety feature so that people can detect the presence of the gas. Per NFPA 58, the odorant must be detectable when the fuel gas is at a concentration of not less than one-fifth of its lower explosive limit (LEL). Butyl mercaptan is most often used in natural gas, whereas ethyl mercaptan or thiophane is most often used in LP gas. Typically, natural gas is not odorized while being transferred in a transmission pipeline system. Odorization generally only occurs when natural gas is introduced into a distribution system to be used by the end consumer. In the case of LP gas, odorant is added prior to being delivered to a bulk distribution facility.

If the investigation reveals that fuel gas may be involved, the investigator should verify the presence of an odorant. Often the investigator can smell the odorant during the investigation.

Propane should be tested by obtaining a liquid sample for laboratory analysis, as gas samples will not provide an accurate measurement. LP gas samples should be obtained using stain tubes and approved sampling techniques. Because there is no universal stain tube for all odorants, the investigator must identify the odorant and use the appropriate stain tube. It is recommended that the investigator seek assistance from someone who is experienced in this field to obtain the sample(s). Once collected, the sample(s) should be submitted for analysis as soon as possible to a laboratory that is able to analyze the sample for odorization. The laboratory's capabilities should be determined by the investigator prior to submitting the sample(s).

Natural gas samples, on the other hand, should be collected using Tedlar bags and tested at the scene with field detection equipment.

Gas Systems

The systems that transport various fuel gases are similar in that they all pipe the gas directly into consumers' buildings and appliances. One common difference, however, is that natural gas is supplied from a central location via underground supply service lines (a type of distribution line). LP gas or other types of storage gas are contained in a bulk supply, often at the consumer's location.

Natural Gas Systems

Natural gas is supplied through a transmission pipeline, then through main pipelines (mains), and then through service mains or service laterals before it is connected to the meter at the consumer's location. TABLE 8-2 provides a description of the natural gas supply system and typical pressures found.

LP Gas Systems

LP gas distribution systems are similar in operation to natural gas systems except that the LP gas storage supply is often located at the consumer's site. LP gas storage containers can also be housed in bulk storage locations and piped underground, similar to natural gas systems.

Table 8-2 Natural Gas Supply System from Supplier to Consumer

Type of System	Typical Use	Pressure
Transmission pipeline	Used to transport the natural gas from the main facilities to the local utility	1200 psi (8275 kPa) gas from storage/production
Main pipelines	Used to distribute the gas (1000 kPa); typically 60 psi	Varies; seldom exceeds 150 psi; a centralized grid system: 400 kPa or less
Service lines (service laterals)	Used to connect customers to the main pipelines	Ranges from 4- to 10-inch water column (1.0 to 2.5 kPa)
Metering	Measures the volume of gas passing through a pipe	Depends on the consumer's needs, from 4-inch water column (2.74 kPa) to several psi

LP Gas Storage Containers

The differences among portable tanks, cargo tanks, cylinders, and containers are discussed in NFPA 58. The following discussion provides an introduction to those systems.

Tanks

A tank is defined as a storage container with greater than 1000 gal (4000 L) water capacity TABLE 8-3. The design and construction of LP storage tanks are governed by regulations of the American Society of Mechanical Engineers (ASME) FIGURE 8-2. The typical working pressure in the storage tanks

Table 8-3 LP Gas Tank Chart

	150 Gallons	250 Gallons	500 Gallons	1000 Gallons
Length	85″	92″	120″	190″
Diameter	25″	30″	37″	41″
Weight	320 lbs	485 lbs	950 lbs	1750 lbs

FIGURE 8-2 LP gas storage tank.

varies depending on the temperature of the LP, but can range as high as 200 to 250 psi (1400 to 1700 kPa) and may not be transported with more than 5 percent liquid capacity.

Cargo tanks are those containers permanently mounted on a chassis and are used for transporting LP gas. Portable tanks are used for transporting LP gas, but they are not mounted on a chassis. Their quantities exceed a 1000 lb (450 kg) water capacity.

Cylinders

Cylinders are considered upright containers. They have a water capacity of 1000 lbs (450 kg) or less. The design and construction of cylinders are governed by regulations of the U.S. Department of Transportation. The pressure in the cylinders can be the same as that in tanks or containers, ranging up to 200 to 250 psi (1400 to 1700 kPa). Cylinders are most frequently used in rural homes and businesses, mobile homes, and recreational vehicles and for outdoor barbecue grills and motor fuel FIGURE 8-3.

■ Container Appurtenances

The devices that are connected to the openings in tanks and other containers—such as pressure relief devices, control valves, and gauges—are called container appurtenances FIGURE 8-4. The following is a discussion of these devices.

FIGURE 8-3 LP gas cylinder.

FIGURE 8-4 LP gas tank gauge.

Pressure Relief Devices

Containers are governed by U.S. Department of Transportation regulations and are required to have pressure relief valves or fusible plugs. These pressure relief devices are pressure- or temperature-activated to prevent pressure from exceeding a predetermined maximum and rupturing a container.

Pressure relief valves are designed to open at a specific pressure, usually around 250 psi (1700 kPa). The pressure relief valve is generally placed in the container where it releases the vapor, although there are some exceptions. The pressure in a container is directly related to the temperature of the LP gas. As the temperature rises, the pressure rises. NFPA 58 provides methods to determine the pressure in a container if the temperature of the gas is known.

Fusible plugs are thermally activated devices that open and vent the contents of a container. Once activated, they cannot be reused. Above-ground storage tanks with less than 1200 lbs (540 kg) of water capacity may have fusible plugs. The melting temperature of these plugs is between 208°F (98°C) and 220°F (104°C).

Connections for Flow Control

Some container appurtenances control the flow of gas from the container. The most common is a shutoff valve. In some installations, there are excess-flow check valves that monitor the flow from the container; these valves activate if the flow exceeds a set amount. There are also valves called backflow valves that prevent the gas from re-entering the container or distribution system.

Liquid Level Gauging Devices

Often there are appurtenances on containers that indicate the liquid level of the propane. The types of liquid level gauging devices include fixed level gauges, which are fixed at a maximum level, and variable gauges. Types of variable liquid level gauges include float, magnetic, rotary, and slip tube.

Pressure Gauges

Pressure gauges are another type of container appurtenance, in this case depicting the internal pressure of the tank. The gauges are connected directly to the tank or sometimes through the valve. Pressure gauges do not indicate the quantity of liquid propane in the tank.

Pressure Regulators

Pressure regulators are attached through the gas supply system to reduce the pressure so that it can be used by the appliance or utilization equipment. In some propane systems, there are multiple (two) regulators that reduce the propane pressure in two stages. Often, when there are two stage regulators, one of the regulators will be located at the tank, and the other will be located near the appliances or where the fuel delivery system enters the structure. It is common to see the LP pressure reduced for use in the appliances to approximately 11-inch to 14-inch water column (w.c.) (2.74 to 3.5 kPa).

The propane pressure should be reduced from tank pressure to where it can be utilized by the equipment. The tank pressure will vary depending on the temperature of the liquid TABLE 8-4.

Vaporizers

Equipment is sometimes used to assist in vaporizing the liquid propane. Vaporizers are frequently used when there is a demand for large quantities of propane, such as for industrial uses or in cold weather environments FIGURE 8-5. Vaporizers heat the liquid propane, converting it to a gas.

Table 8-4 Typical Pressure Inside Propane Tanks as It Relates to the Temperature of the Propane

Temperature	Pressure
0°F (–18°C)	28 psi (190 kPa)
70°F (21°C)	127 psi (876 kPa)
130°F (54°C)	286 psi (1970 kPa)

FIGURE 8-5 A vaporizer.

FIGURE 8-6 A large natural gas meter.

Common Fuel Gas System Components

Whether the building is a single-family home, a commercial restaurant, or a large manufacturing facility, the fuel gas systems of all buildings will contain various distribution piping, valves, burners, and regulating devices. These various components are designed, installed, and used to deliver the fuel gas to the appliances safely. The investigator must examine these components to determine whether they caused or contributed to the fire or explosion event.

■ Pressure Regulation (Reduction)

The most common types of pressure regulators are the diaphragm type, which has a spring that is set to control the pressure, and the lever type. The amount of gas used may also be measured through a pressure regulator commonly known as a gas meter FIGURE 8-6. The vents on the regulators must be clear for the device to operate properly. If the vent becomes plugged or obstructed, the pressure regulator might not function properly. In cold or flood-prone environments, it is important to place regulators where water or ice accumulations cannot obstruct the vent openings. TABLE 8-5 gives a description of the general utilization pressures for appliances.

■ Service Piping Systems

Service piping provides for delivery of the fuel gas from the main lines to the user FIGURE 8-7.

Table 8-5 Utilization Pressures for Appliances

Use	Pressure
Nonindustrial natural gas applications	4 to 10 in. w.c. (1 to 2.5 kPa)
Nonindustrial propane applications	11 to 14 in. w.c. (2.7 to 3.5 kPa)

FIGURE 8-7 A 20-inch gas main pipeline.

The materials for these gas mains and services are most often constructed of wrought iron, copper, brass, aluminum alloy, or plastic. Service lines are often buried underground and are regulated by NFPA 58. Underground piping should be buried deep enough to prevent physical damage and should be protected from corrosion. If the piping runs under a building or areas that are utilized by the public, it should be encased in an approved conduit.

Valves

Valves will be installed on the piping system to control the flow of the gas. These valves are often installed where the gas piping exits to ground, prior to the meter or regulator. Valves are also located prior to each appliance and in appliances. There are several types of valves used in a common fuel gas system. Some of the valves are automatic and are controlled by the appliance to be closed, open, or partially open. Some of these automatic valves are configured to shut off automatically in an emergency condition, such as fire or excess flow. Some of the common valves in appliances are individual burner valves, main burner control valves, and manual reset valves TABLE 8-6.

Table 8-6 Valve Types and Their Uses

Valve Type	Use
Automatic valve	Includes a valve and an operating mechanism that controls the gas supply to a burner during operation of an appliance. The operating mechanism may be activated by gas pressure, electrical means, or mechanical means.
Automatic gas shutoff valve	Used in connection with an automatic gas shutoff device to shut off the gas supply to a fuel gas-burning appliance.
Individual burner valve	A valve that controls the gas supply to an individual burner.
Main burner control valve	A valve that controls the gas supply to a main burner manifold.
Manual reset valve	An automatic shutoff valve installed in the gas supply piping and set to shut off when unsafe conditions occur. It remains closed until manually reset.
Relief valve	A safety valve designed to prevent the rupture of a pressure vessel by relieving excess pressure.
Service shutoff valve	A valve (usually installed by the utility gas supplier between the service meter or source of supply and the customer piping system) used to shut off gas to the entire piping system.
Shutoff valve	A valve (located in the piping system and readily accessible and operable by the consumer) used to shut off individual appliances or equipment.

Gas Burners

Most gas utilization equipment uses some type of gas burner. The gas burners and/or orifices on appliances may not be interchangeable between natural gas and propane. Ignition of the gas from a burner can occur by several methods, including manual ignition, pilot lights, and pilotless igniters. When a burner is ignited by manual ignition, a person provides the spark or flame. Gas burners may also be ignited from pilot lights, small flames in the appliance near the burners. The pilot light must be of sufficient size and close enough to ignite the gas as it escapes from the burner. In some designs, the pilot light burns constantly, and in other designs, the burner is ignited electronically. In electronically lighted piloted appliances, electric arcs ignite the pilots, which in turn ignite the burners. Pilotless burners do not have pilots but are ignited by electronic arc or resistance heating elements. Many systems stop the flow of gas in the event that the burner does not ignite.

FIGURE 8-8 shows burners and the gas control valve for a typical furnace. FIGURE 8-9 shows the pilot line and gas control valve on a water heater.

© Jones & Bartlett Learning. Photographed by Susan Schultz
FIGURE 8-8 Furnace burners and gas control valve.

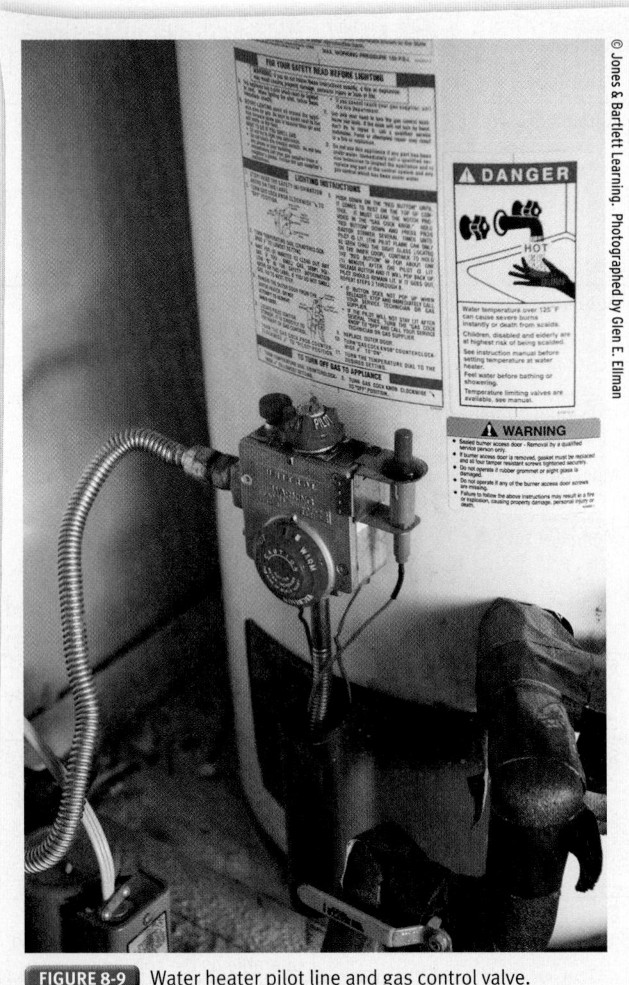

© Jones & Bartlett Learning. Photographed by Glen E. Ellman

FIGURE 8-9 Water heater pilot line and gas control valve.

Common Piping in Buildings

This section discusses the common piping components in buildings and the utilization equipment for fuel gas piping systems. The size of the piping is determined by the maximum flow required by the equipment that is connected to that pipe. Common piping in buildings is often similar to that described earlier in the section on service lines. Piping material can include wrought iron (black pipe), copper, or brass. Corrugated stainless steel tubing (CSST) is also used in commercial and residential applications and can be installed along floor joists, along ceiling joists, or inside wall cavities with more flexibility and fewer joints. Aluminum alloy may be used, but is not suitable for underground applications. Plastic may not be used indoors, although it may be used for exterior or underground applications. It is important to maintain the evidentiary value of piping and its components when conducting an investigation. Parts of the piping system should not be removed, and if the pipe needs to be cut, the investigator must note all of the changes made for reconstruction.

> **Safety Tip**
>
> Do not begin to test or disassemble fuel gas systems without proper training. A trained gas expert or investigator experienced with the investigation of a fuel gas system-related event should be obtained to check the system for leaks. Prematurely removing system components or altering the system prior to proper analysis and testing may lead to a claim of spoliation that can adversely affect court proceedings.

> **Safety Tip**
>
> Over the years, corrugated stainless steel tubing (CSST) has been found to be susceptible to leaks following lightning strikes or over electric current events resulting in a fuel-fed fire. Like all other gas distribution systems, CSST must be installed by qualified personnel and the system must be appropriately bonded and grounded to reduce these risks.

Joints and Fittings

Joints and fittings may be screwed, flanged, or welded. Screwed fittings are fittings with threads. Flanged fittings are connected with the use of a special device that flanges the end of the pipe and requires a fitting to complete the connection, or fittings may be welded. Under some circumstances, compression fittings—which include a device that compresses the fitting to the pipe for a seal—may be used. Plastic piping, which may be used outdoors only, may not be threaded.

Piping Installation

Piping installation is regulated by NFPA 54 and NFPA 58. Piping installations should not weaken the building structure. The piping must be supported with the proper devices and not by other pipes or appliances. The piping should be equipped with drip legs where appropriate, usually near appliances, to capture foreign material. All unused outlets or openings in piping should be capped, not simply controlled by a valve, to prevent gas from escaping.

Main Shutoff Valves

Main shutoff valves should be located on the exterior of the structure, although in older installations they may be located inside the building FIGURE 8-10. They should be placed upstream of each service regulator in order to provide total shutdown if needed. If the investigator suspects the main shutoff valve has leaked or failed, he or she should have a trained technician examine the appliance.

> **Fire Investigator Tip**
>
> Locating the main shutoff valve may be difficult. It may be hidden in bushes, plants, or other concealing structures.

FIGURE 8-10 The main shutoff valve controls the flow of gas throughout the entire building.

Prohibited Locations

As required by NFPA 54 and other recognized gas installation codes, natural gas piping cannot be run through air ducts, clothes chutes, chimneys, gas vents, ventilating ducts, dumbwaiters, or elevator shafts. Furthermore, gas piping should never be installed in areas that are subject to damage or exposed to corrosion.

Electrical Bonding and Grounding

Each aboveground portion of a piping system is required to be electrically bonded and the system grounded.

Common Appliance and Equipment Requirements

The appliance must be compatible with the type of gas it is using. When appliances are designed for natural gas but are used with propane, or when appliances are designed for propane but are used with natural gas, the results can be catastrophic, including fire or explosion.

Appliances should be placed where they will not be subject to damage. They should also not be placed where flammable vapors may accumulate because the appliance may ignite the gas vapors. In certain conditions, gas appliances can be installed in locations such as an automobile garage if they are located above floor level.

Appliance Installation

Where the system pressure is greater than the appliance is designed to handle, an additional regulator should be installed. (This situation was described earlier in this chapter in the discussion of multiple regulators.)

A gas appliance should be located where there is easy access for service or shutdown; however, this is not always the case. For example, commercial cooking appliances such as deep fryers and grill tops are usually supplied from the rear of the appliance, requiring the entire unit to be pulled away from the wall to shut down the supply of gas.

The location should also have sufficient clearance between the appliance and combustibles, including building construction, to ensure that the heat of the appliance does not ignite the combustible materials or items. The minimum required clearance is often displayed on the appliance itself. Appliances and accessories should also be approved as acceptable by the authority having jurisdiction and in accordance with NFPA 54. The electrical connections to the appliances should be in accordance with NFPA 70, *National Electrical Code*.

Venting and Air Supply

Exhaust venting is required to prevent buildup of products of combustion inside the building. Codes require that fuel-burning appliances be vented. Some equipment, however, does not require venting—including ranges, ovens, and small space heaters—because if they are used properly, they will not produce significant products of combustion; however, if these appliances are used improperly, they can allow products of combustion to build up enough to cause harm.

Fresh air supply or combustion air supply venting from the exterior of the building is often necessary for proper operation of gas-fired appliances. If the volume of the room is sufficient, sometimes the installation of combustion air is not required. Many new appliances are designed so that the combustion air directly enters the appliance. When appliance burners begin to operate inefficiently because of an improper fuel/air mixture, carbon may begin to accumulate within the combustion chamber and flue FIGURE 8-11. With time, these deposits can block the flow of combustion products through the flue or possibly start a fire.

Appliance Controls

Fuel gas appliances have controls that are generally categorized as the following:

- Temperature controls
- Ignition and shutoff devices
- Gas appliance pressure regulators
- Gas flow control accessories

FIGURE 8-11 Carbon deposits inside an improperly operating water heater burner.

Fuel Gas Utilization Equipment

NFPA 921, *Guide for Fire and Explosion Investigations,* provides a list of common fuel gas utilization equipment and a description of that equipment TABLE 8-7.

The investigator may encounter other types of gas utilization equipment, but all have similar features in their requirement to regulate the gas, control the flow of the gas, and burn the gas. The most common forms of gas utilization equipment include air heating, water heating, cooking, engines, illumination, and incinerators.

Several different types of appliances use fuel gas in their operation. These appliances include water heaters, furnaces, clothes dryers, and ranges FIGURE 8-12. The investigator should

Table 8-7 Common Fuel Gas Utilization Equipment

Equipment	Description
Air heating	Boilers, forced air furnaces, space heaters, floor furnaces, radiant heaters, duct furnaces, clothing dryers
Water heating	Direct-flame burners for the heating of potable or industrial-process water
Cooking	Ranges, stove-top burners, broilers, and cooking ovens
Refrigeration and cooling	Absorption system refrigeration and cooling systems
Engines	Stationary and motor vehicle engines and auxiliary power units on service vehicles (such as pumps on tank trucks or electrical power generation)
Illumination	Gas lamps used for outdoor lighting; residential examples include driveways, patios, porches, and swimming pools; commercial examples include streets, shopping centers, and hotels
Incinerators, toilets, exhaust afterburners	Burn rubbish, refuse, garbage, animal solids, organic waste, and industrial-process waste

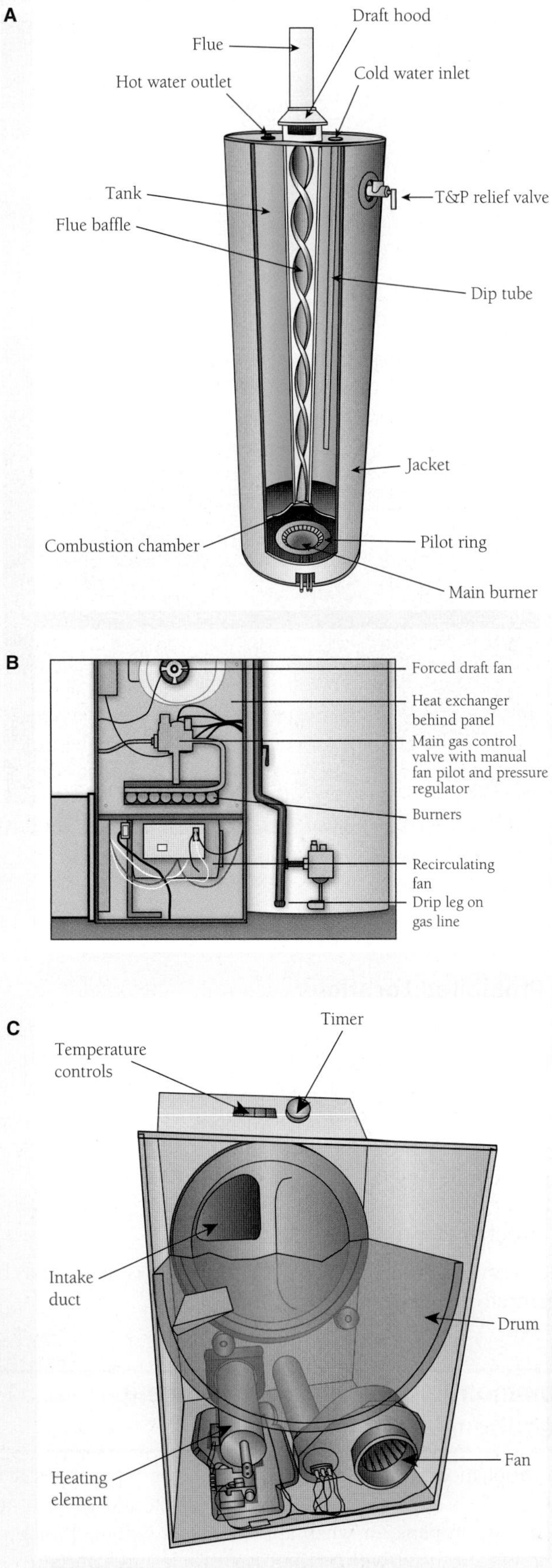

A
Draft hood
Flue
Hot water outlet
Cold water inlet
Tank
T&P relief valve
Flue baffle
Dip tube
Jacket
Combustion chamber
Pilot ring
Main burner

B
Forced draft fan
Heat exchanger behind panel
Main gas control valve with manual fan pilot and pressure regulator
Burners
Recirculating fan
Drip leg on gas line

C
Timer
Temperature controls
Intake duct
Drum
Fan
Heating element

FIGURE 8-12 **A.** Water heater components. **B.** Furnace components. **C.** Clothes dryer components.

be familiar with the design features and operation of these four common fuel gas appliances (also see the Appliances chapter in this textbook).

Investigating Fuel Gas Systems

Analysis of the fuel gas system can provide information if it was involved in the origin or cause of the explosion or fire. As with all parts of a fire investigation, the analysis of the fuel gas system and each component of that system should be done in a systematic manner to ensure thoroughness. This systematic examination should include the proper documentation of the fuel gas system—measuring and diagramming the system, and identifying and measuring the piping and distribution system materials, including valves, couplings, or connectors. This documentation should include detailed information regarding the system from the appliance back to and including the fuel gas source. Notations should be made of the position of valves or other controls (opened or closed). The flow and pressure rate should be measured or documented as appropriate, and any faults in the system such as breaks, cracks, or holes should be noted and documented. The goal is to determine whether, and to what extent, it operated or failed. NFPA 921 provides guidance in conducting the investigation. The investigator should interview all witnesses, property owners, fire fighters, and so forth to determine what condition and activities occurred prior to the fire or explosion that may have involved the fuel gas system of the building. Certain activities such as moving equipment, servicing appliances, and the installation of new fixtures may be factors that the investigator should consider.

Compliance with Codes and Standards

The investigator should evaluate the design, manufacture, construction, and installation of the equipment and system to determine compliance with accepted codes and standards, calling on additional expertise or resources if necessary. There are differences in installation requirements for the various types of gas systems. The design and construction of the appliance itself are often regulated by an approval agency such as the American Gas Association. The installation manual for the appliance often provides information as to its installation requirements.

Leakage Causes

The main causes of gas-fueled fires and explosions involve leakage in the fuel gas delivery system or in the appliance itself. The most common leakage occurs at pipe junctions, unlit pilot lights or burners, uncapped pipes, malfunctioning appliances and controls, areas of corrosion in pipes, and points of physical pipe damage FIGURE 8-13.

FIGURE 8-13 This water heater was damaged by a fire that initiated with leakage and then involved combustibles stored next to the appliance.

Pipe Junctions

Leaks can occur in pipe junctions if the threading is inadequate for a good connection, if the junction between the pipe and coupling is improperly joined or threaded, or if there is an improper use of pipe joint compound.

Unlit Pilot Lights or Burners

Most modern gas utilization equipment stops the flow of the gas to the burners if the pilot light or the burner is not lit. It is possible to have gas leakage in these systems if the sensing system malfunctions or if the automatic shutoff valve does not operate. The amount of gas that flows from an unlit pilot light is generally not sufficient to cause an explosion or fuel fire unless the gas is somehow confined to a space or leaks for an extended period of time. Burners that have been turned on but not lit can provide sufficient gas leakage to fuel an explosion or fire. Some equipment contains a safety device that monitors the flame from the burner, and if there is no flame, it shuts off the flow of gas. This equipment can stop the flow of gas to an unlit burner; however, if there is a failure of the sensing device or a gas valve fails, gas will continue to flow. It is common to see such burners without this sensing equipment, especially in cooking appliances.

Uncapped Pipes and Outlets

Appliances may have been disconnected from the fuel gas line without a cap on the end of the gas pipe and outlets. If the shutoff valve is open, gas escapes from the uncapped pipe.

Fire Investigator Tip

Investigators should not overlook that misuse of the equipment or improper storage of combustible items near these pieces of equipment may also cause a fire.

Malfunctioning Appliances and Controls

Fugitive gas may leak around the controls, valves, fittings, and pipe junctions within the appliance control FIGURE 8-14. For example, dirt or debris could cause valves to allow gas to pass through when they should be closed. Additional expertise or resources may be needed to analyze the equipment.

Gas Pressure Regulators

A failure can occur within the internal diaphragm of a gas regulator, which would allow a leak from the regulator and potentially allow excess pressure to the system. Failures can also occur on the rubber-like seals that control the input of the gas, which would allow a leak around the regulator and potentially allow excess pressure into the system. The vents to the regulator can become plugged or fail, allowing the system to become overpressurized.

Corrosion

Statistics indicate that corrosion may cause as many as 30 percent of all known gas leaks. Corrosion can take place above or below ground. Generally, this type of failure takes a long time to develop. Corrosion can typically be caused by rust, electrolysis, or microbiological organisms. Flexible brass appliance connectors often crack from corrosion and stress and are a factor in many fires and explosions.

Physical Damage to the System

The investigator should keep in mind that the point of the leak may be remote from the actual damage. The strain may appear at pipe junctions or unions. Damage often occurs at the threaded portions of the pipe. Hidden or underground pipes are often damaged by construction FIGURE 8-15.

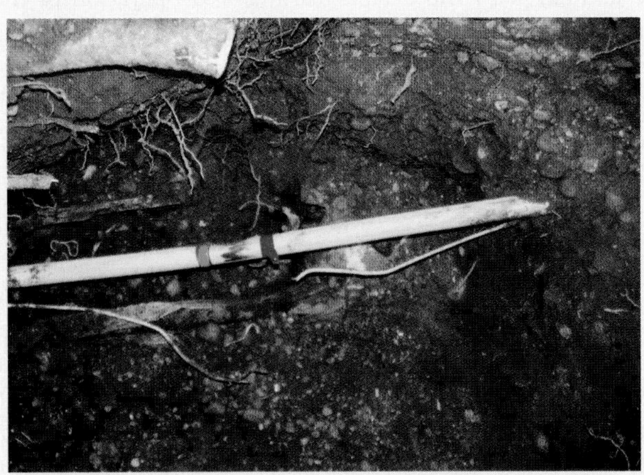

FIGURE 8-15 A gas line that has passed through a broken waste line. The auger of a "drain snake" that cut the plastic yellow gas line is visible behind the red tape.

If the investigator suspects that fuel gas may be a factor in the fire investigation, he or she should test for leaks of the gas supply system prior to its disassembly. Expertise in conducting such evaluations is needed.

Pressure Testing

It is recommended that the gas supply system be pressurized, with the pressure not exceeding the pressure contained within that system. Sometimes the pressure may be as little as 1- to 4-in. w.c. (248.8 to 996.4 kPa). Damaged sections of the piping should be isolated so that the investigator can test undamaged sections. Care should be taken not to unscrew or disassemble components before testing, as evidence can be inadvertently destroyed at the point where the leak occurred. The investigator should attempt to isolate the piping for testing in an area where the leak did not occur and where the proper fittings can be attached.

Gas Meter Testing

Testing can also be done with a gas meter, but only if it is safe to reintroduce gas into the system. The meter can detect the gas escaping from the system FIGURE 8-16.

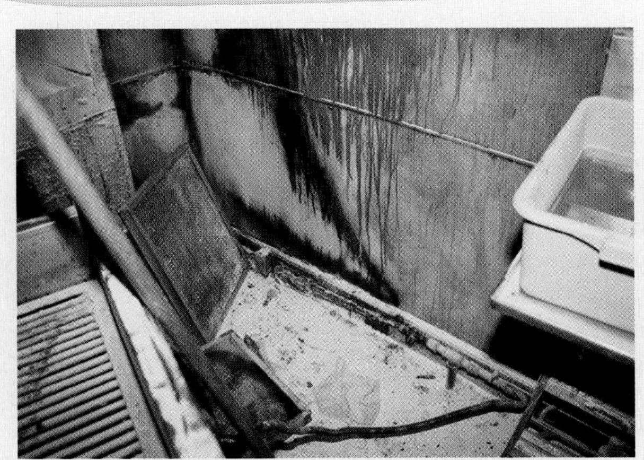

FIGURE 8-14 Photograph of a natural gas line that ignited behind a commercial fryer that was not properly connected.

FIGURE 8-16 Gas regulator test.

Pressure Drop Method

To check gas piping, the system is pressurized with air or an inert gas. This type of test is appropriate for operating systems of 0.5 psi (3.4 kPa) or less. The system is pressurized, the air supply capped, and the pressure monitored by a pressure gauge.

Locating Leaks

There are a number of methods used to locate leaks in fuel gas piping—for example, the soap bubble test or the use of a gas detector survey.

Soap Bubble Solution

The soap bubble solution is a simple method to locate leaks in a pressurized system and entails the application of soap to the piping in question. If bubbling is observed, then there is a leak at that location FIGURE 8-17.

Gas Detector Surveys

Survey devices such as flammable gas indicators, combustible gas indicators, explosion meters, or "sniffers" may be used to detect the presence of gas, hydrocarbon gas, or vapors.

A combustible gas indicator is used in gas detector surveys and may also detect the presence of other gases. When using this method on the exterior of the building, the investigator should test for fugitive gases at every possible ground opening, including all drains or pipes that exit the ground. In gas detector surveys performed on the interior, the investigator should test junction boxes and unions of gas piping.

If the investigator suspects an underground gas leak, an appropriate surveying procedure is the use of bar holes. After a series of underground holes is drilled in the vicinity of the underground piping, readings are taken at each hole and are graphed or charted. This type of survey can potentially lead to the area of the leak. The utility company uses an electronic locator device to locate the underground gas lines.

Exterior surveys also include vegetation surveys. Vegetation situated in the vicinity of a long-term gas leak will die, turn brown, or be stunted.

Testing Flow Rates and Pressures

Any time a fuel gas system is suspected to be involved in the ignition of a fire, components of the system that survive should be tested to determine whether they were operating properly. Testing may be conducted either in the field or at a laboratory to analyze regulator operating pressures and safety devices and to locate leaks within a system. As previously mentioned, testing should be conducted only by those trained and knowledgeable to conduct these tests, and if a flammable gas is being tested, all potential ignition sources should be removed.

If an investigator determines that section(s) of gas piping should be collected, care should be taken to ensure the evidentiary value of the components. It is preferred that the section of interest be removed from the system by cutting the pipes outside the area to be isolated FIGURE 8-18. Screw junctions, unions, tees, or elbows should not be unscrewed because unscrewing them may inadvertently damage or destroy evidence of a loose or improper connection. When removing the section of interest, the orientation and configuration of the isolated area related to the overall system is preserved.

Underground Migration of Fuel Gases

It is possible for gases to migrate great distances underground before entering a structure or exiting the ground. The gas can enter underground gas lines, underground sewer lines, underground electrical and telephone conduits, or underground drain tiles. Gas can enter structures by migrating through cracks or holes in foundations. Weather changes (such as frozen ground) can cause the gas to be routed in a different direction or to migrate underground for greater distances.

When gas passes through a filter medium such as soil, the medium can "scrub" the odorant out of the gas. The gas then becomes odorless and might not be detected. NFPA 54 requires that gas meters be installed at least 3 feet (0.9 meters) from sources of ignition and protected from damage. Odorant

FIGURE 8-17 Soap bubble solution test.

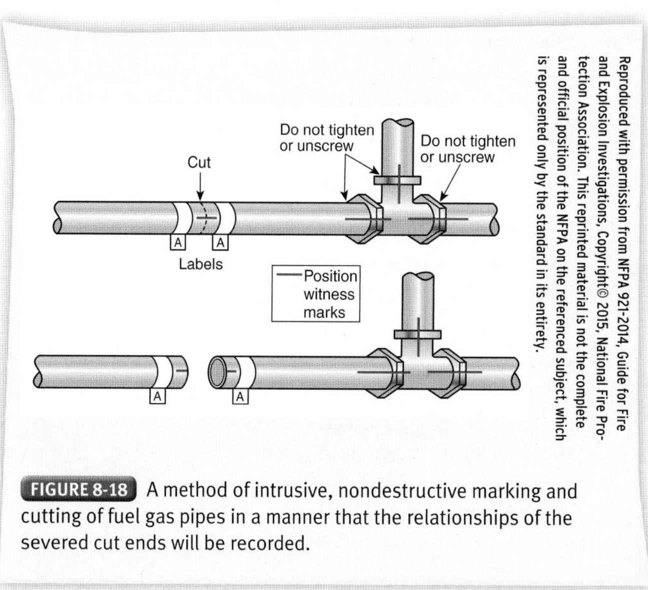

Reproduced with permission from NFPA 921-2014, Guide for Fire and Explosion Investigations, Copyright© 2015, National Fire Protection Association. This reprinted material is not the complete and official position of the NFPA on the referenced subject, which is represented only by the standard in its entirety.

FIGURE 8-18 A method of intrusive, nondestructive marking and cutting of fuel gas pipes in a manner that the relationships of the severed cut ends will be recorded.

can also be lost through adsorption into the piping or container system. This can occur with both new steel or plastic piping or containers or components that have been in use for an extended period of time. Loss of odorant can also occur as a result of reduced flow rates during seasonal changes as usage levels decrease, or when mercaptan odorants oxidize as a result of contact with ferrous-metal containers. Odorant may also be trapped in water or other liquids that may collect within the container or piping system as a result of contamination or condensation.

Wrap-Up

© Greg Henry/ShutterStock, Inc.

Ready for Review

- During every fire scene examination, the presence and condition of fuel gas systems need to be examined and documented.
- Fuel gas systems will likely be either natural gas or propane.
- An understanding of each fuel gas's properties will allow the investigator to interpret fire and damage patterns accurately.
- The systems that transport various fuel gases are similar in that they all pipe the gas directly into consumers' buildings and appliances.
- Natural gas is supplied through a transmission pipeline, then through main pipelines (mains), and then through service mains or service laterals before it is connected to the meter at the consumer's location.
- LP gas distribution systems are similar in operation to natural gas systems except that the LP gas storage supply is often located at the consumer's site.
- The fuel gas systems of all buildings will contain various distribution piping, valves, burners, and regulating devices, which are designed, installed, and used to deliver the fuel gas to the appliances safely.
- It is important to maintain the evidentiary value of piping and its components when conducting an investigation.
- When appliances are designed for natural gas but are used with propane, or are designed for propane but are used with natural gas, the results can be catastrophic, including fire or explosion.
- The most common forms of gas utilization equipment include air heating, water heating, cooking, engines, illumination, and incinerators.
- As with all parts of a fire investigation, the analysis of the fuel gas system and each component of that system should be done in a systematic manner to ensure thoroughness.

Hot Terms

Backflow valve Prevents gas from re-entering a container or distribution system.

Bar holes Holes driven into the surface of the ground or pavement with either weighted metal bars or drills. Gas detectors are then inserted into the holes in order to detect the location of a gas leak.

Combustible gas indicator An instrument that samples air and indicates whether combustible vapors are present. Some units may indicate the percentage of the lower explosive limit of the air/gas mixture.

Commercial propane Derived from the refining of petroleum, a liquefied gas comprising 95 percent propane and propylene and 5 percent other gases.

Container appurtenances The devices that are connected to the openings in tanks and other containers—such as pressure relief devices, control valves, and gauges.

Fixed level gauge Primarily used to indicate when the filling of a tank or cylinder has reached its maximum allowable fill volume. They do not indicate liquid levels above or below their fixed lengths.

Fuel gas Includes natural gas, liquefied petroleum gas in the vapor phase only, liquefied petroleum gas–air mixtures, manufactured gases, and mixtures of these gases, plus gas–air mixtures within the flammable range, with the fuel gas or the flammable component of a mixture being a commercially distributed product.

Fugitive gas Fuel gases that escape from their piping, storage, or utilization systems and serve as easily ignited fuels for fires and explosions.

Fusible plugs Thermally activated devices that open and vent the contents of a container.

Gas burner A device that allows fuel gases and air to properly mix and produce a flame.

Liquefied petroleum (LP gas) (propane) Petroleum gases condensed to a liquid state with moderate pressure and normal temperatures to allow for more efficient distribution.

Lower explosive limit (LEL) The lowest concentration of fuel in a specified oxidant in which combustion can occur; also known as the lower flammable limit (LFL).

Natural gas A naturally occurring, largely hydrocarbon gas product recovered by drilling wells into underground pockets, often in association with crude petroleum.

Pressure gauge A type of container appurtenance depicting the internal pressure of a tank. The gauges are connected directly to the tank or sometimes through the valve. Pressure gauges do not indicate the quantity of liquid propane in the tank.

Pressure relief valve A valve designed to open at a specific pressure, usually around 250 psi (1700 kPa). It is generally placed in the container where it releases the vapor.

Shutoff valve A valve (located in the piping system and readily accessible and operable by the consumer) used to shut off individual appliances or equipment.

Vaporizers Heaters that are used to heat and vaporize propane where larger quantities of propane are required, such as for industrial application.

Variable gauge Gauge that gives readings of the liquid contents of containers, primarily tanks or large cylinders. It gives readings at virtually any level of liquid volume.

Venting The removal of combustion products as well as process fumes (e.g., flue gases) to the outer air.

© Greg Henry/ShutterStock, Inc.

FIRE INVESTIGATOR *in action*

You respond to a report of a small fire in a utility closet at an apartment complex. When you arrive, a tenant advises she was looking for something when she opened the utility closet and discovered a small flame on her natural gas line. When you inspect the utility closet, you observe a flame emanating from a corrugated stainless steel (CSST) gas line that provides natural gas to a furnace. Closer inspection of the CSST reveals it is in contact with the rigid, black pipe supply line.

While interviewing the tenant, she states the building was struck by lightning 4 days ago and portions of the electric service had been damaged in her unit; however, the utility closet was never inspected. She states she did not notice the flame or any odors of gas prior to opening the closet.

1. What is the simplest way to detect a leak in the fuel gas system?
   - **A.** Pressure testing
   - **B.** Soap bubble solution
   - **C.** Visual inspection
   - **D.** Gas detector survey
2. Which of the following may cause the fuel gas to escape?
   - **A.** Mechanical damage to the delivery system
   - **B.** Loose fittings and/or joints
   - **C.** Malfunctioning appliances or devices
   - **D.** All of the above
3. What is the common odorant used for propane gas?
   - **A.** Ethyl mercaptan
   - **B.** Butyl mercaptan
   - **C.** Nitrile mercaptan
   - **D.** Methyl mercaptan
4. Which of the following locations is not approved for natural gas installations?
   - **A.** Utility spaces
   - **B.** Out buildings
   - **C.** Basements
   - **D.** Clothes chutes

Fire-Related Human Behavior

© Photos.com

© Matthew J. Lee/The Boston Globe/Getty Images

Knowledge Objectives

After studying this chapter, you should be able to:

- Discuss the general considerations of human response at a fire incident. (pp 158–160)
- Identify factors related to fire initiation. (pp 161–162)
- Describe the three recognized age categories of youth firesetters and the reasons that they are drawn to setting fires. (p 162)
- Describe the ways in which an occupant may react once a threat is identified. (pp 162–163)

Skills Objectives

After studying this chapter, you should be able to:

- Analyze fire-related human behavior. (pp 158–163)
- Integrate human behavior into the total investigation. (pp 158–163)
- Reference standards, guidelines, and regulations dealing with safety warnings and product design. (pp 162–163)

Additional NFPA Reference

NFPA 921, *Guide for Fire and Explosion Investigations*

CHAPTER 9

FESHE Course Outcomes

Fire Investigation I

There are no Fire Investigation I (FESHE) course outcomes for this chapter.

Fire Investigation II

There are no Fire Investigation II (FESHE) course outcomes for this chapter.

You Are the Fire Investigator

© Jones and Bartlett Publishers. Photographed by Glen E. Ellman

Fire personnel have requested that you conduct an investigation at a local restaurant where three people have died and several others were injured. You conduct interviews with several of the patrons who were at the scene and experienced injuries due to smoke inhalation. They reported seeing four or five people in the kitchen area when a small fire occurred in a fryer. When you meet with the incident commander, he leads you to a cooler in the rear of the building where three deceased people were located. The incident commander states the victims were located after the fire was extinguished inside the cooler with the door closed.

1. Why would there be victims located within the cooler?
2. Why were there injuries when witnesses reported the fire as initially being small?
3. What individual and/or group characteristics would you consider about the victims?
4. How would the presence of fire and smoke detection and suppression systems impact your investigation?

Introduction

This chapter is based on Chapter 11, "Fire-Related Human Behavior," of NFPA 921, *Guide for Fire and Explosion Investigations*. Fire-related human behavior is a distinct field of study that emerged in the 1970s. A fire's origin, development, and consequences are all related—either directly or indirectly—to the actions and/or omissions of human beings. Understanding the behavior of witnesses or occupants is integral to the fire scene examination. By understanding why the occupants or witnesses behaved in a particular manner, the investigator can better evaluate the effects of a particular fire.

General Considerations of Human Response

Research conducted within the past several decades indicates that an individual's or group's behavior before, during, and after a fire can provide valuable insight into the role of the fire-related human behavior. Factors that affect this behavior include characteristics of the individual, characteristics of the group or population to which that individual belongs, characteristics of the physical setting where the fire occurs, and characteristics of the fire itself.

■ Characteristics of the Individual

An individual's actions are shaped by physiological factors, including his or her physical limitations, limitations of cognitive comprehension, and knowledge of the physical setting. These factors all affect an individual's ability to recognize and assess a threat and respond to it.

Physical Limitations

An individual's age (as it relates to mobility), physical disabilities, incapacitating or limiting injuries, medical conditions, and/or chemical impairment can adversely affect the ability of that person to take appropriate actions before and during a fire. The very young and very old are most susceptible to these limitations.

Cognitive Comprehension Limitations

Along with physical limitations, cognitive/mental limitations can affect an individual's ability to recognize and react appropriately to a fire or explosion. Some factors that can limit a person's cognitive ability are age (as it relates to mental comprehension, such as a child hiding instead of escaping), level of rest, alcohol use, drug use (legal or illegal), developmental disability, mental illness, and inhalation of smoke and toxic gases. Cognitive comprehension limitations can cause a significant delay in an individual's response time, thereby placing that person at increased risk for injury or death.

Familiarity with Physical Setting

An individual's familiarity with the setting can make escape more likely, although physical limitations and cognitive impairments can minimize the advantages of such familiarity, such as a person's becoming lost in his or her own home. In larger, unfamiliar structures, individuals tend to leave by the same route they took to enter—perhaps because of their lack of knowledge of their surroundings or perhaps because during extreme situations the brain fixates on the most recent method of entrance.

■ Characteristics of the Group or Population

An individual's response to a threat is affected not only by his or her own circumstances. When interacting with others, other significant factors are the size and structure of the group that the individual is with, the permanence of the group, and the group's roles and norms.

Group Size

An individual's response to a threat or purported threat is tempered by the size of the group with which the individual is associated. Smaller groups aid in the response time of the individuals associated with it. Larger groups have a tendency not to respond within appropriate time frames. This fact can be attributed to the individual's desire not to be first or to a disinclination to disrupt the workings of the group.

Group Structure

When a group has a formalized structure, individuals who perceive themselves as members but not in leadership or authoritarian positions delay their response until those who are in the leadership or authoritarian positions respond to the situation. Examples of this type of group are school, hospital, nursing home, and religious facility populations.

Group Permanence

Research has indicated that the degree of familiarity among the individuals in a group also affects response times. If the group is established and its members know one another well, the individuals react and notify each other in a more timely manner than they do if the group is newly formed or its members are unfamiliar with each other. Examples of groups with a high degree of permanence are families, sports teams, choirs, and clubs.

Roles and Norms

A group's roles and norms in terms of gender, social class, occupation, education, and so forth can also affect its response to threats. Gender roles may provide good examples. In a fire or explosion situation, studies have shown that males are more likely to engage in activities to suppress or defuse the threat, whereas females are more likely to engage in reporting of the threat.

© Jones & Bartlett Learning. Courtesy of MIEMSS

FIGURE 9-1 A fire alarm system can include lights and horns to notify occupants.

■ Characteristics of the Physical Setting

The characteristics of a burning structure will affect the response of the individuals within it. Some factors that affect response are the location of exits, the number of exits, the structure's height, and the operation of its fire warning systems and fire suppression systems.

Location of Exits

It is important for fire exits to be clearly marked. If occupants are unfamiliar with the building, they often react with increased levels of stress, which in turn can result in unpredictable behavior. Furthermore, as was mentioned previously, occupants who are unfamiliar with a building try to exit by the door that they used to enter, even if that behavior moves them in the direction of the threat.

Number of Exits

The number of exits available for escape drastically affects the occupants' behavior. If there are too few exit routes or if the exits are blocked or restricted, occupants are exposed to additional danger.

Height of the Structure

Some people believe that they are less safe in a tall building during a fire. This misconception can result in unpredictable behavior, such as jumping out of windows, during an emergency.

Fire Alarm Systems

Fire alarm systems can help occupants recognize a threat FIGURE 9-1. Research has also shown that voice or directive messages may generate better response than the use of lights and horns alone. Unfortunately, if there have been a number of false alarms, people tend to delay their response until the actual fire emergency is confirmed. This delay can result in the occupants being trapped and unable to escape.

Fire Suppression Systems

The presence of fire suppression systems can positively and negatively affect the response of the occupants FIGURE 9-2. With fire suppression systems in place, a positive effect is that

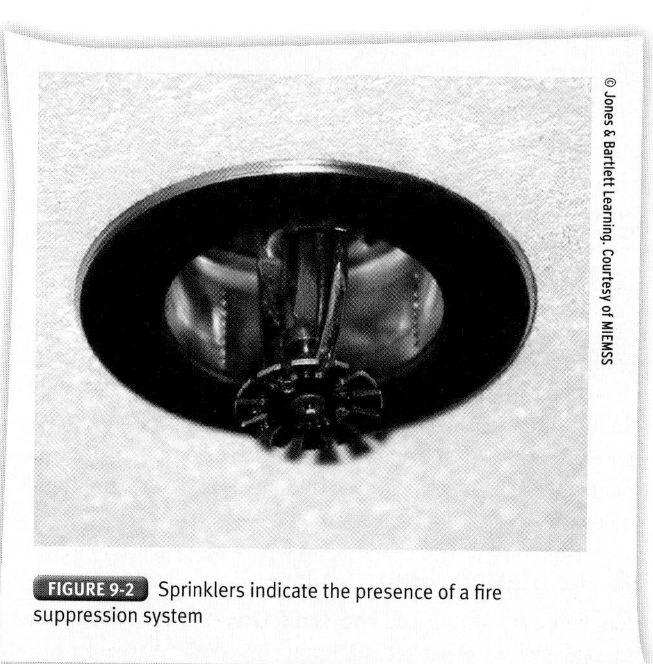
© Jones & Bartlett Learning. Courtesy of MIEMSS

FIGURE 9-2 Sprinklers indicate the presence of a fire suppression system

occupants traditionally have more time to react to a threat. Unfortunately, the amount of time can become exaggerated in the minds of the occupants to the extent that they remain in the structure far too long. A negative effect of discharged fire suppression systems is that they could reduce the visibility of occupants trying to escape.

Characteristics of the Fire

An individual's or group's response to a threat is tempered by their perception of the hazard or threat. Visual and olfactory indicators shape responses. Characteristics of a fire that can influence people's behavior include the presence of flames, the presence of smoke, and the effects of toxic gases and oxygen depletion.

Presence of Flames

Most individuals do not understand the threat presented by flames. They have limited exposure to fire and incorrectly may believe that the presence of small flames is not a hazard. They do not understand that the fire can grow exponentially and will release toxic gases and products into the air. Because they do not understand these principles, individuals tend to disregard small flames as a source of danger.

Presence of Smoke

The presence of smoke in a structure is also misunderstood. There is a general lack of knowledge about the toxic content of smoke and its incapacitating effects on humans. There is also a misperception that lighter-colored smoke is not as dangerous as darker-colored smoke, when in reality both have the potential to be fatal.

Effects of Toxic Gases and Oxygen Depletion

Many people do not understand fire behavior. As a result, they do not understand that fire consumes oxygen during the burning process. In an enclosed space, this fact equates to diminishment of the available oxygen. Depletion of oxygen below 15 percent results in impairment of motor and mental skills and can be fatal to occupants. In addition, the combustion process releases toxic by-products that, when inhaled, adversely affect the occupant and can also result in death. (For more information on fire behavior, see the "Basic Fire Science" chapter of this text.)

Examples of Human Behavior Contributing to Fire Fatalities

Over the course of history, many catastrophic fire events have resulted in significant loss of life. Although there are usually many factors that contribute to the scale of the fire event—including arson, inappropriate use of heat-producing devices, and uncorrected code violations—the response of the occupants also plays a factor. Several examples are highlighted below.

The Station Nightclub Fire

On February 20, 2003, the band Great White was engaged to play before a packed nightclub in West Warwick, Rhode Island. At approximately 11:07 pm, the band began to play its opening song, *Desert Moon*. As part of the performance, the band set off three pyrotechnic "gerbs," which shot fountains of sparks against the walls of the stage area. The ignition of foam material on the walls of the stage area led to a fast-developing fire that took the lives of 100 people. That night, a TV news crew was filming the event to produce a segment on nightclub safety. Ironically, their video captured the tragic event that is one of America's deadliest fires.

Watching the video footage, viewers are offered a horrific accounting of fire dynamics and human behavior. Spectators and even band members failed to see the wall covering ignite or realize its danger. First recognition of the fire by patrons occurred 24 seconds after ignition, with most patrons beginning to evacuate the building 30 seconds after ignition when the band stopped playing. Some of the patrons had been drinking alcoholic beverages. The majority of the patrons attempted to exit the nightclub through the main entrance—the entry point they were familiar with—even though there were three other exits for the structure.

The resulting mass exodus through the main doors led to congestion of people. The doorway soon filled with a tangled mass of victims. In the end, 96 people died in the fire, and another 4 died in local hospitals. An additional 230 people were injured, with only 132 escaping unharmed.

MGM Grand Hotel Casino Fire

On November 21, 1980, a fire broke out in a vacant restaurant in the MGM Grand Hotel and Casino in Las Vegas, Nevada, resulting in the death of 85 people and the injury of 650 others. The fire originated in a wall soffit and is suspected of being caused by an electrical fault that occurred after the installation of copper refrigeration lines for a pastry display case in the same concealed space. Vibrations from an improperly secured evaporator fan sent vibrations through the length of the refrigeration lines installed in direct contact with the electrical conduit. Over time, these vibrations resulted in faults, leading to this catastrophic fire.

The fire spread was aided by wallpaper, PVC piping, and plastic mirrors, and eventually traveled through the lobby, sending a fireball out the front entrance along the strip. The fire progression was compounded by the fact that the fire originated in an area of the casino/hotel that had received a sprinkler exemption. Most of the fatalities occurred as a result of toxic fumes that were circulated through the hotel's heating, ventilation, and air conditioning (HVAC) system as a result of faulty dampers. A significant number of the fatalities occurred in the upper floors of the hotel as a result of smoke inhalation and carbon monoxide poisoning, and many of these occurred in the stairwells that served as chimneys transporting the toxic smoke and gases to the upper areas of the structure. Once in the stairwells, the occupants became trapped as the doors behind them locked and the only open exits were on the ground floor and roof.

A review by the National Fire Protection Association (NFPA) indicated that the hotel occupants did not panic and many took rational steps to preserve their lives, including warning other occupants, placing towels at the base of their doors, and covering their faces with wet cloths.

Several months later, on February 10, 1981, an arson fire occurred at the Las Vegas Hilton. Using the knowledge they gained from the investigation of the MGM Grand Fire, fire fighters used local television networks to notify people to stay in their rooms and out of the hallways and stairwells. As a result, only eight people perished in this fire.

Kiss Night Club Fire

On January 27, 2013, a fire caused by the inappropriate use of an outdoor pyrotechnic device inside the Kiss Night Club in Santa Maria, Rio Grande do Sul, Brazil, resulted in the death of 242 people and the injury of 168 others. Much like the Station Night Club fire, a pyrotechnic device ignited acoustic foam lining the ceiling of the stage area. Following the breakout of the fire, a stampede occurred as people were trying to escape. These efforts were significantly hampered by the lack of exit signs and emergency exits. Most of the victims succumbed to smoke inhalation, and many people died as they either attempted to hide in bathrooms or mistook them as exits. More than 150 were injured by the crush at the front door and the rapidly accumulating smoke within the nightclub.

Other factors contributing to the significance of this event include the furnishing of false information regarding the number of emergency exits during the permit issuing process and the lack of or false information provided regarding functioning fire extinguishers.

Following this catastrophic event, Brazilian authorities conducted safety inspections of similar night clubs, resulting in the closure of 58 facilities for safety concerns.

Factors Related to Fire Initiation

Fire and explosion incidents frequently occur as a result of an act or omission by one or more individuals that could have occurred before or during the fire. These actions can encourage or prevent the spread of fire. The investigator will need to assess these actions, which may include the opening or closing of doors, operation of protection systems, and rescue efforts. Primary areas relating to fire initiation are improper maintenance and operation of equipment or appliances; careless housekeeping; failure to follow product labels, instructions, warnings, and recalls; and violations of fire safety codes and standards.

Improper Maintenance and Operation

During the lifetime of most equipment, there is a prescribed maintenance and cleaning schedule that is usually provided by the manufacturer and should be followed to prevent malfunction. When the equipment is capable of explosion or starting a fire, the required maintenance becomes critically important. Likewise, the operating procedures for equipment or appliances are designed to ensure safety. If either of these two areas is neglected or improperly performed, the lapse can lead to a fire or explosion. The investigator should examine all maintenance records and operating instructions carefully. If a fire can be traced to improper maintenance or operation, the person or persons responsible might not provide these records voluntarily.

Housekeeping

A majority of household equipment—specifically, equipment capable of initiating an explosion or fire—has instructions identifying clearance distances for combustibles. For example, paper products or liquid accelerants should not be stored adjacent to the pilot light of a water heater. Carelessly discarded smoking materials, such as cigarettes or matches, can ignite a fire. Grease buildup in cooking areas or improperly stored cleaning solutions are other hazards. The investigator should note any housekeeping irregularities that may have contributed to the fire or explosion.

Product Labels, Instructions, and Warnings

Labels, instructions, and warnings are placed on products to prevent their misuse or abuse FIGURE 9-3. Manufacturers' labels inform the user of the product's capabilities. Instructions provide the user with all pertinent information about how the product was intended to be used. Warnings alert the user about the dangers that can occur if the product is not used as intended and remind the user about the hazards of a product. A proper warning contains four key elements: an alert word to signal danger, a statement of the danger, a statement of how to avoid the danger, and an explanation of the consequences of the danger.

Alert words are used to draw the user's attention to the information. The words "Caution," "Warning," and "Danger" are common alert words.

A statement of the danger identifies the nature and extent of the danger and the gravity of the risk of injury. Warnings

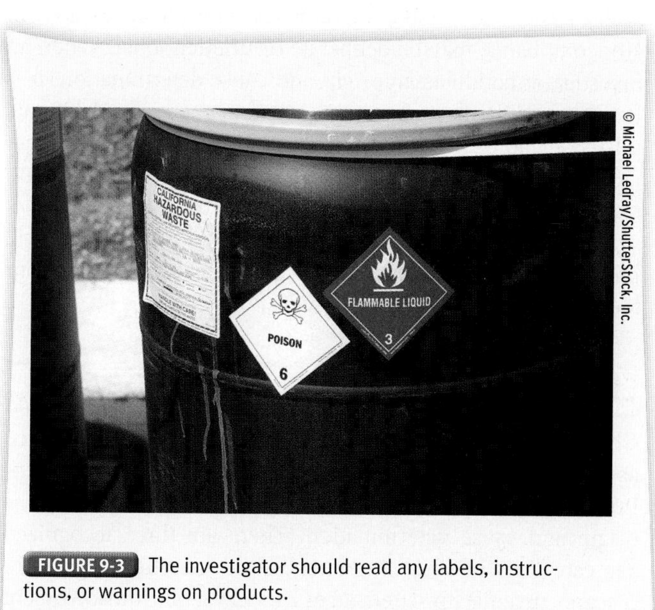

© Michael Ledray/ShutterStock, Inc.

FIGURE 9-3 The investigator should read any labels, instructions, or warnings on products.

Fire Investigator Tip

According to ANSI Z535.4, *Product Safety Signs and Labels*:

- CAUTION: Indicates a potentially hazardous situation that, if not avoided, **may** result in minor or moderate injury.
- WARNING: Indicates a potentially hazardous situation that, if not avoided, **could** result in death or serious injury.
- DANGER: Indicates an imminently hazardous situation that, if not avoided, **will** result in death or serious injury.

provide information on how to avoid the risk and consequences of the danger.

Government and industry have over the years developed accepted guidelines and standards for labels, warnings, and product instructions. Institutions such as American National Standards Institute (ANSI), Underwriters Laboratory (UL), FM Global, and various United States Codes and Regulations have addressed these requirements.

Recalls

Recall notices are a method of notifying consumers of a product defect that was identified after the product was released for consumption. For the most part, recalls are the result of identifying a dangerous situation that could arise even when the product is used in its intended manner. If an individual disregards a recall notice and continues to use the product, the result can be a fire, explosion, or other catastrophic event. Recall notices can be found through the Consumer Product Safety Commission.

Violations of Fire Safety Codes and Standards

The investigator must be aware that noncompliance with fire safety codes and standards can result in a fire or explosion. Noncompliance may be deliberate or unintentional. When an investigator conducts an origin and cause determination, it is frequently difficult to determine whether there was deliberate misuse or abuse of the product, carelessness, or some other factor that contributed to the event. The investigator should examine training records, maintenance records, and other documentation to determine whether there is a pattern that will point toward the potential cause of the event. Check with your local fire prevention bureau for violations issued.

Children and Fire

Children are drawn to setting fires for several reasons, such as curiosity, frustration, anger, revenge, or a need for attention. The investigator should be aware of this fact during the origin and cause determination. There are three recognized age categories of youth firesetters: child firesetters (ages 2 to 6 years), juvenile firesetters (ages 7 to 13 years), and adolescent firesetters (ages 14 to 16 years).

Fire Investigator Tip

Child firesetters: ages 2 to 6 years
Juvenile firesetters: ages 7 to 13 years
Adolescent firesetters: ages 14 to 16 years

Child Firesetters

Child firesetters are children in the 2- to 6-year-old age group who are traditionally curious and set fires in hidden locations, out of sight of adults. Children at this age are not usually able to form the intent of causing damage with fire.

Juvenile Firesetters

Fires set by juvenile firesetters (children aged 7 to 13 years) are typically a symptom or indicator of a psychological or emotional problem, often caused by a broken family environment or physical or emotional trauma or abuse. These fires are often set in and around the home or sometimes in an educational setting.

Adolescent Firesetters

Fires set by adolescent firesetters (teens aged 14 to 16 years) are usually symptomatic of stress, anxiety, anger, or another psychological or emotional problem. The most usual targets of these fires are schools, churches, vacant buildings, fields, and vacant lots. The fires are frequently associated with disruptive behavior, a broken home life, or a poor social environment. (See the "Incendiary Fires" chapter of this text for a discussion of the motives associated with such fires.)

Recognition and Response to Fires

In a fire situation, the occupant's ability to recognize the danger is critical to his or her survival. The occupant must be able to respond appropriately to the perceived and actual dangers associated with the event. Sensory perception can be affected by physical or mental state or chemical ingestion (alcohol, drugs).

The occupant's actions are based on the following four factors:

1. Sight: Direct view of flames, smoke and visual alarms, or flicker
2. Sound: Crackling of flames, failure of windows, audible alarms, a dog barking, children crying, voices, or shouts
3. Feel: Temperature rise or structural failure
4. Smell: Smoke odor

When the occupant identifies a threat (whether fire or explosion), the person must make a decision about how he or she is going to react. The occupant has several choices, including ignoring the problem, investigating, fighting the fire, signaling an alarm, rescuing or giving aid to others, reentering the structure (after successfully escaping), fleeing the fire, or remaining in place. The final decision will be affected by the occupant's state of mind.

The ability to escape is affected by the identifiability of escape routes, distance to the escape routes, fire conditions (such as smoke, heat, or flames), the presence of dead-end corridors, the presence of obstacles or people blocking the escape path, and the individual's physical disabilities or impairments.

Interviews with event survivors can provide the investigator with information that will be beneficial in determining how people actually behaved before and during the fire or explosion FIGURE 9-4.

Personal interviews can help to establish the following:

- Prefire conditions
- Fire and smoke development
- Fuel packages and their location and orientation
- Victims' activities before, during, and after discovery of the fire or explosion
- Actions taken by individuals that resulted in their survival (e.g., escaping or taking refuge)
- Decisions made by survivors and reasons for those decisions
- Critical fire events such as flashover, structural failure, window breakage, alarm sounding, first observation of smoke, first observation of flame, fire department arrival, and contact with others in the building

© Neil Barris/MLIVE.COM/Landov

FIGURE 9-4 Interviews can lead investigators to important information.

Wrap-Up

Ready for Review

- Investigators can better evaluate the effects of a particular fire if they understand why the occupants or witnesses behaved in a particular manner.
- Factors that affect behavior include characteristics of the individual, characteristics of the group or population to which that individual belongs, characteristics of the physical setting where the fire occurs, and characteristics of the fire itself.
- Fire and explosion incidents frequently occur as a result of an act or omission by one or more individuals that could have occurred before or during the fire.
- Children are drawn to setting fires for several reasons, such as curiosity, frustration, anger, revenge, or a need for attention. Age categories of youth firesetters are: child firesetters (ages 2 to 6 years), juvenile firesetters (ages 7 to 13 years), and adolescent firesetters (ages 14 to 16 years).
- In a fire situation, the occupant's ability to recognize the danger is critical to his or her survival. The occupant must be able to respond appropriately to the perceived and actual dangers associated with the event.

Hot Terms

Adolescent firesetters Adolescents (ages 14 to 16 years) who are often responsible for fires that occur at places other than their homes.

Child firesetters Children (ages 2 to 6 years) who are often responsible for fires in their homes or in the immediate area.

Juvenile firesetters Juveniles (ages 7 to 13 years) who are often responsible for fires that start in their homes or in the immediate environment.

Recall notices A method of notifying consumers of a product defect that was identified after the product was released for consumption.

Statement of the danger A warning that identifies the nature and extent of the danger of a product and the gravity of the risk of injury.

© Greg Henry/ShutterStock, Inc.

FIRE INVESTIGATOR *in action*

While conducting a fire investigation that resulted in the deaths of an elderly couple, you interview several neighbors who state that the husband had exited the home and reported that his wife was bedridden and could not escape. They tell you that after he attempted to extinguish the fire located on the rear exterior, he then re-entered the home through the front door in an attempt to rescue his wife.

The husband is found lying face-down on the floor in the dining room, which is located in the opposite direction of his wife's bedroom from the front door.

1. What factor played the biggest role in the inability of the wife to escape?
   A. Physical setting
   B. Physical limitations
   C. Presence of smoke
   D. Presence of flames
2. Below what percentage of oxygen will impairment of motor and mental skills begin?
   A. 21 percent
   B. 19 percent
   C. 15 percent
   D. 12 percent
3. What likely caused the husband to travel in the opposite direction from where his wife was located?
   A. His age
   B. Visible flames in the dining room
   C. Toxic gases present inside
   D. Unfamiliarity with the physical setting
4. You have determined this fire was the result of a 6-year-old child who lives next door who had ignited leaves and sticks in a bucket next to the rear exterior of the elderly couple's home. In which group of firesetters would the child be classified?
   A. Child firesetters
   B. Juvenile firesetters
   C. Pre-teen firesetters
   D. Adolescent firesetters

Legal Considerations

© Photos.com

© ZUMA Press, Inc/Alamy

Knowledge Objectives

After studying this chapter, you should be able to:

- Identify legal considerations for fire and explosion investigations NFPA 4.1.5 NFPA 4.2.1 NFPA 4.6. (pp 168–173)
- Describe the role of an investigator as a trial witness NFPA 4.1.4 NFPA 4.1.5 NFPA 4.7 NFPA 4.7.1 NFPA 4.7.2 NFPA 4.7.3. (pp 173–175)
- Identify and describe the types of evidence in an investigation NFPA 4.1.5 NFPA 4.2.1 NFPA 4.6 NFPA 4.6.1 NFPA 4.7.3. (pp 175–180)
- Discuss arson, its forms, and its legal repercussions NFPA 4.1.5 NFPA 4.6 NFPA 4.7.2 NFPA 4.7.3. (pp 180–181)
- Identify fire-related criminal acts NFPA 4.1.5 NFPA 4.6 NFPA 4.7.2 NFPA 4.7.3. (p 181)
- Describe the role of civil litigation in fire investigation cases NFPA 4.1.5 NFPA 4.7.2 NFPA 4.7.3. (pp 181–183)

Skills Objectives

After studying this chapter, you should be able to:

- Prepare and revise your curriculum vitae so that it is suitable for your use as an expert witness. (pp 177–180)
- Demonstrate effective techniques for delivering testimony NFPA 4.7.3. (pp 177–180)

Additional NFPA Reference

NFPA 921, *Guide for Fire and Explosion Investigations*

CHAPTER 10

FESHE Course Outcomes

Fire Investigation I

2. Describe the implications of constitutional amendments as they apply to fire investigations. (pp 168–172, 175)
3. Identify key case law decisions that have affected fire investigations. (pp 170–171, 178–179)

Fire Investigation II

1. Explain the rule of law as it pertains to arrest, search, and seizure. (pp 169–172)
12. Explain the role of the fire investigator in courtroom demeanor and testifying. (pp 173–175)

You Are the Fire Investigator

© Jones and Bartlett Publishers. Photographed by Glen E. Ellman.

The day after you conducted an investigation at a single-family home, you are contacted by a friend of the homeowners who believes the fire was purposely started because the homeowners were unable to continue making mortgage payments. This person states the homeowners had made several comments in the past about "selling the home back" to the insurance company and had conducted research online about how to start a fire. You and your partner return to the fire scene the following day to retrieve the home computer, but the residence has already been boarded up and is now secure. Your partner begins to remove one of the sheets of plywood being used to secure the residence when the homeowners arrive and question why you are there.

1. What are the possible legal issues that affect your right to re-enter the residence?
2. How will rules related to evidence affect your investigation?
3. While interviewing the homeowner, what legal rights must you be aware of?
4. How might your investigation impact both criminal and civil actions against the homeowner?

Introduction

Most qualified fire investigators receive training based on NFPA 1033, *Standard for Professional Qualifications for Fire Investigator*, and NFPA 921, *Guide for Fire and Explosion Investigations*. Nevertheless, many are surprised to learn the potential impact of these documents when used in litigation.

NFPA 1033 requires investigators to have basic knowledge on a wide variety of legal topics. It also requires investigators to develop skills relating to report writing and testifying for legal proceedings. NFPA 1033 is the national professional qualification standard for anyone wanting to serve as a fire investigator. It sets forth the minimum job performance requirements (JPRs) to be a qualified fire investigator in either the public or private sector. As such, it serves as a benchmark for all fire investigators. It does this in four ways:

1. By specifying 16 broad-ranging topics about which investigators must have and maintain a basic knowledge beyond the high school level
2. By identifying seven main responsibilities of fire investigators
3. By pinpointing the job performance requirements that encompass the job of a fire investigator
4. By listing the knowledge and skills necessary to do each task involved in an investigator's job

NFPA 921 and NFPA 1033 are closely related, and it is useful to read them together. NFPA 1033 specifically cross-references NFPA 921 as a source of the information NFPA 1033 requires investigators to have on the list of 16 topics that form the foundation of an investigator's knowledge base. Further, NFPA 1033 recommends that investigators consult NFPA 921 for the investigator's basic methodology and procedural matters, which are relevant to the skills listed in NFPA 1033 as mandatory for fire investigators. Although NFPA 921 is defined as a "guide" and NFPA 1033 is defined as a "standard," it is NFPA 921 that is most frequently cited in court cases as a source used to evaluate the qualifications of investigators as expert witnesses and to assess the reliability of their opinions. NFPA 921 also has a Legal Considerations chapter touching on most of the legal topics listed in NFPA 1033.

Legal Considerations During Investigation

All aspects of the investigation can be scrutinized during legal proceedings. The legal proceedings may involve criminal prosecution or civil actions in tort or contract. Laws are in a constant state of flux and can vary by jurisdiction. The investigator is advised to be familiar with applicable laws or seek guidance from local counsel. Adhering to the appropriate legal requirements will help ensure that the investigator's testimony is admissible and reliable.

Authority to Conduct the Investigation

The legal authority to conduct a fire scene investigation can be granted by law or contract. For law enforcement and public sector investigations, the authority is conferred by statute, ordinance, relevant case law, or agency rule. The scope and limitations on the power of the investigator are contained in those laws and the court decisions that construe those laws. There can also be constitutional limitations on public sector investigators' authority under applicable state or federal law.

In the private sector, the investigator's authority is a contractual right that is derived from the person or organization authorizing the investigation. One example is where a property owner or the attorney, acting as agent for the owner, hires an investigator and defines the scope and purpose of the investigation.

Sometimes the property owner is not the one directly responsible for instigating an investigation. In cases in which the owner has property or liability insurance respecting the property where the fire occurred, the owner's insurance company is usually the one retaining and instructing investigators.

In this situation, the property owner is said to be a third party to the investigation. In such cases, additional legal considerations can be involved. Authority to investigate insurance claims is granted under the terms of the insurance policy, which generally requires that a claimant allow the insurance company to conduct a fire scene investigation before the claimant is eligible to receive payment for the loss (in the case of property insurance) or a defense to liability claims (in the case of liability insurance).

Every state and the District of Columbia have adopted some form of immunity reporting act that requires insurance companies to report to local law enforcement authorities fires that may have been the result of criminal acts. These statutes are known as *arson-reporting/immunity acts*. Most of the statutes are based on the model act created by the National Association of Insurance Commissioners. These acts are intended to facilitate the exchange of information between law enforcement authorities and insurance companies in the investigation of arson fires. These arson-reporting/immunity acts require insurers to report to specified law enforcement agencies any loss due to an incendiary fire. Some acts also permit insurers to report fires of a suspicious origin. Note that NFPA 921 does not recognize a suspicious origin as an accurate or acceptable level of proof for fire cause determinations; however, this is the language used by some of these statutes. A designated public agency can make a written request listing the types of information and documents it requires the insurance company to release. This can include information about the subject fire and may even include information about the insurance policy or the insured's claim history. In most jurisdictions, the acts require that this information remain confidential until it is required for use in civil or criminal litigation. In exchange for providing this information, the insurance companies are granted statutory immunity from civil or criminal liability.

In addition to the requirement for providing information to public agencies, many of the immunity reporting acts contain reciprocal provisions authorizing the exchange of information from those agencies to the insurance companies to assist in their civil investigations. Several of the acts also authorize the exchange of information directly between insurance companies.

Licensing Requirements

In many states, non-public sector and law enforcement fire investigators must be licensed as private investigators or private fire investigators to conduct investigations in the state. This may require successful completion of a written exam and payment of a fee. Some of the statutes have exemptions, for example, for employees of insurance companies or for insurance adjusters investigating insurance claims. Other statutes provide reciprocity with some states. If you plan to conduct investigations in the private sector, check the legal requirements of each state where you will conduct investigations to ensure that you are complying with the legal requirements.

Also, be aware that states regulate the licensing of professional engineers. If you are an engineer planning to conduct origin and cause investigations, you should check the relevant statutes in each state where you intend to work to see how the state defines the respective roles of investigator and engineer and which licensing requirements you must obtain.

Right of Entry

Even if an investigator has the legal authority to conduct a fire or explosion investigation, a right of entry must also exist. This right can be granted by consent. In the case of public sector investigations, the investigator must be legally authorized to enter the property that is the scene of the fire to examine it or to gather evidence.

In other words, although a statute may require that any fire in which a person was injured be investigated, this does not give the public investigator who is authorized by this statute the automatic right to enter the scene. Constitutional protections might require a search warrant to be issued before the investigator has a legal right of entry. Understanding your state and local laws and statutes and consulting with your state and local prosecutors will provide the proper understanding of your right of entry and ability to conduct an investigation and to collect pertinent evidence.

In the private sector, an attorney for a tenant in a burned structure might hire an investigator to conduct a scene investigation of the building; however, that investigator might need consent of the public authorities who have control of the scene, or of the owner of the property, before entering to investigate.

An investigator who enters a scene to conduct an investigation or gather evidence without legal authority or the necessary consent might face criminal charges or an action for civil liability, such as trespassing. Another possible repercussion is that the evidence obtained from an illegal entry onto the scene might not be admissible as evidence at trial.

During both the public sector (government) investigation and the insurance investigation, the security of the scene and the preservation of fire patterns and potential evidence are of great importance. Government investigators should also make an effort to communicate with the party who takes custody of the fire scene after the suppression of the fire. The responsible party should be advised of the need to preserve the scene pending further investigation by other interested parties.

Public Sector Investigator Right of Entry

The authority to investigate a fire does not necessarily grant the investigator the right of entry at the fire scene. Public sector/law enforcement investigators are bound by constitutional limitations under search and seizure law. A fire scene investigation is considered a search and seizure, just as with any other crime scene investigation.

Fire Investigator Tip

A public sector investigator should know and meet the prosecuting attorney in his or her jurisdiction who handles fire cases and should create a protocol for when the prosecutor should be notified of a criminal fire investigation and the circumstances in which a search warrant should be obtained.

Under the Fourth Amendment of the U.S. Constitution and parallel provisions under state constitutional law, every entry onto a fire scene must be justified. There are four general circumstances that justify a search and seizure to conduct a fire scene investigation:

1. Consent
2. Exigent circumstances
3. An administrative search warrant
4. A criminal search warrant

Consent

A consent search is perhaps the most convenient method of entry to ensure compliance with the Fourth Amendment. Whenever there has been a delay in starting the fire scene investigation, consent to search should be obtained to remove all legal concerns about the timing and duration of the investigation; however, there are strict requirements for obtaining lawful consent to search. First and foremost, consent must be obtained from the proper party. Although the owner of the property is often the proper person to give consent, this is not always the case. The test to be applied is "common authority and control" over the premises. Property that has been leased or rented to an individual usually requires consent from the tenant or occupant rather than the owner, even though the owner or landlord may have the right to enter and inspect the premises under the terms of a lease agreement.

The person who grants consent can authorize the search of only those areas under his or her common authority and control. In the case of roommates, one roommate cannot authorize a search of the other's private areas, such as a separate bedroom or personal closet. In all consent searches, only those areas within the common authority and control of the person giving consent may be searched.

The courts consider a properly executed warrant as the safest method of entry, and even though one of the other methods of entry might be appropriate to the situation, consent should always be requested whenever possible. The investigator should bear in mind that when a fire scene investigation is conducted pursuant to consent, the person who gives the consent may limit the scope of the search and may even revoke consent at any time. Consent to search a fire scene should always be obtained in writing and be witnessed.

Exigent Circumstance

Public sector/law enforcement investigators have the authority to conduct a search and seizure whenever an exigent circumstance exists, requiring an immediate response in the interest of public safety. The U.S. Supreme Court decisions in *Michigan v. Tyler* and *Michigan v. Clifford* outline the scope of a fire investigator's right of entry under this rule. Every fire is considered a public safety emergency that requires a response by fire service. When a fire is reported, the exigent circumstance allows entry onto property to suppress the fire. Similarly, determining a fire's cause is part of the government response and is supported by strong public safety justifications. Thus, the investigation of a fire scene is authorized as part of the exigent circumstance surrounding every fire.

There is no open-ended authority to enter a fire scene and conduct an investigation, however; the investigation must take place within a "reasonable time" following suppression and extinguishment of the fire. Whether a fire scene investigation has been conducted within a reasonable time depends on a variety of facts and circumstances and, if the investigator's actions are challenged, whether the investigation was conducted within a "reasonable time" will be determined by the judge.

Generally, the investigation must commence at the first available opportunity following suppression and extinguishment of the fire. Any delay in beginning the investigation must be justified for the investigation to be considered part of the exigent circumstances surrounding the fire. A delay to wait for daylight is generally considered acceptable; however, waiting until later in the day when the investigation could easily have begun earlier may result in an illegal search and seizure. The investigator should always respond immediately when called to investigate a fire.

Additionally, the investigation may continue only for a "reasonable time," so the investigator must ensure that the investigation is progressing at an appropriate pace and that it does not extend unnecessarily long. The investigator relying on the exigent circumstances doctrine must be able to document and prove why an investigation lasted the time that it did, which may include documenting such factors as scene size, weather, safety issues, necessary tasks (including those outlined by NFPA 921 and NFPA 1033), and other applicable information.

When an investigator is in doubt as to what is "reasonable time," legal advice should be sought. Also, rather than relying exclusively on the existence of an exigent circumstance, the investigator should seek consent whenever possible.

Administrative Search Warrant

An administrative search warrant is a third method of entry for the public sector/law enforcement investigator, expressly authorized by the U.S. Supreme Court in the *Michigan v. Tyler* case. It is seldom used by investigators and even less often recognized by courts. An administrative search warrant is intended for situations in which the reasonable time after a fire has already passed and consent to search is refused or is otherwise not available. It requires a sworn affidavit presented to a judge in a court of competent jurisdiction for the limited purpose of gaining entry to a fire scene to determine origin and cause. The applicant must show in the affidavit:

1. That he or she has the administrative authority to conduct fire scene investigations in the jurisdiction, such as the authority granted by an applicable fire code
2. That a fire has occurred (although the investigator cannot lawfully enter the premises at this time)

3. That the court's authorization is needed to fulfill the administrative responsibilities of investigating the fire to determine its origin and cause

An administrative search warrant cannot be employed once there is evidence of arson or some other crime, as outlined in *Michigan v. Clifford*. When a fire scene investigation pursuant to an administrative search warrant uncovers evidence of a crime, any further search of the premises requires a criminal search warrant.

Criminal Search Warrant

A criminal search warrant is the fourth alternative for gaining entry to a fire scene for the purpose of conducting an investigation. This method is a judicial authorization based on the filing of a sworn affidavit by a government agent who must be legally authorized to apply for the warrant.

Unlike an administrative search warrant, a criminal search warrant is issued only upon a showing of probable cause, establishing that a crime has occurred and that evidence of the crime is believed to be located in the premises to be searched. The affidavit must demonstrate reliable information and specific facts known to the applicant who is seeking the warrant. The premises and the specific areas within the premises where the search will be conducted must be precisely identified. It is strongly recommended that a prosecutor or legal advisor assist in the preparation of the affidavit and application for warrant.

Private Sector Investigator Right of Entry

For a fire investigator in the private sector, there are no constitutional issues to confront. The private investigator utilizes consent to entry almost exclusively—either express consent from the owner or occupant or implied consent under the provisions of an insurance policy. The right of entry can also be granted by court order. A private investigator may not gain entry to the fire scene in the absence of a court order if the owner or occupant refuses to allow access; however, if an insured claimant refuses to allow access to the fire scene by his or her own insurance company's fire investigator, the insurance claim may be denied. Once granted access to the scene, the private investigator should be familiar with the scope of the investigation. For instance, an investigator representing the interest of a tenant should know not to impound evidence that pertains to the interest of the building owner.

Witness Interviews and Interrogation

There are special constitutional issues that a public sector investigator must address when conducting witness interviews for the statement of a witness to be legally admissible at trial. These issues involve the Fifth Amendment protection against self-incrimination and the Sixth Amendment right to counsel (i.e., to have an attorney present during questioning) FIGURE 10-1. These issues are the foundation of the interrogation guidelines of the Miranda Rule (*Miranda v. Arizona*).

Under the Fifth Amendment, a person cannot be compelled to testify against himself or herself or to provide incriminating testimony that can be used later in a prosecution. The

© Jones & Bartlett Learning. Photographed by Glen E. Ellman

FIGURE 10-1 A witness has a right to have an attorney present during questioning.

Sixth Amendment contains the constitutional "right to counsel," which is also part of the protections afforded under the Miranda Rule. The Miranda Rule requires that a witness being interrogated or interviewed in a custodial setting receive certain warnings before the statement can legally be taken and used as evidence later. The Miranda Rule has four components:

1. The right to remain silent and refuse to give a statement
2. The warning that any statement made can later be used as evidence against the witness
3. The right to have an attorney present to represent the witness during questioning
4. The right to have an attorney provided free of charge, before any questioning begins, if the person cannot afford to hire an attorney

The Miranda Rule states that if the interviewee is being asked questions while "in custody," the full set of warnings must be given, and the interviewee must agree to waive those rights before speaking to the investigator. Otherwise the statements made by the interviewee will not be admissible at trial. The rule applies to questioning done by an investigator who is representing a governmental agency or acting at the request of one. Mere questioning of a witness does not in itself invoke the Miranda Rule; the interview or interrogation must take place in a custodial setting. In determining whether a custodial setting existed, a court will consider various factors, including these:

- Apparent authority of the interviewer to detain or arrest the witness. A private citizen or a public official who lacks arrest powers will generally not be required to comply with the Miranda Rule.
- Location of the interview. The interview is more likely to be considered custodial if the witness does not have complete freedom of movement, including the ability to walk away from the interview. This could occur if the interview is in a locked room or a locked facility, if the exits are blocked by government representatives, or if the interviewee is handcuffed.

- Length and context of the interview itself. The longer the interview lasts and the more intense it becomes, the more likely that it will be considered a custodial interrogation—even if it began as a simple witness interview.
- Participants (and observers) at the interview. A large number of interviewers and spectators, especially law enforcement officials, can be considered a circumstance intimidating enough to change a mere witness interview into a custodial interrogation.

When there is any doubt about whether the interviewee might be considered to be in custody, the best practice is to advise the individual of all Miranda rights. When an individual has been given a Miranda warning and agrees to make a statement, a written waiver of rights should be signed and witnessed. The interview itself should be tape recorded or videotaped whenever possible. A good general rule is to have no more than two investigators conduct the interview. This is especially important if the interview room is small.

FIGURE 10-2 Spoliation of evidence is an issue at almost every fire scene. This dryer has been dismantled by fire crews.

Arrests and Detentions

The Fourth Amendment prohibits unreasonable searches and seizures. This includes seizures of persons. When a government official such as a law enforcement officer arrests or otherwise detains a person, this amounts to a seizure.

A seizure occurs when the government exercises control over a person. All seizures must comply with the Fourth Amendment safeguards. These safeguards differ depending on whether the seizure amounts to an arrest or is merely a detention. Whether a person is arrested or merely detained is determined by the circumstances of the situation.

Over time, the Supreme Court has developed a body of rules to govern the detentions and arrests. Here is a summary of those rules that are most applicable to fire investigators exercising law enforcement powers.

Arrests

An officer making an arrest must have probable cause to make the arrest, meaning that he or she has acceptably trustworthy information that would lead a reasonable person to believe that the suspect had committed (or is committing) a crime. The information supporting probable cause can be based on the officer's personal knowledge or on another source, such as a police communiqué, a posted bulletin, witness statements, or information from an informant that appears to be reliable.

If a person's Constitutional rights are violated and the arrest is unlawful, a number of remedies are possible. For example, a judge might dismiss the charges against the person, or evidence flowing from the arrest might be excluded from evidence at trial.

Detentions

An officer may briefly detain a person if the officer has a reasonable suspicion that this person has committed (or is committing) a crime. A reasonable suspicion is something less than probable cause, but more than a vague notion. Whether the officer has information that gives him or her a reasonable suspicion will depend on all of the circumstances. The detention must be brief and only as long as is necessary for the officer to conduct a limited investigation to confirm or disprove the suspicion that caused the officer to stop the person.

Spoliation of Evidence

Spoliation of evidence as a legal issue has existed for hundreds of years but has taken on great importance only relatively recently. It is now an issue that arises in almost every fire scene investigation. Although spoliation was traditionally an issue reserved for civil litigants, it can also appear in criminal matters.

Spoliation of evidence is an inevitable by-product of any fire scene investigation **FIGURE 10-2**. In a very real sense, an entire fire scene is critical evidence in a case. When an investigator has conducted a scene examination by moving fire debris and other evidence at the scene, the scene itself is necessarily altered. When a thorough fire scene examination has been conducted with the movement of debris using a methodical layering process, the result can be a scene that is devoid of virtually all evidence, leaving only a bare slab.

Local code provisions often allow a municipality to demolish a fire-damaged building immediately in the interest of public safety. The private sector investigator may resort to a court-ordered relief to protect against the destruction of evidence and allow a documentation of the scene and potential evidence. In the event that demolition is authorized, the investigator should strive to document the scene as thoroughly as possible under the circumstances.

When evidence is collected and removed from the scene, further change may occur. The examination of equipment and appliances found at a fire scene may involve destructive testing, which constitutes another form of change, alteration, or destruction. In each of these situations, the investigator can face claims of spoliation when other parties later find that they are unable to investigate independently the evidence from the fire scene because that evidence is no longer available as it existed immediately after the fire.

When spoliation has been found to occur, a court may take any number of actions against the responsible party as sanctions for the conduct that caused the spoliation. Those sanctions may include the following:

- Prohibiting any testimony about the spoliated evidence by the responsible party
- Prohibiting the responsible party from presenting any testimony at all about the fire scene investigation
- Instructing the members of a jury that they may infer that the spoliated evidence would have been damaging to the responsible party's case
- Striking the responsible party's evidence and entering judgment in favor of the party victimized by the spoliation

Moreover, courts in some states are now recognizing the tort of spoliation, a legal cause of action allowing a spoliator to be sued for his or her actions.

The concept of spoliation is a matter of fundamental fairness. In cases in which one party has lost, destroyed, or altered physical evidence that prejudices another party, the courts will use the doctrine of spoliation to balance the scales. The sanctions that a court may impose are designed to correct the prejudice that the injured party has suffered by being unable to examine and test independently the evidence in order to defend itself against claims relating to the evidence or to use the evidence to assert its own claims against other parties.

For all investigators, claims of spoliation can best be avoided by using a commonsense approach. Evidence should never be needlessly discarded, destroyed, or lost. Whenever evidence is uncovered that could be important to the interests of another party (such as in product liability cases), the evidence should be immediately preserved, and notification should be sent to all parties that have a potential interest in the evidence to allow them the opportunity to examine it. During this portion of the scene examination, the evidence should be secured in place, and the entire fire scene should be clearly marked and protected to prevent unauthorized movement of the evidence. This security should continue until the evidence is removed from the scene. Destructive testing of physical evidence should be preceded by notice to all known interested parties, permitting them an opportunity to have experts present and observe or participate in the testing. NFPA 921 defines an interested party as "any person, entity, or organization, including their representatives, with statutory obligations or whose legal rights or interests may be affected by the investigation of a specific incident (NFPA 921 Section 3.3.110, p 17)." (Notice to interested parties does not necessarily apply to evidence collected during a criminal investigation.)

The inevitable alteration of evidence as fire debris is examined and moved during a fire scene examination generally does not constitute actionable spoliation, especially if the scene has been properly documented and photographed before, during, and after movement of the evidence. Thus, proper fire scene documentation, which should already be employed in every investigation, is the first line of defense against claims of spoliation. When specific evidence is uncovered (such as equipment or an appliance) that is believed to have been a factor in the fire, those items should always be carefully preserved, and notification should be immediately sent to all interested parties. With this as an established investigative protocol, the threat of spoliation is substantially reduced.

Fire Investigator Tip

A fire investigator should not discuss potential findings with anyone not involved in the investigation. "Under investigation" is the proper response to any inquiries.

The nature and content of notice that should be given to all known interested parties at the outset of a scene investigation to allow them the opportunity to retain experts and attend the investigation are as follows:

- Notification to interested parties (via telephone, letter, or email; oral notification should be confirmed in writing)
- Nature of notice to interested parties
- Content of notice
  - Date of the incident
  - Nature of the incident
  - Incident location
  - Nature and extent of loss (damage, death, or injury)
  - The interested party's potential connection to the incident
  - Next action date
  - Circumstances affecting the scene (such as pending demolition orders or environmental conditions)
  - A request to reply by a certain date
  - Contact information as to whom the notified person is to reply
  - The identity of the individual or entity controlling the scene
  - A roster of all parties to whom notice has been provided

NFPA 921 recognizes that public sector investigators may have different notification responsibilities from those in the private sector. Furthermore, the duty to provide such notice will vary based on the jurisdiction, scope, and procedures of the investigators as well as the circumstances surrounding the fire.

Investigator as Trial Witness

To be a successful witness at trial, whether you are a fact witness (such as an evidence technician) or an expert witness, it is very important to become comfortable with the courtroom and with civil and criminal trial proceedings. This learning process begins with knowing the procedural steps that lead up to trial and the types of oral and written testimony that may be given at or before trial.

Reports

Whether expert witnesses prepare a report in a given case will depend on many factors. Some of these are as follows:

- Whether the criminal or civil rules of procedure require witnesses who will testify at trial as experts to prepare a report (as does Federal Rules of Civil Procedure [FRCP] Rule 26[a][2][B]).
- The trial strategies decided on by the expert and the attorney who will call the expert as a witness.
- The advantages and disadvantages of preparing a report as summarized later in this section.
- Considerations of whether the report is admissible in evidence and whether disclosure of the report is required to be given to the opposing party(s). This will be of particular concern in criminal cases.
- The dictates of any internal (company or departmental) procedures or industry standards that you observe, including NFPA 1033, ASTM E 620, and ASTM E 678. For example, public investigators typically prepare reports of every significant investigation performed. If these reports contain opinions and conclusions, they may be considered expert reports in subsequent legal proceedings.

The content of expert reports will be determined by various factors, including procedural rules, your usual practice, and the requirements of the attorney who has retained you or who is calling you as an expert witness. In civil cases, the federal courts and many state courts require reports to be prepared and disclosed to other parties by any expert who may testify at trial. In the federal jurisdiction, this requirement is reflected in FRCP Rule 26(a)(2)(B).

In addition to complying with legal procedural rules and ensuring that your report is complete, it is also very important that every report you prepare be accurate and easy to read and understand. It should detail and describe every step of the scientific method as it was applied during the investigation and analysis. Check and recheck your facts from the records you have reviewed and your investigation documentation. Carefully proofread your report. Your writing style should be professional and strong, couched in good grammar, with no spelling errors. The report should permit a person who is unfamiliar with fire terminology to understand the facts and conclusions being reported. See the Documentation chapter of this text for further information on report writing.

If your writing skills are not strong, obtain some training on proper grammar, through either reading or taking courses. There are many good books on writing at public libraries. Have another investigator in your department or office critically review your report for internal consistency, grammar, and spelling.

The investigation report should be a complete and accurate summary of the actions, observations, and conclusions of the fire investigator. Depending on the scope of practice for the individual or organization, the report may include a narrative, photographs, video, interviews, diagrams, and other created documents as supporting information. With the report alone, the fire investigator should be able to describe and support the determination of origin and cause.

Fire Investigator Tip

In situations in which you are not required by law or by the attorney calling you as an expert to prepare a report, there are a number of considerations to help you decide whether to prepare a report:

- Quite some time may pass between your investigation and trial. If your report was prepared shortly after your investigation was completed, reviewing it can help refresh your memory for trial.
- A well-organized report can be used by you and the attorney who calls your testimony to structure your direct examination at trial.
- The report provides a useful summary of the material you reviewed, the investigations and tests you conducted, and the analysis leading up to each opinion. This summary will help your attorney prepare for trial and coordinate your testimony with that of other experts and fact witnesses.
- To the extent that your report goes beyond your field of expertise or is inaccurate or incomplete in any way, it can be used to challenge the admissibility of your opinion or become an effective tool against you in cross-examination.
- If you prepare a report with your final opinions before all investigations are complete, the weight or admissibility of your opinion may be subject to challenge on the basis that it is not based on sufficient facts or data pursuant to FRE Rule 702.
- If the attorney who calls your evidence works with you in drafting or preparing your report, this information can be discovered by the opponent and may substantially weaken the weight and credibility of your expert evidence, giving the appearance of bias.

Whether an expert report is discoverable by the opponent will affect the decision of whether and when to prepare a report. The best course is to discuss the matter of preparing a report and the nature of its contents with the attorney who will call you to testify before preparing any report (even a draft or preliminary report).

Testifying in Court

Whether testifying as an expert or as a lay witness (described further below), the investigator should stay in close contact with the attorney calling him or her as a witness to make sure there is a full understanding of the subject matter of the testimony and any expectations of the witness. The witness should make sure the attorney calling him or her understands the scene being described, the science and reasoning behind any conclusions, and any technical terminology being used. The investigator should also understand what documentation should be provided and brought to court.

Professional dress and demeanor are critical. The investigator should wear business attire or a uniform, as appropriate, and maintain a professional, discrete demeanor at all times while in the courthouse. Witnesses at trial or other hearings are frequently sequestered, meaning that they are not allowed in court to hear the testimony of other witnesses. To avoid a possible violation of a court sequestration order, witnesses should arrange to stay outside the courtroom until called, and refrain from discussing the case with anyone other than the attorney who called them.

All witnesses are typically sworn to tell the truth before they sit to testify. Questions are asked of the witness by the attorney who called him or her in direct examination, usually in an open-ended format. Cross examination questions are then asked by the opposing attorney(s), who may ask leading questions. Usually the original attorney is provided an opportunity to ask further open-ended questions on re-direct examination, and further cross-examination may be allowed. In some jurisdictions, judges and even jurors may also ask questions of the witness.

The fire investigator's demeanor while testifying should always be professional and objective. Answer questions clearly and succinctly, and make eye contact with the judge and jury. Avoid unnecessary movement, as well as nervous gestures and expressions, and keep your hands in your lap when they are not in use or gesturing appropriately. If you do not understand a question, ask for it to be repeated or explained. Treat all attorneys with equal professionalism, and avoid displays of anger or disdain. When permitted, leave the witness stand to give demonstrations and explain exhibits.

At times, an attorney may interpose an objection to a question or answer during testimony. An investigator should always stop answering if an objection is raised, until ruled upon by the judge. Typically, if the judge sustains an objection to a question asked, the witness should not answer. If the objection is overruled, an answer is permitted.

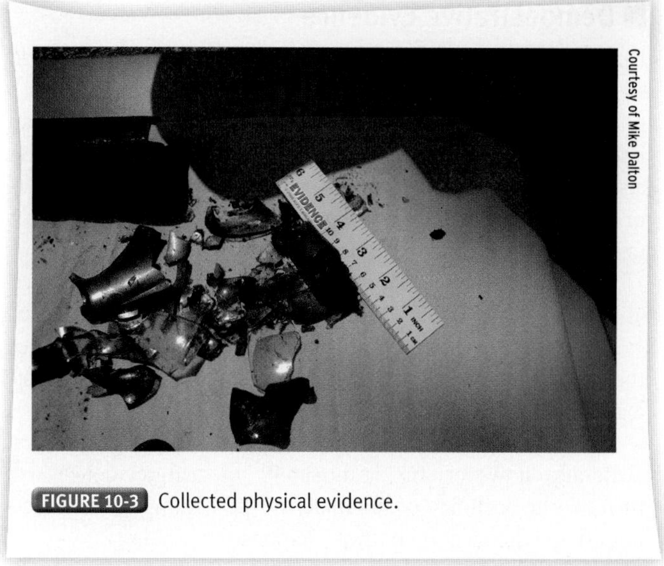

Courtesy of Mike Dalton

FIGURE 10-3 Collected physical evidence.

Types of Evidence

Evidence can be categorized in a number of different ways. Four generally recognized categories of evidence that will form the foundation of any civil or criminal fire trial are:

1. Real or physical evidence
2. Demonstrative evidence
3. Documentary evidence
4. Testimonial evidence

These categories are not distinctive, and some evidence will fall into more than one category.

Real or Physical Evidence

Real evidence refers to physical items that may be produced in court for inspection by the judge and jury. The range of real evidence that might be relevant in a fire case is huge, including everything from minuscule cloth fibers that an arsonist may have left on a victim at the scene to a large section of a wall that is removed from the scene and preserved for later analysis FIGURE 10-3.

Real evidence is commonly examined by experts and may be subjected to testing. It may also be diagramed and photographed and become the subject of demonstrative evidence.

Chapter 17 of NFPA 921 addresses physical evidence that may be relevant to the issues of the origin, cause, spread, or responsibility for a fire, along with recommendations for recognizing, collecting, and preserving this type of evidence at a fire scene. In cases of fire death or injuries, physical evidence may include the deceased's remains and the victim's injuries. Dealing with this type of evidence is covered by NFPA 921, Chapter 25. The actual method of collection of physical evidence requires compliance with technical standards for collecting and handling evidence as outlined in NFPA 921. To be admissible at trial, evidence must be properly transported and stored and must be authenticated by establishing a chain of custody (a record of those who had the evidence in their custody and how it was moved). Even if the chain of custody is proven, some real evidence may be too prejudicial to be allowed into evidence for the jury to see, such as a burned corpse.

In criminal cases, the collection of physical evidence at the fire scene is governed by the constitutional limitations on search and seizure. The right to collection and possession of physical evidence requires compliance with those constitutional standards.

Evidence that has been altered, destroyed, or lost is spoliated evidence. It may be inadmissible at trial, or its existence may result in other evidence being deemed inadmissible as well, such as expert reports that rely on spoliated evidence. In some cases, the investigator who is responsible for the spoliation may be prohibited from testifying. In extreme cases, spoliated evidence can result in outright dismissal of the case.

Demonstrative Evidence

Demonstrative evidence is an important form of evidence used at trial. It includes any type of tangible evidence, such as a photograph, diagram, or chart, that is utilized to demonstrate an issue relevant to the case. The term demonstrative evidence can also refer to items that were gathered from the fire scene (real or physical evidence), as well as other tangible evidence items (as opposed to witness testimony).

The most common forms of demonstrative evidence used at trial are diagrams, charts, timelines, models, and other such evidence that an expert witness or attorney specially creates for the trial to illustrate or demonstrate a point to the jury. The basic legal requirements for such evidence are relevance to the issues at trial and usefulness to the jury in understanding those issues. Generally, however, the admissibility of such evidence will turn on the accuracy or reliability of the assumptions or facts underlying the creation of these demonstrative aids.

Some forms of demonstrative evidence (photographs, film, or digital media) generally do not require authentication by the person who takes the photographs (although it is preferable to do so), nor is there a requirement for proving the chain of custody for such evidence. Photographs, videos, slides, or digital media must be properly authenticated by showing that they are a fair and accurate depiction of the subject. Such evidence is almost always found to be admissible when it has been properly authenticated in this way.

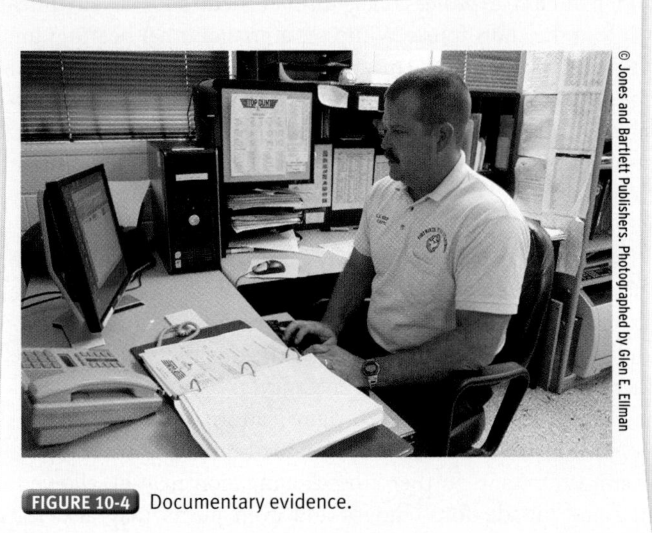

© Jones and Bartlett Publishers. Photographed by Glen E. Ellman

FIGURE 10-4 Documentary evidence.

Documentary Evidence

The third form of evidence used at trial is documentary evidence. This category includes any type of written record or document that is relevant to the case. Business records, incident reports, telephone records, bank records, insurance policies, claim documents, correspondence, written statements from witnesses, investigative notes, and transcripts of recorded interviews are all examples of documentary evidence FIGURE 10-4.

Once again, there are a number of legal issues to consider. First, the evidence must have been properly obtained. With respect to a criminal case, the public sector investigator must consider Fourth Amendment issues of search and seizure law, as well as Fifth Amendment issues of self-incrimination (if, for example, the subject of a witness interview is a criminal suspect).

Both public and private sector investigators should be aware that all documentary evidence must be properly authenticated at trial under the standards of relevance and reliability. In the case of business records, bank records, telephone records, and other such documentary evidence, authentication will usually require the testimony of a records custodian to confirm the authenticity and accuracy of the documents. Documentary evidence may be subject to a legal challenge as hearsay if the information was obtained from an out-of-court source; however, there are many exceptions to the rule against hearsay evidence that may allow the documentary evidence to be used once a proper legal foundation has been laid. To avoid hearsay challenges, investigators should work with the prosecutor or attorney handling the trial to ensure that documentary evidence is properly presented.

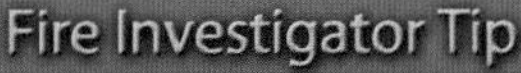

Fire Investigator Tip

Demonstrative evidence can be very powerful because it is interesting and, if properly prepared and presented, will capture the attention of the jury. If you are presenting demonstrative evidence such as a chart, computer model, simulation, or even a PowerPoint presentation that includes photographs and graphical summaries of evidence, here are some suggestions:

- Ensure that you show the evidence or demonstrate the simulation to the attorney who calls you as a witness well in advance of trial. The attorney may need to disclose it to the court or opposing side or make a motion (a request for the court to take action) before trial to deal with the admissibility of such evidence.
- Be prepared to testify that the chart, graph, model, or simulation is accurate in every detail.
- Recognize that one way to challenge demonstrative evidence is to show that the underlying facts or data on which such evidence is based are false or incorrect. Therefore, be prepared to provide details of the underlying facts and where they are found in the trial evidence (i.e., from witness testimony, documents, physical evidence).
- If you are running a simulation or conducting a demonstration, be sure that you have practiced and that it will work as predicted.

FIGURE 10-5 Testimony.

Testimonial Evidence

The fourth form of evidence at trial is testimonial evidence, which is verbal testimony of a witness given under oath or affirmation and subject to cross-examination by the opposing party FIGURE 10-5. Although relevance and reliability are the baseline requirements for admitting testimonial evidence at trial, it is subject to the full range of requirements under the rules of evidence and may be prohibited for a variety of reasons. For instance, a witness may not be competent to testify at all, a witness may not be competent to testify about a specific issue, the testimony may be irrelevant or immaterial to the case, the testimony may include improper hearsay statements from other persons, the testimony may be speculative or unfounded, or the testimony may contain the opinion of a witness who is not qualified to offer opinion testimony (usually only an expert witness is qualified to offer opinion testimony).

In the case of expert testimony, especially the testimony of an origin and cause expert, there are several requirements for qualifying an expert and presenting expert testimony, as addressed in the Expert Witnesses subsection later in this chapter.

During the testimonial process, the investigator may be involved with affidavits, interrogatories, depositions, and trial testimony. An affidavit is a voluntary written statement of fact or opinion. The affidavit is made under oath and is signed by the author. An interrogatory is a set of questions that is served by one party involved in litigation to another involved party. The response, known as answers to interrogatories, is generated by the served party. All forms of interrogatory are made under oath and signed by the respective party.

A deposition is an oral testimony made under oath. The witness, or deponent, is notified of the deposition by subpoena. The proceedings are usually conducted in an office setting but are legal proceedings made under oath with the use of a stenographer. There may be legal counsel for numerous parties involved, and each is given an opportunity to question the deponent. A transcript is generated that is later reviewed and corrected by the deponent. The deposition transcript may be used if the proceeding continues to trial.

If the civil or criminal matter is not resolved through plea or settlement, the matter proceeds to trial. The trial is presided over by a judge and possibly a jury. After the judgment is made, appeals can be made by parties not satisfied with the result.

Fact Witnesses

A lay witness is a person who testifies about matters within his or her own firsthand knowledge. This type of witness is also known as a fact witness because the testimony is about facts that the witness has personally experienced—something the witness has seen, heard, touched, smelled, or tasted. Fact witnesses are limited in their testimony to the types of opinions that they can give.

As with all witnesses, fact witnesses must be competent to testify. This means that they must be able to observe, remember, and communicate their testimony to the court. They must also be able to understand and appreciate the nature of an oath and the obligation to tell the truth. Witnesses are generally presumed to be competent unless their competency is successfully challenged or a federal or state rule provides otherwise. An attorney may challenge the competency of a witness who falls short in any of these areas (e.g., a very young child or a person who is suffering from advanced stages of mental illness).

Even when a witness is competent, an opposing attorney can challenge the testimony by cross-examination to show that it is not credible or reliable, on several grounds, such as:

- The witness is biased or has a vested interest in the outcome of the case (e.g., the witness may have a close personal relationship with the party on whose behalf he or she is testifying, or the witness may have a general prejudice against opposing witnesses, such as the police).
- The witness's powers of perception are not trustworthy (as in a case of eyewitness testimony where the witness was too far away to have heard what he or she alleges to have heard or where the lighting and obstacles at the time make it unlikely that the witness saw what he or she testifies to have seen).
- The witness's testimony at trial contradicts prior oral or written statements made by the witness concerning the same events.
- The witness's testimony contradicts other evidence such as physical evidence, documentary evidence, or the testimony of other witnesses (perhaps where a witness testifies as to having seen the outbreak of fire at 1:00 p.m., yet the 911 reporting the fire call was registered at 12:00 noon; or a witness says all of his or her possessions were in the house when it burned, but investigators sifting the scene find that pictures were missing from walls and clothing was missing from closets and dresser drawers).

Fact witnesses are not limited to bystanders or eyewitnesses to an event. You, as an investigator, will very likely become a fact witness in a criminal or civil trial at some time. If you were present before the fire was extinguished or before the fire crews left, you may testify as a fact witness about what you observed of the fire suppression and overhaul activities. If you testify about having collected, handled, or transported physical evidence, you do so as a lay witness. (You are a lay witness rather than an expert to the extent that you do not render opinions about the meaning of the evidence.)

Evidence technicians (those who are specially trained to document, collect, preserve, and transport evidence) require special training to do their jobs properly; however, they may not have sufficient scientific, technical, or other specialized knowledge to qualify them to give opinion evidence about the implications of the fire or explosion as would an expert.

Fact witnesses are restricted in their testimony by Federal Rules of Evidence (FRE) Rule 701 or a comparable rule in state court. Rule 701 effectively limits the types of opinions that fact witnesses can give. They may give an opinion about things like the speed a vehicle was traveling when they saw it, or the reasons they think that a person appeared drunk or tired, happy or distraught, or that a building was "fully involved" in flames; however, a fact witness, even an evidence technician who has training based on NFPA 921 and NFPA 1033, may not give an opinion about the origin, cause, spread, or responsibility for a fire unless that witness is qualified by the trial court as an expert witness.

Expert Witnesses

An expert witness assumes a special role in the courtroom. Expert witnesses may testify about facts in their own personal knowledge (like lay witnesses), as well as facts made known to them before or at trial, facts that they are asked to assume, or facts that are generally agreed on by experts in the particular field. Most significantly, experts are permitted to render an opinion respecting the fire or explosion FIGURE 10-6.

© Corbis/age fotostock

FIGURE 10-6 Expert witness.

The nature of the opinion that an expert may render will be limited to some extent by the rules in the jurisdiction where the expert is testifying. In some states, an expert may not give an opinion on the ultimate issue that the jury is there to decide. For example, in some states, an expert may give an opinion about the mechanics of how an incendiary fire (defined by NFPA 921 as "a fire that is deliberately set with the intent to cause the fire to occur in an area where the fire should not be [NFPA 921 Section 3.3.108, p 17]") was started, but is not allowed to say that it is an arson fire started by the defendant (the person sued in a civil proceeding or accused in a criminal proceeding). If you are testifying as an expert, speak with the attorney calling you as a witness about any limits on your testimony and how your opinions should be presented.

Another limit to expert opinions in all jurisdictions is that the opinion will have to be relevant to the issues in the case and fall within the field in which the court has recognized the expert as qualified. Therefore, if an origin and cause expert has no specialized experience, knowledge, or training in photocopier machines, the court may not permit that expert to testify about how a design fault in the photocopier was the cause of the fire. Similarly, a trial judge court may not allow a person who is qualified to testify about only the mechanics of propane explosions to give an opinion about why people did not smell the propane or whether it was odorized properly.

Before an expert may testify as such, there are several hurdles to overcome.

Admissibility and Relevance of Expert Testimony

As with all other forms of evidence, testimonial evidence from an expert must be relevant to the issues in the case and must be reliable; however, the measure of reliability for an expert witness can be a complex analysis. In federal court cases and recently in many state court jurisdictions, expert testimony is governed by the Daubert rule (*Daubert v. Merrell Dow Pharmaceuticals*). Under the Daubert standard, the trial judge takes a proactive role in screening the testimony of prospective expert witnesses. The judge must assess the reliability component of an expert's testimony by challenging the expert as to both the conclusions (opinions) to be presented and the techniques and methods used to reach those conclusions. The judge may consider various factors in evaluating an expert's testimony. These factors include, but are not limited to, the following:

- Whether the theory or technique has been or can be tested and whether the hypothesis underlying the theory or technique can be falsified
- Whether the theory or technique has been subjected to peer review or publication
- Whether the theory or technique has a known or potential rate of error
- The existence and maintenance of standards controlling the technique's operation
- Whether there has been general acceptance of the theory or technique in the relevant scientific community

In addition to those general criteria, a trial judge may ask any other questions that are appropriate to the particular facts of the case to ensure that the testimony is reliable enough to be heard by a jury.

Some jurisdictions apply other standards to measure the reliability of expert evidence. Prior to the Daubert decision, to be admissible, expert testimony based on novel scientific theories or techniques had to pass a test that came from a 1923 federal appellate court decision in *Frye v. United States*, known as the Frye "general acceptance" test. Pursuant to the Frye test, evidence was necessary to show that the principle or technique on which the opinion was based "was sufficiently established to have gained general acceptance in the particular field in which it belongs (*Frye v. United States*, 293 F. 1013 [D.C. Cir. 1923])."

Some states continue to apply the Frye test for novel scientific testimony. Others use a modified test, using elements of Daubert and Frye. Regardless of whether Frye or Daubert is the standard, NFPA 921 can be utilized in every jurisdiction in the United States to measure whether a fire investigator's theory or technique has gained general acceptance in the relevant scientific community (i.e., the community of fire investigators).

In jurisdictions where the Daubert rule applies, expert witnesses can expect a particularly strong challenge to their testimony and findings by both the opposing party and the trial judge. In responding to a Daubert challenge, the investigator must be able to demonstrate that both the methods used in the investigation and the conclusions reached by the investigator can be proven to be reliable by reference to objective and documented sources.

Familiarity with the entire NFPA 921 document and the ability to cite readily the appropriate sections addressing any issue in question will enable an investigator to respond effectively to a Daubert challenge. Because NFPA 1033 is the industry standard for fire investigators and because it references NFPA 921 as described earlier in this chapter, NFPA 1033 lends additional weight to employing NFPA 921 as a measure of reliability.

Qualifications of Expert

As we have seen, the first admissibility requirement in FRE Rule 702 is that to testify as an expert, a witness must be "qualified as an expert by knowledge, skill, experience, training, or education." The trial judge is the person who determines whether an expert is qualified. This decision is based on the exercise of the judge's discretion as guided by precedent cases. Counsel may stipulate to the qualifications of an opposing expert, in other words, admit that the expert is qualified, not challenging the admissibility of the expert's testimony on that ground. Alternatively, an attorney may try to prove either that a witness is not sufficiently qualified to give any expert opinion testimony or that a witness is qualified as an expert, but in a more limited field or on more limited issues than those for which the expert was originally proffered. If the qualifications of an expert are not challenged, the trial court will typically recognize the expert as qualified. If the qualifications are challenged, the court will hear evidence on the issue.

Challenges on the qualifications of experts in fire and explosion cases are growing increasingly common. Many of these challenges are framed using NFPA 921 and NFPA 1033 as benchmarks.

Fire Investigator Tip

Basic components of an expert's qualifications include the following:

- Personal background information (name, age, current place of business or employer, duties in current position)
- Formal education (schools, colleges, or universities attended; dates attended; diplomas or degrees received; any specialization obtained)
- Skills and experience (emphasizing those that relate to your qualifications to testify as an expert in fire or explosion cases)
- Professional licenses or certifications (including the date of any license or certification, the licensing/certifying body, the dates of re-certification, and the jurisdictions where the license/certification are recognized—i.e., state, national, international)
- Distinctions and awards (relating to your work in your field of expertise)
- Publications (using formal citation format and including the name[s] of any co-authors, title of the work, title of the journal or collection if yours was part of a larger work, name and city of the publisher, publication date)
- Memberships and offices in professional associations (including committee memberships or chairs and other special appointments. Be careful about listing memberships that were in name only, such as committees to which you were appointed but did not actively participate. Much headway can be made cross-examining you on empty titles.)
- Professional presentations (speaking engagements at seminars, conferences, and the like)
- Academic or teaching appointments

You should also keep track of the following additional information, which might be the subject of questions raised by the attorney calling you as a witness or by the opposing attorney at your deposition or when you are cross-examined:

- Publications to which you subscribe or read regularly
- Books that you have read in your field or that were the basis of courses you have taken, including NFPA 921 and NFPA 1033, and other books like the *NFPA Fire Protection Handbook*
- Other cases in which you testified (including affidavits, depositions, testimony in preliminary motions, and trial testimony)

Disclose to the attorney who will call you as a witness any negative aspects of your background or qualifications so that the attorney can help you plan how to limit the potential damage from this type of information. Negative aspects might include courses that you failed, certifications that you allowed to lapse, successful challenges to the admissibility of your testimony in prior cases, or accusations of ethical improprieties.

The types of information attorneys and judges will consider when evaluating an expert's qualifications should appear on your curriculum vitae (CV). Be scrupulous to ensure that your CV is fair and accurate with respect to each item listed. In preparing to testify in a particular case, review your CV, focusing on the specific aspects that relate to issues in the case. Although your qualifications should always be accurate, you should present them to a judge or jury in a manner that shows how your expertise qualifies you to address specific issues in the case at hand.

Remember that NFPA 1033 requires (and the legal system expects) experts to stay current in their fields. Devote adequate time to reading professional journals and other publications, attending seminars and association meetings, and obtaining further formal education to the extent necessary to maintain and improve your qualifications.

Examination of Witnesses

The questioning of witnesses customarily involves two or more parties following a course of direct and cross-examination. The direct examination is conducted first by the party who produced the witness. After direct examination, the opposing party is able to conduct a cross-examination. The purpose of the cross-examination is to extract more detail from the direct examination or to challenge the testimony.

Arson

Arson is one of the oldest criminal offenses under the law. It dates back to the early common law of England, when it was one of only three capital offenses (along with treason and murder). The common law of England originally defined the crime of arson as the malicious burning of another's home at night. For several centuries, this was the only form of intentional fire setting that was considered to be a crime. Criminal laws have since been codified and modified in the United States.

Today, the crime of arson can be committed in many forms. It can include causing fire or explosion to residences, buildings, vehicles, aircraft, watercraft, personal property, and other structures and materials. The elements of arson crimes vary among jurisdictions. An investigator must be closely familiar with the arson statutes in his or her state and the federal jurisdiction.

Most arson offenses are felony crimes punishable by imprisonment and fines. Arson to an occupied structure is generally considered to be first-degree arson, punishable by substantial prison time, ranging up to a life sentence in some jurisdictions. Acts of second-degree arson or third-degree arson are generally punishable as felony offenses as well. The intentional burning of other types of property that are not designated under the arson statutes may be considered vandalism or criminal mischief.

NFPA 921 Section 12.5.6.3 includes a list of factors for investigators to consider when determining whether the crime of arson has occurred; however, remember that you must consult the arson statutes of each jurisdiction, which contain requirements specific to that jurisdiction.

The District of Columbia and every state jurisdiction have enacted arson-reporting/immunity statutes. These statutes generally require that insurers report suspected acts of arson to local law enforcement authorities and provide a mechanism for the release of insurance claim information to public officials regarding fires that may have been the result of a criminal act. The process is initiated by a written request from the municipal authorities citing the applicable code and the request for information contained in the insurance file. The code requirement protects the insurance company from civil liability. Failure to comply with the request can result in criminal prosecution against the insurance company.

Circumstantial Evidence

A prosecutor often has to prove arson entirely by circumstantial evidence rather than by testimony of an eyewitness to the start of the fire. Circumstantial evidence is evidence based on inference and not on personal observation. Arson crimes will usually require proof of three things:

1. The fire was incendiary.
2. The person who set the fire had the necessary criminal intent.
3. The defendant set the fire or arranged for someone else to do so.

If there was no eyewitness who saw how the fire started, the investigator will rely on circumstantial evidence to determine the cause of the fire and who was at fault.

NFPA 921, Chapter 24, Incendiary Fires, contains a long list of indicators for an investigator to consider that may, together with other evidence, lead to the inference that a fire was incendiary. A few examples are as follows:

- Multiple fires: Two or more separate, nonrelated, simultaneously burning fires
- Trailers: Deliberately introduced fuel or manipulation of existing fuels to aid in the spread of fire from one area to another
- Incendiary device: A mechanism used to initiate an incendiary fire, such as a lit cigarette placed in a book of paper matches and propped over a container of combustible materials, or a Molotov cocktail
- The presence of ignitible liquids in the area of origin that cannot be explained other than as a device to start or spread the fire

Once a fire has been classified as incendiary, the next step is to identify the firesetter. This involves an investigation into who had both the opportunity and a motive to set the fire. Motive is defined by NFPA 921 as "an inner drive or impulse that is the cause, reason, or incentive that induces or prompts a specific behavior (NFPA 921 Section 24.4.9.1.1 p 241)." Distinguishing motive from criminal intent is necessary to prove intent as an element of a crime. Criminal intent is the state of mind to commit the criminal act (or sometimes omission) without any justification, excuse, or other defense. In contrast, motive is the reason behind an act. A person may have a motive to commit arson but had nothing to do with the fire.

For example, in difficult financial times, an insured might have the financial motive to burn her home. She might also have fallen asleep while cooking deep fried potatoes on the stove and accidentally caused a fire. In this situation, the fire is accidental, even though she had a motive for burning her own home. In other words, a person can have a motive for causing a fire, without any criminal intent. Without criminal intent, she did not commit arson.

Although motive is generally not a required element of the crime of arson, a prosecutor will want evidence of motive to present to the jury. Without an explanation of why a person would commit such a serious crime as arson, jurors may be reluctant to convict. It is human nature to want to know why someone committed arson. NFPA 921, Chapter 24, contains a summary of the traditional classifications of motives for fire setting.

In addition to classifying a fire as incendiary and determining motive, another aspect of building a circumstantial case of arson is to identify those who had the opportunity to set the fire. Opportunity involves the investigation of those who were present at the time the fire started and who had access to the site. Witness reports as to the whereabouts of suspects at the time of the fire, and paper trails such as gas or other receipts and phone records, may shed light on this issue. For fires that occurred in locked premises with no evidence of a break and enter, the persons with keys or other access are a critical consideration. The strongest circumstantial case will come down to proof of the person who had exclusive opportunity to set the fire.

Other Fire-Related Criminal Acts

In addition to arson, other criminal offenses may be related to a fire scene investigation. These fire-related crimes are numerous, including vandalism at one end of the spectrum and terrorism at the other. These are only some of the potential offenses that may be relevant in a fire investigation:

- Burning to defraud
- Insurance fraud
- Unauthorized burning
- Manufacture or possession of fire bombs and incendiary devices
- Wildfire arson
- Domestic violence
- Child endangerment/child abuse
- Homicide and attempted homicide
- Endangerment/injury to fire fighters
- Obstruction/interference with fire fighters
- Disabling of fire suppression/fire alarm systems
- Reckless endangerment
- Failure to report a fire
- Vandalism or malicious mischief
- Burglary/trespassing

In addition to these offenses, virtually any other type of criminal activity may be uncovered in the course of the investigation. A fire may have been set to cover evidence of other crimes.

Burden of Proof

In a criminal matter, the prosecution must prove the defendant's guilt beyond a reasonable doubt. This is a heavier burden than the burden of proof in a civil case such as a claim based in tort or contract. In most civil cases, the plaintiff must prove his or her claim by a preponderance of the evidence, sometimes also termed "more likely than not," or proof on a "balance of probabilities."

However, in a civil claim where the defense alleges that the plaintiff was involved in wrongdoing comparable to a criminal act, the defendant must establish the defense by "clear and convincing" evidence. This is a greater degree of proof than in ordinary civil cases. For example, an insurance company might defend a claim made by someone it insures, asserting that the insured burned his or her own property or committed insurance fraud by inflating the claim made on the loss. In this situation, the defendant insurance company must prove its defense by clear and convincing evidence, meaning it must show that it is highly probable this defense is true. This "clear and convincing" evidence burden is less than the "beyond a reasonable doubt" burden prosecutors must meet in criminal cases such as arson or fraud, but greater than the burden on the plaintiff in most civil cases.

Civil Litigation

Civil litigation in fire investigation cases is far more prevalent than criminal prosecution. An investigator for a public agency who has conducted a fire scene examination may become involved in civil litigation related to that fire, regardless of the outcome of his or her own investigation and the results of any subsequent prosecution. For private sector investigators, the prospect of becoming involved in civil litigation at some point is a virtual certainty.

Civil litigation may be based on one or more causes of action. A cause of action is a situation in which one person is entitled to obtain legal redress from another person. It is usually described by the legal theory on which the lawsuit is based. Examples of some causes of action that can be commenced as a result of a fire or explosion are breach of contract, product liability, negligence, a civil action based on fire code violations, or a civil claim for an intentional wrong such as arson. Every fire scene investigator must have an understanding of civil litigation principles in order to respond effectively when called to testify in a civil proceeding.

There are, of course, fundamental differences between the criminal and civil courts. First among these differences is the fact that the consequences of the proceeding are far more severe in a criminal case. An individual's life and liberty are at stake in a criminal case; a criminal defendant who is convicted at trial faces the prospect of imprisonment. Civil cases and their outcomes are usually concerned only with the award of money damages, but although life and personal freedom may not be at stake in a civil case, the individual's financial freedom may be at risk. Fire litigation cases in civil court are usually high-stakes cases. For this reason, they are seriously and aggressively litigated by the parties and their counsel.

A fire litigation case in civil court can directly involve the actions of the public sector investigator and may ultimately rest on his or her findings. Frequently, one of the parties seeks to use a public investigator as an expert witness for its side of the case to have a presumably impartial witness support the case. Private sector investigators, on the other hand, are always challenged on the fact that they have been paid for their testimony and findings. When a public sector investigator can be used in a civil case, the stigma of having been hired to testify for that side of the case is not a factor.

The procedures in civil litigation are essentially the same as the procedures in criminal cases in most respects, yet there are significant differences. The discovery phase of a civil case is usually much longer and much more involved than that in a criminal case. Dozens of witnesses may be deposed. Hundreds of records may be subpoenaed, and numerous procedural motions may be filed by the parties.

The forms of discovery consist primarily of requests to produce, interrogatories, depositions, and reports. The request to produce will specify particular documents or document categories that are to be produced, and must be responded to within a specified time frame. If documents or other items are needed from a non-party to the litigation, subpoenas can be served upon them for the production of relevant documents on a particular date and at a specified location. The interrogatory is a set of questions served on one party by another. The response, called answers to interrogatories, is also provided in writing. Responses to requests for production and interrogatories are made under oath and are signed by the responding party. Depositions are the verbal testimony given by witnesses or experts. The process is done under oath and is documented by a stenographer. The deposition transcript can be used in court if the proceeding goes to trial. Written reports made by experts can be used during all forms of discovery. The written report should address activities conducted, materials reviewed, the opinions the expert will address, and the basis for arriving at such opinions. Publications by the expert, prior expert testimony, and documentation of compensation for work conducted can also be included.

In federal court cases, witnesses who will be testifying as experts must make certain disclosures at the early stages of the litigation under Rule 26 of the Federal Rules of Civil Procedure. At the very outset of the litigation, all expert witnesses must be identified to the other side in a written disclosure. The substance of the expert's anticipated testimony must be disclosed, along with a copy of the expert's reports and analyses. An expert who has not yet prepared a report must do so to comply with Rule 26. The disclosure must include, among other things, the facts or data considered by the witness in reaching the opinions and conclusions in the case. Additionally, the expert must provide a current CV or resume with complete information about his or her educational background, employment history, professional qualifications, fees charged, articles or publications written, and prior testimony as an expert. The information about prior testimony must be provided in detail, including the style (name) of each case, the court in which the case was tried, the attorneys who litigated the case, and the party who hired the witness. This information is required for every case in which the expert has testified as an expert witness within the preceding 4 years.

With increasing frequency, fire investigators are finding themselves more than mere spectators in civil litigation. Lawsuits have been filed against public sector investigators and private sector investigators alike, as well as their agencies and companies. Negligence claims and professional liability (malpractice) claims are asserted against fire scene investigators for improper investigations, incorrect findings, erroneous conclusions, false arrest, and malicious prosecution in criminal cases, as well as for bad faith in insurance claims cases and the spoliation of evidence in both criminal and civil cases.

Negligence

Negligence is conduct that falls within the legal standard established to protect others against harm. A cause of action for negligence from a fire or explosion can develop in more ways than most people can imagine. The essence of negligence is carelessness.

Negligence can range from unplanned accidents caused by mere carelessness to gross disregard for the rights, property, or safety of others. There are a number of defenses to negligence. The conduct of the plaintiff, the one who brings a civil suit to a court of law, is a primary consideration. If the plaintiff's own negligence contributes to his or her injury, the effect will depend on the law in the given state. In most states, the plaintiff's damage award will be reduced by the percentage of the plaintiff's fault. In other states, the plaintiff's claim may be completely barred.

Some examples of damages caused by negligence are as follows:

- Careless disposition of smoking materials (like cigarettes), causing a fire
- Cooking accidents, such as grease fires
- Fires caused by inadequate lightning protection as required by law
- Fires that spread from fireplaces or campground fire pits
- Fires caused by improper use, maintenance, or repair of gas or electrical systems or appliances

As seen in the next section, negligence can sometimes be inferred or proven when a person breaches fire safety laws, causing a fire. Breach of safety codes may also give rise to a negligence action even when the cause of the fire is not known. If a life safety code is breached, causing someone to be injured or killed in a fire, or if the unreasonable spread of a fire can be traced back to the contravention of safety laws, the persons responsible may be liable for negligence.

Product Liability

Manufacturers or sellers of defective products may be liable for persons injured by those products. The defect leading to the injury can arise in manufacturing defects, inadequate warnings, or design defects. This area of law comes from both common law principles and legislation. It is also sometimes called manufacturers' liability.

Fire incidents frequently result in product liability claims. Products giving rise to such claims are many and varied. They include motor vehicles, appliances, and components of building systems like electrical, HVAC, or gas systems. A product liability claim can even result when propane or natural gas for use by consumers is not properly odorized, allowing a leak to go undetected and ultimately cause a fire or explosion.

When any sort of a product is involved in the cause or spread of a fire, it is important for you to explore a number of avenues of investigation. Here is a brief overview of some investigative tasks that may be required:

- Reviewing NFPA 921 subsection 12.3.5 respecting spoliation of evidence before dealing with products at the fire scene. Note the requirements for notification to interested parties and documenting the product prior to altering it or conducting any destructive testing.
- Establishing, as much as possible, the chain of custody of the product as it passed from the manufacturer to the seller and ultimately to the person who owned it at the time of the incident. This means collecting any of the following documents and information that are still available: receipts for the product's purchase; packaging (boxes, warning labels); users' manuals; warranty cards; serial number(s); and information on the make, model, and date of manufacture. In some instances, it may even be necessary to trace how the product was transported and warehoused before reaching the consumer.
- Identifying component parts of the product that may have been defective, resulting in the fire or explosion, and obtaining as many details as possible to determine whether a different company was responsible for manufacturing separate components.
- Tracing the history of the product, including the use (or misuse) by anyone, including the owner, and the service and maintenance history and records.
- Retaining one or more experts who can examine the product and/or its components and provide opinions on any defects in the manufacture, warning labels, or design of the product and/or its components. Experts should also be asked to consider whether repairs or modifications were made to the product that contributed to the loss.
- Obtaining an exemplar of the product so that an undamaged product of the same make, model, and date of manufacture is available for comparison to the fire-damaged product.

Maintaining the chain of custody and the integrity of the product after the fire is also critical. It should be documented in place at the scene, using appropriate methods, including photographing and diagramming and perhaps video recording. If a product will be further damaged if it is moved, it might be necessary to arrange for a portable X-ray to be taken before the product is moved so that the status of its internal components can be recorded. An investigator may need to be creative when packaging and preserving a damaged product. For example, to move a coffee maker when plastic parts have melted, affixing it to a countertop, may require that the entire segment of the countertop holding the appliance be cut away and moved so that the appliance is not damaged further by removing it from the countertop.

Strict Liability

Courts apply the concept of strict liability in product liability cases in which a seller is held liable for any and all defective or hazardous products that unduly threaten a consumer's personal safety. The concept applies to all entities involved in the manufacturing and selling of the product. It is based on the theory that by presenting a product to the public for sale, the manufacturer is representing that the product is safe for its intended use. To prove a product's liability case when strict liability is the standard, the plaintiff typically does not have to prove negligence on the part of the manufacturer or seller.

Wrap-Up

© Greg Henry/ShutterStock, Inc.

Ready for Review

- NFPA 921 and NFPA 1033 can be used in litigation to evaluate the qualifications of experts and to support or challenge the reliability of their opinions.
- All aspects of the investigation can be scrutinized during legal proceedings.
- The legal authority to conduct a fire scene investigation can be granted by law or contract. The scope and limitations on the power of the investigator are contained in those laws and the court decisions that construe those laws.
- Even if an investigator has the legal authority to conduct a fire or explosion investigation, a right of entry must also exist.
- There are special constitutional issues that a public sector investigator must address when conducting witness interviews for the statement of a witness to be legally admissible at trial. These issues involve the Fifth Amendment protection against self-incrimination and the Sixth Amendment right to counsel.

- A seizure occurs when the government exercises control over a person. Arrests and other detentions of persons must comply with the Fourth Amendment safeguards. The circumstances determine whether a person is detained for investigative purposes or is placed under arrest.
- Spoliation of evidence is an inevitable by-product of any fire scene investigation.
- To be a successful witness at trial, whether you are a fact witness or an expert witness, it is very important to become comfortable with the courtroom and with civil and criminal trial proceedings.
- The content of expert reports will be determined by various factors, including procedural rules, your usual practice, and the requirements of the attorney who has retained you or who is calling you as an expert witness.
- Evidence can be categorized in a number of different ways: real or physical evidence, demonstrative evidence, documentary evidence, and testimonial evidence.
- The crime of arson can be committed in many forms, including causing fire or explosion to residences, buildings, vehicles, aircraft, watercraft, personal property, and other structures and materials.
- A prosecutor often has to prove arson entirely by circumstantial evidence rather than by testimony of an eyewitness to the start of the fire.
- In addition to arson, other criminal offenses may be related to a fire scene investigation. These fire-related crimes are numerous, including vandalism at one end of the spectrum and terrorism at the other.
- In a criminal matter, the prosecution must prove the defendant's guilt beyond a reasonable doubt. This is a heavier burden than the burden of proof in a civil case such as a claim based in tort or contract.
- An investigator for a public agency who has conducted a fire scene examination may become involved in civil litigation related to that fire, regardless of the outcome of his or her own investigation and the results of any subsequent prosecution.
- Negligence can range from unplanned accidents caused by mere carelessness to gross disregard for the rights, property, or safety of others.
- Fire incidents frequently result in product liability claims. Products giving rise to such claims are many and varied, but can include motor vehicles, appliances, and components of building systems.

Hot Terms

Affidavit A voluntary written statement of fact or opinion made under oath and signed by the author.

Arson The crime of maliciously and intentionally, or recklessly, starting a fire or causing an explosion. Legal definitions of "arson" are defined by statutes and judicial decisions that vary among jurisdictions.

Chain of custody The trail of accountability that documents the possession of evidence in an investigation.

Circumstantial evidence Proof of a fact indirectly based on logical inference rather than personal knowledge.

Consent Permission from the owner or occupant to examine property.

Constitutional law Law that is based on the U.S. Constitution.

Defendant The entity against whom a claim is brought in a court.

Demonstrative evidence Any type of tangible evidence relevant to a case, for example, diagrams and photographs.

Deposition One type of pre-trial oral testimony made under oath.

Documentary evidence Any type of written record or document that is relevant to the case.

Evidence technicians Individuals who are specially trained to document, collect, preserve, and transport evidence.

Expert witness One who is recognized by a court as qualified by specialized knowledge, skills, experience, or education on an issue.

Fact witness One who testifies about facts from their personal knowledge or experiences; also called a lay witness.

Incendiary fire A fire that is deliberately set with the intent to cause the fire to occur in an area where the fire should not be.

Interested parties One whose legal rights or interests might be affected by the investigation of a particular incident.

Interrogatory A set of questions served by one party involved in litigation to another involved party.

Lay witness One who testifies about facts from his or her personal knowledge or experiences; also called a fact witness.

Motion A request for court action regarding facts, documents, and evidence that are identified during the discovery phase of a legal proceeding. Motions may argue the admissibility of the involved facts, documents, or evidence.

Motive An inner drive or impulse that is the cause, reason, or incentive that induces or prompts a specific behavior.

Negligence Failure to provide the same care that a person with similar training would provide.

Plaintiff The entity who brings a claim to court.

Spoliation Loss, destruction, or material alteration of an object or document that is evidence or potential evidence in a legal proceeding by the one who has the responsibility for its preservation.

Testimonial evidence Verbal testimony of a witness given under oath or affirmation and subject to cross-examination by the opposing party.

Tort A civil wrong leading to a legal claim for damages.

Trial A legal proceeding that is presided over by a judge, who also makes rulings on the admissibility of evidence. Many trials utilize juries as the finder of fact.

© Greg Henry/ShutterStock, Inc.

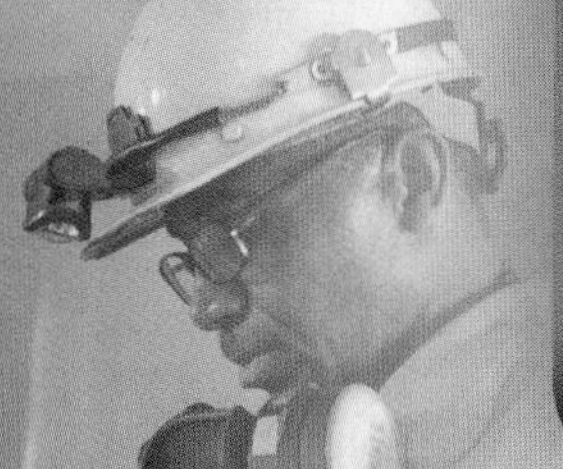

FIRE INVESTIGATOR *in action*

While preparing for a criminal trial, the prosecutor asks to meet with you to review the case. The prosecutor begins by having you tell him about your training and experience and list any cases you have testified in related to fire investigation. He then asks you to explain what your investigation has revealed, specifically the evidence and witness statements you have obtained during the investigation. You advise him of several ignitible liquid samples you had recovered in which a laboratory analysis identified the presence of gasoline and a metal rod with fabric tied to it, which appears to be a homemade torch. Next you explain that several family members had provided verbal statements indicating they had overheard the defendant stating he was going to "burn his house down."

When you complete your overview of the case, the prosecutor seems uneasy and begins to ask specific questions as to what the significance is of each item of evidence you recovered, as well as the credibility of the witnesses. After addressing these questions, the prosecutor states he is uncomfortable with the case because he believes it relies too heavily on circumstantial evidence, and he believes some of the witnesses may recant their statements because they are related to the defendant.

1. The family members' statements would be considered which type of evidence?
   A. Circumstantial
   B. Real
   C. Demonstrative
   D. Testimonial
2. The document used to maintain the accountability of evidence is called:
   A. a consent form.
   B. a chain of custody form.
   C. a curriculum vitae.
   D. Miranda Rights.
3. What type of witness is the person who discovered the fire?
   A. Expert witness
   B. Direct witness
   C. Lay witness
   D. None of the above
4. Which amendment addresses search and seizure during a fire investigation?
   A. First Amendment
   B. Fourth Amendment
   C. Fifth Amendment
   D. Tenth Amendment
5. What is the legal term for the loss, destruction, or altering of physical evidence?
   A. Spoliation of evidence
   B. Destruction of evidence
   C. Elimination of evidence
   D. All of the above

Safety

© Photos.com

© Robert J. Bennett/age fotostock

Knowledge Objectives

After studying this chapter you should be able to:

- Discuss the fire investigator's responsibility for safety at a fire scene NFPA 4.2.1 NFPA 4.2.2 NFPA 4.2.3. (pp 188–190)
- Describe how to conduct a hazard and risk assessment NFPA 4.1.3 NFPA 4.2.1 NFPA 4.2.2 NFPA 4.2.3. (p 190)
- Describe safety clothing and equipment to be used at fire and explosion sites NFPA 4.2.1 NFPA 4.2.2 NFPA 4.2.3. (pp 190–193)
- Discuss the role of personal health and safety for the fire investigator NFPA 4.2.2 NFPA 4.2.3. (pp 193–196)
- Identify and describe factors that influence scene safety NFPA 4.2.2 NFPA 4.2.3. (pp 196–199)
- Discuss the impact of criminal acts or acts of terrorism on fire investigator safety NFPA 4.2.1 NFPA 4.2.2 NFPA 4.2.3. (p 199)
- Discuss safety concerns with materials after they have left the scene NFPA 4.2.2. (p 199)
- Discuss the role of communication in fire investigator safety NFPA 4.2.1 NFPA 4.2.2. (pp 199–200)
- Discuss the role of Occupational Safety and Health Administration (OSHA) standards in keeping fire investigators safe NFPA 4.1.3. (p 200)
- Discuss the possibility of safety concerns related to the public NFPA 4.2.1 NFPA 4.2.2. (pp 200–201)

Skills Objectives

After studying this chapter you should be able to:

- Perform a fire scene hazard assessment so that hazards are identified and removed or mitigated prior to the initiation of the fire scene examination NFPA 4.1.3 NFPA 4.2.1 NFPA 4.2.2 NFPA 4.2.3. (p 190)
- Secure the fire ground and safeguard personnel and bystanders NFPA 4.2.1 NFPA 4.2.2. (pp 189–190, 193–196)
- Determine the best clothing and equipment to use to safeguard the fire investigators NFPA 4.2.2. (pp 190–193)

Additional NFPA References

NFPA 472, *Standard for Competence of Responders to Hazardous Materials/Weapons of Mass Destruction Incidents*

NFPA 473, *Standard for Competencies for EMS Personnel Responding to Hazardous Materials Incidents/Weapons of Mass Destruction Incidents*

NFPA 921, *Guide for Fire and Explosion Investigations*

NFPA 1404, *Standard for Fire Service Respiratory Protection Training*

NFPA 1500, *Standard on Fire Department Occupational Safety and Health Program*

NFPA 1852, *Standard on Selection, Care, and Maintenance of Open-Circuit Self-Contained Breathing Apparatus (SCBA)*

NFPA 1971, *Standard on Protective Ensembles for Structural Fire Fighting and Proximity Fire Fighting*

NFPA 1977, *Standard on Protective Clothing and Equipment for Wildland Fire Fighting*

NFPA 1981, *Standard on Open-Circuit Self-Contained Breathing Apparatus (SCBA) for Emergency Services*

NFPA 1982, *Standard on Personal Alert Safety Systems (PASS)*

CHAPTER 11

FESHE Course Outcomes

Fire Investigation I

9. Recognize potential health and safety hazards. (pp 188–201)

Fire Investigation II

There are no Fire Investigation II (FESHE) course outcomes for this chapter.

You Are the Fire Investigator

© Jones and Bartlett Publishers. Photographed by Glen E. Ellman

Prior to beginning your examination of a basement fire in a 19th-century farmhouse, you notice a large hole in the living room on the first floor. Fire personnel have stated the damage within the basement is extensive, with most of the first floor joists being severely charred. You examine the rooms on the first floor and note several locations where the flooring feels soft and portions of the wood decking begin to break under your weight. At this time, you slowly back away from this area and meet with the incident commander to express your concerns.

1. What safety hazards may exist at the scene that the investigator must consider?
2. How does the suppression of the fire impact your safety during the investigation?
3. What steps can you take to protect yourself and others during the investigation?
4. What safety standards may apply to your investigation?

Introduction

The training that a fire investigator receives provides the necessary foundation for a long, safe, and healthy career; however, in today's rapidly changing society, the fire investigator must always update training and knowledge in the area of safety. Safety should be one of the primary concerns at every fire scene. Fire investigators must follow the safety principles that are utilized during firefighting operations. They must also be trained to recognize and protect themselves and others from the hazards of the fire scene. Management has to address safety standards when creating organizational policies and procedures and should reference documents listed in this chapter among other reference materials.

Fire investigators must be concerned not only with the physical hazards of a fire scene, but also with the chemicals and other toxic substances present at the scene. They must be aware of the long-term effects of repeated exposures to toxins or even one-time exposure to a toxin at harmful levels. There are numerous cases of fire investigators who are fighting serious health problems as the result of investigations. Groups of fire investigators who would normally be expected to be healthy are experiencing cancer at a higher rate than the general population. All too often, fire fighters succumb to illnesses that were caused by their response to toxins or biological compounds present at a fire or created by a fire. The fire departments in Fort Lauderdale (Florida), Newark (New Jersey), and New York City are among those that have experienced fires that caused responders to succumb to similar cancers years after the fire was forgotten. Therefore, the investigator must address the health hazards of fire investigation as well as the physical hazards at each fire scene to ensure the safety of all who operate at the fire scene during the investigative phase of the incident.

Research by the National Institute for Occupational Safety and Health (NIOSH) and the Bureau of Alcohol, Tobacco, Firearms, and Explosives (ATF) in 1998 and 2007 found that many of the compounds present during and after overhaul operations contaminate investigators' clothing and can then impact the health of both the investigator and the investigator's coworkers and families. These findings prompted a recommendation that protective clothing be worn by fire investigators at a scene.

Responsibility for Safety at a Fire Scene

The fire investigator must wear many hats during the course of any origin and cause investigation. NFPA 921, *Guide for Fire and Explosion Investigations,* states that fire scenes, by their nature, are dangerous places that present varied hazards that threaten the immediate and long-term health of the investigator. The fire investigator may be expected to operate in areas that contain toxic atmospheres, compromised structural systems, and sometimes even criminals, so caution must always be exercised. Therefore, investigators have a duty to protect themselves and others (e.g., laborers working at the scene) from hazards that are present at the fire scene, be cognizant of the hazards present at each scene, and stay in control of the scene for safety as well as scene preservation concerns.

The investigative phase is not considered to be part of the emergency phase of the incident, and it is extremely rare for a fire investigation to be conducted under emergency conditions. Nevertheless, it is important to maintain the same safety standards that were in place during the initial phases of the incident.

Safety Tip

When working a room with compromised structural members, tie fire line banner tape to the ends of protruding studs, pipes, and other high and low hazards. The items that you avoid at the beginning of an investigation can be forgotten and cause injury later as your concentration is on digging or observing the fire patterns.

Investigating the Scene Alone

The fire scene contains many hazards that can affect the health and safety of the fire investigator. Some of these hazards have long-term effects, whereas others present an immediate risk to the investigator's health and safety. Ideally, the investigator should not work alone during any fire scene examination. The presence of at least two fire investigators provides for another set of eyes and hands to assess the dangers and handle problems that might be encountered. If an investigator becomes injured or trapped, the second investigator can provide assistance or call for help. The work area must be deemed safe before a fire investigation can begin.

Unfortunately, because of monetary or staffing restrictions, fire investigators are often required to operate alone at fire scenes, even when an immediate hazard may be present. When this is the situation, the investigator needs to notify a responsible off-site contact person of the location, the expected work area at that scene, and the expected time required to complete the investigation. The investigator should always carry a two-way radio, a cellular telephone, or other means of communication. If there is an unexpected delay, the investigator should notify the off-site contact person of the delay and adjust the estimated completion time and area of operation. To prevent complacency, the investigator should be diligent in communicating with the off-site contact person, and the off-site person should contact the investigator if he or she does not hear from the investigator as planned.

Safeguarding Technical Personnel

The fire investigation often involves technical considerations that the investigator might not be qualified to address. This calls for additional experts to take part in the investigation. These experts, although they may be highly trained in their respective fields, are often untrained in how to operate safely at the fire scene. It is the responsibility of the investigator to ensure that all of the personnel who assist with the investigation are made aware of the hazards and of the precautions that have been taken to make the work area safe.

Safeguarding Bystanders

The fire scene is a scene of destruction that is foreign to most people. For this reason, the fire scene—especially a large-loss fire scene—often attracts crowds of people who come to observe the devastation. Some fire scenes take on almost a circus-like atmosphere, with numerous people loitering in the area. These individuals often attempt to get as close as they can to the incident, exposing themselves to the hazards of the fire scene. The fire investigator should request the assistance of police officers or fire personnel to use marking devices such as rope, fire line, or barrier tape to encourage bystanders to maintain a safe distance from the scene. The fire investigator is responsible for the scene, and this responsibility includes the safety and protection of anyone who may wander into a hazardous area FIGURE 11-1.

All bystanders, including the occupants of the fire-damaged structure and nonessential responders, should be kept at a safe distance from the fire scene. The fire investigator (or the incident commander, if applicable) should establish or adjust and enforce collapse zones in their particular area. The suppression team will have already identified the scene's collapse zones, and the incident commander will have control of them and of the general scene safety. The collapse zone must be identified by markers, barricades, vehicles, or specialized scene tape. The incident commander and/or the investigator in charge should inform all personnel of dangers at the scene. Responders should be made aware of the collapse zones verbally because the presence of ordinary scene tape will not necessarily signal a collapse danger to police officers, fire fighters, or contract personnel who are accustomed to crossing scene tape as part of their duties. Bystanders who wander too close to the fire ground could be injured by any one of the myriad hazards present at the fire scene. The fire investigator should also identify a safe, secure place to conduct interviews, away from the distraction and hazards of the fire scene FIGURE 11-2.

While the scene is under government control, the fire investigator can request that the police department assign an officer to remain stationed at the fire scene to prevent unauthorized entry by bystanders or others. To maintain proper

Courtesy of Rodney J. Pevytoe

FIGURE 11-1 Example of a hazardous area.

Fire Investigator Tip

The person in control at an explosion or fire investigation scene should have the following items and information, at a minimum:

- First aid kit
- Local emergency notification numbers
- Knowledge of the location of emergency medical care
- An emergency notification signal

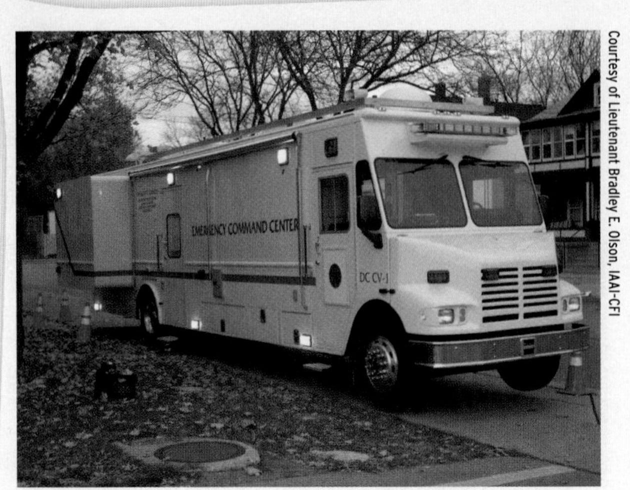

Courtesy of Lieutenant Bradley E. Olson, IAAI-CFI

FIGURE 11-2 Identify a safe, secure place to conduct interviews, away from the distraction and hazards of the fire scene.

security and to preserve the evidence, the investigator should maintain a written log of all authorized people who enter the scene or should ensure that another authorized person maintains such a log. If any unauthorized people enter the scene, the investigator should make note of the identity of such people, the date and time of entry, and the area in which the unauthorized people entered, after which they should be required to leave the scene.

Hazard and Risk Assessment

One of the first tasks the investigator should conduct at a fire scene is the assessment of the hazards and risks. This is a three-step process. This detailed assessment will allow the investigator to:

1. Identify the hazards.
2. Determine the risk of the hazards.
3. Control the hazards.

Identify the Hazards

The hazards identified on scene may be classified as follows:

- Physical hazards: Are there many places that an investigator could slip, trip, and fall?
- Structural hazards: Is the building structurally sound?
- Electrical and gas hazards: Is the electrical and/or gas system still connected/intact?
- Chemical hazards: Is there a material safety data sheet (MSDS) or safety data sheet (SDS), or are there other reference materials available to manage the chemical hazards present?
- Biological hazards: Is there a risk of a hazard caused by bacteria, virus, insects, plants, humans, or animals?
- Mechanical hazards: Is the equipment still operational or functional? Is the machinery specialized, requiring a technical expert? Are all guards in place for all moving equipment?

Determine the Risk of the Hazards

Although all fire scenes pose some form of hazard to the investigator, the level of risk at each will vary. The investigator should access the hazards present and determine the likelihood that contact would be made with the hazards and, if so, in what form. Knowing this will help to address the measures required to control the hazard.

Control the Hazards

With the identification of a hazard, the investigator must then determine the most effective manner to mitigate the risks by comparing the risk level to a suitable benchmark or acceptance criterion. The risks may be controlled with the following:

- Engineering controls, such as shoring to reinforce damaged structural areas, demolition after documentation, or, in complex cases, obtaining the services of a structural engineer.
- Administrative controls, such as the isolation of an area with the use of effective signage, barrier tape informing personnel of the hazards, or restriction of entry to key personnel.
- Proper selection and use of personal protective equipment (PPE), including boots, gloves, helmet, clothing, and eyewear, with a greater level of protection as deemed necessary by the scene assessment. This is generally considered the least effective of the control measures, but it may be a suitable one under the conditions encountered by the investigator.

Safety Clothing and Equipment

Although there is rarely a good reason for a fire investigator to enter the burning building before the fire has been brought under control, fire investigators must protect themselves from physical, biological, chemical, and radiological hazards while conducting an origin and cause investigation. Toxic gases, hazardous chemicals, asbestos, flammable gases, and oxygen-deficient atmospheres are a few of the dangers at almost every fire scene.

In the initial phases of the fire incident, the fire investigator will be busy conducting interviews and documenting the scene as well as searching the surrounding areas for potential demonstrative, documentary, or testimonial evidence. If an occasion arises in which the investigator feels that there is a need to enter the building before the fire has been extinguished, the investigator must be equipped with structural firefighting gear and a self-contained breathing apparatus (SCBA), and must be trained in its proper use **FIGURE 11-3**.

This protective clothing and equipment must meet or exceed the requirements of NFPA. Just as police officers are saved by their bulletproof vests and fire fighters are saved by their turnout gear and SCBA, fire investigators have protective clothing and equipment available to them that should be worn and used at all fire scenes. Fire investigators must be trained in the correct usage, limitations, and decontamination (or disposal) of clothing and equipment used at the scene. Protective

© Jones & Bartlett Learning. Photographed by Glen E. Ellman

FIGURE 11-3 The complete structural firefighting ensemble consists of a helmet, coat, trousers, protective hood, gloves, boots, SCBA, and personal alert safety system (PASS) device.

clothing and equipment should be properly decontaminated prior to the investigator leaving the scene. If this is not possible, equipment and clothing can be removed at the scene, transported in such a way that contamination cannot spread, and later washed according to the manufacturer's instructions.

Safety Tip

Fire investigators often wear fire fighter PPE when working at a scene. The PPE for fire fighters must be manufactured according to exacting standards. Each item must have a permanent label verifying that the particular item meets the requirements of the standard. Usage limitations as well as cleaning and maintenance instructions should also be provided.

The requirements for firefighting PPE are outlined in two standards:

- NFPA 1971, *Standard on Protective Ensembles for Structural Fire Fighting and Proximity Fire Fighting*
- NFPA 1977, *Standard on Protective Clothing and Equipment for Wildland Fire Fighting*

The requirements for SCBA and PASS devices are outlined in two standards:

- NFPA 1981, *Standard on Open-Circuit Self-Contained Breathing Apparatus (SCBA) for Emergency Services*
- NFPA 1982, *Standard on Personal Alert Safety Systems (PASS)*

Other requirements, for training and maintenance, are set forth in two additional standards:

- NFPA 1404, *Standard for Fire Service Respiratory Protection Training*
- NFPA 1852, *Standard on Selection, Care, and Maintenance of Open-Circuit Self-Contained Breathing Apparatus (SCBA)*

The Helmet

The protective ensemble should start with a helmet, which should be worn during all fire investigations. It is important to choose a helmet that is sturdy enough to provide the desired level of protection, but is comfortable so that the investigator will wear it during extended operations. This helmet must protect the wearer from impact and from electrical hazards. An improperly constructed helmet or a helmet with unapproved modifications could cause injury either from being struck or by electrocution. There should also be a suspension system to absorb all but the most severe impact. The helmet should have a chinstrap to hold it in place after an impact. The chinstrap should be used whenever the helmet is worn in potentially hazardous areas, because without it the helmet may fall off and additional falling debris could cause severe head injuries. The helmet may also have some form of built-in eye protection that is easily deployed by the wearer. This eye protection should be constructed of a material that resists scratches.

Protective Clothing

The investigator should wear protective clothing to prevent contamination of street clothes. This clothing could be firefighting turnout gear or coveralls. The fire investigator is exposed to toxins at every fire scene. These toxins become embedded in the investigator's clothing and in anything the investigator contacts after the exposure—including the investigator's vehicle, office furniture, and home—if proper precautions are not taken. This cross-contamination exposes the investigator and his or her coworkers and families to the toxins from the fire scene. The coveralls or turnout gear should be washed after every fire scene examination in accordance with the manufacturer's instructions. This procedure often calls for machine washing with a mild soap and water—but not in a washing machine that is normally used to clean other clothing. A separate washing machine or a company that specializes in cleaning firefighting equipment should be used. Certain toxins that are present at every fire scene (such as benzene and hydrogen cyanide) are absorbable through the skin. The improper cleaning of protective clothing can cause repeated exposure to these carcinogens, as well as others. The use of disposable coveralls would eliminate having to wash your protective clothing.

Respiratory Protection

Today's fire investigator also needs to address respiratory protection at every fire scene. This respiratory protection could come in the form of SCBA, cartridge filter masks, or engineering controls such as mechanical ventilation. No matter which form of respiratory protection is chosen, the work area should be monitored at all times during the investigation. The ideal protection can be obtained with the use of engineering controls to remove the hazard from the work area. Positive-pressure ventilation will provide the work area with fresh air efficiently and effectively. There may be times when positive-pressure ventilation is not practical. During these investigations, the fire investigator can use a combination of positive-pressure ventilation and negative-pressure ventilation to help remove the toxic gases from the area or reduce their concentrations.

The fire investigator should have a good grasp of the theory behind mechanical ventilation to be able to ventilate any area where the fire investigation may take place. Atmospheric monitoring must be performed before the investigator may be in a fire-affected building without respiratory protection, and this monitoring should be repeated periodically during the investigation. The investigator must understand that even if an atmosphere is found to be acceptable in a specific working area in a building, hazardous atmospheres could still exist in closed or hidden spaces, basements, and other below-grade areas.

When respiratory protection is utilized, the employer must have a written respiratory protection program in place. All employees who will be utilizing respiratory protection should be trained in the selection, use, and maintenance of the proper equipment for the hazards encountered. The training should also address emergency procedures for any equipment failures. Medical surveillance is also required to determine the physical abilities and limitations of the user. Periodic medical examinations may be necessary to determine whether any employee has been exposed to a toxin that has the potential of causing a health issue.

Self-Contained Breathing Apparatus

There may be times when ventilation alone cannot decrease the levels of the toxins to a safe level. During these investigations, respiratory protection must be utilized. SCBA provides the highest level of respiratory protection FIGURE 11-4.

© Jones and Bartlett Learning

FIGURE 11-4 Because SCBA carries its own air supply in a pressurized cylinder, its use is limited by the amount of air in the cylinder.

Unfortunately, SCBA limits the amount of time that the user can be protected because of the limited amount of breathing air in the compressed air cylinder. Usually, a compressed air cylinder is rated to provide 30 to 60 minutes of breathing air, although this time varies depending on the physical condition of the wearer and the level of physical and mental strain the user is experiencing. The SCBA face piece also restricts the visual abilities of the user, thereby limiting the investigator's ability to observe the scene. An alternative, the supplied air breathing apparatus (SABA), provides breathing air for longer periods, but the hose is limited to 300 feet. The investigator who wears any breathing apparatus should work in 30-minute periods to allow for rehydration, rest, and monitoring of physical condition caused by the stress and strain of wearing a breathing apparatus. If the investigator is utilizing SABA, it is tempting to extend the 30-minute work period, exposing the investigator to dangerous strains and exposures to heat stress, which should be avoided.

Air-Purifying Respirator

The air-purifying respirator (APR) can be utilized at certain fire scenes; however, this type of respiratory protection has severe limitations. Air monitoring is a must when using an APR because the user is not protected from oxygen-deficient atmospheres (oxygen levels below 19.5 percent) or from areas where toxins exist at a level that is deemed immediately dangerous to life and health (IDLH). The APR is toxin-specific, so the user must know all the hazards that are present at the scene to be able to select the most effective filter. Unfortunately, there is no possible way to identify all the hazards present at any fire scene. Whenever unknown quantities of two or more molecular compounds break down because of their exposure to an unidentified amount of heat, as in a fire, it is impossible to determine exactly what material is ultimately created and what its level of toxicity might be. Investigators can determine general hazards, but the unknown is always present at any fire scene.

APRs are provided with and without eye protection. FIGURE 11-5 shows a form of APR without built-in eye protection. For those APRs that do not have built-in eye protection, the investigator should wear vented goggles to protect the eyes.

Eye Protection

The fire investigator should wear eye protection that provides the maximum level of safety. Snug-fitting goggles are more effective than the face shield on a helmet when fire debris becomes airborne and could contact the user's eyes. To protect the eyes fully, safety goggles should be used along with the eye protection provided by the helmet. The investigator should remember that beyond obvious eye injuries, any toxins that enter the eyes will quickly enter the circulatory and nervous systems. Eye protection should also shield the sun's ultraviolet rays.

Foot Protection

Proper foot protection can prevent major and minor health-threatening injuries. A safety shoe protects the foot from

FIGURE 11-5 Air-purifying respirators can be used only when there is a sufficient amount of oxygen in the atmosphere.

the hazards associated with stepping on nails or sharp objects, as well as from heavy objects falling on the toes. A properly waterproofed shoe protects against exposure to toxins that are present at fire scenes. Rubber boots provide a high level of protection, as do the traditional thigh-high firefighting boots when they are pulled up over the knees and used in association with hard kneepads. Boots utilized at the investigation should conform to NFPA 1971.

■ Gloves

The fire investigator should wear gloves at every fire scene. Traditional firefighting gloves provide limited protection from heat and physical injuries, but they do not provide any chemical protection. Therefore, the fire investigator should consider double-gloving during the investigation, wearing an inner glove that provides a limited level of chemical protection. The inner glove should be made of a material that has been chosen after conducting an assessment of the hazards at the fire scene. For instance, latex gloves provide protection from biological agents but are not effective during extended periods of wear under wet conditions. For the average fire scene, nitrile gloves with some puncture resistance are commonly used and provide a higher level of protection. The inner glove should be worn from the beginning of the investigation. To prevent cross-contamination, the investigator should don a new pair of inner gloves before collecting each sample.

■ Safety Equipment

Each fire scene is different, and at times, additional equipment may be needed to provide a safe work environment. All work areas should be well lit to prevent trip-and-fall injuries and to enable the investigator to assess the hazards effectively and complete the investigation. Portable lighting is required at almost every fire scene, but at no time should a generator be brought into an enclosed work area; the buildup of carbon monoxide would quickly become a serious health issue. Furthermore, when an internal combustion engine is brought into the building, the fire investigator can no longer state that the investigation team did not bring an ignitible liquid into the scene.

Some investigations require lifelines and fall protection. This equipment should be maintained and stored in accordance with the manufacturer's specifications. The investigator should not depend on any ladders that are present at the fire scene other than those brought by the fire department. Any ladders that were in the fire building may have been exposed to the effects of the fire, or the owner of the building may have abused or improperly maintained the ladders before the fire. If a ladder is required, the investigator should be comfortable with its condition and should ensure that it is the proper height and style for the intended use.

Some safety equipment requires specialized training and experience. Even if the investigator is capable of operating specialized equipment, an operator who routinely works with the equipment is more skillful and should be utilized if available. For instance, shoring equipment or other urban search and rescue techniques may be required to protect the investigator at a fire scene.

Personal Health and Safety

The investigator should guard against complacency, which can lead to injury and/or exposure to the toxins present at every fire scene. The fire investigator cannot afford to take a chance or risk. These risks and exposures over the duration of a career can accumulate and lead to life-threatening health problems.

There are many chemical, biological, and radiological hazards that may threaten personal health and safety. For instance, the investigator who responds to a medical or construction facility can be exposed to radiological materials that are stored at these facilities. In addition, nearly every fatal fire scene contains biological hazards. Rarely does the fire destroy all of the bodily fluids of a fire victim. Blood tests show that nearly all fatal fire victims have toxic levels of cyanide in their systems—levels so high that some medical examiners are questioning whether carbon monoxide (generally the most prevalent toxin at a fire scene) or cyanide is killing fire victims. Even nonhuman biological materials can cause allergic reactions, infection, and diseases. Biologic substances from plants, animals, insects, bacteria, viruses, trash, and so forth all pose potential hazards to those on a scene.

Materials that are used in today's society are causing every fire scene to be volatile and hazardous. Toxins such as benzene, toluene, formaldehyde, and cyanide, among others, can be found at every fire scene. These toxic materials pose a problem when they get to the susceptible target organ at levels above

Fire Investigator Tip

General safety meetings at an investigation scene should occur as often as necessary, but a minimum of twice daily. Ideal times for safety meetings are the start of day, end of day, at organized breaks, or when beginning a new phase of the investigation.

the toxicity limits or in a form that the body cannot overcome. The toxic reaction depends on the dose and the duration of the exposure. The dose can be acute, whereby a single incident exposes the individual to harmful levels of the toxin, or chronic, whereby the individual is exposed at a lower concentration, but the toxin is absorbed over an extended period of time until a harmful accumulation is present in the victim's system.

The investigator must also remember that toxins remain in the air after the fire has been extinguished. During the investigation phase of the fire incident, the toxic gases have cooled and are settling or mixing with the air throughout the room, where they are affecting the fire investigator through inhalation or absorption through the skin. In addition to the chemicals released in a fire, a fire at a drug lab or similar occupancy can greatly increase the risk of harmful chemical contamination. Even a residential fire can expose investigators to pesticides and other chemicals. Knowledge of the effects and routes of exposure of chemicals can help those on scene protect themselves against toxins TABLE 11-1.

The fire investigator also has a risk of inhalation, ingestion, and skin absorption of harmful airborne particulate material. Asbestos, silica dust, heavy metals, and many other substances are present at many fire scenes. Like other contaminants, these products may also adhere to the investigator's clothing and equipment. As part of overall scene safety, proper decontamination of equipment and person should be practiced.

Other hazards that potentially exist at fire scenes occur because fire consumes oxygen and can create an oxygen-deficient atmosphere in confined areas. Chemicals that are present at the fire scene may have been released because of the failure of their containers. Carbon monoxide, a common by-product of fire, can cause chemical asphyxiation. Carbon monoxide joins with hemoglobin 250 times more readily than does oxygen, thereby preventing the body's cells from receiving oxygen. This effect can be cumulative and will build after several exposures, causing a collapse in less-than-toxic atmospheres. Cigarette smoking is an example of the cumulative effect of carbon monoxide, which can raise the carbon monoxide level in an individual to 10 to 15 percent even without the individual's being exposed to a fire. It is important to keep in mind that the atmosphere may change during an investigation. A safe atmosphere may turn toxic, and continual atmospheric monitoring may be required. Hydrogen cyanide itself is considered one of the most dangerous gases at an investigation scene due to the level of exposure from smoldering synthetic materials. Consideration should be given to having a hydrogen cyanide detector present at every fire scene investigation.

The physical hazards that cause trip, fall, or crushing injuries are well known to everyone in the field. These injuries occur because fire scenes are cluttered and dangerous places. The fire investigator must constantly be aware of his or her footing and also take note of what is standing nearby or hanging above the work area. Animals, including pets, may present a hazard to the fire investigator if they are distressed.

The fire investigator should assess the fire scene and choose the proper PPE to block the routes of exposure from the hazards present at each specific fire scene. A fire investigation is not a life-or-death situation. Therefore, the investigator should not risk health and safety to determine the origin and cause of any fire. Utilize engineering controls available to protect the investigative team from harmful materials. An example would be a fire in a building with possible asbestos contamination, which would need to be mitigated before the investigation. This is often easier said than done; however, with the proper vigilance and by resisting complacency, the fire investigator can be protected from the known harmful materials present at every fire scene.

Remember that hazardous substances can also be used as a form of terrorism. Be alert and always consider the threat of terrorism or that a crime has been committed.

The fire investigator should discuss any possible hazards with the site manager and utilize SDS and site safety plans before beginning the investigation FIGURE 11-6. Keep in mind that there may be several safety plans of varying topics and complexities in use at a scene TABLE 11-2.

■ Investigator Fatigue

The reconstruction and investigation of a fire scene are often long, tedious, and physically strenuous processes. Safety measures require the frequent use of heavy clothing and respiratory protection. All of these factors may combine to fatigue the investigator. Fatigue may cause the investigator to tire and lose strength and judgment, all of which could place the investigator in jeopardy. As part of investigator safety, it is important to receive periodic rest, food, and fluids in a safe environment. Sanitation and wash stations are required. Proper washing will also help prevent ingestion of contaminants.

Table 11-1 Chemical Exposure

Effects	Routes	Toxicity
■ Local (site of contact) ■ Systemic (affecting organs)	■ Inhalation (breathing) ■ Cutaneous (absorption through skin or eyes) ■ Ingestion (through the mouth) ■ Injection (penetration from contaminated object)	■ Acute (high exposure over short time) ■ Chronic (repeated low exposure over long time) ■ Cumulative (repeated exposure) ■ Latency (period between exposure and effects)

Courtesy of Tanner Industries, Inc.

MATERIAL SAFETY DATA SHEET
ANHYDROUS AMMONIA

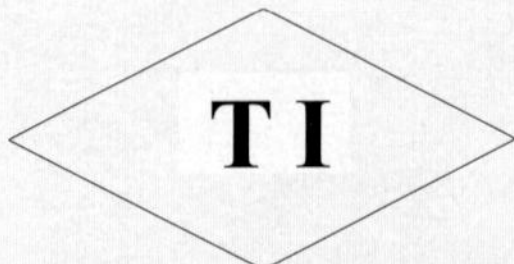

DISTRIBUTORS:
TANNER INDUSTRIES, INC.

DIVISIONS:

NATIONAL AMMONIA	NORTHEASTERN AMMONIA
HAMLER INDUSTRIES	BOWER AMMONIA & CHEMICAL

735 Davisville Road, Third Floor, Southampton, PA 18966; 215-322-1238
CORPORATE EMERGENCY TELEPHONE NUMBER: 800-643-6226 CHEMTREC: 800-424-9300

DESCRIPTION

CHEMICAL NAME: Ammonia, Anhydrous **CAS REGISTRY NO**: 7664-41-7
SYNONYMS: Ammonia **CHEMICAL FAMILY**: Inorganic Nitrogen Compound
FORMULA: NH_3 **MOL. WT**: 17.03 (NH_3) **COMPOSITION**: 99+% Ammonia

STATEMENT OF HEALTH HAZARD

HAZARD DESCRIPTION:
Ammonia is an irritant and corrosive to the skin, eyes, respiratory tract and mucous membranes. Exposure to liquid or rapidly expanding gases may cause severe chemical burns and frostbite to the eyes, lungs and skin. Skin and respiratory related diseases could be aggravated by exposure.

Not recognized by OSHA as a carcinogen.
Not listed in the National Toxicology Program.
Not listed as a carcinogen by the International Agency for Research on Cancer.

EXPOSURE LIMITS FOR AMMONIA: Vapor

OSHA	50 ppm,	35 mg / m³ PEL	8 hour TWA
NIOSH	35 ppm,	27 mg / m³ STEL 15 minutes	
	25 ppm,	18 mg / m³ REL	10 hour TWA
	300 ppm,	IDLH	
ACGIH	25 ppm,	18 mg / m³ TLV	8 hour TWA
	35 ppm,	27 mg / m³ STEL 15 minutes	

TOXICITY: LD 50 (Oral / Rat) 350 mg / kg

PHYSICAL DATA

BOILING POINT: -28°F at 1 atm.
PH: N/A
SPECIFIC GRAVITY OF GAS (air = 1): 0.596 at 32°F
SPECIFIC GRAVITY OF LIQUID (water = 1): 0.682 at 28°F (Compared to water at 39°F).
PERCENT VOLATILE: 100% at 212°F
APPEARANCE AND ODOR: Colorless liquid or gas with pungent odor.
CRITICAL TEMPERATURE: 271.4°F
GAS SPECIFIC VOLUME: 20.78 Ft^3/Lb at 32°F and 1 atm.

VAPOR DENSITY: 0.0481 Lb/Ft^3 at 32°F
LIQUID DENSITY: 38.00 Lb/Ft^3 at 70°F
APPROXIMATE FREEZING POINT: -108°F
WEIGHT (per gallon): 5.15 pounds at 60°F
VAPOR PRESSURE: 114 psig at 70°F
SOLUBILITY IN WATER (per 100 pounds of water): 86.9 pounds at 32°F, 51 pounds at 68°F
SURFACE TENSION: 23.4 dynes/cm at 52°F
CRITICAL PRESSURE: 111.5 atm

Revision: September 2005 Page 1 of 4 Prepared By: JRP

FIGURE 11-6 Material safety data sheets (MSDS) like this one are being transitioned to SDS. Like the MSDS, an SDS will provide all of the information that you will need when dealing with hazardous chemicals.

Table 11-2 Emergency Action Plans

Plan	Description
Emergency evacuation plan	■ Used when the scene suddenly becomes unsafe ■ Includes exit routes, gathering location, and confirmation of complete evacuation
Medical emergency plan	■ Used when someone on-scene requires medical care ■ Includes locations of emergency facilities, phone numbers, and a first-aid kit
Severe weather plan	■ Used when there is severe weather ■ Includes notification method and meeting place for those on-scene
Fire emergency plan	■ Used when a fire or explosion occurs during an investigation ■ Includes fire department location, phone number, and notification method; exit routes; meeting place(s); and confirmation of complete evacuation for those on-scene
Additional emergency action plan	■ Used when scene-specific issues arise ■ Includes information specific to the scene, as determined by the person in control of the scene

© Glen E. Ellman

FIGURE 11-7 Fire suppression is required before the investigator can proceed with a full investigation.

Factors Influencing Scene Safety

There are many factors that can influence the safety of a fire or explosion scene. The stability of the structure as a result of the fire suppression activities is one example. Utilities also pose hazards to the investigator; for example, live wires may be in close proximity to water that was used for suppression. This section discusses these hazards in more detail.

Status of Suppression

An investigator should not enter a burning structure or one that is not completely extinguished without the permission of the incident commander. If such entry is necessary, the investigator should not do so unless he or she is trained to make such entry and is accompanied by fire suppression personnel. The fire investigator must also coordinate all ongoing investigation activities with the suppression crews and keep the incident commander advised of the investigator's progress and of any additional areas being entered by the investigator. The urge to "freelance" should always be resisted. During this suppression phase of the incident, the priority is to extinguish the fire and protect lives, not to conduct an investigation, although the two can be done simultaneously if the circumstances so require FIGURE 11-7.

When an investigator has begun an investigation in what is believed to be an extinguished structure fire, the fire investigator must be aware of the possibility of a "rekindle" occurring in the building. The rekindled fire could grow to the point at which the fire investigator's escape is cut off. For this reason, the means of exit should be at opposite ends of the work area, and hose lines should remain in place during the initial phase of the investigation. When the investigation is complete, overhaul operations must be conducted to ensure that the fire has been completely extinguished.

> **Fire Investigator Tip**
>
> The fire investigation community teaches fire fighters not to destroy the fire scene. Classes on the fire fighter's role in origin and cause investigations stress the need to keep overhaul to a minimum until the scene has been documented and examined by an investigator. Because of this training, overhaul is commonly kept to a minimum, and firefighting operations are suspended when the forward progress of the fire is stopped.

Whenever the investigator enters a fire-damaged building, he or she must remain aware of the fastest or safest means of escape. The investigator should, if possible, identify two separate means of escape from the area of operation. One means of escape could be a ladder being placed to a window or to the roof. If this type of escape is necessary, the ladder must remain in place at all times when the area is occupied by the investigator.

The separate means of escape should be monitored to ensure that it remains free of obstructions during the investigation.

Structural Stability

Before entering any fire scene, the investigator must assess the stability of the structure and eliminate any physical hazards by either the use of shoring materials or demolition. Depending on the nature of the structure, the investigator may need to seek the help of a qualified structural engineer. The fire investigator must not take for granted that any portion of a building that remains standing after the fire is secure and safe. It is important to know that a collapsing wall normally does not make any sound until it hits the ground or something standing between the falling material and the ground. The investigator should also be mindful of hidden holes or weaknesses in the floor caused by the fire itself; by water buildup from the suppression activities; or by weather-related weight factors such as ice, snow, wind, or rain.

Freestanding chimneys pose a substantial risk to both fire fighters and fire investigators. These structures depend on the building structural members for their stability. When the building is compromised, so is the chimney.

The portions of the building that are not supported by trusses are also vulnerable to collapse. Most buildings are built with the minimum amount of materials required by engineering principles and local building codes. It is rare for a building to be constructed with additional materials over those required by code. It is also possible that a building's construction was deficient and not consistent with the applicable code or that a building that conformed to the applicable code when it was built would not be compliant with current codes. The removal of any portion of a building places unintended loads on the remainder of the building, which in turn puts the building in danger of collapse. When the roof collapses, the walls are now largely free standing. If a wall collapses, the roof and other walls are now in danger of succumbing to the forces of gravity. All components of the building's construction are interdependent; the removal of any component alters the system that was put in place to form the building.

Building Load

A building has two different types of building loads: the dead load and the live load. (See the "Building Systems" chapter in this textbook for more information on building loads.)

When the structure is designed, the designer takes into account its intended use and the expected live load. Often, because of changing occupancies, buildings are used in ways that the designer did not intend. Walls are moved when occupancies change from residential to commercial to industrial. These changes normally do not take into account the design of the building, thereby placing the building in stress. Poor maintenance, poor repairs, and aging of the building materials induce further stress on the structure. A fire can be the final straw or the means to set in motion the cause of a later collapse. When wood studs or joists are compromised or destroyed by fire, the surrounding studs or joists must carry the load. The studs may be able to accomplish this for only a short period of time before failure.

Impact Load

Another consideration related to structural stability is the impact load, which is a sudden added load that can be caused by a fire fighter jumping from a ladder onto a roof or a fire investigator jumping down onto a floor. This sudden load may cause the collapse of already weakened or overburdened structural members. A partial building collapse may also provide the impact load that brings down the remainder of the building.

The most stable type of structure is the pyramid, which utilizes a decreasing pattern of masonry building materials. This "pile" will resist collapse. If we take these same building materials and stack them on top of each other, we cause instability. A freestanding wall, whether it is a parapet wall (the portion of a wall that extends above the level of a roof) or any wall that is no longer supported by other structural members, should never be trusted without shoring. The safest way to handle a freestanding wall is to allow gravity to accomplish its goal and knock the wall down in a manner and at a time that the fire investigator determines.

An investigation in a building where the stability of the structure is under question will strain any fire investigation team. The team should make the work area safer by mitigating the hazardous condition. If the stability of the building is a concern, the investigator should utilize a controlled demolition and stabilization of the structure. Demolition is the safest because the hazard is removed. The demolition team should work under the direct supervision of a trained fire investigator who has intimate knowledge of the goals of the investigation. This investigator can then direct the equipment operator to remove the portions of the structure that are unstable without destroying the areas of interest to the investigation team. A good heavy-equipment operator working in conjunction with an experienced fire investigator can remove the hazardous portions of a fire building without destroying evidence or the investigation FIGURE 11-8.

This controlled demolition will make the fire scene safer and allow for a successful completion of the origin and cause investigation.

Utilities

Fire investigators will often come into contact with the fuel and electrical systems in a fire scene and will need to examine these systems. An investigator may incorrectly make the assumption that a system has been disconnected by fire suppression crews. Do not assume. Determine before entering a scene whether an electrical system is energized, the lines of a fuel system are charged, or water mains are operating.

Courtesy of Mike Dalton

FIGURE 11-8 Technical personnel such as heavy-equipment operators work in conjunction with investigators to remove hazardous debris or uncover evidence.

Electrical Hazards

The fire investigator should never assume anything when it comes to electrical safety. The fire department may have disconnected the power only to have it reenergized later. In addition, some structures may have more than one electrical feed, allowing one system to be shut down and another energized within the same structure. Temporary wiring is often not installed, grounded, or insulated correctly. There are many electrical hazard scenarios that could pose a risk to the investigator:

- Assume that all wires are energized, even when the meter has been removed or disconnected.
- Be alert to fallen electrical wires.
- Look out for antennas that have fallen on power lines, for metal siding that has become energized, and for underground wiring.
- Operate ladders and other equipment cautiously when near overhead electric lines.
- Do not depend on rubber footwear as an insulator.

Safety Tip

Put a small AC voltage detector in your pocket at the beginning of an investigation. It will always be handy and is less cumbersome than a multimeter, because it can be used to test for power in circuit breaker panels, outlets, and electrical runs. It can also detect voltage through the insulation of undamaged conductors.

Safety Tip

Be alert for puddles of water near energized electrical wires, appliances, or other equipment.

- Do not enter a flooded basement if the electrical system is energized.
- Shut off electric power remotely if the scene contains an explosive atmosphere.
- Communicate and cooperate with the utility company and make use of their expertise.
- Locate and avoid underground electric supply cables before digging or excavating.
- Be alert for additional electrical services that may not be disconnected, extension cords from neighboring buildings, emergency lighting units, backup generators, and similar installations. These electrical units remain energized after the electrical power has been shut off.
- Use a meter to determine whether the electricity is off.

Although the investigator should not disconnect electrical power, the investigator should ensure that the proper utility company does. The investigator should be knowledgeable about the safe use of testing equipment such as a voltmeter or multimeter. This will allow the investigator to test electrical systems personally for energy. Avoid electrocution by ensuring that the testing device is rated for the voltage supplied to the structure. Once electrical power has been disconnected or interrupted, the use of a lockout/tagout process should be employed by the investigator.

Standing Water

Firefighting operations themselves can add to the stress on a building's structural members. Water weighs 8.35 lb/gal (3.8 kg/L). A $2^1/_2$-inch (6.35-cm) or even a $1^3/_4$-inch (4.45-cm) hose line putting out up to 250 gal (946.4 L) per minute will place an additional live load of 2087.5 lb (946.9 kg) into the building every minute that that hose line is in operation. This additional load must be drained before the investigation can begin. During cold-weather investigations, the removal of this water can be difficult or even impossible. The water may be of unknown depth and may contain unseen hazards. Similarly, standing suppression foam can hide holes and hazards from the investigator's view.

Fire fighters often cut holes in floors to remove water from the structure. The fire investigator must be aware of this possibility and identify the locations of any holes. Often, in urban areas, fire fighters use doors from the building to cover these holes and any holes caused by the fire itself. These doors must not be relied on to hold the weight of the fire investigator.

Criminal Acts or Acts of Terrorism

Fires and explosions are powerful weapons that may be utilized in criminal acts or acts of terrorism. In the event that a device is used as the initiator of the fire, the investigator should consider the potential of additional devices in the area. Devices may have failed to function or been intentionally left to harm personnel. Devices may also utilize hazardous materials in their construction that can leave a residue, creating an additional exposure.

■ Secondary Devices

A commonly used terrorist tactic is to leave secondary devices that are meant to explode or deploy after the initial incident in order to cause damage or injury to first responders and to create additional panic. Personnel on the scene should be wary of any unusual packages or containers. Officials may need to utilize resources from hazardous materials or explosive ordnance experts to address identified or potential risks.

■ Residue Chemicals

Fires and explosions may be initiated or enhanced with the use of various chemicals. Investigators should use appropriate protective clothing and respiratory measures when handling any artifacts of a device or equipment because exposure could harm the investigator.

■ Biological and Radiological Terrorism

As part of a fire or explosion, a terrorist may employ the use of biological or radiological components. Any scene that has been determined to pose a substantial risk of these types of components will require special safety and training measures for all personnel. This can include the safe rendering of the scene prior to the entry of investigators. If the scene cannot be rendered safe, then only those investigators trained and properly equipped to work in such an environment should be allowed to enter the scene.

■ Exposure to Tools and Equipment

Any items worn or utilized in an investigation have a risk of contamination after being used in hazardous environments. All such items should be handled and cleaned with this risk in mind. Equipment that cannot be decontaminated should be disposed of in an appropriate manner.

Safety and Off-Scene Investigation

Safety and potential exposure issues do not end when the investigator leaves the scene. All items removed from the scene, including evidence and equipment, should be handled, labeled, and stored in a safe manner.

Communication

The fire investigator must be able to operate at the fire scene in conjunction with the fire fighters. The investigation must be conducted within the incident command system (ICS) through the incident commander.

The fire investigator has a responsibility to contact the incident commander when he or she arrives at the scene. The incident commander can relay any pertinent information about the fire that could be valuable to the investigation. During this transfer of information, the fire investigator should also determine whether the fire fighters have discovered any hazards. The incident commander can relay information about structural stability, chemicals, or any other hazards present at that scene. A safety plan should be formulated to address such topics as PPE, emergency action plans, hazardous materials, and physical hazards. Depending on scene complexity, the safety plan may include a formal organizational structure. Safety meetings should be scheduled several times per day.

The emergency action plan should address escape routes, medical and fire emergencies, accountability, severe weather, and any other scene-specific potential hazard. When confined spaces are involved, a site program should be developed and implemented by confined space–trained personnel.

While investigating large losses in large buildings or in areas where it is difficult to keep track of the investigation team—especially if suppression activities are still in progress—an ICS, including an accountability system, must be utilized. The investigation team must report to the accountability officer and verify the areas where the investigators will be working and the specific jobs in which they will be involved while operating in that area. When they are finished in that location, the investigators should inform the accountability officer of the completion of that portion of the investigation and then verify the new location where they will be working. This information allows the accountability officer and the incident commander to monitor all personnel operating at the fire scene.

When the fire suppression units clear the scene, there must be a face-to-face transfer of command from the incident commander to the fire investigator in charge of the investigation. The transfer of command should identify any areas of concern that can affect the safety of any personnel remaining at the scene. The investigator at that point becomes responsible for the control of the scene and for the safety of everyone at the scene. Any personnel who are brought to the scene must be briefed on all of the known hazards, PPE requirements, evacuation signals, and emergency medical procedures. The lead investigator is also responsible for the security of the scene. It is essential to prevent the destruction or removal of evidence and to prevent bystanders from taking a "sightseeing trip" through the fire scene and becoming injured or removing evidence.

Occupational Safety and Health Administration

The Occupational Safety and Health Administration (OSHA) requires that private and public employees have a workplace that is safe, and it has identified five particular elements instrumental in keeping these workplaces safe:

1. Management commitment and employee participation
2. Hazard and risk assessment
3. Hazard prevention and control
4. Safety and health training and education
5. Long-term commitment

More information regarding hazard identification, evaluation, and prevention can be found in NFPA 1500, *Standard on Fire Department Occupational Safety and Health Program*.

The following OSHA standards most directly affect the fire investigator. Countries outside of the United States may not use some or any of these standards, although they may have similar regulatory agencies. The student should identify the regulations that apply to his or her departmental or company policy, as well as the safety regulations of the investigator's local agencies, state, province, or country.

29 CFR 1910.120, HAZWOPER

OSHA regulations in the 29 CFR 1910.120, HAZWOPER (HAZardous Waste OPerations and Emergency Response) standard, deal with operations at a hazardous waste site and the emergency response to hazardous materials incidents. By their nature, all fire scenes are potential hazardous waste sites. The HAZWOPER standard, as referenced in 40 CFR, Protection of Environment, Part 311, states that the provisions in 29 CFR 1910.120 apply to state and local employees who are involved in hazardous waste operations. This protects all employees, even public employees operating in states that do not have a state OSHA plan. All fire investigators should be familiar with the following NFPA documents, which can clarify the HAZWOPER standard:

- NFPA 472, *Standard for Competence of Responders to Hazardous Materials/Weapons of Mass Destruction Incidents*
- NFPA 473, *Standard for Competencies for EMS Personnel Responding to Hazardous Materials Incidents/Weapons of Mass Destruction Incidents*

As stated in these documents, the investigator is expected to understand the states of matter and to recognize hazards by occupancy, shape of containers, markings, and odors reported to the responder.

29 CFR 1910.146, Permit-Required Confined Spaces

The Permit-Required Confined Space standard deals with the safety aspects of entering a confined space. Before this standard was enacted, 60 percent of all confined space incident deaths were incurred by would-be rescuers. The fire investigator should be able to identify a confined space and at no time should a confined space be entered without the proper precautions in place. A Permit-Required Confined Space, as defined by OSHA, includes the following criteria:

1. A space where an employee can bodily enter and perform assigned work
2. Limited or restricted means of entrance or exit
3. Not designed for continuous employee occupancy

Entry into a confined space requires special training and further requires the continuous monitoring of the atmosphere, fall protection, ventilating the space, and wearing an SABA. Additional trained support and backup personnel must also be present.

29 CFR 1910.147, The Control of Hazardous Energy (Lockout/Tagout)

Whenever an investigative team works around equipment or wiring where an unexpected energization, startup of machines, or release of stored energy could result in the injury of the investigators, the team must consider OSHA's Control of Hazardous Energy (Lockout/Tagout) standard. This standard utilizes a lockout/tagout device that disables the equipment. Every investigator who works in the area places a padlock on the device, and each investigator retains the only key to his or her own lock. Then, on completing the investigation, each investigator leaving a hazardous area must unlock the device before power can be restored or energy released. This way, energy cannot be restored to the equipment until every investigator has left the hazardous area and has been accounted for. This standard should be followed whenever the investigator is working in or around heavy equipment or energized equipment or where a valve can be operated remotely, releasing a material that can engulf the investigator.

Additional Safety Concerns

One thing that many fire fighters forget when they become involved in fire investigation is that not everyone in the public holds the image of them as public servants who are there to help. They can sometimes be viewed as part of the law enforcement establishment. Like police officers, they must realize that the criminal is not the only threat to their safety. Private and public sector investigators often work alone and must be cognizant of additional hazards, such as the possibility of being attacked or fired on by others who are not necessarily involved in the fire, but simply dislike law enforcement or authority figures. The investigator must realize that a fire may have been set intentionally inside the structure. The person who commits arson does not consider the safety of others before his or her own interests. When an investigator is conducting an investigation, he or she should consider whether a bulletproof vest should be worn and should utilize any known street survival tactics.

The fire investigation field can be a rewarding career, but fire investigators must realize that there are hazards. The origin and cause investigation is not a life-or-death matter, and no lives should be risked to conduct this investigation. Any risks must always be weighed against the gain. If no lives are at risk, the investigator cannot justify risking his or her own life or health to conduct a fire investigation.

Wrap-Up

Ready for Review

- Fire investigators must be concerned not only with the physical hazards of a fire scene, but also with the chemicals and other toxic substances present at the scene.
- Investigators have a duty to protect themselves and others at the scene from hazards that are present at the fire scene, be cognizant of the hazards present at each scene, and stay in control of the scene for safety as well as scene preservation concerns.
- The three steps of the hazard and risk assessment process are:
    - Identify the hazards.
    - Determine the risk of the hazards.
    - Control the hazards.
- If the investigator feels that there is a need to enter the building before the fire has been extinguished, he or she must be equipped with structural firefighting gear and a self-contained breathing apparatus (SCBA) and must be trained in their proper use.
- The investigator should guard against complacency, which can lead to injury and/or exposure to the toxins present at every fire scene.
- There are many factors that can influence the safety of a fire or explosion scene, including structure stability and utilities.
- If a device is used as the initiator, the investigator should consider the potential of additional devices in the area.
- All items removed from the scene, including evidence and equipment, should be handled, labeled, and stored in a safe manner.
- The fire investigator must contact the incident commander when he or she arrives at the scene. The incident commander can relay any pertinent information about the fire that could be valuable to the investigation.
- The Occupational Safety and Health Administration (OSHA) requires that private and public employees have a workplace that is safe, and it has identified five particular elements instrumental in keeping these workplaces safe.
- Private and public sector investigators often work alone and must be cognizant of additional hazards, such as the possibility of being attacked or fired on by others who are not necessarily involved in the fire but simply dislike law enforcement or authority figures.

Hot Terms

Collapse zones Distances around a structure that may be affected by a structural collapse. The area should be identified by markers or specialized scene tape that indicates there is a potential for a structural collapse.

Control of Hazardous Energy (Lockout/Tagout) standard The federal OSHA regulation that governs work around equipment or wiring where an unexpected energization, startup of machines, or release of stored energy could result in the injury of the investigators. This standard utilizes a lockout/tagout device that disables the electrical equipment. Specifics can be found in 29 CFR 1910.147.

HAZWOPER (HAZardous Waste OPerations and Emergency Response) standard The federal OSHA regulation that governs hazardous materials waste site and response training. Specifics can be found in 29 CFR 1910.120.

Impact load A sudden added load to a structure.

Incident command system (ICS) The combination of facilities, equipment, personnel, procedures, and communications under a standard organizational structure to manage assigned resources effectively to accomplish stated objectives for an incident.

Permit-Required Confined Space standard The federal OSHA regulation that governs any space that an employee can bodily enter and perform assigned work, with a limited or restricted means of entrance or exit, and that is not designed for continuous employee occupancy. Specifics can be found in 29 CFR 1910.146.

© Greg Henry/ShutterStock, Inc.

FIRE INVESTIGATOR *in action*

Early one morning, you are requested to assist with investigating a large fire at a metal plating factory. Due to the various chemicals used, fire fighters have not entered the factory and have been applying hundreds of gallons of water per minute from master streams for the past several hours to suppress the fire. As you speak with the business owner, he relates his concern with the amount of water "running off" because many of the chemicals are either corrosive and/or toxic. He then states he has contacted his chemist, who is on his way to assist you with the investigation.

1. Which type of hazard is least likely to be present at this fire?
   - **A.** Electrical
   - **B.** Collapse
   - **C.** Chemical
   - **D.** Terrorism
2. If the investigator must enter a structure before the fire is suppressed, he or she must do which of the following?
   - **A.** Wear an SCBA
   - **B.** Don firefighting protective clothing
   - **C.** Both A and B
   - **D.** None of the above
3. What does HAZWOPER stand for?
   - **A.** Hazardous Waste Operations and Emergency Remediation
   - **B.** Hazardous Waste Operators and Emergency Responders
   - **C.** Hazardous Weapons and Emergency Response
   - **D.** Hazardous Waste Operations and Emergency Response
4. Which of the following should the investigator be most concerned with at the scene of a terrorist event?
   - **A.** Pooling water
   - **B.** Confined spaces
   - **C.** Secondary devices
   - **D.** Oxygen-deficient atmospheres
5. If you are uncertain if an electrical circuit is energized, you should:
   - **A.** conduct a spark test.
   - **B.** use an AC tester or multimeter.
   - **C.** assume the circuit is de-energized.
   - **D.** None of the above

Sources of Information

© Photos.com

© Glen E. Ellman

Knowledge Objectives

After studying this chapter, you should be able to:

- Discuss the reliability of information NFPA 4.6.1 NFPA 4.6.2. (p 206)
- Identify legal considerations when finding sources of information NFPA 4.5.2 NFPA 4.6.1 NFPA 4.6.2. (p 206)
- Identify and describe different forms of information NFPA 4.6 NFPA 4.6.1 NFPA 4.6.2. (pp 206–207)
- Describe how to prepare for and conduct interviews in an investigation NFPA 4.5 NFPA 4.5.1 NFPA 4.5.2 NFPA 4.5.3 NFPA 4.6 NFPA 4.6.1 NFPA 4.6.2 (pp 207–208).
- Identify government sources of information NFPA 4.6 NFPA 4.6.1 NFPA 4.6.2. (pp 208–209)
- Identify private sources of information NFPA 4.6 NFPA 4.6.1 NFPA 4.6.2. (pp 209–211)

Skills Objectives

After studying this chapter, you should be able to:

- Expand a fire investigation beyond the immediate fire scene through a variety of sources of information NFPA 4.6.1 NFPA 4.6.2. (pp 206–211)
- Conduct interviews to gather information pertinent to an investigation NFPA 4.5.2 NFPA 4.6.2. (pp 207–208)

Additional NFPA Reference

NFPA 921, *Guide for Fire and Explosion Investigations*

CHAPTER 12

FESHE Course Outcomes

Fire Investigation I

There are no Fire Investigation I (FESHE) course outcomes for this chapter.

Fire Investigation II

11. Discuss interviewing techniques. (pp 207–208)
13. List the sources and technology available for fire investigations. (pp 206–211)
14. Describe procedures for conducting background investigations. (pp 208–211)

You Are the Fire Investigator

© Jones and Bartlett Publishers. Photographed by Glen E. Ellman

After arriving at your office, you receive an anonymous phone call from someone who states he has information about a house fire you investigated several days prior. This person states he had overheard a friend talking about starting a fire at this house because he was upset with the owner of the house because of an incident that occurred at their workplace. The caller also states that he does not want to be involved and refuses to provide you with his name or address.

1. What other information may be useful from the caller?
2. How can you determine the reliability of the caller?
3. Why did the anonymous caller provide information but not wish to be involved?
4. What legal issues may exist?

Introduction

A thorough fire investigation starts with examination of the fire scene and evaluation of the documentation of the scene. Additional sources of information must be researched and analyzed by the fire investigator. These combined efforts provide the investigator with the opportunity to establish the origin, cause, and, where appropriate, the responsibility for a particular fire.

Advances in technology have greatly enhanced the fire investigator's ability to conduct his or her investigations accurately and efficiently. The Internet has dramatically reduced the amount of time required by investigators to research data and locate information. Investigators can find numerous electronic databases on the Internet related to safety warnings and recalls, lightning strikes and weather information, and property owner data from a municipal/county assessor's website. Improved radio and telecommunications equipment has allowed for nearly instant information exchange among organizations and agencies.

Fire investigation teams, task forces, and strike teams are becoming more popular with many agencies that allow investigators from different communities to exchange information freely among themselves and to identify trends or types of incidents they are experiencing. Additionally, members of these teams often assist each other with investigations, providing technical support and resources as well as experience to the scene.

As with any information obtained during an investigation, investigators must be cautious to ensure that the data and facts obtained are accurate and relevant to their investigation. Similarly, investigators must be cautious as to what information is disseminated to other parties, including fire fighters and fire investigators, to ensure that the investigation is not compromised.

Reliability of Information

Important information is available to the fire investigator through interviews, written records and electronic data, and visual and scientific documentation. It is important that the investigator evaluate the accuracy of the information and the source. The investigator's common sense, knowledge, and experience will help with this evaluation. The investigator should also take into account the source's reputation and/or particular interest in the investigation.

Legal Considerations

The availability of information to the fire investigator is governed by legal considerations. The Freedom of Information Act provides public access to information held by the federal government. Most federal agencies have created procedures for the disclosure of information and appeals processes when a request is denied. Most states have also enacted similar laws that allow for the public disclosure of information concerning government operations and their work products. Investigators must be aware of these laws and understand that the laws, rules, and procedures may vary greatly from state to state. Most states have laws that give similar access to state records. Privileged communications are those statements made by certain persons within a protected relationship, such as husband–wife, priest–penitent, attorney–client, and doctor–patient. These communications are generally defined by state law and vary from state to state. Similar to privileged communications, confidential communications are those statements made under circumstances showing that the speaker intended the statements only for the ears of the person addressed.

Forms of Information

There are many forms of information that can be gathered by the investigator. They include verbal, written, visual, and electronic information.

Verbal information is limited to the spoken word. Verbal information can be gathered a number of ways, including through the interview of witnesses, phone calls, recordings, radio and television broadcasts, and transmissions. For example, a recorded verbal account can provide useful and accurate information, even if the interviewee was not an eyewitness to the event.

Written information is obtained through sources such as written reports or documents, reference materials, and newspapers. The investigator will often receive written information throughout the course of an investigation. The investigator may be required to document and maintain the chain of custody for original documents.

Visual information is limited to information gathered using the sense of sight. Sources include photos, videos, movies, and computer animations. Still photography and video footage provide uniquely informative documentation of a fire incident.

Electronic information is gathered through the use of computers and cell phones. The system maintained by a particular source is essential information for the investigator. For example, if the fire investigator needs information about a particular chemical's container found at the scene of an investigation, the investigator could use a computer to find out more about the substance and then contact the manufacturer if additional information is needed.

The investigator must consider all potential sources of information that may prove useful. Security cameras on adjacent buildings or automated teller machines (ATMs), employee timesheets, alarm companies, pawn shop records, realtors, mechanics, EZ Pass toll records, and other service providers may all provide critical information for the investigator. It is up to the investigator to determine what information is appropriate and useful for his or her investigation.

© Glen E. Ellman

FIGURE 12-1 Interview witnesses, fire fighters, and potential suspects for information about the origin and cause of a fire.

Fire Investigator Tip

Verify any witness statements with known facts. Often, witnesses provide false or inaccurate information either to mislead you or as a result of attempting to help you too much.

Interviews

Information obtained during interviews may prove to be the most valuable information obtained during a fire investigation. Interviews of witnesses, fire fighters, or even potential suspects can often provide critical information as to the origin and cause of a fire **FIGURE 12-1**. Effective interviews are accomplished through proper preparation and documentation. Investigators should attempt to conduct interviews as soon as possible to ensure that witnesses are located, identified, and interviewed in a timely manner. Often, these essential interviews are conducted at the scene or within a vehicle. If needed, follow-up interviews may also occur at various locations, including the witness's home or work or within the investigator's office. The location of these interviews should be considered based on the type of interview being conducted. Whether interviews are conducted by police or fire personnel may depend on jurisdiction or on the nature of the investigation. Interviews regarding fire location, causation, and development should, when possible, be conducted by persons trained in the determination of fire origin and cause.

The credibility of a witness should always be considered by the investigator as part of the evaluation of the witness's statements. Witnesses with a specific interest in the outcome of a fire investigation may present issues with credibility. Their statements should be evaluated with caution and carefully compared to the objective information known by the investigator. Credibility issues may exist whenever a witness has a motive to lie or distort facts in a manner favorable to himself or herself. For example, a witness may hide his or her involvement in causing a fire out of fear of incurring civil or criminal liability.

Purpose of Interviews

Fire investigators may gain crucial information about a fire event during an interview. Witness interviews may reveal possible sources of ignitions, fuel packages and their arrangement within the room of origin, or potential suspects or persons of interest. For example, the fire investigator may wish to interview the last person within a structure (or within the room of origin) to determine any activities or events that may have occurred that could have caused the fire. Even witnesses who were not present at the time of the fire may yield valuable information that can assist the investigator. It is important to note that during an interview, the investigator must not only record the information provided but also determine the quality and usefulness of the information.

Preparation for Interviews

Similar to conducting the scene examination, interviews require a certain amount of skill and training. Investigators must be prepared to conduct an interview and have a thorough understanding of the facts of the incident to that point. These facts allow the investigator to create a plan for conducting the interview, including questions that are crafted to elicit information pertinent to the investigation. Interviews with the incident commander and fire fighters prior to the scene examination can provide the investigator with information as to the size and extent of the fire on their arrival and unusual circumstances they may have encountered. The investigator may choose to prepare questions that are specific to what is believed to be the role of a particular witness in the case, but must remain flexible during the questioning in case the witness is able to provide unexpected information.

Once a witness has been identified, he or she should be interviewed as soon as possible to ensure that the facts are correctly remembered by the witness and not clouded by discussion with other witnesses. These interviews often occur at the scene; however, other locations may be more suitable, such as a designated interview room. If needed, follow-up interviews can be conducted at a later time that may be more practical. Interviews are as dynamic as the scene itself and require the investigator to remain flexible and adaptive to each witness. The investigator should inform the witness who they are and provide appropriate credentials when requested. Likewise, the investigator should also establish the identity of all persons interviewed as well as the information listed in TABLE 12-1.

This information may prove useful at a later date if you need to conduct a follow-up interview or locate the witness for the purpose of testifying in court. It may be several months or years after the fire that an arrest or trial occurs. Key witnesses may have changed jobs or relocated or may become reluctant to participate in that time.

Questions should be meaningful and designed to elicit information from the witness. Open-ended questions work best with most witnesses because the witnesses are allowed to tell their answer almost like a story. Closed-ended questions generally will produce only one- and two-word answers that lack meaning and explanation. For example, the investigator may wish to know what the witness saw. A closed-ended question may be, "Did you see the fire?," whereas an open-ended question might be, "What exactly did you see?"

Table 12-1 Witness Information

- Full name
- Date of birth
- Social Security number
- Driver's license number
- Physical description
- Home address
- Home and cellular telephone numbers
- Place of employment
- Business address
- Business telephone number
- Other information that may be deemed pertinent to establish positive identification

Safety Tip

Be cautious of anyone you approach to interview. The person may mistake your intentions or may dislike those who wear uniforms. Always keep your personal safety the highest priority during an investigation, regardless of the location.

Both question types are useful when appropriate. Open-ended questions work best when attempting to obtain information during initial interviews and in follow-ups when new information is requested. Closed-ended questions are useful for very distinct responses where "yes" or "no" will suffice. In addition to the information provided in the interview, the investigator should also consider the wording chosen by the witness, as well as nonverbal cues such as eye contact and body language (although these cues can vary depending on the witness's age, gender, and cultural background).

Documenting the Interview

Every interview should be documented in a manner that will allow the investigator to review the information provided at a later time. An investigator may speak to numerous witnesses or speak to a witness on more than one occasion during an investigation. Being able to review these statements will allow the investigator to analyze the information obtained and determine whether statements either collaborate or conflict. This analysis may provide the investigator with new facts or sources of information (leads) that can be further investigated.

Although there are numerous methods of documenting an interview, written notes and tape or digital recordings are the most common. Written notes do not require any special equipment other than a writing utensil and paper; however, poor penmanship may make the analysis more difficult while you are attempting to decipher the notes at a later date. Tape recorders allow for a complete record of the interview; however, ambient noise, other people speaking, and equipment malfunctions may cause the investigator to lose his or her interview record.

Another method is to conduct video-recorded statements of interviews. The person speaking can be identified, unlike in a recorded statement; however, on-scene use is limited and not practical. Additionally, state and local laws may prohibit the use of video-recorded statements without the consent of the person being interviewed.

Government Sources of Information

Government records provide a wealth of information that is accessible to the investigator. Street maps, building permits, blueprints, and property ownership records contain some of the vital data that investigators need for a fire scene investigation. Municipal, county, state, and federal governments all maintain records that should be accessed as appropriate for background information.

Table 12-2 Municipal Government

Form of Municipal Government	Description/Benefit
Municipal Clerk	Maintains all records related to municipal licensing and municipal operations
Municipal Assessor	Maintains all public records related to real estate, including plot plans, maps, and taxable real property
Municipal Treasurer	Can provide public records related to names and addresses of property owners, legal descriptions of property, and the amount of paid or owed taxes on a property
Municipal Street or Public Works Department	Maintains records and maps of municipal conduits, drains, sewers, street addresses, and all old and current street names, including alleys and right of ways
Municipal Building Department	Has records related to building, electrical, and plumbing permits and archive building blueprints and files
Municipal Health Department	Maintains records of births and deaths and investigations related to health hazards
Municipal Board of Education	Contains records related to the school system and may assist with identifying and locating school-age offenders
Municipal Police Department	Can provide records related to local criminal investigations and evidence storage and retention
Municipal Fire Department	Maintains records related to fire and EMS incidents and life-safety inspections
Other Municipal Agencies	Varies but may include public works, parks and recreation, and water distribution

Table 12-3 County Government

Form of County Government	Description/Benefit
County Recorder	Responsible for recording legal documents that determine ownership of real property and maintains files of birth, death, and marriage records, as well as bankruptcy documents
County Clerk	Maintains public records related to civil litigation, probate records, and other documents related to county business
County Assessor	Maintains records related to property and plats, including property owners, addresses, and taxable value
County Treasurer	Can provide information related to property owners, tax mailing addresses, legal descriptions, and the amount of either owed or paid taxes on property, and also maintains all county financial records
County Coroner/Medical Examiner	Can provide information related to the identification of victims, manner and cause of death, as well as any items found either near or on the victim
County Sheriff's Department	Can provide both investigative and technical support for county criminal investigations and provides polygraph services, evidence collection, and evidence storage and retention
Other County Agencies	Various departments include parks and recreation, conservancy districts, Homeland Security, and Emergency Management

Municipal Government

Townships, villages, and cities are all forms of local municipal governments. These entities create and enforce laws and regulations specific to their jurisdictional boundaries and provide various services funded through either levies or property taxes approved by their citizens.

TABLE 12-2 provides the positions, departments, and agencies within municipal governments. Not every municipality will have each of the departments listed.

County Government

The next level of government is that of the county, borough, or parish. A county is a subdivision within a state that consists of the various townships, villages, and cities within its boundaries. Like municipal governments, various laws and ordinances are created and enforced throughout the county as approved by the constituents.

TABLE 12-3 provides the positions, departments, and agencies within county governments.

State Government

State governments are subnational entities that function both independently and with the federal government (discussed later). In the United States, state governments operate based on their own constitutions, as well as supporting the U.S. Constitution. Additionally, state governments may provide privileges such as drivers' licenses, hunting and fishing licenses, marriage licenses, and birth certificates.

TABLE 12-4 provides the positions, departments, and agencies within state governments.

Federal Government

Established by the Constitution, the federal government is the highest level of government. The three branches of federal government—the executive branch, legislative branch, and judicial branch—comprise the primary functions of government that are further divided into various departments and agencies.

TABLE 12-5 provides the positions, departments, and agencies within federal governments.

Private Sources of Information

There are numerous groups, professional services, and organizations that provide services that may be useful during an investigation. Test data, insurance records, and various standards may exist that provide the investigator with valuable resources that governmental agencies may not be able to provide. TABLE 12-6 provides descriptions of the private sources of information that are available to the fire investigator.

There are an infinite number of potential information sources the fire investigator may access to assist with his or her

Table 12-4 State Government

Form of State Government	Description/Benefit
Secretary of State	Maintains records related to charters and annual reports of corporations, charters of villages and cities, and trade name and trademark registrations
State Treasurer	Maintains public records related to state financial business
State Department of Vital Statistics	Maintains records for births, deaths, and marriages
State Department of Revenue	May assist with locating tax records of individuals or corporations, both past and present, as well as locating individuals through child support records
State Department of Regulation	Source for information such as professional licenses, results of licensing exams, and regulated businesses
State Department of Transportation	May provide information about highway construction and improvements, motor vehicle accident investigations, and vehicle registration and operator testing and regulations
State Department of Natural Resources	Responsible for conservation and protection of water lands, rivers, lakes, and forest areas. This department may also provide records related to hunting and fishing licenses, waste disposal regulations, and cooperation with the Environmental Protection Agency (EPA).
State Insurance Commissioner's Office	Can provide assistance related to licensed insurance companies, insurance agents (both past and present), and computer complaints
State Police	May provide information related to state criminal investigations. Some state police agencies may also conduct fire investigations or provide assistance to local agencies. Additionally, they may also operate their own forensic laboratory.
State Fire Marshal's Office	Can provide information regarding fire incidents within the state, building inspection records, fireworks and pyrotechnics, and boiler inspections. Most State Fire Marshals also have fire investigators who may provide assistance with conducting origin and cause investigations.
Other State Agencies	May include the Department of Motor Vehicles, Department of Homeland Security, Liquor Enforcement

Table 12-5 Federal Government

Form of Federal Government	Description/Benefit
Department of Agriculture	Maintains records related to food stamps, meat inspections, and dairy products, and oversees the U.S. Forestry service
Department of Commerce	Maintains records related (but not limited) to highway projects and names and addresses of ships fishing in local waters, trade lists, and patents
Department of Defense	May provide public records related to the five military branches. Also, all branches maintain their own investigation units.
Department of Health and Human Services	Maintains records related to Social Security and the Food and Drug Administration and maintains an investigation unit
Department of Housing and Urban Development	Maintains records related to public housing and federal assistance
Department of the Interior	Maintains records related to fish and game activities and Indian Affairs. The National Park Service is maintained within this department.
Department of Labor	Can provide information related to labor and management, including overtime and pay and combats age discrimination
Department of State	Can assist with investigations by locating visas of foreign nationals and companies that operate within the country and abroad
Department of Transportation	Can provide records related to vehicle transportation and hazardous materials
Department of Justice	Assists with records related to antitrust and civil rights violations. This department includes the Civil Rights Division, the Criminal Division, the Drug Enforcement Administration, the Federal Bureau of Investigation, and the Immigration and Naturalization Service. Bureau of Alcohol, Tobacco, Firearms, and Explosives (ATF) may provide technical information on fire science and investigation, as well as information on license holders, manufacturers, and importers of firearms.
U.S. Postal Service	Postal inspectors may provide information that has been routed through the mail system.
Department of Energy	Provides information related to the nation's energy policies and programs
U.S. Fire Administration	Oversees the National Fire Incident Reporting System (NFIRS) and the Arson Information Management System (AIMS), and provides research, reference, and technical information related to fire investigation. USFA also provides numerous fire service-based programs, training, education, and information databases.

Table 12-5 Federal Government (*continued*)

Form of Federal Government	Description/Benefit
National Oceanic and Atmospheric Administration	Maintains records of past and present weather data
Internal Revenue Service	Maintains public records related to the compliance of all federal tax laws. May assist with matters related to the federal income tax.
Department of Homeland Security	U.S. Customs and Border Protection (CBP) regulates importers and exporters, customhouse brokers, and truckers, as well as licensing vessels not licensed by the U.S. Coast Guard.
U.S. Secret Service	Maintains public records related to counterfeiting and forgery of U.S. currency and conducts investigations related to threats against all current and former presidents and their families, as well as foreign heads of state.
Federal Emergency Management Agency	Provides federal planning, response, and assistance related to federally declared disasters, both natural and man-made.
Other federal agencies	Various other federal agencies may provide the investigator with public information useful for their investigation.

investigation. As an investigator continues to seek all potential information sources, he or she may identify individuals or entities that normally would not be sought. The investigator should compile and frequently evaluate his or her investigative file. This will permit the investigator to identify areas for further investigation, evaluate the relationship between the information and documents that have been gathered to date, and identify further corroborating evidence (as well as any discrepancies that may exist within the information that he or she already has).

Table 12-6 Private Sources of Information

Private Sources of Information	Description/Benefit
National Fire Protection Association (NFPA)	Develops, creates, and revises various standards and guides that may assist with an investigation
Society of Fire Protection Engineers (SFPE)	Works to advance fire protection engineering, including publishing various documents that may prove useful to the investigator
ASTM International	Develops voluntary consensus standards that may be used by architects during the design phase as well as procedures for fire tests
National Association of Fire Investigators (NAFI)	Provides training related to fire investigation topics and implemented the National Certification Board
International Association of Arson Investigators (IAAI)	Provides training programs for investigators including the CFITrainer.net online for fire investigators. Provides professional credentials for fire investigators and works to control arson and other related crimes.
American National Standards Institute (ANSI)	Facilitates the development of standards by accrediting the procedures of organizations that develop national consensus standards
Regional fire investigation organizations	May exist at either the state or town level and provide contacts that can be consulted
Real estate industry	Maintains valuable records concerning structures and their owners, which can aid in the detection of fraud and arson
Abstract and title companies	Maintain records related to former and current property owners, as well as escrow account maps and tract books
Financial institutions	Maintain various records of both individuals' and businesses' financial records, as well as information of loan companies, brokers, and transfer agents
Insurance industry	Maintains valuable records concerning structures and their owners, which can aid in the detection of fraud and arson
Educational institutions	Can provide information related to a person's background and personal interests
Utility companies	Maintain databases of customers, as well as documented problems on the status of their distribution equipment
Trade organizations	Act as a clearinghouse of information specific to their discipline and usually create and publish trade magazines
Local television stations	Often provide investigators with copies of videotape related to an incident
Lightning detection networks	Lightning and weather data can play an important part in causal analysis of a fire scene.
Other private sources	Numerous other private sources exist that the investigator may find helpful depending on the specific needs of the investigation.

Wrap-Up

Ready for Review

- A thorough fire investigation includes examination of the fire scene, evaluation of the documentation of the scene, and research and analysis of additional sources of information.
- The investigator should use common sense, knowledge, and experience to evaluate the accuracy of the information that is obtained and the credibility of the source.
- The availability of information to the fire investigator is governed by legal considerations.
- There are many forms of information that can be gathered by the investigator, including verbal, written, visual, and electronic information.
- Information obtained during interviews may prove to be the most valuable information obtained during a fire investigation.
- During an interview, the investigator must not only record the information provided, but also determine the quality and usefulness of the information.
- Investigators must be prepared to conduct an interview and have a thorough understanding of the facts of the incident to that point. These facts allow the investigator to create a plan for conducting the interview, including questions that are crafted to elicit information pertinent to the investigation.
- Every interview should be documented in a manner that will allow the investigator to review the information provided at a later time.
- Government records provide a wealth of information that is accessible to the investigator.
- Municipal governments create and enforce laws and regulations specific to their jurisdictional boundaries and provide various services funded through either levies or income taxes approved by their citizens.
- A county is a subdivision within a state that consists of the various townships, villages, and cities within its boundaries. Laws and ordinances are created and enforced throughout the county as approved by the constituents.
- State governments are subnational entities that function both independently and with the federal government.
- The three branches of federal government—the executive branch, legislative branch, and judicial branch—comprise the primary functions of government that are further divided into various departments and agencies.
- There are numerous groups, services, and organizations that provide services that may be useful during an investigation.

Hot Terms

Confidential communication Those statements made under circumstances showing that the speaker intended the statements only for the ears of the person addressed.

Credibility The quality of being believable or trustworthy.

Investigative file The organized collection of all of the documentary information in a fire case, including verbal, written, and visual information from the scene; the reports of investigators and other professionals; and any other investigation or research that has been conducted.

Privileged communication Those statements made by certain persons within a protected relationship such as a husband–wife, attorney–client, or priest–penitent. These communications are protected by law from forced disclosure on the witness stand at the option of the witness, spouse, client, or penitent.

© Greg Henry/ShutterStock, Inc.

FIRE INVESTIGATOR *in action*

After determining a house fire is incendiary in nature, you interview a neighbor who states the current owner of the property had purchased it from a family member within the past 6 months. This neighbor states she had been told by the previous owner they had purchased a new home in an adjacent county but could not sell this property. After asking several questions, you ask if anything appeared unusual about the current owners. The neighbor's response was that she never saw any moving vehicles or activity to indicate the current owners had ever moved in. She states that the current owners were seldom home and believed she had seen the previous owners at the property the week prior.

After interviewing the neighbor, you conduct a records search and locate several properties in neighboring counties that are owned by this family, several of which were purchased in the past year well below the appraised value. Further research determines that at least two of these properties had suffered fires in the past 3 months.

1. Where would you locate information related to property owner records?
   A. County Recorder
   B. County Clerk
   C. County Engineer
   D. County Assessor
2. Which type of question typically elicits more information from the person being interviewed?
   A. Closed-ended
   B. Open-ended
   C. Rhetorical
   D. All of the above
3. Which piece of information would you typically not need from a witness?
   A. Name
   B. Telephone number
   C. Date of birth
   D. Place of birth
4. Where might you find information related to the other fires you identified in your investigation?
   A. Local fire departments
   B. State Fire Marshal Office
   C. Federal Emergency Management Agency
   D. Both A and B

Planning the Investigation

© Photos.com

© David Ryder/Getty Images

Knowledge Objectives

After studying this chapter, you should be able to:

- Identify the basic information that should be gathered about an incident for use in a fire investigation. (p 217)
- Describe how to organize the investigative functions NFPA 4.1.6. (pp 217–218)
- Discuss the role of the preinvestigation team meeting. (pp 218–219)
- Discuss the role of specialized personnel and technical consultants NFPA 4.1.4 NFPA 4.6.3. (pp 219–220)
- Discuss the role of case management in fire investigations NFPA 4.1.6. (pp 220–221)

Skills Objectives

After studying this chapter, you should be able to:

- Develop a preplan to include resources to be prepared to conduct a fire investigation when the "team approach" is required NFPA 4.1.6 NFPA 4.6.3. (pp 216–221)

Additional NFPA Reference

NFPA 921, *Guide for Fire and Explosion Investigations*

CHAPTER 13

FESHE Course Outcomes

Fire Investigation I

There are no Fire Investigation I (FESHE) course outcomes for this chapter.

Fire Investigation II

There are no Fire Investigation II (FESHE) course outcomes for this chapter.

You Are the Fire Investigator

© Jones and Bartlett Publishers. Photographed by Glen E. Ellman

As part of a fire investigation taskforce, you respond to a neighboring jurisdiction and meet with several investigators who have responded to assist with a large fire involving several residential structures. Fire suppression personnel are continuing to extinguish the fires while you and the other investigators begin to discuss the scope of the investigation. As the team discusses the incident, one of the investigators suggests that an action plan and roles be established to ensure all tasks and activities are completed in a systematic and efficient manner.

1. What information would be useful while conducting the preinvestigation meeting?
2. What investigation activities will likely need to be assigned for this investigation?
3. Where may the team locate specialized personnel to assist with examinations of electrical systems, gas utilities systems, and residential mechanical systems?
4. What type of equipment or supplies may be needed to complete the investigation?

Introduction

For any fire investigation, planning is an important first step. Considerations should include the size and complexity of the fire scene, level of safety at the scene, potential for loss of life, number of investigators, technical or expert support for the type of loss, heavy equipment, staffing, and budget.

It is important to utilize the team concept whenever possible. It is recognized that the investigator may be required to perform all of the functions of an investigation and must preplan for that type of investigation as well. A fire scene investigation includes scene examination, reconstruction, photography, sketching, evidence collection, witness interviews, and other varied tasks that require diverse skills. Through use of the team concept, the investigator in charge can delegate these functions to the individuals best qualified to perform them and thus ensure a thorough and professional investigation **FIGURE 13-1**. The investigator may be required to perform many or all of these functions, depending on the assignment.

© Jones & Bartlett Learning. Courtesy of MIEMSS.

FIGURE 13-1 Part of the investigator's job is to delegate tasks.

A team concept also provides for a level of safety and security in case of injury. If the investigation requires that the investigator act alone, safety issues are still important and must be addressed. This may include having a call-in time or safety checks to ensure safety (see the "Safety" chapter in this text).

Prior to an investigation, the investigator should identify and develop resources that may be needed. For example, the investigator must ensure he or she has the necessary equipment to conduct an investigation, including lighting, cameras, tools, evidence containers, and protective gear. The need for outside equipment or labor resources, such as heavy equipment or debris removal, must be identified as well. The lead investigator must preplan for these functions and identify the resources and funding to carry them out. Experts from government agencies, professional societies, trade groups, consulting firms, or colleges or universities that may be beneficial to the investigation need to be identified.

Preplanning for expertise in various fields will provide resources when needed to respond to a particular type of incident. This may include engineers or scientists of various backgrounds—for example, electrical wiring systems, electrical equipment, heating equipment, forensic debris analysis, forensic body analysis, fire suppression and detection equipment, fire dynamics, and fire spread.

An important part of the investigator's preplan is management of the scene, including (but not limited to) scene security,

Fire Investigator Tip

A fire scene investigation includes data gathering, photography, sketching, evidence collection, witness interviews, scene security and preservation, and other varied tasks that require diverse skills. It is essential that the investigator plan appropriately.

entry, processing, and controlling activities. The investigator must understand the organization and operation of the investigative team within the incident command system, coordinating the team, and acting as the liaison to manage the resources and the team. The team concept should utilize special talents or training that individual team members possess in the areas of electrical, heating and air conditioning, and other engineering fields as needed. Other investigators working under the direction of the team leader perform the functions to complete the investigation, reporting to the lead investigator and the team. The need for outside equipment and labor resources, such as heavy equipment or removing debris, must be identified as well. The lead investigator must preplan for this function and identify the resources (including funding) to carry out such functions.

Each of the concepts and functions described in the following section may be required to be conducted by the investigator.

Basic Incident Information

Although all fires are unique, certain information must be gathered at all fire scenes during the initial assignment and response to the investigation. The investigator must remember that the accuracy of the information that is gathered at this point is paramount to determining many of the procedures and needs that may follow, and without thorough receipt of information, the subsequent planning to handle the event may jeopardize the results.

During the initial assignment and response, basic incident information obtained should include the following:

- The location of the incident
- The date and time of the incident
- The reporting party and discovering party
- The weather conditions at the time of the incident
- The size and complexity of the incident
- The type and use of the structure involved in the incident
- The nature and extent of the damage
- The security of the scene
- The purpose of the investigation

This information provides tools to determine what resources will be needed for the initial response. The underlying conditions of the investigation must be ascertained to respond to the incident properly with personnel or resources. For example, if the stability of the structure is compromised, it may require structural evaluations or shoring to be conducted before the scene investigation can proceed. Or, due to severe structural compromise, a structural engineer or public building official may condemn (red tag) a structure, eliminating the possibility of interior evidence collection. If it is a large loss with the potential for multiple interested parties or if there are fatalities, the response may require heavy equipment, laborers, and environmental monitoring. Complete data are instrumental in providing resources necessary to complete the task.

Weather may also play an important role in the type of equipment and resources needed. Where conditions are expected to be freezing, there may be a need to seal the structure and to provide heat until the compartment is suitable to conduct an investigation without destroying data during the debris removal process. On the opposite side of the spectrum, hot weather conditions may require an early morning investigation with frequent breaks and shelter.

Security will almost always be an issue, both prior to and during the investigation. Potential safety issues should never be overlooked. Fire scenes provide for hazards that are often unique and must be preplanned. Some of the more common safety issues include overhead hazards, foot hazards, slips and falls, structural integrity issues, and electrical issues. However, fire investigations are often in areas of other hazards, such as the hazardous chemicals that may have been present before the fire and those that have developed during the fire. These hazards may require that special personal protective equipment (PPE) be used. Investigators operating alone should also understand these issues and prepare for them. As part of preplanning, an investigator operating alone can set a time with another person for a safety check, with one person calling the other, or with the other stopping by the scene (see the "Safety" chapter of this textbook).

The purpose of the investigation may vary with the investigator's position within a department or organization or the assignment. Understanding the assignment will help the investigator plan for the appropriate response and resources. The assignments may require the investigator to define the origin only, or the fire spread issues, or the function of particular equipment or safety devices. Public sector investigators are often charged with determining whether a crime has been committed. Each type of assignment may bring unique issues requiring different resources. For example, if the assignment requires the understanding of not only origin and cause but also responsibility for the cause, postfire scene investigation resources may become an issue in completion of the assignment.

Organizing the Investigation Functions

The person who is responsible for planning the investigation should identify functions and objectives of investigations to which they may be required to respond. The investigator can then preplan for the investigation and identify potential resources and team members according to their skills and the

© Jones & Bartlett Learning. Photographed by Glen E. Ellman

FIGURE 13-2 Team members assemble prior to an investigation.

investigation's needs FIGURE 13-2. Often the investigation team members are not only required to conduct the scene examination, but also to conduct an investigation away from the scene by gathering documents and interviews.

It is important to keep the team concept in mind at this stage of the planning process. There are six basic functions that are commonly performed in each investigation:

1. Leadership and coordination
2. Safety assessment
3. Photography, note taking, mapping, and diagramming
4. Interviewing witnesses
5. Searching the scene
6. Evidence collection and preservation

Keep in mind that it may also be necessary to fill specialized roles to complete the investigation; for example, the investigator may need personnel qualified to evaluate the fire detection and control, electrical, heating, or air conditioning systems, or personnel experienced in evaluating fire spread or ignition dynamics. The fire investigator should utilize talents or training that are available and should fill all functions with the available personnel from within his or her team; however, this may not always be feasible, and outside resources may be required.

When an investigator is assigned to perform an investigation, the assignment typically includes not only investigative duties on the scene, but also the continuation of the investigation after the scene examination is complete. Therefore, the preplanning must also include resources for any postscene follow-up investigation that may be required. Such resources are often needed to determine responsibility for fire cause, fire spread, or other issues.

This same type of selectivity in choosing resources may be required as an investigation continues through the various stages of data gathering.

The investigator should be familiar with the data, documentation, and resources that may be available to complete the postfire investigation for completion of the assignment. If the scene examination has determined that there is a criminal act, the investigator must understand how to obtain data away from the scene, such as financial records, financial history, building records, and business statements. Similarly, if the investigator's responsibility is to determine spread issues, the investigator needs to be able to obtain information on building systems, fire spread experts, flame spread characteristics of products, and codes.

It is therefore important for the investigator to be aware of his or her responsibility for the investigations assigned and to identify the resources that may be needed to complete that investigation.

> **Fire Investigator Tip**
>
> A team is organized according the needs of the investigation.

Preinvestigation Team Meeting

Under the team concept, the preinvestigation team meeting is the ideal forum for the team leader or lead investigator to explain the goals and objectives of the team and to introduce team members. The team leader should address issues such as jurisdictional boundaries, legal authority to conduct the investigation, assignment of specific responsibilities, and collection of basic information FIGURE 13-3.

The team leader should also ensure that each member of the investigation team is aware of the conditions at the scene and any safety considerations. The team leader should discuss the safety clothing and required equipment that are currently available and other supporting equipment that may be necessary.

The team leader should remind team members that plans made in the preinvestigation team meeting could be revised later, as needed, to achieve incident objectives and goals. Once the investigation begins, there will be briefings for all team members to relay pertinent information that is gathered during the day. Team members will get an updated overview and will have the opportunity to request that additional information be sought by interviewers, to request the assistance of specialized personnel, and so forth.

Personal Safety Equipment

There are a number of pieces of protective equipment that are recommended for use by fire investigators. These include:

- Eye protection
- Flashlight

© Jones & Bartlett Learning. Photographed by Glen E. Ellman

FIGURE 13-3 The preinvestigation team meeting allows for direct communication among team members.

Table 13-1	Recommended Tools and Equipment
Absorbent material	Air blaster (found in camera stores)
Ax	Batteries
Broom	Calcination gauge
Camera and film	Char gauge/digital caliper
Claw hammer	Directional compass
Evidence-collection container	Evidence labels with adhesive
Hand towels	Hatchet
Hydrocarbon detector	Ladder
Lighting	Magnet
Marking pens	Metal detector/probe
Paintbrushes	Paper towels/wiping cloths
Pen knife	Pliers/wire cutters
Pry bar	Rake
Rope	Ruler/straight edge
Saw	Screwdrivers (various types)
Shovel	Sieve/sifting screens
Soap and hand cleaner or moist towelettes	Styrofoam cups
Tape measure	Tape recorder
Tongs	Tweezers
Twine	Voltage detector
Voltmeter/ohmmeter	Water
Writing/drawing equipment	

- Gloves
- Helmet or hard hat
- Respiratory protection (the type will depend on the exposure)
- Safety boots or shoes
- Turnout gear or coveralls

Tools and Equipment

There are a number of tools and/or equipment that are recommended for fire investigation TABLE 13-1.

Specialized Personnel and Technical Consultants

Planning should always allow for the possibility that at some time during the investigation, it will be necessary to bring in additional personnel or consultants—people with specialized expertise to assist in analyzing the incident or specific aspects of the investigation.

To avoid problems later, fire scene investigators must recognize the limitations of their expertise and preplan for resources to fill those limitations. Problems are likely to occur if an investigator goes beyond his or her expertise. For example, most investigators do not have a background or training in evaluating failure modes in the electrical systems of mechanical equipment. Identifying the requirements and the type of investigation assigned will require the investigator to identify his or her own limitations.

Networking with other investigators or associations is of particular benefit to identifying resources to meet these needs and limitations. These investigation networks should be noted during preplanning in the event that the investigator finds the need for additional resources or expertise that has not been previously developed.

As discussed earlier in the chapter, it is advantageous to identify these resources in advance of the incident. The investigator must ensure that there is no conflict of interest with any of the outside experts. There are different types of experts or technical consultants. The title or degree of the consultant does not ensure that the consultant is the exact type of resource you require. There are fields of specialties within experts; for instance, simply because a person is an electrical engineer or mechanical engineer does not necessarily mean that he or she can evaluate a building wiring system issue or a commercial oven fire. The expert may have experience in electrical microcircuit boards but not in building wiring or expertise in residential furnaces but not in commercial ovens. An example of this type of need may be the postscene investigation when the data being analyzed include reports, documents, or financial data. This stage of the investigation may require resources and an expert with competencies to analyze and identify or confirm theories or conclusions for responsibility of the fire or a motive. For example, evaluation of building history, construction records, codes, and flame spread characteristics of materials may provide data to support theories of fire spread and affix responsibility to a building feature or material for the fire spreading and allowing the damage to occur. In a fire determined to be incendiary in origin, a forensic evaluation of the financial history, inventory records, or other such data may affix motive for an incendiary fire.

Various types of specialists to consider include a materials engineer or scientist, a mechanical engineer, an electrical engineer, a chemical engineer/chemist, a structural engineer,

Table 13-2 Specialized Personnel

Position	Description
Materials engineer or scientist	A person with specialized knowledge of how materials react to different conditions, including heat and fire (e.g., someone with a metallurgical background for metals or a polymer scientist or chemist for plastics)
Mechanical engineer	A person qualified to analyze complex mechanical systems or equipment (e.g., heating, ventilation, and air-conditioning [HVAC] systems) and who may also be able to perform strength-of-material tests
Electrical engineer	A person who can provide information regarding building fire alarm systems, energy systems, power supplies, or other electrical systems or components
Chemical engineer/chemist	A person qualified to help identify and analyze possible failure modes where chemicals are concerned
Fire science and engineering specialists	A variety of experts who can provide advice and assistance understanding the dynamics of fire spread from the origin; the energy needed for ignition; issues relating to causation, spread, and fire dynamics; and the reaction of fire protection and detection systems
Fire protection engineer	A person who deals with the relationship of ignition sources to materials to determine what may have started the fire; knows how fire affects materials and structures; is able to assist in the analysis of how a fire detection or suppression system may have failed; and knows crucial information about building and fire codes, fire test methods, fire performance of materials, computer modeling of fires, and failure analysis
Fire engineering technologist	A person with a bachelor of science degree in fire engineering technology, fire and safety engineering technology, or a similar discipline, or recognized equivalent, who has studied various topics related to fire science, investigation, suppression, extinguishment, fire prevention, hazardous materials, fire-related human behavior, safety and loss management, fire and safety codes and standards, and fire science research
Fire engineering technician	A person with an associate of science–level degree in fire and safety engineering technology or similar discipline, or recognized equivalent, who has studied various topics related to fire science, investigation, suppression, fire prevention, hazardous materials, fire-related human behavior, safety and loss management, fire and safety codes and standards, and fire science research
Industry expert	A person who is an expert in a field related to a specialized industry, piece of equipment, or processing system involved in an investigation
Attorney	A person who can provide legal assistance regarding rules of evidence, search and seizure laws, gaining access to a fire scene, obtaining court orders, and dealing with spoliation concerns
Insurance agent/adjuster	A person who can offer information regarding the building and its contents prior to the fire, fire protection systems in the building, and the condition of those systems
Canine teams	Teams of dogs with handlers used to assist in collecting samples for laboratory analysis to identify the presence of ignitible liquids

a fire protection engineer, a fire engineering technologist and technician, an industry expert, an attorney, an insurance agent/adjuster, and an accelerant detection canine team TABLE 13-2.

Case Management

The fire investigator will need to develop a method to organize the information that has been gathered and generated throughout the investigation by all of the various people who might have been involved.

Notes compiled during the investigation must be reviewed, and information from the notes must be provided in reports. NFPA 921 suggests that notes are evidence and need to be retained with the file, even after reports are generated. If your record retention is by electronic means, the notes can be scanned and saved. On investigations that require a team approach, the lead investigator is often asked to author the report. It is important that the empirical data compiled through the team concept are recorded from all participants of the team. All of the notes of the team should be likewise maintained.

The complexity of the investigation will dictate the detail of sketches, drawings, or data collection. A simple, small fire may require only a sketch, whereas a large, complex investigation may require several drawings—fire flow drawings, photo position drawings—or other visual aids to support the investigation and the analysis. The complexity of the report is often dictated by the responsibility of the investigator or mandated by jurisdictional requirements, but in all cases, the empirical data should be noted with the analysis of the conclusions.

Depending on jurisdictional requirements of the investigator, the fire scene investigation report defining the origin and cause may be separate from the postfire investigation, which may involve the analyses of responsibility. The investigator must understand and preplan to ensure that the investigation includes the necessary data and properly presents findings. All data that have been secured should be maintained, regardless of their form (notes, drawings, etc.).

The case management of the entire investigation from the fire scene to postfire investigation will ensure that you obtain all of the empirical data to meet the scope of the investigation. It is the investigator's responsibility, through proper case management, to meet the objectives of his or her assignment. With proper management and data collection, the investigator can then publish the information as required—in reports or in litigation presentations.

Most often the case management will be of an investigation conducted by the lone investigator, not the team. In this type of investigation, the investigator will need to wear many hats to ensure that the investigation meets all applicable requirements for proper investigation, data gathering, and documentation for publication FIGURE 13-4.

© Jones and Bartlett Publishers. Photographed by Glen E. Ellman.

FIGURE 13-4 Organization is a major role of the fire investigation position.

Wrap-Up

Ready for Review

- For any fire investigation, consider the size and complexity of the fire scene, level of safety at the scene, potential for loss of life, number of investigators, technical or expert support for the type of loss, heavy equipment, staffing, and budget.
- The investigator must remember that the accuracy of the information gathered at a fire scene is paramount to determining many of the procedures and needs that may follow, and without thorough receipt of information, the subsequent planning to handle the event may jeopardize the results.
- The person who is responsible for planning the investigation should preplan for the investigation and identify potential resources and team members according to their skills and the investigation's needs.
- Under the team concept, the preinvestigation team meeting is the ideal forum for the team leader or lead investigator to explain the goals of the team and to introduce team members.
- There are a number of pieces of protective equipment that are recommended to be worn by fire investigators, as well as tools and equipment to be carried.
- Planning should always allow for the possibility that at some time during the investigation, it will be necessary to bring in additional personnel or consultants with specialized expertise to assist in analyzing the incident or specific aspects of the investigation.
- The fire investigator will need to develop a method to organize the information that has been gathered and generated throughout the investigation by all of the various people who might have been involved.

Hot Terms

Preinvestigation team meeting A meeting that takes place prior to the on-scene investigation. The team leader or investigator addresses questions of jurisdictional boundaries and assigns specific responsibilities to the team members. Personnel are advised of the condition of the scene and the safety precautions required.

© Greg Henry/ShutterStock, Inc.

FIRE INVESTIGATOR *in action*

A large fire has occurred in a four-story building that contains a woodworking business on the first and second floors and storage for various companies and individuals on the third and fourth floors. During the fire, the majority of the building collapsed, with one wall burying one of your fire engines and damaging several neighboring structures. The following day, you begin to receive phone calls from various insurance companies and private fire investigators representing the various other business and individuals who have sustained damages during the fire. They all indicate they wish to begin their examination; however, you have maintained custody of the scene with the assistance of your local police department.

1. While planning the investigation, the lead investigator should begin by:
   A. taking scene photographs.
   B. obtaining the basic information of the incident.
   C. locating the property owner.
   D. obtaining witness statements.
2. Investigators should identify and develop resources:
   A. before the incident occurs.
   B. after notification of the incident.
   C. upon identifying the need.
   D. All of the above
3. A case management system allows the investigator to:
   A. properly document the cause of the fire.
   B. identify potential sources of ignition.
   C. coordinate only complex investigations.
   D. organize information gathered during the investigation.
4. A specialized person who can assist with identifying deficiencies in a sprinkler or alarm system is a:
   A. mechanical enginner.
   B. structural engineer.
   C. fire engineering technologist.
   D. fire protection engineer.

Documentation of the Investigation

© Photos.com

© Adam Rountree/Bloomberg/Getty Images

Knowledge Objectives

After studying this chapter, you should be able to:

- Describe the use of photography in fire investigation NFPA 4.3 NFPA 4.3.2 NFPA 4.6.2. (pp 226–232)
- Describe the use of diagrams and drawings in fire investigation NFPA 4.3 NFPA 4.3.1 NFPA 4.6.2. (pp 232–235)
- Describe the use of note taking in fire investigation NFPA 4.3 NFPA 4.3.3 NFPA 4.6.2. (p 235)
- Describe the use of reports in fire investigation NFPA 4.7 NFPA 4.7.1. (p 235)

Skills Objectives

After studying this chapter, you should be able to:

- Document conditions at an investigation scene using various media NFPA 4.3 NFPA 4.3.2 NFPA 4.3.3 NFPA 4.6.2 NFPA 4.7 NFPA 4.7.1. (pp 226–235)

Additional NFPA References

NFPA 170, *Standard for Fire Safety and Emergency Symbols*

NFPA 921, *Guide for Fire and Explosion Investigations*

CHAPTER 14

FESHE Course Outcomes

Fire Investigation I

There are no Fire Investigation I (FESHE) course outcomes for this chapter.

Fire Investigation II

9. List the procedures for fire scene documentation. (pp 226–235)
13. List the sources and technology available for fire investigations. (pp 226–227, 230, 234–235)

You Are the Fire Investigator

©Jones and Bartlett Publishers. Photographed by Glen E. Ellman

You are meeting with a private fire investigator at a fire scene you investigated several days ago. After advising the investigator of your determination that gasoline had been applied to the living room and then ignited, the investigator begins to photograph the exterior of the structure, including the electric service and natural gas meter. This investigator then begins to document the interior with photographs, including all the appliances and fixtures, and then begins to create several diagrams of the room of origin, including the size and dimensions of the windows, door, and ceiling height.

1. What potential value may the various diagrams have after completing the scene examination?
2. How would making a video recording of the scene be useful?
3. What documents or reports exist that may assist with your investigation?
4. What is the advantage of using digital photography to document the scene?

Introduction

In recording any fire or explosion scene, the investigator's goal is to record the scene through a medium that will allow the investigator to recall his or her observations at a later date and then document those observations. The compilation of these data allows for the support and verification of opinions and conclusions in a court proceeding. The documentation types that are most often used include photographs, digital video or videotapes, diagrams, maps, overlays, tape recordings, notes, and reports.

Photography

The primary method of recording the scene is through visual media. Photography—whether digital, still, or video—provides the investigator with pictures of the scene that can be used as points of reference when writing the report. These pictures present the investigator and other examiners with the most concise depiction of the condition of the scene, thus improving the process of pattern identification. Furthermore, because the report may be written sometime after the actual site investigation, the visual recordings are an effective way of reminding the investigator of the condition of the fire scene at the time that it was investigated.

Photographs are acceptable for presentation in court, and digital photos are widely used. Typically, courts accept photographs that are objective, have not been altered (e.g., cropped or brightened), and do not inflame or exaggerate. The presenter of the photographs to the court must be able to affirm that the photographs show a true and accurate depiction of the scene as he or she saw it. Although time and expense are important considerations in achieving this, it is always preferable to have too many photos rather than too few.

Before the visual recording process begins, the investigator should have a fundamental understanding of photography, including familiarity with the equipment and accessories, lighting and movement, and common camera settings. These issues are important because the predominant color encountered in a fire or explosion scene is black. In interior photography of a structure that has been subjected to the effects of thermal and mechanical insult, various shades of black must be realistically depicted in the photographs. Proper lighting and exposure allow the investigator's recordings to reflect accurately what he or she observes.

To get the proper representation of what occurred on the scene, it is best to obtain various forms of documentation. A videotape or digital video used in conjunction with still photographs, for example, will be more effective. Some investigators will provide a narration while videoing the scene, starting with the exterior and then transitioning to the interior. The end result will also recall exactly what the investigator is visually seeing, assessing the presentation of fire damage and fire spread patterns and other effects of a fire.

If the investigator would like to become more familiar with cameras or video equipment, there are many courses available. Training in crime scene photography offered at local or state criminal justice training academies, through college courses, or by camera clubs or supply stores may be possible.

Timing

Recording the scene through visual representation should be accomplished as early as possible in the investigation, before the scene becomes altered, disturbed, or destroyed because of suppression or other activities. In addition, by documenting the scene early, the investigator can ensure accurate representation of the condition of the scene in case the scene becomes unsafe later due to building collapse or other environmental or mechanical hazards that could prevent entry and hamper the investigation. An excellent opportunity for fire investigators within the public sector, if possible, is to enter with fire suppression personnel during fire attack and witness the actual fire movement within the structure.

It is also advantageous to photograph the scene continuously through the sifting, reconstruction, and evidence examination phases. Photographing a scene after it has been fully excavated and reconstructed is an important final step in any fire scene examination.

Basics

The easiest way to understand how a camera works is to compare it with the human eye. The human eye and the camera each project an inverted image on a light-sensitive surface. A digital camera has a charge coupled device (CCD) similar to the retina in the eye. The iris (eye) and the diaphragm (camera) regulate the amount of light that is admitted. The camera shutter also controls the time during which the light is admitted.

Cameras

A multitude of cameras are available at a wide range of prices. The camera choice is primarily predicated on financial resources and the investigator's personal skill level. Automatic cameras are the easiest to operate. They determine the primary scope of the photograph and then focus on the most obvious image in the viewfinder. These cameras can provide a sense of comfort to some investigators because they adjust the lens opening (f-stop), control the shutter speed, operate the flash, and focus the lens. These features remove many potential pitfalls for the inexperienced photographer.

Some skilled investigators, however, prefer manual cameras because they allow the user to adjust focus and other settings specifically to suit the immediate circumstances and obtain specialty photographs that the automatic camera's built-in options cannot perform. Most cameras are both manual and automatic, providing a choice for the investigator.

A camera having a resolution of at least 5 megapixels is recommended. Most digital cameras automatically display a digital image of the photo a second or two after the photo is taken. This enables the fire scene photographer to review the image immediately for proper exposure and focus. Digital cameras also give the investigator the ability to enlarge the photo view to ensure that sufficient detail is available. Most of these cameras have the capacity to take hundreds of photos, which can be downloaded very easily into many computers for storage. Depending on agency or company procedures, the media card is either stored and replaced with another or erased and returned to service.

Image Authentication

To be admissible into evidence in court, photographs captured by the investigator must meet the tests of a true and accurate representation and relevance to the testimony. With readily available computer technology, digital images can be enhanced to correct brightness, color, and contrast. If an image has been enhanced, however, it is the responsibility of the investigator to ensure that the original image is preserved and the extent of enhancement is documented.

Investigators should take steps to preserve the original image and establish a methodology that will allow for authentication of the image. Procedures should also be established for image storage, such as placement on the appropriate storage medium or the use of a computer software program that does not allow the original image to be altered and saved with the original file name.

> **Fire Investigator Tip**
>
> Batteries will drain quickly in cold weather. As a backup, always carry extra batteries. Keeping your camera in a warm vehicle until it is needed at a cold fire scene will help to extend battery life.

> **Fire Investigator Tip**
>
> When using digital photography, do not delete any photos, even if they are not of the desired quality (e.g., blurry or a photo taken accidentally). The case file should contain all photos taken to avoid accusation of evidence or information concealment.

Lenses

A camera lens is used to gather light and to focus the image on the camera's detector. Camera lenses can accentuate the quality of the photograph. The investigator should understand that various camera lenses produce different results. Telephoto and zoom lenses allow the photographer to see detail from a distance or to accentuate minute details, although some require faster shutter speeds to provide a sharp image. Other types of camera lenses can distort the appearance of detail. For example, a fish-eye lens can depict a higher percentage of the inside of a small room, but it can also distort the peripheries of the photograph, exaggerating curves in objects.

The focal length of a lens refers to what the camera sees through a given lens. Lenses range in size from wide-angle lenses to telephoto and zoom lenses. With film cameras, the 50-mm lens provided a view similar to that of a human eye. With digital cameras, there is no specific equivalent, but a 35-mm length provides a comparable result in many cameras. One common lens used by many fire scene photographers is an 18–55 mm with close focus capability that provides a range of effective fire scene photographic exposures. Fire investigators should avoid changing camera lenses inside a fire scene to avoid contamination of the lens and camera.

For a given focal length lens, detail is depicted in the size of the aperture opening or f-stop. The f-stop regulates the amount of light received at the lens that is transmitted to the camera sensor (digital photography). A higher f-stop denotes a smaller aperture opening and allows less light to reach the camera's sensor, whereas a lower f-stop allows a greater amount of light.

In general, when a higher f-stop (smaller aperture) is selected, shutter speeds need to be slower to create a photo that has acceptable levels of brightness. If light conditions permit the selection of a higher f-stop, the photo will have greater depth of field, which is the range of distance in which an image will appear acceptably sharp. Otherwise put, a higher depth of field increases the range of objects included within the focused area of the photo.

The investigator should seek the aid of a person knowledgeable in the use of digital cameras when buying or using digital cameras with interchangeable lenses. The focal lengths for lenses used on traditional 35-mm film cameras are different from the focal lengths used on most digital camera lenses.

Filters

The use of a neutral UV filter on a camera is recommended for all fire scene photography. The UV filter does not alter the tone or color of the image, so court admissibility is not affected, and it provides protection for the delicate surface of a lens.

Shutter Speed

Shutter speeds are typically denoted in fractions of a second. The higher the denominator, the faster the shutter (e.g., 1/500 sec is faster than 1/100 sec) and the shorter time the shutter is open. Faster shutter speeds are necessary to freeze fast-moving subjects. Although this is unlikely to be an issue in fire scene photography, faster speeds, when available, also reduce the risk of a photo being blurred due to movement or unsteadiness of the camera or the photographer. However, a good exposure also requires a minimum amount of light. As the aperture decreases (f-stop increases), so does the amount of light admitted per unit time, thus requiring a longer shutter speed. To avoid blurring the image, a tripod is required for shutter speeds below 1/60 sec (60) with ordinary (35-mm) lenses. These minimum speeds increase when zoom or telephoto technology and equipment are used.

Lighting

Lighting plays an integral part in photography. The easiest light source available to the investigator is the sun, although the circumstances of the examination may dictate recording the scene at a time other than during daylight. In those circumstances, an alternative light source should be used. The most popular is the flash attachment, which may be permanently mounted on the camera, temporarily mounted, or separate from the camera. Because of the large dark and light areas and the charred surfaces that are found at fire scenes, these scenes are difficult to illuminate effectively for photography. A high-quality flash that is capable of providing a vast area of light for multiple exposures is a necessary tool in fire scene photography. Other alternative light sources include portable lights, such as floodlights and flashlights. The glare from a flash or floodlight may distort the appearance of an object. Bounce flashes, light diffusers, and other techniques will help with this.

The time a photograph is taken may, in some instances, affect the ability of the viewer to understand what is depicted in the image. Photographs of an identical subject in naturally lighted conditions taken at noon may differ significantly from a photograph taken at dusk. Many investigators use flash units during the daytime to fill in shaded areas with light, creating a clearer photograph.

It is recommended that the investigator utilize a flash unit that can be separated from the camera system, allowing the investigator to angle the light source as needed under the circumstances. This capability enables the photographer to use the built-in flash on the camera, with the separate flash unit increasing the amount of light available for a proper exposure. Multiple flash units and remote operating devices called slaves can illuminate large areas.

A ring flash is often used for close-up work. This specialized flash unit fits on the end of the lens and is often used when photographing a critical piece of evidence such as an arc mark or tool mark.

Special Types of Photography

Advancements in the technology of photography have created new photographic tools for the investigator. Images using infrared, laser, panoramic, and microscopic photography may be useful to the investigator in documentation.

Composition and Technique

Photographs are an integral part of the examination and should reflect the condition of the scene as seen by the investigator. The photographs should be taken in a predetermined manner and in accordance with accepted practices in the fire and explosion investigation field—for example, by illustrating the philosophy of examination from areas of least damage to areas of most damage.

An effective general technique at a fire scene is to begin the documentation on the exterior perimeter, progressing to the interior, and from the least damaged areas to the most damaged areas in a sequential manner. Critical evidence should be documented by photographing the subject from different angles and distance, including downward from a ladder. To show the degree of smoke spread or evidence of undamaged areas, it is important to photograph the entire fire scene, not just the suspected location of origin.

Sequential Photos

The photographic documentation of the scene should depict sequential views of the scene. A photograph of a relatively small subject (e.g., a chair) is taken first from a distance (perhaps from the doorway to the room) to show the position of the subject in relation to other fixed objects such as a door casing or radiator, followed by a shot from a medium distance showing more detail, and ending with a close-up of the subject FIGURE 14-1. Sequential photography allows the observer of the photograph to understand better the totality of the view and the relationship of the subject to the overall surroundings.

Mosaics

Another method for depicting the totality of the scene is to use a mosaic of photographs. Mosaic photographs are a series of photographs that encompass a large area by overlapping the start of one photograph where the previous photograph ended FIGURE 14-2. Mosaics are used when a wide-angle lens is not available and a panoramic view is desirable. To create a mosaic, the investigator should identify the breadth of the photograph according to readily identifiable landmarks. Each ensuing photograph should encompass a portion of that landmark so that the final appearance is of one large photograph encompassing an overall view of a particular area. A tripod will allow for a consistent mosaic pattern and alleviate movement and blurred photos.

Computer programs are available that will digitally stitch mosaic images from a series of digital photos.

Photo Diagram and Photo Log

When recording the scene, the investigator should annotate a diagram of the site, identifying the point from which each photograph was taken, the direction of the photograph, the placement of the item, and the photo number. This is referred to as a photo diagram FIGURE 14-3.

There are occasions when the time that the photograph was taken is important as well, and this should be annotated

Courtesy of Jamie Novak, Novak Investigations Inc. and the St. Paul Fire Department

Courtesy of Jamie Novak, Novak Investigations Inc. and the St. Paul Fire Department

Courtesy of Jamie Novak, Novak Investigations Inc. and the St. Paul Fire Department

FIGURE 14-1 Sequential photographs of a chair.

on the diagram. Many digital cameras can imprint the date and time on the photographic image. The photographer should identify the diagram by affixing his or her initials, the date, location of the scene, and any other pertinent identifiers. A diagram like this is helpful to individuals who did not visit the fire scene. It gives them an overall picture of the condition of the scene and what was observed by the investigation team. Virtually all digital cameras record the date and time of a photo in the data that are saved with the photo.

Scene photos should be accompanied by a photo log that provides a list or table of photographs with a reference to each photo (such as the photo number assigned by the camera) and a brief description of the item or area photographed in each. The photo log should have a caption listing, at a minimum, the date of the photos, the scene involved, a case number if relevant, and the identity of the photographer. A photo log is highly useful for the investigator and others to understand and recall later what was depicted in the photos.

A fire investigator may work with another investigator to create a photo diagram or log while photos are being taken, and can even delegate the photography of the fire scene to others. However, he or she should take care that the photos that are being taken depict the items and views needed to fully document the scene. For example, the investigator may want photos that are taken at a different angle or with special detail

© Jones & Bartlett Learning, Photographed by Glen E. Ellman

FIGURE 14-2 Mosaic of warehouse burn scene from aerial truck.

to convey a particular item. These needs should be clearly conveyed to the assistant taking the photos.

Photography and the Courts

One of the potential uses of photographs is with testimony before a court. Prior to any photo being admitted into the court, there may be several challenges to its use. The photo must be a true and accurate depiction of the content of the photo. A court may exclude photos that may be too inflammatory to jurors, such as photos that depict a gruesome death scene. Photos may also be rejected if they lack clarity or if they lack relevancy to the matter before the court. Most courts permit printed photos as well as photos projected or depicted on screen.

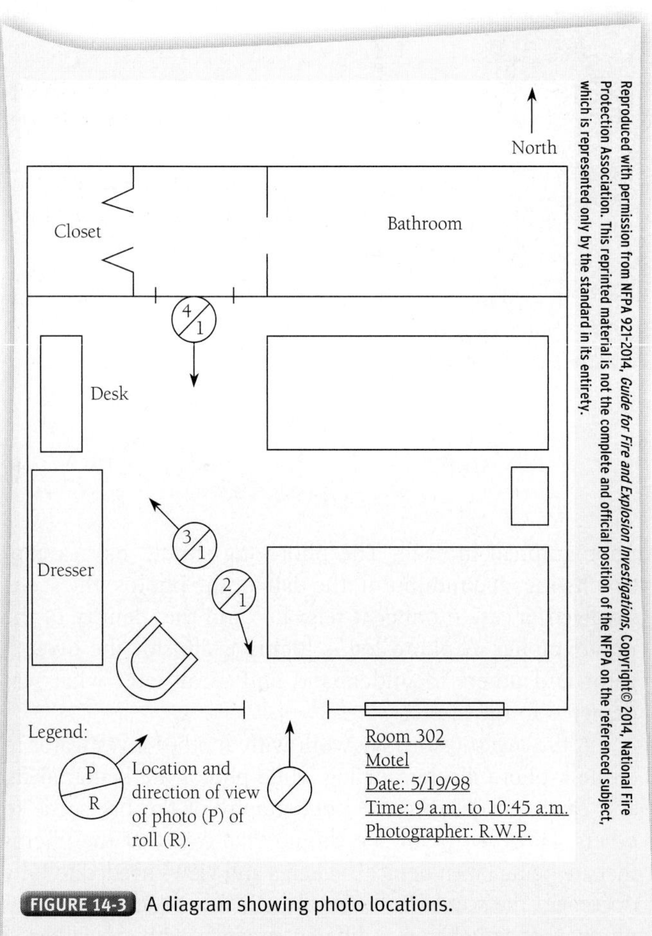

Reproduced with permission from NFPA 921-2014, *Guide for Fire and Explosion Investigations*, Copyright© 2014, National Fire Protection Association. This reprinted material is not the complete and official position of the NFPA on the referenced subject, which is represented only by the standard in its entirety.

FIGURE 14-3 A diagram showing photo locations.

Video Photography

Video photography has become an acceptable medium for orienting viewers to the scene. Newer digital video technology provides important low-light and self-focusing features. A big advantage of using video is the ability to orient viewers to the entire scene by recording complete views of important areas. However, video photography should not be the sole medium used to convey conditions at the scene. Still photography is important for documenting the scene, especially for documenting evidence and scene conditions. When documentation of testimony is needed, video cameras can be used to interview witnesses, suspects, occupants, and owners.

Video recording is a method of documenting and recording the scene. The investigator is able to record the scene and narrate at the same time. The recording, not necessarily used for presentation, can be used by the investigator to refresh his or her memory of the conditions and evidence location. Most entry-level digital video cameras cannot turn off their microphone. Investigators should avoid inappropriate comments or "off the cuff" analysis of cause, which may prove embarrassing later.

Suggested Activities to Be Documented

It is important to document as much of the scene as possible. Some suggested activities to record are conditions on arrival, suppression, overhaul, observers, and origin and cause determination. The progression of the fire, its colors, its reaction to suppression activities, and the overhaul procedures employed

Fire Investigator Tip

Video recording can be used to document the examination, position, and condition of evidence.

are all important in helping the fire investigator to determine the origin and cause. Photographs can also document the extent of damage to the victims or structure.

Photographs of the crowd observing the fire scene activity can help the investigator identify individuals who may have knowledge beneficial to the investigation. These photographs can also help the investigator identify individuals who are seen at multiple fires or are known by the law enforcement or firefighting community.

All fire investigators should routinely determine whether initial witnesses to the fire such as the 911 reporting party documented what they saw with a camera phone or digital camera. Many times media news outlets have staff or stringers respond to large fires. These professionals sometimes arrive on scene prior to apparatus and often have superior quality photographic equipment. Reporters may be busy interviewing victims or witnesses prior to your arrival and should always be contacted.

Photographs of the suppression activities can help the fire investigator understand why the fire reacted in a particular manner. Also, when documenting suppression activities, the location of hydrants, engine companies, apparatus, and hose placement should be noted.

Exterior photographs are important and can be used to establish the location of the fire scene. Exterior photographs should include street signs, house numbers, or other identifiable landmarks that are likely to remain for some time FIGURE 14-4. They should also include surrounding locations of the fire scene, and all angular views of the exterior of the fire scene FIGURE 14-5. Exterior photos could also be taken of the address numbers of affected buildings. This may be useful for documentation and search warrant purposes.

Structural photographs document the extent of damage to the structure after heat and flame exposure. These photographs should be taken from multiple views to record heat and flame damage and fire spread patterns. Structural failures or deficiencies should be captured in photos because these can play a role in the outcome of the fire. The photographs that are taken should be useful to the investigators for explaining their analysis of the fire scene to supervisors, prosecutors, the court system, and insurance representatives.

Interior photographs assist the investigator in documenting the scene and more thoroughly describing the conditions within the structure. All significant points that were accessed or created by the fire should be photographed, along with significant smoke, heat, and burn patterns.

The condition of rooms within the immediate area of the fire should also be photographed to document thermal and smoke conditions. All heat-producing appliances or equipment should be photographed to document their condition with regard to the area of fire origin. The condition of furniture and furnishings should also be documented in photographs depicting heat and smoke travel.

© cindygoff/iStock/Thinkstock

Courtesy of Rodney J. Pevytoe

FIGURE 14-4 **A.** Street signs are identifiable landmarks. **B.** A fire number is one possible address identifier.

The condition of entrance and exit routes, especially doors and windows, should be documented. The photographs should depict their condition at the time of the fire (whether doors were locked or open, whether windows were open or closed, etc.). The condition of interior fire detection and control devices should also be documented (whether sprinklers

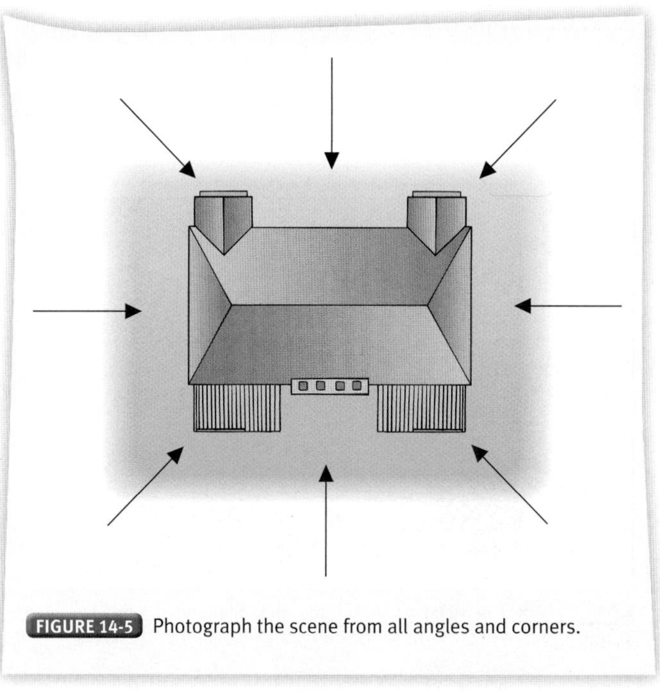

FIGURE 14-5 Photograph the scene from all angles and corners.

activated, whether extinguishers were employed, whether smoke detectors activated). Clocks may indicate the time that power ceased or the time at which fire or heat physically stopped their movement; however, caution must be taken when interpreting battery-operated clocks.

The utilities present at a structure should be thoroughly documented to include the condition of various controls and appliances such as transformers, panels, meters, regulators, and valves. The point at which the utilities enter the structure should be documented to include their location and condition at the time of the fire. The condition of fuses and/or circuit breakers is important, as is the documentation of all potential ignition sources, such as electrical wiring, extension cords, power cords, and appliances. Within the area of fire origin, an investigator should photographically document the condition of appliances and the position of all valves and controls.

During the scene examination, photographs should be taken of all items of evidentiary value. For instance, the photo in FIGURE 14-6 documents the wick from a Molotov cocktail found at a fire scene. Photographs of potential evidence should allow the viewer to locate the item in the scene and its general appearance. More detailed photographs of evidentiary items can be taken after they are moved if the photographer is unable to obtain detailed photos due to scene constraints such as lighting, location, or hazards. A highly visible ruler can be used to provide a scale to identify the size of items in photographs. There should be one photo showing the evidence and another photo of the ruler next to the evidence.

Photographs should also be taken of occupant or victim actions and locations at the scene. These pictures should document any survivors' or victims' location at the time of the fire, any indicators of actions taken by them during the fire, and any end result such as serious injury or death. If the scene includes a fatality, the position of the victim should be thoroughly recorded. If the full condition of the victim cannot be documented at the scene because of lighting, scene hazards, or other obstacles, additional photos should be taken as soon as possible after removal.

When a witness or victim reports that he or she saw a potentially significant event, the photographer should attempt to photograph the view from the position of the witness.

© Jones & Bartlett Learning. Photographed by Jessica Elias

FIGURE 14-6 The wick from a Molotov cocktail found at a fire scene.

Aerial photographs can help the investigator clarify scene arrangement and large evidence items such as a vehicle or body.

Satellite and street-level photography, available in many areas and viewable online, can possibly provide the investigator with both pre- and postincident photos of the scene.

Presentation of Photographs

When displaying or presenting photography of a fire scene examination, the investigator should choose the videos or photographs that present the examination and causal determination with the greatest clarity. The investigator should determine which media are most acceptable to the local court system, law enforcement agencies, fire departments, and insurance companies.

Video presentations are effective means of conveying scene conditions and actions undertaken at the scene. Many law enforcement agencies, fire and police academies, school systems, and courts have the equipment necessary for use in computer or video presentations. Photograph enlargements and projected/screen images are commonly found in court hearings. Computer presentations are an effective and comprehensive medium for exhibiting large quantities of visual material. They allow for integration of multiple formats into a single presentation. Be sure to have backup hard copies in the event of a computer failure for courtroom presentations. Also ensure the screen presentation can be seen from all angles in the courtroom.

Diagrams and Drawings

Diagrams and sketches are used to support the investigator's memory and to document details of the scene. For example, a diagram could be used to conduct an important witness interview or to provide a means for orienting photographs. Sketches can aid the investigator in documenting fire patterns, fire growth, and scene conditions. The accuracy of the data used to create these materials is essential.

Types of Drawings

Diagrams are formal drawings that are completed after the investigation. Sketches are generally freehand diagrams or diagrams drawn with minimal tools that are completed at the scene. The differences among the types of drawings relate to the amount of detail that is incorporated, the type of construction of the structure, features of the structure, equipment, and other factors that are important as to the origin, cause, and spread of the fire. Every investigation should include fire scene sketches. This is especially important for investigations likely to be involved in criminal or civil litigation.

When the fire investigator is determining the type of drawing to use, he or she should decide what needs to be shown on the drawing. If a fire investigator is unable to create a detailed drawing, he or she should create a rough sketch that depicts the area of fire origin, the arrangement of items within this area, and the location of doors and/or windows. This rough sketch should also include accurate taped measurements of walls and ceilings; placement of doors, windows, and other fixed objects; and the position of furniture and appliances within the area of fire origin. FIGURE 14-7 includes an example of a rough sketch

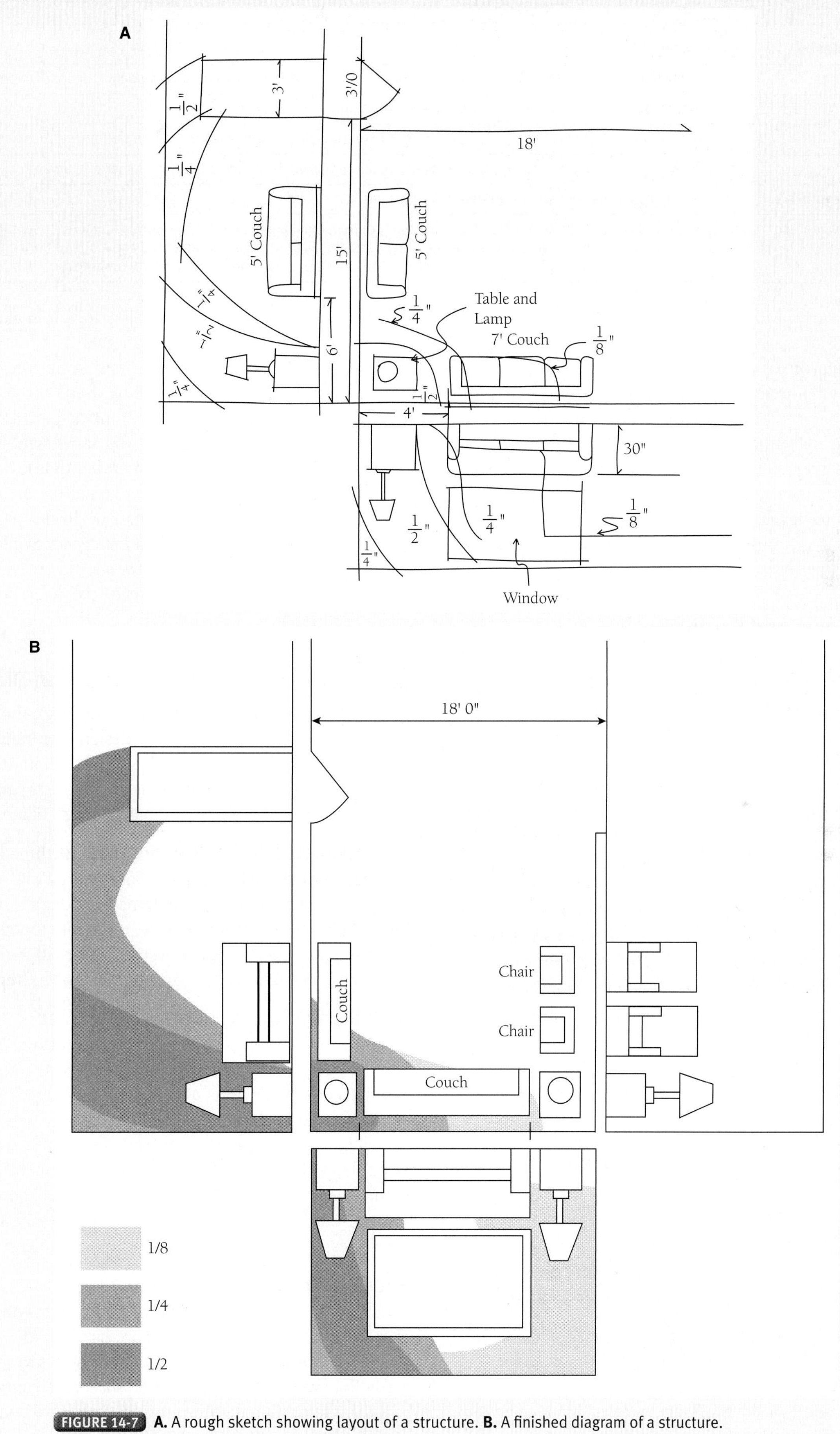

FIGURE 14-7 **A.** A rough sketch showing layout of a structure. **B.** A finished diagram of a structure.

Table 14-1 Detailed Documentation

Type of Documentation	Example
Site or area sketches	Show the location of apparatus, water supplies, or similar information. This may also include inspection reports.
Floor plans	Identify locations of rooms, stairs, windows, doors, or other features.
Elevations	Single-plane diagrams that show an interior or exterior wall and specific information about it.
Details and sections	Show specific features of an item such as the position of switches and controls or damage to an item.
Exploded view diagrams	Used to show assembly of components or parts lists.
Three-dimensional representations	The investigator should measure and document dimensions to develop a three-dimensional representation, regardless of whether these plans already exist. Thickness of walls; air gaps in doors; and the slope of floors, walls, and ceilings are examples of the information that should be measured, obtained, and documented.

that shows basic layout of the structure, as well as the finished diagram made from that sketch.

Several types of drawings can be used to illustrate fire scene conditions. Sometimes a detailed diagram is necessary for proper documentation, but at other times, a rough sketch will suffice when the analysis and conclusions are simple. The type of drawing chosen depends on the degree of detail required.

When the scene or litigation is more complex, the investigator may need to develop or acquire detailed resources such as building plans or construction documentation. TABLE 14-1 outlines details that may be needed.

Drawing Tools and Equipment

There are a variety of options for the investigator to use in the creation of a scene diagram. Often the size and the complexity of the scene and the investigation dictate the type and degree of detail included. In critical cases such as fatal incidents, homicides, incendiary fires, or large or commercial fires, the investigator should obtain architectural plans or diagrams from equipment manufacturers. Surveyors may be useful in accurately mapping large-scale scenes. Many larger municipalities and state and federal law enforcement agencies are able to perform computer-assisted mapping using laser devices.

Recent advances in computers and software have allowed investigators to utilize various types of programs to develop diagrams. Computer-assisted drawing (CAD) programs may be used alone or in combination with computer fire models. A good drawing package will offer many tools to the investigator. Some programs provide for multiple layers in the sketch, which would provide depiction of possible prefire and postfire conditions. No matter what type or brand of program is utilized, the investigator should choose one that offers flexibility of design, automatic dimensioning, and a variety of "libraries" that contain predrawn objects such as furniture, electrical, plumbing, and HVAC components.

Diagram Elements

Depending on the complexity of the investigation, there are many elements that should be included on sketches and diagrams:

- General information: The investigator should identify the name of the person who created the diagram, the diagram title, and date of preparation.
- Identification of compass orientation: Most often the sketch will note north at the top of the page.
- Scale: The investigator should draw to scale and should indicate on the drawing whether it is "not to scale" or "to scale." If the drawing is to scale, the investigator should specify the scale being used.
- Symbols: The investigator must be consistent with the use of symbols, not using the same symbol for multiple purposes. It is recommended that the investigator utilize symbols used in engineering or architecture. Fire protection symbols can be found in NFPA 170, *Standard for Fire Safety and Emergency Symbols*.
- Legend: The fire investigator should create a legend for any drawing, indicating what the referenced symbols represent.

Prepared Design and Construction Drawings

Prepared design and construction diagrams are those that were developed for the construction or design of a building, equipment, or even appliance. Care should be taken in the interpretation of the diagrams because items or buildings may have been modified during the construction and/or use of a building.

Architectural and Engineering Drawings

During the construction phase, various drawings will be used by the contractors. These diagrams may depict only utilities, fire detection and suppression equipment, topographical, or other unique aspects needed in the construction process. It is helpful for the investigator to be familiar with these drawings.

Architectural or Engineering Schedules

In larger buildings, it often becomes necessary to detail the type of equipment used within the structure. A detailed list of this type is known as a schedule. Schedules may be broken down by equipment type and generally detail all of that particular type of equipment used within the structure. For example, a door schedule details the placement of all doors within the structure. Other examples where schedules are used include interior finish schedules, electrical schedules, HVAC schedules, and plumbing and lighting schedules.

Specifications

When an architect or engineer prepares a drawing, the types of materials used in construction are listed on the specification sheet. The specification sheet matches materials to specific placement. For example, the specification sheet could annotate the use of R-17 insulation in the attic area, 5/8-inch plywood in a ground floor bedroom, and so forth. These drawings can usually be obtained from the contractor or the building inspection department in the community in which the structure was built. These drawings also include the architect's elevation drawings. These are drawings of the outside of the building, which are helpful if the building is a total loss. If possible, as-built plans should also be obtained. These are architectural representations of how the structure was actually built and may vary from the original plans.

Note Taking

In addition to visual representation of the scene, the investigator should incorporate investigative note taking to supplement those items that cannot be photographed or sketched. These notes may be referred to as field notes. Note taking may include names and addresses, model or serial numbers, witness statements, photo logs, identification of items, types of materials, or investigator observations such as burn patterns and building conditions. Notes may also contain information that is obtained from witness interviews and document reviews that relate to the scene. Although forms are not required and cannot be used as an incident report, there are forms that have been developed for data collection. The forms may aid the investigator in gathering the necessary data to prepare the report.

Many investigators have found that digital recording devices are very useful in taking recorded interviews of witnesses, victims, suspects, and others as well as easily recording notes while making your scene examination. These electronic statements and notes are easily downloaded into the case file on a computer along with digital photos and other digital information. Additionally, these devices do not require tapes and can hold as much as 24 hours of recorded materials.

The investigator should be careful not to rely solely on one medium to record the scene. For example, the use of a portable recorder or another device should not be the only means of gathering data.

When the investigator is ready to complete a finished report detailing the information about the fire scene, he or she will find the documentation produced and recorded during the investigative process to be invaluable. A review of notes, data forms, photos, and sketches will assist the investigator in providing an accurate and detailed scene report.

Many departments have established a uniform policy on the retention of notes. Any notes maintained are subject to examination or "discovery" by interested parties, and this may occur long after the information is collected or the report is written. Investigators should examine their department/company policy on the retention of notes and adhere to it during every investigation.

Reports

The purpose of a report is to communicate the observations, analysis, and conclusions made during an investigation. Reports generally contain descriptive information, pertinent facts, and opinions and conclusions TABLE 14-2.

Every agency may have a different format for the reporting process. NFPA 921 does not prescribe a particular format of a report. When a case goes to court, rules of procedure may require certain report formats, and/or the inclusion of specific categories of information. All reports should be done in a clear, precise, and accurate manner. The report is often distributed to others outside of the investigator's department or company, and an organized, well-written report creates a favorable impression of the investigator's credibility and professionalism. The report should contain the foundation for any conclusions or opinions that are expressed and should demonstrate use of the scientific method. All relevant information should be included, including all of the data and observations that support the hypotheses considered by the investigator. When information is obtained from witnesses or other sources, those witnesses and sources should be identified along with the information they provided. The investigator should carefully avoid speculation and unsupported opinions. Where appropriate, technical definitions and information can include references to NFPA 921 and other relevant documents.

Reports are most effective when they are written in an organized fashion. The information in a report is easier to understand if it is reported in categories; for example, an investigator may assign separate sections of the report to interviews, physical observations, outside research, conclusions, and other topics. The investigator can develop a report template or utilize reporting software that provides templates. When possible, the investigator should use the same format for each report.

In a team setting, different investigators may be assigned to document different parts of the investigation. For example, an investigator may be tasked with reporting on interviews she conducted (or for which she acted as a scribe), while another is asked to report on the physical findings in a scene, and a third investigator is assigned to report the team's investigative conclusions. In such situations, only one investigator should report on any given topic. This avoids conflict and confusion among reports.

Every report should, at a minimum, be subjected to an administrative or technical review to ensure that the investigator's logic and conclusions can be followed by others and that typographical errors are avoided.

Table 14-2 Report Information

Type of Information	Example
Descriptive information	Date, time, and location of incident; date and location of examination; date the report was prepared; name of the person requesting the report; scope of the investigation; and nature of the report
Pertinent facts	Description of the scene, items examined, and evidence collected; data and observations that form the basis for hypothesis development, testing, and conclusions
Opinions and conclusions	The opinions and conclusions of the investigator; foundation on which the opinion/conclusion is based; and the name, address, and affiliation of each person who expressed an opinion/conclusion

Wrap-Up

Ready for Review

- In recording any fire or explosion scene, the investigator's goal is to record the scene through a medium that will allow the investigator to recall his or her observations at a later date and then document those observations.
- The primary method of recording the scene is through visual media. Photography—whether digital, still, or video—provides the investigator with pictures of the scene that can be used as points of reference when writing the report.
- Recording the scene through visual representation should be accomplished as early as possible in the investigation, before the scene becomes altered, disturbed, or destroyed because of suppression or other activities.
- Photographs should be taken in a predetermined manner and in accordance with accepted practices in the fire and explosion investigation field.
- A big advantage of using video is the ability to orient viewers to the entire scene by recording complete views of important areas.
- It is important to document as much of the scene as possible. Some suggested activities to record are conditions on arrival, suppression, overhaul, observers, and origin and cause determination.
- Computer presentations are becoming an effective and comprehensive medium for exhibiting large quantities of visual material.
- Diagrams and sketches are used to support the investigator's memory and to document details of the scene.
- The differences among the types of drawings relate to the amount of detail that is incorporated, the type of construction of the structure, features of the structure, equipment, and other factors that are important as to the origin, cause, and spread of the fire.
- There are a variety of options for the investigator to use in the creation of a scene diagram. Often the size and the complexity of the scene and the investigation dictate the type and degree of detail included.
- In addition to visual representation of the scene, the investigator should incorporate investigative note taking to supplement those items that cannot be photographed or sketched.
- The purpose of a report is to communicate the observation, analysis, and conclusions made during an investigation. Reports generally contain descriptive information, pertinent facts, and opinions and conclusions.

Hot Terms

Diagrams Formal drawings that are completed after the scene investigation is completed.

Mosaic photographs Used when a sufficiently wide angle lens is not available and a panoramic view is desired. A mosaic is created by assembling a number of photographs in overlay form to give a more-than-peripheral view of an area.

Photo diagram A diagram indicating the direction from which each of the photographs was taken.

Photo log A written chart or list of all photos taken at a fire scene, including photo numbers or other identifiers and a brief description of the item or area photographed.

Ring flash Reduces glare and gives adequate lighting for the subject matter in close-up photography.

Schedule Used on larger projects, this details the types of equipment in lists.

Sequential photography Shows the relationship of a small subject to its relative position in a known area. The small subject is first photographed from a distant position, where it is shown in context with its surroundings. Additional photographs are then taken increasingly closer until the subject is the focus of the entire frame.

Sketches Freehand diagrams or diagrams drawn with minimal tools that are completed at the scene and can be either three- or two-dimensional.

© Greg Henry/ShutterStock, Inc.

FIRE INVESTIGATOR *in action*

Your client has asked you to prepare a final report for a fire you investigated last year. When you open your case file, you remember you did not complete a scene diagram, and the notes you took during your interview with several witnesses seem incomplete and are difficult to read. You review your scene photographs and discover that several photographs you recorded during the scene examination are either blurry or too dark to accurately depict the cause of the fire. Although you remember the investigation, certain facts are unclear and you are having difficulty completing the report.

1. Photographs may only be used in court if:
   A. both parties agree that the photographs were not altered.
   B. the investigator affirms they are a true and accurate representation of the scene.
   C. they are black and white.
   D. All of the above
2. Exterior photographs should include all of the following except:
   A. weather conditions.
   B. house numbers.
   C. street signs.
   D. landmarks.
3. Which of the following is a formal drawing document created after the scene examination has occurred?
   A. A sketch
   B. A diagram
   C. An architectural drawing
   D. An engineering drawing
4. Which of the following is the best medium for note taking during an investigation?
   A. Notepads
   B. Digital audio recording
   C. Pre-made forms
   D. Investigators should not rely on one medium

Physical Evidence

© Photos.com

© John B. Carnett/Bonnier Corporation/Getty Images

Knowledge Objectives

After studying this chapter, you should be able to:

- Identify types of physical evidence found at a fire investigation scene NFPA 1.3.7 NFPA 4.4 NFPA 4.4.2. (pp 240–241)
- Describe how to preserve the fire scene and physical evidence NFPA 1.3.7 NFPA 4.4 NFPA 4.4.2. (pp 241–243)
- Describe how to avoid contaminating physical evidence NFPA 1.3.7 NFPA 4.4 NFPA 4.4.2. (pp 243–244)
- Describe how to collect physical evidence NFPA 1.3.7 NFPA 4.4 NFPA 4.4.2 NFPA 4.4.4. (pp 244–248)
- Describe how to identify collected physical evidence NFPA 1.3.7 NFPA 4.4 NFPA 4.4.2. (pp 248–249)
- Describe how to transport and store physical evidence NFPA 4.4 NFPA 4.4.2 NFPA 4.4.4. (p 249)
- Describe the examination and testing process for physical evidence NFPA 1.3.7 NFPA 4.4 NFPA 4.4.2. (pp 249–251)
- Describe how and when physical evidence should be disposed of NFPA 4.4.2 NFPA 4.4.5. (p 251)

Skills Objectives

After studying this chapter, you should be able to:

- Process physical evidence using recommended and accepted methods NFPA 1.3.7 NFPA 4.4 NFPA 4.4.2. (pp 240–251)

Additional NFPA Reference

NFPA 921, *Guide for Fire and Explosion Investigations*

CHAPTER 15

FESHE Course Outcomes

Fire Investigation I

1. Identify the responsibilities of a fire fighter when responding to the scene of a fire, including scene security and evidence preservation. (pp 240–243)

Fire Investigation II

7. List the procedures for fingerprinting and evidence collection/preservation. (pp 244–249)

You Are the Fire Investigator

© Jones and Bartlett Publishers. Photographed by Glen E. Ellman

While investigating a fatal fire, fire suppression personnel locate a large metal pipe that appears to have blood on it. They bring you the pipe and indicate that it was located within the dining room, which did not sustain any flame or heat damage. Upon entering the structure and examining the victim within the room of origin, you notice injuries that do not appear to be caused by the fire. Additionally, you note there is a fire pattern on the floor that appears to have been created by an ignitible liquid.

1. What influence do fire suppression personnel and their tactics have on physical evidence?
2. How can physical evidence assist in determining the cause of a fire?
3. Where will you locate physical evidence during your scene examination?
4. What is the best method to collect and preserve physical evidence?

Introduction

Physical evidence, also referred to as real evidence, is any physical or tangible item that tends to prove or disprove a particular fact or issue. At a fire scene, the entire scene, including the fire patterns, sources of ignition, security and fire detection equipment, and items associated with the cause of the fire, is considered physical evidence. This type of evidence can be produced in court or other proceedings if it is properly identified, documented, collected, preserved, and analyzed.

The fire investigator is responsible for locating, photographing, properly identifying, collecting, documenting, examining and storing, and arranging for testing of physical evidence from a fire or explosion scene. The first stage of preservation of potential physical evidence on a fire ground begins with the firefighting or suppression operation itself. Fire investigators in a jurisdiction can help themselves by offering training to fire crews and fire officers in how to recognize the importance of evidence preservation and basic ways to avoid inadvertent destruction of physical evidence.

A common occurrence with which a fire investigator must contend is trying to locate actual physical evidence once the fire has been extinguished and overhaul has been completed. Fire investigation and law enforcement associations may offer basic instructional training at the state or county level in the area of preservation of physical evidence at fires, explosions, traffic accidents, and other joint operations.

Safety Tip

Fire investigators should be aware of the hazards associated with their work, especially within fire scenes. Careful assessment of the structural integrity of a building and all utilities is constantly required. Any occupancy may contain dangerous substances such as pesticides, chemicals, or other hazards. Fire investigators are urged to protect themselves at all times by wearing personal protective equipment such as respiration masks, gloves, helmets, and other gear and by routinely performing the due diligence required to maintain their health and safety.

Physical Evidence

Common physical evidence at a fire scene may include traces of ignitible liquid in flooring, a tool mark that is at a point of forcible entry or that indicates adjustment of a critical valve, a faulted electrical circuit, fingerprints, blood, or other physical items or marks that help the investigator establish fact.

Physical evidence can be obvious, such as a melted accelerant container in a retail store lobby, or it can be latent (hidden), such as a fingerprint on a tool that might not even be visible until it is developed. The investigator has to be patient and methodical in the discovery of any possible sources of physical evidence.

Fire patterns, the visual or measurable physical effects that remain after a fire, are artifact evidence. Artifact evidence at a fire scene may also include the remains of the first fuel ignited, the competent source of ignition that ignited the first fuel, and other materials that influenced the fire's growth and development.

Fire investigators must be able to recognize objects that seem to be out of place and identify the evidence that should be collected from a fire scene. The decision may be based on the scope of the investigation, legal requirements, or limitations. The investigator should initially adopt a mindset of inclusiveness when it comes to physical evidence. If the object seems to be in the wrong place, the investigator can assume that it is potential evidence until the investigation process proves otherwise.

Fire investigators can improve their results by getting training in physical evidence recognition and collection and acquiring the needed equipment. Although fire investigators would probably be better at recognizing fire-damaged evidence, certain types of evidence such as tool marks, footwear or tire track impressions, and fingerprints would ordinarily be better left to crime scene technicians. A recommended approach in complex investigations is to establish a close working relationship with law enforcement crime scene evidence technicians who are already skilled in the recognition, collection, analysis, and preservation of trace evidence and who possess all of the needed equipment.

Usually, physical evidence requires laboratory examination and the testimony of an expert witness to establish its significance at a trial or proceeding. Fire investigators should be in contact during the scene examination with the forensic laboratory technician who will be evaluating the evidence to determine the quantity, packaging, storage, and transportation methods appropriate for the category of physical evidence.

Physical evidence is something that can be observed or physically handled by a judge or jury and differs from other forms of trial evidence such as direct evidence (testimony of witnesses who observe acts or detect something through their five senses and surveillance equipment such as CCTV), demonstrative evidence (photographs, maps, X-rays, visible tests, and demonstrations), or circumstantial evidence (facts that usually attend other facts to be proven and are drawn by logical inference from them).

Preservation of the Fire Scene and Physical Evidence

Fire scene evidence preservation is typically initiated by the first responders during the initial attack and overhaul. Ideally, the fire scene, particularly the room of fire origin, should be left as intact and undisturbed as possible. Physical evidence can take many forms and can exist in a variety of locations in a fire or explosion scene examination. The decision of what physical evidence to collect rests with the fire investigator. It is important to understand that although the most important physical evidence in a fire investigation is generally found within the area of origin, important evidence can also be found elsewhere within the fire scene and frequently outside the fire scene as well FIGURE 15-1.

Courtesy of Captain David Jackson, Saginaw Township Fire Department

FIGURE 15-1 Physical evidence is generally found within the area of origin.

The value of evidence preserved at the scene is often not known until the end of the investigation. It is for this reason that any materials thought to be debris or rubbish that are removed from a scene be photographed, contained, and marked or labeled with the location where they were found. For example, if a piece to an appliance is found to be missing, it could later be found in a labeled container containing debris from that specific location.

Skill in recognizing, documenting, collecting, and preserving physical evidence comes from specialized training and experience and is among the most important capabilities of any investigator. Experienced fire investigators understand the need to bring crime scene technicians or other evidence specialists onto the fireground to deal with identification and recovery of physical evidence that may exceed the investigator's own training or equipment capabilities.

Errors in identification, documentation, recovery, or preservation may preclude later analysis and use at trial. Inadvertent or intentional spoliation of physical evidence could potentially expose the investigator to legal sanctions.

Protecting Evidence

After a fire has been brought under control, interior crews may begin to discover potentially important physical evidence. There are a number of effective ways to protect physical evidence on a fireground. The best way is to post a fire fighter at the entrance to the area of fire origin or near a critical piece of evidence to restrict access, create a fire watch, and provide additional suppression as required. Trained investigators should be called to the fire scene as early as possible in this process.

Potential physical evidence, such as a gasoline container, a gun, a tool that may have been used in a burglary, an appliance that shows evidence of electrical faulting, or similar item, can often be "taped off" or identified by an evidence cone. The incident commander (IC) should always be notified when potential evidence is discovered.

A common technique used by crime scene technicians for years has been to "flag, bag, and tag." This process is to identify items that may have evidentiary value and flag with small flags or flagging tape similar to that used by surveyors to identify and protect items during the initial examination of the scene. The location of the items is documented and then the technician returns and collects the items in the appropriate containers. The technician identifies the evidence with tags that note the contents of the container and its location, thus creating the chain of custody. To avoid evidence cross-contamination, a new pair of gloves should be worn for each piece of evidence collected and processed at the scene.

Roles and Responsibilities of Fire Suppression Personnel

As previously discussed, the first stage of preservation of potential physical evidence on a fire ground begins with the firefighting or suppression operation. Although fire fighters are not responsible for determining the origin or cause of fire, they play an integral part in the investigation. Because the specific cause of a fire may not be known until after a full scene examination is conducted, a professional approach would be to consider the entire fire scene as physical evidence to be protected to the extent possible by controlled firefighting and conservative overhaul operations, especially in the area of fire origin. Fire fighters are trained to look for indicators of incendiarism, including multiple fires, incendiary devices, or ignitible liquids. These rules are set forth in the professional qualifications for fire fighter and fire officer in other National Fire Protection Association (NFPA) publications.

In initial fire ground operations, the security of each door and window is an issue. During the size-up, well-trained fire fighters should always take note of existing conditions during their first contact with the scene, including audible alarms, open or unlocked doors or windows, odors, and objects that do not seem to belong where they are located. Once life and safety issues are controlled, the IC should ensure that the fire scene is protected from any further destruction.

Preservation

The investigation of fire cause differs from many other types of investigations, such as investigation of homicide or traffic accidents, because the actual cause of a fire might not be known until the full scene investigation has been completed. It is difficult to predict what the significant physical evidence will be at a given fire scene or where it will be located. For this reason, the investigator should think of the entire scene as potential evidence until the process of investigation can narrow down the key areas.

The general rule is that the best physical evidence is usually located at the point of the most critical action of the person responsible for initiating the fire. In the large majority of incendiary and accidental fires caused by the negligence of a person, the area of fire origin is where one would normally expect to find the most important evidence.

In a majority of fire scenes, the fire service IC can recognize the area of fire origin simply by applying his or her experience and training. The IC may be able to obtain additional information from eyewitnesses who were present at the fire development or from the first public safety official to arrive at the scene.

Early suppression of fire within the area of origin should always be a priority to help preserve evidence there. In cases in which the area of fire origin is known and the fire has extended into other areas of a structure, it is desirable, whenever sufficient assets are present, to keep the area of origin "darkened down" by periodically directing suppression there. Fire attacks by indirect ceiling spray or fog streams are always best whenever possible because these methods are apt to have the least impact on potential evidence.

Caution in Fire Suppression Operations

Old-fashioned firefighting operations, such as directing high-pressure, straight streams of water into potential areas of fire origin or overhauling all room contents out a window and pulling down all walls and ceilings, obviously destroy evidence and are avoided in well-managed fireground operations. High-pressure water streams or positive-pressure ventilation can create confusing fire damage patterns and can destroy physical evidence. They should be used only in conditions in which there is no alternative. Pulling salvage covers on top of the contents within the area of fire origin potentially introduces a source of cross-contamination and should be avoided. Every effort must be made to leave items on the fireground, especially within the area of fire origin, in their prefire position.

Close coordination of fire suppression and fire investigation in a community can minimize the "time out of service" issue for fire suppression crews while maximizing the fire prevention, code enforcement, and law enforcement benefits of an effective fire investigation program. The solution to many of these traditional problems is to call experienced fire investigators to the scene as early in the firefighting process as possible. Inadvertent damage to physical evidence and loss of potential witnesses can be avoided by early intervention.

Management of overhaul is crucial, especially within the area of fire origin. It is at this stage that remaining evidence that was not damaged by fire is susceptible to damage or likely to be displaced. Walls and ceilings that reveal fire movement and intensity pattern evidence and damage patterns to furniture, appliances, and other equipment within the area of origin are critical evidence that must be documented and analyzed before overhaul takes place. Highly effective fire knockdown and overhaul techniques have been developed that can stop fire and minimize inadvertent destruction of evidence.

Fire suppression personnel must avoid adjusting valves or knobs and switches on appliances or circuit breakers FIGURE 15-2. The practice of throwing all breakers in a breaker panel can usually be avoided completely by simply throwing the main breaker to "off."

One of the standard practices in maintaining crime scene integrity within law enforcement agencies is to post personnel at potential entry points and restrict entry to the scene to personnel whose job it is to examine, document, and collect evidence. This could be accomplished on a fireground by posting fire fighters at each entry point to the area of fire origin.

The original position of a switch or valve is often of crucial importance. It is safest for fire crews to interrupt electric or gas service at the street or to shut off the main valve or main breaker

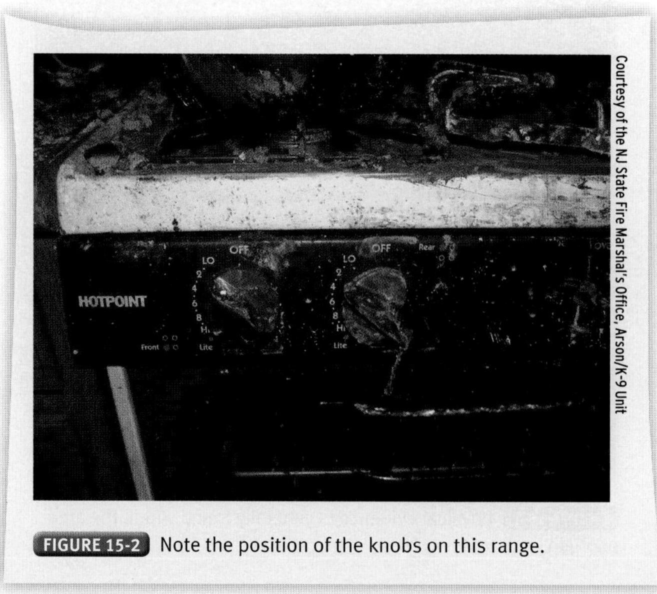

Courtesy of the NJ State Fire Marshal's Office, Arson/K-9 Unit

FIGURE 15-2 Note the position of the knobs on this range.

at the utility entrance. Individual gas appliances always have a shutoff valve behind the appliance, and this should be used to cut the fuel gas flow. Suppression personnel should also fuel any power tools that they may be using outside of the fire scene to avoid spilling fuel into the area under investigation.

In general, physical evidence should be left where it is found until trained investigators can assume responsibility for the scene and guide documentation and recovery.

Roles and Responsibilities of the Fire Investigator

Before any evidence is removed from the area of fire origin, it needs to be analyzed, photographically documented, and fixed in a diagram that indicates its location and position before it is collected. Field notes should document the condition of the evidence when it is discovered and should list the other people who are present. In FIGURE 15-3, the position of a Molotov cocktail is documented in a diagram. In FIGURE 15-4, a photograph is used to document the location and position of the evidence.

Skilled investigators, prosecutors, and medical examiners caution against moving physical evidence, such as containers of a suspected accelerant or dead bodies, from a fire scene before they can be carefully examined and their location documented.

Likewise, in cases in which there is evidence of forcible entry, pieces of an explosive device, or some other potential item of physical evidence, the general rule is that the evidence should be left undisturbed and firefighting operations, whenever possible, should be directed to a secondary position away from potential evidence sites. (See the section on spoliation of evidence in the "Legal Considerations" chapter in this textbook.)

FIGURE 15-4 The position of evidence is documented in this photograph.

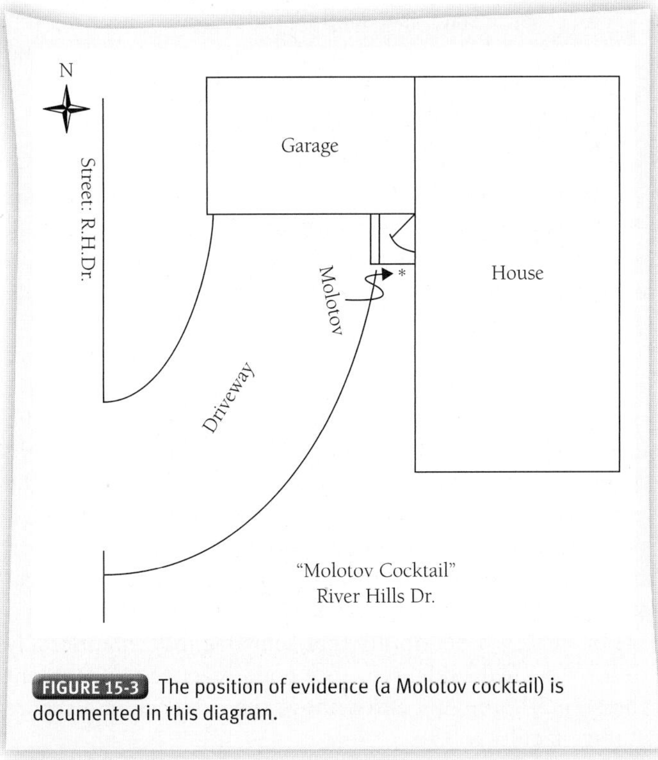

FIGURE 15-3 The position of evidence (a Molotov cocktail) is documented in this diagram.

Contamination of Physical Evidence

Avoiding contamination of physical evidence is of concern in any scene investigation. Opportunities to contaminate evidence occur during firefighting, overhaul and salvage operations, evidence handling, storage, and transportation.

Contamination During Collection

Most contamination of physical evidence occurs during its collection through the use of contaminated tools and protective equipment. This is referred to as cross-contamination, which is the unintentional transfer of a substance from one fire scene or location contaminated with a residue to an evidence collection site. There are four major potential sources of cross-contamination at a fire scene: tools, protective equipment such as gloves, evidence containers, and emergency equipment. Fire fighters and fire investigators should comply with certain housekeeping procedures to help prevent cross-contamination.

Contamination of Evidence Containers

New containers should be used to collect evidence, and clean disposable gloves should be worn by the collector for each new collection. The use of collection tools, such as the collection

Fire Investigator Tip

To limit contamination of evidence, consider using the evidence container itself to collect the evidence.

container itself, may also aid in avoiding cross-contamination. It is recommended that fire investigators carry a supply of new, unused, lined one-pint, one-quart, and one-gallon paint-style evidence cans, or their equivalent, in which to store residue and comparison samples. It is a good practice to seal the cans immediately on receipt from the supplier and before putting them into a response vehicle, to minimize any possibility that the interior of the cans could be contaminated during storage or transport by the vehicle, gear, or other equipment. Other types of physical evidence can be placed in cardboard boxes or, in some cases, plastic bags. It is important to be trained in the collection of physical evidence or to consult with an evidence technician at a fire scene to maximize the potential of collecting valuable evidence and to minimize errors.

Tools

Fire investigators should be equipped with a special tool kit to process fire scenes. These tools should be kept separate from other fire department equipment and must never be coated with any rust-preventive material. After a fire scene examination has been completed, these tools should be cleaned with detergent and rinsed with a strong stream of water.

It is recommended that fire investigators use steel blade tools (shovel, hoe, brick trowel, chisel, razor, etc.) and squeegees with hard rubber blades for excavation and ignitible liquid evidence sampling. Some types of common equipment (bristle brooms) and safety gear (fire fighter gloves) probably cannot be cleaned once they have been contaminated. Disposable gloves should always be used when handling physical evidence, and a new pair should be used each time a new item of evidence is collected.

It is necessary to clean each tool that was used for excavation or evidence sampling before taking it into the fire scene and, again, between evidence collection sites within the scene. Laboratory testing has found concentrated liquid dishwashing detergents that dissolve grease to be effective in dissolving ignitible liquid residue on steel tools when the tools are scrubbed with a clean scrub brush and flushed with clean water. The investigator should submit a sample of the liquid detergent to his or her forensic laboratory to ascertain its properties and ingredients.

If investigators have an accelerant detection canine or a sensitive hydrocarbon detector available, it can be used to double check the tools after cleaning and before use.

Ignitible liquids derived from crude oil are generally not soluble in water alone.

Protective Equipment

It is important to clean boots before entering the area where samples are to be taken and during the scene examination process and to avoid walking through contaminated areas en route to the collection site. Investigators should not handle ignitible liquid residue samples while wearing fire gloves but should instead wear disposable gloves to handle potential residue evidence. Several pairs of new, clean disposable gloves can be carried in an investigator's pocket or kit. Two disposable surgical-type gloves will conveniently fit into an empty 35-mm film container with a snap top. Each time an investigator collects evidence samples, the investigator should change gloves and collection equipment to minimize the chance of cross-contamination.

Emergency Equipment

Fire fighters should not bring tools or generators powered by petroleum distillates into an area of fire origin. An investigator may later test this area for the presence or absence of fuels. All of the housekeeping rules suggested here apply in any situation in which an investigator has handled anything that may contain a petroleum distillate before going into the area of fire origin where arson is suspected. If fuel-powered tools must be brought into the area of fire origin, the investigator should be alerted to this fact.

Collection of Physical Evidence

The collection of physical evidence is governed by the type, form, size, and condition of the evidence. An additional concern is whether handling and packaging a specific piece of evidence will inadvertently degrade fragile forms of evidence such as latent fingerprints or a tool mark. If there is any doubt in the mind of the fire investigator about how to handle and preserve a piece of evidence properly, the investigator should always obtain the assistance of a trained evidence technician.

Collection methods depend on the following factors:

- Physical state: Is it solid, liquid, or gas?
- Physical characteristics: What are its size, shape, and weight?
- Fragility: Will the evidence disintegrate or break?
- Volatility: Will the evidence evaporate?

In many jurisdictions, public safety fire investigators are principally concerned with fires of incendiary origin, whereas civil fire investigators need to be concerned with evidence of arson and also evidence that establishes liability or negligence.

As a general rule, public fire investigators should not remove ordinary appliances, wiring, and the like that have malfunctioned and started a fire. When a noncriminal fire cause is established, public safety investigators should determine the identity of the insurance carrier from the property owner and make appropriate notification. In most cases, the property owner can provide the name of a local insurance agent or broker who can quickly make notification. In cases in which the fire scene is structurally unstable or there is a possibility that someone may intentionally destroy or steal evidence, it is a good practice to document the scene thoroughly and remove the evidence to a secure storage facility.

Documenting the Collection of Physical Evidence

Field notes, reports, sketches, and diagrams should be used to document the collection of physical evidence. This should include photographs and diagrams with measurements. The investigator should make a list that contains each piece of evidence removed from the scene and document the names of the person(s) who collected the evidence. This documentation should begin before the physical evidence is moved. This will help the fire investigator establish the following:

- The origin of the physical evidence, by providing its location, condition, and relationship to the investigation
- That the physical evidence has not been altered or contaminated

After the physical evidence has been documented with field notes and photographs and fixed in a diagram, it is ready to be collected and preserved. Ensure that the documentation of evidence includes information on all parties who took custody of the evidence, even if just for a short time.

Collection of Traditional Forensic Physical Evidence

Traditional forensic physical evidence includes, but is not limited to, finger and palm prints, bodily fluids such as blood and saliva, hair and fibers, footwear impressions, tool marks, soils and sand, woods and sawdust, glass, paint, metals, handwriting, questioned documents, and general types of trace evidence.

Specific procedures for recovery of traditional physical evidence are entirely dependent on the type and condition of the evidence. Fire investigators who are trained and equipped to make recoveries of specific types of evidence can do so. They should bring in trained evidence technicians from local, state, and federal law enforcement agencies when the collection process exceeds their training or equipment capabilities. The investigator should follow ASTM International standards as well as policies and procedures of the laboratory that will look at or test the evidence.

Collection of Evidence for Accelerant Testing

An accelerant is any fuel or oxidizer, often an ignitible liquid, used to initiate a fire or to increase the rate of growth or speed of the spread of a fire. Accelerants can be solids, liquids, or gases. ASTM E 1387, *Standard Test Method for Ignitable Liquid Residues in Extracts from Fire Debris Samples by Gas Chromatography*, or ASTM E 1618, *Standard Test Method for Ignitable Liquid Residues in Extracts from Fire Debris by Gas Chromatography—Mass Spectrometry*, should be followed for this type of collection or testing.

The following sections on liquid evidence are excerpted, with permission of Robert A. Corry, from *A Pocket Guide to Accelerant Evidence Collection, Second Edition,* 2001.

Liquid Accelerant Characteristics

The most commonly encountered arson accelerants are ignitible liquids such as gasoline or kerosene. These liquids share important physical properties that govern their behavior in a fire environment and especially as evidence that can be recovered by a skilled investigator using the proper tools. Liquid accelerants are readily absorbed by structural components, furnishings, and other debris; they generally float when in contact with water (alcohol is an exception); and they are persistent once trapped in porous material.

Liquids have physical properties that cause them to behave differently from most gases or solids. The most common ignitible liquids used as fire accelerants have unique characteristics that manifest themselves in both fire development and evidence from fire scenes. Key properties of common ignitible liquids include the following:

- Liquids flow downgrade and tend to form pools or puddles in low areas.
- Almost all hydrocarbon liquids are lighter than water, are immiscible, and may display "rainbow" coloration (sheen floating on water). Certain other common ignitible liquids (e.g., alcohol and acetone) are water soluble.
- Almost all commonly used ignitible liquid accelerants tend to form flammable or explosive vapors at room temperature.
- The vapors of most commonly used ignitible liquid accelerants are heavier than air and tend to flow downward into stairwells, cellars, drains, pipe chases, elevator shafts, and so on.
- Many ignitible liquids used as fire accelerants are readily absorbed by structural materials and natural or synthetic substances.
- Many ignitible liquids are powerful solvents, which tend to dissolve or stain many floor surfaces, finishes, and adhesives.
- Common ignitible liquids used as fire accelerants do not ignite spontaneously.
- Ignition of a given ignitible liquid vapor requires that the vapor be within its flammable or explosive range at the point where it encounters an ignition source at or above its ignition temperature.
- When an ignitible liquid is poured onto a floor and ignited, two major things occur:
  - Many types of synthetic surfaces (e.g., vinyl) or surface treatments mollify (soften) beneath the liquid.
  - At the edges of the pool, burning vapors adjacent to the liquid edge cause many floor surfaces, such as wood, to char, whereas certain others, such as vinyl, melt and then char. As the liquid pool boils off, its edge recedes. Floor surface charring (or melting and charring) follows the receding liquid edge. The floor area under the ignitible liquid is protected from the effects of burning until the liquid boils off that section.
- Experiments indicate that the greatest temperatures in an ignitible liquid-accelerant fire occur above the center of the burning liquid pool. Some experiments have shown that maximum concentrations of ignitible liquid residues are found at the edges of the fire pattern and

Table 15-1 Ignitible Liquid Collection Areas

Most Desirable Collection Areas	Least Desirable Collection Areas
• Lowest areas and insulated areas within the fire damage pour pattern • Samples taken from porous plastic or synthetic fibers • Cloth, paper, and cardboard in direct contact with the pattern • Inside seams, tears, and cracks • Floor drains and bases of load-bearing columns or walls	• Deeply charred wood • Gray ash • Edge of a hole burned through a floor • Samples from absolutely nonporous surfaces • In general, areas that were exposed to the greatest heat or hose streams

minimum concentrations toward the center, and some have found the opposite. This uncertainty is avoided by taking samples from both the edges and the center.

- Ignitible liquids with high vapor pressure, such as alcohol or acetone, tend to "flash and scorch" a surface, whereas ignitible liquids with higher boiling components, such as kerosene or turpentine, tend to "wick, melt, and burn," leaving stronger patterns. The amount of ventilation available to the fire is a factor in fire pattern appearance.

As a general rule, all fire scenes should be thoroughly photographed inside and outside before any debris excavation is done. As pieces of potential evidence are recovered, the investigator should photograph each in place and fix its location in a crime scene sketch.

The major skill of accelerant residue evidence collection is knowing what to collect and what not to collect. Ignitible liquids used as accelerants burn better than most of the surfaces onto which they are poured. The investigator can expect to find better, stronger samples in protected areas and inside absorbent materials within the pour pattern TABLE 15-1.

A properly trained and disciplined accelerant detection canine (AK-9) team can be of great assistance in identifying areas to collect samples that have a higher probability of laboratory confirmation than samples taken without the assistance of this asset. A second valuable factor is that the AK-9 team can search a scene much faster than any investigator and can help to locate potential evidence areas. The dogs can assist in searching vehicles, clothing worn by suspects, and tools or other instruments. AK-9s can also assist in delayering areas suspected of being sites where an ignitible liquid was used. The alert of an accelerant detection canine is not sufficient evidence of the presence of an ignitible liquid, and any material collected from such an alert must be tested by a competent laboratory before any conclusions may be made about the presence or absence of such a liquid.

Collection of Liquid Samples for Ignitible Liquid Testing

In cases in which a container is suspected of containing an ignitible liquid used as an accelerant or in which such a liquid was a factor in an accidental case, there is a series of recommendations for evidence recovery:

1. Find out whether investigators recognize the odor of the liquid to allow later testimony about odor recognition.
2. Collect a sample of the liquid into a sterile glass container with a hard plastic cap by pouring or drawing the liquid into a sterile pipette or eyedropper. It is good practice to place a piece of aluminum foil on top of the container before screwing down the cap.
3. Remove the container from the fire scene. Pour a small amount onto a safe surface, and attempt to ignite it to allow later testimony about ignition properties.

If an investigator has evidence to suspect that a specific type or brand of ignitible liquid was used to set a fire (from the odor in the debris or an accelerant can left behind), he or she should inform the chemist who is working on the project.

Sterile, 3- or 4-oz (90- to 118-mL) pharmacy bottles with hard plastic caps (not glued cap liners) are recommended for collection of suspected pure ignitible liquid samples. Using a sterile eyedropper or pipette, the investigator should collect about 2 or 3 oz (60 to 90 mL) of the liquid, place a piece of aluminum foil over the bottle top, screw down the cap, and place a label on the side of the bottle.

Sterile cotton balls or gauze bandages are recommended for skimming suspected ignitible liquid residue sheen (rainbow colors) off the surface of suppression water in the pour pattern area.

The types of glass bottles, pipettes, and eyedroppers described can usually be obtained from a forensic supplier, a forensic laboratory, or a pharmacy. A plastic fishing tackle box makes a convenient container in which to keep a supply of pipettes, glass bottles, gauze pads, and evidence labels. Consult with a law enforcement evidence technician to learn where to obtain specialized supplies.

Collection of Liquid Evidence Absorbed by Solid Materials

Ignitible liquid residue sampling at a fire scene can be done in a way that maximizes laboratory identification of ignitible liquid residues. Most of the laboratory procedures for liquid evidence that has been absorbed by solid materials involves testing "headspace" vapor in various ways. Headspace is the zone inside a sealed evidence can between the top of fire debris and the bottom of the lid. Fire chemists generally recommend that evidence containers be filled to two-thirds volume with the debris sample, leaving the top one-third volume as empty air headspace.

New, lined steel paint cans with V-groove lids are recommended for the collection, preservation, and analysis of fire debris that is suspected of containing ignitible liquid residue. Many jurisdictions are successfully using evidence cans lined

Fire Investigator Tip

A pure ignitible liquid sample should never be placed in a metal can. If the can should become heated, internal vapor pressure could pop off the lid.

with Teflon or similar materials, which have the additional benefit of inhibiting rust. Best practice includes sealing and submitting an empty exemplar can to the laboratory along with the samples.

After taking the sample, the investigator should seal it in the can by firmly tapping the V-groove lid onto the can top, trying not to distort the sides of the can, which could lead to seepage of volatile gases. Some agencies require investigators to further secure the lid with tamper-resistant tape. The identification mark or evidence label should be placed on the side of the evidence can, not on the lid.

The latest generation of heat-sealed, plastic evidence pouches designed for use with ignitible liquid residue has eliminated earlier problems, according to a 1991 laboratory test conducted by the Bureau of Alcohol, Tobacco, Firearms and Explosives (ATF) laboratory. This type of evidence container has important advantages but remains puncture prone. The investigator should ask his or her laboratory for its recommendation. Whichever evidence container is chosen, a sample container should be submitted for testing and comparison.

Training for obtaining samples of materials suspected of containing ignitible liquid residue is provided by state and federal fire investigation schools.

The investigator should always keep in mind that ignitible liquids exposed to fire quickly burn away. Sampling for ignitible liquid residues involves identifying and sampling from the base of absorbent materials placed on a floor surface and cutting narrow splinter samples from seams and joints on wood flooring, thresholds, door casings, furniture, the edges of suspected pour patterns on carpet, and so forth. Whenever possible, the investigator should try to pulverize, shred, or splinter sample residue evidence material. Breaking a large solid sample into many smaller pieces dramatically increases the surface area from which to extract residue. Excess water should be drained.

The sample should be loaded vertically (i.e., "chimney roll") whenever the type of sample permits (e.g., carpet or pad samples, linoleum seams, paper, cardboard, or cloth). Sample specimens from seams and joints in wood flooring should be splintered or pulverized and loaded vertically in the container when possible. When collecting sample material that is best stacked vertically, it may be helpful to lay the evidence can on its side while loading it.

A strong sample should be placed in the container that is closest to its size without packing down. The investigator should not dilute the sample by adding material that is not suspected of containing ignitible liquid residue just for the sake of filling the can, but use a smaller container instead.

If a strong odor of an ignitible liquid is present in the sample, a normal-sized sample should be taken whenever possible (at least one quart). When a weak odor or no discernible ignitible liquid odor is present, large-size samples should be taken as required (not more than one gallon each). If a small object such as a matchbook is found to contain an ignitible liquid odor, the object should be sealed in aluminum foil and then further sealed in a suitable evidence container. The chemist should be advised of the odor.

When photographing physical evidence, it is considered good practice to include a permanent feature (e.g., a radiator, wall, valve, or door casing) in each evidence photograph.

The basic rule with line drawings of fire scenes is to measure the location of any movable item of evidence from two or more fixed, permanent objects whenever possible. The investigator should always use a magnetic compass to orient the drawing to north.

Collection of Solid Samples for Accelerant Testing

Solid accelerant may include commercial chemical compositions such as Thermite, Fire Starter paste, or composite fireplace logs available for purchase. Chemical incendiary materials made from common household chemicals and cosmetics can also be used to set fires. These materials require special handling to prevent dangerous injuries and also special packaging to preserve the sample. Call your regional or state crime laboratory for advice if these are encountered.

Comparison Samples and Exemplar Products

Comparison samples should be taken whenever sampling liquid or solid materials are believed to contain ignitible liquid residue. This is especially true when sampling from a surface believed to have a petrochemical base, as in the case of many common floor surfaces such as carpet or vinyl. Comparison samples can often be successfully obtained on the same carpet or surface, but away from the area where the ignitible liquid residue is believed to be by removing sample material from underneath large objects that sit on the floor, such as couches or cabinets. Taking a sample of uncontaminated carpet for laboratory testing helps to establish a baseline of chemical components to which the contaminated sample can be compared.

There are times when a comparison sample will not be available because of damage of the scene. There are also times when a comparison sample is not always necessary. It is up to the analyst to make the determination if one is necessary. It is, however, often difficult for the investigator to return to the scene after the initial investigation to obtain the sample; therefore, a comparison sample should be obtained during the initial investigation if one is available.

In cases in which an appliance or control is believed to have caused a fire, the investigator can often learn the model type and source by simply interviewing the occupants of the area of fire origin. In serious fire cases, obtaining an exemplar unit is often desirable. A good resource for product safety information is the U.S. Consumer Product Safety Commission (CPSC) website, which has information on product recalls by manufacturer and model. This information includes products that have been recalled for fire and electrocution hazards. Mechanical or electrical engineers are usually involved in examination to determine the cause of a failure.

Fire investigators are cautioned not to perform destructive examinations that exceed their training and experience. In these cases, the investigator should always consult with the insurance carrier and the person sustaining the loss about the need to perform any testing or plans to remove any evidence that could be destroyed or further damaged.

Collection of Gaseous Samples

The investigator should not enter any area where a potentially explosive atmosphere is present. There are several methods that can be used to collect samples of gaseous vapors, including commercially available sampling devices and evacuated air sampling devices. Contact state or federal laboratories for advice if confronted by this situation.

Collection of Electrical Equipment and System Components

Before examination of electrical equipment, the first priority is to make certain that there is no electrical current to the equipment. If there is any question, it is advisable not to touch the equipment. A qualified electrician should be called to the scene.

Fire-damaged electrical equipment, such as electrical panels or switchgear, appliances, controls, and the like, needs to be examined, photographed, and fully documented on scene because movement will often degrade its evidentiary value.

Whenever possible, the appliance or suspected defective circuit should be left in place and an expert brought back to examine it. If such a resource is not available and the building has fire insurance coverage, contact the insurance carrier. In most cases, if there are reasonable grounds to believe that an electrical malfunction caused a fire, the carrier will assign a qualified expert to assist fire investigators. Potential evidence of a non-incendiary fire cause should be left in place, pending notification of other interested parties. A group inspection of the scene and potential evidence can then be scheduled.

Any wire that is cut for evidence collection should be identifiable with reference to at least one of the following:

- The device or appliance the wire is from
- The circuit breaker or fuse number or location the wire is from
- The wire's path between the device and circuit protector

If there is an undamaged appliance or control on the fire ground that is the same make and model as the one suspected of causing the fire, the fire investigator should give strong consideration to collection of the exemplar for later comparison.

> **Fire Investigator Tip**
>
> Fingerprints will often remain on surfaces even after a fire. Investigators should always search any location such as a window or door opening where there is evidence of forcible entry or drawers or other objects that have been moved for latent fingerprints and tool marks. Investigators trained in retrieving latent fingerprints and casting tool marks and who have the equipment should recover this evidence. Otherwise, contact a qualified crime scene technician for assistance.

Identification of Collected Physical Evidence

The first stage in identifying collected physical evidence is photographing the evidence where it was found and placing the evidence into a scene drawing by measuring its exact position from fixed objects such as walls or door casings.

It is an accepted practice to take an overall photograph of any important article of evidence to show its relationship to other major objects in the room or vicinity, a second photo to show the object in its immediate surroundings, and then a close-up photo of the item. A fourth photo of the evidence should be taken with a measurement scale in the photo to establish size.

The collection of an item of evidence must include marking the evidence for positive identification FIGURE 15-5.

Crime scene procedure calls for marking, tagging, or bagging of evidence. It is best to mark directly on the evidence itself if this will not be destructive. Otherwise, the item can be tagged or placed in a bag. The investigator should be careful not to place a mark on a piece of evidence in a location where other evidence such as a latent fingerprint or tool mark may exist.

When marking an item of physical evidence, tag, or package for identification, the following data should be included:

- Date and time collected
- Case number
- Location
- Brief description of the evidence
- Where and at what time the item was discovered
- Name of the investigator(s) collecting the evidence

© Jones & Bartlett Learning. Photographed by Glen E. Ellman

FIGURE 15-5 Mark the evidence container.

The materials used for marking the evidence should not be susceptible to removal, damage, or alteration.

A complete documentation and detailed list of all physical evidence collected should be prepared and maintained by the investigator. The list should include the following:

- Identification of the evidence by particular case
- Individual evidence/item number or letter
- Evidence collection location
- A brief description of the evidence item

Transportation and Storage of Physical Evidence

At some point, physical evidence needs to be transported for examination or laboratory testing. The major concern at this point is maintaining the physical integrity of the evidence.

The investigator may use several methods to transport evidence. Personal delivery is the preferred method. When this is not possible, the evidence may be shipped via the post office or a common carrier. Specific requirements vary depending on the type of evidence.

The laboratory should be consulted on the safest way to ship a given category of evidence. Some materials, such as explosive devices or flammable liquids, or electrical evidence such as circuit breakers, relays, or thermostats, cannot be shipped under normal circumstances. Evidence should be stored under conditions that maintain it in the best possible condition, guarding it against excessive heat, humidity, or other sources of contamination. Dry, dark, and cool conditions are preferred for most evidence. Refrigeration or freezing may be recommended depending on the type of evidence. Consult with the postal service or the shipping company for regulations and limitations on shipping hazardous materials.

A form of return receipt is necessary for all shipments of evidence, and a letter of transmittal is generally appropriate as well. The transmittal letter should include a detailed list of the evidence submitted, the nature and scope of the testing requested, and possibly the circumstances surrounding the event in question. The fire investigator should also include his or her name, address, and phone number so the examiner can contact the investigator if necessary. If the package is mailed, the investigator should ship it by registered mail or commercial courier service, and a return receipt and signature surveillance should be requested.

Crime Scene Search Evidence Report

Name of subject ________

Offense ________

Date of incident ________ Time ________ a.m. p.m.

Search officer ________

Evidence description ________

____

Location ________

____

Chain of Possession

Received from ________

By ________

Date ________ Time ________ a.m. p.m.

Received from ________

By ________

Date ________ Time ________ a.m. p.m.

Received from ________

By ________

Date ________ Time ________ a.m. p.m.

Received from ________

By ________

Date ________ Time ________ a.m. p.m.

Reproduced with permission from NFPA 921-2011, *Guide for Fire and Explosion Investigations*, Copyright© 2010, National Fire Protection Association. This reprinted material is not the complete and official position of the NFPA on the referenced subject, which is represented only by the standard in its entirety.

FIGURE 15-6 A chain of custody form.

Chain of Custody

Forensic and legal requirements mandate that evidence be positively identified and maintained in a chain of custody from the point where the evidence is collected right to the courtroom. The integrity of any piece of evidence may be challenged, so the investigator needs to keep an accurate historical accounting of the evidence, which is referred to as chain of custody FIGURE 15-6. This record details all of the parties who have come into possession of the evidence. The inability to show the chain of custody for an evidentiary item can lead to challenges of admission as evidence in court.

The strongest chain of custody has one link: the fire investigator who collected the evidence. If possible, the investigator who recovers the evidence should place it into storage, transport it to and from the laboratory, place it back into storage, and then bring it to court. If this is not possible, then the investigator must be certain to maintain a strict accountability for each person who has custody of the evidence in accordance with local requirements.

Examination and Testing of Physical Evidence

After an item of evidence has been collected, it may be submitted for examination or testing. Testing is often performed to establish chemical composition or for failure analysis. Examination and testing often require that the evidence be altered somehow, without impacting the item's value as evidence. Debris may be removed from the item, as might a case or cover be opened or removed to allow full inspection of the item. The investigator is then able to conduct such tests as X-rays, depth of char, density, and so forth. If it is determined that testing may alter the evidence, parties of interest should be notified prior to testing, in order to afford them an opportunity to object or to have a representative present at the testing.

Laboratory Examination and Testing

Although many fire investigators use the laboratory strictly for analysis of suspected residue of ignitible liquids and materials, modern forensic laboratories can provide a wide variety of tests on many types of physical evidence. Fire investigators are urged to obtain training in crime scene processing and laboratory science to learn the scope of these capabilities. Some of the tests conducted on evidence are listed in TABLE 15-2.

Sufficiency of Samples

The fire investigator should be familiar with the purposes and standards established for each test. Each test requires a minimum amount of evidence to facilitate an accurate test. The investigator should confirm that the testing facility is conducting the appropriate test in accordance with the established standards before the testing is conducted.

Comparative Examination and Testing

Comparative examination is generally a function of an engineering examination performed in a laboratory. In cases where a destructive examination of a complex appliance suspected of causing a fire is to be performed, the investigators will very often attempt to obtain an exemplar for comparison purposes.

Table 15-2 Methods of Testing Evidence

Testing Method
Gas Chromatography (GC)
Mass Spectrometry (MS)
Infrared Spectrophotometer (IR)
Atomic Absorption (AA)
X-Ray Fluorescence
Flash Point by Tag Closed Tester (ASTM D 56)
Flash and Fire Points by Cleveland Open Cup (ASTM D 92)
Flash Point by Pensky–Martens Closed Tester (ASTM D 93)
Flash Point and Fire Point of Liquids by Tag Open-Cup Apparatus (ASTM D 1310)
Flash Point by Setaflash Closed Tester (ASTM D 3828)
Autoignition Temperature of Liquid Chemicals (ASTM E 659)
Heat of Combustion of Hydrocarbon Fuels by Bomb Calorimeter (Precision Method) (ASTM D 4809)
Flammability of Apparel Textiles (ASTM D 1230)
Cigarette Ignition Resistance of Mock-up Upholstered Furniture Assemblies (ASTM E 1352)
Cigarette Ignition Resistance of Components of Upholstered Furniture (ASTM E 1353)
Flammability of Finished Textile Floor-Covering Materials (ASTM D 2859)
Flammability of Aerosol Products (ASTM D 3065)
Surface Burning Characteristics of Building Materials (ASTM E 84)
Fire Tests of Roof Coverings (ASTM E 108)
Critical Radiant Flux of Floor-Covering Systems Using a Radiant Heat Energy Source (ASTM E 648)
Room Fire Experiments (ASTM E 603)
Concentration Limits of Flammability of Chemicals (ASTM E 681)
Measurement of Gases Present or Generated During Fires (ASTM E 800)
Heat and Visible Smoke Release Rates for Materials and Products (ASTM E 906)
Pressure and Rate of Pressure Rise for Combustible Dusts (ASTM E 1226)
Heat and Visible Smoke Release Rates for Materials and Products Using an Oxygen Consumption Calorimeter (ASTM E 1354)
Ignition Properties of Plastics (ASTM D 1929)
Dielectric Withstand Voltage (Mil-Std–202F Method 301)
Insulation Resistance (Mil-Std–202F Method 302)

Fire investigators should be aware that oftentimes an exemplar may be present within the fire scene. Examples are lamps, extension cords, outlets, switches, or circuit breakers.

Disposition of Physical Evidence

Review standard operating procedures on how your agency or firm packages, labels, and stores evidence that it takes from a fire scene. After the evidence has been gathered, tested, and stored, the question arises as to how long the item needs to be maintained. Many factors influence the length of storage. Legally, some cases—such as a murder-related investigation—may have no statute of limitations, and thus, evidence may be brought from storage many years after the fact. Even after a conviction in a criminal case, the defendant is entitled to appeal, and the evidence should be maintained during the time allotted for an appeal. In other cases, the evidence may be returned to the owner.

Civil cases also have lengthy time frames during which evidence may be called into court. In general, the fire investigator should maintain all evidence until proper written authorization to dispose of the evidence has been received from all who are concerned in the investigation. After that, items of evidence should be returned to the owner.

The disposal of evidence should be conducted safely and in a timely manner. Documentation of the evidence and its disposal should be in keeping with regulations in your jurisdiction.

Wrap-Up

© Greg Henry/ShutterStock, Inc.

Ready for Review

- Physical evidence, also referred to as *real evidence*, is any physical or tangible item that tends to prove or disprove a particular fact or issue.
- Common physical evidence at a fire scene may include traces of ignitible liquid in flooring, a tool mark that is at a point of forcible entry or that indicates adjustment of a critical valve, a faulted electrical circuit, fingerprints, blood, or other physical item or mark that helps the investigator establish fact.
- The value of evidence preserved at the scene is often not known until the end of the investigation.
- Opportunities to contaminate evidence occur during firefighting, overhaul and salvage operations, evidence handling, storage, and transportation.
- The collection of physical evidence is governed by the type, form, size, and condition of the evidence.
- The first stage in identifying collected physical evidence is photographing the evidence where it was found and placing the evidence into a scene drawing by measuring its exact position from fixed objects such as walls or door casings.
- The investigator may use several methods to transport evidence. Personal delivery is the preferred method.
- Examination and testing of evidence often require that the evidence be altered somehow, without impacting the item's value as evidence.
- Review standard operating procedures on how your agency or firm packages, labels, and stores evidence that it takes from a fire scene.

Hot Terms

Accelerant Any fuel or oxidizer, often an ignitible liquid, used to initiate a fire or increase the rate of growth or speed the spread of fire.

Circumstantial evidence Facts that usually attend other facts to be proven and are drawn by logical inference from them.

Cross-contamination The unintentional transfer of a substance from one fire scene or location contaminated with a residue to an evidence collection site.

Demonstrative evidence Photographs, maps, X-rays, visible tests, and demonstrations.

Direct evidence Testimony of witnesses who observe acts or detect something through their five senses and surveillance equipment such as CCTV.

Fire patterns The visual or measurable physical effects that remain after a fire.

Headspace The zone inside a sealed evidence can between the top of fire debris and the bottom of the lid.

Physical evidence Any physical or tangible item that tends to prove or disprove a particular fact or issue. Also referred to as *real evidence*.

Real evidence Any physical or tangible item that tends to prove or disprove a particular fact or issue. Also referred to as *physical evidence*.

Traditional forensic physical evidence Includes, but is not limited to, finger and palm prints, bodily fluids such as blood and saliva, hair and fibers, footwear impressions, tool marks, soils and sand, woods and sawdust, glass, paint, metals, handwriting, questioned documents, and general types of trace evidence.

FIRE INVESTIGATOR *in action*

As you are conducting an investigation of a roof fire at a large automotive parts supplier, you identify several pieces of equipment near the point of origin that may have caused the fire. Fire personnel have already removed a large portion of the roofing and one exhaust fan from this area and are beginning to cut out the electrical conduits that provide power to the fan because they are in the way of extinguishing "hot spots."

You advise the incident commander of your need to examine the items on the roof, at which time he directs his fire fighters to stand by until you complete your examination. Once on the roof, you notice several gasoline-powered tools and a fuel can lying on its side and leaking.

1. What is the best way to protect the area of origin and potential physical evidence after the fire is extinguished?

- **A.** Have fire personnel remove the items as they locate them.
- **B.** Post a fire fighter at the entrance or access point.
- **C.** Overhaul the area of origin, beginning at the entry.
- **D.** Direct fire personnel to remove only large objects during overhaul.

2. Which of the following is not a likely cause of cross-contamination?

- **A.** Properly cleaned metal tools
- **B.** Using a gas-powered generator
- **C.** Using the same gloves between samples
- **D.** Leaving evidence containers unsealed

3. Prior to collecting physical evidence, it should be:

- **A.** photographed in place.
- **B.** placed on a scene diagram with exact measurements.
- **C.** Both A and B
- **D.** None of the above

4. Traditional evidence such as tool marks, footwear or tire track impressions, and fingerprints are best collected and preserved by:

- **A.** fire investigators.
- **B.** fire investigation technicians.
- **C.** police detectives.
- **D.** crime scene evidence technicians.

5. Which of the following is the best location for preserving most types of physical evidence?

- **A.** Dark, dry locations with low humidity
- **B.** Well-lit locations with no more than 50 percent humidity
- **C.** Warm locations with low humidity
- **D.** Dark locations with no more than 50 percent humidity

Origin Determination

© Photos.com

© Glen E. Ellman

Knowledge Objectives

After studying this chapter, you should be able to:

- Describe the recommended techniques for determining the origin of a fire NFPA 4.2 NFPA 4.2.6 NFPA 4.2.7. (pp 256–261)
- Describe how to analyze various data collected during a fire investigation NFPA 4.2.6 NFPA 4.2.7. (pp 261–265)
- Describe how to select the final hypothesis NFPA 4.2.7. (p 265)
- Describe the fire investigator's considerations when the origin cannot be sufficiently defined NFPA 4.2.7. (p 265)

Skills Objectives

After studying this chapter, you should be able to:

- Determine a fire's area of origin and point of origin using the scientific method. (pp 256–265)
- Accurately reconstruct the area of origin. (pp 259–261)
- Analyze a fire scene's data. (pp 261–265)

Additional NFPA Reference

NFPA 921, *Guide for Fire and Explosion Investigations*

CHAPTER 16

FESHE Course Outcomes

Fire Investigation I

11. Identify cause and origin and differentiate between accidental and incendiary. (pp 256–265)

Fire Investigation II

There are no Fire Investigation II (FESHE) course outcomes for this chapter.

You Are the Fire Investigator

© Jones and Bartlett Publishers. Photographed by Glen E. Ellman

You are investigating a fire that has caused extensive damage to three individual dorm rooms in a college dormitory. After conducting an exterior examination, you note smoke staining from only one of the second floor windows.

Your interior examination finds significant smoke staining in the second floor hallway with a flame pattern emanating from room 207. The room is severely damaged, but you locate fire patterns that indicate the fire originated near a bed on the east side of the room. In this area, you locate several extension cords, a candle, and an ashtray.

1. What information may the occupants of the room be able to provide to identify the point of origin?
2. How would you document the area of origin?
3. What analytical techniques could you perform to determine the specific cause for this fire?
4. What building utilities or equipment could provide additional information as to the cause of the fire?

Introduction

The fire's area of origin refers to the room, building, or general area in which the point of origin is located. The point of origin is the exact physical location, as far as the investigator can determine, where a heat source and a fuel come into contact with each other and a fire begins, such as where a lighted match comes into contact with a sofa cushion. Determining the area and point of origin of a fire is essential to determining its cause. The task may be easy or difficult, depending on the amount of evidence, facts, or data that are available; the reliability of the evidence; and the amount of destruction.

The goal of determining the origin is to identify in three dimensions the location at which the fire began. This chapter examines the method used to determine the origin and describes various investigative steps, including initial scene assessment, safety assessment, interior and exterior examination, fire scene reconstruction, fire spread scenarios, and total burns as they relate to origin determination in structures. Although this chapter focuses on structure fires, the method is generally applicable to the determination of origin of all fires.

Although they are treated as two separate topics, origin determination and cause determination are closely related. If the investigator cannot reliably identify the area of origin, it is very difficult to determine the fire's cause or ignition scenario reliably.

To determine the origin of a fire, the investigator must be systematic in examining the fire scene. By being systematic and using the scientific method as described in the "Basic Methodology" chapter of this text, the investigator can reduce any potential errors or oversights that could creep into the process. By employing similar techniques on every fire, the investigator develops a standardized approach that should consistently and reliably address the requirements of a complete fire scene investigation.

Acceptable systematic approaches for the investigator include, but are not limited to, working from the area of least damage to the area of greatest damage or from the highest point to the lowest point. Whatever order is selected for inspection, the point is to assess all pertinent areas of the fire scene. This will help to ensure that an objective analysis can be performed.

Fire Investigator Tip

The purpose of determining the origin is to identify in three dimensions where the fire began (point of origin). The ignition source is usually at or near the accurately determined point of origin.

Investigators should establish and adhere to a systematic procedure for each type of incident they are investigating (structure, vehicle, boat, etc.). Following the same procedure repeatedly will allow an investigator to ensure that he or she does not overlook significant information and that he or she avoids jumping to conclusions or succumbing to expectation bias. To this end, utilization of checklists can be helpful, but they are not foolproof.

Relevant information relating to origin can be obtained from one or more of the following sources:

- Witness information
- Fire patterns
- Arc mapping
- Heat and flame vector analysis
- Depth of char/calcination survey
- Application of fundamental principles of fire dynamics

Overall Methodology

The methodology recommended for examination of a fire scene includes an initial scene assessment, development of a preliminary fire spread hypothesis, an in-depth scene examination, reconstruction of the fire scene, and development of a final fire spread hypothesis. These steps should lead to the accurate identification of the fire's origin. This methodology is meant to guide, but not limit, the investigator's activities during the process of origin determination. Investigators should consider all aspects of the fire event (e.g., witness statements, fire suppression operations). An individual investigator's experience and expertise also play an important role in the determination of fire origin.

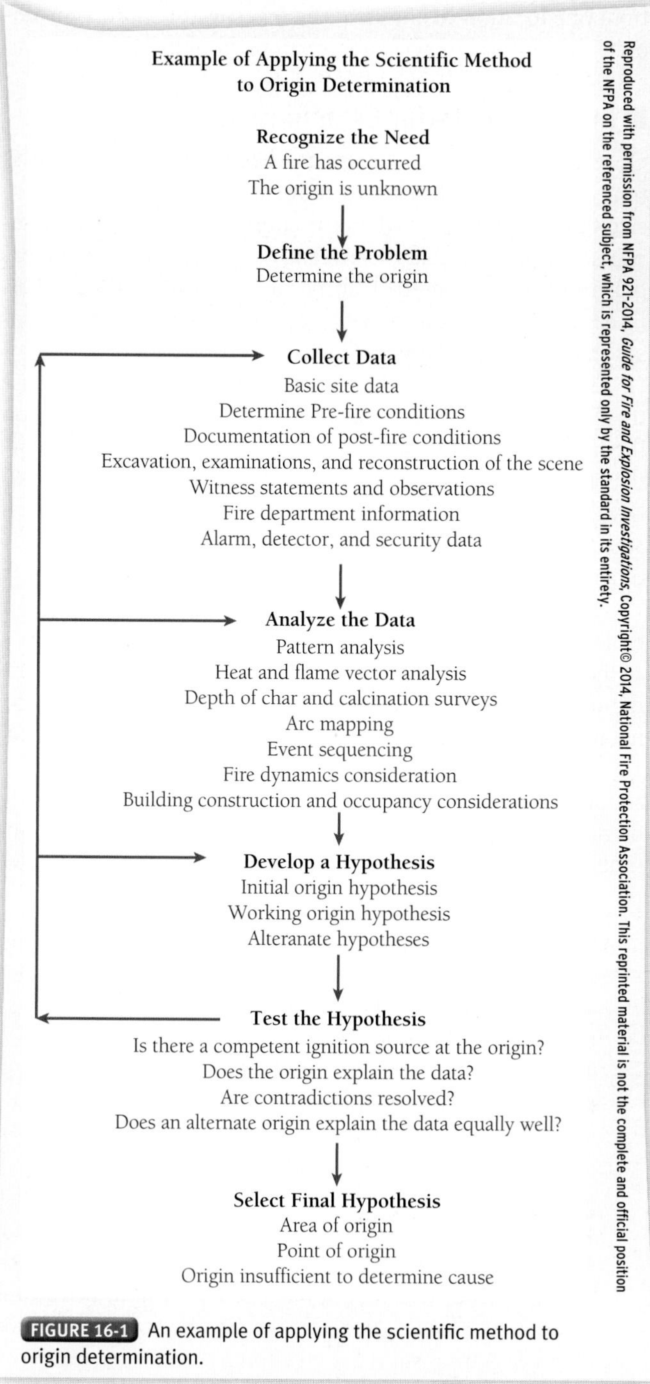

Reproduced with permission from NFPA 921-2014, *Guide for Fire and Explosion Investigations*, Copyright© 2014, National Fire Protection Association. This reprinted material is not the complete and official position of the NFPA on the referenced subject, which is represented only by the standard in its entirety.

FIGURE 16-1 An example of applying the scientific method to origin determination.

The scientific method is utilized by the investigator, as part of a systematic approach, to determine the origin of the fire. **FIGURE 16-1** depicts a flow chart demonstrating the application of the scientific method to origin determination. The analysis of the various data collected by the fire investigator will always yield at least one (e.g., the fire started in the structure) and usually multiple hypotheses regarding the origin. These early hypotheses are considered "working" hypotheses. Testing of hypotheses should continue as new data are collected. Additional testing may result in revision or disregard of a hypothesis.

Testing of origin hypotheses requires inclusion of all available data and application of fundamental principles of fire dynamics. An investigator should not focus exclusively on identifying a first fuel ignited and a competent ignition source in the proposed area of origin. Failure to consider growth and spread can lead to erroneous conclusions. For example, a cigarette in a trash container may be a competent ignition source for paper in the container. Placement of the trash container, however, must be by secondary and subsequent fuels for the fire to spread. If the trash container were next to a noncombustible wall with no other fuel in the area, it is unlikely the fire could spread, but if the trash container were under a wooden desk in an office cubicle, the fire would likely spread to adjacent combustibles. A fire spread analysis must be conducted to determine whether the physical damage and other available data are consistent with the origin hypothesis.

The body of data must be considered collectively to determine which of the hypotheses fit all of the available evidence. When an otherwise plausible hypothesis fails to fit some piece of data, the investigator should try to reconcile and identify whether it is the hypothesis or the data that are wrong. There are, however, instances where a single piece of evidence can be the basis for an accurate determination of origin. Such evidence may include a credible eyewitness, security video, or some piece of irrefutable physical evidence.

Many of the tasks performed by the fire investigator aimed at determining the origin occur simultaneously (photography, witness interviews, etc.). Likewise, application of the scientific method (data collection, analysis, hypotheses development and testing) occurs continuously during the investigation. These are not mutually exclusive processes.

Identification of the point of origin is not always possible; however, the inability to determine the point of origin does not, in and of itself, eliminate the ability to develop a credible and defensible origin and cause hypothesis.

Origin determination necessarily involves identification and interpretation of fire patterns (see the "Fire Patterns" chapter of this textbook). The surfaces in a fire scene record patterns continuously over the life of a fire. The investigator must realize that as a fire progresses, previously generated patterns can be altered or obliterated. Determining the sequence in which these patterns were created, known as sequential pattern analysis, is the key to determining the origin. The investigator must collect and organize the fire pattern data into a sequential format to be able to comprehend and articulate the origin and spread of the fire.

The process for data collection, as it relates to origin determination, includes the initial scene assessment, safety assessment, excavation and reconstruction, and collection of data from nonscene sources. These topics are discussed in greater detail later in this chapter.

Initial Scene Assessment

The determination of origin begins with the initial scene assessment. The initial scene assessment provides an overall look at the structure and surrounding area. In this step, data collection for the determination of origin begins. The purpose of this assessment is to determine the scope of the investigation—for example, equipment, manpower, safety, and security requirements. Furthermore, it helps to identify areas that require detailed inspection.

The first order of business in the initial scene assessment is the safety assessment. During the safety assessment of the scene, the investigator will need to assess each of the hazards discussed in the "Safety" chapter (in this text and in NFPA 921). The overall goal is to determine whether it is safe to enter the scene, what steps are required to make the scene safe to enter, and what personal protective equipment (PPE) will be required when entry is made. Only after appropriate safety measures have been implemented should the initial scene assessment continue.

The initial scene assessment should include examination of the site and of surrounding areas. This involves looking for fire patterns away from the core of the scene. Important witnesses can be identified, among them the reporting party and neighbors. It is important for the investigator to document findings during this phase.

As part of the initial scene assessment, the investigator should walk around the entire exterior to evaluate the possibility of extension from an outside fire, to examine the building construction details and materials, to determine the occupancy or use of the building, and to identify areas that may require further study. During this and all subsequent phases, the investigator should constantly be evaluating the safety of the building and identifying any conditions that need to be corrected. These conditions include securing utilities, shoring unstable areas, removing water, cordoning off unsafe areas, and noting inhalation dangers.

Building Identifiers

Beginning with the initial scene assessment, it may be beneficial to establish building identifiers and to use them throughout the investigation. These identifiers provide consistency in various aspects of an investigation—such as in the description of photographs, diagram preparation, report writing, and testimony. Building identifiers or a building marking system can include any of the following:

- Points on a compass (north side of the building, etc.)
- Front, back, left side, right side
- Similar method such as those used in the incident command (Side A, Side B, etc.) or for urban search and rescue

Weather Considerations

The weather at the time of the fire is an important component of the initial scene assessment. Wind can play a role in the development of the fire and the patterns left behind. Weather information can be obtained from a number of sources, including the local airport, local media outlets, the National Weather Service, and Internet sources. Another source would be the weather observations of witnesses and fire fighters obtained through interviews. It is important to consider the distance between the point at which the weather readings were taken and the fire scene. Weather conditions can vary dramatically from one location to another. The investigator should corroborate official reports with observations from local witnesses.

> **Fire Investigator Tip**
>
> Do not be hasty during the initial scene assessment. Time spent during this phase may save significant time and effort later. This includes determining the safety of the scene, staffing and equipment requirements, and the areas that will require a detailed examination.

In-Depth Exterior Examination

Once the initial scene assessment has been completed, it is time for the investigation to move into a more comprehensive stage. This stage is when the efforts of the first two phases begin to pay off, in that the investigator has identified the areas that need to be examined further and can safely do so. This in-depth examination is made by using observations, photographs, sketches, or other means of documentation.

A detailed exterior surface examination is critical because there are issues that may be brought up later in the investigation that may be important and related to the exterior. For example, the location and condition of a window (open or closed) may have affected the ventilation and the subsequent fire growth and spread inside the structure. If this condition is not documented, then its value in determining the fire spread is lost.

Another critical reason to conduct an in-depth exterior examination is to look for additional fires that may have occurred at the scene. Investigators may mistakenly zero in on the first fire-damaged area they find or the area with the greatest damage, not realizing that there could be other areas where fires occurred.

As with the initial scene assessment, the investigator should adhere to a systematic procedure. Even if the fire clearly originated in the interior, this process should be used to document various building features, including, but not limited to, the following:

- Prefire conditions: The state of repair of the building and its components, the conditions of structural elements, and the condition of the structure's fire suppression and detection systems
- Utilities: Type, location, and meter readings
- Doors, windows, and other openings: Their location, condition (open, closed, or broken, and for how long prior to the fire), and security mechanisms
- Explosions: The presence or absence of evidence that explosions may have affected the exterior components
- Fire damage: Overall damage, damage to natural openings (windows, doors), damage resulting in unnatural openings (holes made by the fire or explosion, holes made during suppression efforts)

The exterior evaluation of fire damage should include consideration of the fire-suppression efforts. These efforts could have a significant impact on the spread of the fire and fire patterns. It is important to interview fire fighters to determine what actions were taken on the fire ground, such as forcible entry, ventilation, overhaul, and other related activities.

Often, the fire origin will be on the interior of a structure. Even if the fire did start on the exterior, the conditions inside the building should be documented, especially any damage that occurred as a result of the fire spreading to the interior.

In-Depth Interior Examination

All interior rooms and other relevant areas should be inspected to identify areas requiring detailed inspection. Additional safety hazards should also be identified at this time. Interior damage should be compared with the location and type of damage on the exterior. Among the things to examine are the following:

- Prefire conditions of the structure
- The contents of the structure
- Storage of contents
- Housekeeping
- Maintenance
- Evidence of explosion damage
- Areas of fire damage
- Building systems (heating, ventilation, and air conditioning [HVAC]; electrical; fuel gas; fire protection; alarms; and security)
- The composition of the surface coverings of walls and floors
- Position of windows, doors, and other openings (ventilation aspects)
- Indicators of smoke and heat movement (i.e., fire patterns) on various surfaces
- Relative extent of damage (severe, moderate, minor, none) in each area

Although sometimes difficult, every effort to identify the prefire conditions of the structure should be made when warranted. Various conditions, including construction detail and/or defects; holes in floors, walls, or the roof; and missing doors or windows may have a significant effect on fire development and spread.

It may also be necessary, when considering fire development and spread, to identify the prefire conditions of various building systems. These systems can include electrical, fuel gas, HVAC, fire protection, and others.

HVAC systems can enhance the development and spread of fire, particularly if in operation during the fire. Inspection of filters for soot accumulation or heat damage can help determine whether the system was in operation during the fire.

The location of HVAC vents, both supply and return, in any areas of interest should be documented. Documentation should include the location and dimensions of vent openings, ducts/plenums and the settings of the thermostat and fan(s), and the power supply.

Some HVAC systems, particularly in commercial structures, are fitted with detectors and manual or automatic dampers designed to change or stop operation of the system in the event of fire. The condition, location, and position of these devices should be documented, and monitoring records (if any) should be sought.

The fuel gas service should be inspected and documented to include or exclude the possibility of contribution to the fire. This system should be documented in detail when an ignition or spread scenario includes involvement of fuel gas. In such cases, inspection should include documentation of the position of valves and testing to determine the supply pressure (if possible) and to locate any leaks. Note that fires frequently compromise the integrity of gas piping, threaded joints, and other connections.

In some structures, there may be oil-fired heaters or other appliances requiring a liquid fuel. These systems may be permanent (large boiler) or portable (kerosene heater). The location of such devices, fuel reservoirs, and the quantity of fuel remaining should be documented. If the ignition hypothesis includes this system, a sample of the liquid fuel should be taken.

The electrical system of the structure should be documented, beginning at the service entrance. Important information to gather includes the main panel amperage and voltage input, condition of circuit breakers or fuses (type, rating, position), and areas of flame or heat damage.

Any appliances plugged into receptacles in the area of origin should be identified. Note the location of the receptacles, what appliances are plugged in and where, and the position of switches. A more detailed inspection and documentation of the electrical system may be required where an electrical ignition scenario is hypothesized or when an arc map survey is warranted. It may become necessary to trace circuits back to the panel, noting the condition of the circuit protection and loads along the circuit.

Fire protection systems and the data recorded by such systems can be of tremendous value in reconstructing the sequence of events in a fire. Detector activation and sprinkler response can be utilized to help identify an area of origin and/or spread scenario. Off-site monitoring services can provide these data. Some systems retain such data on site in the central panel, which can be downloaded and preserved. The data in such systems must be retrieved quickly because the electrical supply to the structure has likely been terminated and backup battery life is limited. Special equipment and a qualified technician are generally required to retrieve these data, as they are easily compromised.

The location and condition of devices associated with the fire protection system should be documented. Particular attention should be paid to the height of wall-mounted devices or, on ceilings, the distance from walls. Additional value can be drawn from the location of activated sprinklers.

Postfire Alterations

The investigator should also note any apparent postfire alterations that may include debris removal or movement, content removal or movement, alterations to electrical service, gas meter removal, and indications of prior investigations. The investigator should identify those responsible for the alterations and interview them to determine, as closely as possible, the conditions before the alterations were made.

Debris Removal and Reconstruction

Adequate debris removal and reconstruction are essential to observe, document, and analyze fire patterns accurately. In this phase of the investigation, the investigator attempts to recreate the conditions that existed prior to the fire FIGURE 16-2. By removing the debris carefully and systematically and then placing the remaining contents back in their prefire location, a great deal can be learned.

To locate and recover evidence, it is necessary to remove the debris in a systematic and controlled manner. This ensures that there is no misinterpretation of patterns that became

Courtesy of Rodney J. Peyfoe

FIGURE 16-2 Fire scene reconstruction recreates the conditions that existed prior to the fire.

obscured by debris or by loss of evidence contained in the debris. The investigator must remember that safety is of paramount concern during this phase. Debris removal can make a structure unstable. Hazardous materials may become apparent, or energized electrical wires may become exposed. When removing debris, the investigation team should remove only as much of the debris as is necessary. (See the "Safety" chapter of this text for more information about safety.)

Adequate removal of debris is essential in leading the investigator to the correct analysis. Fire suppression crews should attempt to minimize content and debris removal during overhaul before the start of the investigation. If this is not possible, the investigator should document the conditions prior to debris removal. This documentation should include photographs and notes.

The debris should be moved only once, digging and exposing materials from the top down, similar to an archaeology dig. It should be removed in a systematic fashion, and the process should be fully documented, including the location, condition, and orientation of any contents that are uncovered. The investigation team should work together, deciding what is important and what is not before beginning the process. Hand tools rather than heavy equipment should be used to remove debris, when possible.

Fire Investigator Tip

Debris removal is a tedious and laborious task. Plan ahead so the debris has to be moved only once.

Safety Tip

Debris removal can weaken a fire-damaged structure and cause collapse.

Heavy Equipment Use

Some fire scenes may require use of heavy equipment to facilitate debris removal or to mitigate safety hazards. The fire scene should be fully documented prior to the use of heavy equipment and frequently during its use.

Fire scene inspections progress at a much slower rate than routine construction demolition. A briefing should be held with the equipment operator and crew prior to the commencement of work. The equipment operator should clearly understand the investigative goals, the specific area to be worked in, and evidence or items of interest to be recovered.

Heavy equipment operation is noisy, can damage evidence or contaminate the scene, and can create safety hazards for the investigator. The operator should be under the supervision of a fire investigator at all times while the equipment is in use. The fire investigator and operator should be in direct contact via radio or through use of hand signals.

If the presence of ignitible liquids in the scene is in question, sampling should be completed before heavy equipment is brought in. The equipment should be inspected for leaks and documented prior to use. Sampling of the equipment fluids can be performed to help address contamination concerns. Refueling or lubrication of the equipment should occur only in a designated location well away from any area(s) of interest. It is preferable that heavy equipment be positioned outside the structure during use. If large amounts of debris or structure need to be removed, this should be done in a way that does not change or add debris to the underlying area. Shoring of structural elements should be considered as an alternative to demolition where possible.

Where heavy equipment must be operated within the scene, the equipment should initially be positioned outside the area of interest. Once an area has been inspected, documented, cleared, and samples taken if indicated, the equipment can move progressively into the scene. Loading and unloading of the debris should be monitored to identify relevant items of evidence.

Large or heavy items can be lifted out of the scene utilizing rigging to minimize potential damage. The items can be further inspected, documented, and even stored on scene (assuming proper security exists) for the duration of the investigation. Circumstances may require removal of evidentiary items from the scene. Removal should be well documented and can complicate ongoing inspection of the scene.

Avoiding Contamination and Spoliation

Potential ignition sources, initial fuels, or other important evidence can be identified during excavation of the scene. Care should be taken to limit damage to and prevent contamination of any such items. Simple decontamination and processing techniques can prevent these problems.

Washing floors after debris removal may help reveal fire patterns. Caution should be exercised when using high-pressure streams, as they can damage evidence. Washing should be done only after adequate debris removal, after samples have been collected, and after proper scene documentation has been completed.

Both public- and private-sector investigators should recognize that civil litigation may result from fire incidents.

Parties that may be interested in such litigation should be identified, placed on notice, and invited to participate in the ongoing inspection. It is important that potentially interested parties be provided the opportunity to view the scene and evidentiary items in place where possible. It is understood that different requirements for criminal investigations exist.

The fire scene reconstruction phase is important because it can allow for a strong visualization of how the fire developed and spread (sequential pattern analysis). During the previous phases of the investigation (initial scene assessment, in-depth exterior examination, and in-depth interior examination), the investigator can generally narrow down an area of origin. The reconstruction efforts can therefore be limited to the probable area of origin as opposed to the entire area of damage.

Contents

Identifying the prefire location and orientation of contents is essential to fire scene reconstruction. Reconstruction may also include replacement of structural components. Detailed documentation of contents, or remains of contents, located during debris removal should be made.

Postfire reconstruction is complicated by displacement of contents during suppression or overhaul. Protected areas may exist that can help identify the prefire location of an item. Witness observations can also aid in reconstruction. Care should be taken to avoid inaccurate reconstruction, as it can result in false data being included in the analysis.

Prefire Conditions

Persons familiar with the involved structure (e.g., owners, occupants, workers, neighbors) can provide information regarding the type and placement of contents, general conditions or alterations to the structure, fire protection systems, etc. Frequently, these persons will be able to sketch the placement of furnishings or fixtures. Sometimes they may have prefire photographs or video; however, the investigators should be cautious with these details and confirm when the individuals were last in the fire location.

Public records (e.g., county assessor or treasurer, fire department inspections, and building department permits and inspections) will also help to determine prefire conditions. Internet searches can sometimes lead to photographs of the interior of the scene from real estate listings. Commercial information sources may be able to provide such things as aerial or satellite images.

Knowledge of prefire conditions can assist in the identification of fuels in the building and area of origin. It is important not only to identify the first fuel ignited but also subsequent fuels that allow the fire to spread.

Mathematical calculations and computer fire models are sometimes used in reconstructing the fire dynamics or fire development; however, these methods require a certain level of expertise. They are also highly dependent on the correct information being entered to produce a valid solution (a situation known in programming as "garbage in, garbage out"). If the investigator is not comfortable using these tools, or does not have someone available who has the proper level of expertise, it is better not to use them.

Fire Investigator Tip

NFPA 921 Appendix A contains a form to assist in collecting the necessary information for use in calculations and computer fire modeling.

Security Cameras

Security cameras may have captured events leading up to and during a fire. The video-recording device may be found distant from the fire-damaged area. Consider also that neighboring buildings, banks, or ATMs may have captured important data. Like fire protection alarm data, camera footage can also assist with preparation of a timeline of events. This is a time-sensitive consideration, however, because many security systems store data for a limited time.

Intrusion Alarm Systems

Security systems can provide helpful information if monitored. These alarms can be activated by flames, heat, or smoke movement. A trouble signal may also be generated if wiring or devices are damaged by heat. The same cautions indicated for fire protection systems apply to burglar alarm systems.

Witness Observations

Witness observations are of considerable importance. Relevant data can be what was seen, heard, or smelled. The investigator should evaluate witness information with physical evidence of origin. Contradictions between witness information and physical evidence should be resolved.

Analyze the Data

Once the data have been collected from the scene and other appropriate sources, it is time for the investigator to determine the point of origin. To do so, by applying the scientific method, the investigator must analyze the collected data to develop any number of reasonable origin hypotheses. Some investigative and analytical techniques that may be useful are heat and flame vector analysis, depth of char surveys, depth of calcination surveys, and arc survey diagrams. The origin hypothesis should explain not only where the fire started, but how it spread throughout the structure. The hypothesis must be supported by the evidence (data) gathered through the fire investigation process.

Any evidence that contradicts the origin hypothesis proposed by the fire investigator must be resolved. If the hypothesis is not supported by the facts, the investigator may have to return to one of the earlier steps in the scientific method and repeat the process from that point to ensure that a sound conclusion is reached (see the "Basic Methodology" chapter of this textbook).

Fire Pattern Analysis

Principles of fire dynamics, heat transfer, and fire pattern generation must be understood and applied during the analysis of the collected data to determine the origin of a fire accurately.

The reader is referred to chapters on basic methodology and basic fire science.

It is essential that all observed patterns be considered in the analysis. Reliance on a single pattern, except in cases of extremely small fires, will likely result in inaccurate origin hypotheses.

Remember that the fire patterns present after suppression represent what is left of all patterns recorded during the life of the fire. As previously stated, patterns can be altered or obliterated by ensuing fire. This is particularly prevalent in cases of full room involvement, post-flashover fires, or structural collapse. Also, pattern interpretation can be significantly complicated by a rekindle.

As discussed, fire patterns are generated by either the spread of the heat/flames or the intensity of burning. It is important to recognize that the factors such as fuel form, type, and geometry; heat release rate; and ventilation may lead to intensity patterns that may not correlate to the origin. Fire movement patterns are better indicators of origin than intensity patterns are. In well-developed compartment fires, it may be difficult to differentiate between movement and intensity patterns.

Many factors affect fire pattern generation. Identical fuel packages will burn differently depending on placement in the room. The size, geometry, and heat release rate of the fuel, as well as the location of ventilation, will have at least as much effect on fire pattern generation as the length of time it burned. The investigator should not assume that the point of greatest damage was where the fire burned longest and therefore is the origin. The possibility that flames spread from a smaller fuel package to a larger package must be considered.

Ventilation can have a dramatic effect on fire behavior and heat release rate. Identification of ventilation factors (e.g., door/window position, fire suppression strategy) is crucial in evaluating fire patterns. Ventilation-controlled fires will burn with greater intensity (and cause greater damage) because of openings that provide access to oxygen. Full room involvement will also change, and may mask or obliterate patterns made during the fuel-controlled (early) phase of the fire.

Analytical Tools

One of the analytical tools available to the investigator is the heat and flame vector analysis, which is a process used to assist with fire spread analysis and origin determination. Arrows representing the investigator's assessment of the direction of heat/flame spread are placed on a detailed diagram of the fire scene, and the patterns can be traced back to the fire origin.

To utilize this method, the investigator must first prepare a diagram of the scene. The diagram should be reasonably detailed and include the location of doors, windows, and pertinent contents showing identifiable patterns. The investigator then draws arrows on the diagram to depict their interpretation of heat and/or flame spread based on the patterns in the scene.

The arrows can point in the direction of fire travel or back to the heat source, as long as the direction is consistent throughout the diagram. The length of the arrow should reflect the actual size or magnitude of the pattern.

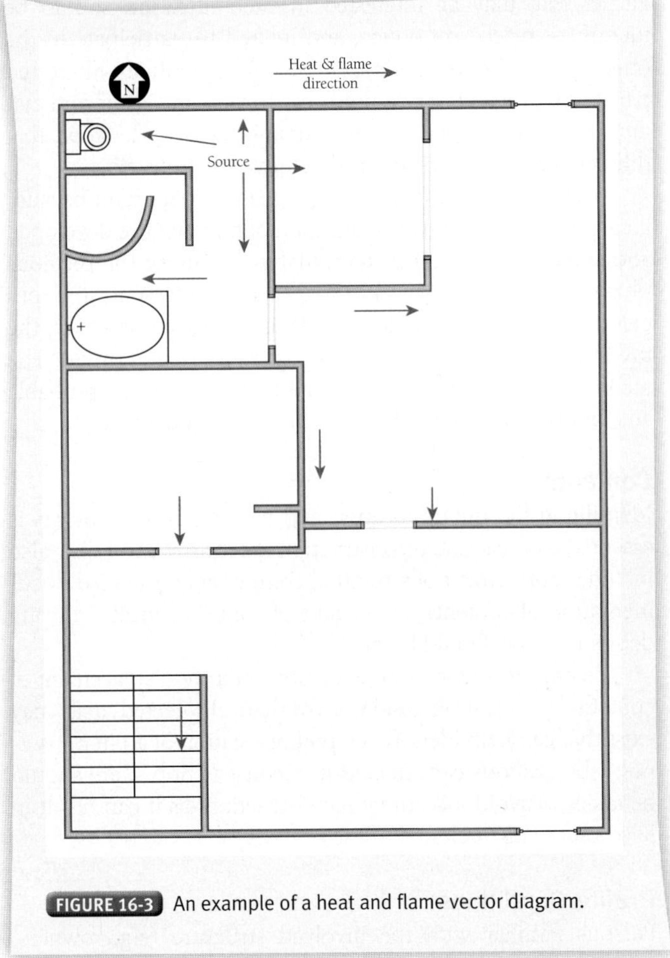

FIGURE 16-3 An example of a heat and flame vector diagram.

In a legend accompanying the diagram, the investigator may provide details of the pattern, for example, pattern geometry, height above the floor, height of the pattern vertex, and the fire effect that constitutes the pattern.

Complementary vectors can be considered together to show actual heat and flame spread directions. When using complementary vectors, the investigator should clearly identify which vectors indicate actual flame patterns and which represent heat flow. The point of the heat and flame vector analysis is to illustrate graphically the investigator's interpretation of the fire patterns. The diagram can also be used to identify conflicting patterns that need to be explained FIGURE 16-3.

Sometimes fire patterns are not visually obvious. Depth of char or depth of calcination surveys may help the investigator identify areas of greater and lesser heat damage and define the associated lines of demarcation defining the patterns. Depth measurements are plotted on a diagram. Points of equal or nearly equal depth (isochar) can be connected by lines, thereby revealing patterns.

Depth of Char Analysis

Measurements of the relative depth of char on identical fuels are plotted on a detailed scene diagram to determine locations within a structure that were exposed to a heat source and were part of the fire spread. The key to appropriate use of this tool is documenting the relative depth of charring from point to point and locating places where damage was most severe due to exposure, ventilation, or fuel placement. The investigator may then be able to deduce the direction of fire spread.

We know fire patterns are created by heat sources. It is important to note that any particular pattern may not have been created by the initial fuel. For example, if a fire originates in one part of a compartment and spreads to involve ignitible liquid stored elsewhere in that compartment, the investigator may be misled by the intensity patterns from the burning ignitible liquid. It is of critical importance that the investigator considers pattern analysis with an accurate understanding of the development of the fire and application of basic fire dynamics.

The investigator is cautioned that several factors can influence the validity of char depth analysis. These factors include:

- Single versus multiple heat sources
- Ventilation factors influencing rate of burning or fire intensity
- Consistency of the measuring technique or method

Comparison of char depths should be made only on like materials. The depth of char analysis is most reliable when applied to fire spread evaluation. It is not intended to determine specific burn times or quantify intensity of heat from any particular fuel package.

Consistency of the method of measuring char depth is critical to generating reliable data. Thin, blunt-edge probes are best. Sharply pointed objects are not suitable. Certain types of calipers with round depth probes and tire tread depth gauges are excellent. The same tool should be utilized on comparable materials. Consistent pressure of insertion is also necessary. Research continues into the development of a mechanized tool that will automatically deliver a consistent pressure for every reading.

When measuring char depth, the investigator must note if any of the material has been consumed or broken off during the fire. The depth of the missing material should be added to the measured remaining depth of char.

Where fugitive fuel gas was the first fuel ignited, char depths are likely to be relatively uniform. Variances in char depth may exist in locations where the gas had pocketed or ignited other fuels. Deeper charring may exist at the point of the gas leak if there was continued burning of a pressurized gas jet.

Depth of Calcination Analysis

Measurements of the relative depth of calcination (observable physical changes in gypsum wallboard) are plotted on a diagram to determine locations within a structure that were exposed longest to a heat source. Relative depth of calcination on gypsum wallboard can indicate differences in total heating. Deeper calcination readings indicate longer exposure to a heat source or exposure to greater intensity/temperature during the fire. The same cautions regarding depth of char measurements apply to depth of calcination measurements. Additionally, the finish on the wallboard (e.g., paint, stucco) may influence the heat effect, particularly if the finish is combustible. Also, water applied during fire suppression can soften gypsum wallboard to the point where no reliable measurements could be obtained.

Two methods for measuring the depth of calcination are available to the investigator: the probe method, as described in the depth of char section, and cross-sectional inspection. When measuring a cross section, small sections (at least 2 inches [5 cm] in diameter) of a full thickness of the wallboard must be removed. The depth of the visually different calcined layer can then be recorded. A depth of calcination diagram can be prepared in the same manner described for depth of char measurements.

Arc Mapping

Arc mapping is a technique in which the investigator uses the identification of locations of electrical arcing to help determine the area of origin. This technique is based on the predictable behavior of energized electrical circuits exposed to a spreading fire. The locations of the arc sites in the building and in relation to each other, plotted on an arc survey diagram, can reveal a pattern, which can be used in the analysis of the sequence in which the affected parts of the electrical system were compromised. This sequential data can be used in combination with other data to define the area of origin. Once the origin is identified, any arc sites within that area can be evaluated as a potential ignition source.

There are circumstances that can make it more difficult or impossible to utilize this technique, such as melting of conductors due to fire exposure or postfire reenergizing of the electrical system (and subsequent arcing).

The following procedure can be utilized to perform an arc mapping survey:

1. Identify the area to be surveyed.
2. Diagram the area as accurately as possible.
3. Establish zones within the area (ceiling, north wall, etc.).
4. Identify all conductors passing through each zone. Document the attached devices (junction boxes, outlets, switches); the loads on the system, if possible; and the direction of power distribution.
5. Systematically and closely inspect each of the conductors in a given zone.
6. Inspect the length of conductors visually and by hand to identify any melt damage or other anomalies. Note that conductors can be damaged during removal from conduit. Also, bear in mind the cautions against spoliation before removing conductors from conduit.
7. Determine whether the damage is the result of arcing, environmental heat, or eutectic (alloying) effect.
8. Plot the location of arcing events on the diagram. Also, document the physical characteristics of the anomaly (e.g., severed, faulted to another conductor in the same cable).

9. Mark the locations of arcing on the conductors with cable ties, flagging tape, or other suitable nondestructive means for photography and documentation.
10. Collect and preserve the items where required.

As previously stated, the conductors should be examined by touch for anomalies. Another method to locate anomalies is to rub a cotton ball along the conductor. The anomalies may be surface imperfections such as dirt accumulations, scale, or oxidation. Arcing sites will be concave or convex in shape. Closely inspect anomalies for evidence of metal loss or deposition of metal on the conductor. There will usually be complementary damage on adjacent conductors (a notch in one conductor and an adhered mass on an adjacent conductor). Do not stop the inspection of a conductor just because a spot of suspected arcing is found. Inspect the entire length to determine whether multiple arcing sites are present. Fire-caused melting is generally widespread and can involve multiple conductors in an area. Arcing events are highly localized. The edges of the melt damage are generally quite distinct.

It is very important to note that differentiating between arcing- and fire-caused (environmental) melting can be difficult. If an investigator relies on incorrect identification of melt damage as an arc site, hypotheses developed from these flawed data will be incorrect. Utilization of subject matter experts and advanced laboratory analysis techniques may be required to validate melt damage as the result of an arcing event.

The visibility of arc damage is related to the duration of the arcing event and the timing of the arc. In alternating current (AC) circuits, the potential difference between two conductors depends on the point in the AC cycle at which the arc occurs. During the typical 120-volt AC cycle, the potential between the "hot" conductor and the neutral or ground conductor ranges between +120 V and –120 V, 50–60 times per second. As the potential approaches 0 volts, physical damage from the arcing event will be less severe and therefore more difficult to see.

Marking arc sites will help illustrate the arc location and patterns in photographs and video. The relationship in space of the arc sites to each other and to those conductors that did not arc is the significant evidence, not the individual arc sites.

Precision in preparing the diagram will aid in reducing errors in the analysis of the data. Both plan and elevation views are recommended where appropriate. It is also important to document type and status of the overcurrent protection device for each involved circuit (fuse, breaker, ground fault, or arc fault circuit interruption) FIGURE 16-4.

When collecting conductors from an arc survey site, ensure that the conductors are properly identified, labeled, and photographed before removal. These conductors can be very brittle and broken easily. Associated devices can fall off the conductors, potentially eliminating the possibility of future circuit tracing.

Like the other analytical tools discussed, arc mapping is useful not only for analyzing the sequence of events, but for developing and testing origin hypotheses. If a conductor has been severed by an arcing event, then it can be correctly concluded that any arcing events "downstream" from the point where the wires were severed occurred before the wires were severed. (Exceptions include prefire wiring errors and backfeeding from an uninterruptible power supply or generator.) If an area of origin is hypothesized then, in the absence of evidence to the contrary, one can reasonably expect the origin to be one of the first areas where the electrical system was compromised.

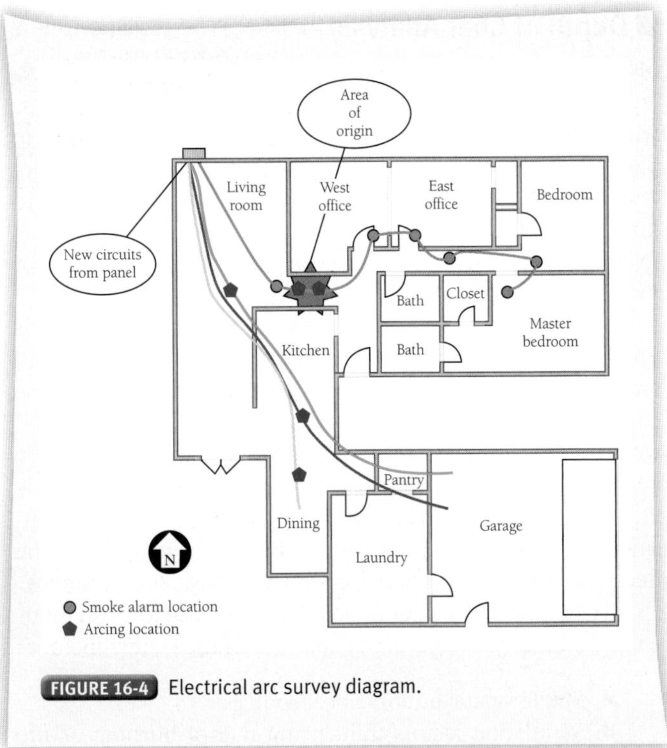

FIGURE 16-4 Electrical arc survey diagram.

Once all data have been collected, they need to be analyzed. Fundamentals of fire dynamics can be applied to the analysis of the collected data and to aid in the development of origin hypotheses. Analysis of the sequence of events during a fire can be useful in determining the origin. Analyses can develop information such as potential first fuels ignited, the sequence of subsequent fuel involvement, recognition of lacking data to be collected, and identification of potential competent ignition sources.

Based on the analysis of all of the data, the investigator can now develop one or more hypotheses to explain the origin and spread of the fire. A hypothesis may be developed early in the investigation and not be disproved by the additional data. Ultimately, the investigator must be able to articulate, no matter the order of the investigative steps taken, how the steps taken conform to the scientific method.

Conformance with the scientific method requires that once a hypothesis is developed, it be tested. Deductive reasoning tests are based on the premise that if the hypothesis is true, then the fire scene should exhibit certain characteristics (assuming the fire did not grow so large the

Fire Investigator Tip

A critical question to be answered is this: "Are there any other origin hypotheses that are consistent with the data?"

characteristics were obliterated). Techniques available for hypothesis testing include timeline analysis, fire modeling, and experimental testing. Application of these techniques to hypothesis testing can help identify gaps or inconsistencies. These techniques can be applied to data analysis and hypothesis development.

Timelines can be used to understand the relationship in time among significant events and circumstances identified or known to have occurred during the fire.

Fundamentals of fire dynamics can be applied to a hypothesis through various computer programs. These fire models use incident specific information (collected by the investigator) to predict fire development at given points in time.

Physical experiments, including live burn exercises, can be used to test a hypothesis. If the experiment produces results consistent with the evidence and data from the fire scene, the experiment would support the hypothesis. Conversely, if the experiment produces different evidence, the hypothesis may need to be discarded or new information developed and considered in the analysis.

A technically valid origin determination is one that is consistent with the available data. The following questions should be answered during the process of hypothesis testing:

- Can a fire starting at the hypothesized origin result in the observed damage?
- Is the growth and development of a fire starting at the hypothesized origin consistent with the established timeline?
- Is there a competent ignition source at the hypothesized origin?

The lack of a competent ignition source in the area of origin may require the investigator to reanalyze the validity of the origin hypothesis. Contradictory witness observations must be explained through physical evidence, reliability, or weight of the information. The origin should not be "determined" just because a readily ignitible fuel and a potential ignition source are found together.

Selecting the Final Hypothesis

Once the origin hypotheses have been tested, only one hypothesis should survive the testing for the investigator to be able to select it as the final hypothesis. The investigator should then review the entire process to ensure that all credible data are accounted for and credible alternate hypotheses have been eliminated. The investigator should document the facts that support the origin hypothesis to the exclusion of all others.

It is unusual for a hypothesis to be totally consistent with all of the data collected. It is important to note that not all data collected have the same value. Data must be evaluated by the investigator to determine their reliability and relative weight. Contradictions must be recognized and resolution attempted. Incomplete data can make this impossible. If contradictions cannot be resolved, the origin hypothesis should be reevaluated.

Case File Review

In large fire losses or where personal injury or death occurs, review of an investigator's work product by other investigators is likely. The original investigator may be required to produce the data that he or she collected and his or her analyses. Another investigator should, theoretically, be able to review this evidence and reach the same conclusion regarding the origin. Differences in opinion may arise because of the weight given to various data by different investigators, application of differing theoretical explanations (fire dynamics) to the underlying facts of the case, or as the result of other outside influences such as expectation bias.

Origin Insufficiently Defined

The inability to identify the point or area of origin does not, in and of itself, preclude a valid determination of cause. Take, for example, a scenario involving ignition of fugitive fuel gas resulting in a total burn. Hypotheses can be developed and tested regarding probable ignition sources for the fuel gas. The result may be multiple potential ignition sources, but the first fuel ignited and the reason for the fuel leak (cause) can be identified.

When the origin cannot be narrowed to a reasonable size (e.g., a room within a structure), the investigator should be able to explain why. A few of the possible reasons for an inability to define the area of origin include the absence of reliable fire patterns due to the extent of damage, building collapse, or near complete consumption of combustible material. Extensive damage may also preclude use of various analytical tools such as heat and flame vector analysis or arc mapping.

In the absence of physical evidence to establish the origin, an eyewitness to the early stages of the fire may be able to provide reliable evidence. An attempt should be made to obtain photographs or video of the fire in its early stages from witnesses or first responding units such as police officers.

A total burn—a fire that has consumed nearly all of the fuel (contents and/or structure)—is a difficult, but not impossible, fire scene to investigate. Although it is usually not possible to determine the origin and cause of a total burn, there are other areas of interest that may need to be considered. Valuable data can still be recovered from a total burn site. The investigator should therefore attempt to follow the procedures outlined in this chapter to gather data.

Wrap-Up

Ready for Review

- The fire's area of origin refers to the room, building, or general area where the point of origin is located. The point of origin is the exact physical location where a heat source and a fuel come into contact with each other and a fire begins.
- The methodology recommended for examination of a fire scene includes an initial scene assessment, development of a preliminary fire spread hypothesis, an in-depth scene examination, reconstruction of the fire scene, and development of a final fire spread hypothesis.
- The determination of origin begins with the initial scene assessment, which provides an overall look at the structure and surrounding area.
- A detailed exterior surface examination is critical because there are issues that may be brought up later in the investigation that may be important and related to the exterior.
- All interior rooms and other relevant areas should be inspected to identify areas requiring detailed inspection.
- Investigators should note any apparent postfire alterations that might include debris removal or movement, content removal or movement, alterations to electrical service, gas meter removal, and indications of prior investigations.
- Adequate debris removal and reconstruction are essential to observe, document, and analyze fire patterns accurately.
- The origin hypothesis should explain not only where the fire started, but how it spread throughout the structure.
- Principles of fire dynamics, heat transfer, and fire pattern generation must be understood and applied during the analysis of the collected data to determine the origin of a fire accurately.
- Measurements of the relative depth of char on identical fuels are plotted on a detailed scene diagram to determine locations within a structure that were exposed to a heat source and were part of the fire spread.
- Arc mapping is a technique in which the investigator uses the identification of locations of electrical arcing to help determine the area of origin.
- Once the origin hypotheses have been tested, only one hypothesis should survive the testing for the investigator to be able to select it as the final hypothesis.
- In large fire losses or where personal injury or death occurs, review of an investigator's work product by other investigators is likely.
- The inability to identify the point or area of origin does not, in and of itself, preclude a valid determination of cause.

Hot Terms

Arc survey diagrams Locations of electrical arcing are identified and plotted on a diagram of the affected area of the structure. The spatial relationship of the arc sites can create a pattern and help establish the sequence of damage. This analysis can be used on building circuits and electrical devices within a compartment to help identify or narrow the area of origin.

Area of origin The room, building, or general area in which the point of origin is located.

Depth of calcination surveys Measurements of the relative depth of calcination (observable physical changes in gypsum wallboard) plotted on a detailed scene diagram to determine locations within a structure that were exposed longest to a heat source.

Depth of char surveys Measurements of the relative depth of char on identical fuels plotted on a detailed scene diagram to determine locations within a structure that were exposed longest to a heat source.

Fire scene reconstruction The process of removal of debris and replacement of contents or structural elements in their prefire positions. Reconstruction can also include recreating the physical scene during fire scene analysis investigation.

Fire spread analysis The process of identifying fire patterns relating to the movement of fire from one place to another and the sequence in which the patterns were produced to trace the fire back to an origin.

Heat and flame vector analysis A process to assist with fire spread analysis and origin determination in which arrows representing the investigator's assessment of the direction of heat/flame spread are placed on a detailed diagram of the fire scene. The patterns can be traced back to the fire origin.

Initial scene assessment A preliminary phase of the investigation to ensure preservation of the scene; to identify safety hazards, manpower, and equipment needs; and to determine areas that require further inspection.

Isochar A line on a diagram connecting equal points of char depth.

Point of origin The exact physical location where a heat source and a fuel come into contact with each other and a fire begins.

Safety assessment Inspection to determine whether the scene is safe to enter and the steps necessary to render the scene safe.

Sequential pattern analysis Application of principles of fire science to the analysis of fire pattern data (including fuel packages and geometry, compartment geometry, ventilation, fire suppression operations, witness information, etc.) to determine the origin, growth, and spread of a fire.

Total burn A fire scene where a fire continued to burn until most combustibles were consumed and the fire self-extinguished because of a lack of fuel or was extinguished when the fuel load was reduced by burning and there was sufficient suppression agent application to extinguish the fire.

© Greg Henry/ShutterStock, Inc.

FIRE INVESTIGATOR *in action*

You are requested to conduct an origin and cause investigation at a large barn located in a rural area of your community. Upon your arrival, you note the structure is severely damaged and partially collapsed. Fire personnel advise that when they arrived, the entire structure was on fire and the west end had partially collapsed. As you start your exterior examination, you note the remains of a passenger vehicle, several metal tool boxes, and what appears to be the electric service panel. You begin your interior examination and identify two electric circuits in the west end of the structure as well as several electrical devices connected to the remains of an outlet.

1. Which of the following individuals may provide the most accurate information related to the origin of this fire?
   A. The owner of the barn
   B. The first-arriving fire personnel
   C. The person who reported the fire
   D. All of the above
2. The owner of the barn states that he had been welding a pipe inside the structure approximately 1 hour prior to the fire being discovered. What information could the owner provide that would be most useful for determining the cause?
   A. The location of items prior to the fire occurring
   B. The building's orientation
   C. The weather conditions at the time of the fire
   D. Information related to the electric service provider
3. What technique could be used to narrow the area of origin related to the electrical system?
   A. Reconstruction
   B. Arc mapping
   C. Fire modeling
   D. Depth of char analysis
4. Which of the following situations may limit your ability to determine the origin of the fire?
   A. Collapse of the structure
   B. Consumption of combustible materials
   C. Absence of reliable fire patterns
   D. All of the above
5. Which tools are most useful for removing debris from the area of origin?
   A. Heavy equipment
   B. Portable vacuum equipment
   C. Portable lighting equipment
   D. Hand tools

Fire Cause Determination

© Photos.com

© Jones & Bartlett Learning. Photographed by Glen E. Ellman

Knowledge Objectives

After studying this chapter, you should be able to:

- Explain the roles that source and form of heat of ignition play in fire cause. (pp 270–271)
- Discuss identification of the first fuel ignited NFPA 4.2 NFPA 4.2.6. (p 271)
- Identify the potential oxidizing agents at a fire scene NFPA 4.2 NFPA 4.2.6. (p 271)
- Discuss identification of the ignition sequence NFPA 4.2 NFPA 4.2.6. (p 271)
- Explain the use of the scientific method in cause determination NFPA 4.1.2 NFPA 4.2 NFPA 4.2.6. (pp 271–273)
- Describe the process of elimination and negative corpus. (p 273)
- Discuss levels of certainty in investigative opinions NFPA 4.6.5. (p 273)

Skills Objectives

After studying this chapter, you should be able to:

- Identify the first fuel ignited NFPA 4.2 NFPA 4.2.6. (p 271)
- Identify the ignition source NFPA 4.2 NFPA 4.2.6. (p 271)
- Use the scientific method in cause determination NFPA 4.1.2 NFPA 4.2 NFPA 4.2.6. (pp 271–273)

Additional NFPA Reference

NFPA 921, *Guide for Fire and Explosion Investigations*

CHAPTER 17

FESHE Course Outcomes

Fire Investigation I

11. Identify cause and origin and differentiate between accidental and incendiary. (pp 271–273)

Fire Investigation II

There are no Fire Investigation II (FESHE) course outcomes for this chapter.

You Are the Fire Investigator

© Jones and Bartlett Publishers. Photographed by Glen E. Ellman.

Early Christmas morning, you are investigating a fire that originated in the living room of the north-facing unit in a duplex. After removing the debris in the room, you locate steel wires consistent with those used for artificial Christmas trees. In reconstructing the scene, you determine the fire originated near the center of the room where the owner's son had placed a beanbag chair the night before. A wood-burning fireplace is located approximately 2 feet south of the area where the owner stated the chair had been located.

1. How would you apply the scientific method to identify the first fuel ignited?
2. What potential sources of ignition likely exist in the living room?
3. What data would you collect to identify the ignition source?
4. How would the size, shape, and construction of the beanbag affect the ignition sequence?

Introduction

Determining cause is not limited to identifying the ignition sources (such as smoking materials), the first fuel ignited (such as an upholstered chair), the oxidizing agent, or the human actions (such as failure to properly dispose of smoking materials). It also involves identification of the circumstances, conditions, or agents that brought the fuel, ignition source, and oxidizer together and the conditions under which the fire was able to start and propagate. At the onset of the investigation, the investigator should identify the scope of practice and authority of the entity requiring the investigation. For example, a municipal investigator may not have the need or resources to identify the event and/or component(s) responsible for an electrical failure, but would be compelled to have available resources to pursue the party responsible for an incendiary fire. Conversely, the private sector investigator would have more responsibility to facilitate a forensic examination of an electrical component while deferring to municipal authorities to conduct a criminal investigation. Understanding the scope of practice will facilitate the course of the investigation and often will protect and preserve the fire scene for the party most responsible to conduct a more comprehensive investigation. This chapter focuses on determining the cause of a fire or explosion. The classification of fire causes is discussed in the "Classification of Fire Cause" chapter in this textbook.

Fire Investigator Tip

Arson is neither a fire cause nor a proper classification. *Incendiary* is the classification for a deliberately set fire; the crime committed is arson.

Ignition Source

The source of the ignition energy will be at or near the point of origin. Many times the source of the ignition or its remains will still be at the point of origin **FIGURE 17-1**. If present, it may be damaged or even unrecognizable. It may have been destroyed by the fire or moved by suppression operations, or it may have been moved or transported away from the origin. In all cases, the investigator should work to identify the ignition source to determine fire cause correctly, and should always consider the possibility of ignition sources that are no longer found at the point of origin.

Any identified ignition source must be "competent," meaning that it has sufficient energy to ignite the first fuel and that it is capable of transferring that energy to the fuel long enough to bring the fuel to its ignition temperature. This process is identified with three steps:

© FirePhoto/Alamy

FIGURE 17-1 Often the source of the ignition will remain at the point of origin.

1. Generation: The competent ignition source must be able to generate energy in the form of heat and must be generating heat at the time of ignition.
2. Transmission: Whether by conduction, convection, radiation, or direct flame contact, the heat source must be able to transmit enough energy to the fuel to reach its ignition temperature.
3. Heating: The target fuel must have heated to the point of ignition. This depends on several factors, including the material's thermal inertia—the product of thermal conductivity, density, and specific heat.

The determined ignition source must be able to generate energy sufficient to raise the fuel to its ignition temperature and transmit that energy to the fuel. The term *thermal inertia* is used to describe the response of fuel to the energy that is impacting it. Energy may reflect off, transmit through, disperse into, and heat the material being impacted. Items with a low thermal inertia will dissipate heat at a slower rate, and therefore have a higher surface temperature than items with a high thermal inertia when exposed to equal heat energy.

When the investigation shows that the fuel ignited was a vapor such as gasoline, liquefied petroleum (LP) gas, or natural gas, the ignition source must be competent and present when the fuel gas is in a proper range of flammable limits. The ignition source may be difficult to determine because it may be an arc from a switch or electric motor or an open flame source such as a pilot light.

The ignition source for ignitible vapors may also be more difficult to identify because of the ability of the vapor cloud to migrate throughout multiple compartments. Within the area of the vapor cloud, there may be multiple competent ignition sources. The ignition source would be the one that was present when that portion of the ignitible vapor cloud was within its flammable limits. A detailed interview with individuals who have sufficient knowledge of the contents of the room of origin and the activities that took place there should be conducted. These interviews will assist the investigator in the identification of ignition sources present. Additionally, reports from police, fire inspectors, insurance companies, and others may assist with the determination of cause.

First Fuel Ignited

The first material ignited that sustains the combustion process beyond the actual ignition source must be identified. Identification of the first fuel is an important part of testing the hypothesis. Although the first fuel ignited may have an ignition temperature that is well within the temperatures produced by the ignition source, the configuration of the first fuel ignited is important to consider. With like fuels, one arranged with a high surface-to-mass ratio and the other with low surface-to-mass, the high surface-to-mass will require less thermal energy to ignite. An example is a 2 × 4 block of wood and the same block of wood shaved into a pile of thin shavings. With the same overall mass but a much higher total surface area, the pile of shavings will ignite at the same temperature but will require less energy to bring the fuel to the ignition temperature. The shavings ignite more easily because they do not have sufficient mass to dissipate the heat, so they more quickly reach ignition temperature. In all instances, the heat source must have sufficient heat and duration of exposure to heat the fuel source to its ignition temperature.

Oxidizing Agent

In most fires, the oxygen in the air will serve as the oxidizing agent, or oxidant. Fire intensity can be enhanced by the introduction of other oxidants into a fire. Commonly encountered supplemental oxidants include medical oxygen (e.g., compressed gas cylinders or oxygen generators) or certain chemicals (e.g., pool sanitizers). Sometimes residue of chemical oxidants will remain after the fire. If such residue is found, it should be collected for laboratory analysis. If an oxidant other than atmospheric oxygen is identified at or near the area of origin, it should be documented, and its role in fire development should be considered.

Ignition Sequence

The mere presence of a fuel and a competent ignition source does not alone result in a fire. One of the elements that the investigator should identify is the ignition sequence or the events that brought the ignition source and the fuel together, thus establishing the fire cause FIGURE 17-2.

As an example, the investigator may conclude that a kitchen fire started in a toaster, but what was the event that caused this to occur? Several possibilities exist for consideration—a design defect, alterations to a component, malfunction of an internal component, or improper usage by the consumer. The investigator should be prepared to address the event, possibly a failure mode, within the investigator's respective scope of expertise. In some instances, an additional expert may be required to identify more thoroughly the event and failure mode of a suspected appliance or component. It may also be useful to develop timelines to organize and analyze collected data.

The Scientific Method in Cause Determination

Once the origin of a fire has been accurately identified, the investigator can work to determine cause. As with the determination of origin, this process requires the use of the scientific method (more fully discussed in the "Basic Methodology" chapter in this text). To determine cause accurately, data must be collected and analyzed; only then can cause hypotheses be developed and tested for each potential ignition source. The appropriate way to apply the scientific method to testing the cause hypothesis is to attempt to disprove the hypothesis. Working to prove a hypothesis can lead to errors resulting from expectation or confirmation bias. Hypotheses that cannot be disproved may be considered either possible or probable.

FIGURE 17-2 The cause of a fire includes three steps: An ignition source, the first fuel ignited, and the cause of the ignition source and first fuel ignited coming together. Here, the heater is the ignition source, the curtain is the first fuel ignited, and the proximity of the two caused the fire.

Data Collection

Essential data to be collected to determine cause include identification of fuels, potential ignition sources, and unusual oxidants in the area of origin. Information from persons with knowledge of the area can be helpful in gathering this information, as can the review of prefire photographs.

Evaluation of the competence of a hypothesized ignition source requires identification of the initial fuel. The initial fuel may not survive the fire in recognizable form. The initial fuel may be building materials, interior furnishings, or a component part of an appliance. When diffuse fuels (i.e., gas, vapor, or dust) are the initial fuel, the point of origin (ignition) may be remote from the location of other fuels that sustain combustion. A flash fire may ignite light combustibles, such as window treatments, in multiple locations, further complicating the analysis.

The form and geometry of the initial fuel must be such that the ignition source is capable of generating and transferring sufficient heat energy over the allotted time to result in ignition. The first fuel ignited must be capable of igniting the subsequent fuels for the fire to spread. Identification of this information is known as a fuel analysis.

All potential ignition sources in the area of origin must be identified. If the room of origin became fully involved, this means examining every potential ignition source in the room to determine whether it is competent, in light of the energy of the ignition source and the physical properties of the fuel ignited. This process is known as the ignition source analysis.

Data may also need to be collected from outside the area of origin, for example if a sample is needed of unburned fuel or if an exemplar ignition source is available. The investigator should also consider ignition sources that do not correspond to a physical device that has been identified on scene, such as open flame and static electricity. In the absence of a physical device, other evidence is needed to establish such an ignition source.

Data Analysis

Once all of the potential initial fuels, potential ignition sources, oxidants, and relevant circumstances/activities have been identified, analysis of the whole of the data can be conducted. The investigator must not jump to a conclusion about a fire cause just because a fuel and an ignition source happen to coincide with the area of origin. A separate hypothesis must be developed and tested for each possible initial fuel/potential ignition source/oxidant combination identified.

The investigator must establish the ignition sequence, which is the sequence of events and circumstances that allowed the initial fuel and the ignition source to come together and result in a fire. In conducting this analysis, it is important to realize that the investigator must establish with evidence the events that occurred or that were logically necessary for the fire to have started. This is an extremely important concept. It is recognized that physical evidence of the ignition source may not exist after the fire and that there are circumstances in which the ignition sequence can be logically inferred when physical evidence of the ignition source does not exist. This information must be established with evidence, not by the absence of it. Through testing of alternate hypotheses of potential ignition sequences, the investigator can identify the one sequence that is consistent with all known facts.

Certain evidence is particularly supportive of the inference of a reliable ignition sequence in the absence of physical evidence of the ignition source. These include, but are not limited to, diffuse fuel ignitions, multiple origins, trailers, witnessed ignitions, and presence of ignitible liquids confirmed by laboratory analysis and without an innocent explanation for their presence.

Factors to be considered in developing an ignition sequence include the following:

- How and when the initial fuel came to be present at the origin
- How and when the oxidant came to be present (if an unusual oxidant is involved)
- How and when the competent ignition source came to be present

- How and when the competent ignition source transferred heat to the initial fuel
- What acts or omissions, and in what sequence, brought together the fuel, oxidizer, and ignition source
- How the initial fuel ignited subsequent fuels to result in fire spread

Cause Hypotheses

As prescribed by the scientific method, the investigator should develop and test alternate hypotheses of cause to identify the one that is consistent with all known facts. Hypothesis testing may include reference to scientific and trade literature, which can contain not only useful information and data, but also may have descriptions of experiments and testing performed by others. The investigator may also choose to do his or her own physical tests or experiments, keeping in mind that such tests or experiments must be performed in a way that will produce reliable and applicable results for the investigation at hand. Further, as discussed in the "Basic Methodology" chapter of this text, cognitive testing of hypotheses is commonly and appropriately used by fire investigators to test ideas and hypotheses against the known data.

Process of Elimination and Negative Corpus

The process of elimination is essential to the analysis of cause and to the investigator's efforts to test and evaluate hypotheses. In many investigations, the investigator will be successful in testing and disproving a range of hypotheses, to the point that one remains that has not been able to be refuted. However, as mentioned above, the remaining hypothesis must not be adopted as the investigator's final opinion simply because it is the last one standing. Rather, it must be supported by, and consistent with, the evidence. Thus, the investigator must not identify an ignition source by simply eliminating all other ignition sources known or believed to have been in the area of origin, then claiming such elimination is proof of the identified source, even though there is no independent evidence to support it. This is an improper technique and is termed *negative corpus*. Negative corpus is inconsistent with the use of the scientific method (which NFPA 1033 requires all investigators to use in their analysis of cause) because it involves making conclusions without the collection and analysis of data. Use of negative corpus has typically been seen in fires classified as incendiary, where an investigator would propose and eliminate a number of accidental ignition sources, then rely on such elimination as proof of an incendiary (human hands) ignition. To avoid any suggestion that he or she is relying on negative corpus, the investigator must base all hypotheses and determinations on the analysis of facts, which are derived from observations, physical and other evidence, science, calculations, and experiments.

Levels of Certainty

The investigator should establish a level of certainty to opinions formed. The investigator is the individual who must assign weight to or confidence in any particular piece of data considered and the final determination of cause. The investigator may establish one of two levels of certainty based on their confidence in the data collected.

- Probable: A level of certainty that would correspond to being more likely true than not. This level of certainty would hold the determination to be at a level greater than 50 percent. This level of certainty is required for a fire cause to be classified as natural, accidental, or incendiary.
- Possible: A level of certainty that would be regarded as feasible but not probable. In the event that two hypotheses are formed of equal level of certainty, they each must be considered possible. When the level of certainty is determined to be possible, the cause of the fire should be considered undetermined.

The investigator should not confuse the terms *possible* and *probable* as used previously with *probable cause* as used in the legal context of a criminal investigation, nor should these terms be equated with burdens of proof required in civil or criminal cases. (See the "Legal Considerations" chapter in this textbook.)

Wrap-Up

Ready for Review

- Determining cause is not limited to identifying the ignition sources, the first fuel ignited, the oxidizing agent, or the human actions. It also involves identification of the circumstances, conditions, or agencies that brought the fuel, ignition source, and oxidizer together and the conditions under which the fire was able to start and propagate.
- Many times the source of the ignition or its remains will still be at the point of origin.
- Identification of the first fuel ignited is an important part of the hypothesis testing.
- Depending on the properties of the ignition source, physical evidence of it may remain at the point of origin.
- In most fires, the oxygen in the air will serve as the oxidizing agent, or oxidant. Fire intensity can be enhanced by introduction of other oxidants into a fire.
- The investigator should identify the ignition sequence or the events that brought the ignition source and the fuel together, thus establishing the fire cause and responsibility.
- The appropriate way to apply the scientific method to testing the cause hypothesis is to attempt to disprove the hypothesis.
- Essential data to be collected to determine cause include identification of fuels, potential ignition sources, and unusual oxidants in the area of origin.
- Once all of the potential initial fuels, potential ignition sources, oxidants, and relevant circumstances/activities have been identified, analysis of the whole of the data can be conducted.
- The investigator should establish a level of certainty to opinions formed.

Hot Terms

Fuel analysis Identification of the first fuel ignited.

Ignition sequence The sequence of events and circumstances that allow the initial fuel and the ignition source to come together and result in a fire.

Ignition source analysis Process through which all potential ignition sources in the area of origin are identified and then considered in light of the physical properties of the first fuel ignited, fundamental scientific principles, and other available data.

Thermal inertia The product of thermal conductivity, density, and specific heat (or heat capacity).

© Greg Henry/ShutterStock, Inc.

FIRE INVESTIGATOR *in action*

You have determined that a fire within a pawn shop originated in the corner of an office where two safes, a microwave, and a dorm-sized refrigerator are located. The fire is relatively small and fire patterns indicate the point of origin to be in front of the two safes. The microwave and refrigerator are each sitting on a safe. The damage to the refrigerator and microwave is only to the sides exposed to the fire, and they do not appear to be involved with the cause of the fire. The remains of a waste can with charred papers are present on the floor where the fire pattern originates.

An interview with the employee present at the time of the fire reveals he had opened the safes approximately 5 minutes before going outside to smoke a cigarette. He stated he was only outside for a few minutes when he noticed white smoke inside the business.

1. What is the process used by the investigator to identify the ignition source?
   - **A.** The scientific method
   - **B.** Ignition sequence assessment
   - **C.** Ignition source analysis
   - **D.** All of the above
2. Which factor most influences the speed at which this fire developed?
   - **A.** The type of business
   - **B.** The time of day
   - **C.** The contents stored within
   - **D.** The size and geometry of the items at the point of origin
3. Which of the following terms refers to the absence or elimination of all possible ignition sources as being evidence?
   - **A.** Preponderance of the evidence
   - **B.** Probable cause
   - **C.** Negative corpus
   - **D.** Circumstantial evidence
4. You have determined the microwave and refrigerator did not cause the fire; however, the employee denies smoking inside the business. What level of certainty would you place on a discarded smoking material as the cause?
   - **A.** Low certainty
   - **B.** Probable
   - **C.** Possible
   - **D.** Undetermined

Classification of Fire Cause

© Photos.com

Courtesy of Mike Dalton

Knowledge Objectives

After studying this chapter, you should be able to:

- Describe the accidental fire cause classification. (p 278)
- Describe the natural fire cause classification. (pp 278–279)
- Describe the incendiary fire cause classification. (p 279)
- Describe the undetermined fire cause classification. (p 279)
- Discuss fire cause classification and how it may change. (p 280)

Skills Objectives

After studying this chapter, you should be able to:

- Classify a fire cause. (pp 278–280)

Additional NFPA Reference

NFPA 921, *Guide for Fire and Explosion Investigations*

CHAPTER 18

FESHE Course Outcomes

Fire Investigation I

11. Identify cause and origin and differentiate between accidental and incendiary. (pp 278–280)

Fire Investigation II

There are no Fire Investigation II (FESHE) course outcomes for this chapter.

You Are the Fire Investigator

© Jones and Bartlett Publishers. Photographed by Glen E. Ellman

You are beginning your scene investigation of a large outdoor fire involving stored plastic pallets. Your have interviewed several witnesses who described seeing a "marine flare" falling into the area of the fire minutes prior to its discovery, and fire personnel described locating the fire at the top of a 20-foot tall stall upon their arrival. One person had been identified in the area as shooting flares the week prior to the fire. As you walk toward this person's home, you are stopped by a resident who states he had put out a small fire in his yard minutes prior to the pallet fire and would like for you to look at the device that caused the fire. The resident then leads you to a small burned area in his yard, and you recognize the device as a Chinese lantern.

1. If it is illegal to use Chinese lanterns, how does this influence your classification of the fire?
2. How would you classify this fire?

Introduction

Cause classification is the culmination of identifying the four elements of the fire (ignition source, first fuel ignited, oxidizer, and ignition factor) and then categorizing them according to the generally accepted definitions. Remember that the *cause* of the fire is related to the above mentioned criteria, whereas classification is more about assigning responsibility, reporting, or gathering statistics. This will vary depending on the agency and sometimes on current state or provincial rules or statutes. The *determination* of the cause and the *classification* of the cause are separate activities. Classification is completed after reviewing all the data (facts) collected and making a well-informed determination as to the classification based on those data (facts). The investigator should not use motive to classify a cause; it should be classified based on facts found at the scene.

There are four general categories to which the cause of an explosion or fire can be classified:

1. Accidental
2. Natural
3. Incendiary
4. Undetermined

Fire Investigator Tip

Suspicious is not an accurate description for a fire cause. Avoid using this term.

Courtesy of Mike Dalton

FIGURE 18-1 At this scene, combustibles were placed carelessly near a wall heating unit and ignited when the unit turned on, damaging the structure.

Accidental Fire Cause

Accidental fires are those that are not the result of a deliberate (intentional) act FIGURE 18-1. This also encompasses friendly fires that are ignited deliberately but become hostile, such as intentionally ignited brush or trash fires that spread beyond the intended confinement. Traditionally, the accidental category has been used to classify fires ignited by juveniles who are below the legal age of responsibility. Careless acts, no matter how they appear, will still fall into this classification; the investigator must remember to use the data (facts) he or she has gathered.

Natural Fire Cause

The natural fire cause classification is for fires that ignite without human intervention. This category includes fires that result from natural phenomena such as lightning, wind, and earthquakes FIGURE 18-2.

© Costazzurra/ShutterStock, Inc.

FIGURE 18-2 Fires resulting from natural phenomena are classified as natural fires.

Incendiary Fire Cause

Incendiary fires result from deliberate acts—that is, fires that result from intentional actions or circumstances for the fire to occur in areas where it should not have occurred. Fires resulting from reckless or negligent acts are not specifically referenced but may be included on the basis of the circumstances in which "the person knows the fire should not be ignited." The mindset or mental state (intent) of the fire setter is therefore a key element of this classification FIGURE 18-3.

Take, for example, the case of a child who used an open flame device to experiment or "play" with setting things on fire. Knowledge that the fire should not have been set can be evidenced by the location in which this behavior takes place (e.g., in the bedroom closet). Such a fire may be properly classified as incendiary even though the child cannot be shown to have acted with malice.

Undetermined Fire Cause

The undetermined fire classification is an appropriate category for fires that have not yet been investigated, for fires that are under investigation, and for fires that have been investigated and the cause is not proven to an acceptable level of certainty FIGURE 18-4. However, the failure to determine the ignition source of the fire does not automatically require the fire investigator to rule the fire undetermined. If sufficient evidence establishes a factor, such as the use of an accelerant, that evidence may be sufficient to establish the fire cause.

There should be no negative stigma attached to the undetermined classification. Fire investigations should be approached without presumption; therefore, all fire investigations theoretically start with an undetermined classification (fires that are under investigation). Approaching a fire investigation without presumption also helps to prevent an investigator from falling prey to expectation or confirmation biases.

Newly discovered information may require an investigator to re-analyze an investigation. In the event that analysis results in a determination of the cause, a previously "undetermined" fire can be classified without issue.

Courtesy of Greg Lampkin

FIGURE 18-3 Containers with ignitible liquids were placed in a structure set up for destruction intentionally by an arsonist.

Courtesy of Mike Dalton

FIGURE 18-4 A complete burn down of a structure with no witnesses. The power was connected, but there was no information as to what the cause could be. The current data support an undetermined cause.

Fire Investigator Tip

Arson is a term that denotes a crime and is therefore determined by judicial process. A fire classified as incendiary can have sufficient probable cause to be called arson by the law enforcement community. Depending on local laws, not all incendiary fires are arson.

Discussion of Classification

The first three classifications (accidental, natural, and incendiary) can be reached when the hypothesis is supported by the data. If the hypothesis is not supported by the facts (data), the investigator returns to one of the earlier steps in the scientific method and repeats the process from that point to ensure that a supportable conclusion is reached (see the "Basic Methodology" chapter in this textbook). If the cause cannot be determined because the data are insufficient to support the hypothesis, then the cause must be classified as *undetermined*. With the development of new data, the investigator should reevaluate the original classification of *undetermined* and change the classification as supported by the data.

Even after a classification has been made, if new evidence (data) should become available that does not support the original hypothesis, then the investigator should retest the original hypothesis with the new data. If appropriate, the investigator should reach a new cause determination.

It is not necessary that all of the elements of the cause be identified for the cause to be classified. In limited circumstances, it may be possible to make a credible cause determination in the absence of the *physical* evidence of the ignition source. With sufficient evidence, an ignition scenario that logically occurred or would need to have occurred for the fire to start can be reliably hypothesized and tested. The investigator cannot use evidence of motive to classify the cause (see the "Incendiary Fires" chapter in this textbook).

There also may be circumstances in which the available evidence allows for the identification of the ignition source and the first fuel, but the cause cannot be classified because the circumstances bringing the essential elements together cannot be established. In these circumstances, the appropriate classification is *undetermined*.

At no time is a fire to be classified as *suspicious*. This is not an acceptable classification for a fire cause. *Suspicious* is not an appropriate term to describe the unexplained and carries a negative connotation. Although the untrained layperson, including news media, might refer to a fire as suspicious, this term is inappropriate and should not be used by the fire investigator. Suspicion refers to a level of proof or level of certainty. Fires where the level of certainty is only a suspicion should be classified as "undetermined" (see Table 2-1, Hypothesis Certainty, in the "Basic Fire Methodology" chapter).

Wrap-Up

Ready for Review

- Cause classification is the culmination of identifying the four elements of the fire (ignition source, first fuel ignited, oxidizer, and ignition factor) and then categorizing them according to the generally accepted definitions.
- Accidental fires are those that are not the result of a deliberate (intentional) act, including fires that are ignited deliberately but become hostile, such as intentionally ignited brush or trash fires that spread beyond the intended confinement.
- The natural fire cause classification is for fires that ignite without human intervention, including fires that result from natural phenomena such as lightning, wind, and earthquakes.
- Incendiary fires result from deliberate actions or intent or circumstances for the fire to occur in areas where it should not have occurred.
- The undetermined fire classification is an appropriate category for fires that have not yet been investigated, are under investigation, or have been investigated but the cause is not proven to an acceptable level of certainty.
- Even after a classification has been made, if new evidence (data) should become available that does not support the original hypothesis, then the investigator should retest the original hypothesis with the new data.

Hot Terms

Accidental fires Fires for which the proven cause does not involve an intentional human act to ignite or spread fire into an area where the fire should not be.

Incendiary fires Fires that are deliberately set with the intent to cause the fire to occur in an area where the fire should not be.

Natural fire Fire caused without direct human intervention or action, such as fire resulting from lightning, earthquake, and wind.

Undetermined fire Classification of fire when the cause cannot be proven to an acceptable level of certainty.

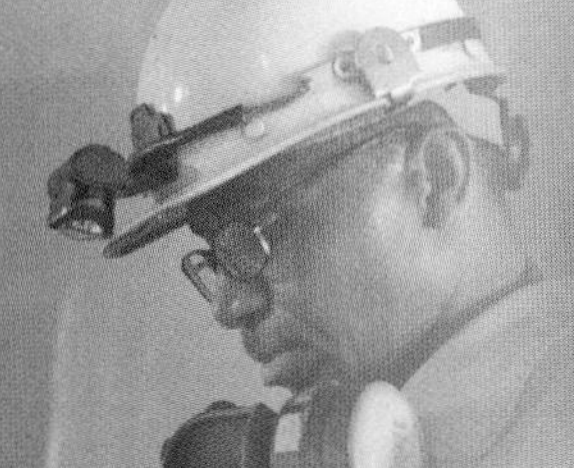

FIRE INVESTIGATOR *in action*

After conducting a lengthy investigation of several fires involving trash cans, dumpsters, and debris piles within a normally quiet neighborhood, you eventually identify a 7-year-old boy who admits to starting these fires because he "likes to watch the fire." The child seems remorseful about the fires and states he "did not want to hurt anyone."

After speaking with his parents, you learn the child has been "very curious" about fire in the past several months and has asked several times to go visit the fire station near his home.

1. What should be the classification of this fire?
   A. Accidental
   B. Incendiary
   C. Child firesetter
   D. None of the above
2. A fire caused by an unattended candle that ignites nearby combustibles is classified as:
   A. accidental.
   B. unintentional.
   C. Both A and B
   D. None of the above
3. A fire caused by a lightning strike would be classified as:
   A. an act of nature.
   B. accidental.
   C. natural.
   D. undetermined.
4. A fire that has "unusual" circumstances, but no specific cause is identified, should be classified as:
   A. incendiary.
   B. accidental.
   C. suspicious.
   D. undetermined.

© Greg Henry/ShutterStock, Inc.

© Glen E. Ellman

Analyzing the Incident for Cause and Responsibility

© Photos.com

© Jones & Bartlett Learning. Photographed by Glen E. Ellman

Knowledge Objectives

After studying this chapter, you should be able to:

- Explain the differences between cause and responsibility NFPA 1.3.7. (p 284)
- List factors that can contribute to fire spread. (pp 284–285)
- Discuss the nature of responsibility in fires and explosions NFPA 4.5.1 NFPA 4.6.3 NFPA 4.6.5. (p 285)
- Identify analytical tools used for failure analysis NFPA 1.3.7 NFPA 4.6.5. (pp 285–286)

Skills Objectives

After studying this chapter, you should be able to:

- Analyze a fire incident for cause and responsibility NFPA 4.6.5. (pp 284–286)
- Analyze an explosion incident for cause and responsibility NFPA 4.6.5. (pp 284–286)

Additional NFPA Reference

NFPA 921, *Guide for Fire and Explosion Investigations*

CHAPTER 19

FESHE Course Outcomes

Fire Investigation I

There are no Fire Investigation I (FESHE) course outcomes for this chapter.

Fire Investigation II

There are no Fire Investigation II (FESHE) course outcomes for this chapter.

You Are the Fire Investigator

© Jones and Bartlett Publishers. Photographed by Glen E. Ellman.

While investigating a fire that has caused near-complete destruction of a large livestock building, fire personnel advise you there was a second fire in an identical building. The building does not appear to have sustained any damage, but as you enter the structure, you notice plastic containers on the floor and a dark-colored residue inside and surrounding each container. A large hole is melted through the plastic grate flooring in one of the holding pens where at least one container had been located.

During your interviews with the owner, he states that an exterminating company had been fumigating the two buildings earlier that day, using a chemical mixture that must be placed on a noncombustible surface to ensure the reaction does not ignite combustible materials.

1. What potential events may have led to the cause of the fires in either building?
2. What role may the exterminating company have played in the cause of these fires?
3. What factors may have influenced the difference in the amount of damage between the two buildings?
4. How might you establish the cause and determine potential responsibility for these fires?

Introduction

Although the focus of an investigation is typically on the fire cause, there are other significant features that are often addressed during an investigation. Some or all of these features may be addressed to varying degrees, depending on the investigator's responsibility. Responsibility for a fire or explosion incident is the accountability of a person or other entity for the event or sequence of events that caused the fire or explosion, spread of the fire, bodily injuries, loss of life, or property damage. In some cases, the investigator's determination of responsibility for an incident may involve more issues than simply isolating the cause of the fire. The investigator may need to address responsibility issues in any or all of four classifications:

1. Cause of the fire or explosion. This area identifies the elements of a cause: the heat source, the first fuel ignited, the oxidizer, and the conditions or circumstances that allowed these components to come together and result in a fire or explosion.
2. Cause of damage to property resulting from the incident. This area considers the factors that are responsible for fire spread from the origin. It also includes factors that might have contributed to the extent of the loss. These factors may include the combustibility of contents or construction materials, the adequacy or inadequacy of passive fire protection systems (e.g., fire walls and fire doors) and active fire protection systems (e.g., sprinklers), and structural compliance to applicable fire or building codes.
3. Cause of bodily injury or loss of life. This area identifies factors related to human injuries or deaths. The factors may include analysis of fire alarm or smoke detection systems, means of egress, the role of materials that emit products that are harmful to humans during a fire, the reasons for any fire fighter injuries, or building codes and code violations.
4. Degree to which human involvement contributed to any of the above. This area concerns factors related to human contribution (i.e., act or omission) to any of the features listed above, and includes negligence and incendiary acts. For example, these could include combustible materials stored too close to a heat source, unattended cooking, lack of maintenance, improper electrical or gas work, or improper installation of appliances.

The cause of a fire or explosion is not the ignition source (e.g., cigarette fire, short circuit). The cause includes the circumstances or actions that caused a competent ignition source to come in contact with a fuel and an oxidizer and transfer enough energy to start a fire. The circumstance or action could be negligence (such as when a pot is left unattended on a stove), a product defect (such as the improper assembly of a component inside an appliance), or an intentional act (such as pouring an ignitible liquid in a house and lighting it). (See the "Classification of Fire Cause" chapter for more information.)

Cause of Property Damage and Injuries

The rate of development and spread of fire/smoke influences the amount of property damage caused by a fire. Development and spread of heat and smoke are of great significance in identifying causal factors of bodily injury or death. Factors contributing to the development and spread of fire and smoke that should be considered by the investigator in all cases include the following:

- Code violations (such as propping open fire doors)
- Compartmentation (construction detail or passive fire protection assemblies)
- Detection and alarm systems (timeliness of activation or notification)
- Fire suppression (both fixed systems and agency operations)

- Fuel load and geometry (See the "Fire Patterns" chapter in this textbook.)
- Housekeeping (poor housekeeping can result in more rapid fire spread and hamper the suppression activities of the fire department)
- Human behavior (acts or omissions)
- Increase in hazard or change in occupancy (rendering original fire protection system design inadequate)
- Structural or system failure (lack of fire stops, smoke separations, or adequate fire and smoke detection; utility failure; or protection systems out of service)
- Ventilation effects (door and window openings, HVAC operation, and firefighting operations)

Issues that commonly arise in large-loss fires include building construction; whether a building was remodeled or modified in a design-inappropriate manner or a manner that violated codes and standards; whether post-construction modifications and penetrations contributed to the spread of fire and smoke; whether a change of occupancy added fire load or changed other risk factors; whether detection and suppression systems were adequate for the occupancy of the building; whether these systems worked properly or were broken, damaged, or restricted; whether code violations occurred with regard to exits, access to areas, product storage, electric connections and cordage, and the like; and factors, if any, that interfered with efficient fire suppression by persons on scene and the fire department.

The investigator may also need to take into account the factors that caused or contributed to any consequential damage, meaning damage that is secondary to the original damage caused by fire, smoke, and firefighting operations. Consequential damage can include damage that results from the loss or unavailability of power, water, or gas; weather damage that happens because a building is compromised; rust and corrosion; damage from water or mold; theft; and contamination.

When investigating bodily injury or deaths, additional factors may also have relevance. These include the following:

- Toxicity (from products of combustion)
- Hazardous materials (exposure to materials released as a result, not the cause, of a fire)
- Means of egress (design, maintenance, visibility, operation or obstruction of exits and areas of refuge)

Issues that commonly arise in fires with injuries or fatalities can include whether adequate exits were available, marked, and known to victims; the degree of intoxication or other mental impairment of any victims; the age of the victims; the amount of time taken by victims and others to recognize the threat of fire or smoke; the behavior of victims and others upon detecting the threat; whether persons involved were trained or knowledgeable on how to behave in emergencies; and whether the occupant load was exceeded.

Fire Investigator Tip

Insufficient preparation by emergency responders may impact the casualties of the incident.

Nature of Responsibility

Determining responsibility is a process that occurs after the fire cause has been determined. It is through determining responsibility that codes and standards can be changed, fire and life safety can be addressed, and civil or criminal litigation can be initiated.

The nature of responsibility in a fire or explosion incident may be in the form of an act or an omission, whether the act or omission is deliberate or unintentional. Responsibility for some aspects of a fire or explosion event may be attributed regardless of the fire cause classification, including undetermined. Although the courts will generally affix a legally binding finding of responsibility (liability), the court relies on evidence presented by those who performed the investigation and analysis. Responsibility may fall on multiple parties in a given incident.

Investigation of responsibility will typically involve conducting interviews and gathering various reports, records, and other data to identify contributing factors. Consultation with technical experts (e.g., electrical or mechanical engineers) may be required. Determining responsibility frequently requires that the investigator conduct a failure analysis.

Failure Analysis

Failure analysis is establishing which factors contributing to any injuries, loss of life, and property damage that may have resulted from the incident. The information gathered through a failure analysis of an incident can prove invaluable in helping to ensure that these losses are not repeated in the future. When looking at an incident from a failure analysis point of view, it is important to understand that it is often a chain of events that contribute to the cause.

Some types of analytical tools for failure analysis are as follows:

- Timelines
- Systems analysis
- Failure mode and effects analysis (FMEA)
- Fault trees
- Mathematical modeling
- Heat transfer analysis
- Flammable gas concentrations
- Hydraulic analysis
- Thermodynamic chemical equilibrium analysis
- Structural analysis
- Egress analysis
- Fire dynamics analysis

Some examples of circumstances to which failure analysis can be applied during an investigation are as follows:

- Did the sprinkler system operate? If so, why did it not control the fire? Was sufficient water provided for the hazard being protected against?
- Were there sufficient exits for people to escape from the fire? How did people react when they tried to escape?

- Did the fire alarm sound? If not, why not? Were the smoke detectors properly placed to react to the fire development?
- If the identified ignition source was an appliance or other equipment, was it designed and installed properly? Was it modified or tampered with?

Some of the most tragic fires have resulted in significant building and fire code changes. One of the significant contributing factors to the loss of life in the Cocoanut Grove nightclub fire in Boston, Massachusetts, in 1942 was the lack of proper exits FIGURE 19-1. The two hotel fires that struck Las Vegas in the 1980s resulted in a widespread movement in the lodging industry to call for the installation of sprinklers. The fire at Seton Hall University that killed three students in 2000 was the impetus for the installation of sprinklers in all student housing in the state of New Jersey.

These incidents and others resulted in dramatic changes because failure analysis identified significant contributing factors. Going beyond the initial determination of the cause of the fire brought out other factors relating to responsibility.

© AP Photos

FIGURE 19-1 Lack of proper exits contributed to the tragedy of the Cocoanut Grove nightclub fire in 1942.

The reader is cautioned that a proper failure analysis often requires specialized education, training, and experience (e.g., electrical, mechanical, or fire protection engineers). Investigators should seek assistance from subject matter experts where appropriate. (See the "Failure Analysis and Analytical Tools" chapter in this textbook.)

Wrap-Up

© Greg Henry/ShutterStock, Inc.

Ready for Review

- Responsibility for a fire or explosion incident is the accountability of a person or other entity for the event or sequence of events that caused the fire or explosion, spread of the fire, bodily injuries, loss of life, or property damage.
- The rate of development and spread of fire/smoke influences the amount of property damage caused by a fire.
- It is through determining responsibility that codes and standards can be changed, fire and life safety can be addressed, and civil or criminal litigation can be initiated.
- Failure analysis is establishing which factors contributing to any injuries, loss of life, and property damage that may have resulted from the incident. The information gathered through a failure analysis of an incident can prove invaluable in helping to ensure that these losses are not repeated in the future.

Hot Terms

Failure analysis A logical, systematic examination of an item, component, assembly, or structure and its place and function within a system, conducted in order to identify and analyze the probability, causes, and consequences of potential and real failures.

Responsibility The accountability of a person or other entity for the event or sequence of events that caused the fire or explosion, spread of the fire, bodily injuries, loss of life, or property damage.

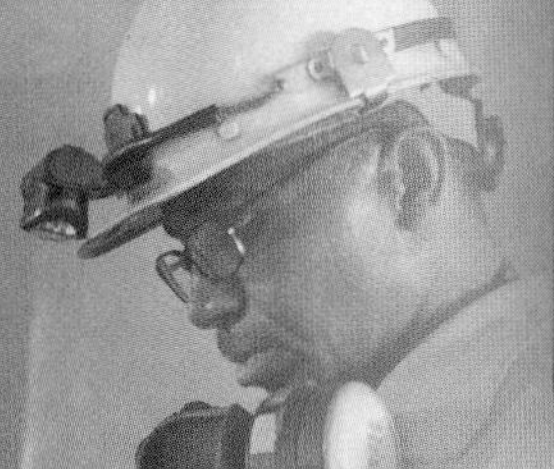

FIRE INVESTIGATOR *in action*

You respond to an explosion that has caused extensive damage to a metal shipping container that was being used as a storage shed. The container is in several pieces, with the contents dispersed over a large area. The doors to the container are located several hundred feet away from the remains of the container; however, the padlock is found still secured.

An interview with the business manager reveals the container was used to store tools, supplies, and equipment for a recycling business. The manager states that 2 days prior to the explosion, an employee had placed two full 5-gallon gasoline cans near the doors to prevent them from being stolen over the weekend.

During your examination of the scene, you determine the only source of ignition was a small refrigerator in the container that was powered by an extension cord from the office. Additionally, employees state that they noticed an odor of gasoline from the area of the container several hours prior to the explosion.

1. What is the likely cause of the explosion?
   - **A.** Undetermined until further information is developed
   - **B.** Accidental explosion
   - **C.** Fugitive vapors ignited by an electrical arc
   - **D.** A possible incendiary act
2. What factors contributed to the explosion?
   - **A.** The storage of gasoline within the container
   - **B.** The container being closed and secure
   - **C.** An electrical arc from the refrigerator's compressor relay switch
   - **D.** All of the above
3. Which of the following is not considered a failure analysis tool?
   - **A.** Photography
   - **B.** Fault tree
   - **C.** Timeline
   - **D.** System analysis
4. Which investigative activity may assist with establishing responsibility for the explosion?
   - **A.** Conducting interviews
   - **B.** Reviewing reports
   - **C.** Both A and B
   - **D.** None of the above

Failure Analysis and Analytical Tools

© Photos.com

© Jones & Bartlett Learning. Photographed by Glen E. Ellman

Knowledge Objectives

After studying this chapter, you should be able to:

- Explain the use of timelines in fire investigation NFPA 4.4.3 NFPA 4.6.2. (pp 290–292)
- Explain the purpose and implementation of system analysis in fire investigation NFPA 4.4.3 NFPA 4.6.2. (pp 292–294)
- Identify the components of mathematical and engineering modeling and how they can be used to investigate fire incidents NFPA 4.4.3 NFPA 4.6.2. (pp 294–297)
- Explain the use of graphic representations, including computer modeling, in fire investigation NFPA 4.4.3 NFPA 4.6.2. (p 297)
- Explain the process of fire testing and its importance in fire investigation NFPA 4.4.3 NFPA 4.6.2. (pp 297–299)

Skills Objectives

After studying this chapter, you should be able to:

- Organize information into a rational and logical format NFPA 4.4.3 NFPA 4.6.2. (pp 290–294)
- Use models to analyze fire incident data NFPA 4.4.3 NFPA 4.6.2. (pp 294–297)

Additional NFPA Reference

NFPA 921, *Guide to Fire and Explosion Investigations*

CHAPTER 20

FESHE Course Outcomes

Fire Investigation I

There are no Fire Investigation I (FESHE) course outcomes for this chapter.

Fire Investigation II

13. List the sources and technology available for fire investigations. (pp 290–299)

You Are the Fire Investigator

© Jones and Bartlett Publishers. Photographed by Glen E. Ellman

You have determined a fire originated at a furnace located in a basement. The furnace had been installed several weeks prior to the fire by a local HVAC company and had been operating continuously since that time.

The property owner states that she had begun to notice a "burning odor" approximately 5 hours prior to seeing smoke emanating from the first floor ventilation ducts. She further states that when she went to the basement to see what was causing the smoke, she noticed the outside of the furnace access panel "glowing orange."

1. What information will you need to obtain to determine why the furnace caught fire?
2. Why might you create a timeline for this investigation?
3. Which investigative tools may assist you with describing and visualizing the events that caused this fire?

Introduction

As part of the investigation of a fire/explosion event, the investigator will often be faced with a multitude of data and facts. The investigator will need to be able to organize and analyze the information to formulate a hypothesis on the event. The investigator has several analytical tools available to assist him or her in interpreting the information obtained, including the following:

- Timelines
- System analysis
- Mathematical/engineering modeling
- Graphic representations
- Scene reconstruction
- Fire testing

Timelines

A timeline is a graphic or narrative representation of events related to the fire incident, arranged in chronological order. Understanding the timeline of any incident is key to creating the failure analysis because it assists the investigator in determining the sequence of events that occurred. A timeline is a comprehensive listing of specific events, no matter how inconsequential, that can be verified to a stated degree.

Timelines may include events that occurred before, during, and after the fire. Estimates of fire size or conditions are often valuable in developing timelines. With the use of fire dynamics, fire conditions can be related to specific events. Information that is gathered from detection and suppression systems may be useful in determining the spread of the fire, as well as establishing a time for the events. Remember that the value of the timeline is directly related to the accuracy of the information it includes, and the reliability of a timeline is a function of the level of confidence that can be placed in specific elements that it includes. Not all information can be related to a specific time and may have to be listed as a time interval during which the event occurred. A variety of components are used to develop timelines. These include incidents described as either *hard times* or *soft times*.

Hard Times (Actual)

When developing a timeline, incidents that can be related to a known exact time are generally referred to as hard times (TABLE 20-1). This means that the time of occurrence is specifically known. For example, a fire department's incident history records can provide the specific times when units were dispatched, arrived on scene, and so forth. Such data serve as benchmarks in developing the timeline.

Soft Times (Estimated)

Other times can be considered soft times. Soft time is either estimated or relative and is generally provided by witnesses.

Table 20-1 Hard Time Sources
Dispatch telephone or radio logs
Emergency medical service (EMS) reports
Alarm system records (on-site, central station, fire dispatch, etc.)
Inspection reports (building, health, fire)
Utility company records (maintenance, emergency, and repair records; cell phone records)
Private videos and photos
Media coverage (newspaper photographer, radio, television, and magazines)
Timers (clocks, time clocks, security timers, water softeners, and lawn sprinkler systems)
Weather reports (NOAA, airports, and lightning tracking services)
Maintenance records (current and/or prior owner/tenant)
Interviews establishing observations and activities of witnesses
Computer-based fire department alarms, communication audio tapes, and transcripts
Building or systems installation permits

For example, a witness may not be able to state precisely the time a certain event such as a flashover occurred or the time when an occupant was sent to leave a building but may be able to relate such observations to another event, such as the arrival time of the fire department or the moment when a bus or train went by, a television program was on, or an event let out. Through interviews and further gathering of data, an investigator may be able to narrow down the range of occurrence for such soft times to relative time periods. Relative time can be subjective in nature and varies with the witness providing the information. Witnesses should refer to their actions and observations in relation to each other, in relation to other events, and they should be as specific as possible.

Estimated time is an approximation based on information or calculations that may or may not be relative to other events or activities.

Benchmark Events

Some events are particularly valuable as a foundation for the timeline or may have significant relation to the cause, spread, detection, or extinguishment of a fire. These events are referred to as benchmark events.

It is important when developing a timeline that includes different hard time sources to note any discrepancies between the clocks and synchronize the times. Such data should be double-checked to ensure that errors were not made in synchronizing all such sources of data.

For example, Mr. Jones ends his work day and punches out of work at 6:00 p.m. He begins his 25-mile drive back home in heavy traffic. As he turns onto a side street, he passes a furniture store and notes that the store is closed and appears quiet. Mr. Jones continues his travel and passes the fire station, seeing fire trucks responding to a fire call at the furniture store. The trucks radioed their departure to the fire at 6:32 p.m. Using the benchmark events established with the work time clock and the fire department dispatch records, an estimated relative time may be established for when Mr. Jones passed the store.

Multiple Timelines

A variety of timelines may be required to evaluate and document sequence events precipitating the fire, events during the actual fire incident, and postfire activities effectively. Depending on complexity, these may be evaluated in a macro or micro timeline.

A macro evaluation of events can cover months or even years and may incorporate activities that occurred a long time before the fire, such as those related to building construction, modification of codes, and/or code enforcement activities. Conversely, a micro evaluation is used to look at small or narrow segments of the macro timeline in detail. Examples of micro incidents may be travel times for a witness to walk between rooms or the dispatch and on-scene times.

Parallel timelines can be used to look at multiple events that occur simultaneously. Graphic timelines can be very helpful in putting together the sequence of events that transpired, and approaches involving matrices may be helpful when multiple events occur simultaneously TABLE 20-2.

The actual creation of the timeline may be accomplished with several methods. A simple paper and pencil timeline may offer all that is needed for the investigation. Various companies provide computer software programs that are used for the construction of timelines and investigative charts. In some larger law enforcement agencies, there may an intelligence section with personnel who will assist with the gathering of the information and then construct reference charts and reports such as timelines for the investigator.

Scaled Timeline

A scaled timeline is useful in displaying the time and event and its chronological relationship to other events. A scaled timeline will show the time of each event with the spacing between the time events being scaled in a manner that would show the elapsed time between each event FIGURE 20-1. The timeline may also list hard times above the line and soft times below the line.

Table 20-2 Sample of an Abbreviated Matrix Used for Developing a Timeline

Time	Engine 1	Engine 2	Ladder 1	Battalion 1
0604	Dispatched	Dispatched	Dispatched	Dispatched
0605	Responding	Responding	Responding	Responding
0608	On scene, assuming command			
0610	Making entry with hose line			
0611		On scene	On scene	
0613	Attacking fire in NW bedroom	Advancing to second floor with hose line for primary search and rescue	Extending ground ladder to roof for ventilation	
0614				On scene, assuming command
0615	Fire knocked down in NW bedroom	Victim located in hallway, second floor	Accessing roof	Requesting ambulance
0616	Conducting primary search, first floor	Victim removal	Making ventilation opening	

The investigator may want to add columns for the police department or for nonfire service witness accounts.

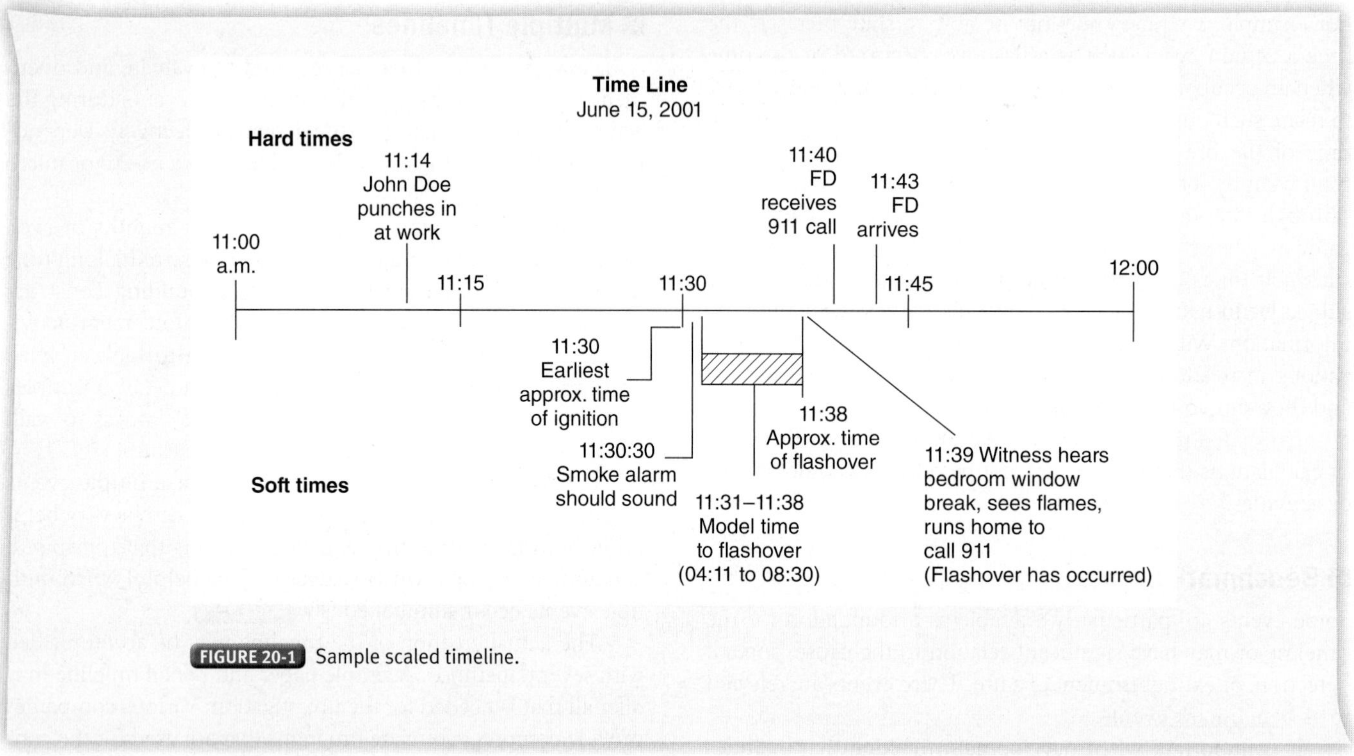

FIGURE 20-1 Sample scaled timeline.

System Analysis

System analysis is an analytical approach that takes into account characteristics, behavior, and performance of a variety of elements—including human activities and mechanical features of equipment—and integrates these to provide as complete a picture of events surrounding an incident as possible.

Simple examples of system analysis include evaluations of incidents that investigators make every day. An informal system analysis can take the form of a properly functioning stove left on by a homeowner, which leads to a kitchen fire: One aspect of such an incident involves human factors, and the second aspect involves the properties of the stove and its surroundings. Together, these make up the system whose properties led to the incident.

A variety of tools are available for the investigator to use in analyzing such incidents. Some of these tools (fault trees and failure mode and effects analysis) are discussed in the next sections, and they provide a systematic method of analyzing systems to determine hazards or faults.

Fault Trees

Fault trees (also known as decision trees) illustrate series of events and decisions that must take place for a specific outcome to occur. When this type of graphic logic or reasoning is applied to a situation, a solution may become more readily apparent, and incorrect solutions may be eliminated from consideration. When developing a fault tree, the investigator should use deductive reasoning. A fault tree diagram places, in logical sequence and position, the conditions and chains of events that are necessary for a given fire or explosion to occur.

Readers may be familiar with similar decision trees used in electronics or computer programming with "and" and "or" gates at the point where decisions must be made FIGURE 20-2.

For an "and" gate, all events or conditions must be present. For an "or" gate, any one of several events or conditions must be present. Fault trees identify the conditions and chain of events of a fire or explosion. If the conditions or events did not happen in an order that would lead to the event, then the proposed scenario is not possible. The investigator may apply probabilities to the events. The end result may produce multiple feasible scenarios for the fire scene.

There is software for developing fault trees. The data to be used in fault tree analysis or failure mode and effect analysis (discussed in the next section) can be obtained from the following places:

- Operations and maintenance manuals
- Maintenance records
- Parts replacement and repair records
- Design documents
- Services of expert with knowledge of the system
- Examination and testing of exemplar equipment or materials
- Component reliability databases
- Building plans and specifications
- Fire department reports
- Incident scene documentation
- Witness statements
- Medical records of victims
- Human behavior information

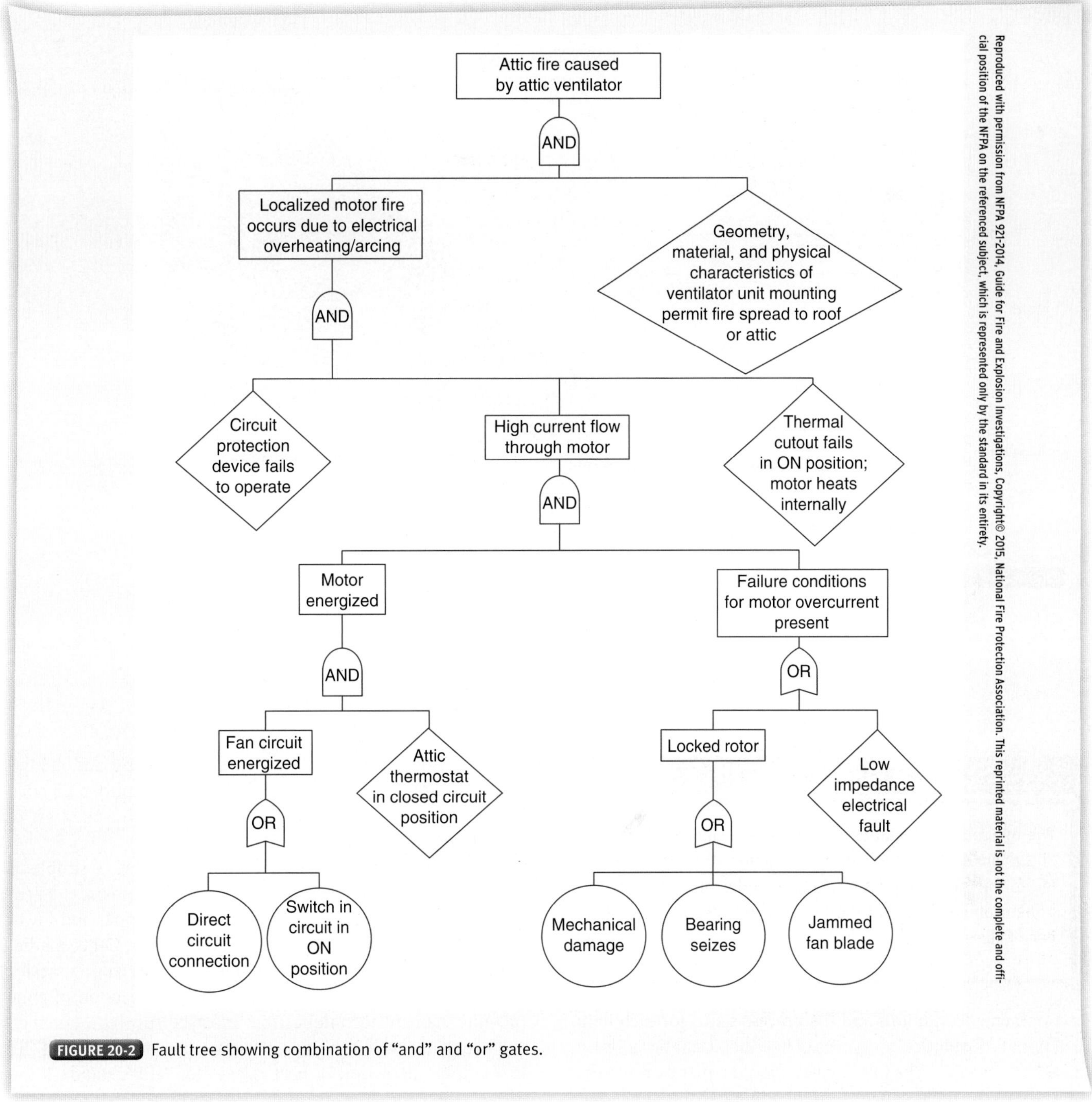

Reproduced with permission from NFPA 921-2014, Guide for Fire and Explosion Investigations, Copyright© 2015, National Fire Protection Association. This reprinted material is not the complete and official position of the NFPA on the referenced subject, which is represented only by the standard in its entirety.

FIGURE 20-2 Fault tree showing combination of "and" and "or" gates.

Failure Mode and Effects Analysis

Failure mode and effects analysis (FMEA) is another graphical method or technique used to determine the causes and effects involved with an event or subevent leading to a fire. By identifying specific components associated with potential ignition sources or fire spread, it may be possible to identify specific predecessor events occurring or activities preceding an incident.

FMEA is prepared by filling in a table with various column headings to address the needs of the particular investigation. The column headings are flexible; however, each FMEA contains the following items:

- Item or action being analyzed
- Basic fault (failure) or error that created the hazard
- Consequence of the failure

The development of an FMEA table can be involved or quite simple, depending on the complexity of the incident or the depth of analysis required. When compiling the information on an FMEA, the investigator should consider the

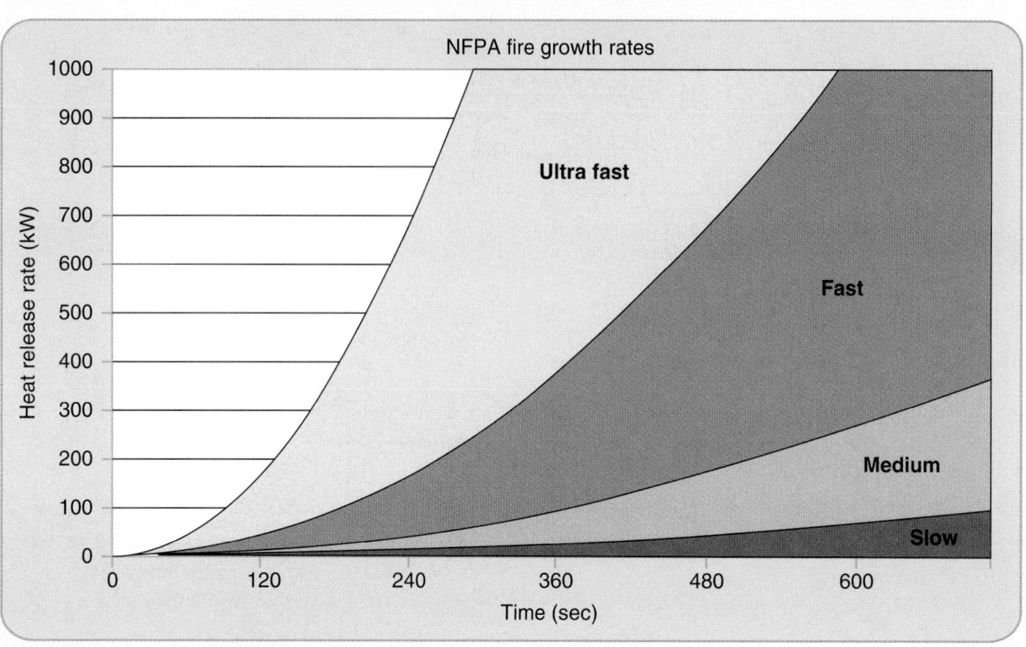

FIGURE 20-3 Fire growth curves.

Fire Investigator Tip

The minimum information for an FMEA includes a list of all system components and human actions that may have led to the incident, possible failure modes for each component and action, and the immediate consequences of each failure.

environmental conditions and the process status for each item or action. Probabilities or degrees of likelihood can be assigned to each occurrence. The investigator should remember that the accuracy of the table depends on identifying the system components and human actions that are relevant to the incident.

Mathematical/Engineering Modeling

Mathematical (engineering) modeling of fire incidents is generally conducted using hand calculations or formulas to evaluate specific issues or computer-aided analysis to examine more complex fire dynamics and fire progression issues.

A widely used group of modeling techniques involves the application of engineering models. Data are developed that incorporate both known and approximated properties of materials and systems, specific features and components of a fire incident, and physical property estimates defined to a stated degree of certainty. To these assumed "fact sets" or "modeling input data," accepted engineering and analytical techniques are applied to develop descriptions of what may or may not have occurred under a clearly stated set of conditions.

Mathematical (engineering) models can be as simple as those used when flammable gas concentrations are calculated on a hand calculator to determine the upper and lower explosive limits for a given gas in a given space. Other mathematical models may be quite complex, such as those used for fire growth modeling. These involve the application of zone or finite element techniques that must be run by a qualified modeler on a well-equipped personal computer. FIGURE 20-3 is a graphic depiction of heat release for NFPA standard fire growth rates for slow, medium, fast, and ultrafast fire growth rates.

It is important to understand that these techniques are only tools used to test a hypothesis and that repeated calculations should be carried out to bracket the conditions that most likely existed with a given incident. This type of modeling can be used to determine a probable scenario given a specific set of boundary conditions. Models do not necessarily provide a definitive solution. As with all evidence, modeling results should be considered in light of other data that have been gathered during the fire investigation. Models can frequently be used to help support or disprove a particular hypothesis. Even in investigations in which the origin and the cause of the fire are not of issue, the fire model may be useful in the understanding of fire damage or injury.

The use of these analytical tools may depend on the scope of the investigator's role in the investigation and its practical purpose in the question at hand. The services of a special expert may be needed to complete the analysis.

Various disciplines use forms of mathematical models to simulate or predict events using established scientific principles and empirical data. In some investigations, these models have also provided useful information in the fire or explosion investigation.

Limitations of Mathematical/Engineering Modeling

Mathematical (engineering) modeling, whether that is hand calculations or computer fire models, works with some limitations and assumptions that need to be considered. These models are useful to the investigator in testing his or her hypotheses but should never be used as the sole basis of a fire origin and cause determination.

The results of mathematical models may be affected by several factors. Inputs to the model are subject to uncertainties. The use of standardized methods of input for the selected model will narrow the uncertainties. Often, generic data from fire science literature or exemplar models are used in the models and may also contribute to the uncertainties. Results from the mathematical models are subject to uncertainties from the approximations made within the model itself.

When using a model, the investigator should retain the original input and output files as part of the investigative record. This process will allow other interested parties to examine the full content of the work product.

Mathematical modeling must be conducted by an individual who has sufficient training and experience to carry out the type of analysis required. Examples of typical sorts of engineering modeling techniques that can be used effectively in fire investigation work are discussed later.

Heat Transfer Analysis

Heat transfer models allow the investigator to determine how heat was transferred from a source to a target by one or more of the common heat transfer modes: conduction, convection, or radiation.

Heat transfer models can be used to test hypotheses such as the competency of a given heat source to act as an ignition source in a given fire causation scenario. Other useful applications of heat transfer modeling are to explain damage or ignition to adjacent buildings via fire spread, ignition of secondary fuel items, and transmission of heat through building elements.

Flammable Gas Concentrations

By determining what the concentration of gas was within a given space, the investigator can support or disprove the theory of whether the presence of a flammable gas during an incident was a contributing factor to the sequence of events that occurred. Flame concentration results can support or refute theories about the location and size of a leak or the competency of an ignition source.

Hydraulic Analysis

When a fire is not controlled by an operating sprinkler system, it is important to determine whether the sprinkler system functioned as designed and intended and whether the design was adequate for the particular occupancy and furnishings present.

The analysis of a sprinkler system and its water supply should determine whether the system and water supply were matched to the hazard being protected. Questions should be asked to determine whether there were any faults in the system, such as closed valves or faulty sprinklers. It may be necessary to evaluate why an apparently functional system did not control a given fire. This involves an analysis of the water supply characteristics, water flow through the piping network, and sprinkler discharge characteristics. It may also involve developing a fire growth or heat release model to assess the potential effectiveness of the sprinkler system under various circumstances.

Hydraulic analysis can also be used on systems such as carbon dioxide, gaseous suppression agents, dry chemicals, and fuel distribution systems.

Thermodynamic Chemical Equilibrium Analysis

Fires and explosions that are believed to be caused by reactions of known or suspected chemical mixtures can be investigated by a thermodynamics analysis of the probable chemical mixtures and potential contaminants. Thermodynamics analysis can be used in incidents involving chemical reactions and to determine the feasibility of a given scenario. This analysis may be useful in evaluating a hypothesis about chemical reactions, role of contamination, role of ambient conditions, potential of overheating, and other related scenarios. Computer programs can be used to make these predictions using input data detailing the properties of the chemicals in question. This analysis may show that the chemicals have no thermodynamic reaction and can be eliminated from the hypothesis. Analysis may also show the chemicals are capable of being thermodynamically favored, and further investigation would be required to determine whether they are sufficient to cause ignition.

Structural Analysis

Structural analysis can provide the investigator with clues to why a building failed by collapse at a given point during a fire. This type of analysis can be critical in helping to point toward where the fire was able to have the most significant impact on the strength of the building.

Egress Analysis

The reason why a fire victim did not escape from a given fire scene is a critical question for a fire investigator to answer. The cause for such a failure may be related to egress design features (or lack thereof) of a building or to maintenance issues. The investigator should obtain data such as the location of the exits, egress routes, travel distance, and egress route widths to help in this analysis. Computer-based fire models such as Consolidated Model of Fire and Smoke Transport (CFAST) have modules that can provide egress data for the specific fire situations they can be used to model.

Fire Dynamics Analysis

A number of methods are available to assist the investigator in the analysis of fire dynamics and fire growth, including specialized fire dynamics routines (hand calculations) and computer models (zone models and field CFD models). These analytical methodologies can be used to test and evaluate origin and cause hypotheses and can assist in the evaluation of physical and eyewitness evidence used as part of the hypothesis development process.

Computer models can assist the investigator in attempting to determine the growth and development of the fire. These models, which are based on established mathematical equations, may be useful in the prediction of several fire-related factors and/or events such as:

- Time to flashover
- Gas temperatures or concentrations
- Flow rates of fire-related gases
- Temperatures of interior surfaces
- Time of activation of fire detection and suppression devices
- Effects of events such as opening doors and windows

Computer models are a useful tool in the evaluation of a hypothesis formed by the investigator. Utilizing known input data, the model can be compared with witness accounts and physical evidence as a test of a developed hypothesis or account.

The investigator is reminded and cautioned that these models are only tools and are subject to certain limitations based on uncertainties that must be assessed and analyzed. Although they can provide additional data on a fire, modeling results must be weighed in relation to other information gathered during the investigation and the reliability of the input data and assumptions made.

A number of variables or uncertainties that can influence fire modeling results include the following:

- Fire load characteristics
- Ventilation openings (size and open or closed)
- HVAC flow rates
- Heat release rates

The two primary types of models currently used to assess fire growth are zone models and field CFD models.

Specialized Fire Dynamics Routines

Specialized fire dynamics routines are simplified procedures that require minimal data to run a computer model and can often answer a narrowly focused question. These procedures often involve the use of specific algebraic hand calculation equations and generally require much less data and information. These equations have been derived through research and experimentation and require the input of variables derived from information developed from the fire scene and obtained from appropriate reference materials.

These equations and hand calculations can be utilized to evaluate specific issues, including the following:

- Time to flashover
- Heat flux
- Heat release rate
- Time to ignition
- Flame height
- Detector activation
- Gas concentration
- Flow rates of smoke, gas, and unburned fuels

Computer Fire Models

The two primary types of computer models currently used to assess fire growth are zone models and CFD models. Computer models allow for a more complex and detailed analysis of fire growth and behavior that is useful in both predictive analysis and testing and post-incident investigation and hypothesis testing. These models also incorporate certain assumptions that must be considered when analyzing results and are subject to limitations.

Zone models usually divide a compartment into two zones: a hot upper zone and a cooler lower zone. These models assume universal conditions throughout the zone and allow for the expansion of the zones during the course of the fire. These applications can routinely be run on personal computers and are generally well accepted and validated by peer review.

CFD models are more complex than zone models and divide the compartment into many small cells. Numerous calculations occur in each cell, and activity in one cell affects the surrounding cells. These models allow for a more detailed analysis of a fire event but also require a much greater level of expertise. These models usually require large capacity computer workstations and can be time consuming and very

Fire Investigator Tip

Fire dynamics and fire growth can be analyzed using several methods, including the following:

- Specialized fire dynamics routines, such as hand calculations
- Computer models, such as zone models and field computational fluid dynamics [CFD] models

expensive to operate. CFD models are useful in evaluating fire progression in irregular geometry or where very fine detail is required. The use of these models in fire investigation and litigation is increasing.

Graphic Representations

Another useful tool to represent what has occurred during a given fire incident is graphic representations. These include drawings of all sorts, as well as physical models and computer animations, which are being used increasingly today. Some computer models also work with a graphic interface that allows for a visual representation or interpretation of the data derived from the model results. For instance, the NIST Fire Dynamics Simulator is a CFD model that works with NIST SmokeView to provide a visual animation of the fire progression. Graphic representations are frequently used by both the investigation team and forensic evaluation personnel.

Among the reasons that an investigation team uses graphics are to understand an incident location better, to assist in interviewing witnesses, and to define and identify materials and systems and their involvement in an incident.

In forensic applications, graphics can be used to help a judge and/or jury to better understand important features of a fire scene and underlying scientific and engineering principles that caused a particular outcome in a given fire. Graphic representations should not be confused with mathematical models, described earlier. Mathematical models are based on calculations, whereas graphic representations are based on geometric representations of a scene or a set of facts.

Guidelines for Selection and Use of a Fire Model

Numerous factors should be taken into consideration when selecting the fire model to use on a particular fire incident. First, the investigator should identify the hypothesis(es) to be tested or other question(s) to be answered by the modeling procedure. The model that is chosen preliminarily should be validated (this information should be available from the model developer and available in documentation). If possible, the user should consider the degree of uncertainty that is expected to result from the use of the model, both from the model itself and from the data that is chosen to be entered.

Fire Testing

Fire testing is a process that can help check data collected or test a specific hypothesis. Testing can be conducted in the field and/or in a controlled environment and may range from bench tests to full-size re-creations of the event. These tests are useful in determining several factors or events in the fire, such as origin and cause, fire spread, combustion characteristics, and effects of fire on materials.

As with fire modeling, when used in a proper manner, fire testing can be useful in the evaluation of a hypothesis. However, the investigator should be mindful of potential differences and inaccuracies between test conditions and the conditions at the time of the fire being investigated. For example, differences may exist in weather conditions, missing windows or loss of glass, fuel loads associated with contents, and ventilation effects. Fire testing may be used as a tool to test a hypothesis, but the data obtained should not be relied on as an absolute. Fire testing can provide useful information, although it is not possible to re-create all conditions of a given fire perfectly. To the extent possible, the methods, procedures, and instruments used in testing should follow accepted norms of practice. Following these established practices will help ensure credibility of the results.

Examples of fire testing may include determining the burn-through time of a wall or door using recognized fire endurance testing techniques such as those found in the ASTM E-119 standard or determining the heat release characteristics of a piece of cushioned furniture to assess whether it could have been responsible for fire growth and the damage patterns noted. Application of the types of data that are developed in testing can provide invaluable information to support or refute various hypotheses being evaluated. Fire testing does have limitations in that it is not possible to re-create all aspects of a specific fire.

FIGURE 20-4 shows a series of images of 4-minute fire growth progression prior to extinguishment. The fire was initiated with a single match igniting a single sheet of newspaper.

FIGURE 20-5 shows temperature development during ASTM E-1537-99 fire testing of a couch. Measurements were made at locations 4 ft (1.2 m) above the couch and directly above the couch back.

Data Required for Modeling and Testing

To conduct valid modeling and testing, it is important that the investigator gather data that are as accurate and complete as possible. As in all other endeavors, the "garbage in, garbage out" concept applies here. It should be anticipated that during court proceedings, test results will be subjected to a Daubert review, and the results that are presented will be only as valid as the accuracy of the information from which they were derived and the care that was taken to develop them. (See the "Legal Considerations" chapter for more information on the Daubert Rule.)

Important information to be obtained includes structural information, materials and contents, and ventilation information.

Structural information, such as the room dimensions and size of structural components as they apply to the fire information, should include the following:

- Length, width, and height of rooms and buildings
- Wall thickness
- Slopes of floors and/or ceilings
- Construction materials
- Construction features present

© Jones & Bartlett Learning. Photographed by Glen E. Ellman

© Jones & Bartlett Learning. Photographed by Glen E. Ellman

© Jones & Bartlett Learning. Photographed by Glen E. Ellman

© Jones & Bartlett Learning. Photographed by Glen E. Ellman

© Jones & Bartlett Learning. Photographed by Glen E. Ellman

FIGURE 20-4 A fire's progression over a 4-minute period.

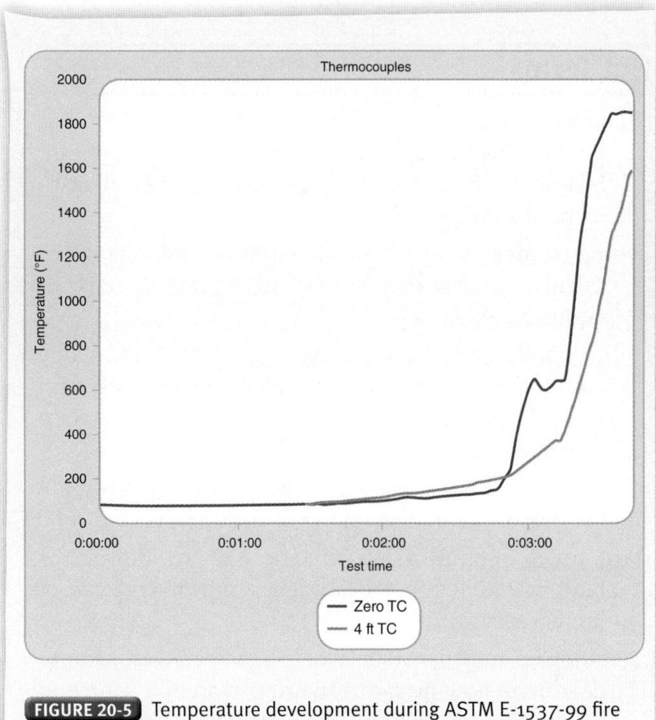

FIGURE 20-5 Temperature development during ASTM E-1537-99 fire testing of a couch.

Materials and Contents

A meaningful analysis of a fire requires understanding of the heat release rate, the fire growth rate, and total heat released. The analysis should be documented and include the following:

- Type of contents, including materials
- Location of contents
- Configuration and condition of contents

Ventilation

Understanding ventilation conditions is important to the validity of a fire test or model. The position and conditions of openings should be included, as should the following:

- Location of openings
- Size of openings
- Status (open or closed)
- Area of usable opening (fully open, partially open, closed)
- Ventilation effects, including wind and HVAC
- Fire department operations

Wrap-Up

© Greg Henry/ShutterStock, Inc.

Ready for Review

- The investigator has several analytical tools available to assist him or her in interpreting the information obtained during an investigation.
- Understanding the timeline of any incident is key to creating the failure analysis because it assists the investigator in determining the sequence of events that occurred.
- When developing a timeline, incidents that can be related to a known exact time are generally referred to as hard times.
- Soft time is either estimated or relative and is generally provided by witnesses.
- Benchmark events are particularly valuable as a foundation for the timeline or may have significant relation to the cause, spread, detection, or extinguishment of a fire.
- A variety of timelines may be required to evaluate and document effectively sequence events precipitating the fire, events during the actual fire incident, and postfire activities.
- Analytical approaches involving systems analysis take into account characteristics, behavior, and performance of a variety of elements—including human activities and mechanical features of equipment—and integrate these to provide as complete a picture of events surrounding an incident as possible.
- Various disciplines use forms of mathematical models to simulate or predict events using established scientific principles and empirical data.
- Graphic models include drawings of all sorts as well as physical models and computer animations. This class of models is frequently used by both the investigation team and forensic evaluation personnel.
- Fire testing can help check data collected or test a specific hypothesis. It can be conducted in the field and/or in a controlled environment and may range from bench tests to full-size re-creations of the event.

Hot Terms

Benchmark events Events that are particularly valuable as a foundation for the timeline or may have significant relation to the cause, spread, detection, or extinguishment of a fire.

Estimated time An approximation based on information or calculations that may or may not be relative to other events or activities.

Failure mode and effects analysis (FMEA) A technique used to identify basic sources of failure within a system and to follow the consequences of these failures in a systematic fashion.

Fault trees Logic diagrams that can be used to analyze a fire or explosion. Also known as decision trees.

Hard times Specific points in time that are directly or indirectly linked to a reliable clock or timing device of known accuracy.

Heat transfer models Models that allow the investigator to determine how heat was transferred from a source to a target by one or more of the common heat transfer modes: conduction, convection, or radiation.

Relative time The chronological order of events or activities that can be identified in relation to other events or activities.

Soft times Estimated or relative points in time.

System analysis An analytical approach that takes into account characteristics, behavior, and performance of a variety of elements.

Timeline A graphic or narrative representation of events related to the fire incident, arranged in chronological order.

FIRE INVESTIGATOR *in action*

Several people have been injured in a four-story hotel fire that you are investigating. The fire was discovered by first floor residents after smoke was observed emanating from a second floor balcony. Shortly afterwards, flames were noticed in the second floor hallway, and then the smoke detectors for the building began to sound. Those injured were located on the third and fourth floors of the hotel at the time of the fire and were rescued by fire personnel. Their injuries range from minor scrapes and bruises to severe smoke inhalation; however, everyone is expected to make a full recovery.

As you examine the loss structure, you notice several fire code violations, including open junction boxes, missing fire extinguishers, debris and stored materials located within the stairwell, and nonfunctioning exit signage.

1. Which tool might you use to determine why those injured could not exit the structure?
   - **A.** Engineering model
   - **B.** Smoke migration model
   - **C.** Egress analysis
   - **D.** Heat transfer analysis
2. Which of the following is considered a benchmark event for this fire?
   - **A.** Time of checkout for guests
   - **B.** Time of day the fire occurred
   - **C.** Time when the continental breakfast was ready
   - **D.** Time recorded when the alarm system activated
3. Which analytical tool illustrates a series of events and decisions that must take place for a specific outcome to occur?
   - **A.** Systems analysis
   - **B.** Fault tree
   - **C.** Fire dynamics model
   - **D.** Timeline
4. Timelines should be represented:
   - **A.** as a graphic.
   - **B.** in narrative form.
   - **C.** Both A and B
   - **D.** None of the above

Explosions

© Photos.com

© Jones & Bartlett Learning. Photographed by Glen E. Ellman

Knowledge Objectives

After studying this chapter, you should be able to:

- Describe the types of explosions NFPA 4.2.9. (pp 304–306)
- Discuss the characteristics of explosion damage NFPA 4.2.9. (p 306)
- Discuss the effects of explosions NFPA 4.2.9. (pp 306–308)
- Discuss the factors that control explosions NFPA 4.2.9. (pp 308–309)
- Discuss the characteristics of seated, nonseated, gas and vapor, dust, backdraft or smoke, and outdoor vapor cloud explosions NFPA 4.2.9. (pp 309–312)
- Describe the two main types of explosives NFPA 4.2.9. (pp 312–313)
- Explain how to investigate an explosion scene NFPA 4.2.9. (pp 313–317)

Skills Objectives

After studying this chapter, you should be able to:

- Conduct an explosion investigation NFPA 4.2.9. (pp 304–317)

Additional NFPA References

NFPA 68, *Standard on Explosion Protection by Deflagration Venting*

NFPA 921, *Guide for Fire and Explosion Investigations*

CHAPTER 21

FESHE Course Outcomes

Fire Investigation I

There are no Fire Investigation I (FESHE) course outcomes for this chapter.

Fire Investigation II

8. Evaluate the use of incendiary devices, explosives, and bombs. (pp 312–313)

You Are the Fire Investigator

© Jones and Bartlett Publishers. Photographed by Glen E. Ellman

Your department is dispatched to a house fire with a possible explosion. The first-arriving apparatus reports that the front half of a two-story home has collapsed and numerous people appear injured. You arrive several minutes later and observe a large excavator in the front yard covered by debris. A trench dug from the street has exposed the natural gas line to the structure. The gas line is intact up to the excavator, where the line has been pulled out of the ground.

1. What safety issues may exist at the scene?
2. How will an assessment of the damage assist with determining the point of origin?
3. How will witnesses to the explosion assist you with your investigation?
4. How will you identify the source of the fuel and the ignition source?

Introduction

An explosion is determined by the evidence, which includes damage or change to a container caused by the confinement of a blast overpressure (evidenced by damage or changes to structures or objects, or by injury to persons). In fire and explosion investigations, an explosion is a sudden conversion of potential energy, either chemical or mechanical, into kinetic energy (the energy possessed by a system or object as a result of its motion). The conversion into kinetic energy produces and releases gas(es) under pressure, which then cause mechanical damage to materials, including movement, shattering, and changing. Although there is usually a loud noise with an explosion, it is not an element of primary criteria as the production and violent release of the gas(es).

The ignition of a flammable vapor/air mix in a closed container that pops off the lid or damages the can is an explosion, whereas the same ignition occurring in an open field is a deflagration. A failure and bursting of a container caused by hydrostatic pressure (from the fluid itself) is not an explosion because it is not created by gas. Factors controlling the explosion effects and damages include the type, configuration, and amount of fuel, as well as the size, shape, and material of the container or construction and the type and amount of venting present.

Flash fires are a unique form of burning that can lead to investigative confusion because of the fire patterns they create. A flash fire is a fire that spreads rapidly by means of a flame front through a diffuse fuel such as dust, gas, or ignitible liquid vapor and without production of damaging pressure. These patterns can include full consumption or only burning of diffuse materials away from the initial fire or explosion.

This chapter is based on investigation of diffuse (not concentrated) fuels in lightweight construction structures. The practices discussed may not readily transfer to other types of fuels and construction. Additional research and specialized expertise beyond what is presented may be required; this is especially true with condensed fuels and high explosives. Be aware of the possibility of secondary explosions and/or devices. If you are not sure, do not proceed; request qualified personnel to conduct the investigation. It is best to have mutual aid or specialized resources identified prior to the time of need. These resources will vary depending on your jurisdiction but are available from local, county, state, regional, and federal law enforcement, fire, and other agencies.

Types of Explosions

The two major types of explosions are mechanical and chemical, which are distinguished by the source and the mechanism that produces the explosion FIGURE 21-1. There are also several subtypes of explosions that fall under these major types.

Mechanical Explosions

A mechanical explosion does not involve changes in the basic chemical nature of the substance(s) in the vessel; it is a purely physical reaction. A mechanical explosion is the rupture of a vessel or container (such as a cylinder, tank, or boiler) that results in the release of pressurized gas or vapor. The pressure leading to the mechanical explosion is not due to a chemical reaction or change in chemical composition of the substances involved.

A boiling liquid expanding vapor explosion (BLEVE) is the most frequent type of mechanical explosion that the user of this text will encounter. FIGURE 21-2 shows the damage from a BLEVE.

In short, a BLEVE is a subtype of mechanical explosion involving containers of liquids that are under pressure and are at temperatures above their atmospheric boiling point. The vessel(s) may range from a small butane lighter to a railroad tanker.

A BLEVE may occur as a result of the internal temperature of the liquid or vapor being increased by exposure to a fire. Exposure to the fire increases pressure within the vessel, leading to rupture. The release of the liquid results in instant

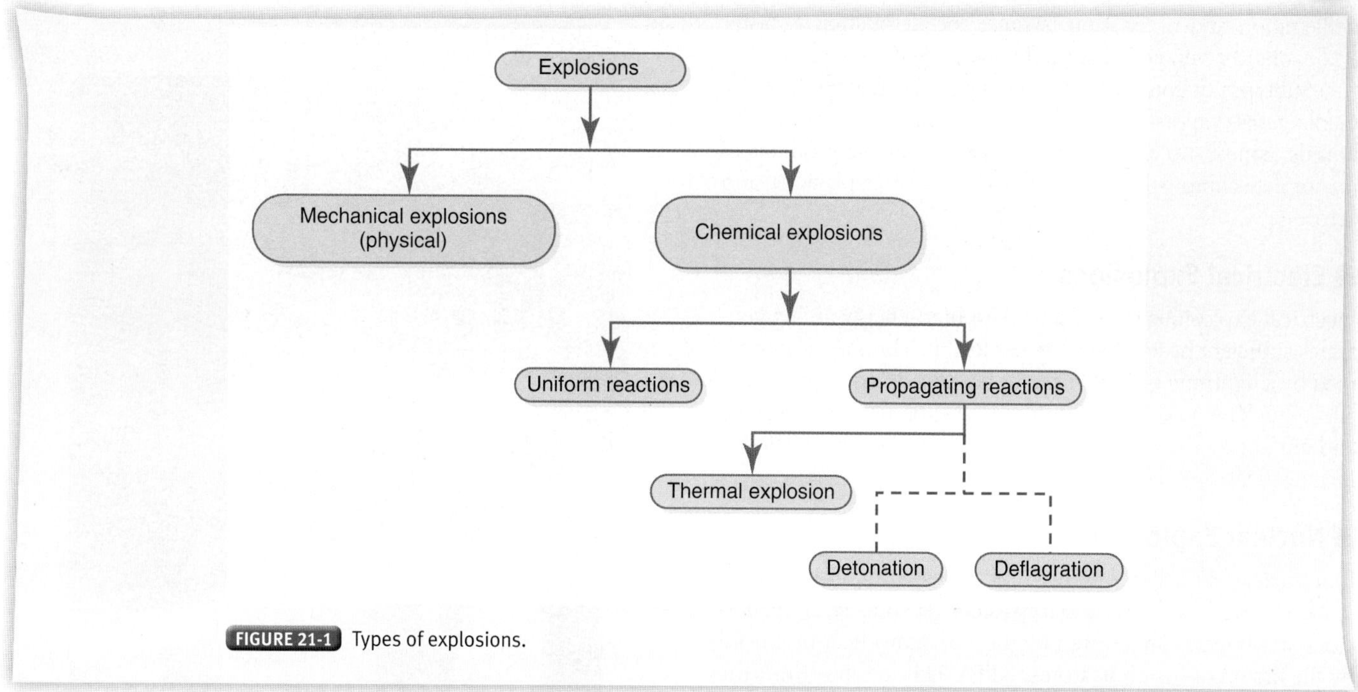

FIGURE 21-1 Types of explosions.

vaporization. If the contents of the vessel are ignitible, then a fire will likely result. The ignition of the vapor usually occurs from the original heat that caused the BLEVE or from another electrical or mechanical source. A common example of a BLEVE that does not involve an ignitible liquid is the rupture of a steam boiler, where the energy source is the pressurized steam. Additionally, a BLEVE may result from overfilling, runaway reaction, vapor explosion, or mechanical failure.

Chemical Explosions

A chemical explosion is one in which an exothermic chemical reaction is the source of the high-pressure gas and the fundamental chemical nature of the fuel is changed. Although chemical explosions can involve solid combustibles or explosive mixtures of fuel and an oxidizer, most typical are the propagating reactions that involve gases, vapors, or dust mixed with air.

FIGURE 21-2 BLEVE damage.

Combustible Explosions

The most common chemical explosions are those caused by the burning of combustible hydrocarbon fuels. A combustion explosion is characterized by the presence of a fuel (such as dust) with air as an oxidizer. The elevated pressures are created by the rapid burning of the fuel and production of combustion by-products and gases. The velocity of the flame front propagation through the fuel determines whether the combustion reaction is classified as a deflagration (less than the speed of sound) or detonation (faster than the speed of sound).

Uniform reactions occur more or less equally throughout the material and include ordinary chemical reactions that form gaseous products at a rate faster than they can be vented. Propagating reactions initiate at a specific point in the material and propagate as a reaction front through the unreacted material.

Thermal explosions are the result of exothermic reactions occurring within confinement without provisions for dissipating the heat of reactions. This can accelerate to the point at which high-pressure gases are generated and an explosion occurs.

A deflagration is a reaction that travels through the air (propagates) at subsonic velocities. A deflagration can be vented successfully. A detonation is defined by NFPA 921 as a combustion reaction that propagates at supersonic velocities, more than 1100 feet per second (340 meters per second). Other authorities define the minimum detonation speed at 1000 m/sec (3250 ft/sec) or higher, and explosives typically have propagation speeds many times the speed of sound.

The difference between a detonation and a deflagration is the magnitude of pressure versus time for the system involved in the combustion reaction. A significant difference between deflagration and detonation is time; the detonation is faster and cannot be vented because of the speed of the reaction.

Subtypes of combustion explosions are classified as flammable gases, vapors of ignitible (flammable and combustible) liquids, combustible dusts, smoke and flammable products of incomplete combustion (such as in a backdraft explosion), and aerosols.

Electrical Explosions

Electrical explosions are caused when high-energy arcing generates sufficient heat to cause an explosion. Thunder accompanied by a lightning bolt is an example of an electrical explosion effect. Such explosions require a special expertise to investigate and are not covered in NFPA 921, *Guide for Fire and Explosion Investigations*.

Nuclear Explosions

In a nuclear explosion, the high pressures within the primary system as well as the secondary system (such as steam generators and boilers) are created by the enormous heat produced by the fission or fusion of atoms. NFPA 921 does not cover the investigation of nuclear explosions, which should be referred to a special expert.

Characterization of Explosion Damage

The terms *low-order damage* and *high-order damage* are preferred to characterize explosion damage. The differences in damage are more a factor of the rate of pressure rise and the strength of the confining vessel or structure than of the maximum pressures within the system.

Low-Order Damage

Low-order damage is produced by pressure rising at a slow rate. The characteristics of low-order damage include walls bulged out or laid down, roofs lifted slightly, windows dislodged with the glass intact, and thrown-out debris that is generally large and found within a short distance from the structure FIGURE 21-3.

High-Order Damage

High-order damage results from rapid rate of pressure rise. High-order damage is characterized by walls, roofs, and structural members shattering. This results in small, pulverized debris that is often thrown great distances from the structure (hundreds of feet) FIGURE 21-4.

Effects of Explosions

An explosion is a gas dynamic phenomenon that, under ideal theoretical circumstances, will manifest itself as an expanding spherical heat and pressure wave front. The heat and pressure waves produce the damage characteristic of explosions. The effects of explosions can be observed in four major groups:

- Blast overpressure and wave effect
- Projected fragments effect (shrapnel)

© Chris Pole/ShutterStock, Inc.

FIGURE 21-3 The windows are dislodged as a result of low-order damage to this dwelling.

© urbancow/iStockphoto.com

FIGURE 21-4 Shattered remains as a result of high-order damage are thrown a great distance from the structure.

- Thermal effect
- Seismic effect (ground shock)

Blast Overpressure and Wave Effect

Blast overpressure and wave effect is a result of large quantities of gas being produced by the explosion of the material. The gases move outward at high speed from the point of origin and then return. The blast overpressure and wave effect creates a positive-pressure phase that is the outward movement of gases and displaced air, whereas the negative-pressure phase is the movement of air rushing toward the area of origin.

Positive- and Negative-Pressure Phases

The positive-pressure phase, with the expanding gases moving outward from the point of origin, is more powerful than the negative-pressure phase and is responsible for most of the pressure damage. The negative-pressure phase results from the rapid outward movement of the positive-pressure phase, which leaves low air pressure behind it FIGURE 21-5. Air then moves into the low-pressure area toward the point of origin.

The positive-pressure phase is more powerful than the negative, but the negative phase may still cause secondary damage by propelling artifacts toward the point of origin.

Shape of Blast Wave Front

The shape of the blast front from an idealized explosion would be spherical, expanding evenly in all directions from the epicenter. In the real world, confinement, obstruction, ignition position, cloud shape, or concentration distribution at the source of the blast pressure wave changes and modifies the direction, shape, and force of the front.

When the containers, structures, or vessels that contain or restrict the blast overpressures are ruptured, they are often broken into fragments that may be thrown great distances depending on their size and shape. These fragments are also called missiles, shrapnel, debris, or projectiles and can cause significant damage.

The positive-pressure phase is responsible for the majority of pressure damage, including weakening of the structure such that the structure can be further damaged by the negative-pressure phase. The negative-pressure phase can cause secondary damage and move items of physical evidence toward the point of origin, thus concealing the point of origin. It is normally less powerful than the positive phase but still has strength to accomplish significant damage, especially where components have been preweakened by the positive phase.

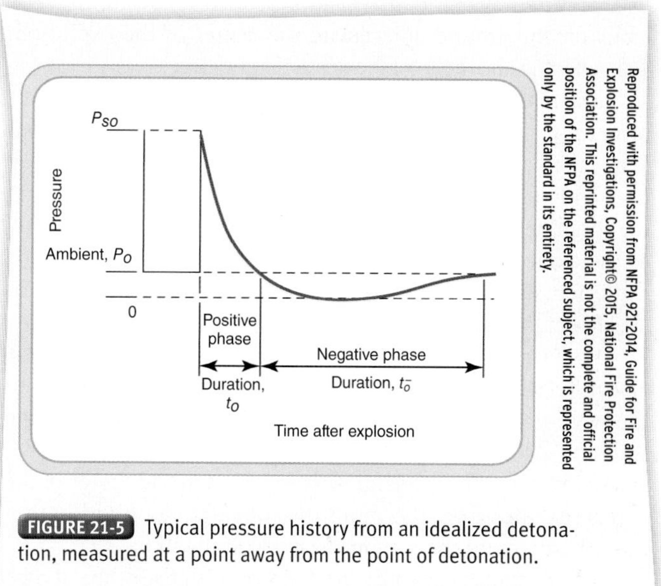

Reproduced with permission from NFPA 921-2014, Guide for Fire and Explosion Investigations, Copyright© 2015, National Fire Protection Association. This reprinted material is not the complete and official position of the NFPA on the referenced subject, which is represented only by the standard in its entirety.

FIGURE 21-5 Typical pressure history from an idealized detonation, measured at a point away from the point of detonation.

Table 21-1 Rate of Pressure Rise Versus Damage

Rate of Pressure Rise	Damage
Slow	■ Pushing or bulging type of damage ■ Weaker parts of the structure will rupture first ■ Characteristic of low-order damage
Rapid	■ Shattering of confining vessel or container ■ Debris will be thrown great distances ■ Characteristic of high-order damage

Confinement changes the shape and force of the front. The direction of the blast wave front may be changed as a result of either the venting path or redirection when the front is reflected off a solid object. The force decreases with distance from the epicenter, assuming no propagating reactions. The correlation between the rate of pressure rise and the damage effects of the explosion is shown in TABLE 21-1.

Where the pressure is less rapid, the venting effect has an important impact on the maximum pressure that develops. For example, in fugitive gas explosions in residential or commercial buildings, the maximum pressure is limited to a pressure that is slightly higher than the pressure that the components of the building enclosure can sustain without rupture. In a well-built residence, this pressure seldom exceeds 3 psi (21 kPa).

Rate of Pressure Rise Versus Maximum Pressure

Explosion blast pressure front damage is not only a product of the total amount of energy but is also, and sometimes more significantly, dependent on the rate of energy release and the resulting rate of pressure rise.

Damage from slower rates of pressure rise is similar to that of low-order damage; windows and other weaker structural members will fail, allowing the blast pressure to vent directly out, thus mitigating the explosive effects. Damage from rapid pressure rise is similar to high-order damage because venting is not developed, there is more shattering of the confining structure components, and debris is thrown great distances. Essentially, a slow rate of pressure rise decreases the maximum pressure through venting. (See NFPA 68, *Standard on Explosion Protection by Deflagration Venting*, for a discussion of calculating the theoretical effects of venting during a deflagration.)

Projected Fragment Effect

Projected fragment effect, also referred to as shrapnel effect, is the pieces of debris that result from the rupture of vessels or containers as a result of the blast pressure fronts. The shrapnel may cause personal injury or damage a great distance from the source of the explosion. The shrapnel may also sever electric

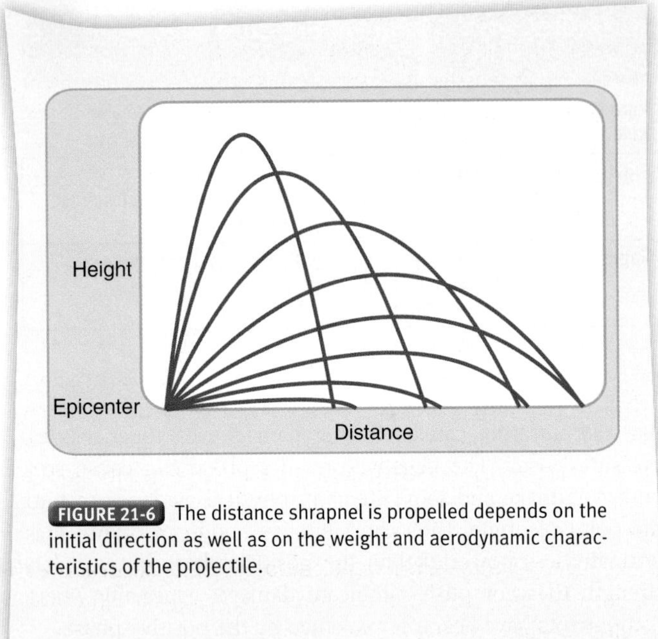

FIGURE 21-6 The distance shrapnel is propelled depends on the initial direction as well as on the weight and aerodynamic characteristics of the projectile.

utility lines, fuel gas lines, or storage containers. The distance that the shrapnel will be propelled is greatly dependent on the initial direction as well as on the weight and aerodynamic characteristics of the projectile FIGURE 21-6.

Thermal Effect

Thermal effect may occur because of the explosions' release of quantities of energy that can ignite nearby combustibles. As noted in TABLE 21-2, detonation produces extremely high temperatures of limited duration, whereas deflagration produces lower temperatures of longer duration. Fireballs (momentary balls of flame during or after an explosion, which may present high-intensity short-term radiation) and firebrands (hot or burning fragments propelled from the explosion) are possible thermal effects, especially with BLEVEs of flammable vapors. The secondary fires may increase the damage as well as injuries from the explosion. Unfortunately, it is often difficult to determine which occurred first, the fire or the explosion. The thermal damage from a chemical explosion depends not only on the nature of the explosive fuel, but also on the duration of the high temperature. Table 21-2 provides an analysis of explosion type and resulting thermal effect. The user should note the part that time plays in thermal effects. Because these effects may ignite fires away from the explosion center, they must be investigated thoroughly to determine the reason for them and to prevent them from misleading the investigation.

Table 21-2 Thermal Effects from Various Explosions

Explosion Type	Effect
Combustion	■ Releases heat energy ■ Can cause secondary fires
Chemical	■ Releases great quantities of heat
Detonating	■ Produces extremely high temperatures of short duration
Deflagration	■ Produces lower temperatures of longer duration

Seismic Effect

The transmission of tremors through the ground is known as seismic effect. The effect is a result of the blast wave expansion, causing structures to be knocked to the ground. Small explosions result in negligible seismic effects, whereas larger explosions may produce damage to structures, underground utility services, pipelines, tanks, and cables.

Factors Controlling Explosion Effects

The factors that control the effects of an explosion are as follows:

- Type and configuration of fuel
- Nature, size, volume, and shape of containment vessel
- Level of congestion and obstacles within the vessel
- Location and magnitude of ignition source
- Venting of the containment vessel
- Relative maximum pressure
- Rate of pressure rise

The nature of the noted factors and their variables can produce a variety of physical effects. The variables that may affect the characteristics of a blast pressure front as it travels from the source include blast pressure front modification by reflection, blast pressure front modification by refraction, and blast focusing. The blast pressure front may amplify because of its reflection from objects in its path. The reflection can cause the overpressure to increase as much as eight times at the surface of reflection. This phenomenon is negligible with deflagrations.

A blast pressure front that encounters a layer of air at a significantly different temperature may cause the front to bend or to refract. This occurs because the speed of sound is proportional to the square root of temperature and air. A low-level temperature inversion may cause refraction, resulting in a focus on the ground adjacent to the center of the explosion. This effect is also negligible with deflagrations.

The makeup of the fuel has a significant effect on the results of an explosion. When the fuel mixes with air and creates turbulence, the flame speed, rate of combustion, and rate of pressure rise all increase. The shape, size, and location of obstacles within the container affect the turbulence and consequently the severity of the explosion. The fuel's container has a great effect on the explosion as well. Its size, shape, construction, volume, materials, design, and internal obstacles all impact the fuel's reaction.

Both the location and the magnitude of the ignition source also influence an explosion. The closer the ignition source is to the wall of the container, the sooner the flame will reach the wall and extinguish, leading to lower pressure and a smaller explosion. The closer the ignition source is to the center of the container, therefore, the more the pressure can build within the container and the greater the explosion will be.

The damage created by the explosion is greatly affected by the ventilation of the fuel prior to the explosion, particularly if the fuel is diffuse, such as gas, vapor, or dust. The more ventilation, and the closer that ventilation is to the ignition source, the weaker is the explosion. A weaker explosion creates less damage, although the damage will be greatest in the path of the venting—for example, a window or a door. Ventilation has less of an effect when the explosion is caused by a detonation because there is not enough time for the venting to relieve sufficient pressure.

Blast Pressure Front

The blast pressure front can be modified by either reflection or refraction (with blast focusing). When a blast pressure front hits an object, the pressure may reflect off the object, thus amplifying the blast pressure front and overpressure. This reaction is especially severe in corners, where pressure may be locally focused.

When a blast pressure front hits a layer of air that is either a different temperature or a different density, the effect is refraction rather than reflection. The relationship between the speed of sound and the temperature can cause the blast front to refract, focusing the blast on the ground around the center of the explosion.

Seated Explosions

The crater or area of greatest damage may be characterized as the seat of the explosion. The seat may be of any size, depending on the amount and strength of the explosive material. The seated explosion is generally characterized by high pressure and rapid rates of pressure rise. The types of fuels that may generate seated explosions include explosives, steam boilers, highly confined fuel gases and liquid fuel vapors, and BLEVEs in small containers.

Explosives generally have a highly centralized epicenter or seat. Accordingly, because explosives have a high-velocity positive-pressure phase, they usually produce craters or localized areas of great damage. Boiler and pressure vessel explosions exhibit effects similar to explosives but with lesser localized overpressure adjacent to the source. Fuel gases are ignitible vapors that, when combined with small vessels, may also produce seated explosions. Finally, a BLEVE produces a seated explosion if the confining vessel is of a small size (such as a can or barrel) and the rate of pressure release is sufficiently rapid.

Nonseated Explosions

A nonseated explosion occurs when the fuels are dispersed or diffused at the time of explosion with moderate rates of pressure rise and subsonic explosive velocities. Supersonic detonations may also produce nonseated explosions, depending on the conditions. Fuel gases, such as natural gas and liquefied petroleum (LP), usually produce nonseated explosions because of the confinement of the gases in large containers such as rooms. Explosions from the vapors of flammable fuel or combustible liquids are usually nonseated explosions. A subsonic explosion speed, as well as the magnitude of the area that they cover, means that small, high-damage seats will not occur. Dust explosions most often occur in confined areas with wide dispersal, such as grain elevators, processing plants, and coal mines. The large areas of origin also preclude the development of explosion seats. Finally, smoke explosions or backdrafts usually involve a widely diffused volume of combustible gases and particulate matter. Accordingly, the explosive velocities are subsonic, thus limiting the production of pronounced seats.

Gas and Vapor Explosions

Fuel gases or the vapors of ignitible liquids are the most commonly encountered explosion. The more violent explosions involving lighter-than-air gases are less frequently encountered than are explosions involving gases or vapors with vapor density greater than 1.0 (heavier than air). Ignition temperatures in the range of 700°F to 1100°F (370°C to 590°C) are common. Thus, fuel gas–air mixtures are the most easily ignitible fuels that may result in an explosion. Minimum ignition energies begin at 0.25 mJ (an extremely low level of energy). FIGURE 21-7 shows the damage resulting from a vapor explosion.

Interpreting Explosion Damage

The explosion damage to structures is related to several factors:

- Fuel–air ratio
- Vapor density of the fuel
- Turbulence effects
- Volume of the confining space
- Location and magnitude of the ignition source
- Venting
- Strength of the structure

Fuel–Air Ratio

The entire volume of air need not be occupied by a flammable mixture of gas and air for there to be an explosion. Relatively small volumes of explosive mixtures capable of causing damage may result from gases or vapors collecting in a given area. Mixtures at or near the lower explosive limit (LEL) or the

Courtesy of Rodney J. Pevytoe

FIGURE 21-7 Low-order damage caused by a propane vapor explosion.

upper explosive limit (UEL) of a gas or vapor usually produce less violent explosions than do those near the optimum concentration. The optimum concentration is usually just slightly rich of stoichiometric (the optimum ratio at which point the combustion will be most efficient). This is a result of lower flame speeds and lower maximum pressures in a less-than-optimum ratio of fuel to air. Thus, explosions of mixtures near the LEL do not tend to produce large quantities of postexplosion fire in that most of the available fuel is consumed during the explosive propagation. It follows that explosions of mixtures near the UEL tend to produce postexplosion fires because of the fuel-rich mixtures. The delayed combustion of the remaining fuel may produce a postexplosion fire. In short, the stoichiometric mixture (slightly fuel-rich) produces the most efficient combustion and therefore the highest flame speeds, rates of pressure rise, and maximum pressures, and consequently the most damage.

Flame speed may vary widely, depending on temperature, pressure, confining volume, confining configuration, combustible concentration, and turbulence. Burning velocity is the rate of flame propagation relative to the velocity of the unburned gas ahead of it. Fundamental burning velocity is an inherent characteristic of a combustible as a fixed value. It is the velocity at which a flame reaction front moves into the unburned mixture as it chemically transforms the fuel and oxidant into combustion products. It is only a fraction of the flame speed. The flame speed is the product of the velocity of the flame front (created by the volume expansion of the combustion products caused by the increase in temperature), and any increase in the number of moles (a measurement of the amount of a substance in terms of the number of particles it contains) and any flow velocity caused by the motion of the gas mixture prior to ignition. The burning velocity of the flame front can be calculated from the fundamental burning velocity as reported in NFPA 68, at standardized conditions of temperature, pressure, and composition of unburned gas.

While conducting an explosion investigation, especially when assessing damage by determining the flame speeds, rate of pressure rise, and maximum pressure, it is essential to understand the explosive limits of mixtures: the concentration of an explosive gas or vapor and air. The optimum mixture is at or above the stoichiometric mixture (slightly fuel-rich), which results in an ideal combustion process where fuel is burned completely and produces the optimum (most violent) explosions, which then produce the greatest flame speeds, rates of pressure rise, and maximum pressure, and thus usually the most damage. Mixtures close to the LEL or UEL produce less violent explosions because the less-than-optimal mixture of fuel and air produce lower flame speeds, rates of pressure rise, and maximum pressure. UEL mixtures will produce secondary explosions because all the fuel is not initially consumed and later mixes with air.

An explosion can occur even if the fuel has not spread throughout the entire container. The fuel's properties help determine the reaction. Similarly, it cannot be assumed that because there was not an explosion, the container is not full of fuel. A delicate balance exists among the various elements involved.

The fuel–air ratio is of further importance in investigating an explosion when the following concepts are applied:

- Adiabatic flame temperature: the theoretical temperature of a complete combustion process that loses neither energy nor heat to its outside environment.
- Laminar burning velocity: The flame propagation rate relative to the unburned gas and the movement of the front into the unburned gas as it chemically combines the fuel and oxidizer into a combustible.
- Expansion ratio: The rate of expansion of a fuel–air mixture's products of combustion after ignition behind the expanding flame front.
- Laminar flame speed: The speed of a freely burning flame relative to a fixed point without turbulent effect.
- Turbulent flame speed: The speed of a turbulent combustion, which is generally relevant to a real explosion. Turbulent flame speeds may exceed the speed of sound in the unreacted mixture.

Vapor Density

Air movement from both natural and forced convection is a dominant mechanism for moving gases in a structure. Vapor density is the ratio of the average molecular weight of a given volume of gas or vapor to that of air at the same temperature and pressure and is also referred to as specific gravity. The vapor density of the gas or vapor may affect the movement of the fugitive gas as it escapes from its container or system. Heavier-than-air gases and vapors (vapor density greater than 1.0, such as ignitible liquids and LP gases) have a tendency to flow to lower areas. In contrast, lighter-than-air gases (such as natural gas) tend to rise and flow to upper areas. Where lighter- or heavier-than-air gases are involved, the operation of heating and air-conditioning systems and temperature radiance may cause mixing and movement that can reduce the effects of vapor density. Field tests have confirmed that vapor density effects are minimized by most air exchange rates that exist in older homes. Thus, vapor density effects are greatest in still air conditions.

Full-scale testing of flammable gas concentrations has shown that near-stoichiometric concentrations of gas develop between the location of the leak and either the ceiling (for lighter-than-air gases) or the floor (for heavier-than-air gases). A heavier-than-air gas that leaks at floor level may create a greater concentration at floor level, but with time, the gas will slowly diffuse upward. The inverse relationship is true for a lighter-than-air gas leak at ceiling height.

Fuels are mixed with air to a mixture within their flammable/explosive ranges, which are usually more air than fuel; thus, the vapor density is closer to that of air than that of the fuel. (Hydrogen, acetylene, and ethylene are known exceptions to this.)

Turbulence

Turbulence within a fuel–air mixture increases the flame speed and thus increases the rate of combustion and the rate of pressure rise. The turbulence may vary according to the size and shape of the combining vessel and thus affect the severity of the explosion. Be especially aware of any mixing or turbulent sources such as fans and forced-air ventilation and watch for chevrons or other structural components that could create turbulence.

Nature of Confining Space

The size, shape, construction, volume, materials, and design of the confining space greatly affect the nature of the explosion damage. In a confining space, a smaller space for the same amount of fuel will affect the rate of pressure rise by causing it to increase greatly, thus causing a more optimal explosion. The same fuel amount in a steel drum will result in a more violent explosion as compared with an explosion in a house. Similarly, obstructions such as walls, columns, and/or other large objects increase flame speed and thus also increase the rate of pressure rise.

Location and Magnitude of Ignition Source

The highest rate of pressure rise will occur if the ignition source is in the center of the confining structure. Although the energy of the ignition source has a minimal effect on the course of the explosion, unusually large ignition sources such as explosive devices can significantly increase the speed of pressure development and thereby cause transition from deflagration to detonation.

Venting

The nature of the venting of the containment vessel for gas, vapor–gas, or dust-fueled explosions has varying effects on explosion damage. A length of pipe may rupture in the center if it is sufficiently long. A room may experience destruction or merely movement of the walls and ceiling depending on the number, size, and location of doors and windows FIGURE 21-8. Furthermore, the venting of a vessel structure can cause damage to materials in the path of the venting. The user should note that if the explosion is typed as a detonation, venting effects are minimal because of the high speeds of the blast pressure fronts. In short, the blast pressure fronts are too fast to allow any venting to relieve the pressure.

FIGURE 21-8 Damage to a wall from a gas explosion.

Courtesy of Nina Scotti, NMS Investigations, Inc.

Strength of the Structure

A structure, for the purpose of this chapter, is anything that can be used to contain something else. The characteristics of a structure impact its strength against an explosion. The strength of the structure will depend on what type of structure it is, what it is made of, and potentially, in the case of a building, the services or utilities it is equipped with. The characteristics of a structure impact its strength against an explosion. For example, a glass jar (simple structure) would react differently from a five-story building made of steel and glass and serviced with natural gas and electrical service (complex structure).

Underground Migration of Fuel Gases

Lighter-than-air and heavier-than-air fuel gases that have escaped from underground piping systems can migrate underground to enter structures, resulting in fires or explosions. The fugitive gases may permeate the soil, migrate upward, and dissipate harmlessly in the air, but if the surface of the ground is obstructed by rain, snow, freezing, or paving, the gases may migrate laterally and enter structures through means of disturbed soils.

Fugitive gases may enter buildings through any opening into the structure, such as sewer lines, electrical and telephone conduits, or drain tiles, or through basement and foundation walls. The user should note that natural gas and propane have no natural odors of their own. Foul-smelling compounds (such as ethyl mercaptan) are added to gases. Odorant verification should be part of any explosion investigation. Gas detectors may be helpful in locating sources of gas leaks.

Multiple Explosions

Secondary explosions or cascade explosions may result when gas and vapors have migrated to adjacent stories or rooms, resulting in pockets. Thus, when ignition or explosion takes place in one story or room, subsequent explosions may occur in adjoining areas or stories. The migration or pocketing of gases may produce areas with different air–fuel mixtures. Accordingly, the dynamics of air–fuel mixtures may result in a series of vapor–gas explosions depending on the ratio of the air–fuel mixture in each area.

Dust Explosions

Violent explosions can be fueled by dust that is dispersed within the air. This is true for combustible materials as well as for materials not normally considered to be combustible. For example, dust explosions may occur from the following materials: agricultural products, carbonaceous materials, chemicals, dyes and pigments, metals, plastics, and resin.

Particle Size

The rate of pressure rise generated by combustion is largely dependent on the surface area of the dispersed dust particles. The finer the dust particles, the more violent the explosion is. The total surface area, and consequently the violence of the explosion, increase as the particle size decreases. In general, the explosion hazard can exist when the particles of the dust are 500 microns or less in diameter.

Concentration

The user must note that minimum explosive concentrations (MEC) may vary with the specific dust, but unlike most gases and vapors, there is generally no reliable maximum limited concentration. The reaction rate is controlled more by the surface-to-air mass ratio than by a maximum concentration. The minimum explosive concentration may be from as low as 0.02 oz/ft^3 to 2 oz/ft^3 (20 g/m^3 to 2000 g/m^3), the most common concentrations being less than 1 oz/ft^3 (1000 g/m^3).

The combustion rate and maximum pressure decrease if the mixture is fuel rich or fuel lean. The rate of pressure rise and the explosion pressure are low at the lower explosive limit and at the high fuel-rich concentration. Thus, as in gases and vapors, the pressure rise and the maximum pressure that occur in dust explosions are high if the dust concentration is at or close to the optimum mixture. The rate of combustion and thereby the rate of pressure rise are greatly increased when there is turbulence within the suspended dust–air mixture. Furthermore, the size and shape of the confining vessel affect the nature of turbulence and thereby can have a great effect on the severity of the dust explosion.

Moisture

Increasing the moisture content of the dust particles increases the minimum energy required for ignition and the ignition temperature of the dust suspension. The initial increase in ignition energy and temperature is low, but as the limiting value of moisture concentration is approached, the rate of increase in ignition energy and temperature becomes high. The user should note that the moisture content of the surrounding air has little effect on the propagation reaction once ignition has occurred.

Sources of Ignition

The variables that affect the ignition sensitivity of a dust include the ignition temperature, the minimum energy, and the minimum concentration. The Bureau of Mines has classified thousands of samples to identify ignition sensitivity as well as the explosion severity by which the index of explosivity can be established. Sources of ignition include the following: open flames, smoking materials, lightbulb filaments, welding operations, electric arcs, static electricity discharge, friction sparks, heated surfaces, and spontaneous heating.

The actual ignition temperature for most material dust ranges from 600°F to 1100°F (320°C to 590°C). Minimum ignition energies are higher for dust than for gas or vapor fields and are generally within the range of 10 to 40 mJ, significantly more than those for most flammable gas vapors.

Multiple Explosions

Dust explosions usually occur in a series within industrial and agribusiness operations. The initial explosion is usually less severe than the secondary explosion. The user should note that the first explosion puts additional dust into suspension, which then results in additional explosions. The structural vibrations in the blast front from the first explosion will propagate faster than the flame front, thus placing the dust ahead of it into suspension. The secondary explosion may progress from one area to another or from one building to another building.

Backdraft or Smoke Explosions

It is common for fires in airtight rooms to become oxygen depleted. This results in the generation of flammable gases due to incomplete combustion. The heated fuels accumulate in areas within the structure where there are insufficient oxygen and insufficient ventilation. The opening of a window or door results in the mixture of the fuels with air. The fuels can then ignite and burn sufficiently fast to produce low-order damage (less than 2 psi or 13.8 kPa). These are called backdrafts and smoke explosions.

Outdoor Vapor Cloud Explosions

The release of gas, vapor, or mist into the atmosphere may result in a cloud forming within the fuel's flammable limits and, subsequently, ignition culminating in an outdoor vapor cloud explosion. The phenomenon is referred to as *unconfined vapor–air explosion* or *unconfined vapor cloud explosion*. It is most frequently related to catastrophic failure of vessels and tankers, large amounts of fuel, and low-lying areas.

Similar to backdrafts and smoke explosions, outdoor vapor cloud explosions occur usually in partial restrictions of a natural or manmade structure. Unconfined environs, previously thought to foster outdoor vapor cloud explosions, will result only in a flash fire. Outdoor vapor cloud explosions are most common in chemical processing plants because significant amounts of fuel are involved; however, it is possible for congestion from natural sources (e.g., trees, vegetation) to accelerate flames and cause significant overpressures.

Explosives

Explosives are categorized into two main types: low explosives and high explosives, which should not be confused with low-order and high-order damage. Remember that there are several caveats in NFPA 921 cautioning to use specialists where explosives are concerned.

Low explosives are characterized by deflagration (subsonic blast pressure wave), a slow rate of reaction with the development of low pressure FIGURE 21-9A. Examples of low explosives are smokeless gunpowder, flash powders, and black powder. Low explosives are designed to work by pushing or heaving effects.

High explosives are designed to produce shattering effects because of their high rate of pressure rise and extremely high detonation pressure FIGURE 21-9B. Thus, a characteristic of high explosives would be the detonation propagation mechanism.

Safety Tip

If there are explosive devices still present on scene, it is not yet time for a fire investigation.

FIGURE 21-9 **A.** Flash powders are one type of low explosive. **B.** Dynamite is one type of high explosive.

The high, localized pressures are responsible for creating localized damage near the center of the explosion.

The effects seen from a diffuse phase (fuel–air) explosion are typically very different from those seen from solid explosives. A fuel–air explosion, most often a slow deflagration, will usually result in structural damage that is uniform and omnidirectional with relatively widespread evidence of burning, scorching, and blistering. The rate of combustion of a solid explosive is extremely fast in comparison to the speed of sound. Thus, pressure does not equalize throughout the explosion volume, and high pressures are generated near the explosive. The location of the explosion should be evidenced by the crushing, splintering, and shattering that are produced by the higher pressure; however, major distance from the source of the explosion may leave little evidence of intense burning or scorching except where shrapnel may have landed on combustible materials. Common examples of high explosives are dynamite, water gel, TNT, ANFO, RDX, and PETN. The extremely high detonation pressure may reach 1,000,000 psi (6,900,000 kPa).

TABLE 21-3 shows some explosion investigation tips. When on the scene of an explosion of unknown origin, the investigator should always be on the lookout for evidence that explosives were involved, such as signs of high-order explosion and high detonation pressure, and should involve explosives specialists when such evidence is present. Such a situation could include the presence of homemade explosives and/or improvised devices of varying levels of sophistication. The investigator should also recognize that secondary devices and explosions are possible.

Only investigators with appropriate training should conduct an investigation relating to possible explosive instances. Investigators who do not have this training should coordinate their investigation with appropriate experts.

Table 21-3 Fuel–Air Explosions Distinguished from Solid Explosive Explosions

Damage	Fuel–Air	Solid Explosive
Structural	Uniform and omnidirectional	Nonuniform Crushing and shattering near the location of the explosion
Fire	Widespread burning, scorching, and blistering	Localized around the source of the explosion

Investigating the Explosion Scene

The objectives of an explosion scene investigation are no different from those of a fire investigation: Determine the origin, identify the fuel and ignition source, determine the cause, and establish the responsibility for the incident.

A systematic approach is important in an explosion investigation. The following investigative procedures are quite comprehensive for the large incident, but the same principles should be applied to small incidents. Again, extensive damage would require the coordination of additional experts, including, but not limited to, an explosion expert and a structural expert. The investigator should note the necessity for absolute control of the scene to eliminate, as much as possible, the potential for contamination or loss of evidence and artifacts.

Securing the Scene

First responders to the explosion should establish and maintain physical control of the structure and surrounding areas. No unauthorized person should enter the scene or have contact with any blast debris, regardless of how remote it is from the scene. Evidence may be small and easily disturbed or removed by traffic in and around the scene. Caution must be used to prevent cross-contamination of the scene by investigators or authorized personnel wearing clothing or footwear that may contain explosive residue from other scenes or sources such as an explosives firing range. Furthermore, securing of the scene may minimize injuries to unauthorized personnel.

Safety Tip

Safety should be constantly in mind. This includes proper equipment, looking for additional hidden devices, stability of the structure (including what you are walking on), weather conditions, status of utilities, standing water, air quality, and biological and chemical exposure risks. All found hazards should be clearly marked to warn others.

Establishing the Scene

The outer perimeter of the incident scene (or blast zone) should be established at one and one-half times the distance of the farthest piece of debris found. Blast debris may be propelled great distances, in which case the scene perimeter should be widened accordingly.

Obtaining Background Information

All investigations should have sufficient background data to establish a timeline for the purpose of analysis.

First, all information should be obtained relating to the incident itself. The information would include a description of the incident site, systems and operations involved, conditions, and sequence of events that led to the incident, including material safety data sheets. Most important, the investigator must identify not only the locations of any combustibles and oxidants that were present and what conditions existed at the time of the incident, but also information relating to which combustibles, oxidants, and hazardous conditions currently exist at the site.

Investigators should examine witness accounts, maintenance records, operational logs, manuals, weather reports, previous incident reports, and other relevant records. As in any failure analysis, the most recent changes to equipment, procedures, and operating conditions may be especially significant. Blueprints of the building and drawings and prints of the process can assist in the proper documentation of the scene.

Establishing a Search Pattern

The scene should be searched from the outer perimeter inward toward the area of greatest damage. The final determination as to the explosion's epicenter should be made only after the entire scene has been examined. The search pattern may be spiral, circular, or grid shaped. The scene itself, with its particular circumstances, often dictates the nature of the pattern. The search pattern should overlap so that no evidence is lost at the edge of any search area, regardless of the methodology. The number of actual searchers needed depends on the size and complexity of the scene. Consistent procedures as to the identifying, logging, photographing, marking, and mapping of evidence must be maintained. The location of evidence may be marked with chalk, spray paint, flags, stakes, or any other marking means. The evidence should be photographed and may be secured or collected once spoliation issues have been addressed.

All fire investigation safety recommendations listed in NFPA 921, Chapter 13, also apply to an investigation of explosions.

Safety at the Scene

Structures that have been involved in an explosion are often structurally damaged. Accordingly, the possibility of floor, wall, ceiling, roof, or entire building collapse is great. Involvement of a structural engineer or construction engineer should be considered. Construction equipment may provide a temporary support mechanism, minimizing risk to the investigators. All toxic materials in the air need to be neutralized. Accordingly, material safety data sheets should be consulted to identify the appropriate personal protective equipment to be used.

A thorough search of the scene should be conducted for any secondary devices before the investigation is initiated. If undetonated devices and explosives are found, the area should be evacuated and isolated, and explosives disposal should be notified.

Initial Scene Assessment

The investigator should make an initial assessment of the type of incident. Safety is paramount, and the investigation must not proceed if safety has not been established. If the investigator determines that the explosion was fueled by explosives or an explosive device, the investigator should discontinue the scene investigation, secure the area, and contact the appropriate entities, such as bomb technicians, to check for hazards and secondary and unexploded devices. The area should also be checked with a gas detector to ensure that explosive or hazardous atmospheres no longer exist. Once safety has been addressed, the following tasks are sequenced to assist the investigator in the initial scene assessment:

1. Identify whether the incident was an explosion or fire.
2. Determine high- or low-order damage.
3. Identify seated or nonseated explosion.
4. Identify the type of explosion.
5. Identify the potential general fuel type.
6. Establish the origin.
7. Establish the ignition source.

Identifying Explosion or Fire

The first task is to determine whether the incident was a fire, explosion, or both, and, if both, which came first. Look for signs of overpressure such as displacement or bulging of walls, floors, ceilings, doors and windows, roofs, structural members, nails, screws, utility service lines, panels, and boxes. Also analyze the extent of heat damage to the structure and its components to determine whether the damage can be attributed to fire alone.

Determining High- or Low-Order Damage

The investigator should determine whether the damage indicates high-order or low-order damage. Review Section 23.3 of NFPA 921 to assist in the classification of the type, quantity, and mixture of the fuel involved.

Identifying Seated or Nonseated Explosion

The investigator should determine whether the explosion was seated or nonseated. Review Sections 23.6 and 23.7 of NFPA 921, which will assist in classifying of the possible fuel involved.

Identifying the Type of Explosion

Identify the type of explosion involved: mechanical (such as BLEVE), chemical, other chemical reaction (such as a combustible explosion or other chemical explosion), or electrical. Often, the presence of damaged equipment or fuel sources can provide evidence of explosion type.

Identifying Potential General Fuel Type

Identify the types of fuels that were potentially available at the explosion scene by identifying the condition and location of utility services (fuel gases) and sources of ignitible dusts or liquids. Analyze the nature of the damage in comparison to the damage patterns consistent with the following: lighter-than-air gases, heavier-than-air gases, liquid vapors, dust, explosives, backdrafts, and BLEVEs.

Establishing the Origin

Attempt to establish the origin of the explosion as soon as possible. The origin is usually identified as the area of most damage—a crater or localized area of severe damage in the case of a seated explosion. If it is a diffused fuel–air explosion, the origin is consistent with the confining volume or room of origin.

Establishing the Ignition Source

Attempt to identify the ignition source by looking for potential sources, including hot surfaces, electrical arcing, static electricity, open flames, sparks, and chemicals in which fuel–air mixtures are involved. If explosives are involved, the ignition source may be a blasting cap or pyrotechnic device. Be sure to note artifacts from the ignition sources that may have survived the explosion.

Vapors—including natural gas, propane, and gasoline vapors—are capable of being ignited by a variety of ignition sources, including electrical arcs from equipment and appliances in normal operation, pilot lights, and sparks. When a gas explosion is under consideration, the investigator should locate all such ignition sources, keeping in mind that they may not be in the area with the most explosion damage. Gaseous fuel released in a structure can migrate among rooms and be ignited remotely from the heaviest fuel concentration, with the flame front then traveling back to areas of greater fuel.

Detailed Scene Assessment

After obtaining the general information from the initial scene assessment, a more detailed study of the blast damage is recommended using the components that are listed here (the investigator should record all findings).

Effects of Explosion

A detailed analysis of the explosion overpressure damage should be made. The articles that are damaged should be identified as having been affected by one or more of the following forces: blast pressure wave—positive phase, blast pressure wave—negative phase, shrapnel impact, thermal energy, and seismic energy.

The investigator should examine the type of damage as to whether the debris was shattered, bent, broken, or flattened, as well as a change in the pattern. At a distance from a detonation explosion center, the pressure rise is moderate, and the artifacts resemble those of a deflagration explosion. Items in the immediate vicinity of the detonation center exhibit splintering and shattering.

The scene should be examined carefully to identify any fragments of foreign material. Estimation of damage from an explosion includes the maximum pressure of the explosion compared with the construction of the structure. A light-framed structure can be damaged with much less overpressure than a reinforced structure, for example.

Damage to personnel from explosion blast pressures is usually a result of acceleration in the high-velocity air stream with subsequent impact against a rigid surface, rather than compression in the airwave itself. More information on blast injuries can be found in the "Fire and Explosion Deaths and Injuries" chapter in this textbook.

Preblast and Postblast Damage

Debris that has been burned and propelled away from the point of origin may indicate that a fire preceded the explosion. Glass fragments with smoke residue and soot found some distance from the structure may indicate a fire of some duration followed by an explosion. Glass fragments that are clean and debris that is not burned but found some distance from the structure may indicate an explosion prior to the fire.

Articles of Evidence

The method used to document scene artifacts may include locating, identifying, noting, logging, photographing, and mapping physical evidence.

The probability of physical evidence being propelled both inside and outside of the structure may result in the evidence being found embedded in walls, resting in adjacent vegetation, inside adjacent structures, and within the body and clothing of victims. Photographs must be taken of the injuries to the victims as well as any materials removed from them during medical treatment. Hardhats, gloves, boots, and respirators, as well as clothing and materials removed from the victims, should be preserved for further examination.

The condition and position of damaged structural components—walls, ceilings, floors, roofs, foundations, support columns, doors, windows, sidewalks, driveways, and patios—should be noted. The condition and position of building contents such as furnishings, appliances, heating or cooking equipment, manufacturing equipment, clothing, and personal effects should also be noted.

The condition and position of the utility equipment, such as fuel gas meters, regulators, fuel gas piping and tanks, electrical boxes and meters, electrical conduits and conductors, heating oil tanks, parts of explosive devices, and fuel vessels should be examined.

Force Vectors

Document the debris that has been propelled away from the area of origin as well as the following parameters: the direction of travel; the distance of travel; the material propelled; and the material's size, weight, and configuration. Items in the flame front path, including victims' clothing and skin, may exhibit damages. They may also show directional patterns. Dust explosions may additionally leave burned and unburned particles on the affected items. This process assists the investigator in identifying the trajectories of the artifacts involved.

Analyze the Origin (Epicenter)

In explosion dynamics analysis, the general path of the explosion force vectors is followed from the least to the most damaged area. The process may require more than one explosion dynamics diagram to identify the debris movement: a large-scale analysis indicates the general area or room for further analysis as to the origin, and a small-scale diagram analyzes the explosion dynamics of the area of origin itself FIGURE 21-10.

It is necessary to plot the directions of debris movement and the relative force necessary for the movement of each major piece of debris. The explosion dynamics analysis may be complicated by secondary explosions. It will help to recall that secondary explosions (dust explosions) are often greater than the primary explosion and thus cause more damage. The analysis of the explosion dynamics is based on debris movement away from the epicenter of explosion in a roughly spherical pattern. The farther an object is from the epicenter, the less force the object is subjected to.

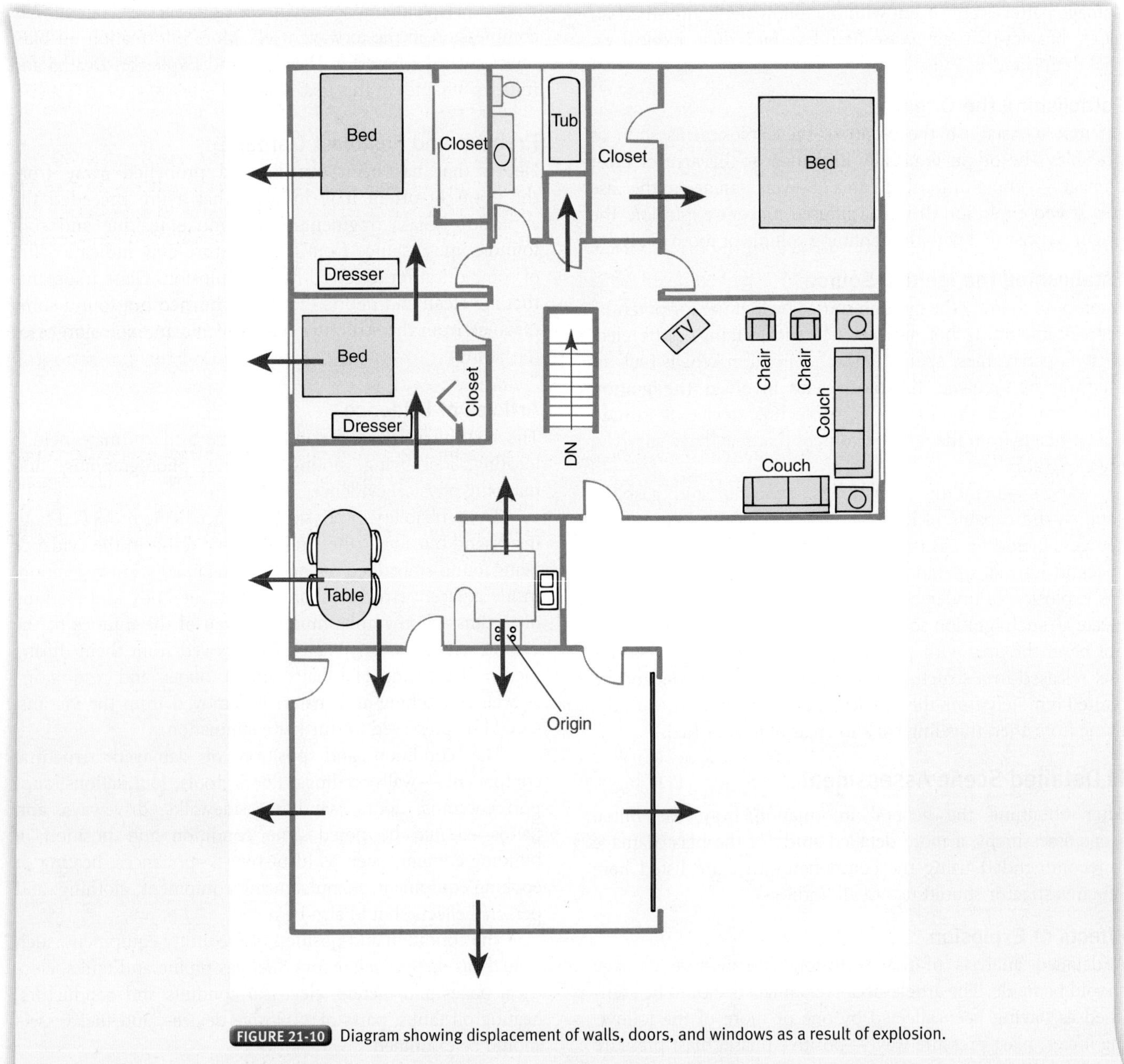

FIGURE 21-10 Diagram showing displacement of walls, doors, and windows as a result of explosion.

Analyze the Fuel Source

All available fuel sources should be considered and then limited to one fuel source that meets all the physical damage criteria. Clinical analysis of debris, soot, soil, and air may assist in identifying the fuel. Gas chromatography, spectrography, and other chemical tests of samples may identify the fuel. Once the fuel has been identified, the investigator can determine the source. All gas piping should be examined, and leak testing should be performed. Furthermore, as has been noted, odor verification should be part of any explosion investigation involving flammable gas. Although stain tubes may be used, the collection of a sample for submission to a lab is the most accurate means.

Analyze the Ignition Source

A careful evaluation of every possible ignition source should be made. The factors to consider include minimum ignition energy of the fuel, ignition energy of the potential ignition source, ignition temperature of the fuel, temperature of the ignition source, location of the ignition source in reference to the fuel, presence of both fuel and ignition source at the time of ignition, and witness accounts of conditions prior to and at the time of the explosion.

Analyze to Establish Cause

An analysis to determine the simultaneous presence of the fuel and the ignition source can include the following techniques: timeline analysis, damage pattern analysis, debris analysis, relative structural damage analysis, correlation of blast yield with damage, analysis of damage items in structure(s), and the correlation of thermal effects.

Timeline Analysis

After gathering information, create a timeline of events prior to and during the explosion using hard and soft times. Consistencies and inconsistencies with cause theories can be inferred to establish a "best fit" theory. (See NFPA 921 Section 22.2 for detailed information on this process.)

Fire Investigator Tip

When releasing the scene, be sure that you document it, including the proper authority taking custody of the scene, with time and date; that you have identified and disclosed all existing health and safety hazards; and that you have identified any other steps that may be needed, such as a structural report, utility company work needed, and cleanup/safety and salvage steps.

Damage Pattern and Debris Analysis

Damage pattern analysis is the documentation of damage patterns, primarily of debris and structural damage, for further analysis. Debris analysis includes the identification, diagramming (location found), photographing, and noting of debris pieces that indicate the direction and G-force of the explosion, which may allow reconstruction of components.

Correlation of Blast Yield with Damage Incurred

Correlation of damage and projectile distance with the type and amount of fuel used should be done to see whether they agree—that is, whether the yield of the hypothesized fuel could have done the existing damage.

Analysis of Damaged Items and Structures

The relation of the type of fuel used and the damage caused may require specialized experts to examine the damage to items and structures.

Correlation of Thermal Effects

Heat damage on a collection of articles from an explosion may be evidence of a fireball or fire during the event and may assist in identification of the type of explosion and/or the fuel. This also may require a specialist in the field to conduct the analysis of the articles.

Wrap-Up

© Greg Henry/ShutterStock, Inc.

Ready for Review

- In fire and explosion investigations, an explosion is a sudden conversion of potential energy, either chemical or mechanical, into kinetic energy (the energy possessed by a system or object as a result of its motion).
- The two major types of explosions are mechanical and chemical and are differentiated by the source and mechanism that produce them.
- The characteristics of explosion damage are low-order and high-order damage.
- The effects of an explosion include blast pressure front effect, shrapnel effect, thermal effect, and seismic effect.
- Factors controlling the explosion effects are type and shape of fuel; nature, size, volume, and shape of containment vehicle; venting of the containment vessel; relative maximum pressure achieved; and rate of pressure rise.
- Analyze the scene and evidence to establish the origin of the explosion. This is done through determining whether the explosion was seated or nonseated, the area of most damage, and the potential fuel. An explosion dynamics analysis is conducted to trace backward from the least to the most damaged area.
- Fuel gases or the vapors of ignitible liquids are the most commonly encountered explosion.
- Dust explosions may occur from the following materials: agricultural products, carbonaceous materials, chemicals, dyes and pigments, metals, plastics, and resin.
- It is common for fires in airtight rooms to become oxygen depleted. This results in the generation of flammable gases due to incomplete combustion and, if air is introduced, the potential for backdrafts and smoke explosions.
- The release of gas, vapor, or mist into the atmosphere may result in a cloud forming within the fuel's flammable limits and, subsequently, ignition, culminating in an outdoor vapor cloud explosion.
- Explosives are categorized into two main types: low explosives and high explosives, which should not be confused with low-order and high-order damage.
- The objectives of an explosion scene investigation are no different from those of a fire investigation: Determine the origin, identify the fuel and ignition source, determine the cause, and establish the responsibility for the incident.

Hot Terms

Blast overpressure Large quantities of gas being produced by the explosion of the material.

BLEVE Boiling liquid expanding vapor explosion.

Burning velocity The rate of flame propagation relative to the velocity of the unburned gas ahead of it.

Chemical explosion Explosion in which a chemical reaction is the source of the high-pressure fuel gas. The fundamental nature of the fuel is changed.

Combustion explosion Explosion caused by the burning of combustible hydrocarbon fuels and characterized by the presence of a fuel with air as an oxidizer.

Deflagration A reaction that propagates at a subsonic velocity through an unreacted medium, less than 1100 feet per second, and can be successfully vented.

Detonation A combustion reaction that propagates at supersonic velocities, greater than 1100 feet (340 m) per second, and cannot be vented because of its speed.

Explosion The sudden conversion of potential energy (chemical or mechanical) into kinetic energy with the production and release of gases under pressure. These high-pressure gases then do mechanical work such as moving, changing, or shattering nearby materials.

Explosion dynamics analysis The process of using force vectors to trace backward from the least to the most areas of damage following the general path of the explosion force vectors.

Explosives Any chemical compounds, mixtures, or devices that function by explosion.

Fireballs Momentary balls of flame during or after an explosion that may present high-intensity, short-term radiation.

Firebrands Hot or burning fragments propelled from an explosion.

Flame speed The local velocity of a freely propagating flame relative to a fixed point.

High-order damage A rapid pressure rise or high-force explosion characterized by a shattering effect on the confining structure or container and long missile distances.

Kinetic energy The energy possessed by a system or object as a result of its motion.

Low-order damage A slow rate of pressure rise or low-force explosion characterized by a pushing or dislodging effect on the confining structure or container and by short missile distances.

Mechanical explosion Rupture of a vessel or container such as a cylinder, tank, or boiler, resulting in the release of pressurized gas or vapor. The pressure leading to the mechanical explosion is not due to a chemical reaction or change in chemical composition of the substances involved.

Nonseated explosion When the fuels in the explosion are dispersed or diffused with moderate rates of pressure rise and subsonic explosive velocities.

Seat of the explosion A craterlike indentation created at the point of origin of an explosion.

Seismic effect The transmission of tremors through the ground as a result of the blast wave expansion causing structures to be knocked down.

Stoichiometric The optimum ratio at which point the combustion will be most efficient.

FIRE INVESTIGATOR *in action*

An explosion has occurred in a suburban neighborhood, causing extensive damage to numerous structures. When you arrive at the scene, the structure of origin is completely destroyed with structural components and contents shattered and dispersed over a large area. Neighboring structures closest to the origin structure have displaced exterior walls and broken windows. After examining the origin structure, you determine the seat of the explosion was located in the basement. The debris from the basement is removed, and the remains of a propane-fired furnace are present. The furnace is largely intact, and the attached piping from the fuel tank is still attached. Closer inspection of the furnace piping reveals the moisture trap had been removed prior to the explosion.

1. The description of damage to the structure of origin is consistent with a(n):
   A. mechanical explosion.
   B. low-order explosion.
   C. electrical explosion.
   D. high-order explosion.
2. Which of the following is not a typical search pattern?
   A. Circular
   B. Grid
   C. Spiral
   D. Sequential
3. The damage to the neighboring structures is likely the result of which explosion effect?
   A. Blast overpressure and wave effect
   B. Projected fragments effect (shrapnel)
   C. Thermal effect
   D. Seismic effect
4. What does the acronym BLEVE stand for?
   A. Boiling Liquid Expanding Violent Explosion
   B. Boiling Liquid Expanding Vapor Explosion
   C. Boiling Liquid Exploding Vapor Event
   D. Boiling Liquid Exploded Vessel Event

Incendiary Fires

© Photos.com

© Guy Corbishley/Alamy

Knowledge Objectives

After studying this chapter, you should be able to:

- Identify and describe incendiary fire indicators. (pp 322–325)
- Describe the assessment of fire growth and damage in incendiary fires. (p 326)
- Identify and describe potential incendiary fire indicators not directly related to combustion. (pp 326–328)
- Record and examine other evidentiary factors NFPA 4.6.4. (pp 328–331)

Skills Objectives

After studying this chapter, you should be able to:

- Examine fuel loads and their composition and location to understand fire growth. (pp 322–326)
- Recognize the motive(s) for a fire based on the recognized list of motives for fire-setters NFPA 4.6.4. (pp 329–331)
- Examine fire protection systems to determine whether they have been altered or sabotaged by an arsonist. (pp 327–328)

Additional NFPA Reference

NFPA 921, *Guide for Fire and Explosion Investigations*

CHAPTER 22

FESHE Course Outcomes

Fire Investigation I

11. Identify cause and origin and differentiate between accidental and incendiary. (pp 322–329)
13. Identify the characteristics of an incendiary fire and common motives of the fire-setter. (pp 322–331)

Fire Investigation II

8. Evaluate the use of incendiary devices, explosives, and bombs. (pp 324–325)

You Are the Fire Investigator

© Jones and Bartlett Publishers. Photographed by Glen E. Ellman

Over the weekend, you and your partner investigate several fires that occurred in vacant structures. Several of these structures are located in the same neighborhoods, with the fires being reported within minutes of each other. The damage to each structure is relatively the same, and all of the fires appear to have been started by setting fires to abandoned pieces of furniture or debris located within each structure. After comparing notes with your partner, you realize each one of the structures had been placed on a demolition list and several are owned by the same property management group. This group owns additional structures in your jurisdiction and had recently been fined for violations of the property maintenance code.

1. What is the significance of these structures being on the demolition list?
2. Why might the property management group be involved with starting any of the fires?
3. Which factors allowed for this fire to grow so large in the middle of the afternoon in a densely populated, high-traffic community?
4. What are the other investigative activities you will need to complete to determine who is responsible for these fires?

Introduction

An incendiary fire is a fire that has been deliberately set with the intent to cause the fire to occur in an area where the fire should not be. This chapter covers the indicators that point to incendiary fires, including indicators not directly related to combustion, and other evidentiary factors such as suspect development and identification.

Incendiary Fire Indicators

The indicators that are discussed in this section should be studied for their possible support of the hypothesis that the fire is incendiary.

Multiple Fires

Multiple fires are fires with no obvious connection that would have allowed one fire to ignite the fuel in (or spread to) another area. A fire in the basement that spread through the walls of a balloon-frame house to the attic would not be classified as an incident with multiple fires, because the fire spread naturally through a vertical opening and ignited combustible fuels in a location that was remote from the initial fire. The investigator must determine that a separate fire was not the natural result of the growth and spread of the initial fire.

Other "natural" means of fire spread that could cause multiple fires include the following:

- Conduction, convection, or radiation
- Flying brands
- Direct flame impingement
- Falling flaming materials (drop down), such as curtains
- Fire spread through shafts, such as pipe chases or air conditioning ducts
- Fire spread within wall or floor cavities within balloon construction
- Overloaded electrical wiring
- Utility system failures
- Fuel gas or dust explosions
- Lightning
- Rupture and launching of aerosol containers

The investigator should not confuse the site of a previous fire with that of a more recent fire. This could cause the investigator to develop the hypothesis that there were multiple simultaneous fires. The fire history of the structure should be obtained and examined. Apparent multiple points can result from sustained burning or smoldering during or after fire suppression or overhaul. Full-room involvement or postflashover conditions can make identifying multiple fires more difficult or impossible. It is important to conduct a full scene examination to determine whether there were separate fires and to establish areas of lesser or greater damage and fire spread patterns. If this full scene examination is not conducted, valuable evidence may not be located and the investigator may not be able to determine accurately whether separate fires occurred. Any investigation should include, where permitted, examination of areas in a subject structure that do not appear to have been damaged by fire.

Trailers

Arsonists attempt to spread fire to other areas by deliberately linking them together with combustible fuels or ignitible liquids called trailers. One fire will, in turn, ignite other areas via these trailers FIGURE 22-1.

Trailers can leave distinctive patterns on horizontal surfaces such as floors. When the floor is cleared of debris, the pattern may be easily discernible; however, it is important to determine that these patterns are not the result of other mechanisms or materials such as open areas bordered by protected areas or the effects of flashover and full-room involvement. At full-room involvement, radiant heat can be expected to create burn patterns on floors that can be misinterpreted as trailer burn patterns (see the "Basic Fire Science" and "Fire Patterns" chapters of this text for additional information).

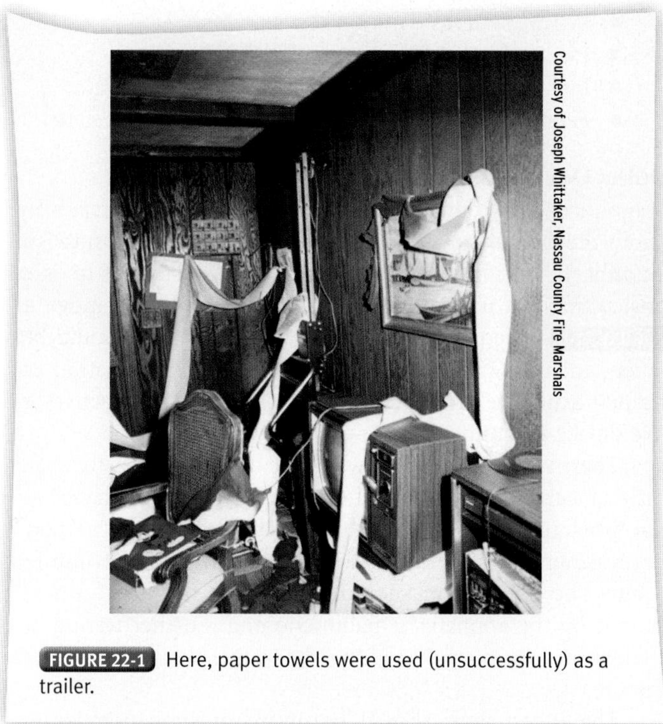

FIGURE 22-1 Here, paper towels were used (unsuccessfully) as a trailer.

Materials that can be used as trailers include the following:

- Ignitible liquids
- Clothing
- Paper
- Straw

Many common household goods (cleaning fluids, gasoline, etc.) are ignitible liquids, and their presence on the fire scene is not necessarily indicative of use as a trailer. It is not the fuel that constitutes a trailer but the manner and location in which the fuel was used.

■ Lack of Expected Fuel Load

When the observable fire damage is not consistent with the observable fuel load, further investigation is warranted. The investigator should attempt to quantify the fire damage that would be caused by the known or reported fuel load in relation to the observable fuel load. However, the absence of expected fuel loads is not enough to classify the fire cause as incendiary.

Examples of areas and spaces that routinely have low or limited fuel loads include corridors, stairways, hallways, and vacant homes. If the origin for a fire is in a low- or limited-fuel area, the investigator should look for physical evidence of fuels such as an ignitible liquid and take samples; however, burning in these areas may not be unusual if the burning represents fire extension or movement from another area, particularly if the adjacent space has developed past flashover.

■ Lack of Expected Ignition Sources

Another indicator that deserves further investigation is when a fire origin does not have a readily apparent competent ignition source. The investigator may have to look closely through the debris in the search for ignition sources that have burned, melted, or been consumed. Closets, crawl spaces, and attics are typical areas, rooms, and spaces in which a limited number of heat sources are present. The fire investigator should be careful to examine the fire-burned area in its totality and not focus on one particular area where the fire damage appears to be greatest. Areas of heavy burn can be influenced by high heat release–rate products, delayed extinguishment, and ventilation effects in that particular area and not be the specific area of origin.

■ Exotic Accelerants

Mixtures of fuels with Class 3 or Class 4 oxidizers and thermite mixtures may be considered exotic accelerants. Some oxidizers are capable of self-ignition. These types of accelerants can cause exceedingly hot fires and generally leave residues that may be visually or chemically identifiable. Indicators of high-temperature accelerants (HTAs) include the following:

- Rapid rate of growth
- Brilliant flares
- Melted steel or concrete

Other reasons for dramatic fire effects and patterns should always be considered. These might include ventilation effects, delays in fire suppression, effects of particular fire suppression tactics, or type and configuration of fuels.

■ Unusual Fuel Load or Configuration

A fire-setter may hope to create a fire that will burn more aggressively or effectively by moving contents or materials into a configuration to allow for more rapid fire growth or fire spread than would be expected if the contents were spaced farther apart. This could also be done in an attempt to provide more complete burning of the fuels. Witnesses may be able to provide information as to the position or location of contents prior to the fire.

The types of fuels can be evaluated to determine whether they are the types expected in a given occupancy, as fuels may be added to an area to assist in fire growth or spread.

An unusual fuel load or configuration should not be assumed to be related to the fire cause. The investigator should seek an explanation from the homeowner or occupant and/or first-arriving engine company for the configuration if it is truly unusual.

■ Burn Injuries

For all classifications of fires, all known burn injuries to persons should be analyzed to determine their relationship to fire ignition and the investigative hypothesis. The investigator should also attempt to determine the type of burn injury (e.g., one resulting from a hot object or open flame) FIGURE 22-2. Because burn injuries may be sustained while setting an incendiary fire, local hospitals should be contacted for identification of recent burn victims. Some jurisdictions require the reporting of burn injuries. A detailed interview of the burn victim may be helpful to determine the origin, cause, or spread of the fire, as well as the victim's activities prior to and during the fire. This information can then be compared to physical findings at the scene to determine

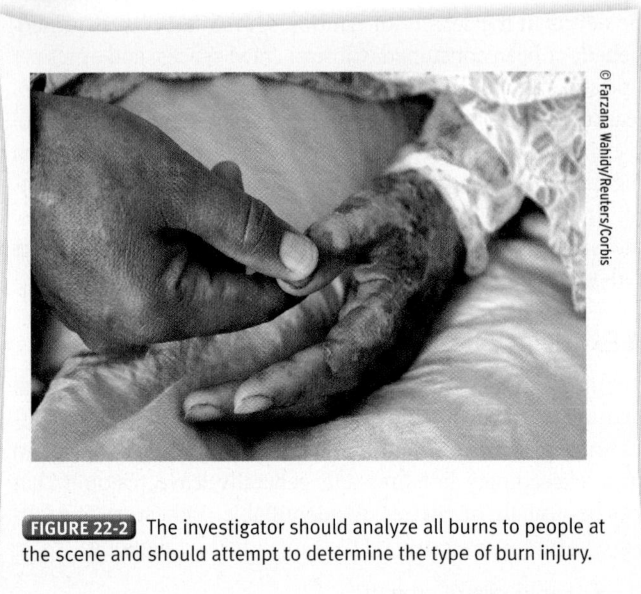
© Farzana Wahidy/Reuters/Corbis

FIGURE 22-2 The investigator should analyze all burns to people at the scene and should attempt to determine the type of burn injury.

whether there is consistency between the two. All burn injuries should be documented, and samples of the victim's clothing may be taken in appropriate cases for laboratory testing for ignitible liquids.

Incendiary Devices

Incendiary devices include a wide range of mechanisms used to initiate an incendiary fire. If incendiary devices are used, remains of the device can often be found.

Almost any appliance or heat-producing device can be used as an incendiary device. If there are no other obvious ignition sources, then efforts should be made to determine whether a device was used. Some examples include the following:

- Combination of cigarette and matchbook
- Candles
- Wiring systems
- Electric heating appliances FIGURE 22-3
- Fire bombs/Molotov cocktails FIGURE 22-4
- Paraffin wax–sawdust incendiary device (fireplace starters)

Delay Devices

Some incendiary devices are constructed as delay devices to allow the fire-setter to leave the area safely or to establish an alibi. If, during the investigation, the investigator finds a device that has not activated, he or she should not move it FIGURE 22-5. Adequate precautions and safeguards should be taken, including evacuation of the area and notification of trained explosive ordnance disposal personnel if an active or live device is found.

There may be occasions when the fire-setter tries to mask the true cause of the fire by attempting to use an appliance as the "obvious" cause of the fire. For example, a fire-setter may pour an ignitible liquid into a coffeemaker, causing a fire to occur. The investigator should not assume that the fire was caused by the appliance's malfunctioning. Further testing or evaluation may be warranted to determine the true cause of the fire.

The investigator should be aware of spoliation issues when conducting this type of investigation, and if the investigator does not have the specific engineering training, he or she should seek qualified assistance. As an example, a pot-on-the-stove fire may cause heavy fire damage to some or all of the control mechanisms on the panel at the upper rear portion of the cook-top surface. Because the components largely comprise a phenolic plastic material, much of it may be heavily damaged, and removing the back panel of the control panel could cause the location of the contacts to collapse and fall apart. Caution should be taken to protect these mechanisms until all interested parties have been put on notice and are present before any destructive testing is done. An alternative to destructive testing is the use of an X-ray machine to document the contact positions of the controllers without taking the panel apart.

© Jones & Bartlett Learning. Photographed by Glen E. Ellman

FIGURE 22-3 This heater was used as an incendiary device.

© Jones & Bartlett Learning. Photographed by Jessica Elias

FIGURE 22-4 Fire bombs are also referred to as Molotov cocktails.

© Jones & Bartlett Learning. Photographed by Glen E. Ellman

FIGURE 22-5 An example of a time-delayed incendiary device.

Because arsonists may set multiple fires, the investigator should inspect the building to determine whether other fires occurred that are noncommunicating. The arsonist may have used similar devices in these other fires that can provide valuable clues. Examine these areas for debris that may contain delay or other incendiary devices or methods that would contribute to the ignition sequence. In cases of multiple fires, often at least one incendiary device has failed to operate, leaving the investigator with valuable evidence. Look for trailers from one burn area to the other(s).

Presence of Ignitible Liquids in Area of Origin

Commonly referred to as an accelerant or liquid accelerant, the presence of ignitible liquids may indicate that the fire was incendiary. Care should be taken to note the location where the ignitible liquid was found. Obviously, the presence of a perceived ignitible liquid found on the floor area of a garage may not be unusual considering that vehicles, lawn equipment, and fueling cans may be stored there. The investigator will have to determine the reason for the presence. Conversely, an ignitible liquid found on the floor area in a random linear pattern in

© Jones & Bartlett Learning. Photographed by Glen E. Ellman

FIGURE 22-6 An accelerant-sniffing canine may be a useful addition to a fire investigation.

a kitchen may be something to consider and should be further examined. It is the investigator's responsibility to determine whether the presence is an intentional placement to start, accelerate, and/or spread the fire. Samples of the debris should always be taken for laboratory analysis. Lab confirmation as to the presence of an ignitible liquid will help substantiate the investigator's case for an intentional, illegal act.

Another avenue to pursue as a means to assist in the identification of a possible ignitible liquid is the use of an accelerant-sniffing canine (AK-9) FIGURE 22-6. These dogs are specially trained in the detection of various ignitible liquids. It should be noted that an AK-9 is only as good as the training it receives from its handler. The investigator should be familiar with the AK-9 handler and the training the dog has had and continues to receive. The AK-9 should only be used as an additional tool to assist in the building of an investigator's case for a suspected criminal act. Because of the nature of the dog alert, and the need to confirm it with laboratory analysis, the AK-9 should never be used as a primary means of identifying the potentially laced areas. Rather, these areas should be identified by investigators based on the data and information gathered during the investigation, as well as their training and experience.

The investigator should perform his or her normal investigation and take samples based on the burn patterns and other identifying characteristics that lead the investigator to believe this to be a suspected area of ignitible liquid use. The AK-9 can then be sent in to examine the room(s) and identify (or not) the areas where there may have been an ignitible liquid. Samples should be lined up on the exterior of the structure and interspersed with comparison sample cans. The dog can then examine the samples in the can and see if it alerts. The dog can be another tool to help the investigator corroborate evidence samples prior to laboratory testing. All substances collected off an AK-9 alert must be validated through laboratory analysis.

Assessment of Fire Growth and Damage

If a fire spreads more quickly than can be explained by the expected fuel load or beyond the area where it would normally be expected to be confined, then the investigator should look more closely at what contributing factors could have caused this to happen. Fire growth is related to a large number of variables, including the volume of the compartment, the height of the ceiling, the heat release rates of fuels, the location of the initial fuel package within the compartment, and ventilation. In the absence of physical evidence, the investigator is cautioned against using subjective terms such as *excessive*, *abnormal*, *unusual*, or *suspicious* to support an incendiary fire cause determination. The proper use of a fire model can provide assistance to the investigator in determining the accuracy of their observations.

Potential Indicators Not Directly Related to Combustion

Certain indicators can assist the investigator in developing an ignition hypothesis, questions for witnesses, or avenues for further investigation. These indicators typically are not related to the determination of the fire cause but tend to show that somebody had prior knowledge of the fire.

Remote Locations with Blocked or Obstructed View

A fire-setter may start a fire in a remote location or one obscured from public view; however, accidental fires can also start in remote locations. Therefore, no cause determination conclusions should be made based on the location of origin. More important to the investigator are obscurations that occur just prior to the fire, such as paper-covered/painted windows or furniture strategically placed in front of exterior openings. These can provide the investigator with information for establishing a timeline, as well as for questioning witnesses.

Fires Near Service Equipment and Appliances

To make a fire appear to be accidental, a fire-setter may set a fire near appliances, hoping that the appliance will appear to be the ignition source for the fire. Each appliance should be carefully evaluated to determine whether it was the ignition source. The investigator should remember issues related to spoliation before any destructive evaluation.

Removal, Replacement, or Absence of Contents Prior to the Fire

Contents are sometimes removed or replaced with items of lesser value prior to a fire. Careful documentation of the remains and the debris may be useful in establishing a fraudulent insurance claim, even in situations in which the fire cause cannot be determined.

The determination that contents have been removed or replaced requires verification that the items were present some time prior to the fire. Assessment concerning the absence, removal, or replacement of any item requires verification from corroborated witness statements, inventory, or sales receipts. Consider documenting all of the occupancy's contents, not just those near the origin.

Personal items or irreplaceable items may be removed prior to a fire. The items that were removed may provide the investigator with avenues for questioning and the identification of suspects. Examples include jewelry, photographs, pets, tax records, business records, and firearms.

Items and contents that may be replaced depend on the occupancy of the building. Some examples are as follows:

- Residential: furniture, clothing, appliances, jewelry, guns
- Industrial/commercial: machinery, equipment, stock, merchandise
- Vehicles: tires, batteries

If the contents of the building are abnormal to the occupancy, this could be an indication that further investigation is necessary.

Blocked or Obstructed Entry

To allow the fire more time to grow, the fire-setter may place obstructions to hinder or slow the firefighting operations. Any unusual obstructions that deny fire vehicle access should be noted and evaluated. Obstructions may include fallen trees, barricades, or larger obstructions that have the potential to block the access of emergency vehicles. To prevent access to the structure, doors and windows may be obstructed as well.

Fire Investigator Tip

Always strive to perform your investigations in a systematic manner. A normal routine should be established for each fire you investigate so that, eventually, you will automatically perform all of your investigations in that same manner consistent with NFPA 921 Sections 4.2 and 4.3. Consistency is key.

Fire Investigator Tip

Be sure to document each step of your investigation photographically. You can never take too many photos. The one photo you will need will invariably be the one that you did not take.

Sabotage

The term sabotage refers to intentional damage or destruction. Fire-setters often develop conditions that lead to rapid and complete destruction of a building and its contents. To accomplish this goal, the fire-setter may sabotage the fire protection systems so notification to occupants and the fire department is delayed FIGURE 22-7.

Damage to Fire-Resistive Assemblies

Fire-resistive assemblies are implemented to separate a structure into compartments and thereby confine fire and smoke from spreading to other parts of the structure. The most common method of fire travel through a structure is open doors, with open stairwell doors having an even greater impact on fire spread. Fire-setters may penetrate fire-resistive assemblies (walls, floors, or ceilings) or prop open doors to facilitate fire and smoke movement; however, penetrations in these assemblies may also be the result of firefighting activities or poor construction, and doors may be propped open by occupants to improve access or ventilation. It is important to conduct a thorough investigation and not to assume sabotage.

© DWImages/Alamy

FIGURE 22-7 Fire-setters may sabotage fire protection systems to delay notification of a fire.

Damage to Fire Protection Systems

Common fire protection systems are listed in TABLE 22-1 (see also the "Fire Protection Systems" chapter).

If the fire suppression or detection systems are operating normally, they should detect or suppress the fire before it has a chance to cause any significant damage. In addition, they could serve to alert the fire department. If the fire suppression system or fire detection system failed, the investigator should inspect the systems for any signs that may indicate the cause of the failure, such as the following:

- Improper installation
- Tampering
- Lack of maintenance
- System shutdown
- Equipment or structural assembly failure

It is also important to determine whether these conditions existed prior to the fire. Inspection and maintenance records could provide valuable clues regarding the prefire status of the systems.

Some methods of disabling these systems include the following:

- Removing or covering smoke detectors
- Obstructing sprinklers
- Shutting off control valves
- Damaging threads on standpipes, hose connections, and fire hydrants
- Placing debris in Siamese connections and fire hydrants or disconnecting alarm bells or sirens
- Starting multiple fires to overload the suppression system

The investigator should always examine the fire protection systems to determine whether any tampering or disabling of the system occurred. Investigators should always check with alarm-monitoring facilities to determine the times that trouble or alarm signals are received. This information may assist in the investigation by further defining a timeline for the fire.

Table 22-1 Common Fire Protection Systems

System Type	Examples
Detection systems	Heat Smoke Flame Security Video
Suppression systems	Sprinkler system Standpipe system Fire department connection
Special extinguishing systems	Carbon dioxide Foam Halon
Water mains and hydrants	

Safety Tip

Some fire-setter actions can add safety concerns for fire fighters and investigators. For example, a fire-setter may intentionally weaken supportive structures in the hope of delaying firefighting efforts and obscuring evidence. Investigators must be attuned to the potential for sabotage to a building and take the necessary safety precautions.

Damage to Buildings

A fire-setter may intentionally damage the building for several reasons, including (1) to hinder the ability of the fire fighters to fight the fire effectively and (2) to provide avenues for the fire to spread beyond its area(s) of origin.

Examples of the first could include the following:

- Cutting openings in the floors that fire fighters could fall through or breaching other fire-rated assemblies such as walls or doors to permit the spread of fire
- Jamming or barricading doors and windows to make it difficult for fire fighters to enter
- Sabotaging fire-rated doors, fire dampers, and the like so that they are in the open position during a fire and will not close automatically

The investigator should exercise due care during the fire scene examination to determine that the failure of any particular fire protection system or components was from an intentional act and not through lack of maintenance of compartmentation areas (fire/smoke doors, etc.) that are commonly propped open or disabled by employees as a normal routine. Other reasons for failure should also be examined, such as manufacturer or installation defect.

Opening Windows and Doors

A fire needs oxygen to burn. To provide ventilation, the fire-setter may open exterior windows and doors that would not normally be open (e.g., during cold weather). The objective of the fire-setter could be to have the fire spread outside of the compartment of the area of origin. To accomplish this effectively, artificial avenues of fire spread may have to be created. It is important for the investigator to determine whether the occupants normally propped the doors open or whether this may have been done to spread the fire. Wind direction may also be a factor that the fire-setter attempted to take advantage of in spreading the fire.

Fire Investigator Tip

You should always examine all available court, real estate, business, maintenance, code enforcement, and any other records and documents for the vehicle, property, structure, or person you are investigating. This may help establish motive for the fire.

Other Evidentiary Factors

Other evidentiary factors are those indicators analyzed after the fire has been classified as incendiary to develop suspect profiles. It is through this analysis that trends or patterns can be detected. The key to this analysis is whether the fire setting is repetitive. Serial fire-setter and serial arsonist are the terms used to identify individuals or groups who are involved in three or more fire-sets. There are three principal trends that may be identified through this analysis:

1. Geographical or cluster: Fire-setting tendencies are within the same geographical location or neighborhood.
2. Temporal frequency: A serial arsonist can choose the same time period or day of the week.
3. Materials and methods: Repetitive fire-setting behavior not only remains in the same geographical location but also uses similar fire-setting materials and methods.

Fire-setters may start an incendiary fire to conceal another crime, such as a homicide or burglary. The issue of which crime occurred first has more to do with motive than with the cause of the fire. Determining motive, however, can aid investigators in their approach to the investigation. The investigator has a number of different mechanisms at his or her disposal to aid in his or her investigation. The examination of bank records, insurance policies, code enforcement or fire inspection complaints or sanctions, National Insurance Crime Bureau (NICB) records, tax records, and so forth can uncover potential motives for the fire. Some indicators revealed by an investigator include the following:

- Financial stress
- History of code violations
- Fires at additional properties owned by a single individual or group
- Overinsured property

Liens, attachments, unpaid taxes, mortgage payments in arrears, real estate for sale, poor business location or competition, economic decline, outdated or overstocked product, loss of jobs, and overinsurance or recent changes in policies to inflate the values all establish potential motives for the fire. The investigator should also examine past insurance histories to determine whether the claimant may have had previous claims for fires in other locations or vehicles and so forth.

Timed Opportunity

Fire-setters sometimes take advantage of conditions or circumstances that add to the chances of successful destruction of the property. A timed opportunity can also increase their chances of not being apprehended. Examples of a few timed opportunities are as follows:

- Natural conditions
  - Floods
  - Snowstorms
  - Hurricanes
  - Earthquakes
  - Electrical storms
  - High winds
  - Low humidity
  - Extreme temperatures
- Civil unrest
- Fire department unavailable
  - Calling in a false alarm
  - Parades
  - At the scene of another working fire

Motives for Fire-Setting Behavior

Motive is defined as an inner drive or impulse that is the cause, reason, or incentive that induces or prompts a specific behavior. Motive indicators in the fire investigation process should be used only to help identify potential suspects. These indicators should not be used to determine or classify the fire cause. Crime analysis is a method of identifying personality traits and characteristics exhibited by an unknown offender. It is the identification and analysis of the personality traits that eventually lead to the classification of a motive. It is possible that more than one motive may have led to the reason for the fire.

Victimology is one of the most important aspects of classifying an offense and determining the motive. It is an understanding of the offender activity with the victim (or targeted property). The investigator should evaluate why this particular victim was targeted for the crime. Very often this will lead to the motive for the crime, which will lead to the offender. Three manifestations of offender behavior at a crime scene are as follows:

- Modus operandi (MO): The method of operation used by the offender. This predominantly applies to serial arsonists. The offender's actions during the perpetration of a crime form the MO. The offender develops and uses the MO over time because it works, but it also continuously evolves. MOs are a learned behavior and may be modified over time as the offender becomes more sophisticated and confident.
- Personation (the signature): The unusual behavior by an offender, beyond that necessary to commit the crime. Most violent crimes such as arson begin within the offender's imagination. For example, the serial arsonist daydreams about building bombs, setting fires, and so forth. When the offender translates these daydreams into action, his or her needs drive him or her to exhibit unusual behavior during the crime. This unusual behavior leaves behind commonalities that can identify that person as the offender.
- Staging (alteration of the crime scene): When someone purposely alters the crime scene prior to the arrival of police. There are generally two reasons associated with staging:
  - To redirect the investigation away from the most logical suspect.
  - To protect the victim or the victim's family. As an example, the offender of a sexual homicide frequently leaves the body in a degrading position. Modesty staging by the family can certainly be understood, but the investigator needs to obtain an accurate description of exactly how the body was situated when found and exactly what that person did to alter the crime scene.

Motive Versus Intent

Intent is generally necessary to show proof of a crime and refers to the state of mind that exists at the time a person acts or fails to act. Motive is the reason that an individual or group may do something and is not generally a required element of a crime.

The classifications discussed in this chapter are based on Douglas et al., *Crime Classification Manual* (*CCM*). This manual helps investigators gain as much information as possible by identifying essential elements of analytical factors used to classify the motive of a murder, arson, or sexual offense.

The behaviors listed in the *CCM* may identify a possible motive, leading investigators to possible suspects. These behaviors apply whether the fire is the result of a one-time occurrence or the action of a serial fire-setter. There are three classifications of repetitive fire-setting behavior:

1. Serial arson involves an offender who sets three or more fires with a cooling-off period between fires.
2. Spree arson involves an arsonist who sets three or more fires at separate locations with no emotional cooling-off period between fires.
3. Mass arson involves an arsonist who sets three or more fires at the same site or location during a limited period of time.

The National Center for the Analysis of Violent Crime (NCAVC) has identified the following six motive classifications as the most effective in identifying offender characteristics for fire-setting behavior:

1. Vandalism
2. Excitement
3. Revenge
4. Crime concealment
5. Profit
6. Extremism

Vandalism Vandalism is defined as mischievous or malicious fire-setting that results in damage to property. Common targets include educational facilities and abandoned structures but also include trash fires and grass fires. Vandalism falls into two categories:

- Willful and malicious mischief: Incendiary fires are those that have no apparent motive or those that are seemingly set at random. These fires are often attributed to juveniles or adolescents.
- Peer or group pressure: Recognition or pressure from peers is another category of vandalism and is predominant among juveniles. Juveniles often are swayed more by those they are with than by thinking about the consequences of their actions.

Excitement People may also set fires because they seek excitement. The perpetrator generally does not intend to hurt anyone in his or her fires; however, unplanned injuries and deaths can occur. There are four subclassifications to this motive.

- Thrill seeking: Offenders are filled with feelings of excitement and power and are often repetitive fire-setters. A psychological need or desire drives him or her to set these fires.
- Attention seeking: Fire-setters in this classification have a need to feel important.
- Recognition: Offenders in this subclass are sometimes referred to as vanity or hero fire-setters. Fire fighters, particularly volunteer fire fighters, and security guards may be likely candidates under this classification. The need for recognition, praise, and/or reward can be a driving motive for their behavior. If they are successful, they will most likely repeat the offense multiple times. These offenders often are still on scene and provide intimate details as to the location of the fire and how to get to its location. They are usually very forthcoming with information and helpful to fire fighters and police.
- Sexual gratification or perversion: Although considered rare, there are documented cases of offenders in this category who set fires as a means for sexual release. They are often still on scene in a remote location that hides them from view.

Revenge An important factor for the offender who sets fires as a means of revenge is his or her perception of a real or imagined injustice done to him or her or to someone he or she cares about. These fires are sometimes premeditated and well planned. These are usually one-time events, but in the case of serial arsonists will occur multiple times.

- Personal retaliation: Something commonly occurs that triggers a retaliatory response. The event may have been a fight, an argument, or a feeling of being taken advantage of, against himself or herself or against others close to the offender. Common targets are the victim's home, personal possessions, or vehicle.
- Societal retaliation: A serial offender may fall into this subclassification. The perpetrator may have feelings of loneliness, rejection, persecution, abuse, or inadequacy that drive him or her to set fires. As these feelings recur, the offender may continue to set fires as a release and for a feeling of gratification. Look for a similar MO in these fires.
- Institutional retaliation: This usually involves offenders who have a grudge against the institution. The offender may be a former employee, customer, patient, or student. Common targets are religious, medical, governmental, and educational institutions or corporations.
- Group retaliation: Targets may involve fraternal, religious, racial, or other groups, including gangs. Evidence of graffiti, symbols or markings, and other vandalism may be found at these scenes.

Crime Concealment In crime concealment, arson is generally a secondary or collateral criminal activity. The purpose is to conceal the initial criminal activity that occurred.

- Murder concealment: When a murder has been committed, the fire is usually set to disguise the fact that there was a death prior to the fire or to destroy forensic evidence that could identify the victim. The fire may also be set with the intent to destroy evidence that could link the offender to the crime.
- Burglary concealment: These fires are usually set to hide the fact that a burglary occurred or to destroy any evidence that could help identify the burglar/fire-setter.
- Destruction of records or documents: The general target in this category is records or documents associated with the business, institution, or corporation. Fires may be started in the file cabinets, which often are found with the drawers left open, or folders of the files or documents. Investigators should focus on employees or owners of the location as potential suspects.

Profit Fires set for profit involve those set for material or monetary gain. The monetary issue may be either directly or indirectly linked. Profit-motivated arson is a commercial crime that involves the least amount of passion of any other motives for arson. Direct gain may come from insurance fraud, elimination or intimidation of business competition, extortion, removal of unwanted structures to increase property values, or escaping financial obligations. Further subclassifications are fraud to collect insurance, to liquidate property, to dissolve a business, to conceal a loss, or to liquidate inventory. Other categories include employment, parcel/property clearance, and competition.

Extremism Extremism as a motive comes into play to further a political, social, or religious cause. These fires may be set by individuals or groups. The fire-setters generally have a great degree of organization, as reflected in their use of more elaborate ignition or incendiary devices. Terrorism and riot/civil disturbance fall into the extremist category.

- Terrorism: The targets chosen by terrorists are rarely random structures. They are most often chosen based on a specific significance that will provide the greatest impact for the message that they want to deliver. Common objectives may be of political or economic significance. Political targets include government offices, newspapers, universities, political party headquarters, animal research facilities, abortion clinics, and military or law enforcement installations. Economic targets may include business offices, distribution facilities, banks, savings and loan facilities, or companies thought to have an adverse impact on the economy. Fire, explosives, and other weapons may be used in these assaults.

- Riot/civil disturbance: Riots and civil disturbances have occurred for a number of reasons. Displeasure with a political election, public outcry for a court finding or police action, and celebration after winning major sporting events are examples. Fires that occur during these events are usually intentional and are generally accompanied by looting and vandalism. Investigators need to ascertain whether the fire was the act of a rioting crowd or building owners who want to benefit in the form of a financial advantage and the fire attributed to the ongoing riots.

Wrap-Up

Ready for Review

- Indicators of incendiary fire include multiple fires, trailers, lack of expected fuel load or ignition sources, exotic accelerants, unusual fuel load or configuration, burn injuries, and incendiary devices.
- If a fire spread more quickly than can be explained by the expected fuel load or beyond the area where it would normally be expected to be confined, then the investigator should look more closely at what contributing factors could have caused this to happen.
- Certain indicators not related to determining fire cause can assist the investigator in developing an ignition hypothesis, questions for witnesses, or avenues for further investigation:
  - Remote locations with obstructed views
  - Fires near service equipment and appliances
  - Removal or replacement of contents prior to the fire
  - Obstructed entry
  - Sabotage
  - Open windows and doors
- Other evidentiary factors are those indicators analyzed after the fire has been classified as incendiary to develop suspect profiles. It is through this analysis that trends or patterns can be detected.
- Fire-setters sometimes take advantage of conditions or circumstances that add to the chances of successful destruction of the property. A timed opportunity can also increase their chances of not being apprehended.
- The NCAVC has identified six motive classifications as the most effective in identifying offender characteristics for fire-setting behavior:
  - Vandalism
  - Excitement
  - Revenge
  - Crime concealment
  - Profit
  - Extremism

Hot Terms

High-temperature accelerants (HTA) Mixtures of fuels with Class 3 or Class 4 oxidizers and thermite mixtures.

Incendiary devices A wide range of mechanisms used to initiate an incendiary fire.

Incendiary fire A fire that is deliberately set with the intent to cause the fire to occur in an area where it should not be.

Intent Necessary to show proof of a crime and refers to the state of mind that exists at the time a person acts or fails to act.

Mass arson The setting of three or more fires at the same site or location during a limited period of time.

Modus operandi (MO) The method of operation used by the offender.

Motive An inner drive or impulse that is the cause, reason, or incentive that induces or prompts a specific behavior.

Personation The unusual behavior by an offender, beyond that necessary to commit the crime.

Sabotage Intentional damage or destruction.

Serial arson Involving an offender who sets three or more fires with a cooling-off period between fires.

Serial arsonist See *serial fire-setter.*

Serial fire-setter Individual or group involved in three or more fire-sets.

Spree arson The setting of three or more fires at separate locations with no emotional cooling-off period between fires.

Staging Purposeful alteration of the crime scene prior to the arrival of police.

Trailers The method to spread fire to other areas by deliberately linking them together with combustible fuels or ignitible liquids.

Vandalism Mischievous or malicious fire-setting that results in damage to property.

Victimology A thorough understanding of the offender activity with the victim (or targeted property).

© Greg Henry/ShutterStock, Inc.

FIRE INVESTIGATOR *in action*

Multiple fire departments have responded to a fire at a nightclub, which you have been requested to investigate. While the fire is being extinguished, you begin to interview several bystanders who state that they first noticed the fire emanating from the front door. The bystanders state that they are not sure if the door was open or closed at that time but are certain the fire extended from floor level to the top of the doorway. Additionally, they state that they had seen a vehicle at the building approximately 1 hour prior to the fire, and they believe the vehicle belongs to the business owner.

After the fire is extinguished and it is safe to enter, you identify three separate fires: one in an office space, a second in a rear utility room, and the third in the foyer of the front door. After excavating and reconstructing each area of origin, you identify fire patterns on the floor that may have been caused by an ignitible liquid. Additionally, the patterns on the floor in the foyer extend to the exterior through the doorway. An interview with the business owner reveals that he had recently closed the business because it had become unprofitable and he was experiencing too many thefts from employees. He states that he had been at the business during the day to remove items of value while he prepared to put the building up for sale. The owner also states that he does not know the cause of the fire but would not be surprised if a former employee had tried to burn his business down.

1. What may have been used to allow the fire-setter to leave the area before the fire started?
   A. Trailer
   B. Delay device
   C. Secondary device
   D. None of the above
2. Based on the information you have obtained, what would be the most likely intent for this fire?
   A. Excitement
   B. Vandalism
   C. Profit
   D. Extremism
3. Which of the following could be a natural reason for multiple fires in this building?
   A. Falling curtains
   B. Flying brands
   C. Overloaded electrical circuits
   D. All of the above
4. Which of the following findings in this fire may be considered a possible indicator of an incendiary fire not related to combustion?
   A. The removal of valuables prior to the fire by the owner
   B. The business being closed at the time of the fire
   C. Patterns that may have been created by ignitible liquids
   D. All of the above

Fire and Explosion Deaths and Injuries

© Photos.com

© David L. Ryan/The Boston Globe/Getty Images

Knowledge Objectives

After studying this chapter, you should be able to:

- Describe the investigative considerations necessary at a fatal fire scene NFPA 4.4.1. (pp 336–338)
- Describe the death-related pathological and toxicological examination of fire and explosion victims NFPA 4.4.1. (pp 338–340)
- Identify and describe the fundamental issues of death investigation NFPA 4.4.1. (pp 340–341)
- Describe how to determine the cause and manner of death NFPA 4.4.1. (pp 341–342)
- Identify and describe postmortem tests and documentation NFPA 4.4.1. (pp 342–343)
- Describe the investigative considerations necessary for a fire that caused injury NFPA 4.4.1. (pp 343–345)

Skills Objectives

After studying this chapter, you should be able to:

- Evaluate a body for details about the fire scene NFPA 4.4.1. (pp 336–343)
- Secure a body as evidence NFPA 4.4.1. (pp 337–338)
- Evaluate an injured victim for details about the fire scene NFPA 4.4.1. (pp 343–345)

Additional NFPA Reference

NFPA 921, *Guide for Fire and Explosion Investigations*

CHAPTER 23

FESHE Course Outcomes

Fire Investigation I

1. Identify the responsibilities of a fire fighter when responding to the scene of a fire, including scene security and evidence preservation. (p 4)

Fire Investigation II

10. Analyze fire-related deaths and injuries. (pp 4–14)

You Are the Fire Investigator

© Jones and Bartlett Publishers. Photographed by Glen E. Ellman.

Fire fighters have located a deceased young man during a fire at a mobile home and have asked you to conduct an origin and cause investigation. Upon your arrival, you meet with the incident commander, who states that the fire was contained to the kitchen, located in the front of the mobile home; however, the victim was located in a rear bedroom.

You find the victim is on the floor, within 2 feet of a door that leads to the exterior. He has sustained second-degree burns to his head, arms, and hands and is lying face-down with a pillow covering his face.

1. What would you consider in documenting the scene?
2. Why is it important for fire personnel not to disturb the victim and area around him?
3. What events or conditions may have kept the victim from escaping?
4. How does an analysis of the pattern of injuries assist with your investigation?

Introduction

Fires and explosions often involve a complex investigation to put together the events of the incident. These investigations are even more complex when the fire or explosion involves serious injury or death. Injury can lead to death, sometimes days or even weeks following an event. As with any investigation, it is important that the correct procedures be followed from the beginning and that every fire and explosion involving serious injury receives a full investigation.

Death Scene Considerations

The investigation of a fatal fire or explosion is a two-part investigation, involving the origin and cause of the fire or explosion and the cause and manner of the death. The body of the victim is the most important piece of evidence in determining the cause and manner of the death. Any time a fire or explosion occurs and a body is discovered, there should be an autopsy performed by a competent forensic pathologist.

Fire Suppression

Without question, first responders to a scene are there to preserve life and property. They also need to be aware of the importance of preserving the scene as much as possible while still effecting fire suppression and essential rescue. Fire suppression personnel need to be aware that their actions with powerful hose streams and overhaul could have an impact on the body and other fragile evidence that may be present. If, on discovery of a victim, there is absolutely no question that the victim is deceased, then every effort should be made to leave the body in its original location. At a crime scene, it would be unthinkable to move a murder victim; the fire or explosion victim should not be treated any differently. An exception to this would be if it is determined that the body of the victim would be damaged further by allowing it to remain in its discovered position. It is preferable to have the body preserved than to have it lie under several floors of debris. If it is determined necessary to remove the body, it should be fully photographed first, along with everything underneath it and around it. The body's physical location should be marked in a manner that allows the exact location to be recalled during debris removal, permitting the area to be further examined for evidence. Always remember that the body is a piece of evidence and should be treated as such.

Notification

When a death occurs, there are legal and procedural requirements for the notification of various parties, including law enforcement, medical examiners, coroners, forensic laboratory staff, and possibly others. Often, these same parties are part of the investigative team.

Documentation

One of the most effective and quickest ways to document the scene is by photography, particularly if the body must be moved early in the investigation because of the environment. It is helpful to take a quick series of photos. Digital color photography in a 5 megapixel or larger format should be used whenever possible. Instant photos and video are useful but often lack the detail and flexibility that may be needed later.

To ensure complete documentation, remember that the body and its surrounding area should be photographed before the scene is disturbed and throughout the process of debris and body removal. Attention should be given to photographing the body with respect to such items as exits and their condition, alarm systems, fire or explosion patterns, running water in a sink, and other physical characteristics of the scene. Possible fire patterns and blast effects on the body should be well documented by photography. The body is one of the pieces of evidence that is part of a fire or explosion scene. It should be photographed prior to being moved, once it is placed in the body bag, when it is removed from the body bag, and during the time that clothing is being removed. Close-up and scale

Safety Tips

All fire scenes pose unique safety concerns that need to be assessed prior to the start of the investigation. In addition to the concerns that a fire poses, the investigators will run the risk of exposure to various biohazard materials. Appropriate personal protective equipment, decontamination, and disposal procedures will need to be addressed.

photos of burns and other injuries should be taken. The location in which the body was resting at the fire scene should also be photographed. If the victim has already been removed, the investigator can reconstruct its location by knotting one corner of a sheet and positioning the sheet (with the knot representing the head) as the body was positioned.

Documentation of the scene should include sketches and diagrams. Diagrams should detail and record the physical dimensions of the scene, the contents of the scene, and the measurements of the body location. The sketch should include the outline of the body for reference purposes **FIGURE 23-1**. Often these sketches are useful in court when photos of a victim are not admissible out of consideration for jury members. The investigative process may include the creation of a victim sketch. This sketch is used to detail burn and other injuries to the victim.

Recovery of Bodies and Evidence

All fire or explosion investigations require a team effort. In cases of death, the investigation often expands the team to include a homicide detective, the medical examiner, the coroner, forensic laboratory personnel, and a forensic pathologist. If the body is badly burned, the expertise of a forensic anthropologist and a forensic dentist (odontologist) is likely to be required.

The investigation often initially focuses on the area where the body is located, but investigators need to remember that important evidence can often be found some distance from the body. Fire suppression, the actions of the victim while alive, and the environment may position evidence away from the body. In many cases, the size of the fire scene may require more than one investigator to conduct the scene investigation–one investigator concentrating on the death investigation while the other concentrates on the determination of area of fire origin, point of origin, and the cause of the fire.

To aid in a thorough examination, the investigators can use a grid system to divide the scene into sections **FIGURE 23-2**. Each grid section needs to be examined and documented, and evidence needs to be identified. If the scene does not lend itself to a grid search, the investigators may wish to consider a spiral pattern. All methods that are employed should have some overlap for complete coverage.

The investigation of the debris and its location may reveal a sequence of events for the fire or explosion. Often, the area that must be searched is covered with multiple layers of debris, which must be removed in a systematic way. The examination of the fire or explosion scene can be compared with an archeological dig, in which each layer is removed and interpreted in relationship to the events. At times, the size of the evidence can require different examination techniques. Small pieces of evidence can be located by removal of debris from a section and through careful screening of the debris. The use of multiple screens of different-sized mesh is helpful in this process.

The area and pathway to the body should be examined and assessed prior to the removal of the body as a means of identifying potential evidence and the preparation for its effective preservation. During the removal of the body, items such as clothing and fire debris could be attached or drop away. Items should be left adhered to the body for later identification; if items drop off, they should be collected and placed with the body for forensic examination. The area where the body was resting should be examined for potential evidentiary items. Anything in the victim's hands should be noted. Because the body is a piece of evidence, it is important that it be treated as such, with proper consideration for chain of custody and cross-contamination. The body should be placed in a new, unused, sealed body bag. During all steps of the excavation and recovery process, photos should be taken of the scene and items uncovered.

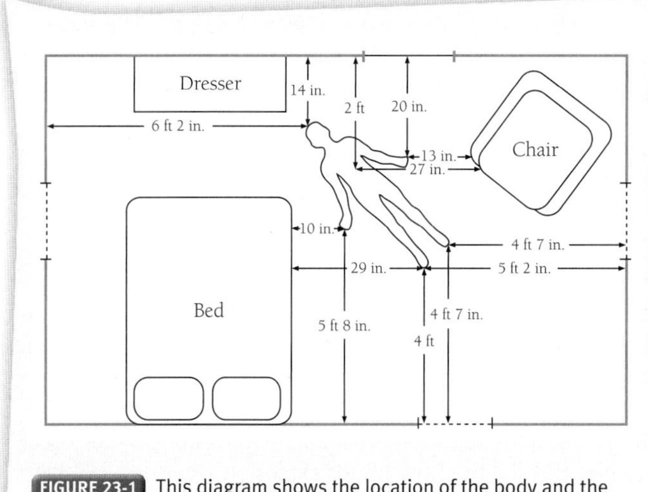

FIGURE 23-1 This diagram shows the location of the body and the dimensions of the surrounding area.

© Jones & Bartlett Learning. Photographed by Glen E. Ellman

FIGURE 23-2 This room has been marked off into sectors to ensure complete coverage.

Fire Investigator Tips

Remember that a body found at a fire or explosion scene must be treated as evidence. Not only do you need to maintain a chain of custody and avoid cross-contamination, but you also have to leave any material attached to the body intact for future examination.

It is possible that the victim's clothing may contain ignitible liquid residue. When possible, the clothing should be preserved for its evidentiary value. In instances where the body was transported away from the scene prior to the arrival of investigators, there should be efforts to obtain and preserve clothing and other potential evidence using proper collection methods.

When the body has been badly burned or fragmented by the fire or explosion, great care should be taken to search for all remains, no matter how small. Burned bones and tissue can blend in with fire debris and may be overlooked. An experienced fatal-fire investigator can be helpful to the team in identification of these items.

The team might wish to bring a skilled forensic anthropologist directly to the scene to assist in skeletal identification as needed. Sometimes it is necessary to determine whether bones recovered at a scene are animal or human FIGURE 23-3. In instances in which animals perish in a fire, it may be useful to conduct a postmortem examination on them to gather additional information.

© National Geographic Image Collection/Alamy

FIGURE 23-3 When examining remains at a scene, it is sometimes difficult to determine whether they are animal or human.

Death-Related Pathological and Toxicological Examination

All victims of a fire or explosion should receive a full autopsy to determine the cause of the death. This examination must take into account several concerns.

One of the first steps in examination of the body should be full-body X-rays. X-rays are useful in identification of any foreign matter in the body, such as a bullet or knife tip FIGURE 23-4. Also, X-rays may be helpful in identifying the victim by determining any past injuries such as a broken bone, surgical metal implant, or dental remains that can be compared with a known set of X-rays.

Testing for the carbon monoxide (CO) levels in blood and tissue is one of the easiest and most common medical tests performed on the fire or explosion victim. This test often reveals much about the cause and sequence of events and should be performed whenever possible. Carbon monoxide is absorbed into the blood and tissue by breathing. It may cause a cherry-pink coloration of the skin, which may not be visible in dark-skinned victims or in victims who are heavily covered in soot. The cherry-pink coloration could be visible in several areas on the body, including the lips and nipples, as well as in the blood that has pooled in the body (caused by postmortem lividity).

Lividity happens after death and is the pooling of the blood in the lower elevations of the body caused by the effects of gravity. Lividity becomes fixed in the body 6 to 9 hours after death FIGURE 23-5.

In addition to the presence of CO, the examination should look for soot or smoke in various areas of the body, including the lungs, trachea, and bronchi. The exterior of the body may show signs of soot around the nose and mouth. The investigator must be aware that these deposits are not always proof that the victim was breathing during the fire. A medical examination of the internal airways and the lungs often shows very clearly black soot deposits that are consistent with the fire victim's breathing in a smoke environment. Soot may also be discovered in the stomach of the fire victim.

The medical examination process should also include a search for other toxic products. With all fires, various combustion products are present. Such compounds as hydrogen cyanide and hydrogen chloride may be discovered in the blood and tissue. The same body elements may also reveal various drugs—both medical and illicit—along with alcohol levels for the victim. Knowledge of these levels is useful to investigators as they try to interpret the ability of the fire or explosion victim to deal with the events as they took place. Gas and liquid chromatography are two techniques that are used to determine these levels.

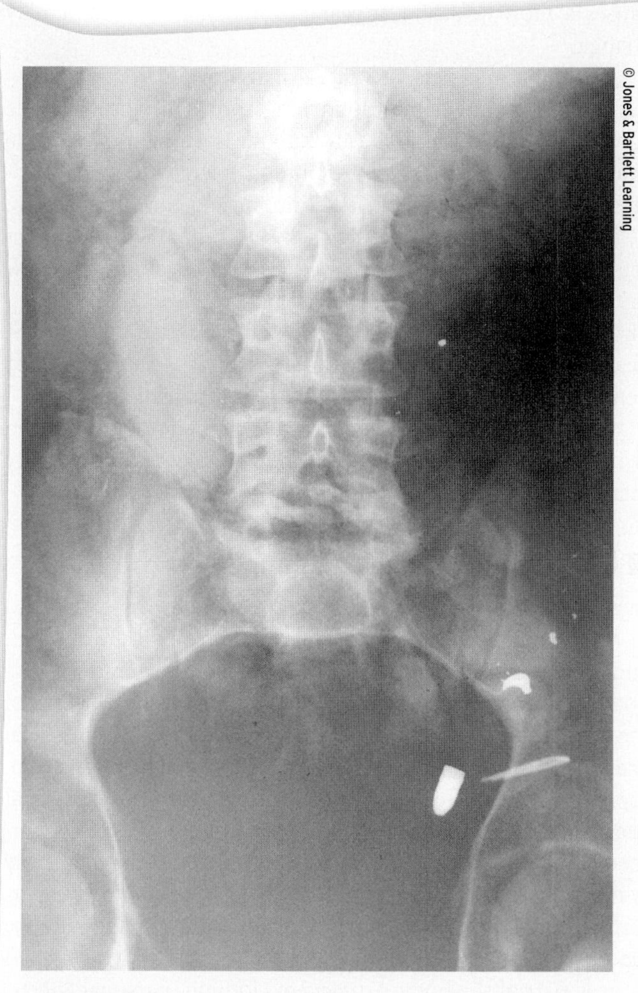
© Jones & Bartlett Learning

FIGURE 23-4 This full-body X-ray shows a bullet in the pelvic area of the fire victim.

Burns

The direct effects of the fire are often noted in burns to the body. This process takes place both before and after death of the victim. Blistering of the skin (second-degree burn) can happen to a more limited degree after death. The effects of the heat after death begin to dehydrate the muscle tissue. As this tissue loses hydration, there is a noticeable shrinkage of the muscle, which in turn causes a constriction or tightening of the muscle. This is noticeable in the facial features of the victim and in the pugilistic attitude. The pugilistic attitude is a boxer-type stance observed in the hands and arms. This constriction of the muscles will also take place in the legs in the later stages of exposure. Often, investigators suspect that the victim was in a defensive stance as if warding off blows to the chest and head when instead it was a stance caused by the fire. This constriction of the muscles can even fracture bones in the arms or legs.

During the examination of the body, special attention needs to be paid to the presence of any blood. Often, blood located on the body is an indication of non-fire-related trauma. The heat of the fire causes dehydration of the tissue, and thus rarely is blood present; one exception to this is small amounts of blood in the nose, ears, or mouth. Any blood that is found should be documented thoroughly.

Consumption of the Body by Fire

Like all things in the fire scene, the body is part of the fuel load. Investigators at a fire scene need to remember to examine the burn patterns and extent of burns on a fire victim as they would any other part of the fuel load FIGURE 23-6. The investigators

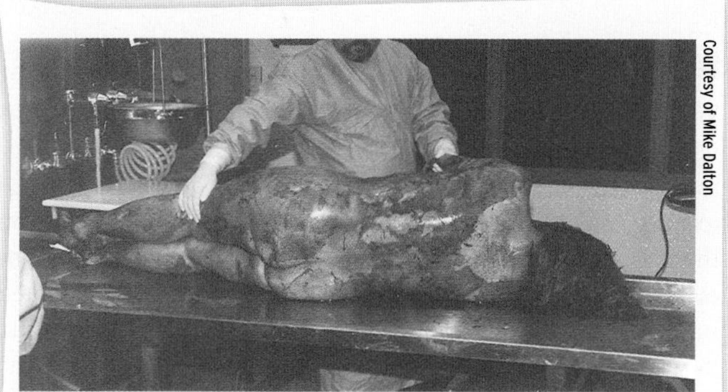
Courtesy of Mike Dalton

FIGURE 23-5 In the hours after death, blood pools in the lower elevations of a body. This can help indicate how long the person has been deceased.

FIGURE 23-6 The body is both part of the fuel load and a piece of evidence and must be examined accordingly.

should make sure that the patterns that are present are consistent with the other patterns in the area. When exposed to sufficient heat, the body is consumed to some degree. The skin and muscle tissues of the body are considered poor fuels, but dehydrate and are consumed with time. Body fat is a combustible component of the body. The bones shrink and then fracture. There is a coloration change to the bone as it is heated. A darkened or black color evolves to a gray-blue and then to chalk white as it is heated.

The skull may fracture or crack from the effects of the fire. The thermal fracture most often occurs along the suture lines of the skull. However, the investigator is warned not to assume that the fire caused any skull fracturing, as the effects of a gunshot or some form of trauma to the skull can also cause fractures. The causation of the fractures may need to be determined by a skilled forensic pathologist or anthropologist.

Much attention has been directed over the years to the phenomenon called spontaneous human combustion. Simply put, the human body does not spontaneously combust. Under the right conditions, the body fat of a fire victim may be absorbed by a wicking material, such as a cotton shirt, and serve as a fuel source for a small but concentrated flame. The energy from the flame may be insufficient to ignite adjacent combustible materials and will, with time, leave the body of the victim consumed in the torso area but may allow the limbs and head (with their lower body fat concentration) to remain.

Fundamental Issues of Death Investigation

The investigator faces a number of fundamental issues with regard to the investigation. These include victim identification, cause of death, manner of death, victim activity, and postmortem changes.

Remains and Victim Identification

One of the critical elements in the investigation related to the body is identification of the victim. As the body is discovered in the fire scene, it may be difficult for the investigator to determine whether the tissue mass is that of an animal or a human. The remains of large animals such as pigs, bears, and deer may resemble human remains. The skeletal remains of a bear paw compared with a human hand may be identifiable only to a trained anthropologist.

The bodies of children and infants pose special problems, as their skeletal structure is less developed than that of an adult. It may be impossible to determine the gender of the prepubertal victim based on examination of skeletal remains. In cremation, the body is often exposed to temperatures over 1800°F (982°C) for 1 to 2 hours. Even after this exposure, there are remains of the skeleton. With a child, there is less development of the skeleton and less tissue mass, so it is therefore possible in extreme cases to have remains that are not identifiable. Where there is a possibility of child remains, the investigator should be especially cautious when examining the scene.

The effects of the fire on the body are varied, depending on many factors. In cases of moderate exposure, it may be possible to make a visual identification of the victim. The investigator needs to bear in mind that the facial skin tightens and causes a more youthful appearance and that hair color may change.

Clothing and personal effects may provide some circumstantial evidence of identification. Fingerprints may be a possible source of identification, as can tattoos. The pugilistic attitude causes the victim's fingers to curl into the palm, which may protect the fingertips sufficiently for printing and comparison.

X-rays of the dental remains and a comparison to known sets are some of the surest ways of identification. The low tissue mass of the head and the hardness of the teeth often help in their preservation in the fire. When the destruction to the head is extensive, the investigator needs to take care to recover dental remains for reconstruction and comparison.

With the recent advances in DNA, its use for identification is enhanced. A comparison of the DNA of the victim to a family member can be helpful.

Coroner and Medical Examiner

In most jurisdictions, the task of determining the cause and manner of death rests with the coroner or medical examiner, and this individual may also be responsible to removing the body.. The fatal-fire investigator should have knowledge of the system in his or her jurisdiction. The coroner or medical examiner will be an important part of the investigation team, especially as it relates to information gained from the medical examination of the deceased. An important role of the coroner/medical examiner is the facilitation of the autopsy. An autopsy is recommended in all fire-related deaths and should be performed by a skilled forensic pathologist. A complete autopsy would include collection of physical evidence that is attached to or is part of the body. Artifacts of clothing, personal effects, bullet remains, chemical residues from ignitible liquids, various tissue and body fluids, and other items may be recovered for future analysis. When possible, it is helpful for the fire investigator to be present during this examination to be able to share information with the medical staff. The fire investigator should become acquainted with the coroner's staff and medical examiners in his or her area and establish an investigative understanding with them, allowing the investigator permission to be present during examinations and other procedures such as photographic documentation. This will improve and facilitate the investigation process for the fire investigator.

Victim Activity

When trying to determine the cause and manner of death, it is helpful for the investigator to examine and interpret evidence to determine the victim's activity prior to the fire. The physical location of the body (such as a bed), clothing, or items found on the body (such as a fire extinguisher) are telltale signs. Patterns of damage to clothing and/or the body should be considered in context with the total scene. Any inconsistencies should be closely examined. Burn patterns to body or clothing may be from attempts to extinguish the fire or to escape. Evidence found in the area may be a helpful indicator of past activity, such as the discovery of smoking material or food left out for cooking. Knowledge of the victim's prefire physical abilities is helpful in understanding the victim's movement or lack thereof.

It is important to remember that not all fire-related deaths are caused by flame, heat, or smoke.

Postmortem Changes

After death, several changes begin to occur to the body. As was mentioned earlier, lividity sets in 6 to 9 hours after death. This pooling of the blood in its lower regions is caused by the effects of gravity. If the body is moved before lividity sets, the pooling of the blood changes. Rigor mortis begins to stiffen the joints of the body a few hours after the death. The effects of the rigor mortis begin to leave the body after about 12 to 24 hours. This loss of rigor proceeds from the body's extremities back to the torso and head of the body. Extreme muscular activities prior to death, as well as elevated surrounding temperatures, will speed the rigor onset. Experienced forensic pathologists may use the status of the rigor on a body to estimate a time of death. Rigor should not be confused with the rigidity of the body's muscles caused by exposure to heat.

The rate of development of changes to the body after death may be affected by several factors, such as time and climate conditions. When possible, a victim timeline should be created. This timeline would include data as to the times of body discovery, removal, transportation, and examination, as well as environmental conditions such as temperature and manner of storage. All of this information could be used later in the interpretation of medical data from pathological and toxicological findings.

Cause and Manner of Death

The ultimate questions related to the body are the cause and manner of the death. The *cause* of the death is regarded as the actual event or injury that brings about the cessation of life, such as smoke inhalation, burns, and gunshot. The *manner* of death is best regarded as the course of events that led up to the accidental, homicidal, suicidal, natural, or undetermined death.

Combustion Products and Their Effects

There are many products from the combustion process that can affect the victim of a fire or explosion. These products may include such items as carbon monoxide, carbon dioxide, nitrogen oxides, halogen acids, hydrogen cyanide, acrolein, benzene, particulates such as soot and ash, and aerosols. Breathing these products or direct skin contact can cause a variety of effects on the human body.

Carbon Monoxide

Carbon monoxide is a product of all fires and is caused by incomplete combustion of the fuel. Thus, the level of CO produced by each fire varies on the basis of the completeness and type of the fuel package. Carbon monoxide is an anesthetic and an asphyxiant. When CO is inhaled, it binds with the hemoglobin in the blood to form carboxyhemoglobin (COHb). Carbon monoxide binds with the hemoglobin in the blood much more readily than oxygen does; therefore, it is possible for the blood to form dangerous levels of COHb with exposure to low concentrations of CO. When the fire survivor is removed from the CO environment, the levels of COHb in the blood begin to decrease; however, this process can take many hours.

The stability of CO in the body is such that it can be tested for hours after the death. As a general rule, levels of COHb in the body of 50 percent or greater are considered fatal. The actual levels can vary, with death possible in some cases at concentrations as low as 20 percent. The level of the COHb in the body may be helpful to the investigator in making determinations about the death of the victim. When levels are below 20 percent, the death most likely was caused by other factors, such as thermal injuries or physical trauma. When levels are 40 percent or greater, the toxic effects of the CO alone or in combination with other factors are likely a contributing cause of death, with most victims dying remote from the room of origin. Studies have shown that at least 60 percent of fire victims die from carbon monoxide poisoning.

There are other, nonfire factors that can cause concentrations of CO in the body. Smokers or people who have been exposed to automobile exhaust may have background CO levels of 4 to 10 percent.

Through the use of the Coburn-Forster-Kane equation (CFK) or the Stewart equation, an estimate of the quantity of CO the victim inhaled can be determined. Using the concentration of CO produced by the fire over time, as well as the weight of the victim and the volume of air exchanged by breathing per minute, the CFK equation can assist in a range for the average exposure to CO to compare to values of COHb determined at autopsy. The Stewart equation estimates the CO inhaled through the relationship among COHb, inhaled CO concentration, respiratory minute volume, and exposure duration. (For a detailed explanation of the CFK and Stewart equations, refer to NFPA 921, Section 25.10.2.)

Cyanide

Hydrogen cyanide (HCN) is another toxicant produced during combustion. Hydrogen cyanide gases are a by-product of the combustion of such common household items as wool, nylon, and various plastics. By itself or in combination with other toxic gases, HCN can incapacitate or cause death. It is more rapidly absorbed into the blood by the respiration process than CO is, which explains HCN's rapid lethal effects.

Cyanide affects the use of oxygen by the body, rather than binding the oxygen with carbon monoxide. Cyanide is most commonly measured in the blood but can also be measured from various tissue samples of the liver, brain, lungs, and kidneys. Samples containing cyanide have not proven the same stability as carbon monoxide. The distribution of cyanide to the various organs may vary; its stability postmortem is generally related to time, storage (temperature), and preservation methods. A victim timeline, as discussed in the Victim Activity section, will assist in the correct determination of the value of the testing done for cyanide.

Other Toxic Gases

With the wide array of fuels in any common setting, there is an equally wide range of gases produced during the fire. Many of these gases cause irritation and swelling or are toxic in nature. Hydrogen chloride and acrolein are two gases commonly found in fires. Hydrogen chloride is released during the burning of polyvinyl plastics and acrolein from the combustion of wood and cellulosic products. These and other by-products should be considered when assessing the ability of a victim to act and move in a fire setting.

Soot and Smoke

Most fuel products will produce some form of smoke and soot material when burned, some more than others. Hot soot particles, when inhaled, may cause thermal injuries leading to edema. Soot may also transport toxins into the body or, in some extreme cases, block the airway of the victim. The liquid mists of the pyrolysis are often acidic in nature and often cause systemic failures on inhalation.

Hypoxia

Hypoxia is caused by a victim breathing in a reduced oxygen environment. As a fire burns in a confined area, it depletes the oxygen in the air. As the level of oxygen is reduced from its starting point of 21 percent to the area of 15 to 10 percent, a gradual increase in respiration occurs, followed by disorientation. Oxygen levels below 10 percent cause unconsciousness and subsequent cessation of breathing. Testing conducted of the blood of a hypoxia victim after death will not provide reliable information because levels of oxygen and carbon dioxide begin to change after death.

Sublethal Inhalation Exposure

Narcotic gas such as CO and HCN, or the effects of hypoxia, affect the response of the person who is exposed by limiting both mental and psychomotor abilities. Additionally, some of these same by-products and smoke can interfere with respiration, irritate eyes, and hinder the ability to see. The levels of these by-products begin to reduce when the person withdraws from the affected area.

Irritant gases can alert people to a fire even in low concentrations. These same irritant gases, when more concentrated, will cause irritation to the level that it may obscure vision or alter behavior. Postfire effects may produce respiratory edema or inflammation.

Smoke will obstruct vision and reduce the speed of travel during escape. The extent to which this is a factor in the escape movement may be changed by several factors, including prior knowledge of the building/area.

Thermal Effects

The thermal effects of the fire can result in death or injury. Hyperthermia is caused when the temperature of the body is greatly elevated. Depending on the time and exposure, hyperthermia may be classified as simple or acute in nature.

Simple hyperthermia is caused by extended exposure (15 minutes or longer) to hot environments. Over time, the body temperature increases, with internal temperatures above 109°F (43°C) being fatal in a few minutes. Humidity and moisture also compound the body's ability to shed excessive heat.

Acute hyperthermia is caused by exposure to high heat levels for a short period of time. Thermal burns also result from this exposure, but the cause of death is related to the elevated body temperature.

The inhalation of hot gases and various toxic gases causes edema (swelling) and inflammation of the airway. The effects of hot gases are generally accompanied by facial burns or singed facial hair. The inhaling of soot may produce thermal injury and may carry toxic compounds or physical blockage to the airway. The fire may consume the oxygen to the point at which it poses a possible risk. Oxygen levels below 15 percent can cause disorientation and loss of judgment. At levels below 10 percent, unconsciousness occurs, followed by cessation of breathing.

Skin Burns

When the temperature of the skin reaches about 110°F (45°C), pain will result. The transfer of thermal energy via conduction, convection, and radiation causes burn injury. Conductive heat transfer to the skin through clothing can occur even though the clothing may not show visible signs of heat. Skin exposed to convective heat with air temperatures above 120°F (49°C) will cause pain and injury. With radiant heat, the greater the heat flux, the faster the damage occurs to the tissue.

Inhalation of Hot Gases

The inhalation of hot gases may result in death or injury. Assessment of this is hampered because of similar effects caused by chemical irritants. The inhalation of hot gases will often be accompanied by burns to skin and facial hair. Researchers have determined that in animals hot gases of 932°F (500°C) resulted in larynx and trachea damage, and steam of 212°F (100°C) resulted in burns deep into the lungs.

Postmortem Tests and Documentation

Various tests and documentation during the postmortem examination provide the investigator with valuable information to aid in identifying the victim as well as establishing the cause and manner of the death. These tests may include the following:

- Blood: For level of COHb, HCN concentration, drugs, alcohol, or poisons
- Internal tissue: For level of volatile hydrocarbons, drugs, or poisons
- Stomach: For contents, including the presence or absence of soot
- Airways: For effects of the fire
- Internal body temperature: To assist in establishing a time and mechanism of death
- X-rays: To assist in identification or location of foreign objects in the body
- Clothing or personal effects: To assist in identification or check for the presence of ignitible liquids
- Recovery of foreign objects in or on the body: Possibly to include bullets, knife parts, explosive device components, and other items
- Sexual assault evidence: To provide possible evidence of other crimes, identification of aggressor, and possible motivation for setting a fire

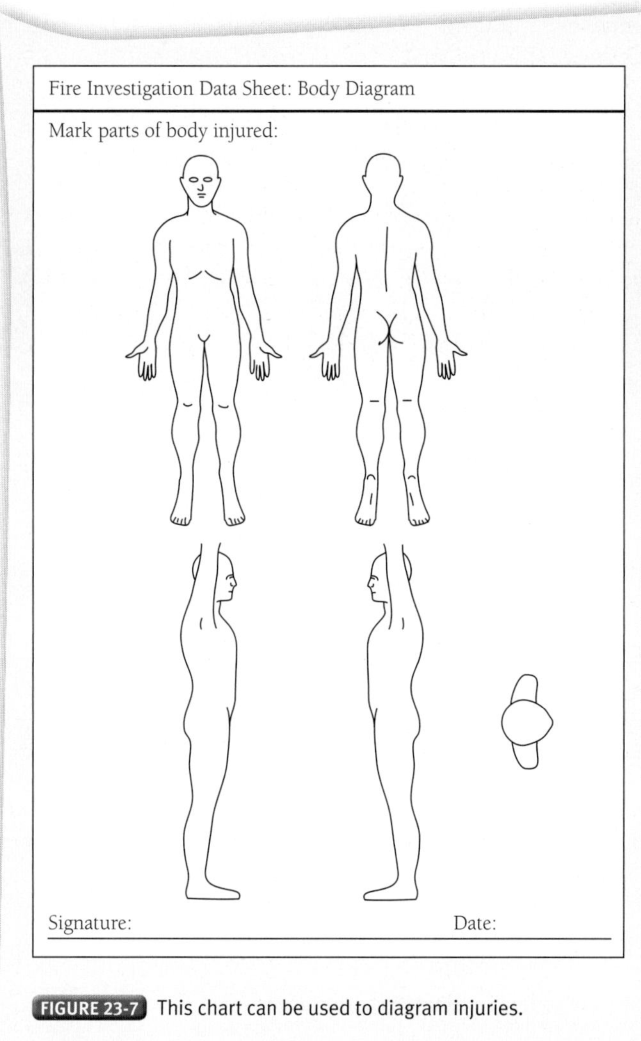
Fire Investigation Data Sheet: Body Diagram

Mark parts of body injured:

Signature: Date:

FIGURE 23-7 This chart can be used to diagram injuries.

- Documentation by photography and sketching: To record injuries and burns to the body and evidence recovered from the body FIGURE 23-7

During the postmortem exam (autopsy), it is helpful for the investigator to be present. The investigator may point out specific evidence and take custody of evidence as it is recovered. The investigator may also share important investigative information with the medical personnel who are performing the tests.

Fire and Explosion Injuries

Deaths from fires and explosions can occur well after the event. Serious injury fires and explosions should be investigated as if they were fatal investigations to ensure completeness. During the investigation of the injury scene, many items should be examined and documented, including physical evidence and medical evidence related to burns and inhalation.

Because of the victim's injuries, statements may not be obtainable for an extended period. All steps to gather information should be used and not delayed in hopes of an interview. Documentation of the nature of injuries should be conducted in a timely manner because, with time, the injury may begin to heal and change in appearance.

Some jurisdictions have laws requiring medical personnel to report burn injuries much in the same manner as gunshot wounds. The investigator should be aware of the laws that apply within his or her jurisdiction.

Physical Evidence

Often, physical evidence, both apparent and microscopic, is located in areas beyond the body. Clothing should be collected as soon as possible to prevent its loss. Clothing may contain physical evidence (in pockets) or may have been a fuel source for the fire. Likewise, the furnishings of the area may be evaluated for burning properties. The type of containers used to collect physical evidence associated with the body is dictated by the purpose of collection, such as blood samples, ignitible liquid presence testing, and such.

As with any fire, potential ignition sources for the fire will be examined and documented.

Some areas require by law that notification be made to governmental agencies when serious fire or explosion injuries occur. These laws are similar to gunshot reporting laws and have been successful in identifying both victims of assault and abuse, as well as perpetrators of arson who are burned in the execution of their crime.

Medical Evidence

Burns

Many terms are used in medical reports to document the evidence of burns. The investigator should be familiar with each one of them.

Degree of Burn Burn injuries need to be documented and assessed to the following degree of burns:

- First degree: Reddening of the skin; also called *superficial* burn.
- Second degree: Blistering of the skin; also called *partial-thickness* burn.
- Third degree: Full-thickness damage to the skin, also called *full-thickness* burn.
- Fourth degree: Damage to the underlying tissue and charring of the tissue.

Body Area (Distribution) The medical community estimates burn damage by using the "rule of nines." This is based on dividing the body into 9-percent segments and adding the portions of the body to obtain a total of the body area affected by the burn injury FIGURE 23-8. Another resource is Table 25.8.1.3 in NFPA 921, *Guide for Fire and Explosion Investigations*. The percentage of body area burned is sometimes used to predict survivability.

Documentation As with all elements in the investigation, thermal injuries should be well documented. This documentation may include sketches and color photography, preferably before treatment or healing, which could change the appearance of the injury.

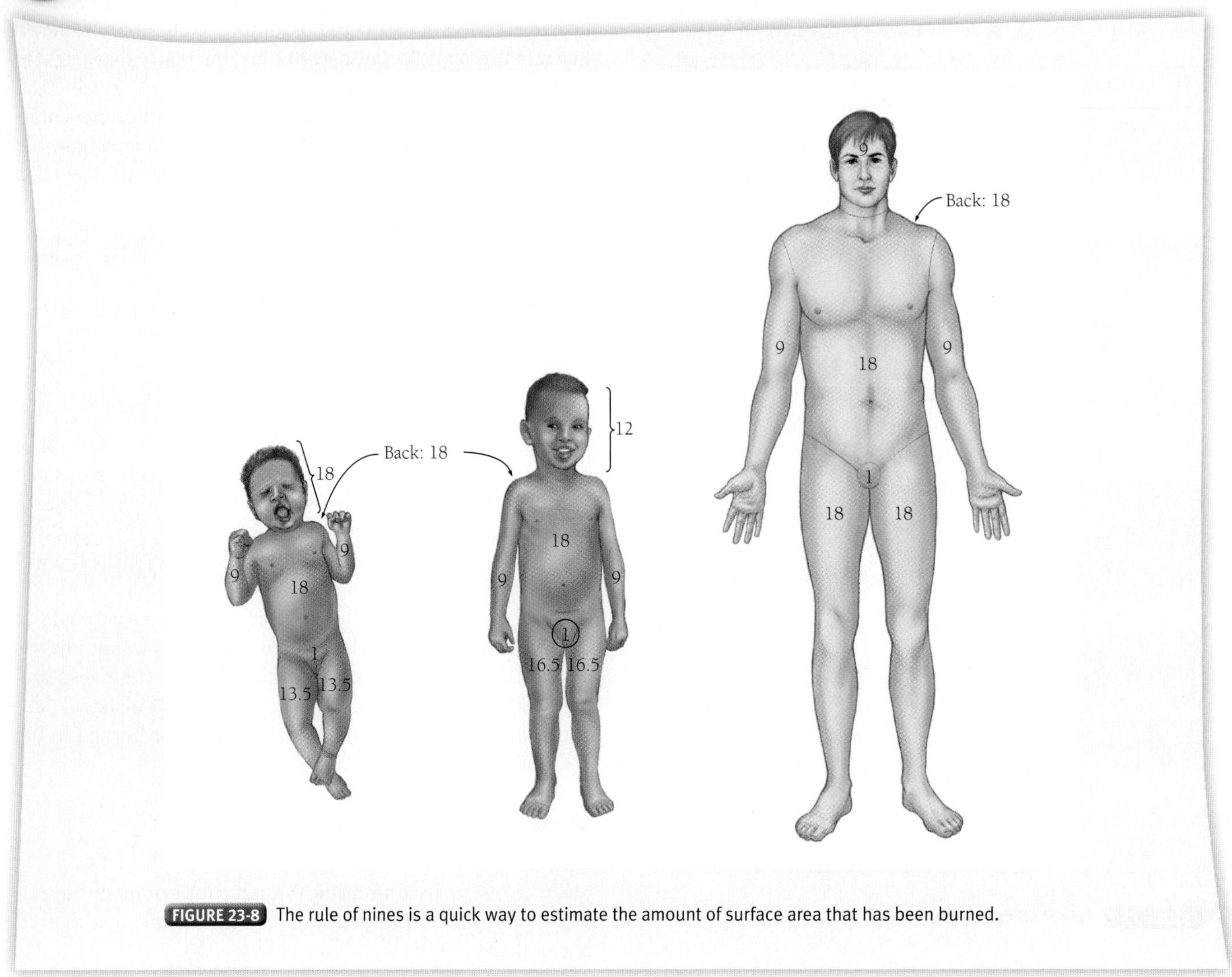

FIGURE 23-8 The rule of nines is a quick way to estimate the amount of surface area that has been burned.

Mechanism of Burn Injury The cause of a burn, whether by scalding, chemical exposure, or hot gases or flames, may not be distinguishable by appearance alone. There is a direct relationship to the formation of burns and the radiant heat flux. As an example, a radiant heat flux of 2 kW/m^2 will cause pain after 30 seconds with no blistering of the skin, whereas a heat flux of 10 kW/m^2 after 5 seconds will cause pain and blister the skin in 12 seconds. Because conducted heat brings the skin in direct contact with the source, it is more dangerous than heat transferred by radiation or convection. This conduction can also occur from heavy clothing that transfers enough heat to burn skin but not burn the fabric.

Inhalation

Fire produces various by-products that affect the person exposed. Testing is needed to determine the levels of these by-products and to understand the actions of the injured individuals and environment to which the victim was exposed.

The percentage of CO eliminated by the body is in direct relationship to the concentration of oxygen afforded the patient. The CO level of a patient in normal atmosphere will reduce by half in about 5 hours. This is greatly accelerated when the patient receives oxygen during treatment. The half-life of COHb in a normal environment is about 250 to 320 minutes, but when the patient receives 100 percent O_2, the half-life is reduced to 60 to 90 minutes. Knowledge and application of this information are useful when COHb levels of people who received treatment are known and interpolated.

Hospital Tests and Documentation On admittance to a hospital, a patient with fire-related injury should have a sample of blood taken for analysis for such items as level of COHb, HCN concentrations, blood alcohol level, and presence of drugs. The blood should be removed as soon as possible because the levels of most items begin to decrease with time and treatment.

Access to Medical Evidence

Many federal and state laws affect the ability of the investigator to obtain medical records. Medical records may in some cases be obtained only with consent of the patient or through legal means, such as court order. Knowledge of laws related to medical records is helpful to the investigator in this process.

Obtaining past medical history may reveal conditions that would affect an individual's ability to comprehend, move about, or detect the dangers of a fire. It may also provide clues to deaths that were determined to have occurred prior to the fire, such as heart attack.

■ Explosion-Related Injuries

The location and distribution of explosion injuries may indicate location and activity of the victim at the time of the explosion and may help to establish the location, orientation, energy, and function of the exploding mechanism or device. The injuries related to an explosion scene are classified into four groups based on the explosion effect that caused them: blast pressure, shrapnel, thermal, and seismic.

Blast pressure injuries are caused from the concussion effects of the explosion. Damage to the internal organs is not uncommon, depending on the level of the blast pressure. The pressure wave may also propel the victim into objects and cause blunt trauma injuries, fractures, lacerations, contusions, and abrasions. With detonations, it may be possible for the body to experience severe injury or amputation. Small particles are also harmful if they are blasted into unprotected skin.

Shrapnel (solid fragment) injuries from fragments from the blast center may cause serious effects on the body, such as amputation, lacerations, or blunt trauma.

Thermal injuries may be caused by the explosive flame fronts, usually causing first- and second-degree burns. Thermal injuries may result from clothing: synthetic fabrics may melt on the victim and cotton fabrics may scorch. Thermal burns are mostly limited to first and second degree; third-degree burns can occur and may be fatal in nature. The brief exposure of the skin to the thermal event may cause skin damage to exposed surfaces, especially those surfaces in direct exposure to the source.

Seismic effects of the explosion may cause injury with the collapse of structures or of their structural elements. These injuries are often seen as blunt trauma, lacerations, fractures, amputation, contusions, and abrasions.

Wrap-Up

Ready for Review

- As with any investigation, it is important that the correct procedures be followed from the beginning and that every fire involving serious injury receive a full investigation.
- The investigation of a fatal fire or explosion is a two-part investigation, involving the origin and cause of the fire or explosion and the cause and manner of the death.
- Documentation of the nature of injuries should be conducted in a timely manner because with time the injury may begin to heal and change in appearance.
- All victims of a fire or explosion should receive a full autopsy to determine the cause of the death.
- The investigator faces a number of fundamental issues with regard to the investigation, including victim identification, cause of death, manner of death, victim activity, and postmortem changes.
- There are many products from the combustion process that can affect the victim of a fire or explosion. Breathing these products or direct skin contact can cause a variety of effects on the human body.
- Acute hyperthermia is caused by exposure to high heat levels for a short period of time. Thermal burns also result from this exposure, but the cause of death is related to the elevated body temperature.
- Various tests and documentation during the postmortem examination provide the investigator with valuable information to aid in identifying the victim as well as establishing the cause and manner of the death.
- Deaths from fires and explosions can occur well after the event. Serious injury fires and explosions should be investigated as if they were fatal investigations to ensure completeness.

Hot Terms

Carboxyhemoglobin (COHb) The carbon monoxide saturation in the blood.

Hypoxia Condition caused by a victim breathing in a reduced-oxygen environment.

Lividity Occurs after death and is the pooling of the blood in the lower elevations of the body caused by the effects of gravity.

Pugilistic attitude A crouching stance with flexed arms, legs, and fingers.

© Greg Henry/ShutterStock, Inc.

FIRE INVESTIGATOR *in action*

You have been requested to investigate a fatal fire in a rural area of your county. Upon your arrival, you are advised that several victims have been located in a second-floor bathroom. The victims include two small children and a younger woman who is believed to be their mother.

Fire fighters have stated that all three victims were found in the bathtub in a position that appeared as if they were hugging each other when they died. They also stated the shower head was flowing water when the victims were found.

As you begin to document the scene, you determine the fire originated in the first-floor living room and ventilated out the front door. Two additional exits are present in the rear of the structure and are intact; however, they are locked from the inside with key-operated deadbolts.

1. What is the likely cause of death for the victims?

A. Cyanide poisoning
B. Smoke inhalation
C. Thermal injuries
D. Drowning

2. Which of the following observations would be significant during your investigation of this fire?

A. The location of the fire
B. The key-locked deadbolts
C. All of the victims being located in the shower
D. All of the above

3. When examining areas around victims for potential evidence, which of the following techniques may be useful?

A. Sifting screens
B. Shoulder-to-shoulder search
C. Grid system
D. Both A and C

4. Which of the following is considered a manner of death?

A. Gunshot
B. Electrocution
C. Drowning
D. Accidental

5. As a general rule, a COHb level of greater than _________ is considered lethal.

A. 25 percent
B. 50 percent
C. 75 percent
D. 90 percent

Appliances

© Photos.com

© Jones and Bartlett Publishers. Photographed by Kimberly Potvin.

Knowledge Objectives

After studying this chapter, you should be able to:

- Describe how to record the fire scene when an appliance is involved. (pp 350–352)
- Describe how to determine the origin of a fire involving an appliance NFPA 4.2 NFPA 4.2.4 NFPA 4.2.6. (p 352)
- Describe how to determine the cause of a fire involving an appliance NFPA 4.2 NFPA 4.2.6. (pp 352–353)
- Identify and describe common appliance components. (pp 353–360)
- Identify and describe common residential appliances and their operation. (pp 360–365)

Skills Objectives

After studying this chapter, you should be able to:

- Evaluate appliances at a fire or explosion scene. (pp 350–365)

Additional NFPA Reference

NFPA 921, *Guide for Fire and Explosion Investigations*

CHAPTER 24

FESHE Course Outcomes

Fire Investigation I

There are no Fire Investigation I (FESHE) course outcomes for this chapter.

Fire Investigation II

6. Analyze electrical causes of fires. (pp 350–365)
13. List the sources and technology available for fire investigations. (pp 352–353)

You Are the Fire Investigator

© Jones and Bartlett Publishers. Photographed by Glen E. Ellman.

You have determined that a fire within a single-family home originated in a utility closet beneath the second floor stairwell, where a gas-fired furnace and an electric water heater were located. Both appliances display severe flame and heat damage, with many of the plastic components being consumed.

As you begin to examine both appliances, you locate fire patterns on the side of the water heater that indicate that the fire originated at one of the heating elements. Additionally, you note the remains of other items such as a broom, mop, and several plastic containers, which appear to be cleaning supplies, near the access panel to the heating element.

1. What may have caused the heating element to fail?
2. What information may you obtain by researching the make and model of the water heater?
3. How would you preserve the remains of the water heater for further testing?
4. Why would having an exemplar appliance be beneficial?

Introduction

This chapter focuses on appliances as ignition sources for fires and explosions. Because appliances should not be considered an ignition source until an adequate origin determination has been completed, this chapter assumes that the origin of the fire has been determined and that an appliance is the suspected ignition source.

Recording the Fire Scene When an Appliance Is Involved

The material in this chapter expands on the information in Chapter 16, Documentation of the Investigation, and Chapter 26, Appliances, of NFPA 921, *Guide for Fire and Explosion Investigations*. The documentation methods outlined in those chapters should be used to record investigations of fires involving appliances. In a fire scene, the most convincing indicators that an appliance caused the fire are as follows:

- Fire patterns that point to the area of origin being near the appliance
- Severe fire damage to the appliance

In addition, if the appliance operates on electrical power:

- Arcing or melting found on conductors either inside or near the appliance

Because of construction crews working, occupants returning, adverse weather conditions, and general deterioration of fire-damaged materials, evidence is vulnerable to damage or alteration. The investigators themselves should not disturb the appliance until it has been thoroughly documented. When it becomes necessary to alter the scene or to move evidence, the investigator is advised to consult NFPA 921 regarding handling of evidence. If evidence is handled improperly, a claim of spoliation can be made. Investigators are advised to avoid this indiscretion and possible resulting legal remedies, even prosecution. There is a need, therefore, to become familiar with and practice the required methods of notifying other parties and documentation prior to scene or evidence alteration, as discussed in Sections 12.3.5.4 and 12.3.5.5 of NFPA 921. NFPA 921 also outlines further considerations to make when conducting destructive testing. See the "Legal Considerations" chapter of this textbook for more information on legal concerns when moving or altering evidence.

After an origin area has been identified and one or more appliances have been found in the origin area, the appliance should be documented by photographs, diagrams, and measurements. Ascertaining the position or setting of any controls, securing the appliance nameplate data, and gathering all component parts of the appliance must also be accomplished as part of the documentation of the scene.

Photographs

The investigator should photograph the entire scene as outlined in the "Documentation of the Investigation" chapter in this textbook, taking care to photograph the appliance from many different angles, including any electrical and fuel lines supplying the appliance. The investigator should record the entire area, including the appliance, to establish its location in relation to combustibles and landmarks in the room. Close-up photographs should also be taken to provide detailed information such as positions of controls and thermal protection devices. The appliance should not be moved until all other fire scene documentation and appliance documentation have been completed. Witness information is helpful to establish position and proximity to combustibles as well as the power state and any recent problems with operation of the appliance.

Diagrams and Measurements

The investigator should diagram the fire scene and locate the appliance on the diagram, being sure to include measurements locating the appliance in relation to fixed landmarks, combustible fuels, power sources, and other elements in the fire scene.

Documenting the Appliance

On a sketch or photograph of the appliance, with descriptive annotation if warranted for clarity, the information in TABLE 24-1 should be documented. This information may require product research on the appliance. Some appliance manufacturers have manuals or online databases that include helpful diagrams and schematics of their appliances. An exemplar of the component(s) in question will provide guidance by similarity to understand function or placement of components in the appliance and will often include descriptive design or layout information. Because this analysis cannot usually occur at the scene, the investigator should still document any of the items in Table 24-1 before moving and transporting the appliance for storage or laboratory inspection.

The investigator should obtain the following identifying information from the appliance. This information is sometimes found on labels or plates located behind service panels or hidden behind or under the appliance. Often the owner will keep all appliance manuals together. An instruction manual may be stored inside the appliance, often in a separate location from the identifying information:

- Manufacturer
- Model number
- Serial number
- Date of manufacture
- Name of product
- Warnings and caution notes
- Recommendations and ratings
- Additional data located on the appliance, such as installation diagrams

Fire Investigator Tip

It is essential to document the status and condition of key control and protection components given in Table 24-1, found on or in the vicinity of the appliance, before moving it. Although moving a control knob or switch is highly discouraged, if one needs to be moved, mark the position first so it can be returned to its original position.

Recovery and Reconstruction of Appliance Components

The investigator should gather together the various components that may have been moved during the fire or firefighting operations and reconstruct the appliance in its prefire location,

Table 24-1 Appliance Documentation

Item	Description	Examples
Primary power source type	Describe the type of *primary* energy used in the appliance.	Electric, gas (natural or propane), fuel oil, solar, or wind
Power source, energy storage	Describe the service origin or storage or generation site of the power.	Utility electrical power supply, battery, propane storage tank, natural gas utility, generator with uninterruptable power supply
Energy source adapter	If one is present, describe the adaptation method (converts source to usable form).	Transformer, power supply, pressure regulator
Power/fuel source feed and connections	Describe feed wire and interface connections to the power source.	Electrical wire distribution = 14/2 AWG solid wire with standard duplex outlets, extension cord = 12 ft 12/2 AWG stranded wire with 4 tap power strip, natural gas 3/8" ID quick connect male plug with 3/8" ID copper pipe
External overload protection	Describe the method of *external* protection from energy overload.	GFCI protection, circuit protection (breakers or fuses), regulator
Bonding and grounding	Describe how the appliance is grounded.	Two or three wire with chassis connected ground, two wire insulated, line-line with earth ground, transformer isolated
Power for controls for appliance	Describe the energy source for the controls used.	Electrical controls—110-V AC, mechanical switches, pneumatic switches, hydraulic servo valve
Internal circuit protection and disconnects	Describe mechanisms used to protect *internal* functions of the appliance, and describe their "as found" state.	GFCI, surge protectors, in-line fuses/circuit breakers, disconnect switches, thermal cutout switches
Operational controls	Describe the appliance control components and their settings as found.	Dials, switches, power settings, thermostat settings
Feedback devices and sensors	Describe the control feedback condition as found.	Temperature sensors, pressure/flow sensors, level sensors, position feedback
Movable parts	Describe the condition and position of important moving parts.	Doors, vents, valves
Cleaning/cooling/heating components	Describe the condition of filters and cooling/warming portions of the appliance.	Filters, cooling fans/pumps, heaters
Clocks and timers	Describe the state of timing mechanisms (if available) in the appliance.	Clock hand position, timer position

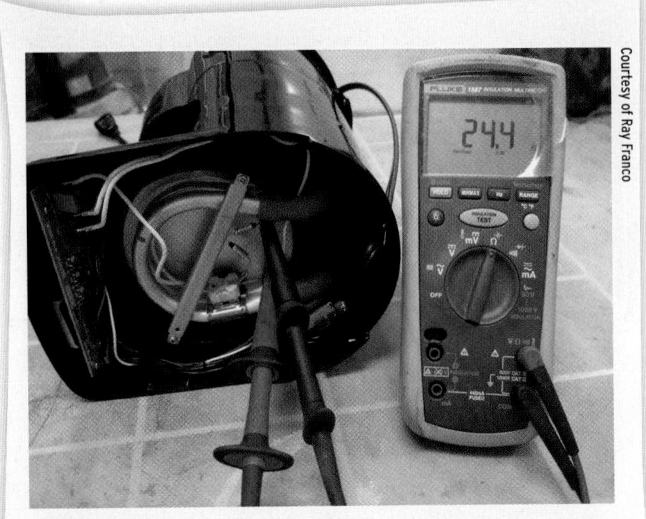

Courtesy of Ray Franco

FIGURE 24-1 This ohmmeter is being used to check the electrical continuity of a coffeemaker.

if possible. Make every effort to recover the property in its original state (as found), with the least damage possible. If necessary, the investigator should plan for a truck or trailer to transport the appliance to a secure location. This transportation should include covers, padding, chains, bags, securing devices, and so forth. Care should be used to keep the appliance upright if possible and to avoid wind or vibration from disturbing sensitive portions of the appliance as the vehicle is driven.

The appliance should not be tested or operated at the fire scene, as this could further damage the appliance and it may be unsafe to do. At this point, all testing must be nondestructive, such as testing for electrical continuity or resistance using a volt-ohmmeter FIGURE 24-1. X-rays may be useful to determine continuity of fuses, thermal protectors, heating elements, and relay contacts. Transformer and motor windings can sometimes be examined with X-rays.

Determining the Origin of a Fire Involving an Appliance

This section applies to and expands on the material in Chapter 6, Fire Patterns, and Chapter 18, Origin Determination, of NFPA 921. Once the origin has been determined to be that of an appliance, the techniques and methodology discussed in this section should be applied to the investigation.

Fire Patterns

Fire patterns should be used to establish that an appliance is located at the point of origin. The investigator should evaluate the fire patterns on the appliance in relation to the remainder of the fire scene. If the appliance shows more severe damage than surrounding items, this could indicate that the fire originated at the appliance. The investigator should keep in mind that plastic appliance parts may have severe damage or be missing, but that this is not necessarily an indicator that the appliance is the point of origin. All other causes for the fire damage that occurred at the appliance must be considered, including debris drop-down or other ignition sources that happen to be next to or on the appliance. The type of enclosure on the appliance must be considered in determining the origin area. Appliances with non-fire-resistant plastic housings can suffer severe fire damage as a result of exposure to fire. An X-ray may be required to see inside melted or collapsed masses to determine condition of key components as mentioned previously.

The investigator should verify that the appliance was connected to an electrical power supply and whether it was energized at the time of the fire. It is important for the investigator to trace and document the power source all the way back to the service panel. Circuit feed conductors should be recovered along with switches and, if possible, circuit protection devices, disconnects, and fuses. Requirements may exist for ground fault circuit interrupter (GFCI) or arc flash protection.

The reconstruction of the fire scene and replacement of the appliance in its prefire position may be necessary to document the fire patterns and provide supporting indication that the fire originated at the appliance. After the fire scene has been reconstructed, the appliance and scene documentation can proceed as previously discussed.

Determining the Cause of a Fire Involving an Appliance

This section applies to and expands on the material in Chapter 19, Fire Cause Determination, of NFPA 921. To determine how the appliance could generate sufficient heat energy to start a fire, the investigator should answer the following questions:

- Was the appliance attached to a power source or fuel source at the time of the fire?
- Was it energized (and/or operating) at the time of the fire?
- Was the appliance operating poorly or abnormally recently?
- Were there recent repairs or maintenance performed on the appliance?
- Was the appliance moved recently?
- Have there been any recalls on the appliance?
- Does the appliance appear to have been modified in any way?
- Could significant heat be generated under recent operating conditions?
- Was moisture present or a factor?
- Were nearby combustible materials coming into close contact with a heat-emitting part of an appliance?

The goal of these questions is to determine whether sufficient heating energy existed, what material the heat ignited, and how the material was ignited by this heat.

It is helpful to consider how the specific type of appliance being examined could most likely have contributed to the fire. For example:

- Water-using appliances, such as dishwashers, washing machines, hot water heaters, and outdoor appliances: Consider fire caused by water/moisture or the lack thereof.

- Heat-emitting appliances, such as heaters, transformers/motors, and clothes dryers: Consider fire caused by ignition of a nearby combustible material, such as clothes or lint/dust that has built up.
- Appliances that draw a large amount of current for an extended period of time, such as heaters and air conditioners: Consider fire being caused by a poor connection or an overloaded circuit.
- Heavy appliances, such as freezers or refrigerators: Consider a short circuit at the power cord caused by some recent moving that left the appliance sitting on top of the cord or by crimping or fraying it.
- Appliances with controls, timers, or electronics and interconnect cabling in them: Consider poor power or circuit board connection problems, relay, power switching or control component failures, or contamination of sensor connections.

Appliance Operation

The investigator should thoroughly understand how the appliance operated and its safeguards. The importance of this cannot be overstated. To understand the operational details sufficiently to assign fault for the loss could require destructive testing, which may alter appliance operation. A good reference source for the operation and design of appliances can be found in home appliance repair books. Sometimes the appliance manufacturer will provide, on request or online, design documentation such as schematics, part lists, operational guides, or software listings that are not normally available to the average consumer.

The investigator should document any modifications made by the manufacturer, end user, installer, or service personnel, especially when power connections or components are involved.

Evaluate and determine whether the electricity was connected, the lines to the appliance were energized, and the appliance was turned on at the time the fire occurred. These conditions must be met for determination of whether an electrical fault or overload occurred and whether sufficient heat was generated to cause ignition of the known surrounding materials (see the "Electricity and Fire" chapter in this textbook).

Disassembly

Usually, disassembly of an appliance should not occur in the field. Disassembly can result in claims of spoliation of evidence. Before beginning any disassembly, the investigator should identify the specific reason disassembly is needed and should create a protocol with the objectives, ground rules, and extent of disassembly identified based on the specific reason. Trying to reveal or clarify the condition or status of some aspect of the evidence is generally the objective sought. For example, some disassembly may be needed to identify the manufacturer of a product. Going beyond what is necessary in order to do this is not recommended. The investigator should disassemble only if he or she has proper expertise, has studied the appliance information, has considered the implications of evidence spoliation, and has given notification to interested parties, including the appliance manufacturer and the insured. Each step of the disassembly process should be documented. The investigator should take notes and photographs or videos documenting the entire disassembly process. X-rays may be considered if disassembly is not possible. Fuses, heating elements, relay contacts, and windings can be examined in this manner. Whenever handling appliances, the investigator should be aware of the possibility of sharp edges, moving parts, and even chemicals or other harmful substances associated with the appliances, and should wear appropriate personal protective equipment. Further, many appliances continue to draw electricity even when shut off, and some electric and electronic parts may remain charged with electricity for some time after disconnection of power.

Exemplar Appliances

An exemplar is an exact duplicate of the appliance in question. An exemplar can be used to help understand how an appliance operates and can be used in testing a proposed ignition scenario. The model and serial numbers on the appliance in question can help the investigator obtain an exact duplicate. If the exemplar is not exactly the same, the investigator must determine whether it is suitable to use. Testing should verify that the appliance could not only generate the necessary heat or electrical discharge, but could also ignite the fuel load. The Consumer Product Safety Commission (CPSC) should be contacted regarding any recall notices on the appliance.

Appliance Components

There are many appliances, and each appliance has a different use and is constructed differently. The following sections describe the common components of appliances.

Housings

The housing of an appliance is the outer shell of the appliance that contains the working components. Although most housings are made of metal or plastic, they may also be made of wood, glass, or ceramics.

Steel

Because of its strength and durability, most metal housings are fabricated from steel or stainless steel. Steel housings may be coated with enamel or plastic, which may be a factor when the appliance is exposed to fire. An example of this would be a refrigerator or washing machine.

Steel melts at high temperatures not normally found in structure fires. Ordinary unfinished steel oxidizes in fire and turns dull blue-gray. If the steel has been deeply oxidized, it may be possible to flake off the oxidation, or oxidation may have been severe enough to penetrate completely through the steel.

Postfire patterns and colors depend on many factors. Ordinary steel housings can have a mottled appearance with colors that include blue-gray, white, black, and reddish brown. Bare galvanized steel could have a whitish coating. Protective

Safety Tip

A steel housing does not shield internal components of the appliance from reaching high temperatures.

coatings on the steel can cause many varied colors when exposed to heat or flames.

Aluminum

Aluminum housings are generally made from formed sheets or castings. Pure aluminum has a low melting temperature of 1220°F (660°C). Aluminum alloys have slightly lower melting temperatures. Aluminum housing will reveal the extent of damage.

Other Metals

Other metals, such as zinc or brass, are used for appliance housings, often for decorative purposes. Zinc melts at 786°F (419°C) and appears as a lump of gray metal when it melts. Brass is an alloy and softens over a range of temperatures; however, it generally has a melting temperature of 1740°F (950°C). Although brass is sometimes used as a housing material, its primary use is for electric terminals.

Plastic

Plastic housings, or housings made from carbon and other elements, are used in appliances that do not normally operate at high temperatures. Plastics can melt at low temperatures and char and decompose at high temperatures. Some plastics will continue to burn on their own once ignited, whereas others may not due to added fire retardants or the chemical composition of the plastic.

Following a brief fire, the plastic housing could be melted and partially charred. The fire investigator must determine whether the heat source was inside or outside the appliance. X-rays provide a view of the internal metal components. In a severe fire, the entire plastic housing may be consumed, but this does not necessarily indicate that the fire started with the appliance.

Phenolic plastics are highly resistant to heat and are often used to make coffeepot handles and circuit breaker cases. These plastics do not melt and do not support combustion. Phenolic plastics form a thin, gray ash layer when moderately heated and turn to a gray ash in a sustained fire. A thin, gray ash layer on the inside, and not the outside, of a phenolic plastic component may indicate that the heating was internal.

Wood

Wood is occasionally used for appliance housings. It may be completely consumed in a fire or may have indicative fire patterns. The fire pattern will help identify whether the fire originated inside or outside the appliance.

Glass

Glass is most commonly used for transparent covers and doors and for decorative purposes on appliances. It readily cracks and may soften and drip under fire conditions.

Ceramics

Ceramics are generally employed for use as a novelty housing and do not melt in fire; however, the decorative glaze that often coats ceramics could possibly melt. Ceramic is also used to support or house the electrical components.

■ Power Sources

The power source for common appliances is generally the alternating current (AC) supplied by the electrical utility company. Information in this chapter is limited to single-phase power that is 240 V AC or less. Power in the United States is supplied at 60 Hz, 120/240 V AC. Most appliances operate on 120 V AC. However, appliances that require 240 V AC (e.g., dryers, ranges, hot tubs) can also work on the same system.

Electrical Cords

Electrical cords can comprise two or three conductors. These conductors are stranded to provide flexibility. Cords with two conductors are found on appliances made before 1962 or on some newer double-insulated appliances. Cords with three conductors are found on newer appliances on which the third conductor is used as a ground for the appliance. These cords can be used for 120- and 240-V AC appliances.

Stranded conductors usually survive fires and may be brittle following exposure to the fire. Electric cords are sometimes trampled on or end up under furniture or under the appliance itself, resulting in partial or poor connections or overheating. With high enough load current, poor connections increase the risk of fire due to resistive heating at the point of connection. Contact clearance spacings may be compromised due to mechanical damage, especially with higher voltages, resulting in faulting. Moisture or contamination can cause short-circuit current. Improper use of indoor cords outdoors or in damp locations can expose contacts to moisture and contamination, creating short-circuit current that can cause heating and even a fire at the contamination point or back at the outlet.

Plugs

Plugs made prior to 1987 and rated for 20 A or less have two straight prongs of equal width. Plugs made after 1987 for 20 A or less have the neutral prong wider than the "hot" prong. This is known as a polarized plug. Some newer plugs have a third prong for grounding. Plugs may have the conductors attached to the prongs inside a molded plastic housing. Some plugs have serial number information on the blades, which is useful for identifying manufacturing information such as date and location of manufacture.

Loose-fitting plugs can create a resistive heating connection, potentially causing a fire that can be identified by missing or partially burned away prongs and receptacles. Whereas the conductors and prongs usually survive a fire, other brass parts may melt. The receptacle does offer some protection to the face of the plug, which can then support a hypothesis regarding whether the appliance was plugged in.

Step-Down Transformer, Power Supply (Adapters)

Some appliances operate at lower voltages such as 6, 12, or 24 V AC or DC. A step-down transformer reduces the 120 V AC provided at the receptacle to the required voltage. The transformer may be a part of the appliance or may be separate from the appliance and plug into the wall receptacle (adapter) or be in-line with the power cord. The wire running from the adapter to the appliance is usually a thinner, two-conductor wire. Shorting of the lower voltage wire is not likely to cause a fire; however, some low-voltage transformers can produce significant energy and high enough current to make resistive heating connections a real problem. Low-voltage (12–48 V DC/V AC) loads are especially vulnerable because the power of a resistive connection varies with the *square* of the current (the current multiplied by itself). For a 12-V AC or DC load, the current is thus 100 times higher than for a 120-V AC load.

For example, to provide 120 W of power to a low-voltage lamp, 1 amp of current is required, but at 12 V AC or DC, this becomes 10 amps of current. If the load connection has a 0.3-ohm resistive connection, the 120-V AC connection dissipates 0.3 watts of power, whereas the 12-V AC or DC connection dissipates 30 watts of power. If embedded with quality combustible material, unprotected connections can cause a fire. This is why, if used outdoors, splices and connections need to be sealed and approved for outdoor use to avoid moisture contamination.

Appliance power supplies typically have protection for loads as described previously. They can use internal transformers or power switching/transforming electronics to isolate the load from the line voltage. These devices can become hot if covered in insulating material. Power supplies should be considered ignition sources because they typically convert high-energy power sources to low-energy loads, but the circuits to do this may malfunction and overheat. Many appliances and electronic devices contain unfused internal power supplies, which can overheat if an internal fault occurs. Some larger entertainment equipment power supplies feed remote devices such as speakers or monitors, which should also be examined for possible faulting. Cables for connecting peripheral components could be suspected if too thin or not the right type.

Batteries

Batteries are energy storage devices and possible ignition sources. Batteries are used for remote or portable devices. Devices that have larger batteries include computer backup systems; security panels; hand tools; and some larger equipment such as floor cleaners, lawn tools, and electric scooters. The style of batteries used with larger devices includes auto, marine, and motorcycle batteries, and the types of cells used are typically lead-acid, although lithium and NiMh cell packs are increasingly becoming available to power these devices. Laptop computer batteries and power tools typically use NiMh or lithium-ion cell packs because of their lightweight construction. Smaller AA/AAA/C/D-sized lithium and NiMh cells contain enough energy to cause a fire if a direct short occurs.

The remains of batteries are usually found after the fire, and it is important to determine what they were connected to. Cable connections from batteries to loads should be examined for faulting and heating (sleeving or melting of insulation). On systems with large batteries, like cars or vans, cables should be fused close to the battery. Fuses and/or breakers should be found and recovered to determine proper sizing.

Battery chargers are also possible ignition sources and are discussed later.

Under normal conditions, small battery-powered circuits generally do not allow sufficient heat buildup to cause ignition. There are, however, certain conditions in which even one battery can provide sufficient power to ignite some materials. Lithium batteries have been known to cause fires due to faulty wiring in electronics. In some cases, the solution in the lithium cells will physically leak, indicating a possible ignition source.

The voltage and energy density for a battery is thus important and is determined by its chemistry. TABLE 24-2 shows several different rechargeable battery types and their characteristics.

Lithium batteries have significantly greater energy density per volume (W hr/L) than lead-acid batteries. Compared with lead-acid, the lithium-metal-polymer battery has over three times the energy density.

FIGURE 24-2 shows the discharge curve for alkaline versus lithium batteries. This curve shows that lithium is a more potent energy source than alkaline in its ability to maintain full voltage during discharge. In a direct short situation, it is possible, but unlikely, that a few alkaline cells could ignite very fine combustible material.

Battery chargers are also possible ignition sources. Cables running from the battery to the charger must be examined for faulting or signs of overheating.

Battery chargers control and limit current required to charge the battery. Typically, they protect the load with fuses or circuit breakers. Some chargers use timers to control charging rate and time; others use the voltage and/or current supplied to the battery to control charging rate. Sometimes they control the charging power path based on battery presence.

These devices are similar to low-voltage power supplies, which have line voltage (120 V AC) input and a controlled DC output. Misconnection or poor connection of the charger

Table 24-2 Battery Energy Characteristics

Manufacturer	Type	Chemistry	Nominal Cell Voltage	Nominal Capacity (Amp-hrs)	Watt-hr/L
Mfgr 1	AA	LiIon	3.6	0.75	365
Mfgr 2	AA	Li Metal	3.0	0.8	324
Mfgr 3	AA	NiCad	1.2	1.0	200
Mfgr 4	AA	NiCad	1.2	1.0	200
Mfgr 5	AA	NiMH	1.2	1.5	246
Mfgr 4	AA	NiMH	1.2	1.5	246
Mfgr 6	AA	Rechargeable alkaline	1.4 to 0.9	1.6	220
Mfgr 7	Box	Sealed lead acid	12	12	103
Mfgr 8	Box	Sealed lead acid	12	12	99

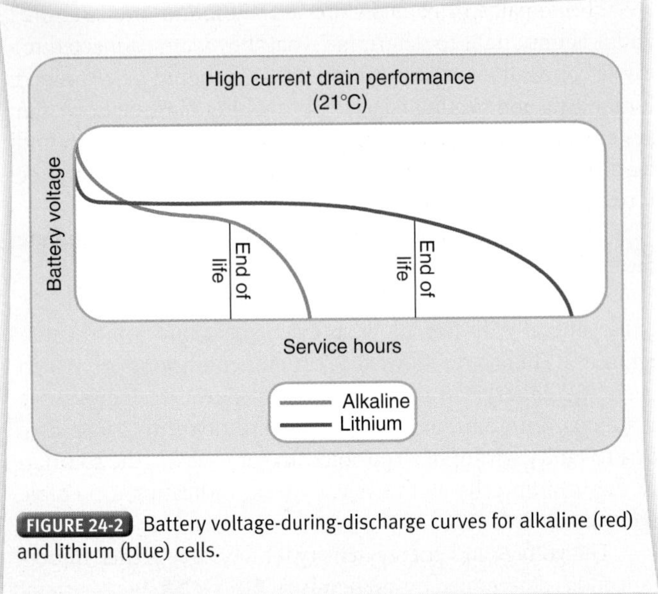

FIGURE 24-2 Battery voltage-during-discharge curves for alkaline (red) and lithium (blue) cells.

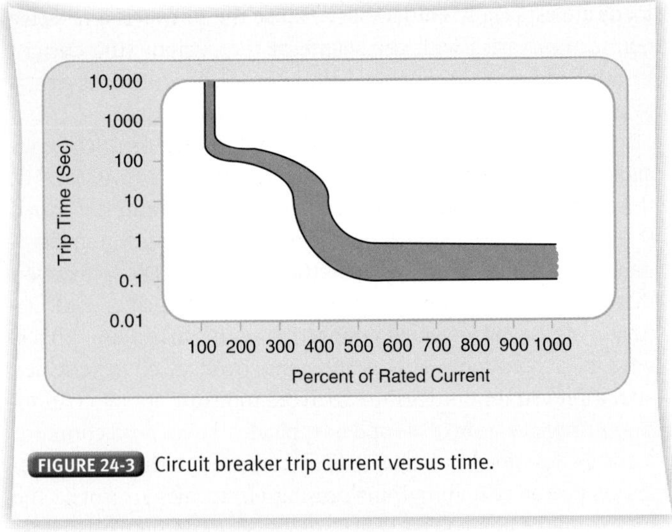

FIGURE 24-3 Circuit breaker trip current versus time.

to battery terminals is a common error to look for. Because of resistive heating connections, nearby combustible materials can be heated to ignition temperatures. Some batteries, if improperly charged, can themselves become overcharged and explode or violently expel hot electrolytic contents, becoming an ignition source. While charging, batteries are often (but not always) disconnected from the equipment that they power during normal operation. This may eliminate the equipment from being a candidate ignition source.

Protective Devices

Overcurrent protection devices employed in appliances and electrical service panels are often fuses and circuit breakers. Remains of the device after a minor fire may show whether the device was being operated. Remains of the device after a major fire may show only that the protective device was present.

If the current is moderate (less than twice the rating), the fuse metal simply melts, thus breaking the circuit. If the current is excessive, the metal vaporizes, leaving an opaque deposit on the glass tube or window.

Circuit breakers may operate thermally or magnetically. A circuit breaker in a fire can trip either as a result of being heated from an external fire or when the current flow exceeds the rating of the circuit breaker. The higher the overcurrent, the faster the breaker trips. Circuit breakers fill a safety, rather than a control, function because they are usually not accurate. The trip time is expressed by plotting current required to open the circuit versus time. It has a maximum and a minimum, indicating that there is a range of times at each current. These values can be different by 100 percent or more, indicating that the trip point is not accurate. The value of current that the breaker must be able to carry and still operate is called the interrupt current. This is typically 1000 A or more. Circuit breakers in appliances may be reset by a lever or a button.

Circuit breakers are usually employed to protect the wiring in the downstream circuit and will not protect upstream wiring or circuits. FIGURE 24-3 shows an appliance circuit breaker trip current versus time plot.

Switches

Many different switch designs are used in common appliances. Switches are used to turn the appliance on or off or to change operating conditions FIGURE 24-4. Postfire examination of a switch can sometimes determine its state and thus whether the appliance was energized at the time of the fire. The investigator should not operate the switch. The electrical

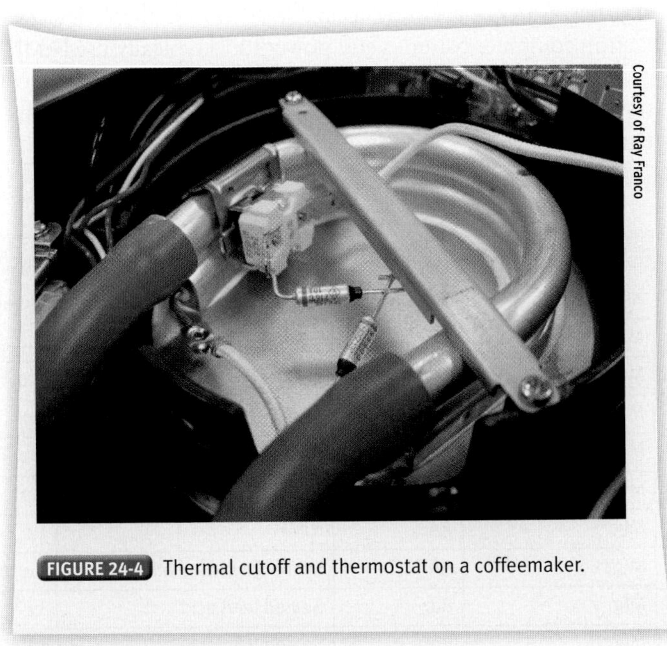

Courtesy of Ray Franco

FIGURE 24-4 Thermal cutoff and thermostat on a coffeemaker.

continuity of the switch should be checked while it is in place. Consider that temperatures, arcing, or mechanical damage could affect the resistance reading. The remains of the switch can be very delicate. Careful documentation of the position of the knobs, levers, or shafts should be performed while the switch is in place. The best situation is where the switch is located in a non-fire-affected area of the appliance or electrical circuit. It is important to make sure that the power to the appliance is off or that the appliance has been unplugged before testing. The investigator should not disassemble the switch unless he or she is qualified to do so. Fire impinging on a switch may destroy the housing, but the switch contacts can still show abnormal operation and should be preserved as valuable evidence.

TABLE 24-3 shows several different switches used in appliances and how they operate. Many of these switches also do double duty and act as sensors; appliances are cost-sensitive products.

Fluid Pressure (Capillary Tube)

A fluid pressure switch operates by fluid in a sensing bulb that is located in an area that can get hot. The fluid expands in the bulb when heated and applies pressure to a bellows device, which in turn opens a set of contacts, breaking the flow of electricity. High temperature can cause the fluid to undergo a phase change, which forces it out of the bulb. The resulting pressure increase of the fluid may crack the bulb or split the tube.

Bimetal

Bimetal switches are much more common. They are composed of two dissimilar pieces of metal that are joined together to form a flat piece. One type of metal expands to a different extent (more or less) than the other type, causing the combined metal structure to bend. This bending motion can actuate a set of electrical contacts blocking or permitting the flow of electricity. After a severe fire, the bimetal device can be distorted far beyond its operating position. This can be a result of heat from the fire and does not necessarily indicate a defective component.

Expanding Metal

The expanding rod switch employs a long rod that is exposed in the heated area. As the temperature rises, the rod expands, opening a set of electrical contacts.

Melting

Some cutoff devices operate by means of a material that melts when the normal operating temperature is exceeded. When the material melts, it releases a spring that opens a set of contacts that stops the flow of electricity. Consider that these devices (and, in general, fuses) may be deliberately bypassed to allow the appliance to be operated without any thermal protection.

Motion Switch

Appliances such as a portable electric heater usually employ a motion (tip-over) switch to ensure that the appliance is operated in its correct designed position. A typical motion switch uses a weighted arm that hangs down and would open the contacts should the heater be tipped over. Another method is a mercury (or other conductive fluid) vial that flows away from contacts when the appliance is tipped.

Contact Damage

Mechanical switches can fail because of an overload that can overheat internal parts or weld the contacts. Poor internal

Table 24-3 Common Switch/Sensors Used in Appliances

Switch Type	Property Sensed	Actuation Method	Mechanism
Manual switch	None	Pushbutton	Mechanical contacts
	None	Throw lever	Mechanical contacts
	None	Rotate knob	Mechanical contacts
Circuit breaker	Overcurrent	Current heats bimetal element and energizes coil	Mechanical contacts
Fuse	Overcurrent	Temperature melts fuse	Fuse
Relay	None (computer control input)	Current energizes coil	Mechanical contacts
Solid state relay	None (computer control input)	Low-voltage command from computer	Solid-state transistor, MOSFET, Triac
Bimetal thermostat	Temperature	Temperature coefficient of expansion of different metals actuates switch	Mechanical contacts
Thermostat	Temperature	Temperature expansion of single metal actuates switch	Mechanical contacts
Capillary thermostat	Temperature	Temperature expansion of liquid actuates switch	Mechanical contacts
Thermal cutoff	Temperature	Temperature melts fuse, actuates switch	Mechanical contacts
Tip-over switch	Orientation	Switch contacts displaced by gravity	Mechanical contacts
Door close switch	Open/close position	Switch contacts displaced by door in place	Mechanical contacts
Pressure switch	Water level or air pressure	Switch contacts enabled/displaced by pressure in line	Mechanical contacts

connections can cause destructive heating and failure. Switches normally show damage created by a parting arc on their metal contact surfaces. Contacts designed to be operated in normal use, such as in a thermostat, are normally pitted because of frequent opening and closing. Safety cutoff switches, however, should not have pitting on the contact faces. If they do, this is a sign of frequent abnormal operation and possibly evidence of a defect. Most switches are designed to snap open or closed to avoid surface pitting, erosion, or possible welding. Damaged or welded contacts do not necessarily prove that the switch caused the fire. Electrically welded contacts typically have normal shapes and have their faces stuck together. Contacts that are melted together in one lump were probably exposed to external heating.

Contact misalignment may be a function of fire heat damage rather than operational failure of the switch.

Solenoids and Relays

Electromechanical solenoids and relays are used to control high-power circuits with a low-power circuit. The remains of these devices normally exist after a fire, and their contacts should be inspected to determine whether they were stuck together during the fire.

Electronic Relays/Switches

Some appliances use electronic switches to turn on loads. These are typically semiconductor components called transistors, MOSFETs, triacs, SCRs, or IGBTs. These devices conduct electricity by semiconductor action and are called *solid state* because there are no physical moving parts and thus no contacts (reduction of contact bounce or chatter). Therefore, it is not possible to determine their switch status at the time of the fire. To be as electrically robust as mechanical switches, they need to be specified as "protected" switches or solid-state relays, which typically have current limit and over-temperature protection, and sometimes overvoltage protection. A small current or a voltage is applied to the gate of the switch, which causes it to allow current flow to a load in the appliance. This type of switch is often used where control of the switches occurs via computer or a special algorithm from low voltage logic. A datasheet describing the operation of the switch can be obtained from the switch manufacturer.

Additional elements are sometimes added ahead of the switch. Examples include a surge protection device such as a metal oxide varistor (MOV) or a series PTC (positive temperature coefficient) resistance to control overcurrent. Some equipment, such as motor drives, may use soft start circuits to reduce inrush current to loads sensitive to current surges. Solid-state relays usually have added features such as zero voltage or zero current tiurn-off to reduce further electrical noise caused by switching. An advantage of electronic switching is that, depending on how the switch is used in the control circuit, contact arcing and switch bounce can be eliminated. Unfortunately, this makes discovery of abnormal operation and safety issues difficult because the evidence of failure is not visible by normal X-ray scan. It is often possible to have a failure analysis (FA) report generated by the manufacturer for the switch, which may give information about the failure mechanism of the switch.

Electronically controlled loads, such as motors, are controlled by semiconductor switches, as described previously here. These require power switching electronics called *drives*. There are several different types of motor drives, with different types and ratings used for their electronic switches. A common control method used by all, however, is the controlled timing of electronic switches to deliver current to the motor windings in the proper sequence to produce the torque required to move the motor rotor. Motors with greater torque require greater current and those with faster speed require greater voltage amplitude, and thus, it is possible to calculate approximately the amount of power consumed by the motor (and thus delivered by the drive). Depending on the amount of power consumed during operation, if a fault occurs in the motor winding or in the electronic switch feeding it, the resulting energy discharge can be sufficient to ignite combustibles nearby, inside the motor, or on the drive circuit board. Various types of switches and switch topologies are used to control motors. A description of these is beyond the scope of this textbook; however, this information is available in application manuals from various power semiconductor and motor drive manufacturers.

Other types of electronically controlled loads require timed electronic switches, which act in a similar manner to motor drives. These may be lighting, heating, or cooling devices or subassemblies within an appliance controlled by precise application of power to the load. A common example is a phase-fired triac controlling a lighting or heating element. The triac is a semiconductor switch that conducts current when the gate is turned ON and blocks current when the gate is turned OFF. Exact control and operation of the triac are beyond the scope of this textbook, but can be explored further by examining circuit diagrams and datasheets.

Transformers

Transformers are devices that reduce AC voltage, usually 120 V AC or 240 V AC, to a lower voltage and can be used to isolate the appliance from its power source. The winding insulation in the transformer may deteriorate after extended use at high ambient or internal temperature. As the winding insulation deteriorates, the impedance drops and more current flows, which in turn generates more heat. This can lead to severe heating, which can cause the windings to fail by melting or create a ground fault. The heat that is generated may ignite the winding insulation or combustibles in the vicinity of the transformer. Components of a transformer often survive a fire. Internal damage of the windings may be shown by a pattern of internal heating, arcing from turn to turn. A thermal cut-off switch may be mounted on the windings to remove power when maximum rated temperature is exceeded. If the transformer is encased in steel, it is unlikely that temperatures could be hot enough to ignite nearby combustibles. However, if there is not a steel covering and if there are paper or plastic winding forms, these can themselves ignite and cause flames to impinge nearby combustibles.

Motors

Motors range from 1/3 to 1 horsepower in major appliances and to small fractional horsepower motors in smaller appliances. Small motors that drive cooling fans are generally not sources of ignition. Typically installed in bathrooms, small shaded pole motors are open frame construction, exposed to moisture and thus accumulating dust and lint on the motor, which can cause overheating, resulting in a fire. These motors are required to have thermal cutoffs to prevent overheating. AC (induction) motors in major appliances usually have starting windings with a centrifugal switch to disengage these windings after startup. The starting windings contain a start capacitor that should be examined because it may have failed or decreased in value with age, causing the motor to take longer or even stall during startup and resulting in overload current in the windings. This can cause the windings to heat sufficiently to ignite the insulation and plastic materials often found around the motor. Protection for the motor is often a fuse link or a thermal cutoff switch that may be mounted on the windings. Motor bearings can overheat and ignite nearby combustibles because of loss of lubrication or increase in friction.

A reduction in the price of magnets has made the permanent magnet motor popular for appliances. These motors replace the rotor windings with permanent magnets and sometimes remove the commutation brushes, which can be a source of arcing. This arcing, although it is controlled, may ignite nearby combustibles.

Stepping motors are used in small movement devices, usually electronics like printers, clocks, and copiers. Because these are typically low-voltage devices, they are not likely to produce the energy needed to ignite a fire.

Heating Elements

Heating elements can ignite combustibles that are in contact with the element itself. Appliances with heating elements are designed to maintain a distance between the element and surrounding combustibles. Sheathed elements are found in ovens and ranges and are made of a high-temperature, high-resistance wire, usually nickel-chromium-iron, which is sometimes called nichrome wire. This wire is surrounded by an electrical insulator and encased in a metal, possibly steel, sheath. Baseboard and space heaters have sheaths made of aluminum. Aluminum sheaths generally melt from external fire exposure. Open elements are composed of wires or ribbons constructed from nichrome. Some appliances use a fan to remove heat from the element and to distribute it. Heat tape, used to warm pipes near appliances, consists of two specified lengths of a known high-resistance element that are electrically isolated except where they are connected at the end of the wires. If a resistive heating connection occurs at the crimp connection point, ignition temperatures can result.

One element method uses steel wire, and another uses conductive plastic. The investigator should know the required resistance of the heating element, which can be tested for resistance with the use of a volt-ohmmeter.

Consider that due to the high amount of current required for electric heating appliances, high-resistance heating can occur if poor connections are found. Also refer to the installation guidelines to see whether electric power cords should be used with heating appliances.

Lighting

Appliances often employ lighting to illuminate work areas, dials, or internal cavities. This lighting is normally of low wattage and is not prone to ignite combustibles. The lighting types normally found are incandescent and fluorescent. Light-emitting diode (LED) lighting is becoming more popular with the advent of high-intensity LEDs, but the power level is still usually lower than what is required for ignition. Some types of higher wattage incandescent lighting may ignite combustibles if they are in contact, such as if an incandescent droplight were taken into an attic and fell into cellulose insulation. Fluorescent lighting operates at a higher voltage; however, the tubes normally do not get hot enough to ignite combustibles.

Recessed lights also generate significant heat, and if the top of the fixture is covered by insulation, the insulation could ignite. Recent building codes specify the use of thermal cutoff switches to interrupt the circuit above a certain temperature. Fixtures with thermal cutoffs and with protective separation of hot bulbs from insulation are marked with an "IC," which means *insulation contact*. Consider that any lights installed in bathrooms are susceptible to moisture and dust accumulating at the terminals, which can lead to short circuits.

Fluorescent Lighting Systems

Fluorescent lighting systems are commonly employed in office settings. They use one or more glass tubes filled with a starting gas and low-pressure mercury. An electrical discharge is sent down the length of the glass tube, exciting the mercury gas. Two methods of starting fluorescent lights are used: the ballast discharge system, which is more common, and the preheated filament system. Ballasts are typically transformers (sometimes called magnetic), but electronic ballasts are more popular, especially in compact fluorescent lamps (CFLs) and work on a switched transistor principle. The ballast is used for creating the high startup discharge voltage required to establish a starting arc across the lamp electrodes. This process is enabled by the starting gas, which results in the mercury vapor creating ultraviolet light that is converted to visible light by the coating on the inside of the tube, known as the phosphor or fluorescent powder. Once the current has started flowing and the lamp is operating, the ballast acts to limit the current through the tube.

There are two main types of fluorescent light ballasts: magnetic and electronic ballasts. The ballast in fluorescent lights can overheat because of internal short circuiting and ignite combustible ceiling materials. Magnetic ballasts incorporate either a reactor or a transformer. Interior fluorescent light fixtures manufactured after 1968 are required to have thermal protection in the ballast.

A "P" on most metal housings indicates that thermal protection exists. All fixtures, indoor and outdoor, manufactured after 1990 are required to have thermal protection in the ballast; however, thermal protection does not ensure that a failure cannot occur. Both the electronic and magnetic fluorescent light ballasts contain pitch or potting compound within the ballast. This provides better heat transfer, reduces noise, and holds the internal parts in place. This pitch can ooze out because of either internal heating or fire exposure. This pitch will normally not ignite other materials unless it is already burning. The electronic ballasts in CFLs use semiconductor switches (transistor/MOSFET/triac) and relatively small passive circuits (capacitor/inductor) switching at high frequency to create the electrical arc required. Some electronic ballasts employ self-resetting thermal protectors, whereas others use fuses for thermal protection. The ballast control ICs usually contain overcurrent protection, which must work correctly for the switching action to be present and for these ballasts to work correctly. For an unsafe condition to occur, the IC must malfunction (many ICs have protection against this as well), and the semiconductor switches must short circuit in order to generate significant heat.

Common failures of magnetic ballasts that initiate fires include arc penetrations into combustible ceiling materials or nearby combustibles and extreme coil overheating that conducts heat into nearby combustibles. If the investigator suspects a ballast failure as the cause of a fire, the ballast, along with the fixture and the wiring, should be preserved for examination by qualified experts. The fixture itself, along with conductors feeding power to it, should be examined for failures, such as arcing nearby or inside the fixture and failure of the lamp holders.

High-Intensity Discharge Lighting Systems

High-intensity discharge (HID) lighting systems are often employed in warehouses, manufacturing facilities, or "big box" retail stores. They utilize a lamp that has a short tube filled with various metal vapors, such as sodium, mercury vapor, or metal halide. An electrical discharge is created along the length of the tube, exciting the metal vapors in the tube, which, in turn, creates light. HID lights operate at higher pressures than common fluorescent lighting systems. HID systems also employ a ballast and a capacitor for starting voltage and for limiting the current flowing through the lamp. These ballasts may be electronic or magnetic, and some of them are protected by fuses or thermal protectors. The ballasts are typically mounted inside the fixture enclosure above the lamp holder and reflector assembly.

Most mercury and metal halide lamps use a cylinder of fused silica/quartz with electrodes at either end. This cylinder is called an arc tube. The arc tubes of metal halide lamps are designed to operate under high pressure, from 5 to 30 atmospheres, and at temperatures up to 2012°F (1100°C), and can unexpectedly rupture due to internal or external factors such as a ballast failure or misapplication or if finger oils are present. The mercury vapor arc tubes can operate in the range of 1112°F to 1472°F (600°C to 800°C) and at pressures from 3 to 5 atmospheres. High-pressure sodium lamps operate at lower pressures, typically close to 1 atmosphere. If the arc tube ruptures for any reason, the outer bulb may break and pieces of extremely hot glass and quartz arc tube may be discharged into the surrounding environment, igniting combustibles where they land.

If the investigator suspects an HID lighting fixture as the cause of the fire, the whole fixture must be preserved for laboratory examination. The investigation of a suspected HID fixture requires the evaluation of the fixture's power supply, wiring, ballast, capacitors, lamp, and any lens present. The fixtures should not be disassembled at the fire scene, but carefully preserved for laboratory examination. It is important to find as much of the lamp remains as possible, preserving all pieces of the arc tube and shielding them from additional damage. The edges of the glass jacket and of the arc tube remains should not be handled, disturbed, or cleaned prior to the laboratory examination. The lamp frame and lamp base pieces should also be collected and preserved. Because of building collapse, firefighting, and overhaul, the presence of a fractured arc tube is not proof that an arc tube ruptured and ignited the fire. It is helpful to gather additional exemplar lamps and fixtures from the scene for comparison and to create a usage history of the lamps in that structure.

Miscellaneous Components

Dimmers and speed controllers are examples of miscellaneous components of appliances. Older appliances may contain rheostats or wire resistors. Components in newer appliances are solid state and are often destroyed in a fire unless the fire is of brief duration. Remaining components can be displaced from printed circuit boards because most solders melt by 400°F (204°C).

Timers can be built into the appliance or stand alone as separate devices. They can be driven by small motors and be badly damaged after a fire. Failure of the timer usually results from gears wearing out or losing teeth and is generally not the cause of the fire.

Thermocouples and thermistors measure temperature and, if they are damaged or if their connections are contaminated, could compromise the operation of the appliance.

A thermopile is a series of thermocouples. They are used in gas appliances to keep a valve open when the pilot is burning. Newer appliances use electric igniters instead of standing pilots.

TABLE 24-4 shows some sensors commonly used in appliances. The sensors already mentioned in Table 24-3 are not included here.

Common Residential Appliances

The components and operation of common appliances are discussed in the following sections. This should help the investigator understand how common appliances work.

Ranges and Ovens

In the common household range or oven, heat is provided by electricity passing through resistance heating coils or by the burning of natural gas or propane FIGURE 24-5. The temperature in the oven is controlled by a thermostat and valve or

Table 24-4 Common Sensors Used in Appliances

Sensor Name	Property Sensed	Sensor Type, Output	Typically Controls
Thermocouple	Temperature	Dissimilar metal contact produces millivolt signal	Heat-producing appliances
Thermistor	Temperature	Thermal-affected change in resistance	Heat-producing appliances
Pressure sensor	Differential or absolute pressure	Strain measurement, change in resistance	Appliances controlling water level or air flow
Flame sensor	Presence of flame	Infrared detector, microvolt signal	Gas valve, blower motor
Current sensor	Current	Current passes through series resistance causing voltage drop, or magnetic measurement causes change in resistance	Current-controlled heaters or cooking appliances, variable speed motors
Proximity sensor	Presence or absence	Capacitance sensor, magnetic sensor, or optical sensor	Door sensors, touchpad, motor speed sensor

switch on the fuel or power supply. In a gas range, a burner fuel supply valve that is manually operated controls fuel flow rate. Ignition is by a standing pilot flame or by an electric igniter. In an electric range, a timing device (temperature control) cycles the flow of electricity through the heating coil to maintain the selected temperature of the electric range. The timing device controls the ON versus OFF period of electric current and is typically a knob, dial, or digital keypad that is manually adjustable. Thus, selecting a higher setting will leave the current ON longer and OFF less.

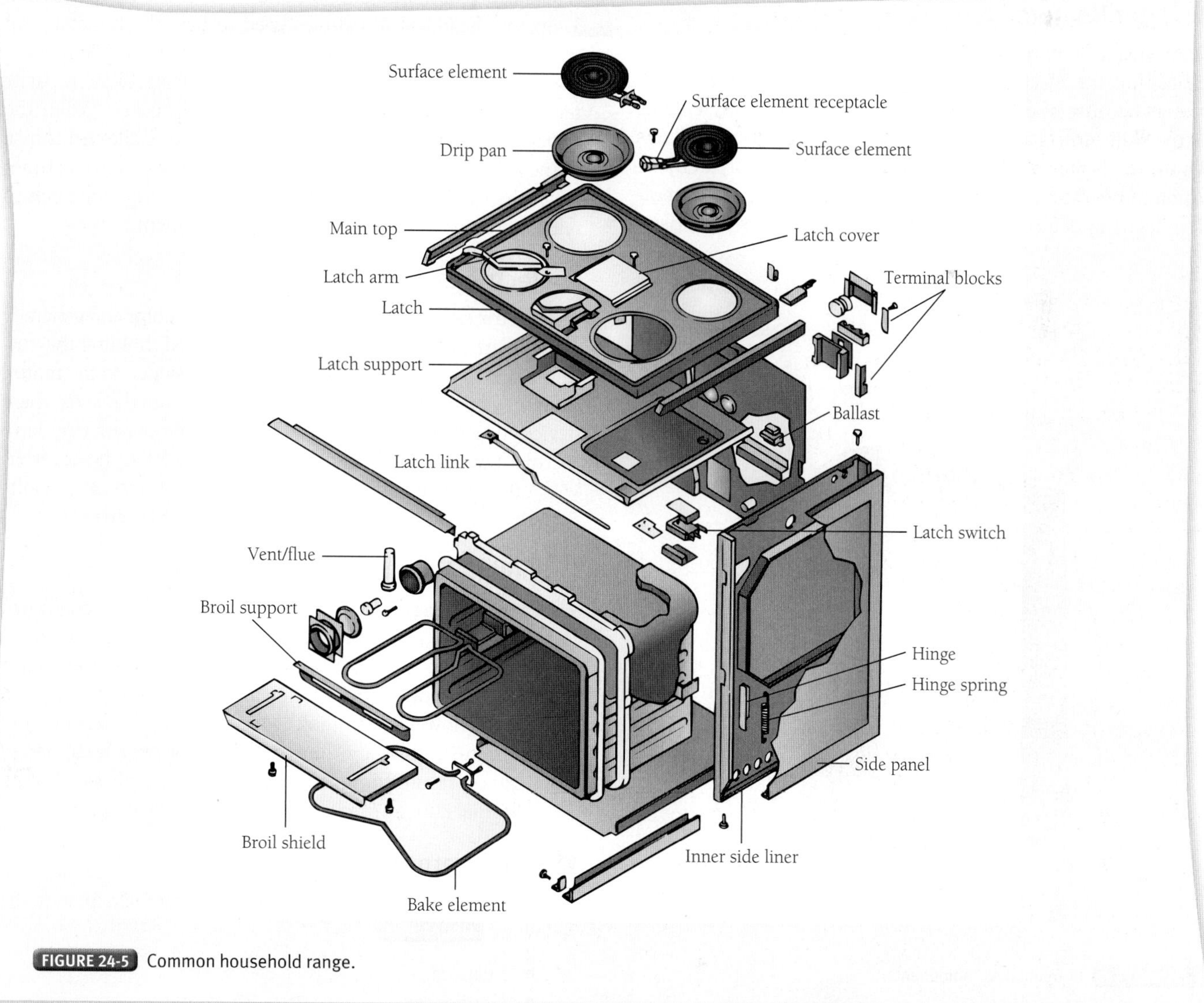

FIGURE 24-5 Common household range.

Although stoves and ovens are typically made of noncombustible materials, any combustible material left on the stove or in the oven while the stove or oven is turned on can catch fire. The postfire settings of the controls are thus important to document.

Air Conditioners and Heat Recovery Ventilators

Air conditioners and heat recovery ventilators (sometimes called HRVs or ERVs) are often operated continuously. Over years of operating in a dusty environment, the dust or contamination in the airflow path can settle over electrical components such as capacitors and cause tracking faults, leading to a high-resistance fault that could ignite nearby combustible material. Compressor and fan motors in these units can overheat because of bearing friction, increase in load, or capacitor degradation, causing excessive temperature rise. If properly designed, the motor's thermal protector will open and the motor will stop running before it reaches dangerously high temperatures; however, this may not prevent a fire if this condition is allowed to persist for a long time. Motor capacitors can develop internal resistance over time, which affects their performance and, if encased in plastic, can cause sufficient heating to self-ignite.

Water Heaters

One cause of fires from electric hot water heaters is moisture contacting the electrical controls. Leaks can occur in water heaters because of corrosion, particularly around the elements or the water inlets and outlets. Additionally, electrical circuits within the heater may come in contact with the polyurethane insulation in the heater jacket, allowing a fire to start and smolder, delaying detection. A lack of water also poses a problem of overheating and damaging control circuitry, which can malfunction and cause a fire.

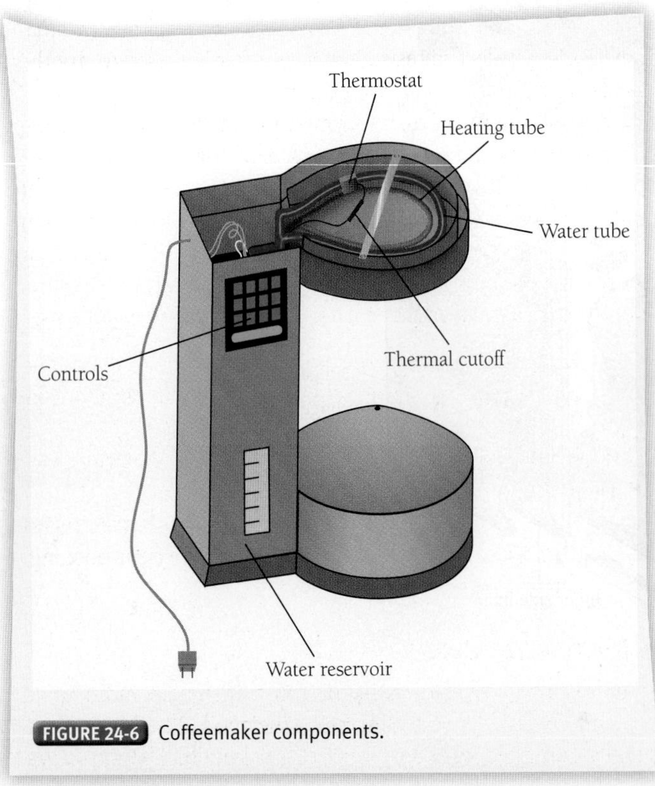

FIGURE 24-6 Coffeemaker components.

Coffeemakers

Standard coffeemaker components consist of a water reservoir, heating tube, carafe, and housing FIGURE 24-6. The heating tube heats the water flowing through it from the water reservoir and forces the water over the ground coffee. The coffee then drips into the carafe. The carafe sits on a warming plate that is heated by the same resistance heating element that heats the heating tube. The resistance heater is controlled by a thermostat that cycles the flow of electricity as needed. A thermal cutoff may be employed, located adjacent to the heating element, as a protection device. Some coffeemakers include automatic timers that turn off the coffeemaker after a selected period of time and turn the appliance on at a predetermined time.

Toasters

The common toaster uses resistance heaters with a sensor that controls the toasting time. The sensor is typically a bimetal strip that is adjustable for different levels of toast darkness. In some cases, the bimetal strip has been blocked by a piece of toast, causing the toaster to stay on and overheat. For this reason, a separate heater is sometimes used to heat the bimetal strip and remove it from the area where food can affect it. When the toasting timer indicates completion, a mechanical latch partly releases the food and turns off the bimetal heater. When the bimetal switch opens, after cooling, the food is allowed to rise to the release point. Newer designs employ an electronic timer to set the toasting time and utilize an electromagnetic latch to release the tray and turn off the heating element.

Electric Can Openers

The electric can opener uses a geared electric motor and generally can operate only when the lever is depressed, holding the cutting wheel in place and acting as the power switch to the motor. Because they operate only when the user presses the lever, these devices usually do not cause fires. Some can openers may have thermal protectors for motor overheat protection; however, as with other appliances, electrical arcing or high resistance faults can occur due to poor connections or damaged cords.

Cooling Fans

The capacitors in box fans and oscillating fans have been known to overheat and cause the plastic casing near the capacitor to ignite; however, the ventilation available around the motor inhibits the motors from experiencing a major fire. Installed in bathrooms, however, fan motors are exposed to moisture and can accumulate dust and lint on the motor, which can cause overheating, resulting in a fire. These motors are required to have thermal cutoff switches to prevent overheating.

Refrigerators

Components contained in a household refrigerator include the following FIGURE 24-7:

- Evaporator
- Condenser

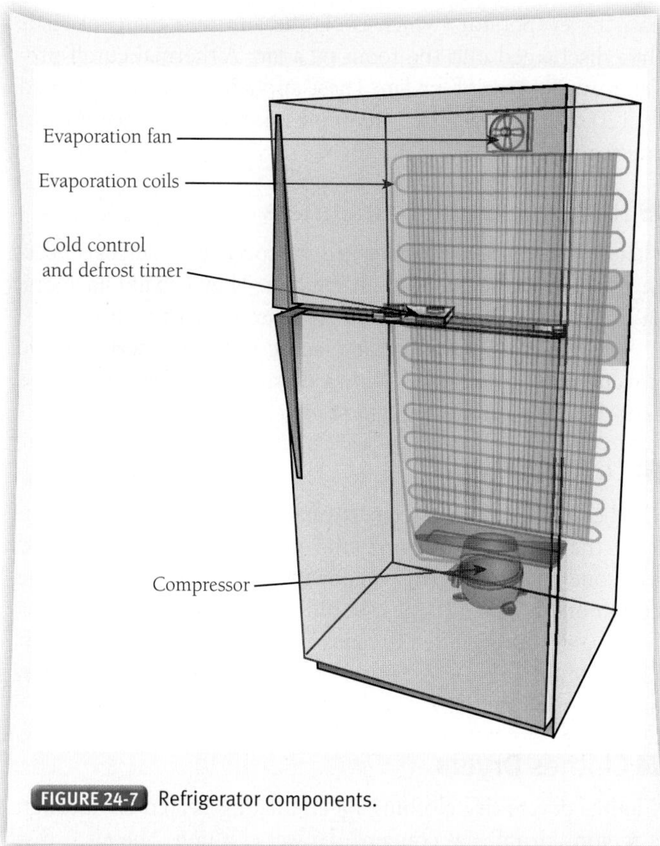

FIGURE 24-7 Refrigerator components.

- Compressor (motor driven)
- Heat exchange medium (fluorocarbon or Freon)
- Tubing to connect the components

Warm air inside older appliances evaporates the heat exchange medium (Freon), turning it to a vapor and transferring the heat to the Freon from the enclosure. The coolant vapor moves to the compressor, and the compressor compresses the vapor. As the coolant condenses to a liquid in the condensing coil, it gives off the heat picked up in the enclosure. The air around the evaporator is cooled, and the air around the condenser is heated. The cool air is circulated in the refrigerator, and the warm air is dissipated into the room. The cycle is controlled by a thermostat or a timing device. The compressor is typically powered by an electric motor equipped with a thermal cutoff switch contained in a sealed container that functions as a heat sink. Possible additional systems include the following:

- Lighting
- Ice maker/water fill valve
- Ice and water dispenser/water fill valve
- Fan for the condenser and possibly one for the evaporator
- Heating coils for:
  - Automatic defrosters
  - Prevention of water condensation
  - Drain pans and ice makers
  - Antisweat heaters

Freezers and refrigerators can cause fire if their power cords or plugs are damaged. For this reason it is important to ask the occupants if they recently moved these appliances, possibly rolling over and resting on the cords.

Recalls have been issued for refrigerators that have experienced fires caused by electrical compressor relay failures and faulty wiring.

Some older refrigerators have a plastic tray with elements to evaporate the defrost water. If the drain valve becomes blocked, this tray can overheat, crack, and expose the elements. When the blockage clears, water fills the tray and a short circuit and fire can occur because of the exposed elements.

Electric Water Dispensers

Failure of the thermal cutoff device or thermostat on water dispensers is a possible cause of fire with these appliances. If the reservoir empties, water can reach high temperatures and evaporate. If there is a leak, water may reach the electronics located underneath the water reservoir, possibly causing a fire.

Deep Fat Fryers

Deep fat fryers are normally found in restaurants and fast food outlets, though smaller models can be found in the home. The thermostat controls the temperature of the cooking oil; however, a thermostat could fail (most commonly when the capillary line connecting the temperature probe in the vat to the thermostat has broken), allowing the oil to overheat and ignite. When this happens, a charred or blackened residue is often found in the cooking vat.

Thermostat contacts have also been known to weld closed with these appliances.

Dishwashers

The dishwasher is a household appliance that employs a pump and a motor. The pump motor may or may not have thermal protection. An electric resistance heater may be used to dry the dishes after the water has been drained.

A major cause of dishwasher fires is moisture contacting the conductors, especially at the top of the door where the controls are. The plastic components that release the detergent could weaken over time, becoming brittle, cracking, and then leaking. When the door is opened, the water inside the door can reach the controller at the top of the door and cause a fire.

Other potential causes of dishwasher fires are as follows:

- Repeated opening and closing of the door over time can stress the wiring harness. The insulation breaks down and a short circuit can result.
- A combustible material contacting the electrical elements could also cause a fire.
- Some control modules in dishwashers contain relays that have failed and self-ignited the plastic housing, causing a fire.

Microwave Ovens

A microwave oven uses a magnetron to generate radio wave (i.e., microwave) energy that heats the food. Microwaves are usually equipped with a thermal cutoff switch. The appliance is usually equipped with internal lighting and possibly a food rotation tray, and it may have thermal cutoff switches above the enclosure. Radiation that is emitted will excite water molecules; if there is no moisture available, other, less volatile molecules will absorb the radiation and start to heat. Therefore, any material that has dried out can overheat and ignite in a microwave oven. Also, any metal objects inside a microwave oven can cause severe sparking, which could lead to a fire.

Space Heaters

There are generally two kinds of space heaters. A convection heater employs a fan to move air across a hot surface, heating the air and dispersing it throughout the room. A radiation heater transfers heat by radiation only. These appliances often are equipped with tip-over switches or other protective devices.

Although heaters are one of the greatest causes of fires, that has more to do with the large electrical currents involved than with the safety of the appliance. Any use of increased current increases the fire risks as well. When designed and used properly, electrical heaters are a safe form of heating.

The following conditions are known to cause fire in fan-forced space heaters:

- Combustible material getting caught
- The heating element detaching because of rough handling and later igniting the plastic casing
- Restriction of the inlet air, but not the over-temperature sensor, causing the heating element to overheat and ignite the plastic casing

The fire risks associated with radiant heaters are that a combustible material is left too close to the heater or the heater falls over and ignites the combustible items on the floor.

Electric Blankets

An electric blanket employs an electric heating element inside the blanket. The controls are typically located on the blanket power cord. Thermal cutoffs are located near (and usually sewn onto) the elements in the blanket. There could be as many as 12 to 15 thermal cutoffs. The blanket is designed to be laid flat and could char and ignite if operated while folded or bunched up.

Window Air Conditioner Units

Window air conditioner units are designed to be placed in a window to cool a room. They function and operate similarly to refrigerators and use similar components and principles for cooling. The air from the room is circulated through the unit, past the evaporator (which cools the air), and the cool air is then discharged into the room by a fan. A thermal cutoff prevents overheating of the fan. These appliances may be powered by 120 or 240 V AC. Fan or compressor motor overheating can cause a fire.

> **Fire Investigator Tip**
>
> It is important to interview the property owner about how the appliance was used or possibly misused.

Hair Dryers and Hair Curlers

Hair dryers employ a high-speed fan to force air across a heating element to heat air for drying hair. They usually have one or more thermal cutoff switches that are resettable.

Hair curling irons use an electric resistance heater inside a wand that is equipped with a thermal cutoff switch. Some models allow water to be added to generate steam.

Clothes Irons

Clothes irons use an electric resistance heater that can heat in both the vertical and horizontal positions and are provided with one or more thermostats or thermal cutoff switches. The condition of these switches should be checked if suspected of a fire. Water can be added to many models for generating steam. Controls on these appliances include temperature selection and an on/off switch.

Clothes Dryers

Clothes dryers dry clothing by circulating heated air through a rotating drum that contains the wet clothing. The air is discharged from the drum through a filter (lint trap) and into a duct that is typically discharged to the exterior of the house. The dryer uses either electricity in 120 or 240 V AC or the combustion of fuel gas to heat the air. The components in a clothes dryer are timing controls, humidity sensors, heat source selectors, intensity selectors, thermal cutoffs, blower motor, and heating elements. FIGURE 24-8 shows the internal components of a clothes dryer.

A clogged lint filter can blow lint back into the interior of the dryer, which can then settle across the base, eventually reaching the element and igniting. Frictional heating from a piece of clothing caught between the moving parts may cause a fire, and the heating of contents such as vegetable oil–soaked rags and plastic bags may cause them to spontaneously combust. Check the contents of a dryer after a fire has occurred.

Clothes dryer design includes a flexible bearing, usually made of a composite of PVC and other synthetics or cotton, on which the drying drum rests. If this bearing comes out of position, it can contact the element and ignite. Because of PVC's fire-retardant properties, these bearings can smolder for very long periods undetected before igniting the clothes inside the drum. A fire from a clothes dryer can therefore occur hours after the dryer has been disconnected from power.

Consumer Electronics

Consumer electronics are very common household appliances. Some examples are radios, compact disc (CD) players, BluRay or DVD players, video cameras, gaming systems, and personal computers, including cell phones and tablets. All of these appliances have a power supply, circuit boards, and a housing.

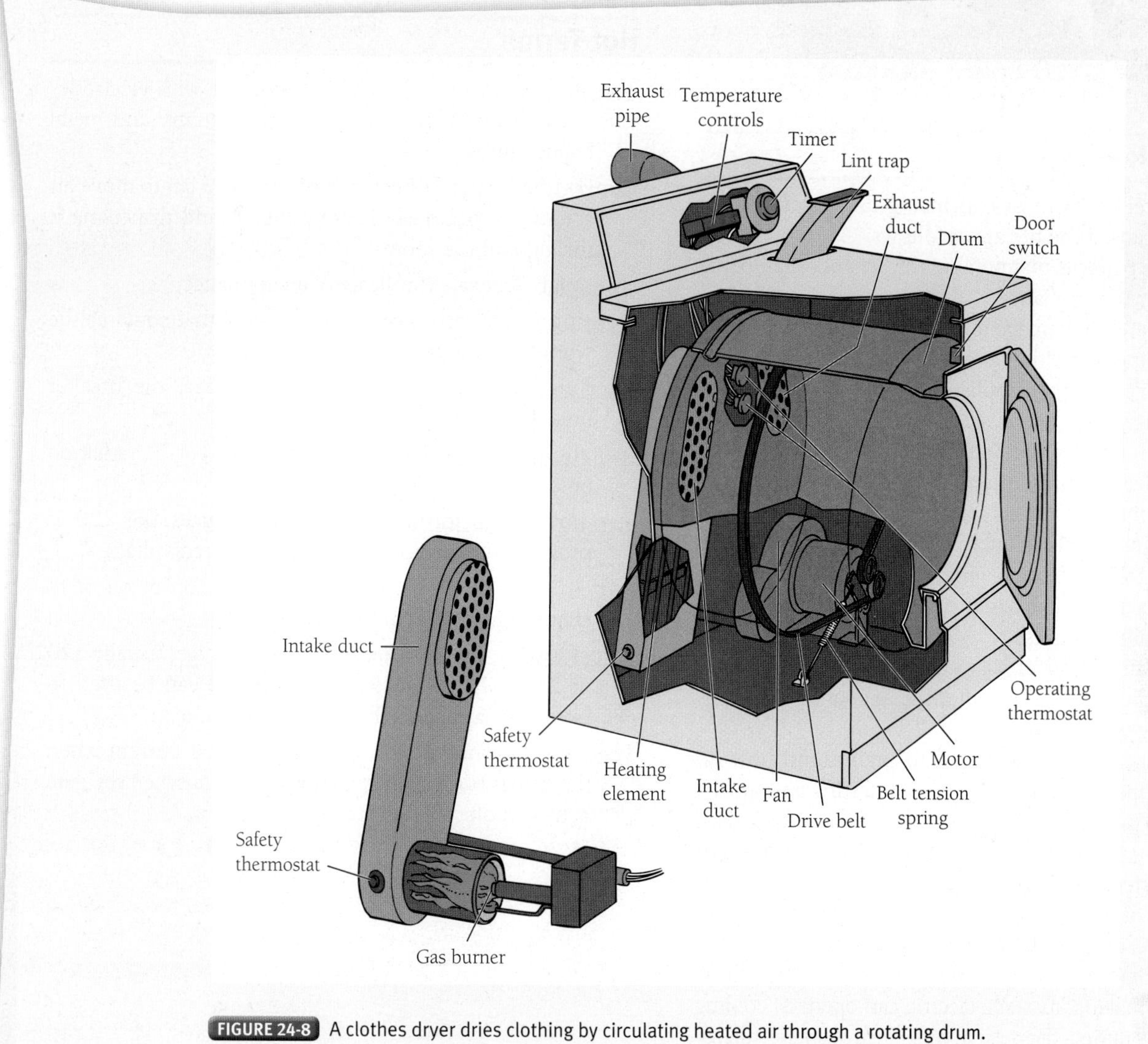

FIGURE 24-8 A clothes dryer dries clothing by circulating heated air through a rotating drum.

Electronics with PC boards (including television sets, stereo components, and computers) have been recalled in recent years, mainly because of poor solder joints and faults in the transformers. When components in a high-energy circuit (power supply) are not soldered well enough, connections worsen with time and use, resulting in carbon tracking on the board. Over time, resistive heating can cause a fire.

Solid-state switches or MOSFETs are used to drive speakers, displays, and motors. If these devices come loose from tight bonding to their heat sinks, they can short circuit and explode. Resulting sparks can ignite nearby combustible material.

Portable equipment, such as laptop computers, uses lithium batteries that, if overcharged, can discharge a flammable liquid that is sufficiently hot to start a fire in nearby combustible material.

Entertainment system/television enclosures are manufactured using fire-rated plastics that do not easily ignite; however, because so many different plastics are used in the manufacturing process and their fire ratings vary, it is still possible for electrical faults to cause a television set to catch fire.

Older fax machines can generate sufficient heat to start a fire if covered with combustible material because they have a thermal print head.

Wrap-Up

© Greg Henry/ShutterStock, Inc.

Ready for Review

- In a fire scene, the most convincing indicators that an appliance caused the fire are as follows:
  - Fire patterns that point to the area of origin being near the appliance
  - Severe fire damage to the appliance
  - Arcing or melting found on conductors either inside or near the appliance
- The investigator should evaluate the fire patterns on the appliance in relation to the remainder of the fire scene.
- The investigator should thoroughly understand how the appliance operated and its safeguards.
- Before beginning disassembly, a protocol should be created with the objectives, ground rules, and extent of disassembly identified based on the specific reason for carrying out the process.
- Postfire examination of a switch can sometimes determine its state and thus whether the appliance was energized at the time of the fire.
- An exemplar can be used to help understand how an appliance operates and can be used in testing a proposed ignition scenario.
- Common appliance components include housings, power sources, switches, solenoids and relays, transformers, motors, heating elements, and lighting.
- Common residential appliances include ranges and ovens, air conditioners and heat recovery ventilators, water heaters, coffee makers, toasters, electric can openers, cooling fans, refrigerators, deep fat fryers, dishwashers, microwave ovens, space heaters, electric blankets, window air-conditioner units, hair dryers and hair curlers, clothes irons, clothes dryers, and consumer electronics.

Hot Terms

Arc tube A cylinder of fused silica/quartz with electrodes at either end that is used in most mercury and metal halide lamps.

Convection heater A heater that employs a fan to move air across a hot surface, heating the air and dispersing it throughout the room.

Exemplar An exact duplicate of an appliance.

Housing In an appliance, the outer shell that contains the working components.

Interrupt current The value of current that the breaker must be able to carry and still operate.

Radiation heater A heater that transfers heat by radiation only.

Step-down transformer Device that reduces the 120 V provided at the receptacle to the required voltage.

Switch Device used to turn the appliance on or off or to change operating conditions.

Transformers Devices that reduce AC voltage, usually 120 or 240 V AC, to a lower voltage and can be used to isolate the appliance from its power source.

Triac A semiconductor switch that conducts current when the gate is turned on and blocks current when the gate is turned off.

FIRE INVESTIGATOR *in action*

You have determined that a fire in a large apartment building originated in the kitchen of one of the units. The kitchen is severely damaged and the remains of a refrigerator, an electric stove, a microwave, and a toaster are present. All of these appliances are severely oxidized with a large amount of debris on them.

An interview with the tenant reveals that the appliances were all new, having been purchased within the past 6 months from a local retail store. She states that she had experienced problems with the stove recently, indicating that the burners would stay on despite the control knobs indicating the burners were off.

After removing the debris and partially reconstructing the area of origin, you determine that the fire originated at the stove. The remains of a metal sauce pan have been located at the left-front burner; however, the burner controls are severely damaged with no indication as to what position they were in at the time of the fire. As you are visually examining the controls, a fire fighter states that he can determine whether the stove was on by taking off the rear panel and disassembling the burner controls to view the switch contacts.

1. If you remove the back of the stove and disassemble the burner controls:

A. a charge of spoliation may be made by other interested parties.

B. you should photograph the controls prior to disassembly.

C. you should conduct a recall search prior to disassembly.

D. Both B and C

2. Which of the following is not a consideration when determining that an appliance caused a fire?

A. The make, model, and serial number of the appliance

B. Whether any moisture was present

C. Whether the appliance was operating properly prior to the fire

D. Whether the appliance was energized at the time of the fire

3. Who should be contacted regarding any recalls related to appliances?

A. The local building official

B. The State Fire Marshal

C. The Consumer Product Safety Commission

D. The appliance manufacturer

4. Which device is used to control high-power circuits with a low-power circuit?

A. A step-up transformer

B. An electronic relay

C. A relay switch

D. A solenoid

Motor Vehicle Fires

© Photos.com

© Nicholas Burke/iStock/Thinkstock

Knowledge Objectives

After studying this chapter, you should be able to:

- Describe the investigative techniques used to analyze a vehicle fire. (pp 370–371)
- Discuss safety issues surrounding vehicle fire investigation. (p 371)
- Identify the fire fuels present in vehicles. (pp 371–372)
- Identify potential ignition sources present in vehicles. (pp 373–374)
- Discuss the various vehicle systems and components as related to fire cause. (pp 375–379)
- Discuss the body systems in a vehicle and how they may affect a vehicle fire. (pp 379–380)
- Explain how to document a vehicle fire scene. (pp 380–381)
- Discuss the process used in the examination of a vehicle fire. (pp 381–382)
- Discuss the investigation of hybrid vehicle fires. (pp 382–383)
- Discuss special considerations for vehicle fires. (pp 383–384)
- Discuss considerations for various types of specialized vehicles and equipment. (pp 384–387)

Skills Objectives

After studying this chapter, you should be able to:

- Conduct a vehicle fire investigation. (pp 370–387)

Additional NFPA References

NFPA 54, *National Fuel Gas Code*

NFPA 58, *Liquefied Petroleum Gas Code*

NFPA 921, *Guide for Fire and Explosion Investigations*

NFPA 1192, *Standard on Recreational Vehicles*

CHAPTER 25

FESHE Course Outcomes

Fire Investigation I

9. Recognize potential health and safety hazards. (p 371)
12. Explain the procedures used for investigating vehicle fires. (pp 370–387)

Fire Investigation II

6. Analyze electrical causes of fires. (pp 373–374)

You Are the Fire Investigator

© Jones and Bartlett Publishers. Photographed by Glen E. Ellman

You have been requested to investigate a fire involving a newer passenger vehicle. When you arrive, fire personnel state the vehicle was fully involved upon their arrival. They state that the owner of the vehicle was very upset because he had just picked up the vehicle from a repair shop.

During your interview with the owner, he states he had recently had repairs made on the vehicle after driving through a flooded roadway. Repairs included replacing all the wiring harnesses, an alternator, cooling fan, and the engine. He states that the vehicle had been driving for several miles when it began to lose power with a loud knocking noise and smoke began to enter the passenger compartment.

As you document the engine compartment, you note that the wiring harnesses are severely damaged and that most of the plastic components have been consumed. In examining the exhaust system, you note a large hole in the bottom of the engine block directly behind the exhaust manifold.

1. What safety concerns should the investigator consider while investigating a vehicle fire?
2. What fuel sources exist within the engine compartment that may have been ignited?
3. Why is the interview with the owner and driver important when investigating a vehicle fire?
4. What should be documented during your vehicle examination?

Introduction

Although investigation of vehicle fires is often considered complex and difficult, fires in automobiles, trucks, heavy equipment, farm and logging equipment, and recreational vehicles or motor homes have the same basic requirements for successful ignition and propagation as structural, marine, or wildland fires. The fire investigator should be able to recognize the burn and char indicators, which may vary from vehicle to vehicle, although patterns may be hard to identify with the tight spaces and excessive fuel loads found in vehicle compartments.. The investigator should also have a good understanding of various fuel characteristics within the vehicle, fire behavior associated with those fuels, and a general knowledge of the working mechanical components. This chapter discusses the various aspects of vehicle systems and methods that assist the investigator in arriving at a successful conclusion.

Investigative Techniques

Before conducting any vehicle fire investigation, investigators should prepare themselves by learning the general systems and characteristics of powered vehicles and equipment in general. Most mobile equipment uses mechanical systems that are similar. For example, cooling systems or fuel systems in automobiles share many design aspects of those employed in trucks and heavy equipment.

The investigator should be aware of the vehicle's safety features; have a general knowledge of the type and use of the vehicle, including its motor and components; and have a plan for conducting the investigation. It is helpful to gather, when possible, information regarding the use and maintenance of the vehicle and circumstances surrounding the fire prior to examination of the site and vehicle. Internet sites may provide information regarding failures and/or recalls for specific vehicles. TABLE 25-1 provides a suggested method in planning the investigation.

Table 25-1 Planning the Investigation of a Motor Vehicle Fire

Safety Concerns	Qualifications	Vehicle Examination
Before viewing the vehicle, in any condition: ■ Ensure its stability. ■ Check that the electrical system on the vehicle is disconnected. ■ Be aware of undeployed air bags and bumpers. (Sodium azide, the expelling agent for the air bag, is a serious safety hazard.) ■ Check for fuel and other fluid leaks that may pose a fire hazard. ■ Take the necessary steps to neutralize hazards. ■ Use proper protective clothing and equipment.	Perform an analysis of the vehicle systems: ■ Know the specific function of each system. ■ Determine whether a system malfunctioned. ■ Determine whether a system has been altered. ■ Determine whether a malfunction or alteration could be responsible for the fire. ■ Preserve evidence for future laboratory examination. If you are not qualified to perform an analysis of the vehicle systems, seek guidance from a qualified individual.	Differentiate three major compartments: ■ Engine compartment ■ Passenger compartment or interior ■ Cargo compartment Differentiate ignition scenarios: ■ Electrical ■ Mechanical ■ Human intervention (actions or inactions) Determine an area of origin.

Fire Investigator Tip

Prior to conducting a vehicle fire investigation, you should be aware of the vehicle's safety features; have a general knowledge of the type and use of the vehicle, including its motor and components; and have a plan for conducting the investigation.

During the initial phase of a vehicle examination, the investigator will look for fire or damage patterns inside or outside the vehicle. These patterns should be used to direct the investigator toward the area, and then point, of origin.

Vehicle Investigation Safety

When conducting an inspection of the vehicle's undercarriage, care should be taken to support and stabilize the vehicle properly with blocking or stands to prevent movement, which may cause injuries. Lifting devices such as jacks, forklifts, tow trucks, or hydraulic/pneumatic devices must never be solely relied upon for support during the inspection.

Undeployed airbags (supplemental restraint systems) in fire- or accident-damaged vehicles may pose a serious safety hazard because they may deploy if sensors are further disturbed. Disconnection of the battery or batteries prior to the inspection may prevent actuation of these systems and deter inadvertent electrical faulting (and possible initiation of another fire) during the examination.

Electrical systems in hybrid or electric vehicles may present an increased shock hazard because of the higher voltages in the systems. Recreational vehicles (RVs) may be connected to 120-V AC or 240-V AC power through a power cord or "shore power" connection.

Other hazards to be aware of include the possibility of fuel spillage from tanks or lines; contamination with coolants, lubricants, or other fluids; cut and puncture hazards from contact with broken glass or sharp metals; and release of energy from damaged spring devices, such as truck parking brake assemblies.

Vehicle Fire Fuels

Fuels in vehicle fires fall into three categories: liquid fuels, gaseous fuels, and solid fuels. Once a fire has begun to develop, any of these fuels may become involved secondarily, creating damage patterns that may be difficult to separate from those developed from burning of the initially ignited materials. The relatively small size of typically noncombustible compartment shells may blur fire patterns, complicating the task of separating primary from secondary damage patterns.

Liquid Fuels

Available liquid fuels include engine fuels (gasoline, diesel), engine lubricants, transmission fluid, power steering fluid, coolant, and brake fluid, and may also include hydraulic fluids or cargo fluids when present TABLE 25-2. Many of these fuels are present in most vehicles. The ignition potential of these fuels depends on the properties of each fuel, the physical state of the fuel (liquid, atomized, or spray form), and the nature of the ignition source. Gasoline is often blended with ethanol for use as motor fuel. This will change the properties of the blended fuel slightly, raising the autoignition temperature and slightly lowering the hot surface ignition temperature.

Gaseous Fuels

Gaseous fuels for motor vehicles are commonly propane or natural gas. These gaseous fuels not only represent alternative motor fuels, but they are also used in recreational vehicles for heating, cooking, and refrigeration. Hydrogen gas is often present and may be released in the area of batteries during charging. As alternative fuels and vehicle power systems evolve, other gaseous and liquid fuels, including hydrogen and compressed natural gas (CNG) or liquefied natural gas (LNG), may become more common. Greater quantities of the various gaseous fuels can be found in larger vehicles, either as fuels or as cargo. TABLE 25-3 outlines the properties of gaseous fuels in motor vehicles.

Solid Fuels

Solid fuels encompass any fuel within or on a vehicle that is not liquid or gaseous in its normal state. Although this is less common in accidental fires, solid fuels may be the first material ignited. Examples include wiring insulation or plastic conduit reaching its ignition temperature due to short circuits or resistive heating of the conductors, and heat from friction causing drive belts, bearings, and tires to ignite.

Solid fuels may contribute significantly to the fire spread. Solid fuels in vehicles include plastics that have heat release rates similar to those of ignitible liquids. Plastics can melt, ignite, and drop flaming pieces, which may spread the fire downward.

Aluminum and magnesium or their alloys, found in engine and vehicle components, can ignite and provide an additional fuel load. Most metals need to be in powder or melted form to burn; however, solid magnesium, which is present in some vehicles, burns vigorously once it is ignited by a competent external heat source.

The presence of melted metals in a vehicle is not necessarily indicative of the presence of an ignitible liquid accelerant. Many of the alloys or metals found in vehicles are not in a pure form or state and therefore have melting temperatures lower than those given on charts and graphs that show such metal in their pure form. When one type of metal falls or melts onto another, the process of alloying can occur. For instance, melted aluminum or zinc could run or drip onto another metal, causing it to melt at a lower-than-normal temperature. Molten metals may also spread the fire as they contact other combustible materials or may cause short circuits between energized cables and grounded components.

A pattern of melted metals may provide additional information to assist the investigator in determining the area of

Table 25-2 Properties of Ignitable Liquids (NFPA 921 Table 27.3.1)

Liquid	Flash Point[a] °C	Flash Point[a] °F	Autoignition Temperature[b] °C	Autoignition Temperature[b] °F	Flammability Limits[c] LFL %	Flammability Limits[c] UFL %	Boiling Point[d] IBP °C	Boiling Point[d] IBP °F	Boiling Point[d] FBP °C	Boiling Point[d] FBP °F	Vapor Density[e] (Air = 1)
Gasoline	−45 to −40	−49 to −40	257–280	4495–536	1.4	7.6	26–49	78–120	171–233	339–452	3–4
Diesel fuel (fuel oil #2)	38–62	100–145	254–260	489–500	0.4	7	127–232	260–450	357–404	675–760	5–6
Brake fluid	110–171	230–340	300–319	572–606	1.2	8.5	232–288	111–142	460–550	238–288	5–6
Power steering fluid	175–180	347–356	360–382	680–720	1	7	309–348	588–658	507–523	945–973	>1
Motor oil	200–280	392–536	340–360	644–680	1	7	299–333	570–631	472–513	882–955	>1
Gear oil	150–270	302–510	>382	>716	1	7	316–371	601–700	>525	>977	>1
Automatic transmission fluid	150–280	302–536	330–382	626–716	1	7	239–242	462–468	507–523	945–973	>1
Ethylene glycol (antifreeze)	110–127	230–261	398–410	748–770	3.2	15.3	196–198	385–388			2.1
Propylene glycol (antifreeze)	93–107	199–225	371–421	700–790	2.6	12.5	187–188	369–370			2.6
Methanol (washer fluid)	11–15	52–55	464–484	867–903	6	36	65	149			1.1

[a]Flash point data were obtained from Technical Data Sheets and Material Safety Data Sheets from manufacturers and suppliers of the major brands of each type of fluid available in the United States. The flash points of gasolines reported in these sources were determined by ASTM D 56. The flash points for diesel fuels, brake fluids, power steering fluids, motor oils, transmission fluids, gear oils, ethylene glycol (antifreeze), propylene glycol (antifreeze), and methanol were determined by ASTM D 56, ASTM D 92, or ASTM D 93.

[b]Autoignition temperature data for gasoline, diesel fuel, brake fluid, ethylene glycol, propylene glycol, and methanol were obtained from Technical Data Sheets and Material Safety Data Sheets from manufacturers and suppliers of the major brands of each type of fluid available in the United States. These sources generally did not report the test method used to determine autoignition temperature; however, ASTM E 659 is the laboratory test method typically used to determine autoignition temperature. Autoignition temperature data for power steering fluid, motor oil, gear oil, and automatic transmission fluid were obtained using ASTM E 659.

[c]Flammability limit data were obtained from Technical Data Sheets and Material Safety Data Sheets from manufacturers and suppliers of the major brands of each in the United States. These sources generally did not specify the laboratory test method used to determine the reported flammability limits; however, ASTM E 681 is a laboratory test method typically used to determine the lower flammability limit (LFL) and the upper flammability limit (UFL).

[d]Boiling range data for gasolines were obtained from the Alliance of Automobile Manufacturers annual North American survey of gasoline properties for 2003. The boiling ranges reported in this survey were determined by ASTM D 86. Boiling range data for diesel fuel were obtained from Technical Data Sheets and Material Safety Data Sheets from manufacturers and suppliers of the major brands of diesel fuel in the United States. These sources generally did not report the laboratory test method used to determine the boiling range of diesel fuel. Boiling range data for brake fluid, power steering fluid, motor oil, gear oil, and automatic transmission fluid were determined by ASTM D 2887. Boiling point data for ethylene glycol, propylene glycol, and methanol were obtained from Material Safety Data Sheets from manufacturers and suppliers of these chemicals. These sources did not report the laboratory test method used to determine boiling point. In the table, IBP and FBP are initial boiling point and final boiling point, respectively.

[e]Vapor density data were obtained from Material Safety Data Sheets from manufacturers and suppliers of these materials.

*Studies include the following:

1. Arndt, S. M., Stevens, D. C., and Arndt, M. W., "The Motor Vehicle in the Post-Crash Environment, An Understanding of Ignition Properties of Spilled Fuels," SAE 1999-01-0086, International Congress and Exposition, Detroit, MI, March 1–4, 1999.
2. LaPointe, N. R., Adams, C. T., and Washington, J., "Autoignition of Gasoline on Hot Surfaces" *Fire and Arson Investigator*, Oct. 2005: pp. 18–21.
3. Colwell, J. D., and Reze, A. "Hot Surface Ignition of Automotive and Aviation Fluids," *Fire Technology*, Second Quarter 2005, pp. 105–123.
4. API PUBL 2216, *Ignition Risk of Hydrocarbon Vapors by Hot Surfaces in the Open Air.*

Reproduced with permission from NFPA 921-2014, Guide for Fire and Explosion Investigations, Copyright© 2015, National Fire Protection Association. This reprinted material is not the complete and official position of the NFPA on the referenced subject, which is represented only by the standard in its entirety.

Table 25-3 Properties of Gaseous Fuels in Motor Vehicles (NFPA 921 Table 27.3.2)

Gas	Autoignition Temperature °C	Autoignition Temperature °F	Flammability Limits (Vol. % fuel in air) LFL	Flammability Limits (Vol. % fuel in air) UFL	Boiling Point °C	Boiling Point °F	Specific Gravity (Air) Vapor Density (air = 1)	Specific Gravity (Air) Min. Ignition Energy (mJ)
Hydrogen	40–572	753–1061	4.0	75.0	−253	−422	0.07	0.018
Natural gas (methane)	632–650	1169–1202	5.3	15.0	−162	−259	0.60	0.280
Propane	450–493	842–919	2.2	9.5	−42	−44	1.56	0.250

Note: The data provided in this table are for generic or typical products and may not represent the values for a specific product. When possible, values specific to the product involved should be obtained from a material safety sheet, product specifications, or by standard test methods.

Reproduced with permission from NFPA 921-2014, Guide for Fire and Explosion Investigations, Copyright© 2015, National Fire Protection Association. This reprinted material is not the complete and official position of the NFPA on the referenced subject, which is represented only by the standard in its entirety.

origin, much as a char pattern analysis is frequently used in structural fire investigations. Melted metals used as indicators should be identified, and their actual melting temperatures should be determined. Many common metals may be identified from their visible characteristics. A precise determination of the composition of metals may require laboratory analysis using technical processes such as Energy Dispersive Spectroscopy (EDS).

Fire Investigator Tip

Investigators should not interpret the presence of melted metals to be an indicator of the use of an ignitible liquid in the belief that only an ignitible liquid can produce temperatures high enough to melt metals.

Ignition Sources

In most cases, the sources of ignition energy in vehicles are the same as those associated with structure fires. Open flames, mechanical and electrical failures, and discarded smoking materials can often lead to a vehicle fire. Vehicles do have some unique ignition sources that should be considered and evaluated by the investigator, including heated exhaust components, various types of bearings, and braking systems.

Open Flames

Although open flames were a common source of fires in a carbureted vehicle when the engine backfired through the carburetor, it is important to realize that fuel injection systems have replaced carburetors in most of today's modern vehicles. The throttle body in most modern vehicles does not handle fuel, which is commonly injected directly into intake manifolds or combustion chambers.

Open flames from smoking materials can occur in ashtrays and could ignite combustibles. Open flames in recreational vehicles may also include appliance pilot lights, burners, and ovens. Exposure to open flames from outside sources, such as wildland or ground cover fires, sometimes results in vehicle fires whose origin may be very difficult to determine, particularly if the vehicle is examined in a location away from the fire scene.

Electrical Sources

The primary source of electrical energy in a vehicle that is not running is the battery. During the investigation, it is important to remember that some vehicles have multiple batteries. Some components that can remain energized when the vehicle is not operating include headlights, power seats, interior lights, alternator, starter solenoid, and relays serving various vehicle systems. Information specific to the vehicle may be obtained from service manuals and/or wiring diagrams.

The investigator should determine whether the vehicle was running at the time of the fire to determine what potential electrical ignition sources were energized.

Fuses, circuit breakers, or fusible links are designed to provide protection for electrical circuits in most vehicles FIGURE 25-1. Most vehicle electrical systems are direct current (DC) with the negative (ground) side of the system connected to the body, frame, and engine. The positive side of the electrical system provides current via the electrical wiring to devices such as starters, alternators, and radios. Any time a positively charged lead goes to ground, a short circuit will occur, with potential for a resulting fire. This may occur if battery cables or other conductors chafe against grounded portions of the vehicle, such as the frame or engine, wearing through their insulation.

The inside of the cover of most circuit breaker panels lists the equipment protected and the size of the fuse. If illegible, this information may often be obtained from the owner's manual or a shop service manual for the vehicle. It is important to view the fuse panel to determine whether a problem exists with an electrical component or circuit and to verify that proper fuses or circuit breakers are in place.

RVs may have alternating current (AC) circuits as well as DC circuits. They may have a converter to convert AC to DC, an inverter to convert DC to AC, and/or a generator to provide AC power. Two or more batteries or up to four auxiliary batteries as a secondary power source for "coach" systems are common FIGURE 25-2. It is important to determine whether the RV was connected to an outside power source through a shore power connection.

Another means by which electrical energy may be an ignition source is through resistance heating, which may be associated with overloaded wiring. Improperly installed or utilized aftermarket equipment can overload the vehicle's electrical systems.

Aftermarket items include audio or video systems and associated hardware, as well as enhanced vehicle lighting (running lights, spotlights, fog lights), navigation systems, and many other types of accessories. Improperly installed aftermarket equipment may lack overcurrent protection.

On commercial vehicles, aftermarket equipment may include electrically operated body systems and communication systems. A high-resistance fault in the electrical wiring can raise the temperature of the wiring insulation to its ignition point without activating the circuit protection. Faults in

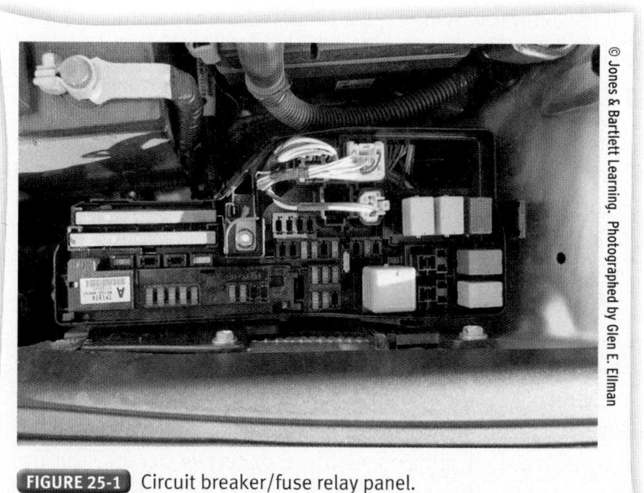
© Jones & Bartlett Learning. Photographed by Glen E. Ellman

FIGURE 25-1 Circuit breaker/fuse relay panel.

© Jones & Bartlett Learning. Photographed by Glen E. Ellman

FIGURE 25-2 This RV has more than one battery.

high-current devices can ignite readily available combustible materials.

Damaged insulation can result in electrical arcing when a charged wire comes into contact with a grounded surface. Crushed or cut wires and crushed batteries can be a source of electrical arcing. It is important to remember that the battery and starter cables are unlikely to be electrically protected (by fuses or overcurrent protection) and that they can carry a large amount of current. If these lines are severed during a crash, arcing can occur and result in a fire. Wiring that is subject to chafing may become devoid of insulation, allowing unintended contact with other metals, thereby creating sparks or resistive heating.

Lamp filaments of broken bulbs are a potential ignition source if the fuel is in a gas, vapor, or liquid spray form. Most lamp filaments are designed to function in a vacuum. The operating temperature of the exposed filament can be as high as 2550°F (1400°C), but when a filament is broken and exposed to air, it operates for only a few seconds and then burns away, removing the ignition source. The glass envelope of a modern headlight, while lit, operates at a temperature well above the ignition temperature of many plastics. If the bulb is displaced and rests against plastic trim components, a fire may result.

During the investigation of a vehicle fire, the investigator should look for evidence of external electrical ignition sources. This could include ports for recreational connections, engine heaters, and battery chargers. Battery chargers can provide an additional electrical source to consider or eliminate during the investigation.

Hot Surfaces

Exhaust system components can provide sufficient temperature to ignite diesel spray and to vaporize gasoline. Slippage of transmission or torque converter components may result in sufficient heat being generated to vaporize fluid, causing a discharge of fluid that can ignite on contact with heated exhaust system components.

Engine oil, power steering fluid, and brake fluid can also ignite when in contact with heated exhaust system components. These liquids may ignite during operation or soon after the engine is shut off, when the temperature of some exhaust system components is still sufficiently elevated. The ignition temperatures for fluids on hot surfaces are generally up to several hundred degrees Fahrenheit above their established autoignition temperatures. The following factors all affect ignition of liquids by a hot surface:

- Ventilation
- Environmental factors
- Autoignition point
- Liquid flash point
- Liquid boiling point
- Liquid vaporization rate
- Atomization of the liquid (effective fuel surface area)
- Latent heat of vaporization of the liquid
- Length of exposure of the liquid to the heated surface and configuration of the hot surface

These factors may all influence the air–fuel ratio at the location of the potential ignition source, creating mixtures that are either too rich or too lean for ignition and/or propagation of the fire.

Other combustible materials, such as plastics, papers, and vegetation, can reach their ignition temperature when in contact with heated vehicle components.

Catalytic converters normally operate with an internal temperature approaching 1300°F (700°C) and a surface temperature in the vicinity of 600°F (315°C), but may reach higher temperatures under heavy loads if air circulation is restricted or if unburned fuel enters the converter because of overfueling or engine misfiring. Under such extreme circumstances, the ceramic matrix material inside the converter may melt and may be ejected from the tailpipe in an incandescent spray.

Mechanical Sparks

Mechanical sparks occur from metal-to-metal contact and can be associated with rotating equipment (alternators, pumps, pulleys, etc.) or bearings. Metal-to-pavement contact (e.g., a broken drive shaft, tire rim, or exhaust pipe) can also produce sparks. The vehicle must be running or moving to produce such an ignition source. Vehicle speeds of 5 mph (8 kph) have been determined to create orange sparks with a temperature of 1470°F (800°C), and greater speeds have created white sparks with a temperature of 2190°F (1200°C). Aluminum-to-pavement sparks are not considered a competent ignition source for most materials. Usually, these types of sparks will not ignite solids. Consideration has to be given to the fuels around the area from which the sparks are emanating and the amount of time the sparks are in contact with these fuels.

Farm, logging, construction, or highway maintenance equipment may have mechanical attachments come in contact with cellulosic crop or plant materials. If damaged or not properly maintained, friction may produce heat capable of igniting those materials. When humidity is low and fuels are dry, mechanical sparks can occur when metal blades or implements hit a rock. The mechanically produced sparks may have the potential energy to ignite surrounding fuels under these conditions. The ability of such sparks to ignite other materials is regulated by their mass–surface area ratio. Smaller sparks will cool more rapidly with exposure to ambient air.

Smoking Materials

Another ignition source identified in vehicle fires is improperly discarded or misused smoking materials. In the hypothesis for this ignition scenario, the ignition sequence for discarded smoking materials centers around the vehicle's ability to confine and retain the heat of the smoking material, and the ventilation aspects of the vehicle. In assessing the possibility of such an ignition sequence, be aware of the Federal Motor Vehicle Safety Standard (FMVSS) #302, which deals with the ignitability of vehicle interior components. Ignition of other fuels such as papers or clothing that are in the vehicle should be considered. If ignition of these materials occurs, the fire may spread to other fuels, including foam and other materials in the vehicle interior.

Systems and Their Function in a Vehicle

The importance of knowing how the various systems in vehicles operate is critical for the fire investigator and can be enhanced by viewing various technical manuals available at the library, car dealerships, and parts suppliers. Quite often, familiarity with a particular vehicle system can be gained by visiting a dealership or automotive repair shop and asking for specific information about the vehicle from a service manager or mechanic. The parts department may also have schematics and views of particular systems.

■ Fuel Systems

Two basic fuel systems are used in gasoline-powered vehicles: vacuum/low-pressure carbureted systems and high-pressure fuel-injection systems. Both take fuel from the gas tank through the use of a fuel pump and deliver the fuel through a fuel line and filter system to the engine, where the fuel is atomized, compressed, and burned.

Vacuum/Low-Pressure Carbureted Systems

The vacuum/low-pressure carbureted systems are most often seen on older automobiles, gasoline-powered farm equipment, lawn maintenance vehicles, and small gasoline engines powering stationary equipment.

Vacuum/low-pressure carbureted systems draw fuel from the fuel tank, usually by means of a mechanical pump attached to the engine, and operate only when the engine is operating. Electric fuel pumps are sometimes used. The fuel is then pumped into the carburetor at a pressure of 3 to 5 psi (20 to 35 kPa), where it is mixed with air in the carburetor (usually at a ratio of 15:1) and drawn into the engine via the intake manifold and into the cylinder, where combustion takes place. FIGURE 25-3 shows a generic view of a vehicle fuel system and its associated components.

Potential problems with vacuum/low-pressure carbureted systems are outlined in TABLE 25-4.

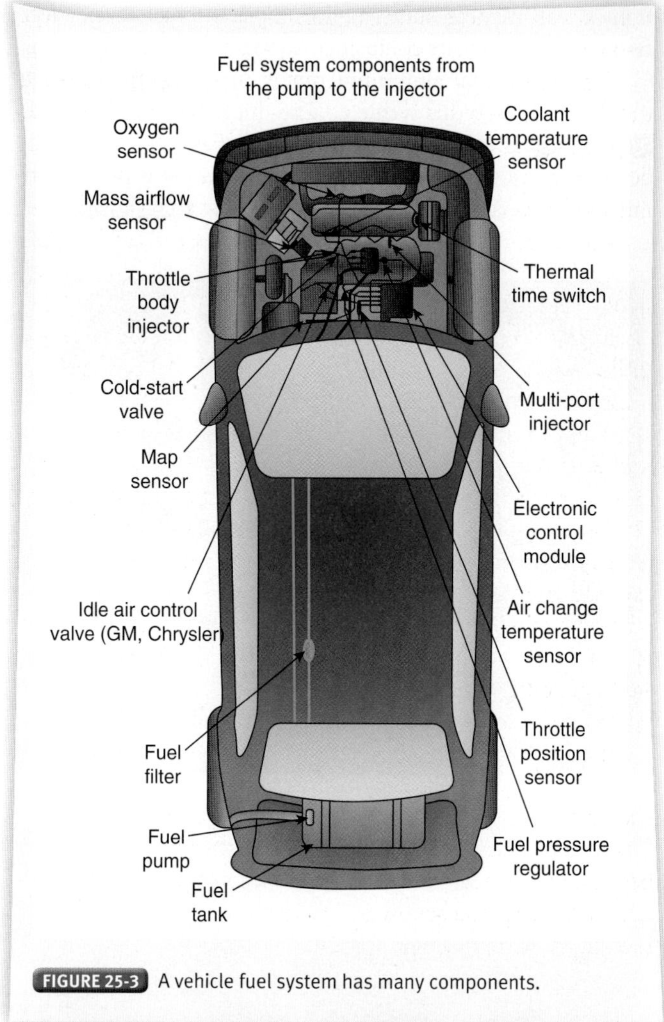

FIGURE 25-3 A vehicle fuel system has many components.

Table 25-4 Vacuum/Low-Pressure Carbureted Systems

Location of a Leak	Problem
Vacuum side	Air is drawn into the system, and the engine will not run.
Pressure side	Fluid can leak as a fine mist or a heavy stream. Ignition is possible if there is an ignition source. If the fuel line leaks as a result of a fire, the leaking fuel vapors can be ignited.

High-Pressure Fuel-Injected Systems

High-pressure fuel-injected systems also retrieve the fuel from the fuel tank and pump the fuel to the engine via an electric fuel pump, which is energized whenever there is a key in the ignition in the "on" or "run" position.

Some of the fuel pumps on modern cars are integrated with the onboard computer, which can be triggered by an inertia switch in a crash that could disable the fuel pump. The fuel is pumped at 35 to 70 psi (240 to 480 kPa) in most vehicles with standard electronic or mechanical fuel injection systems, but may be well over 1700 psi (11.72 mPa) in newer gasoline direct injection (GDI) systems and may have spike pressures of several thousands of pounds per square inch in some modern heavy-duty diesel systems. FIGURE 25-4 depicts an electric fuel pump used in most new vehicles with fuel injection systems. The fuel is pumped to either a single venturi-mounted fuel injector (throttle body) or, more commonly, to a fuel rail assembly on the engine. In both systems, the excess fuel is pumped back to the fuel tank.

Potential problems with high-pressure fuel-injected systems may include leaks under pressure at various fittings, lines, or fuel pressure regulators. In some vehicles, the pressure developed

© Jones & Bartlett Learning. Photographed by Glen E. Ellman

FIGURE 25-4 An electric fuel pump.

from a leak may be sufficient to propel the fuel several feet, resulting in potentially confusing patterns when that fuel burns.

The horizontal aluminum piping in FIGURE 25-5A depicts the fuel rail on this vehicle. The fuel injector shown in FIGURE 25-5B is attached to the fuel rail. Failures can and do occur because of age and types of materials used for the connection between the rail and the supply line FIGURE 25-5C.

© Jones & Bartlett Learning. Photographed by Glen E. Ellman

© Jones & Bartlett Learning. Photographed by Glen E. Ellman

© Jones & Bartlett Learning. Photographed by Glen E. Ellman

FIGURE 25-5 **A.** A fuel rail. **B.** A fuel injector. **C.** The connection between rail and supply line.

Table 25-5 Fuel-Related Problems

Location of a Leak	Problem Noted
Supply side	Operating problems will be noted. Unintended ignition may occur.
Return side	No operating problems will be noted. Unintended ignition may occur.

When being interviewed, the vehicle owner may report fuel-related problems that may assist the investigator in determining an event that led to a fire. A leak on the supply side of the system or problems with the operation of the vehicle should be noted by the operator. Starting difficulty, erratic operation, and stalling all may be problems that were noticed. A leak on the return side of the system can go undetected with no operational problems noted. Fuel lines that enter the tank on the bottom can feed fuel to a fire by gravity, and lines that enter at the top of the tank can feed fuel by siphon effect if they are compromised. Residual pressure in the tank or pressure that is created by the fire could force fuel out of a compromised fuel line. TABLE 25-5 can assist in testing a hypothesis in a suspected fuel-related fire. The investigator should be aware that fuel lines may be constructed of plastic (typically Nylon 12) and may easily be breached by exposure to the fire. Trauma from a motor vehicle accident can compromise either plastic or steel fuel lines. Fuel transported or delivered in such a manner may result in the creation of very confusing patterns of intense fire damage. The condition of these lines, therefore, should be carefully identified and documented.

Diesel Fuel Systems

Diesel fuel systems often use two pumps to deliver the fuel from the tank to the engine with lift pumps that operate under high volume but with low pressure. The diesel fuel is pumped from the tank to an engine-driven fuel injector pump, and the higher-pressure fuel injector pump delivers the fuel to the individual injectors. Combustion air is sometimes provided by natural aspiration, or more commonly through a turbocharger.

Fuel leaks can occur over time because of the vibrations of a diesel engine. If any diesel fuel comes into contact with a hot surface, the diesel fuel may ignite. Another, although rare, potential problem associated with diesel engines is a "runaway" engine in which fuel or oil is drawn in through the air intake system, causing the engine to maintain a very high RPM level. This high RPM level may cause the engine to fail catastrophically.

Natural Gas and Propane Fuel Systems

Natural gas and propane fuels may be stored as CNG or as liquids under very high pressures (about 3000 psi, 21 mPa), which are reduced by a regulator at or near the engine. The fuel flows through a regulator and into the carburetor, where it is mixed with air before entering the engine. Leaks found postfire may not be indicative of prefire leaks, partly because of the different coefficients of expansion between the fittings and piping materials. If a leak does occur, fugitive gases

© Jones & Bartlett Learning. Photographed by Glen E. Ellman

FIGURE 25-6 A turbocharger.

can find an ignition source and flash back to that source. In some cases, the burning fuel or heat from other burning components can cause the fuel tank to rupture during a fire, although these tanks employ pressure relief valves designed to release excess pressure in the event of exposure to outside sources of heat. RVs often have natural gas or LP-gas systems for heating, refrigeration, and cooking. Additional information about natural gas and LP gas systems can be found in NFPA 54, *National Fuel Gas Code*, and NFPA 58, *Liquefied Petroleum Gas Code*.

Turbochargers

Turbochargers can be found on gasoline and diesel-powered engines FIGURE 25-6. The turbocharger increases the power of the engine by forcing pressurized air into the cylinders. The turbocharger uses exhaust gases to rotate the impellor up to 100,000 RPM. Turbochargers and the exhaust manifolds that supply them are the hottest points on the surface of the engine (approaching 1500°F [815°C] under load). If a fuel leak or oil leak occurs and the fluid contacts these surfaces, the fuel can ignite readily. The center bearing of most turbochargers is lubricated by engine oil under pressure. If the turbocharger suffers a mechanical failure during operation, engine oil may be delivered into the hot exhaust stream, potentially resulting in a fire.

> **Fire Investigator Tip**
>
> It is unlikely that a fire investigator will know everything about the systems in all vehicles he or she could encounter. The key is to know where to find the needed information when investigating a fire involving a motor vehicle. Technical manuals can be obtained online or from a library, car dealership, or parts supplier. The parts department of an automotive repair shop may also have schematics and views of particular systems.

Emission Control Systems

Emission control systems are found on most vehicles today. Automotive systems are designed to reduce or control exhaust gas emissions and to collect gasoline vapors while the engine is operating. The emission control systems regulate the fuel input, timing adjustments, and recirculation of exhaust gases through the exhaust gas recirculation (EGR) valve. A charcoal canister collects excess fuel vapors.

Problems that can occur in the emission control system include vapor leakage from the hoses or charcoal canister or overfilling of the fuel tank, which forces liquid into the charcoal canister. This could result in rough idling, stalls, backfires, and overheating of the catalytic converter. If vapors leak from the charcoal canister or associated hoses, a fire may result.

Diesel trucks, under Environmental Protection Agency (EPA) guidelines, incorporate systems to reduce oxides of nitrogen (NOx), carbon dioxide (CO_2), carbon monoxide (CO), and particulate emissions. Trucks built after 2007 typically incorporate software-controlled cooled EGR systems coupled with diesel particulate filters (DPF), which use various methods of "regeneration," or burning of trapped carbon in the DPF. Fuel delivery and, in gasoline engines, ignition timing are computer-controlled in modern engines to optimize both power delivery and emissions control.

Exhaust Systems

Starting at the engine, exhaust system components include an exhaust manifold that is connected directly to the engine, exhaust piping, a catalytic converter, and a muffler. During normal operation of a properly maintained and operating vehicle, the temperatures that enter the catalytic converter can measure 650°F (343°C), and it is normally the hottest part of the exhaust system. An improperly operating engine can raise the temperature of the exhaust system at the catalytic converter enough to ignite undercoating and interior carpeting. A hot exhaust system can ignite grass and other external combustibles.

Internal combustion engines, particularly diesels, tend to eject carbon particles from their exhaust. Under load, some of these particles may ignite or glow and present an ignition hazard if they contact dry grass or duff. Many states require construction tractors and agriculture vehicles to be equipped with spark arrestors.

Motor Vehicle Electrical Systems

The storage battery is the primary energy source in a vehicle. Once the ignition key is turned on, the starter draws energy from the battery to start the engine. Once the engine is running, the alternator provides electrical power to the vehicle. The vehicle battery is normally a 12-V DC negative ground system. The conductor size determines the amount of electrical energy the electrical system can safely carry. Simply put, the bigger the conductor, the more electrical energy the system can carry. The largest conductors extend from the battery to the starter and alternator and to a grounding location. They are rarely protected by overcurrent protection equipment. Arcing on these conductors is commonly found following a fire in the vehicle because of damage to insulation and cable supports.

The battery can also be a fuel source. The battery can generate hydrogen gas and ignite easily if an ignition source is present. Small amounts of hydrogen and oxygen gases may be found in sealed "no maintenance" batteries, and these gases could be released if the battery case is damaged.

12-Volt Electrical Systems

Electrical energy throughout the vehicle comes through the positive side of the electrical system and is routed through a fuse block (sometimes called a load center or power distribution module) with overcurrent protection in the form of circuit breakers or fuses. The electrical system for the vehicle is most often grounded through the frame of the vehicle. Most of the electrical devices in a vehicle have only one conductor (positive). The rest of the circuit is completed through the car frame or body. The exception to this design occurs with trucks and diesel-powered heavy equipment, which often employ positive and negative cables between the batteries, starter, and alternator to reduce power losses in their systems.

Only a few of the electrical circuits have power when the engine is shut off. Those with power may include circuits from the battery to the starter, from the battery to the ignition switch and fuse block, and to power seats. Circuits from the ignition switch to the accessories, such as the clock and cigarette lighter, onboard computers, and aftermarket accessories, may also be powered with the ignition off. Battery power may also be supplied to relays for other accessories.

Wiring diagrams typically show the color of conductor insulation, the device location, and its function. Most vehicle wiring is stranded and can become brittle when heated. The investigator should document the condition and identity of wiring before moving it.

Other Electrical Systems

Electrical systems on some older vehicles have 6-V DC systems. Some can have positive grounding systems. Some trucks, buses, and heavy equipment may have 24-V systems.

Recreational vehicles can have electrical systems that have a combination 12 V DC and 120 V AC. Converters are used to change voltage from 120 V AC to 12 V DC. Inverters change power from 12 V DC to 120 V AC. Many motor homes may have onboard generators that produce 120 V AC.

Event Data Recorders

Many modern vehicles are equipped with an event data recorder (EDR) as part of the airbag actuation system. Vehicle parameters recorded for seconds before and after a crash may include vehicle speed, engine RPM, brake actuation, airbag deployment, crash pulse, and so forth. Retrieval of this information requires special skills and/or equipment and in some cases may be done only by the manufacturer. Damage to the EDR because of exposure to the fire or trauma of a crash may make information retrieval impossible.

Most modern electronically controlled diesel engines also feature data recording capability. If the electronic (or engine) control module (ECM) has not been damaged by the fire, the data can be downloaded in a readable form by an engine technician. Another form of event recording often seen in over-the-road (OTR) trucks is the wireless communications and tracking system developed by Qualcomm and installed in well over half a million OTR tractors. This system can generate a report containing location, operation, and communications information. The OnStar system installed in many newer General Motors vehicles has the capability of recording and reporting on many parameters, including location, operation, systems diagnosis, crash data, and communications.

Many of these systems record and report on information that is considered proprietary, either to a manufacturer, a company, or a vehicle owner. As a result, there may be legal concerns regarding data ownership and/or release.

Mechanical Power Systems

Mechanical failures associated with internal combustion engines can, and frequently do, cause fires FIGURE 25-7. Failure of bearings, rings, or pistons can generate internal heat preliminary to catastrophic engine failure. Such failures can project fragments of connecting rods through the engine block, resulting in the release of a spray of heated engine oil, which ignites on contact with sufficiently heated surfaces. A loss of engine oil from incompletely sealing gaskets can allow the combustible fluid to contact exhaust manifolds and ignite. Coolant loss may allow the glycol component of the coolant to reach ignition temperature atop manifolds or turbochargers. Leaks from extremely high-pressure fuel injection systems (as in modern gasoline or diesel direct injection systems) may supply readily ignitible, atomized fuel into the engine compartment environment.

Lubrication Systems

Most vehicle engines use hydrocarbon-based oils for lubrication. The oil is pumped through the engine at 35 to 60 psi (240 to 415 kPa). An oil leak can occur in a number of locations within the engine compartment. If an oil leak occurs, the oil vapors are susceptible to ignite given the proper circumstances. A lack of oil in the engine can also cause a fire if the engine fails. In diesel-powered vehicles, a turbocharger failure may release lubricating oil into either the exhaust or intake systems, or both. This may result in a runaway engine or a fire.

© AlexKalashnikov/ShutterStock, Inc.

FIGURE 25-7 Mechanical failures associated with internal combustion engines are frequent fire causes.

Liquid Cooling Systems

Most vehicle engines contain a liquid coolant, which is circulated through the engine by a water pump. Generally, the coolant is 50 percent ethylene glycol and 50 percent water. Operating temperatures of the coolant generally range from 180°F to 195°F (80°C to 90°C), with an operating pressure of 16 psi (110 kPa). As the water component evaporates below the boiling temperature of ethylene glycol, a loss of coolant that allows contact with heated exhaust components may result in initiation of a fire.

Air-Cooled Systems

Air-cooled systems are uncommon in most modern vehicles. An air-cooled system circulates the air through the engine by a fan and ductwork. Normally, the fans are belt driven, and if the belt breaks, engine failure can occur.

■ Mechanical Power Distribution

Mechanical power produced by the engine must be delivered to the drive wheels for the vehicle to be put in motion. This power distribution or transfer is accomplished through a transmission and drive axles, both of which require lubrication.

Mechanically Geared Transmissions

Mechanically geared transmissions (manual transmissions) are the means of power transfer from the engine through the clutch assembly (which is located between the engine and the transmission) to the vehicle wheels. Most transmissions contain a transmission fluid for lubrication and the movement of internal valves. If a leak occurs in the vicinity of heated exhaust system components or another suitable ignition source, a fire can occur.

Hydraulically Actuated Transmissions

Hydraulically actuated transmissions (automatic transmissions) have a transmission fluid that is typically cooled by routing through a cooler in the radiator or by auxiliary heat exchangers that may be installed aftermarket. The fluid can be added to the transmission through the tube provided for the transmission fluid level indicator (dipstick). Internal heat developed in the transmission or torque converter (often due to slippage of friction elements) may vaporize transmission fluid, resulting in expulsion of fluid through vents, seals, or the dipstick tube. If transmission fluid contacts heated exhaust components, a fire may result. Transmission fluid that has been overheated may have a brown color and distinctive burned aroma, and may be cloudy because of particulates suspended in the fluid. Laboratory examination of a fluid sample may assist in identifying the nature of any transmission or torque converter failure. A disassembly inspection of both the transmission and torque converter may be necessary to pinpoint the precise failure.

■ Accessories and Braking Systems

Other accessories attached to the engine that can fail include alternators, air conditioning compressors, power steering pumps, air pumps, and vacuum pumps. Mechanical failures that produce friction may present viable ignition sources.

Hydraulic brake systems operate under high pressure, and if a leak occurs, it can generate a spray of brake fluid in the form of ignitible vapors. Fluid spilled from a master cylinder reservoir (sometimes occurring when the cap is left ajar on the reservoir) may contact heated exhaust components, resulting in a fire. Braking systems in all vehicles convert kinetic energy to heat energy, dissipating the heat to the atmosphere. Malfunctions in the braking system may produce more heat than can be safely transferred to the air, resulting in overheating of drums, wheels, tires, or fluids and possibly a fire.

■ Windshield Washer Systems

Windshield washer fluid mixed with water typically contains methyl alcohol or methanol, sometimes at concentrations that are strong enough to form an ignitible vapor when sprayed onto a hot surface. Some washer systems use nozzles that include heaters to keep the fluid from freezing; and the heaters, if they malfunction, can provide a heat source to cause a fire. Additionally, the fluid is typically kept in a plastic container, which may be consumed in a fire. It is important to note whether another part of the vehicle has penetrated the area typically occupied by the washer fluid container.

Body Systems

Many panels in modern vehicles are made of combustible materials. These panels can contribute to the fuel load once a fire has started. Combustible metals are used in some panels, and these can burn intensely. The engine compartment bulkhead can have a number of penetrations for plastic ductwork that burns in a fire and provides an avenue of fire spread from one compartment to another.

■ Interior Finishes and Accessories

Most of the interior seats and padding provide a significant fuel load in a vehicle fire FIGURE 25-8. During investigation of a vehicle fire, the investigator should document any existing or missing accessories within the vehicle passenger compartment and should inventory and photograph contents in passenger areas.

FIGURE 25-8 Interior seats and padding provide a significant fuel load in a vehicle fire.

Cargo Areas

It is important to determine whether the fire started in a cargo area or spread to it from another location. The contents of the cargo area should be inventoried. The spare tire should be examined and the depth of the tire tread noted.

Documenting Motor Vehicle Fire Scenes

It is advantageous to document the vehicle at the fire scene using the same procedures used for documenting a structural fire scene. The investigator should start with the documentation of the area surrounding the vehicle and then focus on the vehicle itself. When possible, the position and location of the vehicle and its relationship to buildings, streets, and other site features should be documented photographically and by measuring and diagramming the site.

Vehicle Identification

Vehicle identification begins with the make, model, and year as well as other identifying features, including the vehicle identification number (VIN) **FIGURE 25-9**. The VIN is commonly found on the dashboard in front of the driver's position, but in some vehicles may be stamped on the engine side of the bulkhead (also known as the firewall). Sometimes the VIN will be located on a label affixed to the driver's door pillar or on the engine block. With a VIN in hand, the investigator can obtain information about the manufacturer, country of origin, body style, engine type, model year, assembly plant, and production number. If standard VIN locations are obscured or destroyed, other derivative VINs may be retrievable in the form of "confidential" VINs. The investigator may need to contact his or her local auto theft unit, the National Insurance Crime Bureau (NICB), or the Canadian Police Information Centre (CPIC) investigator for assistance in locating these hidden VINs.

Vehicle Fire Scene History

It is advisable to determine the actions and events that led up to the fire by conducting interviews with the person(s) who first observed the fire. The owner, passengers, and operator of the vehicle at the time of the fire or the last person to drive the vehicle; nearby residents; fire department personnel; and police officers might also provide pertinent information. Information that may be helpful to obtain includes the following:

- Last use of the vehicle and by whom
- Mileage at the time of the fire
- Operation or problems
- Service or maintenance history
- Fuel level and type, when last fueled and where
- Equipment
- Personal effects in each area of the vehicle
- Photos or videos prior to, during, or after the fire

Courtesy of the NJ State Fire Marshal's Office, Arson/K-9 Unit

FIGURE 25-9 The vehicle identification number provides important information about the production of a vehicle.

Vehicle Particulars

For assistance in identifying systems and component configurations, it is often worthwhile to inspect a similar (exemplar) vehicle. During your inspection of the vehicle, checklists in Annex A of NFPA 921 can be used for assistance in gathering information. Notes and diagrams taken during the inspection are important to document details observed that might otherwise be overlooked, and may help later to clarify details of photographs taken during the examination.

Information about fire causes in vehicles of the same make, model, and year could be important to the investigation. This information can be obtained from the National Highway Traffic Safety Administration, the Insurance Institute for Highway Safety, the Center for Auto Safety, and, in Canada, Transport Canada. Online automotive or fire investigation forums may also present a resource for such information.

Documenting the Scene

When documenting the scene, the entire fire scene should be defined, including the vehicle and the area surrounding the vehicle. The scene itself should be considered as evidence that needs to be protected and preserved until it can be thoroughly documented. When the vehicle has been removed from the scene, both the vehicle and the scene should be documented. The investigator should record the fire scene by making a scene diagram showing reference points and distances. The scene should be photographed, as should surrounding buildings, highway structures, vegetation, other vehicles, tire and foot impressions, fire damage, signs of fuel discharge, and any parts or debris.

The vehicle should be photographed at the scene in a systematic manner similar to that for a structure fire. The investigator should begin on the outside, documenting all surfaces (including top and underside, if possible), damaged and undamaged areas, tires, and tire tread depth. Document the engine compartment, taking overview photographs of all sides and focusing on specific engine areas and components. The investigator should determine and document the positions of windows and doors, inspecting window lift mechanisms for their positions and side window channels for evidence of remaining glass. The investigator should document the paths of fire spread into or out of compartments and cargo spaces. Photos of the cargo space should include photographs of the spare tire and any special equipment such as stereo gear or add-on devices as appropriate. The investigator should photograph any changes or alteration made to the vehicle during the fire suppression effort and/or the initial inspection and, when possible, should also document the vehicle removal process.

Towing Considerations

As noted earlier, the vehicle is often best examined in detail in a controlled setting, away from the fire scene. Removal of the vehicle from the scene, however, may dislodge and potentially lose valuable evidence. The investigator should carefully document the scene surrounding and beneath the vehicle prior to moving the vehicle and as the vehicle is being moved. It may be helpful to wrap the damaged portions of the vehicle prior to transport, preventing evidence from dropping en route.

Documenting Away from the Scene

It is advisable to visit the scene even if the vehicle has been removed. The investigator should secure any background information, including the date and time of loss, the location of loss, the name of the operator, names of any passengers, witness statements, police and fire department reports, the vehicle's current location, and the method of transportation. Documentation should be made of any parts that might be missing or damaged when the vehicle is inspected after it has been moved from the scene. If possible, the vehicle should be protected from the elements.

Motor Vehicle Examinations

The examination of a vehicle fire is a tedious and often complex task that can take longer than the investigation of a simple house fire. A more complete examination of the vehicle can often be better accomplished after removal from the fire scene. As is the case with all fire evidence examinations, it is important to recognize that other entities (such as insurers, manufacturers, and other injured parties) may have a legitimate and legal interest in the investigation and that interest may be compromised by any alteration or removal of evidence. Prior to any destructive examination, those parties should be identified and notified in order to preserve their access to the examination and to prevent a later claim of spoliation of evidence. Refer to NFPA 921 Sections 12.3 and 29.3; ASTM 860, *Standard Practice for Examining and Testing Items That Are or May Become Involved in Litigation*; and ASTM 1188, *Standard Practice for Collection and Preservation of Information and Physical Items by a Technical Investigator*.

The first step in the investigation is to determine an area of fire origin. As with structural fire investigations, the investigator should work from the area of least damage to the area of greatest damage. The examination of the vehicle should begin by noting and documenting exterior damage patterns as indicators of the most intensely damaged compartments FIGURE 25-10. These patterns will assist in working toward the area of origin.

The investigation process of vehicles can be generally divided into three compartments: the engine compartment, passenger and driver area, and cargo space. On the exterior, paint damage, sheet metal distortion, and glass damage often provide directional patterns to lead the investigator toward areas of early fire development. TABLE 25-6 provides the investigator with an overview of windshield indicators.

© D Russell 78/ShutterStock, Inc.

FIGURE 25-10 Vehicle examinations should begin with documentation of exterior damage patterns.

Examination of Vehicle Systems

The inspection of vehicle fuel systems should be systematic, from outside to inside, starting with the fuel tank and working toward the engine. Fuel supply lines and vapor return lines should be documented in the process. The investigator should note and photograph all signs of fire-related damage or rupture if they are present. In the case of larger diesel engines, the fuel lines may consist of hoses that are reinforced with steel braid or may consist of nylon tubing. These hoses should be carefully examined for signs of breaching, mid-hose, due either to contact with energized electrical cables or to chafing, and the ends of hoses should be examined for evidence of failed connections. Modern fuel lines that are constructed or reinforced with nylon or other plastics do not fare well when exposed to fire. Remaining fuel line materials in those vehicles may be important in defining those areas least affected by fire exposure. Be aware that many gasoline injection systems utilize a fuel pressure regulator near the fuel rail that, due to diaphragm failure, may release gasoline under pressure. When examining areas of suspected failure of pressurized fuel components, be aware of acute patterns of damage as an indicator of either the area of origin or as examples of secondary damage.

Table 25-6 Windshield Indicators of First Involvement

Compartment	Indicators
Passenger	Top of the windshield fails Radial burn patterns on the hood
Engine	Bottom of the windshield fails Radial patterns on the doors

Other hoses that should be examined for breaching or crimp failures include coolant hoses, power steering hoses (particularly on the high-pressure side of the system), and oil lines to turbochargers or other engine systems.

Switches, Handles, and Levers

During inspection of the interior compartment, investigators should inspect the position of switches, handles, and levers if possible. The ignition switch should also be checked to see whether the key is present or absent. If this area is severely damaged, investigators can check the floor below for any signs of keys or the ignition lock assembly, which may be found sufficiently intact to support an examination for evidence of key use or prefire damage. Investigators should ascertain and document the position of the windows prior to the fire and determine the position of the transmission gearshift lever.

Hybrid Vehicles and Inspection Safety

Hybrid vehicles use many differing and proprietary designs, some of which employ potentially dangerously high voltages, ranging from 100 V in one brand of small automotive hybrid to 800 V in hybrid commuter buses. Investigators and fire fighters should exercise extra care when examining these vehicles due to a high potential of serious, if not lethal, electric shock. Hybrids feature safety systems to isolate high battery-pack voltage, including a manual disconnect near the battery pack (often located in the trunk or cargo area, behind or beneath a small flap or door) FIGURE 25-11. These high-voltage power supply systems may be compromised if subjected to fire or the trauma of a motor vehicle accident. The investigator, therefore, should use caution in approaching a fire or accident-damaged hybrid vehicle.

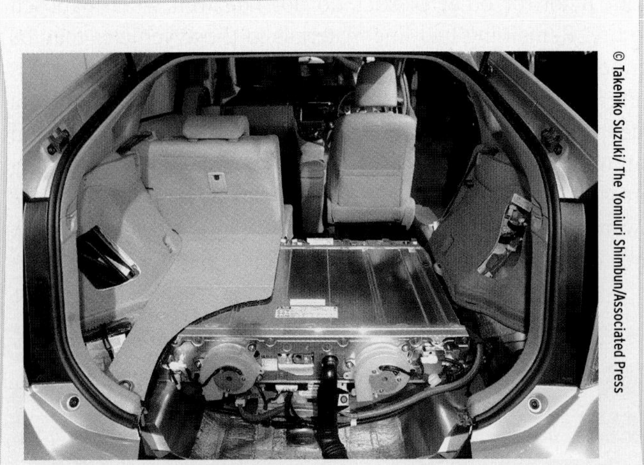

© Takehiko Suzuki/ The Yomiuri Shimbun/Associated Press

FIGURE 25-11 Hybrid vehicles feature safety systems to isolate high battery-pack voltage, including a manual disconnect near the battery pack.

Fire Investigator Tip

Interview owners and operators regarding equipment operation, cleaning, and maintenance schedules. Inspect and document maintenance on the equipment if records are available.

Hybrid Vehicle Technology

Hybrid drive automobiles are a rapidly evolving means of propulsion, thought by many to represent a transitional phase between internal combustion drive systems and non–fossil-fueled systems, possibly of a pure electric, fuel cell variety. As the hybrid designs are relatively new, there is little uniformity in their technical features and specifications. Their general features include a combination of electric power and an internal combustion engine, which is electronically controlled to maximize efficiency. Although the drive system may use up to 600 V DC, supplied by a battery pack, peripheral systems such as lighting, sound, and other convenience systems use a traditional automotive-style 12-V DC system, supplied from a standard 12-V DC battery. High-voltage circuits and harnesses are colored orange for identification. The systems are typically designed to disconnect the battery pack electrically when the key is in the OFF position, in the event of an electrical fault, or upon sensing a crash.

Currently, this technology is being applied to more and larger vehicles, including mid-sized trucks and delivery vehicles, and Class 8 or OTR trucks.

Investigation of Hybrid Vehicle Fires

Hybrid systems differ from previous automotive designs in that they contain battery packs, which can provide potentially lethal electrical shocks and which may, if compromised, provide a unique set of potential ignition sources, as compared with internal combustion-engine vehicles.

The investigator should be aware that automotive wiring containing more than 60 V DC or 30 V AC will be orange in color. This is to warn that the wiring may contain sufficient voltage to be a lethal hazard. The majority of these leads are contained in protective sheathing beneath the vehicle. In addition to the traction motor, higher voltages may be found inside the inverter/converter, which contains the computer control for the system. Because the hybrid's engine does not run at all times, the air conditioning compressor must be electrically driven. Power is drawn from the high-voltage battery pack to run the 300-V DC motor; therefore, the supply wiring harness to the AC compressor motor is colored orange. The steering system requires a pump, but again cannot be driven as in a conventional automobile, so the pump is electrically driven, although by a lower voltage motor (below 60 V DC).

As these systems and vehicles are undergoing rapid developmental changes, the investigator is advised to obtain product-specific information regarding the high-voltage system prior to beginning his or her examination. It may be

Safety Tip

Automotive wiring containing more than 60 V DC or 30 V AC will be orange in color. This is to warn that the wiring may contain sufficient voltage to be a lethal hazard.

prudent to consult with technicians trained in maintenance and repair of the specific vehicle prior to pursuing the vehicle examination. Caution should be exercised with post-crash vehicles or postfire suppression environments that are contaminated with water. The investigator should be aware that the internal combustion equipment of the hybrid vehicle will be much the same as that of a pure gasoline-engined automobile. After positive isolation of the high-voltage system, the investigation may proceed as for other vehicles.

Additional Vehicle Considerations

Total Burns

Total burns are those fires that have consumed all, or nearly all, combustible materials. Total burns are challenging fires to investigate, whether the fire is structural or vehicular. In this type of examination, the investigator's most valuable asset may be the attitude with which he or she approaches the task. It is important to recognize that, although much of the evidence we are seeking may have been altered or destroyed by the fire, there is nearly always remaining information that is retrievable. The investigator should enter the process with a goal of maximizing the collection and documentation of that information.

Investigators should determine the condition of the vehicle prior to the fire and whether any components were missing. They should examine the floorboards for the presence of ignitible liquids or their trace remnants. It is advisable to analyze the fluid levels in the various systems and to take samples. An engine oil analysis or laboratory analysis of the engine oil filter can indicate the condition of the engine prior to the fire, even if the vehicle is severely fire-damaged. Similarly, the prefire condition of the transmission can frequently be determined by analysis of the transmission fluid.

Stolen Vehicles

History has shown that the chances of an accidental fire after a vehicle is stolen are low. Often, vehicles are stolen for parts and are burned to cover the crime. Parts that may have been removed include the wheels, major body panels, engines, transmissions, air bags, stereos, and passenger seats. Sometimes the vehicle was stolen to conceal another crime.

Through debris sifting and inspecting, it may be possible to recover the ignition switch locking tumblers. This is the portion of the ignition switch that holds the key. Forensic laboratory inspection of the tumblers can often indicate whether the ignition switch was defeated and sometimes whether a specific key was used to last operate the vehicle. This examination must be done by a qualified technician, such as a forensic locksmith. The investigator should secure the ignition lock assembly and transport it to the technician, documenting its discovery, removal, and transportation or shipping to maintain the evidence chain properly. If the ignition lock is still in place in the column, removal and transport of the entire column will allow the forensic technician to document the tumbler position and steering lock engagement as well.

Vehicle Ignition Components

When investigating vehicles that were reported stolen and recovered burned, physical evidence of the prefire condition of the ignition lock may be discoverable if the vehicle contains a cut key-style ignition lock. The common location for the ignition lock is on the right side of the steering column; however, it may be mounted in the dash or center console. Vehicle manufacturers utilize only a few styles of ignition lock wafers, and the investigator should be familiar with the construction of the wafer associated with the vehicle that is being investigated.

Typical ignition lock cylinders are constructed of die-cast zinc mounted in a die-cast zinc steering column housing, whereas the lock wafers are typically made of brass. The lower melting temperature of the die-cast zinc steering column housing will typically release the ignition lock from the housing during a fire, and the ignition lock will fall to the floor or into the floorboard debris, typically remaining at least partially intact. The recovery of the ignition lock artifacts can provide the investigator with information regarding the prefire condition of the ignition lock and the time during the fire that the lock collapsed.

The fire investigator should layer the debris on the driver's side floorboard so as to recover the ignition lock artifacts. If the lock was not forcibly removed prior to the fire, some or all of the lock artifacts should be identifiable and recovered from the debris for further analysis. The recovery of the lock remains from within the debris just below the original location of installation would indicate that the lock was likely in the proper position at the time of the fire. The lock remains should show thermal damage, but may also show mechanical damage that may have occurred prior to the fire.

The discovery of an ignition lock on the floor, beneath the fire debris, would indicate that the lock was on the floor prior to the fire and could be an indicator that the ignition lock was removed from its installed position prior to the fire.

When inspecting the ignition lock for evidence of mechanical damage, magnification may be necessary to evaluate the wafer "key way." A detailed examination may show if the lock was forcibly rotated, or "picked." It may also reveal a rotational deformation of the wafer. Sharp gouges or scratches across the inside of the key way may indicate that the lock was picked.

If the remains of the lock are not readily identifiable during the examination of the debris, it may be concealed within component materials and radiographic images may be necessary to identify the ignition lock artifacts.

Some manufacturers use non-brass wafers in ignition locks, which may be melted or destroyed by fire. The investigator may use other ignition lock system components, such as locking lugs, armored lock cylinder caps, tumblers, detent pins, springs, ring antenna, lock core retainer, and other case-hardened parts found in the floor-level debris to determine whether the ignition lock was present in the vehicle at the time of the fire, although these components may not be useful in establishing if the ignition lock was defeated prior to the fire.

Vehicles in Structures

When a vehicle is located in a structure where a fire has occurred, the vehicle should be considered a potential ignition source until it is properly eliminated. The vehicle might be under structural debris that has to be removed before inspection. An external fire could have caused the fuel in the fuel tank or other vehicle fluids to be released and to become involved in the fire. Consider that a fully involved automobile burning inside a garage or other building presents a dramatic fuel load and, as the liquid fuels contained in the vehicle become involved, represents a very high heat-release rate. These factors should be considered in your evaluation of the structural damage, just as the containment of the fire due to the configuration of the building should be considered in your evaluation of the extent of the vehicle damage.

Specialized Vehicles and Equipment

Recreational Vehicles

RVs combine elements of vehicle and residential structures. Their living spaces are similar to houses and mobile homes in terms of contents and materials. Some have plywood flooring, carpets with foam padding, wooden wall paneling, and polyurethane foam furniture. The RV may also be a motorized vehicle, similar to a truck, bus, or automobile, and thereby incorporate many of the same systems seen in those types of vehicles TABLE 25-7. RVs also include trailers such as fifth-wheel trailers and camping trailers.

Resources that may help in determining the origin and cause of an RV fire include sales brochures, owner's manuals, and websites for the manufacturers of the chassis, coach, and other components.

Safety Concerns

The RV may, through its design, comprise a confined space, requiring the investigator to be aware of entry/egress and atmospheric issues. Investigators should ensure that such spaces do not contain hazardous levels of explosive or toxic vapors or gases prior to entering them. Toxic, carcinogenic, or irritating materials such as ammonia, sodium chromate, plastic residues, and fiberglass particles may be present in the thermally damaged RV. Some such materials may become airborne. Their presence should be evaluated prior to entry, and appropriate personal protective equipment (PPE) should be used.

Investigators should also be aware of electrical hazards in the form of 12 V DC from battery-supplied circuits and 120- and 240-V AC power supplied by generators or shore power connections. Liquid (gasoline, diesel) or gaseous (propane, hydrogen) fuels may be onboard. The investigator should perform a proper evaluation and mitigate such potential hazards before undertaking the on-board examination.

The RV should be evaluated for stability hazards prior to the examination. Tires, plywood flooring, sidewalls, and roof structures compromised by the fire may fail during the exam, dropping the RV or its components. These elements must be blocked or shored prior to allowing any personnel within or beneath the RV.

Systems and Components of the RV

RVs may be equipped with equipment and systems unlike those found in other vehicles, including:

- Shore power connections: These consist of a cord, sometimes with a collection of adapters, for connection to a nearby receptacle, as may be available at an RV park or at the owner's home.
- Auxiliary power generator systems: These are either fixed or portable and are powered by an internal combustion engine fueled by gasoline, diesel, or propane. An automatic generator starting system (AGS) that automatically starts and stops the generator under preset conditions of load, demand, battery condition, or time may be incorporated into the system.
- Electrical converters and inverters: These allow power to be interchanged between 12-V DC and 120-V AC systems. The converter transforms 120-V AC power to 12-V DC, while the inverter performs the opposite function. Often, these functions are combined in one electronically controlled unit, which also may be connected to the coach and chassis batteries as an automatic charger.
- Battery systems: Most often, the self-propelled RV will use two separate battery systems. The chassis will have one or more 12-V DC batteries for starting the engine and operating other vehicle systems such as exterior lights and heating systems. The coach will typically employ a separate set of two to four deep-cycle batteries. The two sets of batteries are often connected through an electronic isolator system to allow the batteries to be evenly charged by the engine generator, the auxiliary generator, or the converter system.
- Holding tanks: The RV may be equipped with multiple holding tanks for toilet waste (black water), sink and shower waste (gray water), and fresh or potable water.
- Propane systems: Compressed propane stored in one or more tanks on the RV supplies fuel to power the range or cooktop, refrigerator, water heater, and furnace if so

Table 25-7 Types of Recreational Vehicles

Type	Description
Fifth-wheel travel trailer	Mounted on wheels and towed via a towing mechanism mounted above or forward of a motor vehicle's rear axle
Folding camping trailer	Mounted on wheels and towed by a motor vehicle, it features collapsible side walls and roof that can be unfolded for use and refolded for travel
Travel trailer	Mounted on wheels and towed by a motor vehicle, with roof and sidewalls made of rigid material

equipped. The pressure is modified by a two-stage regulator, and fuel is delivered through pipes or tubing to the individual appliances. The propane system is subject to conditions as specified in NFPA 1192, *Standard on Recreational Vehicles*, NFPA 58, 49 CFR, and ASME *Boiler and Pressure Vessel Code*, Section VIII.

- Stove: A stove may be either a cook top, which drops into a countertop, or a range, which includes an attached oven. In either case, the stove is fueled by propane and is equipped with a nonadjustable propane regulator.
- Water heaters: Powered by propane, 120 V AC, or a combination of both. Typically 12-V DC power is used for control circuits, employing an internal thermostat to regulate water temperature.
- Furnaces: Typically forced-air, using propane as the fuel, with controls and blowers operating on 12 V DC.
- Refrigerators: RV refrigerators do not use a compressor, like home refrigerators; rather, their design involves a closed system of tubes containing water, anhydrous ammonia, sodium chromate, and hydrogen gas under pressure. The ammonia/water mixture is heated by a small propane burner, a 120-V AC heating element, or a 12-V DC heating element. Automatic controls may switch between fuel sources, or the operator may manually select one mode of operation. Historically, fires involving RV refrigerators have involved loss of coolant containment (ammonia and/or hydrogen gas), electrical problems with control circuits, or venting restrictions.

Investigation

Investigation of recreational vehicle fires is similar to that of other motor vehicles. The RV's systems should be examined to determine whether they played a role in ignition or as fuel, and how they may have been involved in the spread of the fire. If appliances are involved in the fire development, the investigator should determine their types, brands, models, and conditions. Care should be used in removing components because of close installation tolerances. Assistance from an electrical or mechanical engineer could be indicated in the evaluation of an RV's heating and electrical systems if a particular system is suspected of having been the ignition source or there is a need to eliminate a system in a suspected arson.

As with the vehicle itself, if individual appliances are potentially involved in the fire cause, their service and recall history should be researched, and the RV's users should be interviewed for any history of malfunction or replacement.

Much information regarding a particular RV's layout and specifications may be found on the manufacturers' websites or through local RV dealers. Most new RVs come from the manufacturer with a packet of owner's manuals, including the RV's specific information as well as that pertaining to installed appliances or equipment. These packets, if available, will assist the investigator in identifying originally installed equipment and appliances. Many manufacturers maintain specific records of installed parts and equipment for their products by serial number. Some manufacturers also maintain a telephone "help line" to provide technical assistance to owners and RV technicians.

Additional Investigation Considerations

Investigation of RV fires should include examination and analysis of all systems to determine their potential for involvement in the fire, as a source of ignition or fire spread. A failure analysis of these systems includes checking for poor installation or function, prefire damage, fuel leaks, and function of safety and control features. Ignition sources may include the same types of sources as those involved in structural fires; those common to motor vehicles; and those that could be related to movement, towing, and mechanical damage.

As with other motor vehicles, RVs are required by the National Highway Traffic Safety Administration (NHTSA) to have a unique VIN. In the case of RVs, the manufacturer also applies a VIN, which can be found in the owner's documents, a label on the exterior of the coach, and often a placard on a galley cabinet. Numbers stamped on the chassis may not be directly related to the VIN.

When investigating an RV fire within a structure, such as a garage, care must be taken to differentiate between the structure's systems and combustibles and the RV's. When the fire occurs in a park or campground, all services provided by the park or campground need to be examined, in addition to the RV's systems. NFPA 1194, *Standard for Recreational Vehicle Parks and Campgrounds*, is a valuable resource in these cases.

Heavy Equipment

Heavy equipment includes earth-moving equipment as well as construction, mining, forestry, landfill, and agricultural equipment. These types of equipment normally serve one or more specific materials-handling functions. Large equipment is often diesel powered with a hydraulic transmission. These systems are sometimes susceptible to failure due to overloading the engine or transmission, a failure of the hydraulic or electrical systems, bearing or engine failures, or the unintended ignition of the materials being handled. Some of this equipment is equipped with fire suppression systems, which the investigator should inspect for function and maintenance history.

Medium- and Heavy-Duty Trucks and Buses

With the exception of OTR tractors, most medium- and heavy-duty trucks are custom built for specific applications. The original manufacturer typically delivers these trucks as a bare cab and chassis, and then beds, bodies, and other equipment are built and installed by one or more secondary shops, manufacturers, or "body companies." Buses frequently are delivered from the original manufacturer as a bare chassis, sometimes with minimal front-end body parts. The complete body is then built and installed by a second manufacturer. Often, a separate company will install other body equipment and systems, such as wheelchair lifts, specialized seating, and so forth. Thus, after a fire in relatively new equipment, several companies may have a warranty or liability interest in the investigation and should be notified prior to any disassembly or other alteration of evidence.

Mass Transit Vehicles

Many transit districts are converting their fleets to alternative fuels. Although diesel engines are still standard in most buses, some districts are now using CNG, LNG, diesel-electric hybrid, or pure electric drive systems, the latter often seen in light rail or streetcar services. Most transit districts perform their own maintenance in-house. The transit district shop may be a valuable reference or technical resource for the investigator.

Earth-Moving, Forestry/Logging, Landfill, and Construction Equipment

Earth-moving, forestry/logging, landfill, and agricultural equipment typically all employ hydraulic systems, providing a reservoir of additional fuel as well as hoses, fittings, and valves that may wear, chafe, or rupture. Chafing between electrical conductors and hydraulic hoses is a frequent cause of fires in these types of equipment. If a conductor chafes through harness coverings and insulation, contact with a grounded steel-reinforced hydraulic hose may provide not only a source of fuel under pressure, but a competent ignition source as well. Logging equipment may accumulate large quantities of "duff," consisting of needles, bark, and other organic material. This material may absorb hydraulic fluid, oil, or diesel fuel and is easily ignited by contact with heated exhaust components, electrical events, or friction-related heat sources. With all equipment operating in an environment in which loose particulate matter is present, radiators and oil coolers may clog, causing overheating of coolant, oils, or other fluids.

Agricultural Equipment

Agricultural equipment is nearly always subject to accumulations of cellulosic materials such as grain dust, wheat or corn chaff, cotton fibers, hay, or straw. Also, harvesting operations are most frequently performed during periods of high ambient temperatures and low humidity. These cellulosic materials are predictably subject to ignition when exposed to high levels of friction, sparks, or contact with sufficiently heated surfaces.

Fires may also result from failure of electrical, fuel, or hydraulic systems; friction developed from bearing or other mechanical component failures; hot surfaces within the equipment; sparks from foreign metals picked up in the field; or friction and sparks from filed mechanical parts that are transported through the crop processing areas of combines, balers, or pickers.

Agricultural equipment requires routine maintenance on a schedule as specified in the operator's manual supplied by the manufacturer. Investigators should interview owners and/or operators with regard to equipment operating, cleaning, and maintenance schedules, and should inspect and document maintenance if records are available. Aftermarket accessories are often installed on agricultural equipment and are sometimes a factor in the initiation of fires. Such accessories should be identified through interviews and/or records and documented by the investigator.

There are many kinds of agricultural equipment, and within these groupings are several makes and models. When investigating a fire involving such equipment, refer to the manufacturer's information for specific details on the particular equipment involved. TABLE 25-8 lists common types of agricultural equipment.

Table 25-8 Agricultural Equipment

Type	Description
Farm tractors	■ They are used for pulling or pushing machinery or trailers, lifting, loading, plowing, tilling, and similar tasks. ■ Engines are usually forward of the operator and run on gasoline or diesel fuel. ■ Transmission and brake types vary, but the transmission includes a power take off (PTO). ■ They often include hydraulic components.
Combines	■ They are used to harvest grain crops: the crop is cut by the header, carried through a threshing mechanism, cleaned in a hopper or tank, then offloaded. ■ Most have turbocharged diesel engines mounted to the rear of the grain tank. ■ Most have hydrostatic drive transmissions.
Forage harvesters	■ They are used to harvest field crops to make silage. ■ Cutting heads cut plants and feed them into processing drums, then to an accelerator device and through a chute to a trailer. ■ Most have turbocharged diesel engines mounted in the rear. ■ They are usually equipped with electronically assisted hydraulic transmissions.
Cotton pickers	■ They are used to harvest cotton. ■ Barbed spindles separate the cotton from the plant, which is then blown into a large basket. ■ Most have turbocharged diesel engines mounted near the center of the equipment, below and to the rear of the operator. ■ They are usually equipped with hydraulic drive transmissions. ■ Additional components include row units attached to a header, a moistener system, a hydraulic system, and a lubrication system.
Sprayers	■ They are similar to tractors but are fitted with solution tanks containing chemicals to treat growing crops. ■ Most do not have PTO attachments. ■ Similar equipment includes windrowers, floaters, spreaders, and fertilizer applicators.
Baling equipment	■ This equipment is used to compress harvested crops into round or square bales that can then be wrapped. ■ In round balers, the crop is rolled inside the equipment and is bound and released through a hydraulic gate when it has reached a certain size. ■ In square balers, a mechanical plunger compresses the crop into thin sheets. When it has reached a certain size, it is bound and released. ■ Additional considerations include the pickups used to pick up crop material, the chain drive systems used to drive the pickup and belt and roller systems, the hydraulic system, and the electrical system.

Safety Tip

The following hazards are common with agricultural equipment, and the fire investigator should take care to ensure safety with these hazards in mind:

- Particulate from burned composite material and vegetation
- Sharp edges from glass or metals
- Overhead hazards due to the size of equipment
- Slippery or uneven surfaces on the equipment or unstable equipment
- Fuels, herbicides, pesticides, and other agricultural chemicals
- Failure of mechanical parts, leading to collapse or spills
- In agricultural fields, presence of mud, snow, plants, snakes, rodents, and insects

The fuels used in agricultural equipment are similar to those used in other motor vehicles, including diesel and biodiesel. The body of the equipment can include plastics, composite materials, and rubber. The materials used will greatly impact the investigation. For example, at high enough temperatures, plastic components may melt to the point where any fire patterns or other evidence are lost.

Safety Concerns

Although investigating fires in agricultural equipment involves many of the same safety concerns as investigating fires in other motor vehicles, there are additional safety considerations to keep in mind:

- Due to the size of the equipment, there is the potential for more fluid (e.g., fuel) to be stored in the equipment. This can affect the fire spread and intensity and can eliminate early fire patterns.
- Special tools or equipment may be necessary to access internal areas of agricultural equipment. In these cases, it is best to confer with someone who has experience with the particular type of equipment to avoid spoliation issues.
- The height of the equipment may interfere with overhead electrical wires, either at the scene or during equipment removal from the scene. If the equipment is potentially affected by high voltage, investigators and others should stay at least 10 ft (3.1 m) away from the equipment until it has been de-energized by qualified personnel.
- A fire in an elevated cotton picker basket may cause instability and accessibility issues if the operator was unable to release the basket and drive away from it.
- The gate on round balers can create severe crushing and entrapment hazards if the gate's hydraulic system fails.

Fire Investigator Tip

A common cause of fire is a buildup of combustible materials, such as vegetation. This debris can collect in all parts of the machinery, leading to ignition through friction. One example is the practice of scrapping, which involves harvest of scrap cotton from dead plants. The conditions for this practice are very dry, and crop residue can build up in the machinery, leading to malfunction and possible ignition.

Wrap-Up

Ready for Review

- Although investigation of vehicle fires is often considered complex and difficult, fires in vehicles and machinery have the same basic requirements for successful ignition and propagation as structural, marine, or wildland fires.
- Well before conducting any vehicle fire investigation, investigators should prepare themselves by learning the general systems and characteristics of powered vehicles and equipment.
- When conducting an inspection of the vehicle's undercarriage, care should be taken to support and stabilize the vehicle properly with blocking or stands to prevent movement, which may cause injuries.
- Fuels in vehicle fires fall into three categories: liquid fuels, gaseous fuels, and solid fuels.
- Vehicles have some unique ignition sources, including heated exhaust components, various types of bearings, and braking systems.
- Knowing how the various systems in vehicles operate is critical for the fire investigator and can be enhanced by viewing technical manuals available at the library, car dealerships, and parts suppliers.
- Vehicle components can influence the spread of fire, including panels and bulkheads.
- It is advantageous to document the vehicle at the fire scene using the same procedures used for documenting a structural fire scene.
- A more complete examination of the vehicle can often be better accomplished after removal from the fire scene.
- Although much of the evidence investigators are seeking may have been altered or destroyed by a total burn fire, there is always remaining information that is retrievable.
- Large equipment is often diesel powered with a hydraulic transmission. These systems are sometimes susceptible to failure due to overloading the engine or transmission; a failure of the fuel, hydraulic, or electrical systems; bearing or engine failures; or the unintended ignition of the materials being handled.
- Agricultural equipment is nearly always subject to accumulations of cellulosic materials such as grain dust, wheat or corn chaff, cotton fibers, hay, or straw. These cellulosic materials are predictably subject to ignition when exposed to high levels of friction, sparks, or contact with sufficiently heated surfaces.
- Hybrid vehicles use many differing and proprietary designs, some of which employ potentially dangerously high voltages, ranging from 100 V in one brand of small automotive hybrid to 800 V in hybrid commuter buses.

Hot Terms

Catalytic converter A device in the exhaust system that exposes exhaust gases to a catalyst metal to promote oxidation of hydrocarbon materials in the exhaust gas.

Electronic (or engine) control module (ECM) An electronic device that controls engine operation parameters, including fuel delivery, throttle control, and safety systems operations.

Event data recorder (EDR) A device to record data before and after a crash event.

Fuel rail An internal passage or external tube connecting a pressurized fuel line to individual fuel injectors.

Hybrid vehicles Vehicles that utilize a combination of internal combustion and electric motors for propulsion.

Total burns Fires that have consumed all, or nearly all, combustible materials.

Turbocharger An exhaust-driven device that compresses intake air to increase engine power.

© Greg Henry/ShutterStock, Inc.

FIRE INVESTIGATOR *in action*

An elderly couple has contacted your department stating they had been on vacation for several weeks and, upon returning home, have discovered their newer model passenger vehicle caught fire. They state that when they opened their garage door, they observed smoke and soot in the garage and a large hole in the hood of their vehicle.

Your examination of the scene indicates that the fire had developed and self-extinguished within the garage. Significant damage is present to the front half of the vehicle, with a large hole present to the right side of the hood, the bottom of the windshield failing, and significant damage to the dashboard and front passenger seat. After documenting the scene and the exterior of the vehicle, you raise the hood to discover extensive damage to the passenger side of the engine compartment.

After examining and documenting the damage to the engine compartment, you begin to inspect the battery cables, which are routed along the passenger side of the vehicle. On top of the passenger fender well, you determine that the positive battery cable has arced and has welded to one of the strut mounting bolts.

1. What does the damage to the windshield indicate?
   - **A.** A fire developed within the rear of the passenger compartment.
   - **B.** A fire developed from the front of the passenger compartment.
   - **C.** A fire developed within the engine compartment.
   - **D.** A fire developed from the dashboard.
2. Which of the following safety hazards should you consider while inspecting the passenger compartment?
   - **A.** Ignition due to a fluid leak
   - **B.** Accidental air bag deployment
   - **C.** Deployment of the bumpers
   - **D.** An electrical shock
3. A fluid leak may be ignited by which of the following components?
   - **A.** Exhaust manifold
   - **B.** Catalytic converter
   - **C.** Brake master cylinder
   - **D.** Both A and B
4. What is the investigator's most valuable asset when investigating a total burn vehicle?
   - **A.** The owner's statement
   - **B.** Documentation of the prefire condition
   - **C.** Identification of missing components
   - **D.** The investigator's attitude toward the investigation

Wildfire Investigations

© Photos.com

© AbleStock

Knowledge Objectives

After studying this chapter, you should be able to:

- Describe the differences among ground fuels, surface fuels, and aerial fuels and their effects on fire spread. (pp 392–393)
- Describe the effects of wind, fuels, topography, weather, fire suppression, and other natural mechanisms on fire spread NFPA 4.2.4 NFPA 4.2.5. (pp 393–395)
- Describe indicators a fire leaves behind that can lead the investigator to the origin of the fire NFPA 4.2 NFPA 4.2.4 NFPA 4.2.5. (pp 395–398)
- List special safety considerations associated with wildland fire investigation. (pp 398–399)
- Describe methods of conducting a wildfire origin investigation NFPA 4.2 NFPA 4.4.2 NFPA 4.6.5. (pp 399–400)
- Describe fire cause categories used in wildland fire investigation NFPA 4.2 NFPA 4.4.2 NFPA 4.6.5. (pp 400–403)
- Identify methods of evidence collection for wildfire investigations NFPA 4.4.2. (p 403)

Skills Objectives

After studying this chapter, you should be able to:

- Employ special safety considerations associated with wildland fire investigation. (pp 398–399)
- Protect the general area of origin. (p 399)
- Conduct an origin investigation NFPA 4.2 NFPA 4.4.2 NFPA 4.6.5. (pp 399–400)
- Conduct a cause investigation NFPA 4.2 NFPA 4.4.2 NFPA 4.6.5. (pp 400–403)
- Collect evidence for wildfire investigations NFPA 4.4.2. (p 403)

Additional NFPA References

NFPA 921, *Guide for Fire and Explosion Investigations*

NFPA 1144, *Standard for Reducing Structure Ignition Hazards from Wildland Fire*

NFPA 1500, *Standard on Fire Department Occupational Safety and Health Program*

NFPA 1977, *Standard on Protective Clothing and Equipment for Wildland Fire Fighting*

CHAPTER 26

FESHE Course Outcomes

Fire Investigation I

9. Recognize potential health and safety hazards. (pp 398–399)

Fire Investigation II

There are no Fire Investigation II (FESHE) course outcomes for this chapter.

You Are the Fire Investigator

© Jones and Bartlett Publishers. Photographed by Glen E. Ellman.

You are at the scene of a large fire that has consumed several acres of farm fields near an interstate highway. The fire is partially contained; however, fire fighters advise that a storm with predicted high winds is entering the area and may change the direction of the fire toward a newly developed residential neighborhood.

During your examination of the scene, you locate the remains of several oxidized aluminum cans and a small ring of rocks in a tree line near the interstate.

1. What types of fuels are present and how will they influence the development of this fire?
2. How will the arriving storm impact your investigation?
3. What fire movement indicators will be present to assist with determining the origin of the fire?
4. What hazards need to be considered during your investigation?

Introduction

Wildland fire investigations involve special techniques, practices, equipment, and terminology and vary greatly from structural fire investigations. Along with these particular skills, a thorough understanding of wildfire behavior is necessary. This chapter describes topography, fuel arrangement, wildfire fuel types, factors that affect fire spread, indicators of directional patterns, conducting an origin and cause investigation, and special safety considerations. As in other specialized investigation situations, consider enlisting the assistance of personnel who specialize in wildland firefighting and/or wildfire investigations.

Wildfire Fuels

Fuels are classified as ground fuels, which include all flammable materials lying on or in the ground; surface fuels, which include all flammable materials just above the ground; and aerial fuels, which include all green and dead materials located in the upper forest canopy. Although aerial fuels may be important to use as indicators to narrow the fire's start to a general area of origin, ground fuels are critical in determining the point of origin.

Fuel Condition Analysis

The physical characteristics of the fuel must be classified by the burning characteristics of individual materials and by the combined effects of the various types of flammable materials present. This then allows for predictions of the rate of spread and general fire behavior.

Ground Fuels

Ground fuels include all flammable materials located between the mineral soil layer and the ground surface. Duff (decomposing organic material above the soil) is not a major influence on the fire spread rate because it is typically moist and tightly compressed with little surface exposure. Roots restrict air supply; however, they may provide an avenue for a fire to spread.

Surface Fuels

Surface fuels are those flammable materials located from the ground surface to approximately 6 ft (2 m) above the ground. Dead leaves and coniferous litter are highly flammable materials and should be considered separately in evaluating surface fuels FIGURE 26-1. Examples include needles dropped from trees. Dead needles attached to trees are especially flammable because they are exposed to air and do not touch the ground, which is often moist.

Grass, weeds, and other small plants are surface fuels that influence the rate of fire spread, based primarily on their

© MrIncredible/iStock/Thinkstock

FIGURE 26-1 Dead leaves and needles are highly flammable surface fuels.

degree of curing. Cured grass is dry grass and is extremely flammable.

Fine fuels and dead wood, which consist of twigs, small limbs, needles, leaves, bark, and rotting material with a diameter of less than 1 inch (2.5 cm), ignite easily and often carry fire from one area to another. They are usually found in areas where logging has taken place. Dry, rotten wood lying on the ground may be ignited by the main fire, which will burn hotter and longer than fine fuels.

Downed logs, stumps, and large limbs require long periods of hot, dry weather before they become highly flammable. These heavy fuels are more of a factor for fire duration than an aid to spread. Low brush and reproduction vegetation may either accelerate or slow the spread rate of a fire because understory vegetation can shade fuels, preventing them from drying rapidly. The understory vegetation often provides the link between ground and aerial fuels.

Aerial Fuels

Aerial fuels are those located from approximately 6 ft (2 m) above the ground surface to the crowns of the canopy. Crowns (twigs and needles or leaves of a tree) are a highly flammable fuel with arrangements that allow free circulation of air and exposure to wind and sun. Snags (standing dead trees) are important aerial fuels that influence fire behavior. Fires start in snags because they are drier and much easier to ignite.

Moss hanging on trees is the most apt to ignite of all aerial fuels and provides a means of spreading fires from ground fuels to other aerial fuels. Moss reacts quickly to changes in relative humidity.

Crowns of high brush are aerial fuels because they are separated by distance from ground fuels. The key factors are volume; live fuel moisture; oil content; arrangement; general condition of ground fuels; and the presence of fine, dead aerial fuels.

Fuel Matrix

Fuels are broken down into four major fuel groups: grass, shrub, timber litter, and logging debris. Each group is further divided into fuel-type models based on predictable behavior under specified weather conditions. These models group fuel elements based on species, fuel form, size, and arrangement.

Factors Affecting Fire Spread

The primary factors that affect fire spread are heat transfer, lateral confinement, weather influence, fuel influence, suppression efforts, and topography. A variety of other factors, both natural and manmade, also influence fire spread.

Heat Transfer

In a wildfire, fire spread is influenced by convective and radiant heat. Convective heat allows the fire to spread from the lower-level grasses and duff to the upper-level branches and crowns. Radiant heat then becomes dominant as the primary heat transfer method as the fire progresses laterally. Radiant heat is the dominant heat transfer method in lower-level fuels, comprising brush and grass on level surfaces.

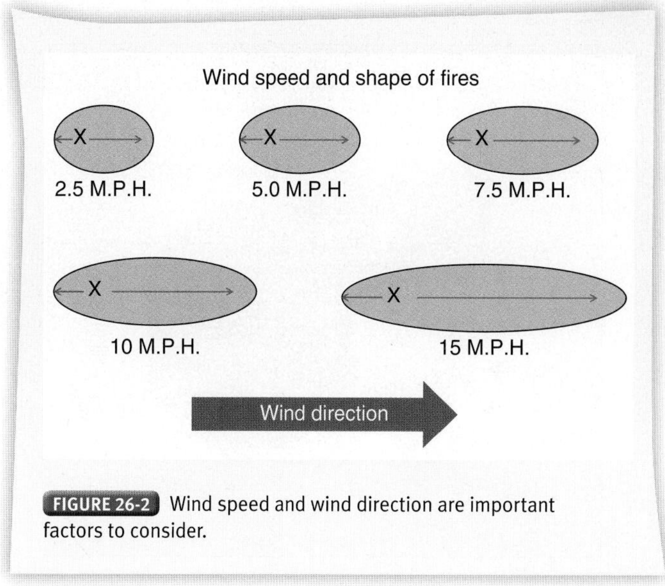

FIGURE 26-2 Wind speed and wind direction are important factors to consider.

Lateral Confinement

Lateral confinement occurs when the fire cannot spread laterally because of inflammable terrain features such as roads, rocky areas, or streams. These confinements can alter the speed, intensity, and direction of spread. Burn patterns may be deceptive as a result of the fire flanking these obstacles.

Wind Influence

Wind influences a fire by pushing the flames ahead and preheating fuel; providing more oxygen; drying out vegetation; and blowing embers, sparks, and airborne firebrands (pieces of burning material) ahead of the fire FIGURE 26-2.

There are four classifications of winds:

1. Meteorological winds are caused by pressure differences in the atmosphere that create weather patterns. They are greatly impacted by the rotation and topography of the Earth.
2. Diurnal winds are the effect of the air being heated by the sun during the day and then cooled during the night. As the temperature of the air increases, the air rises; as the temperature of the air decreases, the air sinks. The fluctuation between the two creates the diurnal winds.
3. Fire winds are created when air is pulled into the bottom of a powerful convection column to replace the rising warmer air. This movement pulls oxygen into the rising plume and encourages fire spread.
4. Foehn winds are the result of air flowing between pressure gradients. As gravity increases the air's speed and the air is compressed, the temperature of the air increases and its humidity decreases, thus improving conditions for fire spread.

Fire Head, Flanks, and Heel (Rear)

The direction in which the local wind is blowing determines the primary route of the fire's advance, led by the fire head,

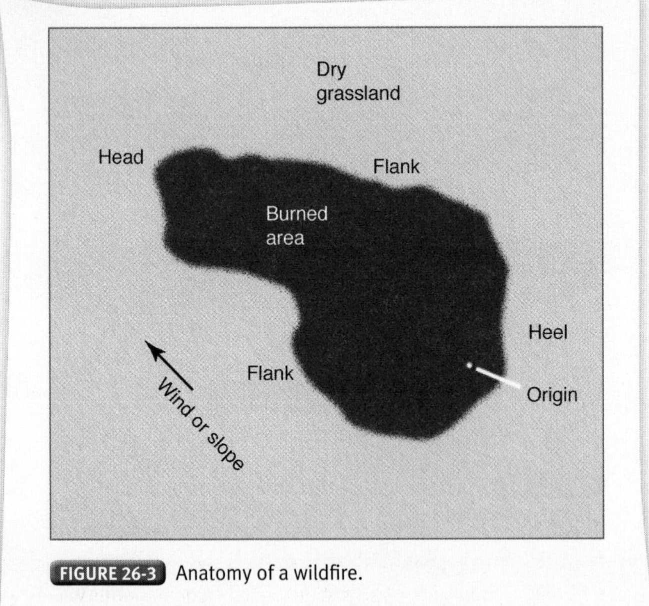

FIGURE 26-3 Anatomy of a wildfire.

or the area of greatest fire intensity. The fire flanks are located on either side of the head. Fire progression on the flanks of the fire is characterized by less intense fire behavior than the head of the fire. The fire heel is located on the opposite end of the fire from the fire head and is less intense. It will generally be backing or burning slowly against the wind or downhill FIGURE 26-3.

Fuel Influence

Ignitibility, rate of burning, and fire spread are influenced by several fuel characteristics. The species of the vegetation determines its moisture content, shape, and density. The moisture present in the fuel is further influenced by the condition of the vegetation, solar exposure, weather, and geographic location. Oil or resin content within the vegetation is also an influence.

The smaller the fuel, the easier the fuel will be to ignite and the faster it will be consumed TABLE 26-1. Ground fuels, such as roots, can burn along their entire length underground and can ignite a surface fire in a different location much later. Surface fuels, including grasses, twigs, needles, or low brush, are commonly involved with the spread of fire. Crown fuels include tall brush, hanging mosses, limbs, leaves, and tree crowns. If fire spreads in crown fuels, the results can often be devastating because all flammables are normally consumed.

Dead fuels are often classified in terms of the time taken for them to reach a percentage of equilibrium with their environment, and are referred to as 1-hour, 10-hour, 100-hour, and 1000-hour fuels depending on their composition, thickness, and location.

Table 26-1 Fuel Size

Size	Burning Characteristics	Examples
Small (fine) fuels	■ Easiest to ignite ■ Rapidly consumed	■ Seedlings and small trees ■ Twigs and leaves ■ Dry grass ■ Brush ■ Dry field crops ■ Pine needles and cones
Large (heavy) fuels	■ Harder to ignite ■ Burns slower and longer	■ Large-diameter trees and brush ■ Large limbs ■ Logs ■ Stumps

Topography

Topography relates to the form of natural and human-made earth surfaces. It affects the intensity and spread of the fire and can greatly influence winds.

Slope is the change in elevation over a given distance, measured by the rise over the run (the differential in height divided by the total horizontal distance, normally measured in feet, then multiplied by 100, equals percent slope). It allows the fuel on the uphill side to be preheated more rapidly than if it were on level ground and is an important factor in fire spread. Wind currents moving uphill during the day also accelerate the fire spread. Burning fuels can roll downhill, igniting other fuels.

Aspect is the direction the slope faces. If a slope faces the sun, it is typically drier and may have a more combustible fuel type than slopes without this solar heating. This results in a greater ease of ignition and faster spread. TABLE 26-2 summarizes these relationships.

Weather

Weather plays an important role in wildfire behavior because it influences the atmospheric stability, temperature, relative humidity, wind velocity, cloud cover, and precipitation.

Weather history is a description of atmospheric conditions over the preceding days or weeks and should be analyzed to determine what influences it might have had on the fire's ignition and burning characteristics. This information can be obtained through wildland fire dispatch centers, the National Weather Service website, and a variety of other Internet sites.

The ambient temperature (air temperature of the surrounding environment) influences the temperature of the fuel. As the radiant energy of the sun heats the atmosphere, the

Fire Investigator Tip

Temperature differences of more than 18°F (8°C) can occur between shaded areas and areas exposed to sunlight, and 2°F to 5°F (–17 to –15 °C) per thousand feet of altitude, depending on air moisture.

Table 26-2 Effect of Slope Aspect on Fire*

Direction	Conditions	Effect
Facing toward the sun	■ Has less vegetation ■ Drier	■ Greater ease of ignition and fire spread
Facing away from the sun	■ Has more vegetation ■ More moist	■ Less ease of ignition and fire spread

*Conditions and effects are opposite in a high desert environment.

ambient temperature increases, in turn increasing the temperature of the fuel and reducing the moisture content therein.

Humidity is the measure of water vapor suspended in air. Humidity is usually expressed as relative humidity. The moisture in the air directly affects the amount of fuel moisture. Warm air can hold more water than cool air. If the air is dry, it will try to pull moisture out of the surrounding vegetation, thus drying out the vegetation and making it more susceptible to fire. If the air is more humid, however, the moisture will transfer to the surrounding vegetation, thus reducing the chance of fire and slowing any fire spread.

Fire Suppression

Fire suppression comprises all activities that lead to the extinguishment of a fire. Fire crews should attempt to protect the potential areas of origin so that the fire's origin and cause can be accurately determined. Common fire suppression activities include the following:

- Fire breaks: Any natural or human-made barriers that are used to stop the spread or reroute the direction of the fire by separating the fuel from the fire. They are also called fire lines or control lines.
- Air drop: The aerial application of water, foam, or retardant mixture directly onto the fire or threatened area or along a strategic position ahead of the fire FIGURE 26-4.
- Firing out: The process of burning the fuel between a fire break and the approaching fire to extend the width of the fire barrier.

Class A Foam

Air mixed with water and Class A foam produces a sudsy coating on combustibles. On wood, it reduces the surface tension of water, allowing it to penetrate much more easily rather than shedding off the surface. It also suspends water for a period of time, allowing for a longer cooling effect, a reduction of oxygen to the fuel source, and prolonged protection to unburned fuels.

© photocdn6/iStock/Thinkstock

FIGURE 26-4 An air drop is the application of water, foam, or retardant from the air.

> **Fire Investigator Tip**
>
> Suppression crews should carefully place wet lines or construct fire lines a short distance from the fire's edge in the area of origin. This will allow the fire to burn undisturbed to the lines, resulting in protection of the fuels, burn patterns, and the origin.

> **Fire Investigator Tip**
>
> Wildland–urban interface fires may result from fire starting in a structure and spreading to the wildlands or starting in the wildlands and spreading to a structure. A wildland fire investigator must clearly make this determination in the event that the legal system becomes involved to address responsibility and financial liability.

Other Natural Mechanisms of Fire Spread

Wind-borne firebrands are hot embers picked up by wind and blown into unburned fuel great distances from the original fire. They can start spot fires remotely from the main fire and could be mistakenly interpreted as deliberately set fires.

A fire storm is a natural phenomenon that attains such intensity that it creates and sustains its own wind system. This system can be so strong that it is capable of producing small dust devil–type circulations called fire whirls, which dart around in unpredictable directions, damage homes and other structures, and quickly spread the fire.

Animals and birds can also cause fire to spread if their fur or feathers catch fire (e.g., from a power line). Once the animal is on fire, wherever that animal goes it can spread the fire.

Indicators of Directional Pattern

Analysis of the directional pattern shown by multiple indicators in a specific area can identify the path of fire spread through the area. By using a systematic approach of back tracking the progress of the fire, the investigator can retrace the path of the fire to the point of origin. This procedure is the accepted and standard technique in wildfire investigation. Visual indicators include differential damage; char patterns; discoloration; carbon staining; and the shape, location, and condition of residual, unburned fuel.

Wildfire V-Shaped Patterns

Wildfire V-shaped patterns are ground surface burn patterns created by the fire spread. When viewed from above, they are generally shaped like the letter V. These are not to be confused with the traditional plume-generated vertical V patterns associated with structural fire investigations.

Wildfire V-shaped patterns have the following characteristics:

- Horizontal, not vertical, patterns
- Affected by wind direction and slope
- Fire flanks (perimeter of fire parallel to direction of spread) widen up the slope in the direction of the wind
- Origin that is normally near the base of the V

Degree of Damage

The degree and type of damage to fuels indicate the intensity, duration, and direction of fire passage TABLE 26-3. Leaves, branches, and large woody material sustain greater damage on the side from which the fire approached. Grass that grows in clumps may not be completely consumed, resulting in protected areas on the side opposite the fire's advance FIGURE 26-5.

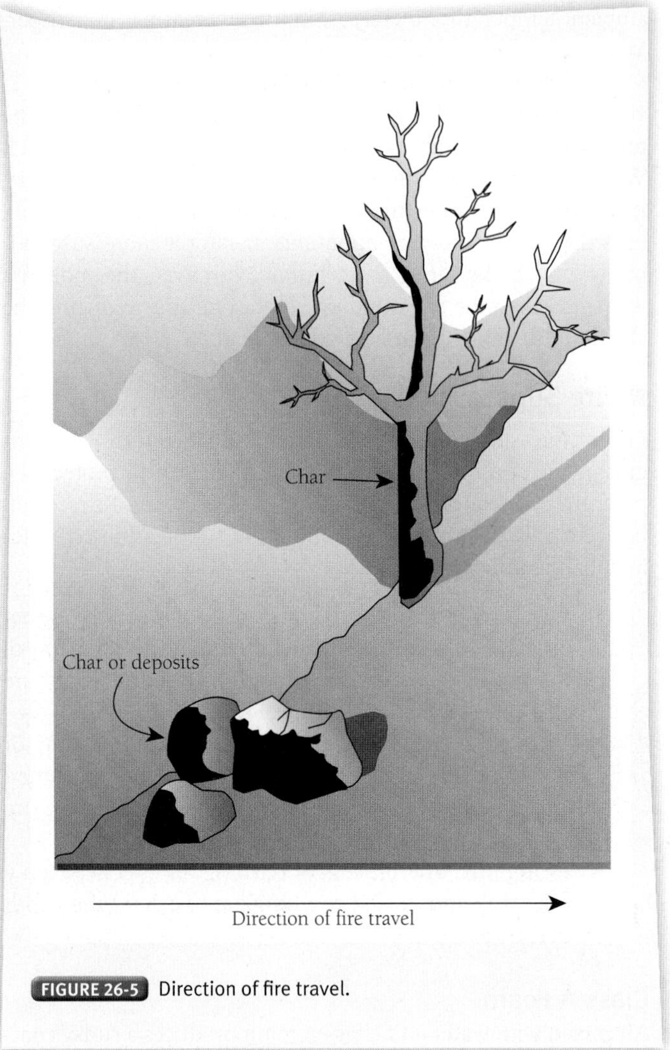

FIGURE 26-5 Direction of fire travel.

Fire Investigator Tip

Place a stem of low growing brush, normally 1 to 3 feet (0.3 to 0.9 meters) in height, between two fingers near the mid-point of the stem. Squeeze the stem slightly, and move your fingers upward. Sooting often appears heavier on one finger than the other. This shows the direction from which the fire advanced. Several samples should be taken from various stems for confirmation. This technique can also be useful on wire fencing.

Table 26-3 Damage Indicators of Fire Passage

Fuel	Appearance	Interpretation
General	Fuel is more damaged on one side than another	Shows direction of spread due to protection
Grass stems	Unburned grass stalks or seed heads lying on the ground	Grass heads typically fall in the opposite direction of the fire's travel
Brush	More upper foliage is burned	Exit side
	Some upper branch tips fall unburned to the ground	Entry side
	Ash deposits not found on fuels	Fuels still burning when the ash was deposited
	Brush cupping Tips of burned stubs are blunt or cupped	Upwind side of fire
	Tips of burned stubs are sharp	Downwind side of fire
	Brush die-out pattern Decreasing fire intensity, charring, and burned branch size	Fire died out after entering brush growth

Grass Stems

As a low-intensity fire burns the bottoms of grass stems, the stems fall back into the fire. If they fall into a burned area behind the fire head, they may remain unburned. Unburned stalks of grass on the ground generally point in the direction of the fire approach. The investigator should examine several grass stems to increase the reliability of this method. He or she should also view the stems outside the burned area to ensure that no other influences, such as wind or heavy rains, have had an effect on their vertical arrangement.

Brush

Brush can be very valuable as a fire direction indicator. It is often damaged by the fire but not fully consumed, leaving the investigator a reliable indicator that has not been moved or destroyed, even during light mop up. Brush can display several indicators such as freezing, degree of damage, depth of char, angle of char, curling of leaves, sooting, white ash deposits, cupping, and die-out patterns. Not many objects can display as many indicators as brush.

Trees

Trees are significant indicators of fire direction, particularly in areas of frontal fire damage FIGURE 26-6. Fire movement is

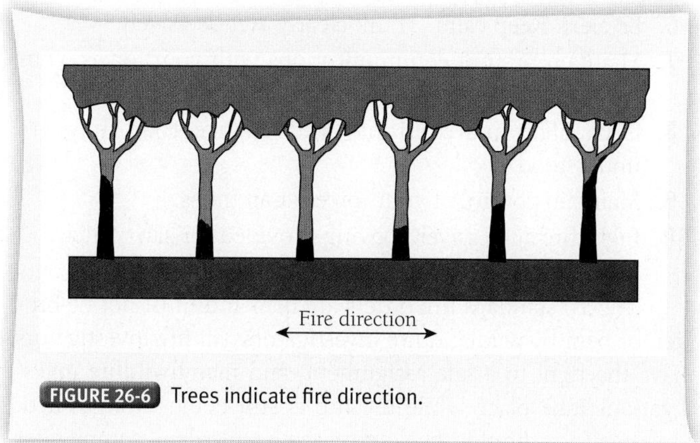

FIGURE 26-6 Trees indicate fire direction.

recorded at ground level around the root base and tree trunk and at flame height by the lower foliage and crown canopy. The char to the trunk surface of a tree is affected by the topography of the land surface. A fire burning uphill or with the wind creates a char pattern that slopes to a greater degree than the ground slope.

Crown damage can also be used to interpret fire direction. Convection and radiant heat ignite lower limbs and then spread upward into the rest of the tree. This action progresses in intensity as the wind action drives the fire away from the windward foliage and branches TABLE 26-4.

As an example, if you observe a triangular unburned area on the side of the crown, your interpretation would be that the unburned area is on the windward, or approaching, side of the fire.

Angle of Char

Angle of char indicators are divided into two groups based on the types of fuels in which they occur. Angle of char can occur in pole-type fuels (e.g., tree trunks, utility poles, fence posts) or in the foliage crowns of brush or timber-type fuels. Angles of char on pole-type fuels are described in more detail in Table 26-4. Angles in foliage crowns are similar but are often not as obvious.

White Ash Deposit

White ash can be the by-product of combustion. More white ash will be created on the sides of objects exposed to greater amounts of heat and flame. Ash is often dispersed downwind

Table 26-4 Trunk Char

Fire Direction	Wind Direction	Observation
Up slope	Up slope	Angle of char on the tree will be greater than the slope angle of the hill
Down slope	Up slope	Char line on the tree is nearly parallel to the slope of the hill
Down slope	Down slope	Angle of tree char will be higher on the downhill side
Up slope	Down slope	Char angle is level with the slope of the hill with only light upslope damage to the tree trunk

and deposited on the windward sides of objects. Fuels facing the advancing fire will appear lighter on the side facing the oncoming fire and darker on the side opposite.

Cupping

Cupping is a concave or cup-shaped char pattern on grass stem ends, small stumps, and the terminal ends of brush and tree limbs. The exposure of the surface to the windward side of the fire leaves a charred surface on the fuel. This side is charred deeper than the opposite side, which is protected from exposure.

Die-Out Pattern

As a fire progresses into different fuel types, the spread may slow, or the fire may extinguish in areas where there is increased fuel moisture, lack of fuels, or other conditions causing a decrease in rate of spread and intensity.

Exposed and Protected Fuels

A noncombustible object or a fuel itself shields the unexposed side of a fuel from heat damage. Fuels will be unburned or show less damage on the side shielded from the advancing fire.

Staining and Sooting

Staining is caused by hot gases, resins, and oils condensing on the surface of objects. This occurs most commonly with noncombustible objects such as metal cans, glass bottles, or rocks. Stains will appear on the side of the object exposed to the flames.

Soot is caused by incomplete combustion and the natural fatty oil content in some vegetation. Carbon soot is typically more heavily deposited on the side facing the approaching fire. Soot deposits can be noticed by rubbing your hand or finger across the surface, comparing opposite sides. Sampling soot on small brush or weed stems in multiple locations can be an effective technique.

Depth of Char

Char on branches or logs exhibits a scale-like appearance. Wood materials lose mass and shrink as they burn, forming this scale-like surface. The side with the deepest charring will typically be on the side facing the advancing fire.

Spalling

Spalling is caused by a breakdown in the tensile strength of the rock's surface that has been exposed to heat, and it will appear as shallow, light-colored craters or chips in the surface of rocks within the fire area. Spalling will usually be accompanied by slabs or flakes exfoliated from the surface of the rock. It is generally associated with advancing fire areas and will appear on the side of the rock exposed to the flames.

Curling

Curling occurs when green leaves curl inward toward the heat source. They fold in the direction from which the fire is coming. This usually occurs with slower moving, lighter burns associated with backing and lateral fire movement.

FIGURE 26-7 Look for sooting and/or staining to determine the direction of spread.

Foliage Freezing

Freezing can occur as fire passes the needles or leaves of trees. Especially in an advancing fire, they tend to become soft and pliable and are easily bent in the direction of prevailing winds. As they cool, they will often remain pointing in the direction of the prevailing wind.

Noncombustibles

A protected area in fuels immediately adjacent to a noncombustible object indicates that the fire direction was from the opposite side of the protected area. Staining and sooting on one side of the object indicate that the fire direction was from the stained side FIGURE 26-7.

Loss of Material

As materials heat, they undergo physical change. Typically, when wood or other combustible surfaces burn, they lose material and mass. The shapes and quantities of remaining combustibles can themselves produce lines of demarcation and ultimately fire patterns to be analyzed by the investigator.

Special Safety Considerations

The basic principles of safety apply for wildfire incidents. These principles include the following 10 standard firefighting orders:

1. Keep informed on fire weather conditions and forecasts.
2. Know what your fire is doing at all times.
3. Base all actions on current and expected behavior of the fire.
4. Identify escape routes and safety zones, and make them known.
5. Post lookouts when there is possible danger.
6. Be alert. Keep calm. Think clearly. Act decisively.
7. Maintain prompt communications with your forces, your supervisor, and adjoining forces.
8. Give clear instructions, and ensure that they are understood.
9. Maintain control of your forces at all times.
10. Fight fire aggressively, having provided for safety first.

The 10 standard fire orders are firm and must not be broken or bent by wildland fire investigators. All fire investigators have the right to a safe assignment, and many wildfire investigations take place while the fire is still being extinguished. Investigators should check in with the incident commander and should ensure that firefighting personnel are always aware of their location and that there is always a clear escape route. Investigators should participate in any daily safety messaging or meetings and should have the ability to contact incident command in the event resources are needed or fire conditions are spotted.

All appropriate personal protective equipment (PPE) for wildland firefighting for your agency should be worn as required. Information on the proper type of protective equipment can be found in NFPA 1977, *Standard on Protective Clothing and Equipment for Wildland Fire Fighting*, and NFPA 1500, *Standard on Fire Department Occupational Safety and Health Program*.

Safety is a major concern on all fires. Investigating wildfires carries its own specific hazards. The following are some considerations that must be addressed while on the fire scene:

Safety Tip

The following conditions require more attention and care:

- Fire not scouted and sized up
- In country not seen in daylight
- Safety zones and escape routes not identified
- Unfamiliar with weather and local factors influencing fire behavior
- Uninformed on strategy, tactics, and hazards
- Instructions and assignments not clear
- No communication link between crew members and supervisors
- Constructing line without safe anchor point
- Building line downhill with fire below
- Attempting frontal assault on fire
- Unburned fuel between you and the fire
- Cannot see main fire and not in contact with anyone who can
- On a hillside where rolling material can ignite fuel below
- Weather gets hotter and drier
- Wind increases and/or changes direction
- Frequent spot fires crossing line
- Terrain or fuels make escape to safety zones difficult
- Feel like taking a nap near fireline

- Maintenance of personnel accountability
- Underground burning hazards
- Falling or rolling debris
- Current and changing weather
- Area still burning or having reburn potential
- Ongoing suppression efforts such as air operations and equipment use
- Downed live power lines
- Hazardous materials
- Wildland fire PPE

Conducting an Origin Investigation

In identifying the area of origin, considering the factors of wind, topography, and fuels, the origin is normally located close to the heel, or rear, of the fire.

Interviews with various parties can help the investigator narrow down the area where the fire first started. Observations of reporting parties can help because they may have observed the fire at a relatively small stage.

Observations of the initial attack crew are extremely useful. Crew members may have observed people or vehicles, can report weather conditions, and can detail initial fire conditions and location. They may have taken photos. Origins are often located in the coolest part of the fire, as it seeks fuels and builds heat to establish advancement. Identify and secure any potential evidence as soon as possible so that it is protected during any additional firefighting and investigation. Airborne personnel can also make observations, but they have a broader perspective on the location of the fire.

Vital information in the investigation of a wildfire can often be provided by other witnesses as well. They tend to be familiar with the area, can give information about the area of origin, and may be able to direct the investigator to a possible cause by describing the condition of the fire (e.g., smoke conditions, intensity, rate of spread, weather).

Satellite or imaging tools that are used primarily to establish fire suppression tactics can also be used to assist in the establishment of the area of origin. Based on the direction of fire spread and data indicating the fire location when first detected, these tools can provide a broader perspective on the fire and fire spread. If vehicles are seen leaving the area, consider a vehicle line-up when appropriate. A vehicle line-up can be used as effectively as a suspect line-up. Witnesses may be able to identify a vehicle seen at a fire location. If the investigator has a vehicle of interest, it can be confirmed or eliminated by this process. The witness should sign and date the photo if one is selected, but only if absolutely sure.

Security of the Area or Point of Origin

The integrity of the fire scene needs to be preserved. The area of origin should be secured by flagging off the area and by posting personnel at the scene to restrict access and to protect evidence such as tire tracks, footprints, and potential ignition sources. If the area has been tampered with, it could mislead the investigation and affect evidence credibility.

Search Techniques

When completing a search for a fire's origin, the following principles should be used:

- Walk the exterior of the fire.
- Look for burn patterns.
- Identify the general area of origin.
- Enter the general area of origin.
- Identify the specific area of origin.
- Grid the specific area of origin.
- Identify the origin.

The search area should be walked twice in opposite directions to view burn patterns from different angles. While examining burn patterns, identify clusters or groups of indicators to ensure your finding of spread direction. Once you are sure of the general area of origin, the flagged-off area can be reduced in size to allow suppression crews to work the area. The general area of origin is the area of the fire that the investigator can narrow down based on macroscale indicators, witness statements, and analysis of fire behavior. It may be a limited area on a small fire or several acres on a large fire. The general area of origin can now be entered from the advancing side and worked to the specific area of origin. The specific origin area is a smaller area, within the general origin area, where the fire's direction of spread was first influenced by wind, fuel, or slope. Generally, this area is characterized by subtle and microscale fire direction indicators. It will usually be no smaller than about 5 × 5 feet (1.5 × 1.5 meters) and may be substantially larger, depending on fire spread indicators and other factors. It is typically characterized by less intense burning. If the identified specific area of origin is small, the area can be searched in its entirety from its perimeter. If the area is large, the site should be broken into smaller areas for systematic close-up examination **FIGURE 26-8**.

The investigator should use metal flagging stakes or labeled flags to identify evidence **FIGURE 26-9**. Labeled flags can also be used to mark the original position of evidence that has been removed from the scene. Some flag staking may take place widely in the general area of origin but will be used much more in the specific area of origin as the true origin is approached. Red-colored flagging stakes are often used to indicate advancing fire, whereas yellow stakes are used for flanking fire and blue stakes for backing fire. White stakes are reserved for evidence and the point of origin.

Fire Investigator Tip

Have security (fire fighter, investigator, or law enforcement) on scene at all times during the investigation. This may require multiple shifts of security if the investigation is lengthy. Ensure that each shift is documented.

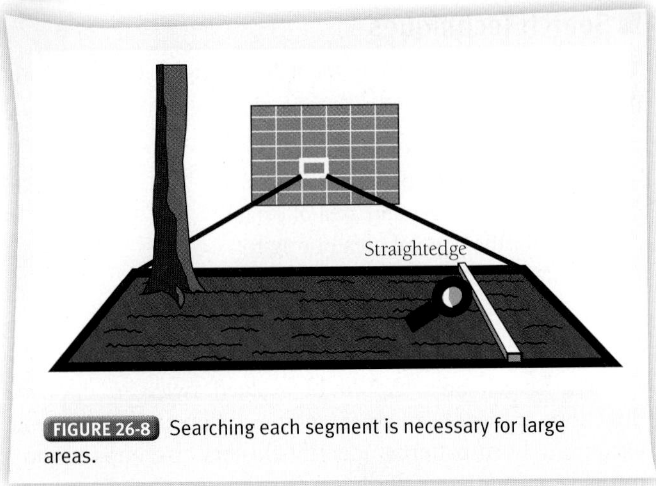

FIGURE 26-8 Searching each segment is necessary for large areas.

FIGURE 26-9 Mark off the general area of origin to protect evidence.

Analyzing fire spread may involve many steps to establish and plot the location of the head, flanks, and heel of the fire when it was discovered. Wind direction and available weather information should be documented and plotted based on the observations of those who discovered the fire. Weather information involves plotting fire head, fire flank, fire heel, and wind direction and speed.

Care must be exercised continually not to destroy evidence or other signs, such as footprints or vehicle tracks. If the area is large, it will be helpful to break the area into segments. This is called "segment division" and allows the investigator to look more closely at specific areas of the fire to determine which may be eliminated. This prevents the need for re-entry and reduces potential for destruction of evidence.

The loop or spiral technique is effective in small areas. Using this method, the investigator approaches from the head of the fire working from flank to flank, back and forth, closing on the origin. A danger to this technique is that if the loops are too large, indicators and evidence can be overlooked and/or destroyed. The distance between lines should get tighter as you approach the area of origin.

The grid, or strip, technique is used for inspecting a large area with more than one investigator. This method can narrow the origin search area very quickly.

The lane technique is one of the best techniques for inspecting an area in great detail. String lines are established, creating lanes to be individually inspected. It is relatively quick and simple to implement and most commonly used when indicators show the origin is near.

The investigator should conduct the examination of the origin area with the minimum amount of disturbance, seeking to expose the evidence of fire spread at the site of the origin rather than destroying or removing it during the investigation. A photographic record sketch with commonly used symbols, along with field notes, should be maintained throughout the investigative process. Each agency should consider writing a standard operating procedure for the use of photography if the evidence is to be used in court.

Search Equipment

The following is a list of some of the equipment that can be used in a wildfire investigation:

- Magnifying glass or reading glasses to enhance small details or identify evidence that was not visible without it.
- Magnets to locate ferrous metal fragments or particles. They should be used on every origin for discovery and exclusion purposes.
- Straightedge to segment the origin area.
- Probe to uncover small pieces of evidence from the surrounding vegetation.
- Comb to separate evidence from debris or pick up evidence to avoid damage.
- Hand-held lights to locate items in low-light areas.
- Air blower (nasal aspirator) to separate light ash from items of interest.
- Metal detector to locate metals that may be of evidentiary value and that could not be easily located with a magnet.
- Sifting screen to separate a suspected item of evidence from the surrounding dirt and vegetation.
- Global positioning satellite (GPS) recorder to obtain the accurate longitudinal and latitudinal position of the fire origin, location of evidence, mapping purposes, and reference points. GPS data can be cross-referenced against onsite survey information, lightning strike data, aerial photography, or satellite imagery for fire cause determination (see NFPA 921, Section 28.8, Fire Cause Determination).

Fire Cause Determination

The objective of every origin and cause investigation is to establish and confirm the cause of the fire. In doing so, the investigator must also determine what did not cause the fire. All findings must be well documented. The heat source or ignition device should also be recovered, if possible, and processed properly as evidence. Testing the hypothesis of how the

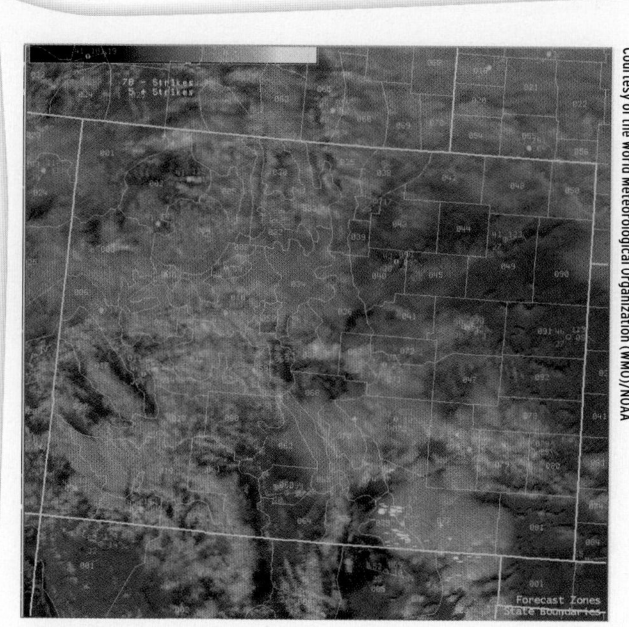

Courtesy of the World Meteorological Organization (WMO)/NOAA

FIGURE 26-10 Lightning data maps are often available from wildland fire agencies.

fire started is required. This can be done by the use of literature or by live testing under controlled conditions. To live test your hypothesis, collect fuels similar to those at the origin to recreate the ignition of the fire while being filmed. Again, this must be performed under controlled conditions, at a safe site away from the fire area. Investigators must follow agency policies and local, state, and federal laws when doing so. Simulate the same weather and topographic conditions if possible. This method can be very useful in court in explaining how the fire started and how fire burns in wildland fuels.

Natural Causes

Wildfires are not always started by the actions of people. Many are ignited by natural causes, such as lightning. Lightning maps are available from several sources and may be used for verification or elimination purposes FIGURE 26-10.

The action of a lightning strike causes fulgurites—slender, usually tubular bodies of glassy rock produced by electrical current striking and then fusing dry, sandy soil—by melting sand in the soil near the base of the tree or area of contact with the ground. Fresh scars may be noticeable on the vegetation. Other physical evidence at the origin may include pieces of wood and bark shredded from the tree or shrub, multiple strikes, holes in the ground, or strange burn patterns on the ground.

Human Causes

Humans can cause fires through actions or by failing to act. Accidental fires are those for which the cause does not involve a deliberate human act to ignite or spread fire into an area where the fire should not be. An incendiary fire is one that is intentionally ignited by a person who knows that the fire should not be ignited.

Campfires

To determine a campfire as cause of a wildfire, look for a ring of rocks and/or soil, along with mounds with the presence of ash, coals, wood, and garbage. Disturbed grasses and soil may indicate sleeping areas. Wildfires caused by campfires are normally accidental and are a result of improper extinguishment.

Smoking

Discarded ignited smoking materials, such as cigarettes, cigars, pipe tobacco, and matches, can start wildfires. Certain conditions must exist for a fire to be started by smoking materials:

- The fuel moisture must be dry, below about 25 percent (although other environmental conditions can override the humidity).
- The fuel source must be fine or powdery (litter or punky wood) or otherwise easily ignited.
- The cigarette or other material must be in contact with the fuel bed and at a particular angle and orientation depending on the weather and fuel conditions.

Debris Burning

Fires occur at dumpsites, timber harvesting operations, and land-clearing operations as well as at residences from garbage and other debris set on fire. These fires can spread to the neighboring vegetation. Burn barrels or incinerators may be a consideration as a fire cause. In windy conditions, hot ash and debris can blow away from the debris burn and start a fire some distance away. Large, woody debris piles, especially if mixed with soil, have been known to hold long-term thermal residency for many months, including over winter, before escaping into adjacent wildland. Witnesses are often useful in determining whether debris burning was the cause of the fire. Check for burn permits, when applicable, for possession and condition compliance.

Incendiary

Incendiary wildfires are deliberately and/or maliciously set with the intent to damage. These fires are normally set in frequently travelled routes and in multiple locations. Evidence such as matches, paper, cigarettes, rubber bands, or candles may be present. The absence of evidence at the origin may also be a strong indicator because the fire was likely started with a device that was then taken from the scene (e.g., a butane lighter). Such a method is referred to as a hot set.

Prescribed Fire (Controlled Burn)

A prescribed fire is the result of an intentional ignition by a person or a naturally caused fire that is allowed to continue to burn according to approved plans to achieve resource management objectives. These fires will at times escape the control of the responsible party because of weather conditions or negligence.

Machinery and Vehicles

Any power or motorized equipment that uses electricity or flammable products in its operation is capable of starting a fire when in proximity to combustible vegetation. Fires can be started by exhaust systems, faulty spark arresters, vehicle fires, and friction. The potential for fire increases when the machinery or vehicle has defective or failed parts; for example, friction heating of bearings, worn brakes, "frozen" shafts, and abrasion may lead to a fire.

Railroad

Along a railroad, fires can be caused by a train, maintenance equipment, or personnel. Common examples include wheel wear on corners, carbon emissions from the exhaust, hot brakes, rail grinding, and personnel smoking. Witnesses are important to the investigation. Trains can sometimes be stopped through dispatch for inspection and to review maintenance reports to determine fire cause, to provide safety to fire fighters in the fire area, and to stop further fire starts.

Fire Play

Wildfires started by children younger than 13 years old are considered juvenile fires. Matches, lighters, cigarettes, or other ignition devices may be found in the area of origin. Other evidence may include bicycle tracks, snack wrappers, toys, or drink containers. Authorities should determine whether the fires were started out of curiosity or whether there is an underlying pathological issue with the child. The child may benefit from referral to a juvenile fire-setter intervention program.

Fireworks

Fireworks are classified as being ground based, hand held, aerial, and explosive. Charred cardboard, paper, colored sticks, plastics, or packaging may be found at or near the origin, and witnesses to the fireworks can be very helpful in providing additional information. Fireworks can produce hot sparks that reach surface temperatures as high as 1200°F (648°C), which is three times hotter than necessary to ignite wildland fuels.

Firearms

Using black powder firearms and firearms discharging tracer, incendiary, and steel core ammunition can cause wildfires. With black powder weapons, normally the burning patch material rather than the burning black powder has ignited the wildland fuels. The burning chemical compounds within tracer and incendiary ammunition can ignite wildland fuels, too. Ammunition with a steel core can strike a rock or other material hard enough to cause sparks and ignite wildland fuels as well.

■ Utilities

Fires can result from wind-damaged trees coming in contact with power line equipment or from the equipment experiencing hardware failure; bird, small animal, or balloon interference; or heavy dust buildup. Sparks or molten metal can ignite flammables, and fulgurites or indentations in the ground are often found if the wire contacts the ground FIGURE 26-11. Pits are often found where the power line comes in contact with a tree. Electric fences may also start fires by transferring rapid electric pulse charges through wires. These pulses ground against dry fuels, causing ignition FIGURE 26-12.

> **Fire Investigator Tip**
>
> There are several reference materials that give more detail that may be helpful. Ensure that these do not conflict with, but instead enhance, NFPA standards.

FIGURE 26-11 This fulgurite was created by a downed power line contacting the ground in sandy soil.

Oil and Gas Drilling

Many fire ignitions related to oil and gas drilling are discussed in previous sections under smoking and equipment use. A well blow-out, flare pit, or stack operation is designed to dispose of excess or unwanted petroleum by-products. These operations

FIGURE 26-12 This electric fence is the source of the fire.

can start fires by direct flame contact or by emitting sparks into flammable fuels.

Miscellaneous

Wildfire causes that cannot be properly classified under other cause groupings are usually classified as *miscellaneous*. This is not an all-inclusive list, but is presented to provide examples of other wildland fire causes.

Spontaneous Heating

Spontaneous heating can be caused by a biological or chemical action and is more likely on warm, humid days where green or moist grass, hay, wood chips, or oily cloth is piled.

Sunlight and Glass Refraction

Clear glass or reflective objects found at or near the point of origin may be the cause of the fire if environmental conditions exist to support ignition. Conditions include prolonged direct sunlight exposure on the reflective object, focusing the sun's rays into small, dry fuels such as grass or duff. Fires started by these items are rare.

Evidence

Evidence collection procedures for wildfire investigations are the same as those for other fires, as outlined in the "Physical Evidence" chapter in this text. These include the following procedures:

- Identifying and gathering
- Preventing contamination
- Collection
- Detailed identification
- Transporting and storage
- Testing

Wrap-Up

© Greg Henry/ShutterStock, Inc.

Ready for Review

- Wildland fire investigations involve special techniques, practices, equipment, and terminology and vary greatly from structural fire investigations.
- Fuels are classified as ground fuels, which include all flammable materials lying on or in the ground; surface fuels, which include all flammable materials just above the ground; and aerial fuels, which include all green and dead materials located in the upper forest canopy.
- The primary factors that affect fire spread are heat transfer, lateral confinement, weather influence, fuel influence, suppression efforts, and topography.
- Analysis of the directional pattern shown by multiple indicators in a specific area can identify the path of fire spread through the area.
- All fire investigators have the right to a safe assignment, and many wildfire investigations take place while the fire is still being extinguished, so the 10 standard fire orders must be followed.
- In identifying the area of origin, considering the factors of wind, topography, and fuels, the origin is normally located close to the heel, or rear, of the fire. Interviews with various parties can help the investigator narrow down the area where the fire first started.
- To live test your fire cause hypothesis, collect fuels similar to those at the origin to recreate the ignition of the fire while being filmed. This must be performed under controlled conditions, at a safe site away from the fire area.
- Investigators must preserve the origin and all evidence correctly, conduct interviews, and prepare professional investigation reports to prepare for cost recovery, where applicable.

Hot Terms

Accidental fires Fires for which the proven cause does not involve an intentional human act to ignite or spread fire into an area where the fire should not be.

Aerial fuels All green or dead materials located in the forest canopy.

© Greg Henry/ShutterStock, Inc.

Air drop The aerial application of water, foam, or retardant mixture directly onto the fire or threatened area or along a strategic position ahead of the fire.

Ambient temperature Air temperature of the surrounding environment.

Aspect The direction the slope faces (N, E, S, W).

Class A foam A mechanically generated aggregation of bubbles that have a lower density than water.

Coniferous litter Primarily needles dropped from coniferous trees but may also include branches, bark, and cones.

Crowns Twigs, needles, or leaves of a tree or bush.

Cupping A charred surface on the fuel that is caused by the exposure of the surface to the windward side of the fire.

Duff The layer of decomposing organic materials lying below the litter layer of freshly fallen twigs, needles, and leaves and immediately above the mineral soil.

Fire breaks Natural or human-made barriers used to stop the spread or reroute the direction of the fire by separating the fuel from the fire.

Fire flanks The parts of a fire's perimeter that are roughly parallel to the main direction of spread.

Fire head The portion of a fire that is moving most rapidly, subject to influences of slope and other topographic features. Large fires burning in more than one drainage of fuel type can develop additional heads.

Fire heel Located at the opposite side of the fire from the head, this part of the fire is less intense and is easier to control.

Fire storm A natural phenomenon where a fire attains such intensity that it creates and sustains its own wind system.

Firebrands A piece of burning material normally transported by wind or convection.

Firing out The process of burning the fuel between a fire break and the approaching fire to extend the width of the fire barrier.

Freezing When leaves, needles, and small branches are heated by a passing fire front, they become very soft and pliable. Prevailing winds or drafts created by the fire bend them in the direction the fire advanced. Once cooled, they remain pointing in this direction.

Fulgurites Slender, usually tubular, bodies of glassy rock produced by electrical current striking and then fusing dry sandy soil.

Ground fuels All combustible materials below the surface litter, including duff, tree or shrub roots, punky wood, peat, and sawdust, that normally support a glowing combustion without flame.

Hot set Fire started in which the device is removed by the individual starting the fire.

Incendiary fire Wildfires intentionally ignited under circumstances in which the person knows that the fire should not be ignited, often with the intent to damage or defraud.

Prescribed fire A fire that resulted from intentional ignition by a person or a naturally caused fire that is allowed to continue to burn according to approved plans to achieve resource management objectives.

Relative humidity The amount of moisture in a given volume of air compared with how much moisture the air could hold at that same temperature.

Slope The steepness of land or a geographical feature that can greatly influence fire behavior.

Snag A standing dead tree, or part of a dead tree, from which at least the smaller branches have fallen.

Spot fires Projections of flaming or burning particles (firebrands) that are found ahead of the flame front.

Surface fuels All combustible materials from the surface of the ground up to about 6 ft (2 m).

Understory vegetation The area under a forest or brush canopy that grows at the lowest height level. Plants in the understory consist of a mixture of seedlings, saplings, shrubs, grasses, and herbs.

Weather history A description of atmospheric conditions over the preceding few days or several weeks.

Wildfire V-shaped patterns Horizontal ground surface burn patterns generated by the fire spread.

FIRE INVESTIGATOR *in action*

During the Fourth of July weekend, you have been requested to investigate a large fire that has damaged several multimillion dollar homes at a beachside community. Fire personnel are still extinguishing spot fires within vegetation between the homes and beach when you arrive.

The incident commander states that a large wildfire was discovered in the sand dunes and had already consumed several wooden walkways before extending to the rear of the residential structures. He states that several neighbors had approached him reporting that there had been individuals shooting fireworks on the beach during the evening.

1. The origin of a wildfire is typically located near the:
   A. flanks.
   B. head.
   C. heel.
   D. perimeter.
2. What is the term for the concave, cup-shaped char pattern on grass stem ends, small stumps, and the terminal ends of brush and tree limbs?
   A. Cupping
   B. Angle of char
   C. V-pattern
   D. Die-out pattern
3. What are green leaves known to do when exposed to a fire?
   A. Freeze in position
   B. Char and ignite
   C. Brown and fall
   D. Curl toward the direction of fire travel
4. Which of the following is the best technique for searching an area in detail?
   A. Lane
   B. Spiral
   C. Strip
   D. Grid

Management of Complex Investigations

© Photos.com

© Jessica Rinaldi/The Boston Globe/Getty Images

Knowledge Objectives

After studying this chapter, you should be able to:

- Discuss site and scene safety considerations for complex investigations NFPA 4.2.1. (pp 408–409)
- Identify interested parties in a complex investigation NFPA 4.1.4. (p 409)
- Describe the basic information and documents needed to manage a complex investigation NFPA 4.1.4 NFPA 4.1.6. (pp 409–410)
- Discuss the importance of communication among interested parties NFPA 4.1.4. (pp 410–412)
- Discuss the considerations required for governmental inquiry into fire investigation NFPA 4.1.4 NFPA 4.4.2. (pp 412–413)
- Describe the organization and management of a major loss investigation NFPA 4.1.4 NFPA 4.2.1. (pp 413–414)
- Discuss evidence control at complex investigations NFPA 4.2.1 NFPA 4.4.2 NFPA 4.4.4. (p 414)
- Describe the logistics involved in complex investigations NFPA 4.1.4 NFPA 4.4.2. (pp 414–415)

Skills Objectives

After studying this chapter, you should be able to:

- Draft an agreement between parties when given incident and investigative information NFPA 4.1.4. (pp 409–412)
- Develop an investigative flow chart for conducting a large loss investigation when given incident information NFPA 4.1.4. (p 412)

Additional NFPA Reference

NFPA 921, *Guide for Fire and Explosion Investigations*

CHAPTER 27

FESHE Course Outcomes

Fire Investigation I

9. Recognize potential health and safety hazards. (pp 408–409, 414–415)

Fire Investigation II

There are no Fire Investigation II (FESHE) course outcomes for this chapter.

You Are the Fire Investigator

© Jones and Bartlett Publishers. Photographed by Glen E. Ellman.

You are conducting an investigation of a large fire in a commercial strip mall. Numerous businesses have been affected, including a nail salon, an electronics store, and a large commercial grocery. You have been informed by the owner of the nail salon that a contractor had been working in the electronics store the day prior. As you begin to document the scene, you meet with several insurance adjusters representing the various businesses and the building owner. They all indicate that they are retaining their own origin and cause investigators and would like a copy of the fire report as soon as it is ready.

The adjuster for the building owner has asked if you have determined where the fire originated and what you believe to be the cause of the fire. You advise the adjuster that you have not been able to enter the structure due to the collapse of the roof and will need to utilize heavy equipment to access the area of origin.

1. What site safety issues should be considered prior to entering the structure?
2. What basic information should be obtained to assist with determining the cause?
3. How would you protect the scene until other interested parties arrive?
4. What logistics and planning need to occur prior to initiating a joint investigation?

Introduction

This chapter is a resource to assist the investigator in the organization and management of investigations in which more than one (usually several) parties are interested in participating and conducting an investigation. An investigation may be complex because of size, scope, and duration of an incident. A complex investigation generally includes multiple simultaneous investigations and involves a number of interested parties. An interested party is any person, entity, or organization, including their representatives, with statutory obligations or whose legal rights or interests may be affected by the investigation of a specific incident. This chapter provides tools for scene management or coordination of multiple interested parties at complex investigations that usually require several days to complete. This often occurs in high-rise buildings, large complexes, or facilities that have multiple occupancies; however, a complex investigation can also occur with small fires in which several parties have an interest.

This chapter does not describe how the investigation should be conducted to gather data to reach an origin or cause hypothesis, nor does it describe how the evidence should be handled. These issues are addressed in other chapters of this text. Rather, it provides tools to organize and manage the function of interested parties to participate in the process. The management structure outlined in this chapter is recommended to coordinate the operations and the different parties involved; however, investigators should always follow the policies and protocols in their own jurisdictions. The manager of the investigation should keep in mind that the goal in conducting a major investigation is usually to properly coordinate so that all parties can conduct their own investigations.

Site and Scene Safety

Site and scene safety are in the interest of all interested parties and needs to be addressed early in the investigation and planning process. The site should be evaluated for safety considerations and the information disseminated to all interested parties. The parties need to understand that this is a fire scene; the structure stability and site are not the same as they were prefire. The parties must bear some responsibility in safety considerations and training in working with these hazards.

Hazardous materials provide unique considerations. It may be necessary to appoint a site safety person who has the authority to stop activities if safety becomes an issue and until the issue can be addressed.

Safety concerns may be of such a magnitude that the retention of a company specializing in the evaluation of site hazards, the implementation of safety plans, and the provision of safety officers should be considered FIGURE 27-1. Site

© John Sartin/ShutterStock, Inc.

FIGURE 27-1 A high level of site hazards (e.g., hazardous materials) may require specialized resources to ensure safety during site investigation.

personnel must follow the safety officer's directives but may express objections and request reentry. A preliminary safety briefing specific to the particular site and scene may be required for anyone entering the investigation site. A record of those having attended such a safety briefing should be maintained. Unresolved objections, as with disagreements, should be recorded and handled.

The site may need to undergo continued site monitoring for concerns such as structural conditions and airborne contaminants. Safety issues should be addressed in the agreement between the parties. (See the "Safety" chapter in this textbook for additional discussion on safety issues relating to fire investigation.)

Fire Investigator Tip

Interested parties are not limited to the owners and occupants of a property and their legal representatives and insurance providers. Interested parties may also include building component, appliance, and equipment manufacturers and installers; gas and electric utilities; suppression or detection equipment representatives; and code enforcers or inspectors. The interested parties likely will retain their own experts, insurance companies, and lawyers that may participate in some aspect of the investigation.

Interested Parties

Often, one of the most difficult responsibilities borne by the manager of the complex investigation is to identify the interested parties who should have access to the site. This is usually coordinated with a legal advisor who is then likely to send out the notices. The owner/occupant, the insurance carrier, and the legal representatives should be part of the identification and notification process. Generally, an interested party is one who may be involved in future litigation and may have an interest as the artifacts and/or evidence are removed or altered.

Interested parties may be identified early in the investigation process. During the course of an investigation, however, other parties may be identified. These parties should be notified of the incident; they may have some responsibility, and they may participate in the investigation. Interested parties that have become aware of the incident through their own sources and wish to participate in the investigation can also approach the investigator in charge.

Interested parties are companies, persons, or other entities that may have an interest in the outcome in the investigation. This, of course, will include the owner, representatives of injured parties or fatalities, and parties to potential civil or criminal prosecution. This may also include building component manufacturers, appliance systems providers, protection and detection system installers, equipment manufacturers, and construction material and finish manufacturers. Other parties that should not be overlooked may include service companies that serviced the equipment in the area of origin, gas and electric utilities, suppression or detection equipment representatives, and even some public agencies such as code enforcers or inspectors. The interested parties likely will retain their own experts, insurance companies, and lawyers that may participate in some aspect of the investigation.

Basic Information and Documents

A plan should be developed for the fire scene investigation. The plan is most often referred to as a protocol and lists the procedures to be followed during the scene investigation. It may be agreed upon by the parties, or drafted by a party in charge (such as a property owner), and it may be provided to all interested parties prior to the scene investigation. Along with the procedures for the investigation, the protocol may designate responsibility for specific tasks, such as keeping the evidence that is collected, how disagreements will be handled, and concerns and procedures surrounding scene safety. The protocol may include a work plan, which is an outline of the tasks to be completed during the investigation, including the order of tasks and the timeline for completion.

There may be more than one protocol developed for the fire investigation; for example, one protocol may be developed for the scene examination, and another for subsequent investigations, including lab examinations or destructive testing.

Joint investigations help to avoid accusations of evidence altering or other indiscretions. In fact, public and private investigators can work jointly, particularly in situations involving technical expertise, assuming that interested parties are given the opportunity to participate in the private investigation. With proper planning and communication, criminal investigations can be conducted parallel with civil investigations. Private investigators can provide various types of support to public investigators.

Meetings

A preliminary meeting may be held before starting the on-scene investigation to discuss safety considerations, address concerns, and set forth the ground rules for conducting activities. Preliminary information and discussions may also occur via telephone calls or emails. The following are some of the concerns that should be addressed during the preliminary meetings:

- Safety
- Protocols
- Planning and timing
- Access
- Organizing the investigative team
- Regular meetings
- Resources
- Preliminary information
- Lighting
- Securing the scene
- Sanitary and comfort needs
- Communications

- Interviews
- Plans and drawings
- Search patterns
- Evidence

Meetings may be held as the investigation progresses to ensure that everyone is signed in, reminded of safety considerations, and informed about changes or scheduling. Only one person from each interested party should be the spokesperson for that group. That person should also have the authority to make decisions on behalf of the group that he or she represents.

The manager of the complex investigation will need to ensure that a site or scene evaluation is conducted to identify potential safety issues and then to address those issues. Information relating to the handling of those issues is found in the "Safety" chapter of this textbook. The investigator who manages the investigation site also needs to coordinate access to the site to ensure that only those who have been authorized gain access to the site. This access may depend on whether certain safety training has occurred or environmental issues are maintained.

Communication Among Interested Parties

Communication among parties plays an important role in major loss investigations. The communication should be maintained to keep the parties informed of the activities and schedule of the scene work and scene examination. Lack of communication may leave parties with a sense that evidence is being hidden from them or that they are not being provided with information to protect their interests. This, of course, can lead to difficulties among the parties. Good communication and sharing of information can create cooperation among the parties and an uneventful, successful site investigation.

Dissemination of Information

Disseminating information among the parties needs to be addressed early; otherwise, a lack of information could lead to difficulties among the parties or even legal action. Without proper communication there is more opportunity for attitudes of distrust, disagreements concerning the activities of the investigation, issues with timing of activities, and management issues. Standard communications may be conducted in various forms, such as through regular meetings, websites, email, or bulletin boards.

Website

On cases where there are many parties and the scene work may take a period of time, a secure website may be beneficial to keep the parties informed. A password is often used to access the website. You may need to consult with someone familiar with setting up such a site. Websites allow interested parties to be informed during times when they are not required to be at the site. They can also include a schedule of activities. During complex investigations, there may be times when no investigation activities are being performed because safety or environmental concerns need to be addressed prior to conducting the scene investigation. There also may be parties that are not interested in participating in parts of the investigation where they have a lower level of interest. The website is also a location where information or data can be disseminated.

Understandings and Agreements

It is important that there are understandings and agreements among the parties regarding how the investigation will be conducted, who has control of the investigation, and the rules or parameters for participating in the investigation.

This can be a complex issue during an investigation when there are multiple parties and each party has multiple investigators who want to have access to certain areas of the site. The understanding between the parties should include an agreement regarding safety, access to the site, costs, scheduling, communication, logistics, protocols, the handling of evidence, documentation, and interviews.

Often, public fire investigators have control of the site at the initiation of the investigation. The public officials are usually charged by statute to determine whether a crime has been committed, and they may need to control the site until they have made that determination. There is an understanding that the owner and other private parties also have an interest in the investigation even if a crime has not been committed. The public officials may provide access for other parties to observe or participate while their investigation is proceeding. Public officials may also determine that participation in the investigation by all of the interested parties will ensure the most efficient and effective investigation without negatively impacting the interests they represent. Proper planning and discussion can often eliminate concerns or issues between the public and private investigator.

Protocols or agreements should address most aspects of the activities during the management of major investigations FIGURE 27-2. Protocols may include the agreements of activities at the scene or away from the scene, such as interviews and evidence storage and examinations. The protocols need to be flexible enough to allow changes as the investigation moves forward and new challenges or issues are discovered. Protocols should be developed with input from interested parties. There may be objections or disputes to the protocol. The handling of objections to the protocol should be addressed in the protocol, such as that the suggestion will be evaluated and implemented if agreed by the parties but that the owner has the right to the final say. The protocol may not be a true consensus because the scene manager, owner, or public authorities may need to override a consensus on some issues. This may occur when there is a concern on criminal issues or proprietary concerns of the owner.

Scheduling

It may not be feasible to allow access to all parties during all phases of an investigation because there would be too many people for the allowable area, safety issues, or lack of space to view the activities properly. A consensus or compromise needs to address these issues, which may include allowing access to only one person per interested party to view the activity or setting up audio and video feeds so that the investigation can be viewed from a safe location. Often, it is best if one person from each party is allowed to be present during

Courtesy of Fire Cause Analysis

Fire Cause Analysis
A DIVISION OF IFT INC.

September 22, 2008
FCA #07=777777

SITE EXAM PROTOCOL

7777 North 77th Street, Phoenix, Arizona

This document sets forth the protocol and timetable for the site examination of the remains of specific equipment, utilities and defined portions of the property defined above. These protocols are written to facilitate the efficient site inspection, and identification of materials needing to be evaluated and tested to identify those factors and circumstances that contributed to the cause of the "event".

1. The fire scene examination will be conducted jointly with all in attendance. The investigation will be conducted on September 22, 2008.

2. A construction crew is on site and will make the site safe to enter. If you observe an unsafe condition it is your responsibility to notify the lead site investigator immediately.

3. The investigative team will conduct the following examination activities:

   A visual inspection of the property. The inspection is to familiarize each of the interests with the building, the contents of the building, and the scope of the site investigation. Each interest is encouraged to identify any locations, items, or material that may have evidentiary value in this investigation. If such is found, it will be documented, and if appropriate, secured as evidence.

   In addition, those locations on the property that hold an interest to any of the investigative team should be made known to the lead investigator. The lead investigator will evaluate and coordinate any requests for scene excavation.

   a. A visual, non-destructive examination will be conducted of any area of interest. As such, if materials warrant future examination and laboratory testing, the components will be marked, documented, and secured as evidence. Any data generated regarding the various materials in the building will be shared with all interested parties upon written notification.

   b. Identification by the parties, of the evidence they deem necessary to assist in future testing and analysis. The components identified by all parties will be appropriately marked and based on the locations of these items, and collection will be made. During the removal process full documentation will be afforded all parties with photos or video as they deem appropriate and in accordance with these protocols

4. All parties must sign in and provide a business card.
5. All parties must sign a "Hold Harmless" agreement prior to entering the site.
6. The use of safety equipment (hard hat / gloves / eye protection / respiratory protection) is left to the discretion of the individuals participating in this site inspection. Particulate dust masks are available.
7. This is a site exam only. No destructive or engineer testing will be conducted at this time.
8. Previously collected evidence will be made available for visual inspection and photographs only.
9. All evidence collected may be subject to further testing. All parties will be notified of the protocol for the exam, test times, dates, and locations.
10. It is the responsibility of each individual to ensure that any evidence that they deem should be collected be brought to the attention of the lead investigator at this inspection. This site will be released to a demolition company immediately following this inspection.
11. Question or amendments to this protocol must be submitted to the lead investigator and will be considered.
12. By signature of the sign in sheet, the parties have read and agreed to this protocol. The sign in sheet is attached, Appendix A.

Offices in San Francisco Bay Area, Sacramento Valley, S. California & Phoenix, AZ

935 PARDEE STREET
BERKELEY, CA. 94710-2623 USA
510.649.1300
800.726.5939
Fax: 510.649.3099

FIGURE 27-2 A sample protocol.

important removal and documentation if space and safety issues can be overcome.

Scheduling for a smaller number of interested parties is often not as difficult as with a large number of interested parties. Not all parties may be available for the time scheduled for the activities. The scene investigation will likely have to move forward, and compromise on the schedule should be achieved if possible. The parties with the scheduling conflict must understand that holding the site for extended times is difficult and may not be feasible. A compromise may be reached by thoroughly documenting the activity or having another person representing that party attend the activities. Scheduling activities that occur after the scene examination is usually not as time sensitive, such as examination of artifacts that have been secured from the scene, and often a mutually agreeable time to perform the examination can be reached.

Good communication on scheduling is important because if parties have sufficient notice of the activities and the schedule, they can arrange their own calendars to accommodate the investigation schedule. Such communication should include the flow chart and activities that are occurring and are scheduled to occur with the best estimates of the timing and as far in advance as can be made available. Such communication can be conducted through frequent exchanges of written communication such as emails, letters, and faxes. Password-protected websites have been used and can provide up-to-date information as changes occur.

The protocol and flow chart developed at the initiation of the investigation are important tools in scheduling tasks and should address the scheduling issues. Scheduling of the parties is a difficult issue, and parties need to accommodate the scheduling effort so that the investigation may move forward.

Cost Sharing

Costs of some portions of the investigation activities can sometimes be agreed to be shared among the parties. This may be true if each of the parties will have an interest in the result of the activity or needs copies of the result of the activities. Such cost sharing may include professional photography or videography during the investigative activities, cost of developing drawings, or site safety personal protective equipment (PPE). Sharing of cost may also include evidence storage, debris removal, personal comfort items, and specialized tests of the scene environment or evidence; however, some of the parties may not agree to cost sharing, especially if they think that they may have little involvement in responsibility for the fire. An agreement should be obtained for any cost sharing and should be put in writing, with the parties' signatures. Ideally, it should happen early in the investigative process.

Information Sharing

The issue of sharing information and interviews should also be addressed early. There could be a need for confidentiality agreements before interviews can be conducted or before information is exchanged to ensure that proprietary or secret information is held in confidence. As has been described in other chapters of this text, interviews are an important part of an investigation so that a final hypothesis can be developed.

Sharing of information among the parties can expedite the investigation process and also provide others with data to develop an accurate analysis and hypotheses. There may be some information that cannot be shared or will have stipulations with providing the information. Such information may be proprietary information on a manufacturing process or information that the owner is not willing to share with competitors. There also may be interviews conducted by a party who thinks they cannot share for concerns of liability or concern for others' safety. Public authorities conduct interviews and examinations that are often not available to be shared with others because of legal considerations. This concern for sharing this information can sometimes be eliminated by having any party who receives the information sign a nondisclosure agreement stating that the information cannot be shared with anyone else. Information sharing should be encouraged so that complete data are available for analysis.

It is usually best to share information that is related to data needed to conduct a complete analysis for hypothesis development, such as interviews with the first witnesses or the people who last used or maintained a suspect piece of equipment. Legal considerations may be an issue to releasing some information.

Amendments to Agreements

The agreements or protocols need to be flexible because it may be necessary to amend an agreement as the scene or activity progresses or changes. If the original protocol is in writing, any changes should also be in writing. All parties must be notified of any suggested changes to the original written agreement. If the protocol being used is verbal, the change should be noted and made known to all parties; this type of verbal agreement often occurs while the scene activities are occurring and where the parties agree to changes on the method of how an activity can be conducted. Such verbal agreements are among the parties present, but because other parties may not be present, the scope of the change should be within the original scope of any written agreement.

Disagreements

A method for handling disputes should be discussed. Disagreements should be noted in writing, with suggested methods to handle the disagreement and a resolution provided. Management of large losses requires a good-faith effort to handle disagreements, and it is possible the matter may end up before the courts. Amendments may be made to the agreement, but they must be in writing.

Governmental Inquiry

It is recognized that the public or law enforcement investigation may have concerns of maintaining evidence and the investigation when allowing access by private investigators; however, with proper coordination and management, private-sector interests may be preserved so that private investigators have the opportunity to conduct separate, independent investigations and to view or examine the evidence before it is preserved or altered. Such investigations have been conducted at sites with the public sector taking the lead and doing the scene work while the private investigators observe the process. The private investigator may be allowed the opportunity to photograph and observe evidence removal while the scene work is in progress. The evidence gathered at the scene may also be very important to the private interests such as owner/resident or insurance company and their ability to recover from the incident.

Public investigators should attempt to avoid altering evidence without consideration of other parties, but may find it necessary to alter evidence during their examination. If such destruction is necessary, consultation with interested parties should be considered. The public investigators should document their alteration or removal of any evidence according to recognized standards and processes. When public investigators identify that there likely are no longer any public issues or they have determined that there is not a criminal issue, the interest of private parties should be consulted. The public investigators may wish to continue to participate in the investigation with the private investigators.

The issue of spoliation of evidence, which is discussed in the "Legal Considerations" chapter in this textbook, continues to evolve through the courts. The courts have generally recognized that other parties may have the right to participate in the scene investigation before it is altered. The tools provided in this chapter help the investigator in charge of the scene coordinate the scene activities.

Management of the Investigation

Proper site management of a major investigation will provide the tools to limit safety-related issues, organize proper facility and site activities, handle evidence issues, and assist in issues that may arise during the investigation. The implementation of a management system prior to the start of the investigation will benefit the investigation process. The site manager's duty is to manage the site and personnel; the site manager may not be the person who is responsible for conducting the investigation. Management of major losses often provides enough issues and activities that need to be addressed without the added responsibility for conducting the investigation; however, there are times when the site manager is also the lead investigator and must wear both hats. Incident management models are available through various fire management courses.

Organizational Models

Management of complex investigations may involve many different parties performing joint, yet independent, investigations simultaneously. There are organizational models available from various organizations, including the National Fire Protection Association (NFPA). Persons familiar with the National Incident Management System (NIMS), commonly used on the fireground, may be able to make use of the NIMS command model in the management of an investigation.

Control of the Site and Scene

Control of the site may change during the course of the investigation. The public sector may initially be in control. Once the public sector has completed its examination, it will often release the scene to the owner, other responsible person, or insurance company. Following the public sector examination of the site, the site owner or occupant will often have the authority to determine who has control.

Securing the Site and Scene

Access to the site and monitoring access may be one of the more challenging managerial issues on a complex investigation. It is for this reason that security personnel may be called in to manage this task FIGURE 27-3. Access should be limited and monitored with sign-in sheets when people enter and exit. Coordination of access may involve name tags or escorts. Escorts are not intended to spy on the parties, but to ensure that safety concerns are met.

Site and Scene Access

Security of the site and scene access are important considerations in preventing scene contamination and safety issues. The entity in control is usually responsible for security and access. The owner should be consulted because there may be issues with proprietary information and valuable equipment remaining at the site. Fences and security guards can often be valuable tools. Often, a sign-in sheet for entry and departure is used to account for those accessing the site. Confirmation of their exit is important to ensure that all are out at the end of the day.

Site-Specific Restrictions or Requirements

There may be site-specific requirements to enter the fire scene. Some common requirements to maintain scene security and safety include no smoking and no food, beverages, containers, or wrappers brought onto the scene. Scene safety often

© BRIAN SNYDER/Reuters/Corbis

FIGURE 27-3 Security personnel may be needed to monitor access to the investigation site.

requires that food not be consumed while on the site because of the potential for ingestion of contaminants. Entering the site usually requires some form of PPE, such as the use of hard hats, safety shoes, and safety glasses. This should be addressed in the protocol and site safety plan.

The scene integrity, including not allowing public access and not allowing debris or other contaminants on site, must be maintained until the investigation is complete and the evidence is removed. The scene security may be required to be maintained until the site has been made safe to passersby and other noninterested parties.

Delegation of Control and Transfer of Control

The choice of the person in control of the site is usually dictated by the circumstances at hand and often changes as the investigation moves forward. The party responsible immediately following suppression may be the public sector investigator. When the public sector investigator determines that no crime has occurred, they often turn the site and investigation over to others, likely the owner or the owner's representative. The owner likely has the legal authority to take control but may not have the resources or expertise; thus, the insurance company and their representatives may take control of the site. This may not always be the case, especially if there are uninsured responsible parties, tenants, fatalities, or injuries involved. There are issues of control, which may require legal consultation depending on the lease agreement, when there are tenant-occupied spaces.

The transfer of control should be planned ahead of time to ensure an orderly transition. The party relinquishing control should provide information on the changes that have been made to the site, evidence secured, and safety issues. The party that is relinquishing control still may participate in the investigation, but no longer is in control. This often occurs when the public official determines that the fire is likely accidental but wishes to continue to participate in the investigation.

Release of Information

Release of information to the public or media should be addressed early, especially in large losses. The activity is usually left to the public authorities and their public information officer. On large losses with multiple public jurisdictions, this can be a sensitive issue. Private parties are unlikely to release information unless through an owner representative. It is often in the best interest of the investigator not to become involved in the release of information to the public.

Evidence Control

Handling of evidence is another key issue in management of complex investigations. An evidence custodian is designated to manage all aspects of this process. Agreements should be made on how the evidence is identified, documented, collected, preserved, transported, and stored. Other factors to consider include identifying who will have access to the evidence once it is stored and how information will be distributed on proposed examination or testing of the evidence.

Evidence Removal from the Scene

The handling of the evidence should be addressed in the protocol and early in the investigation. All interested parties should be allowed to document and see the evidence prior to its being changed or removed from the site. The evidence should be removed in an agreed-upon manner, usually following recognized practices identified in NFPA 921.

The storage of the evidence should also be addressed in the agreement. The evidence collected should be secured in a locked facility with fire-protection features. The cost for the storage should be agreed on by the parties, but usually rests with the owner's insurance carrier. The keys or security usually resides with the owner's representative, but in some cases, multiple locks are allowed so that no one party has access to the evidence.

The interested parties should agree on access to view the evidence. When such viewing occurs, the activity should be available to all parties. Nondestructive exams should be permitted to allow viewing and documention of the evidence, and other interested parties should be notified of such an exam but may not be present; however, when there is to be disassembly or destructive examination of evidence, a protocol needs to be developed prior to any such exam. The protocol should be circulated among the parties for consideration, a laboratory needs to be identified, and an agreeable date for the exam should be identified. Each interested party may have its own expert at such an examination.

Spoliation issues generally dictate that all interested parties participate in any destructive examination of the evidence. Agreements usually provide that everyone involved be notified and have the opportunity to be present whenever access to the evidence is provided.

Logistics

In coordinating the actions of the different parties, the manager of the complex investigation needs to coordinate the parties so that they can safely conduct their activities, avoid conflicts, protect evidence, and identify and address the individual needs of investigators and other personnel on scene.

The logistics of each activity and the coordination of the investigation may not be in the protocol, but need to be in the planning and possibly in the work plan. The logistical considerations are items to provide an orderly flow to allow the scene investigation to progress, including the comfort needs of those participating in the investigations. Such logistics may include outlining in detail the scene investigation activities and environmental issues and obtaining trash collection, toilets, equipment, and security personnel.

Equipment needed to conduct the investigation is part of the preplanning as addressed in the "Planning the Investigation" chapter in this textbook. This may include large equipment to lift ceilings or walls or to remove debris or evidence. Laborers and small hand tools may be needed to conduct the site investigation. Investigators are usually prepared for equipment needs for smaller fire losses, but large losses may bring unique or large equipment issues. Manpower in conducting such an investigation may be obtained by retaining a construction company; there are companies that are equipped to provide such assistance.

In some large sites, transportation can be an issue. For example, if the parking area is remote from where the investigation activities are taking place, the parties may need transportation to the site from the parking lot. Transportation may be needed on both sides of a security checkpoint—one to transport from a parking area to security and one to transport from security to the investigation site, if remote. Transportation of the evidence to the evidence storage facility should also be given consideration, especially if there is a need to transport large items.

Site safety is usually an issue on major investigations. It may be necessary to handle safety-related issues prior to conducting the scene investigation. The investigator may be required to develop a site safety plan. Site safety requires the identification of hazards and then a discussion of the hazards with the parties. Hazards that are common to major losses are structural integrity, overhead hazards, electrical hazards, tripping, slips, and falls. Hazardous material may also be present. Handling hazardous materials requires special considerations and additional resources. The Occupational Safety and Health Administration (OSHA) and NFPA provide guidance in handling such issues. Consideration of an outside vendor may be required. The "Safety" chapter in this textbook provides additional information on this subject.

The site may require an area of cleanup or decontamination, especially if hazardous materials may be present. If materials are present that would pose a contamination hazard if brought outside the scene into public areas, the removal or decontamination of those who have accessed the site may be required. This may include a changing area, washing facilities, or even showers. Personal modesty should be considered. Hazardous material sites provide challenges that must be identified and rectified to prevent contamination from leaving the site.

Another consideration while conducting the scene examination is communications. Some sites are large or noisy, and communication devices, such as two-way radios, may be required. Overhead lighting or spotlights may be brought in instead of using flashlights, which often do not provide enough lighting. There may be an issue of snow and ice that must be removed prior to, or during, the scene examination. The area of interest may need to be covered and heated prior to conducting the examination. Sanitary and comfort needs, such as water, shelter, seating for a rest area, toilet facilities, and an area to get out of the weather (hot or cold), should also be addressed.

Wrap-Up

Ready for Review

- An investigation may be complex because of size, scope, and duration of an incident. A complex investigation generally includes multiple simultaneous investigations and involves a number of interested parties.
- Safety is a key issue in major investigations and must be addressed early in the process. Some safety-related issues may need to be addressed at the site prior to conducting the scene investigation.
- Interested parties should be identified early in the investigation process, but may also be discovered throughout the course of an investigation.
- A protocol should be developed and provided to the interested parties. The protocol may involve building and site plans, manuals, and schematics to help investigators become familiar with the site.
- Communication among parties involved should be maintained to keep the parties informed of the activities and schedule of the scene work and scene examination.
- The public or law enforcement investigation may have concerns of maintaining evidence and the investigation by allowing access of private investigators; however, with proper coordination and management, private-sector interests may be preserved so that private investigators have the opportunity to conduct separate, independent investigations and to view or examine the evidence before it is preserved or altered.
- Proper site management of a major investigation will provide the tools to limit safety-related issues, organize proper facility and site activities, handle evidence issues, and assist in issues that may arise during the investigation.
- Agreements should be made among interested parties on how the evidence is identified, documented, collected, preserved, transported, and stored.
- In coordinating the actions of the different parties, the manager of the complex investigation needs to coordinate the parties so that they can safely conduct their activities, avoid conflicts, protect evidence, and identify and address the personal needs of investigators and other personnel on scene.

Hot Terms

Complex investigation Generally includes multiple simultaneous investigations and involves a number of interested parties.

Entity in control The interested party who has or represents ownership of the scene or is in effective management of the site, scene, or evidence, and is organizing, directing, or controlling the joint actions of the other interested parties.

Evidence custodian Person who is responsible for managing all aspects of evidence control.

Interested party Any person, entity, or organization, including their representatives, with statutory obligations or whose legal rights or interests may be affected by the investigation of a specific incident.

Protocol A description of the specific procedures and methods by which a task or tasks are to be accomplished.

Work plan An outline of the tasks to be completed as part of the investigation, including the order or timeline for completion.

© Greg Henry/ShutterStock, Inc.

FIRE INVESTIGATOR *in action*

You are at the scene of a large fire at a chemical production facility. At the time of the fire, the facility was shut down to allow for the installation of a new chemical processing machine. Several contractors—including the machine manufacturer, electricians, plumbers, and the facility's own maintenance personnel—were working on installing the new machine when a fire occurred within the machine. The fire has caused extensive damage to the facility, and waste chemicals have contaminated the area where the fire originated.

After notifying all of the known interested parties, a joint scene examination is scheduled. Two days prior to the joint examination, you are contacted by an adjuster representing a contractor who was not previously identified. The adjuster states that his insured was a subcontractor of one of the already identified parties and that they were just advised of the joint examination. The adjuster also states that he would like to move the joint scene examination to allow their investigator to attend.

Upon informing the other parties of this new subcontractor and his request to move the inspection date, the attorney representing the chemical facility advises the date cannot be moved because the company is losing money every day they are shut down.

1. Due to the potential of hazardous materials being involved, what should be considered to ensure the safety of all those participating in the investigation?
   - **A.** Obtaining specialized resources to assist with the chemical hazards
   - **B.** Requiring pre- and post-entry medical screening of those who enter the scene
   - **C.** Appointing a safety officer
   - **D.** Both A and C
2. Which of the following methods may be used to communicate information to all interested parties during a complex investigation?
   - **A.** Letters
   - **B.** Emails
   - **C.** Faxes
   - **D.** All of the above
3. Which document provides a description of the procedures and methods to be used in an investigation as well as responsibilities for various tasks, and may include a method for resolving disputes between parties?
   - **A.** A work plan
   - **B.** A sign-in sheet
   - **C.** A site plan
   - **D.** A protocol
4. If the destruction of evidence may occur or is necessary during the public investigator's investigation, the investigator should:
   - **A.** continue his or her investigation because it is separate from the private investigation.
   - **B.** consult the interested parties prior to continuing the investigation.
   - **C.** note where the evidence is located and move it to a safe location.
   - **D.** None of the above

Marine Fire Investigations

(c) Photos.com

© Ian Hainsworth/Alamy

Knowledge Objectives

After studying this chapter, you should be able to:

- Identify types of marine vessels. (p 421)
- List safety considerations related to marine fire investigations. (pp 421–423)
- Identify and describe the systems used on marine vessels. (pp 423–426)
- Explain design and construction features associated with marine vessels. (pp 426–427)
- Describe the various types of propulsion systems on marine vessels and their uses. (pp 427–428)
- List potential ignition sources on marine vessels. (pp 428–429)
- Identify issues related to cargo on marine vessels. (pp 429–430)
- Describe how to document marine vessel fire scenes. (pp 430–431)
- Describe the process for examining a marine vessel fire scene. (pp 431–433)
- Identify considerations for marine vessel fires in structures. (p 433)
- Explain legal considerations related to marine fire investigations. (p 433)

Skills Objectives

After studying this chapter, you should be able to:

- Ensure the safety of the investigator conducting both shore- and water-based investigations. (pp 421–423)
- Conduct a fire scene examination of a marine fire. (p 431)
- Properly document a marine fire. (pp 430–431)
- Locate and identify the area of origin associated with a marine fire. (pp 431–432)

Additional NFPA References

NFPA 302, *Fire Protection Standard for Pleasure and Commercial Motor Craft*

NFPA 921, *Guide for Fire and Explosion Investigations*

CHAPTER 28

FESHE Course Outcomes

Fire Investigation I

There are no Fire Investigation I (FESHE) course outcomes in this chapter.

Fire Investigation II

6. Analyze electrical causes of fires. (pp 428–429)

You Are the Fire Investigator

© Jones and Bartlett Publishers. Photographed by Glen E. Ellman

You are investigating a fire aboard a large houseboat that was tied to a pier on a lake. The vessel is still afloat; however, it broke free from the pier during the fire. The fire also caused extensive damage to the interior compartments.

After having the vessel towed back to the pier, you begin your investigation and determine that the fire originated in the kitchenette area. An interview conducted with the boat owners reveals it had been unoccupied for several weeks but was connected to a shore line at the time of the fire. After systematically removing the debris and reconstructing the area of origin, you have determined the fire originated at a dual-fuel refrigerator.

1. What unique safety hazards must be considered while you conduct your investigation?
2. How will you document the vessel?
3. What types of materials could be part of the vessel's construction?
4. Who would you consider interviewing during your investigation?

Introduction

The field of marine fire investigation is one of the most challenging areas facing investigators. A shipboard fire can encompass many facets of fire investigative knowledge, from compartment fires to vehicle fires to industrial fires. Vessels can vary in size from a small recreational boat less than 20 ft (6 m) long to ultralarge crude carriers thousands of times larger. Vessel fires can occur on land, in a stable drydock, or thousands of miles offshore in hurricane-force winds and high seas. The myriad possibilities facing a marine fire investigator require not only a solid understanding of fire dynamics, but also familiarity with marine terminology, ship construction, vessel operations, maritime safety and fire protection regulations, and international admiralty law.

Because of the wide range of vessels in which a fire investigator may be called to examine a fire scene and render an opinion as to origin and cause, NFPA 921, *Guide for Fire and Explosion Investigations*, focuses on boats that are less than about 65 ft (20 m) long **FIGURE 28-1**. Vessels in this category number well into the hundreds of thousands and experience the majority of marine fires in the United States. These boats

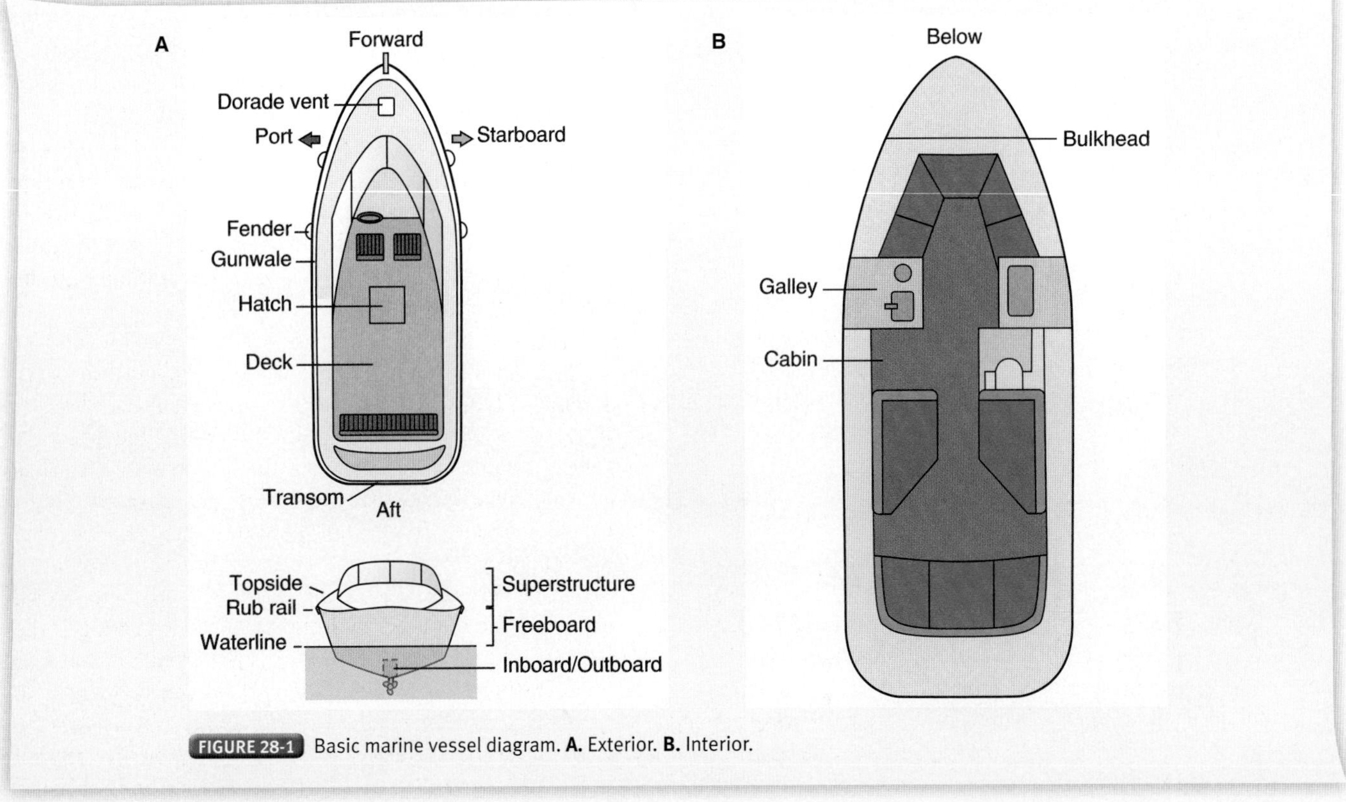

FIGURE 28-1 Basic marine vessel diagram. **A.** Exterior. **B.** Interior.

Table 28-1 Common Marine Vessel Terms

Term	Definition
Aft	Toward the rear of the vessel
Aground	Touching the bottom
Beam	The widest part of a vessel
Below	Beneath the deck
Cabin	A compartment for passengers or crew
Capsize	Turn over
Chain plate	A metal plate used to fasten a shroud or stay to the hull of a sailboat
Dock	A protected water area in which vessels are moored (usually a pier or wharf)
Dorade vent	Deck box ventilation that uses a baffle to keep water out while letting air in below decks
Fender	A cushion placed between marine vessels or between a marine vessel and a pier to prevent damage to the vessel(s)
Forward	Toward the bow of the marine vessel
Freeboard	The vertical distance between the waterline and the gunwale
Galley	The kitchen area of a marine vessel
Gear	Ropes, blocks, tackle, and other equipment
Gunwale	The upper edge of a marine vessel's side
Port	The left side of the marine vessel when looking forward
Rub rail	The rubberized plastic or metal bumper that extends along both sides of the vessel, usually immediately below the gunwales
Sole	The cabin floor, timber extensions under the rudder, or the molded fiberglass deck of a cockpit
Starboard	The right side of the marine vessel when looking forward
Waterline	A line painted on a hull showing the point to which a marine vessel sinks when it is properly trimmed

include most pleasure powerboats, sailboats, and larger luxury yachts. Many of the smaller commercial boats, like fishing boats and passenger craft, also fit into this size range.

Marine vessels can be created using a variety of techniques and materials. Wood has been used for hundreds, if not thousands, of years as the primary construction material; however, as technology has advanced, steel and aluminum, fiberglass, and other materials have become more popular. Vessels may be created using multiple pieces or modules or may be formed as one solid component. Fixtures and furnishings are also made from a variety of materials that have properties similar to those found in a residence or motor vehicle.

There are many terms specific to marine vessels that will be helpful during marine fire investigations. TABLE 28-1 defines several of those terms; additional terms are discussed further in this chapter.

Types of Marine Vessels

There are hundreds of different types of marine vessels, each specifically designed for an intended use. Recreational or personal vessels include jet skis, wave runners, runabouts, catamarans, sailboats, and high-performance speed boats. Commercial vessels include fishing boats, container ships, ferries, recovery vessels, and oil super tankers. Military vessels include destroyers, aircraft carriers, hovercrafts, and submarines.

Marine Vessel Investigation Safety

Because of the threat of taking on too much water, flooding or instability is always a concern and demands the attention of all involved. Ships' crews are generally well trained in such matters and are familiar with dewatering as well as firefighting.

The techniques used by ship- and land-based fire fighters can at times vary greatly. Fire investigators must be cognizant of the potential differences in the ways in which a fire may be attacked at sea as compared with in port and must pay particular attention to how suppression efforts may have altered fire pattern development.

Prior to inspecting a marine vessel, either on land or afloat, the investigator should determine the general safety and stability of the craft and provide for the appropriate level of personal protective equipment (PPE) to conduct the investigation safely. The investigator must ensure that the craft is stable and will not fall over or capsize during the investigation. Crafts still afloat pose a significant hazard because they may be filled with water, which will make them very unstable and prone to capsizing. In these cases, the water should be removed (dewatering) and the craft stabilized prior to continuing. If the investigator is unable to remove the craft from the water, a personal flotation device (PFD) should be worn. In cases of submerged or sunken marine vessels, skilled divers may be required to ensure the safety of the investigator. If electrical service—either shore based (shore power) or from onboard

Fire Investigator Tip

Although some may argue that fire patterns are fire patterns, certain conditions on a vessel may substantially alter fire development and the resulting patterns. The very nature of vessel construction and operation is the source of significant differences in fire behavior and pattern generation. Some of the factors that can affect the interpretation of fire ignition, growth, and pattern development on a marine vessel include the following:

- Common instances of flammable atmospheres
- Effects on ventilation of maintaining watertight integrity at or after a fire's ignition
- High conductivity through steel bulkheads
- Six-way heat transfer to surrounding spaces
- Hidden/rarely accessible spaces
- Subsurface effects of boundary cooling
- Movement of the shipboard platform in heavy seas
- A wide variety of hazardous materials in both the ship's stores and its cargo
- Effects of wind
- Availability of air in the fire compartment
- Use of saltwater for firefighting
- Differences in shipboard versus land-based firefighting

FIGURE 28-2 Conduct air monitoring to ensure that the air in the confined environment is not hazardous.

battery systems—is present, the investigator must de-energize these systems or risk being shocked or electrocuted.

Confined Spaces

Marine vessels are often compartmentalized to allow for containment of water should flooding occur. This often creates small, single-access confined spaces that pose a significant risk to the investigator during his or her investigation. Prior to entering any confined space, the investigator should conduct air monitoring to ensure that hazardous levels of ignitible gases, toxic vapors, or oxygen-enriched/deficient environments are not present FIGURE 28-2. An appropriate level of PPE should be used to protect the investigator, and any lighting or powered equipment used in these spaces should be intrinsically safe.

Airborne Particulates

Because many marine vessels are constructed of fiberglass, the investigator must consider the use of appropriate respiratory protection while conducting the investigation. When fiberglass burns, the resin is consumed, leaving small particulates of fiberglass, which can cause severe irritation to the investigator's respiratory system. A self-contained breathing apparatus (SCBA) or a particulate filter mask should be worn when conducting the examination. In the event that the investigator arrives at the scene soon after the fire is extinguished, air monitoring should be conducted prior to and during the examination.

Energy Sources and Hazardous Materials

Electrical power for a marine vessel may come from a variety of onboard sources. The investigator must locate and disable these systems to protect himself or herself from personal injury and from potentially creating another fire. These electrical sources must be photographed and diagrammed prior to being manipulated and disabled.

Batteries are often used, and more than one may be present in different locations. Atmospheric monitoring must be conducted to ensure that an explosive environment is not present prior to disconnecting the battery cables on each battery. An arc may be produced when the cables are disconnected, creating a risk of ignition if such an environment is present. If an inverter is present, the DC input should be disconnected. Shore lines may also be used, especially if the marine vessel is docked. The shore line should be de-energized and then disconnected from the marine vessel.

Hydrogen gas may be located in battery compartments, and care should be taken to reduce the potential of static discharge or electrical arcs when removing the battery cables.

Various types of liquid and gaseous fuels may be found on a marine vessel. This includes fuels used to propel the marine vessel and fuels for cooking and heating purposes. The investigator must be aware of the potential presence of these systems because a leak may pose a significant risk of fire or explosion should the gas migrate from an uninspected area of the marine vessel.

Safety Tip

Fire investigators should remember that marine vessels will have materials similar to those found in both structural and vehicle fires and will produce similar products of combustion.

The investigator may also encounter various fluids, lubricants, and oils used on the marine vessel that not only pose a risk of fire but are also potentially harmful to the environment. In addition, hydraulic fluids, antifreeze, engine oil, and other lubricants could have leaked from the marine vessel's mechanics, creating a slippery walking surface for the investigator.

Sewage Holding Tank

On marine vessels that contain sewage holding tanks, methane gas may be present. During the fire, the tanks may have been damaged, allowing these gases to escape. The investigator must ensure that proper venting occurs to reduce the risk of fire or explosion. Besides the risk of fire and explosion, biological hazards may also be present.

Structural Concerns

Either during the fire suppression operations or as a result of damage sustained by the fire, water may enter the bilges and lower sections of the marine vessel, causing it to become very unstable. This poses the greatest hazard to the fire investigator while conducting an investigation on a marine vessel that is still floating in the water. The investigator must ensure that the craft is stable prior to conducting the investigation because if the vessel lists or capsizes, the investigator may become trapped within it.

Like in a structure fire, the structural components and finish of the marine vessel can be damaged during the fire, creating additional hazards. The investigator should thoroughly inspect and sound the decking and other walking sources to reduce the risk of collapse FIGURE 28-3. To sound the decking, the investigator should use techniques similar to those of fire fighters sounding floors, using a tool to test the integrity of the decking in front of them.

Both vertical and horizontal openings may be present on a marine vessel to allow access below decks and to other portions of the marine vessel. Although some openings may be visible, hatches, doors, and covers may conceal these openings after a fire. If flooding is present, unprotected openings may also be concealed and pose a significant risk to the investigator while examining the marine vessel.

© Glen E. Ellman

FIGURE 28-3 Inspect any walking surfaces for signs of instability before using them.

> **Safety Tip**
>
> Flammable atmospheres on vessels are common because of the presence of not only hydrocarbon fuels for propulsion, but also chemicals such as paints, solvents, lubricants, and volatile cargos. Vapors from many of these products are often present in compartments such as fuel tanks, fuel day-tank pump rooms, paint lockers, lazarettes, workshops, head spaces of cargo tanks, engine spaces, and void spaces. Most are heavier than air and tend to settle to lower levels. As with ignitible vapors on land, if these vapors collect in pockets within the flammable range of a particular chemical or product and a sufficient ignition source is present, then a fire or explosion may easily result.

Wharves, Docks, and Jetties

Wharves, docks, and jetties are often slippery as a result of their presence near the water. They may also be structurally unstable, especially if they were in contact with a vessel subjected to fire. The investigator should be careful while on these surfaces, especially while transferring to or from a marine vessel.

Submerged Marine Vessels

Ideally, marine vessels that are submerged should be thoroughly documented and photographed prior to being raised and removed from the water. This requires the use of specialized personnel trained to conduct such unique investigations. During the recovery operations, the investigator must monitor and manage the operations from the surface. Care should be taken to reduce the potential of damaging or altering the marine vessel and its appurtenances during these operations. The investigator must also be aware of the potential issue of fuels and other harmful products escaping into the water.

Visual Distress Signals and Pyrotechnics

Pyrotechnic signal flares may be present on marine vessels and should be handled with care if encountered by the investigator. If such a device is discovered, the investigator must secure it to prevent it from accidentally activating.

System Identification and Function

As with any fire investigation, marine vessel fires can be a challenging endeavor. The investigator must be able to recognize and identify properly the various systems and components to determine their role, if any, in the cause of the fire. The extensive use of combustible materials such as foam, plastics, and wood may make the identification of components and systems difficult; however, noncombustible items will often survive and can then be examined.

Fuel Systems: Propulsion and Auxiliary

The fuels used in propulsion systems for marine vessels vary depending on the type of vessel. For example, most personal watercraft and small marine vessels are powered by gasoline and diesel engines that use the combustion energy to turn the propeller shaft, whereas military naval vessels may develop propulsion from direct-drive diesel engines or nuclear power plants that generate heat to operate steam boilers.

The most common types of propulsion systems that the investigator will encounter will be gasoline or diesel engine power plants.

Vacuum/Low Pressure Carbureted

Unlike automobiles, vacuum/low pressure carbureted inboard and outboard (I/O) engines on marine vessels must have backfire flame arrestors and a specialized gasket to prevent the escape of fuel from the venturi opening of the carburetor if flooding occurs.

High Pressure/Marine Fuel Injection Systems

Fuel injection systems, which operate at high pressures, vary among manufacturers. In general, the fuel injection system of all makes and models of engines will include a throttle body, plenum and fuel rail assembly, knock sensor, and engine control module FIGURE 28-4. The fuel storage and delivery systems, however, will still be low-pressure systems with vented fuel tanks.

Occasionally, multiple versions of a specific engine model may exist to allow the manufacturer to meet individual needs and requirements. To determine the specific system design, the investigator must record the serial number, as well as the make and model.

FIGURE 28-4 Marine fuel injection system.

Diesel

A fuel injection system is used for diesel-fueled engines and is specific to each manufacturer. Diesel engines have similar fuel system requirements as gasoline engines and require large air exchanges in the engine room/compartment to maintain adequate combustion air and ambient room temperatures.

Fuel Systems: Cooking and Heating

It is not uncommon to find both cooking and heating systems on a vessel, especially those designed for continuous habitation. These appliances are very similar in size and use to those found in recreational vehicles. During the examination, the investigator may find these devices and should determine whether they potentially are involved with the fire's development and/or spread.

Liquefied Petroleum Gases

Liquefied-petroleum (LP) gas systems may be used for either heating or cooking fuel, which require LP gas cylinders to be stored in compartments that vent at the bottom directly to the outside of the marine vessel. When LP gas is used for cooking, it is necessary to control other potential ignition sources. Appliances that have integrated fuel cylinders of less than 16 ounces are governed by the American Boat and Yacht Council (ABYC) A-30, *Cooking Appliances with Integral LPG Cylinders*. LP gas may be used for either heating or cooking if the appliance meets ABYC A-26, *LPG and CNG Fuels Appliances*.

Compressed Natural Gas

Recently, the use of compressed natural gas (CNG) for cooking has declined in popularity; however, the investigator may still encounter these systems. The CNG cylinders will often be stored in an accommodation space of less than 100 ft^3 (2.8 m^3).

CNG may be used for either heating or cooking if the appliance meets ABYC A-26, *LPG and CNG Fuels Appliances*.

Alcohol

Often used in galley ranges, alcohol may be stored either in tanks integrated with the appliance or in an independent container and pressurized by a hand pump.

Solid Fuels

NFPA 302, *Fire Protection Standard for Pleasure and Commercial Motor Craft*, provides design and installation requirements for solid fuels used on marine vessels. This includes the use of charcoal and wood.

> **Safety Tip**
>
> In the event that an explosion has occurred, it is important that the investigator identify all potential flammable or combustible gases and liquids that may have been present on the vessel.

Diesel and Kerosene

NFPA 302; ABYC A-3, *Galley Stoves*; and ABYC A-7, *Boat Heating Systems*, provide design and installation requirements for the use of diesel fuels for heating and cooking on marine vessels. Diesel and kerosene stoves are often used in place of propane, reducing the risk of migrating vapors. Combustion air is drawn from inside the compartment and is exhausted outside the vessel. Improper installation or use may cause heated gases to enter the compartment, creating a risk of carbon monoxide poisoning or ignition of nearby combustible items.

Turbochargers/Supercharges

Both diesel and gasoline engines may have turbochargers/superchargers installed to improve horsepower for the engine. Although both turbochargers and superchargers send pressurized air into the intake manifold, a turbocharger is powered by hot exhaust gases, whereas a supercharger is powered by the engine via a drive belt.

Turbochargers and superchargers may use the engine oil to lubricate or cool the devices because they operate at very high temperatures. To protect the surrounding materials, water jackets and/or heat blankets or shields may be fitted to these units.

Exhaust Systems

When combustion engines are used for propulsion, the hot gases produced must be safely exhausted. This can be achieved with both wet and dry exhaust systems. A wet system draws water surrounding the marine vessel into the exhaust system, where it travels through the muffler (if provided) and hoses until it exits the marine vessel. The components of these systems must be designed in accordance with ABYC P-1, *Installation of Exhaust Systems*, and NFPA 302. If a dewatered exhaust system is used, the exhaust gases are separated from the water at the muffler. The water is discharged from the transom, while the exhaust is released through the bottom of the marine vessel. These systems are also designed as specified in ABYC P-1.

Dry exhaust systems vent combustion gases through vertical exhaust pipes that are covered with insulating materials. These systems are seldom implemented on recreational boats.

Electrical Systems

In saltwater marine environments, a common difficulty facing vessel owners and operators is protecting electrical and mechanical components from corrosion. Not only is the metal structure of the ship in danger of corroding, but wiring, connectors, and electrical distribution equipment are also at particular risk FIGURE 28-5. Once copper conductors and parts begin to corrode after exposure to saltwater or other corrosive environments, resistance can increase in the electrical circuits. Furthermore, if saltwater is able to infiltrate or wick under wire or cable insulation, such problems can occur at locations in the electrical system other than just at connections. As a result, areas of high resistance can lead to overheating, even in conductors well away from connections where such failures tend to occur most often in electrical systems on land.

© Gari Wyn Williams/Alamy Images

FIGURE 28-5 Corrosion.

Corrosion can be a particular problem for AC power supplied to vessels through shore-tie connections. Conductors exposed to saltwater and not adequately maintained can result in high-resistance connections. In case of partially shorted systems, overcurrent protection devices may not trip, and electrical power may continue to flow. In such instances, fires can result at the plastic or rubberized boots of the shore-to-vessel connection.

Anywhere above decks where there are electrical or electromechanical components, corrosion poses a concern. Although topside (above deck) electrical items are usually protected from sea spray contact and saltwater intrusion, at times it still occurs. As a result, high resistance can occur in the circuitry, causing unequal heating and sometimes fires.

Both AC and DC electric systems may be located on marine vessels and are provided by a variety of sources. A shoreline is often used to provide AC power to marine vessels; however, generators and inverters may also be used as an onboard source. Direct current is provided by onboard battery systems, which power lights and equipment.

Engine Cooling Systems

Marine vessel engines are cooled by either drawing surrounding water into the engine, then discharging the water, or through the use of closed-coolant systems. A seawater system draws the water into the engine via a pump, which then circulates the

Fire Investigator Tip

A review of available statistical information for both large ship and recreational vessel fires suggests that between one half and three quarters of marine fires occur in engine spaces. Many of those involve electrical failures and/or fuel systems.

water through the engine and discharges it through the muffler. A closed system uses a 50/50 blend of glycol antifreeze, similar to an automobile, contained in a heat exchanger. Seawater is circulated into the heat exchanger and then discharged through the exhaust system.

Ventilation

Many marine vessels use a permanently mounted fuel tank that is often concealed within an accommodation space or specifically designed compartment. These tanks are required to have flame arrestors installed and are vented via the top of the tank through a hose, which is then directed overboard (out of the vessel). Gasoline fuel tanks are naturally ventilated (not forced), whereas compartments and associated bilges in these compartments are required to be power ventilated.

Many marine vessels have hatches and portholes that allow for natural ventilation of the accommodation spaces when they are open. Dorade and cowl vents may be present on a marine vessel and allow for ventilation all of the time.

When interpreting fire patterns, the investigator must consider the presence of these ventilation points and their effect on fire movement patterns.

Transmissions

A marine vessel may have either a mechanical or hydraulic-geared transmission. Generally, inboards and some high-performance inboard/outboard (I/O) engines will use a hydraulic-geared transmission, whereas outboards and most I/O will use a mechanical transmission. When a marine vessel uses a hydraulic-geared transmission, it will have its own lubricating oil (SAE 90) and cooler.

Accessories

Like other vehicles, marine vessels may contain various accessories for convenience or comfort. Air conditioners use a heat exchanger that circulates seawater through the coils. Air conditioning compressors will often be found in lower accommodation spaces or the engine room.

A power steering pump may also be present and operates similarly to a motor vehicle, in which a belt-driven pump is present on the propulsion engine.

Refrigeration units such as refrigerators and freezers may be powered by either AC or DC current, LP gas, or a combination of both. If an LP gas system is used in an accommodation space, it must meet the requirements of UL 1500, *Standard for Safety Ignition Protection Test for Marine Products*.

Hydraulic Systems

Hydraulic systems may be used either to steer or to adjust the trim settings on a marine vessel. Trim tabs are powered by a pump motor that is mounted next to a hydraulic tank near the transom. Hydraulic steering systems may be either pump driven or mechanical and will use either rigid or flexible hoses to steer the marine vessel. Although both are closed systems and are relatively safe from ignition, a release of hydraulic fluid on either electrical components nearby or a hot surface may pose a potential ignition risk.

Hydraulic thrusters are generally not used on marine vessels but may be found in a tunnel forward in the marine vessel. These electrically driven units may be driven by either a 12- or 24-V DC motor that is protected by a 200-amp fuse installed near the thruster unit.

Construction

It is imperative that marine fire investigators be familiar with the basics of ship and marine vessel construction. Maintaining adequate strength through well-designed, robust construction allows vessels to operate in heavy weather conditions without failure.

Knowing how these factors apply will assist the investigator in the proper application of fire dynamics principles to explain fire patterns.

Exterior Construction

High-tensile steel is used almost exclusively in the construction of large, seagoing vessels. The hulls and primary construction of smaller commercial vessels or recreational boats may be steel but may also consist of wood, fiberglass-reinforced plastic (FRP), aluminum, or ferro-cement.

The use of FRP and wood in the construction of marine vessels should be considered when the investigator is determining the total fire load. In certain instances, marine vessels may be constructed of FRP that used balsa or high-density foam as a core, which can act as a heat insulator during a fire.

Additionally, the investigator should consider the accessory items either attached or within the marine vessel, which may be combustible and add to the total fire load. These items include communication equipment; low- and medium-density plastic fixtures such as marker and navigation lights, spotlights, and railings; wooden components such as decking, masts, and booms; interior finishes and fixtures; and other combustible items such as sails, PFDs, life rafts, and seat cushions.

Generally, superstructures (cabins and other structures above the main deck) and decks are made of the same material as the hull. Horizontal partitions that form the tops and bottoms of compartments are known as decks. Decks are supported on the underside by beams and longitudinals. They provide strength to the hull by helping resist transverse compressive loads. Bulkheads are the vertical partitions that extend perpendicular to the long axis of the vessel and establish the basic compartmentation of the interior of the vessel. Bulkheads can be either structural or nonstructural. Structural bulkheads connect decks, frames, and hull plating and can withstand severe fluid dynamic pressures of the sea and liquid cargos. Nonstructural bulkheads serve as compartment separations, as in living or accommodation spaces. Structural bulkheads and decks can also function in steel ships to provide various degrees of passive fire protection.

Interior Construction

Depending on the design of the marine vessel, various materials may be used to achieve a lightweight, yet durable, structure similar to those found in structures and automobiles. Generally, interiors of recreational boats are made with FRP, plywood, and veneers. Steel and aluminum are noncombustible and may be used for structural components such as bulkheads and support structures.

Depending on the size and use of a marine vessel, the interior may resemble that of a recreational vehicle, an ordinary home, or a luxury hotel. The interior of accommodation spaces may be finished with wood paneling or painted walls, have carpeting, and contain items such as televisions, lamps, candles, and fans. They often have bedding and furniture constructed of fabric-covered foam padding. Wood items may be coated with organic oils and varnishes as a finish. These items, as well as the potential for human actions in these occupied spaces, must be considered when determining ignition sources, fuel loads, and fire spread issues.

Engine and machinery compartments contain the engine and accessory devices and components needed to propel and operate the marine vessel. These include batteries, generators, invertors, and storage units for waste water, hydraulic fluids, and fuel. Both DC and AC power may be present, with various electrical circuits to accommodate the different electrical needs of the marine vessel. Fuel tanks may be constructed of aluminum, steel, polyethylene, or fiberglass. Fittings and hoses may be constructed of neoprene, synthetic rubber, metal, or nylon. The tanks, fittings, and hoses must meet strict design requirements to prevent the release of fuel and should be examined by the investigator to determine suitability for marine use. Leakage may be the result of damage caused by deterioration, age, heat, vibration, mechanical damage, or corrosion from contact with water.

Because fuel is often stored or transferred in these spaces, flammable/explosive vapor detectors may be present to detect fugitive vapors from either a spill or a leak. These devices should be searched for and examined to determine whether the device was operating properly at the time of the fire.

In some instances, a fire protection and suppression system may be present in these spaces. These systems can displace oxygen and pose a dangerous hazard to the investigator should they discharge. The investigator must ensure these systems are disabled while conducting the investigation.

Storage tanks and holds are often found in the lower portions of the marine vessel, below accommodation spaces. The materials stored in these tanks may vary greatly and are often combustible. The investigator must determine which items were present in storage or holding tanks when conducting his or her investigation. Items improperly stored may either come in contact with an ignition source or, in the case of volatile organics used for maintenance or cleaning, spontaneously ignite.

Propulsion Systems

Propulsion systems are often determined by the use and operation of a particular vessel. This includes electric, liquid fuel, and fuel-gas, or a combination of two or more of these systems. Once the investigator has determined the type of propulsion system(s) present, he or she can begin to identify components that may have contributed to the fire's cause or spread.

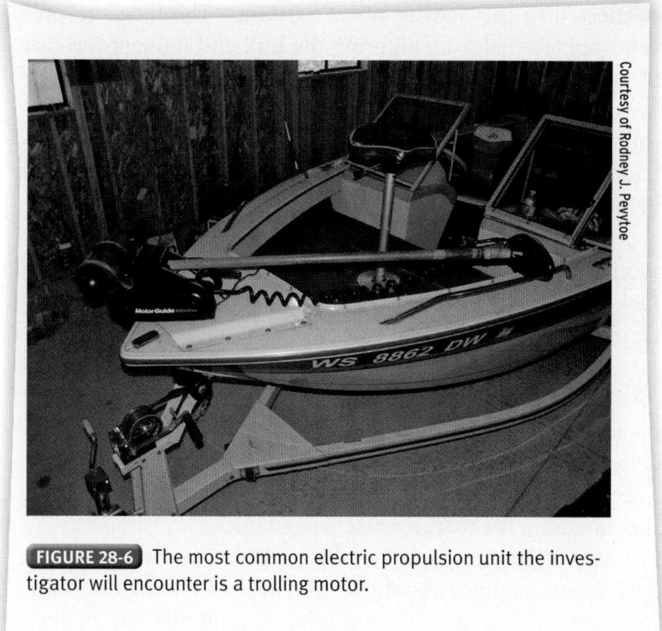

FIGURE 28-6 The most common electric propulsion unit the investigator will encounter is a trolling motor.

Electric Systems

The most common electric propulsion unit the investigator will encounter is a trolling motor: a small electric motor connected to a battery and used primarily for slow-speed maneuvering of recreational boats FIGURE 28-6. Main propulsion from electric motors is not generally used.

Fuels for Marine Vessels with Motorized Propulsion Systems

As discussed earlier, fuel systems may be present for propulsion, appliances, or the generation of electricity. Engines will fall into one of three general categories: outboard engines with self-contained or portable fuel tanks, inboard gasoline engines, or inboard diesel engines.

Outboard motors may be either two- or four-cycle gasoline engines with a carburetor or fuel-injected fuel delivery system. Two-cycle engines use a mixture of gasoline and oil to lubricate the motor, whereas four-cycle engines operate similarly to that of an automobile, where the oil is contained in a separate reservoir. Fuel is delivered by either a low-pressure delivery pump or a high-pressure, engine-mounted fuel pump used in fuel-injected systems.

Inboard arc mostly four-cycle engines that use built-in fuel tanks, fuel lines, fuel filters, and fuel pumps for fuel delivery. The marine vessel may contain more than one tank, which will be connected by an equalizing line known as a manifold to ensure equal distribution of weight. The tanks themselves are vented and all connections, fittings, and valves are located on the top of the tank to prevent draining or leakage. A fill plate is located on top of the fill port and connected to the marine vessel's ground. The fill plate prevents spilled fuel from entering the hull and reduces the risk of a static discharge while fueling. The fill nozzle must be in contact with the fill plate prior to the pump's being activated. Marine vessels constructed after 1977 are required to have antisiphon devices installed on the tanks if the engine is below the level of the tank.

Inboard diesel engines operate similarly to that of an automotive engine in which a high-pressure system will deliver the fuel; however, a low-pressure return line to the tank will likely be present.

Fuels may also be used for cooking, heating, or generators. LP gas is the most common heating, refrigeration, and cooking fuel that the investigator will encounter and it may collect in the lower portions of the marine vessel if vapors leak from the system. Generators are often supplied by a separate fuel line from the marine vessel's fuel tank.

When examining either the propulsion units or accessory items, samples should be obtained of all engine fluids, including fuel, hydraulic, and lubricating fluids, for analysis of prefire condition, just as you would with a motor vehicle fire.

Other Fuel Systems Used for Propulsion

Although uncommon, the investigator may still encounter a marine vessel that is propelled by a steam-driven engine in which a boiler is fired by wood, charcoal, coal, or paraffin. Older ships such as paddle-wheels often used the steam power created by the boilers to drive the pistons that turned the large rear or side-mounted wheels, which were popular in the mid- to late-1800s.

Ignition Sources

The investigator will likely find sources of ignition similar to those in both structural and vehicular fires; however, other unique sources of ignition such as turbochargers and exhaust manifolds may also be located. The investigator should take the time to understand thoroughly the components of the marine vessel's mechanical and electrical systems during his or her investigation to ensure that all potential sources of ignition are identified and examined.

Open Flames

Backfire events through unprotected carburetors are a common source of open flames. To prevent such fires, all inboard and I/O marine vessels are required to have a backfire flame arrestor attached to the air intake with a flame-tight connection. These flame arrestors must be approved for marine use by the United States Coast Guard (USCG) or comply with the Society of Automotive Engineers Standard SAE J-1928. Even with flame arrestors in place, however, fires are still possible if the owner does not keep the arrestor clean and in good condition.

Other sources of open flame ignition include fuel-fired burners used for cooking. Since 1977, however, appliances with pilot flames have been prohibited on gasoline-fueled marine vessels.

Electrical Sources

Most marine vessels will have some type of electrical system used to either start the engine, provide power to running lights, operate auxiliary lighting, or provide other conveniences found in both automobiles and buildings. These systems and devices can all be potential sources of ignition and must be examined and evaluated by the investigator.

Unless installed with a main disconnect switch, marine vessels equipped with batteries may have several electrical circuits energized despite the fact that the engine is not running and the ignition switch is in the "off" position. This includes items such as the bilge pump and the primary wiring circuits to the alternator, ignition switch, and battery cables. These circuits are generally always energized. When the marine vessel is operating, however, the accessory items and fixtures become potential sources of ignition.

The investigator may also find components present that were not rated for watercraft use. Overcurrent devices may have been replaced with oversized fuses and breakers, and accessory items may be added that overload the electrical system.

Marine vessels will most often be found to have both AC and DC electrical systems present for the investigator to examine. Components located below the water line are grounded or bonded to the marine vessel's electrical system to prevent electrical shock as well as serve as a form of lightning protection. Like an automobile, the negative or ground side of the battery is bonded to the engine block, whereas the positive or hot side supplies power to the fuses, accessories, and equipment that operate on the DC electric system. This two-wire system can be damaged or fail and pose a risk of a potential ignition source if the positive side (wiring, connections, or components) becomes grounded to another object. AC systems that generate power from either a generator or inverter may also present potential ignition sources within the circuit, connections, and/or connected appliances.

Overloaded Wiring

When circuits are overloaded because of excessive use of accessory items, the conductor may begin to heat, degrading and igniting the protective insulation. Because most wiring is routed through bulkheads and in the concealed structure of the marine vessel in large bundles and harnesses, heat is not easily dissipated and a significant fire can develop without activating the overcurrent devices. The investigator must interview the appropriate person(s) to determine whether any prefire deficiencies existed and what role they may have played in the ignition of the fire.

Electrical Short Circuiting and Arcs

If wiring becomes worn, brittle, or damaged, an arc or short circuit may occur if exposed conductors come in contact with a grounded surface. The investigator should examine the condition of all electrical wiring to determine whether insulation degradation or damage as a result of charring or other mechanical forces exists. If large current circuits such as the battery cables short or arc, they can easily ignite engine oil, ignitible vapors, or the insulation on the wiring itself.

Electrical Connections

Just like in automobiles, connections may become corroded because of a high-resistance connection in addition to exposure to water and salt. A high-resistance connection can easily ignite surrounding combustible items. The investigator should examine all connections for evidence of resistance heating.

Lightning

If lighting strikes near the marine vessel or the vessel itself, this high-energy discharge can conduct through the marine vessel's structure and electrical system, which can heat and easily ignite combustible items located nearby.

Static Electricity and Incendive Arcs

When loading or unloading cargo, attention must be paid to reducing the possibility of static buildup and discharge. If piping and other mechanical devices used in transferring cargo to or from a ship are not properly grounded, then a static arc could, depending on the type of cargo, result in either a fire or an explosion.

If ignitible vapors are present, static electricity and incendive arcs can pose a significant hazard of fire or explosion. Like in automobiles, refueling operations may produce both ignitible vapors and static electricity and require proper grounding between the fuel tank and fuel nozzle. Gasoline vapors that are heavier than air can collect in the lower portions of the marine vessel, including the accommodation spaces and bilge areas where the generation of static electricity may occur as a result of normal operations.

Care must be taken to ensure that dissimilar cargos, particularly ignitible liquids, are not accidentally mixed. When two liquids with dissimilar volatilities are mixed, the result can be that the overall volatility can increase above that of the higher boiling point liquid. If this occurs, precautions that would normally prevent ignition of the less volatile liquid may no longer be sufficient to avoid a fire or explosion.

■ Hot Surfaces

Exhaust systems and cooking and heating appliances also can ignite combustibles located too close to these objects. Exhaust manifolds generate temperatures that easily ignite engine fluids, including diesel sprays, vaporized gasoline, oils, and hydraulic fluids. Similarly to automobiles, the temperatures of exhaust components may actually increase shortly after the engine is stopped because the cooling system is no longer circulating fluid through the engine. Regardless of the type of fuel used to generate heat, cooking and heating appliances can easily ignite combustibles located near the exposed hot surfaces.

Adequate clearances from combustible items should be maintained. The investigator should conduct interviews with the appropriate person to determine whether any combustible items may have been stored in compartments or locations near one of these hot surfaces.

■ Mechanical Failures

Main bearing failures generally do not result in a fire; however, if a bearing in the alternator, the motor, a pulley, or a pump fails, combustible items nearby may ignite because of either excessive heating or hot metal being expelled from the bearing. If the investigator suspects that a bearing may have failed in one of these devices, he or she should inspect the inner and outer races of the bearing for physical damage.

If a pulley, alternator, or water pump seizes, the drive belt can heat as it passes over these components, generating sufficient heat to ignite the belt and start a fire. An inspection of the engine components can determine whether a pulley wheel is seized or locked up.

■ Smoking Materials

Lit cigarettes are capable of starting a fire, particularly when a lit cigarette meets upholstery and other fabrics. Urethane foam and other materials often used in seating burn easily and can promote spread and intensity of the fire.

Cargo Issues

A concern related to the shipment of cargo is ensuring that incompatible materials are not stored near each other or shipped in the same container or location in a hold. Many thousands of tons of oxidizers and peroxides, for instance, are transported annually aboard ships. It is imperative that such materials are isolated from fuels that might combust in their presence. Even though materials are packaged separately, incompatible cargos must not be placed close to each other in the event that rough seas cause breakage of containers and/or accidental mixing. Once fires do involve spaces containing oxidizers, they can easily grow out of control with little hope of firefighting.

Marine investigators examining the origin and cause of fires that may be cargo-related should review the vessel manifest to identify what materials were being transported at the time of the fire. The International Maritime Dangerous Goods (IMDG) Code provides information regarding classifying various hazardous products and cargo compatibility issues.

In break bulk cargo ships and container vessels, if a fire starts deep in the middle of stacks of cargo, it may be a long time before the fire is detected. In many cases, ships have unloaded cargo containers only to discover that a fire burned in their cargo unnoticed for the entire voyage. On many ships, cargo containers are mounted with only about 18 inches (45.7 cm) of space between them laterally and several containers high. Should a fire occur near the middle of one of these stacks or in a below-deck hold near the center of packaged cargo, the seat of the fire can be extremely difficult to locate and can easily spread before being identified FIGURE 28-7. Because of this, the job of the investigator becomes even more difficult.

© Malcolm Fife/iStock/Thinkstock

FIGURE 28-7 A fire occurring near the middle of a stack of cargo can be extremely difficult to locate and can easily spread before being identified.

Documenting Marine Vessel Fire Scenes

The investigator should approach the documentation of marine vessel fires in the same manner as he or she would a structure or automobile fire. The boat should be inspected at the loss location prior to being moved; however, this is often impractical. Most inspections will be conducted out of the water in a salvage yard, repair facility, or storage lot.

On Land

The investigator should first establish the location of the fire. If the marine vessel was damaged at its current location, then the investigator should determine whether the marine vessel was connected to any shore-based power sources (shore power). The power sources, shore line, and connection to the marine vessel should be photographed and documented. All potential ignition sources and fuels either in or near the marine vessel should be identified, photographed, and documented. The surrounding area, including the manner of storage of the marine vessel, should also be documented. The investigator should also look for evidence of prefire vandalism or mechanical problems with the marine vessel. Vessels that are improperly resting on their trailers, covered with debris, and located in obscure locations with large growing foliage around the vessel may be an indication of the vessel being abandoned.

Fire Investigator Tip

Marine vessels should be removed from the water and secured to allow the investigator an opportunity to examine the entire vessel. Evidence of prefire damage or poor maintenance may be below the water line.

In Water

Similar to land-based investigation, the investigator should determine whether the vessel is floating at the location where the fire occurred. If the vessel has been towed away from the original location, this area as well should be examined by the investigator to locate any possible items of evidentiary value. Interviews with witnesses should reveal whether the vessel was under way, moored, or anchored at the time of the fire. The investigator should also determine whether shore power was present and should examine, photograph, and document the condition of the power source, shore line, and connection at the vessel. Ignition sources and potential fuels in and around the vessel should also be identified, photographed, and documented. Underwater examinations may also be necessary to locate items of potential evidentiary value.

Moored

If a marine vessel is moored at a dock, slip, or seawall, the area should also be examined for potential ignition sources. The construction materials of piers or docks may have contributed to the intensity of the fire or fire spread from another vessel.

Anchored and Under Way

Ignition sources may vary between vessels that are under way, adrift, or at anchor. The investigator should determine which systems would be in use depending on the status of the vessel.

Underwater

Only specialized personnel should conduct underwater surveys of submerged vessels. The vessel and surrounding area should be examined for items of potential evidentiary value, including components of the vessel, dock, or pier or items that are potential ignition sources. Prior to raising the vessel, the orientation and condition of the vessel should be thoroughly documented, as items may be damaged, displaced, or lost during the raising operation.

Marine Vessel Identification

The investigator may locate various types of registrations and markings for the boat after a fire. Generally, a unique identifier will be given to each vessel and will vary depending on local regulations and the vessel's manufacturer. Vessels manufactured or imported into the United States, including homemade vessels, are required to have a hull identification number (HIN) permanently attached (e.g., stamped, engraved) in two separate places on the vessel. The primary is generally located at the starboard side of the transom (right rear), below the

FIGURE 28-8 Boats in the United States are required to have an HIN permanently attached in two separate places on the boat.

rubrail FIGURE 28-8. The secondary location will vary and can be obtained by contacting the manufacturer. If the HIN can be located, the first three letters will indicate the manufacturer code, which can provide further information for the investigator. Registration numbers are normally located underneath the bow of the vessel and can be used to establish ownership. USCG numbers may also be found inside some vessels. A hailing port and name also may be found on the vessel, usually on the transom; however, this is not usually regulated.

Witness Interviews

As with any fire, the investigator should establish a basic scenario in regard to any circumstances related to the fire. The investigator should seek and interview anyone who may have information about the fire, including the owner, operator at the time of the fire, bystanders, and the police and fire department personnel who responded to the incident. The investigator should seek information about the history of the vessel and any specific activities related to the fire, including activities just prior to, during, and after the fire occurred. The questions asked by the investigator will be very similar to questions posed during other types of fire investigation. Questions include, but are not limited to, the following topics:

- Operational condition of the vessel
- Accessory use and locations
- Recent repairs or modifications
- Water and weather conditions
- Activities and operations on the vessel prior to the fire
- How the fire was first discovered
- Whether the boat was working correctly at the time the fire was discovered
- Actions taken after the fire's discovery
- How the fire was extinguished
- An estimated timeline for fire discovery, summoning of help, and extinguishment
- Salvage operations
- Actions taken by the Environmental Protection Agency (EPA) or similar agencies
- Locations of occupants and fire victims
- Actions taken by occupants and any public officials
- Details about any previous investigations involving the vessel
- Any other questions the investigator may develop as a result of the interview itself

Marine Vessel Particulars

After identifying the vessel and obtaining all relevant information, the investigator must begin to determine the construction and design features that would affect the fire's development and growth. This can be achieved by examining an exemplar and reviewing the owner/operator's manual and sales literature associated with the vessel. A recall database search may also prove useful in determining any preexisting safety issues. The Consumer Product Safety Commission and USCG both maintain an extensive database on product recalls related to vessels based off make and model information.

Marine Vessel Examination

As with any fire, the area of origin must be determined by the investigator and may occur in one of four major areas:

- Cockpit/topside
- Engine/fuel compartment
- Accommodation compartment or cabin
- Bilge areas

The investigator should establish a systematic process to evaluate and examine the vessel TABLE 28-2. This may include processes similar to those conducted on structural and vehicular examinations.

Examination of Marine Vessel Systems

The investigator should begin by examining the exterior of the vessel, identifying fire patterns that may allow the investigator to determine the area of origin FIGURE 28-9. The investigator must also consider vents and openings around engine and fuel compartments, because these can affect fire patterns. If a fire occurs in an accommodation space, the fire may consume the

Table 28-2 Examination Outline

Exterior	Hull, outdrive, vents, windows, prefire damage
Top side	Mast, decking, accessory items, application patterns, and evidence of materials being removed or replaced prior to the fire
Interior	Cockpit, sleeping berths, storage areas, galley, and state rooms
Mechanical	Propulsion system, bilge rooms/compartments, and other accessory locations

FIGURE 28-9 Begin by examining the exterior of the boat.

topside or cabin without affecting the other spaces. Because these spaces are usually airtight when closed, the fire will often self-extinguish. Items within these spaces must be examined to determine their role in either the ignition or the spread of the fire. The investigator may find spaces that were used as sleeping areas, lounges, cooking, or storage, which all pose unique ignition and fire spread issues.

The construction of the hull causes accommodation spaces to have sloped surfaces, limited vertical spaces, and offsets that may alter normal fire development and therefore limit the patterns that investigators typically find. By examining the fire patterns present, the investigator can determine whether the fire occurred within or outside an accommodation space. If an engine or fuel compartment fire occurs, it will likely involve fuel vapors and cause extensive damage, spreading to the accommodation spaces. The investigator should thoroughly examine the carburetor, fuel injection systems, fuel delivery, exhaust, and ignition systems to determine their role in the ignition of the fire. Bilge areas are usually located in remote and isolated areas of the vessel and will often collect heavier-than-air fuel gases such as gasoline, diesel, or oil, which may still be present after the fire. Bilge pumps themselves are usually intrinsically safe and not a likely source of ignition if operating properly.

After the compartment of origin is located, a detailed examination of the area should be conducted. Individual systems should be examined to locate evidence of their role in the fire. Fuel tanks should be examined for failures around the edges or bottom or corrosion that may allow fugitive gases to escape. The fuel fill and vent hoses should be examined to determine whether chafing, corrosion, or other damage is present. The ground between the tank fill plate and the tank should also be tested for electrical continuity. If a fuel tank is exposed to heat, a demarcation line may be present where the fuel level inside has autocooled the exterior surface. Plastic tanks may still be intact and contain fuel. The plastic container often softens and fails at the level of the fuel within.

Switches and handles should also be examined and their position documented and photographed. Port lights and hatches should be examined and their position at the time of the fire determined. Generators, battery, and shore-line power transfer switches should also be examined. If the ignition can be identified, determine whether the key is present and whether any damage or tampering occurred. Although many of these switches are made of materials easily consumed, components may still be present and identifiable after the fire.

Marine Vessels in Structures

Marine vessels stored within garages or carports should be examined as though they are potentially the source of ignition; however, the vessel may simply be an additional combustible fuel package. Sources of ignition within the storage space must also be examined by the investigator because a fire originating within the structure can easily spread to the vessel. Vessels afloat that are moored under shelters should be examined with consideration of the shelter itself containing heat and fire gases that can influence the fire development and patterns present on the vessel. The electrical service to and on floating docks must also be examined, as damage from corrosion or movement of the dock itself may be present.

Legal Considerations

The nature of marine vessel fires creates unique legal issues. Admiralty (maritime) law is established in the U.S. Constitution in Article III, Section II, and provides for the federal courts to hold jurisdiction over cases arising around marine commerce, marine navigation, shipping, sailors, and the transportation of passengers and goods by sea. Search and seizure, right of entry, and spoliation of evidence are also potential issues the investigator may encounter and are discussed at length in the "Legal Considerations" chapter in this textbook. There may also be fines and other legal considerations related to contamination if the vessel's fuels or other harmful contaminants located within the vessel escape into the water. The EPA and USCG may be involved with these particular issues. Although federal, state, and local investigators may have authority and jurisdiction over fires occurring on vessels within territorial waters, investigations of fires occurring on foreign-registered ships in international waters will often be the responsibility of that foreign state.

As with other types of fires, marine vessels may be set on fire for various reasons. For example, a recreational boat owner may set his or her boat on fire, believing that it may sink and not be recoverable. In the event that the fire is determined to be incendiary, the investigator may need to coordinate his or her investigation with other agencies such as the USCG or local port authorities.

Wrap-Up

Ready for Review

- Marine vessel fire investigations pose unique safety issues for the fire investigator.
- There are hundreds of different types of marine vessels, each specifically designed for an intended use.
- Fire investigators must be cognizant of the potential differences in the ways in which a fire may be attacked at sea as compared with in port and must pay particular attention to how suppression efforts may have altered fire pattern development.
- The investigator must be able to recognize and properly identify the various systems and components to determine their role, if any, in the cause of the fire.
- It is imperative that marine fire investigators be familiar with the basics of ship and marine vessel construction.
- Once the investigator has determined the type of propulsion system(s) present, he or she can begin to identify components that may have contributed to the fire's cause or spread.
- The investigator should take the time to understand thoroughly the components of the marine vessel's mechanical and electrical systems during his or her investigation to ensure that all potential sources of ignition are identified and examined.
- Marine investigators examining the origin and cause of fires that may be cargo-related should review the vessel manifest to identify what materials were being transported at the time of the fire.
- The investigator should approach the documentation of marine vessel fires in the same manner as he or she would a structure or automobile fire.
- Marine vessels are usually divided into four areas:
  - Topside/cockpit
  - Accommodation space
  - Engine/fuel compartment
  - Bilge area
- Marine vessels stored within garages or carports should be examined as though they are potentially the source of ignition; however, the vessel may simply be an additional combustible fuel package.
- As with any fire, the investigator must be aware of potential legal issues associated with maritime law as well as constitutional rights of the marine vessel owner or operator.

Hot Terms

Accommodation space A space on a vessel designed for people to live in.

Adrift Loose, not on moorings or towline.

Afloat Floating in water; not sinking.

Bilges The lowest, interior part of a vessel's hull; the area where spilled water and oil can collect.

Boat Any vessel that can be placed or stored on another vessel.

Bulkheads The vertical separations in a vessel that form compartments and that correspond to walls in a building. The term does not refer to the vertical sections of a vessel's hull.

Decks Usually continuous, horizontal divisions running the length of a vessel and extending athwart ships. This corresponds to floors in a building on land.

Hatches Openings in a vessel's deck that are fitted with a watertight cover.

Holds Compartments used for carrying cargo below deck in a large vessel.

Hull The outer skin of a vessel including the bottom, sides, and main deck, but not including the superstructure, masts, rigging, and other fittings.

Inboard Toward the interior of a vessel from the outer hull.

Intrinsically safe A protection feature for the safe operation of electronic equipment in explosive atmospheres and/or under irregular operating conditions.

Lazarettes Stowage compartments, often in the aft end of a vessel, sometimes used as a workshop.

Outboard Lying outward from the centerline of a vessel toward or beyond its sides; a type of engine that is attached to and lying aft of a vessel's stern.

Overboard Over the side or out of the vessel.

Races A component of a rolling element bearing that contains the elements and transfers the load to the bearing; generally used in pairs (inner and outer races).

Shore power Electrical power supplied from shore via a cord set.

Starboard The right side of a vessel when looking forward.

Superstructures All vessel structures above the main deck or weather deck.

Topside When below decks, refers to areas on or above the main deck.

Transom The stern cross-section of a square-sterned vessel.

Trolling motor A small electric motor connected to a battery and used primarily for slow-speed maneuvering of recreational boats.

Under way Not attached to land or mooring by tie, anchor, or grounding.

Venturi A narrowed area within a carburetor that causes air to accelerate and create a low pressure area that draws fuel into the intake manifold of the engine.

Vessels A broad grouping of every description of watercraft, other than a seaplane on the water, used or capable of being used as a means of transportation on the water.

© Greg Henry/ShutterStock, Inc.

FIRE INVESTIGATOR *in action*

You have been requested to conduct an investigation of a large, high-performance speed boat that had caught fire overnight and is submerged almost 1 mile from the shore. A recovery team has begun raising the vessel; however, it will be several hours until you are able to begin your inspection.

During your interview with the owner, you are informed that the vessel is approximately 5 years old with numerous upgrades made to increase speed and performance. The owner states that the vessel had been used the previous day during a charity poker run and was anchored overnight near the shore.

After the vessel is raised and placed onshore, you note that the vessel is severely damaged with several holes noted near the center of the hull. An inspection of the engine compartment finds limited flame and heat damage; however, most of the upgraded components the owner reported are not present.

1. Where will the primary hull identification number (HIN) be located?
   - **A.** Port side of the bow
   - **B.** Starboard side of the transom
   - **C.** Directly below the vessel registration
   - **D.** Stamped on the starboard side rail
2. Which of the following is required to be attached to the air intake on vessels with inboard and I/O marine motors?
   - **A.** Backfire flame arrestors
   - **B.** Bird and insect screens
   - **C.** Mass airflow sensors
   - **D.** Purge valves
3. Electronic components utilized in confined spaces, engine compartments, or other locations where explosive atmospheres may exist must be:
   - **A.** bonded.
   - **B.** grounded.
   - **C.** intrinsically safe.
   - **D.** All of the above
4. If you determine this fire is incendiary, with which agency may you need to coordinate?
   - **A.** United States Coast Guard
   - **B.** Local port authority
   - **C.** Both A and B
   - **D.** None of the above

An Extract from: NFPA 1033, *Standard for Professional Qualifications for Fire Investigator*, 2014 Edition

Chapter 1 Administration

1.1* Scope. This standard identifies the minimum job performance requirements (JPRs) for fire investigators.

1.2* Purpose. The purpose of this standard shall be to specify the minimum job performance requirements for serving as a fire investigator in both the private and public sectors.

1.2.1 It is not the intent of this standard to restrict any jurisdiction from exceeding the minimum requirements.

1.2.2 Job performance requirements for each duty are the tasks an individual must be able to perform in order to successfully carry out that duty; however, they are not intended to measure a level of knowledge. Together, the duties and job performance requirements define the parameters of the job of fire investigator.

1.3 General.

1.3.1 The fire investigator shall be at least age 18.

1.3.2 The fire investigator shall have a high school diploma or equivalent.

1.3.3 The authority having jurisdiction shall conduct a thorough background and character investigation prior to accepting an individual as a candidate for certification as a fire investigator.

1.3.4 The job performance requirements for fire investigator shall be completed in accordance with established practices and procedures or as they are defined by law or by the authority having jurisdiction.

1.3.5* The job performance requirements found in this standard are not required to be mastered in the order they appear. Training agencies or authorities shall establish instructional priority and the training program content to prepare individuals to meet the job performance requirements of this standard.

1.3.6* Evaluation of job performance requirements shall be by individuals who are qualified and approved by the authority having jurisdiction.

1.3.7* The investigator shall have and maintain at a minimum an up-to-date basic knowledge of the following topics beyond the high school level:

(1) Fire science
(2) Fire chemistry
(3) Thermodynamics
(4) Thermometry
(5) Fire dynamics
(6) Explosion dynamics
(7) Computer fire modeling
(8) Fire investigation
(9) Fire analysis
(10) Fire investigation methodology
(11) Fire investigation technology
(12) Hazardous materials
(13) Failure analysis and analytical tools
(14) Fire protection systems
(15) Evidence documentation, collection, and preservation
(16) Electricity and electrical systems

1.3.8* The fire investigator shall remain current in the topics listed in 1.3.7 by attending formal education courses, workshops, and seminars and/or through professional publications and journals.

Chapter 4 Fire Investigator

4.1 General.

4.1.1* The fire investigator shall meet the job performance requirements defined in Sections 4.2 through 4.7.

4.1.2* The fire investigator shall employ all elements of the scientific method as the operating analytical process throughout the investigation and for the drawing of conclusions.

4.1.3* Because fire investigators are required to perform activities in adverse conditions, site safety assessments shall be completed on all scenes and regional and national safety standards shall be followed and included in organizational policies and procedures.

4.1.4* The fire investigator shall maintain necessary liaison with other interested professionals and entities.

4.1.5* The fire investigator shall adhere to all applicable legal and regulatory requirements.

4.1.6 The fire investigator shall understand the organization and operation of the investigative team within an incident management system.

4.2* Scene Examination. Duties shall include inspecting and evaluating the fire scene, or evidence of the scene, and/or conducting a comprehensive review of documentation generated during the examination(s) of the scene if the scene is no longer available, so as to determine the area or point of origin, source of ignition, material(s) ignited, and act or activity that brought the ignition source and materials together and to assess the subsequent progression, extinguishment, and containment of the fire.

4.2.1 Secure the fire ground, given marking devices, sufficient personnel, and special tools and equipment, so that unauthorized persons can recognize the perimeters of the investigative scene and are kept from restricted areas and all evidence or potential evidence is protected from damage or destruction.

(A) Requisite Knowledge. Fire ground hazards, types of evidence, and the importance of fire scene security, evidence preservation, and issues relating to spoliation.

(B) Requisite Skills. Use of marking devices.

4.2.2* Conduct an exterior survey, given standard equipment and tools, so that evidence is identified and preserved, fire damage is interpreted, hazards are identified to avoid injuries, accessibility to the property is determined, and all potential means of ingress and egress are discovered.

(A) Requisite Knowledge. The types of building construction and the effects of fire on construction materials, types of evidence commonly found in the perimeter, evidence preservation methods, the effects of fire suppression, fire behavior and spread, fire patterns, and a basic awareness of the dangers of hazardous materials.

(B) Requisite Skills. Ability to assess fire ground and structural condition, observe the damage from and effects of the fire, and interpret fire patterns.

4.2.3 Conduct an interior survey, given standard equipment and tools, so that areas of potential evidentiary value requiring further examination are identified and preserved, the evidentiary value of contents is determined, and hazards are identified in order to avoid injuries.

(A) Requisite Knowledge. The types of building construction and interior finish and the effects of fire on those materials, the effects of fire suppression, fire behavior and spread, evidence preservation methods, fire patterns, effects of building contents on fire growth, the relationship of building contents to the overall investigation, weather conditions at the time of the fire, and fuel moisture.

(B) Requisite Skills. Ability to assess structural conditions, observe the damage and effects of the fire, discover the impact of fire suppression efforts on fire flow and heat propagation, and evaluate protected areas to determine the presence and/or absence of contents.

4.2.4 Interpret fire patterns, given standard equipment and tools and some structural or content remains, so that each individual pattern is evaluated with respect to the burning characteristics of the material involved and in context and relationship with all patterns observed and the mechanisms of heat transfer that led to the formation of the pattern.

(A) Requisite Knowledge. Fire dynamics, fire development, and the interrelationship of heat release rate, form, and ignitibility of materials.

(B) Requisite Skills. Ability to interpret the effects of burning characteristics on different types of materials.

4.2.5 Interpret and analyze fire patterns, given standard equipment and tools and some structural or content remains, so that fire development is determined, methods and effects of suppression are evaluated, false origin area patterns are recognized, and all areas of origin are correctly identified.

(A) Requisite Knowledge. Fire behavior and spread based on fire chemistry, fire dynamics, and physics, fire suppression effects, building construction.

(B) Requisite Skills. Ability to interpret variations of fire patterns on different materials with consideration given to heat release rate, form, and ignitibility; distinguish impact of different types of fuel loads; evaluate fuel trails; and analyze and synthesize information.

4.2.6 Examine and remove fire debris, given standard equipment and tools, so that all debris is checked for fire cause evidence, potential ignition source(s) is identified, and evidence is preserved without investigator-inflicted damage or contamination.

(A) Requisite Knowledge. Basic understanding of ignition processes, characteristics of ignition sources, and ease of ignition of fuels; debris-layering techniques; use of tools and equipment during the debris search; types of fire cause evidence commonly found in various degrees of damage; and evidence-gathering methods and documentation.

(B) Requisite Skills. Ability to employ search techniques that further the discovery of fire cause evidence and ignition sources, use search techniques that incorporate documentation, and collect and preserve evidence.

4.2.7 Reconstruct the area of origin, given standard and, if needed, special equipment and tools as well as sufficient personnel, so that all protected areas and fire patterns are identified and correlated to contents or structural remains, items potentially critical to cause determination and photo documentation are returned to their prefire location, and the area(s) or point(s) of origin is discovered.

(A) Requisite Knowledge. The effects of fire on different types of material and the importance and uses of reconstruction.

(B) Requisite Skills. Ability to examine all materials to determine the effects of fire, identify and distinguish among different types of fire-damaged contents, and return materials to their original position using protected areas and fire patterns.

4.2.8* Inspect the performance of building systems, including detection, suppression, HVAC, utilities, and building compartmentation, given standard and special equipment and tools, so that a determination can be made as to the need for expert resources, an operating system's impact on fire growth and spread is considered in identifying origin areas, defeated and/or failed systems are identified, and the system's potential as a fire cause is recognized.

(A) Requisite Knowledge. Different types of detection, suppression, HVAC, utility, and building compartmentation such as fire walls and fire doors; types of expert resources for building systems; the impact of fire on various systems; common methods used to defeat a system's functional capability; and types of failures.

(B) Requisite Skills. Ability to determine the system's operation and its effect on the fire; identify alterations to, and failure indicators of, building systems; and evaluate the impact of suppression efforts on building systems.

4.2.9 Discriminate the effects of explosions from other types of damage, given standard equipment and tools, so that an explosion is identified and its evidence is preserved.

(A) Requisite Knowledge. Different types of explosions and their causes, characteristics of an explosion, and the difference between low- and high-order explosions.

(B) Requisite Skills. Ability to identify explosive effects on glass, walls, foundations, and other building materials; distinguish between low- and high-order explosion effects; and analyze damage to document the blast zone and origin.

4.3 Documenting the Scene. Duties shall include diagramming the scene, photographing, and taking field notes to be used to compile a final report.

4.3.1 Diagram the scene, given standard tools and equipment, so that the scene is accurately represented and evidence, pertinent contents, significant patterns, and area(s) or point(s) of origin are identified.

(A) Requisite Knowledge. Commonly used symbols and legends that clarify the diagram, types of evidence and patterns that need to be documented, and formats for diagramming the scene.

(B) Requisite Skills. Ability to sketch the scene, basic drafting skills, and evidence recognition and observational skills.

4.3.2* Photographically document the scene, given standard tools and equipment, so that the scene is accurately depicted and the photographs support scene findings.

(A) Requisite Knowledge. Working knowledge of high resolution camera and flash, the types of film, media, and flash available, and the strengths and limitations of each.

(B) Requisite Skills. Ability to use a high-resolution camera, flash, and accessories.

4.3.3 Construct investigative notes, given a fire scene, available documents (e.g., prefire plans and inspection reports), and interview information, so that the notes are accurate, provide further documentation of the scene, and represent complete documentation of the scene findings.

(A) Requisite Knowledge. Relationship between notes, diagrams, and photos, how to reduce scene information into concise notes, and the use of notes during report writing and legal proceedings.

(B) Requisite Skills. Data-reduction skills, note-taking skills, and observational and correlating skills.

4.4 Evidence Collection/Preservation. Duties shall include using proper physical and legal procedures to identify, document, collect, and preserve evidence required within the investigation.

4.4.1 Utilize proper procedures for managing victims and fatalities, given a protocol and appropriate personnel, so that all evidence is discovered and preserved and the protocol procedures are followed.

(A) Requisite Knowledge. Types of evidence associated with fire victims and fatalities and evidence preservation methods.

(B) Requisite Skills. Observational skills and the ability to apply protocols to given situations.

4.4.2* Locate, document, collect, label, package, and store evidence, given standard or special tools and equipment and evidence collection materials, so that it is properly identified, preserved, collected, packaged, and stored for use in testing, legal, or other proceedings and examinations, ensuring cross-contamination and investigator-inflicted damage to evidentiary items is avoided and the chain of custody is established.

(A) Requisite Knowledge. Types of evidence, authority requirements, impact of removing evidentiary items on civil or criminal proceedings (exclusionary or fire-cause supportive evidence), types, capabilities, and limitations of standard and special tools used to locate evidence, types of laboratory tests available, packaging techniques and materials, and impact of evidence collection on the investigation.

(B) Requisite Skills. Ability to recognize different types of evidence and determine whether evidence is critical to the investigation.

4.4.3 Select evidence for analysis, given all information from the investigation, so that items for analysis support specific investigation needs.

(A) Requisite Knowledge. Purposes for submitting items for analysis, types of analytical services available, and capabilities and limitations of the services performing the analysis.

(B) Requisite Skills. Ability to evaluate the fire incident to determine forensic, engineering, or laboratory needs.

4.4.4 Maintain a chain of custody, given standard investigative tools, marking tools, and evidence tags or logs, so that written documentation exists for each piece of evidence and evidence is secured.

(A) Requisite Knowledge. Rules of custody and transfer procedures, types of evidence (e.g., physical evidence obtained at the scene, photos, and documents), and methods of recording the chain of custody.

(B) Requisite Skills. Ability to execute the chain of custody procedures and accurately complete necessary documents.

4.4.5 Dispose of evidence, given jurisdictional or agency regulations and file information, so that the disposal is timely, safely conducted, and in compliance with jurisdictional or agency requirements.

(A) Requisite Knowledge. Disposal services available and common disposal procedures and problems.

(B) Requisite Skills. Documentation skills.

4.5 Interview. Duties shall include obtaining information regarding the overall fire investigation from others through verbal communication.

4.5.1 Develop an interview plan, given no special tools or equipment, so that the plan reflects a strategy to further determine the fire cause and affix responsibility and includes a relevant questioning strategy for each individual to be interviewed that promotes the efficient use of the investigator's time.

(A) Requisite Knowledge. Persons who can provide information that furthers the fire cause determination or the affixing of responsibility, types of questions that are pertinent and efficient to ask of different information sources (first responders, neighbors, witnesses, suspects, and so forth), and pros and cons of interviews versus document gathering.

(B) Requisite Skills. Planning skills, development of focused questions for specific individuals, and evaluation of existing file data to help develop questions and fill investigative gaps.

4.5.2 Conduct interviews, given incident information, so that pertinent information is obtained, follow-up questions are asked, responses to all questions are elicited, and the response to each question is documented accurately.

(A) Requisite Knowledge. Types of interviews, personal information needed for proper documentation or follow-up, documenting methods and tools, and types of nonverbal communications and their meaning.

(B) Requisite Skills. Ability to adjust interviewing strategies based on deductive reasoning, interpret verbal and nonverbal communications, apply legal requirements applicable, and exhibit strong listening skills.

4.5.3 Evaluate interview information, given interview transcripts or notes and incident data, so that all interview data is individually analyzed and correlated with all other interviews, corroborative and conflictive information is documented, and new leads are developed.

(A) Requisite Knowledge. Types of interviews, report evaluation methods, and data correlation methods.

(B) Requisite Skills. Data correlation skills and the ability to evaluate source information (e.g., first responders and other witnesses).

4.6 Post-Incident Investigation. Duties shall include the investigation of all factors beyond the fire scene at the time of the origin and cause determination.

4.6.1 Gather reports and records, given no special tools, equipment, or materials, so that all gathered documents are applicable to the investigation, complete, and authentic; the chain of custody is maintained; and the material is admissible in a legal proceeding.

(A) Requisite Knowledge. Types of reports needed that facilitate determining responsibility for the fire (e.g., police reports, fire reports, insurance policies, financial records, deeds, private investigator reports, outside photos, and videos) and location of these reports.

(B) Requisite Skills. Ability to identify the reports and documents necessary for the investigation, implement the chain of custody, and organizational skills.

4.6.2 Evaluate the investigative file, given all available file information, so that areas for further investigation are identified, the relationship between gathered documents and information is interpreted, and corroborative evidence and information discrepancies are discovered.

(A) Requisite Knowledge. File assessment and/or evaluation methods, including accurate documentation practices, and requisite investigative elements.

(B) Requisite Skills. Information assessment, correlation, and organizational skills.

4.6.3 Coordinate expert resources, given the investigative file, reports, and documents, so that the expert's competencies are matched to the specific investigation needs, financial expenditures are justified, and utilization clearly furthers the investigative goals of determining cause or affixing responsibility.

(A) Requisite Knowledge. How to assess one's own expertise, qualification to be called for expert testimony, types of expert resources (e.g., forensic, CPA, polygraph, financial, human behavior disorders, and engineering), and methods to identify expert resources.

(B) Requisite Skills. Ability to apply expert resources to further the investigation by networking with other investigators to identify experts, questioning experts relative to their qualifications, and developing a utilization plan for use of expert resources.

4.6.4 Establish evidence as to motive and/or opportunity, given an incendiary fire, so that the evidence is supported by documentation and meets the evidentiary requirements of the jurisdiction.

(A) Requisite Knowledge. Types of motives common to incendiary fires, methods used to discover opportunity, and human behavioral patterns relative to fire-setting.

(B) Requisite Skills. Financial analysis, records gathering and analysis, interviewing, and interpreting fire scene information and evidence for relationship to motive and/or opportunity.

4.6.5* Formulate an opinion concerning origin, cause, or responsibility for the fire, given all investigative findings, so that the opinion regarding origin, cause, or responsibility for a fire is supported by the data, facts, records, reports, documents, and evidence.

(A) Requisite Knowledge. Analytical methods and procedures (e.g., hypothesis development and testing, systems analysis, time lines, link analysis, fault tree analysis, and data reduction matrixing).

(B) Requisite Skills. Analytical and assimilation skills.

4.7 Presentations. Duties shall include the presentation of findings to those individuals not involved in the actual investigations.

4.7.1* Prepare a written report, given investigative findings, documentation, and a specific audience, so that the report accurately reflects the investigative findings, is concise, expresses the investigator's opinion, contains facts and data that the investigator relies on in rendering an opinion, contains the reasoning of the investigator by which each opinion was reached, and meets the needs or requirements of the intended audience(s).

(A) Requisite Knowledge. Elements of writing, typical components of a written report, and types of audiences and their respective needs or requirements.

(B) Requisite Skills. Writing skills, ability to analyze information and determine the reader's needs or requirements.

4.7.2 Express investigative findings verbally, given investigative findings, notes, a time allotment, and a specific audience, so that the information is accurate, the presentation is completed within the allotted time, and the presentation includes only need-to-know information for the intended audience.

(A) Requisite Knowledge. Types of investigative findings, the informational needs of various types of audiences, and the impact of releasing information.

(B) Requisite Skills. Communication skills and ability to determine audience needs and correlate findings.

4.7.3 Testify during legal proceedings, given investigative findings, contents of reports, and consultation with legal counsel, so that all pertinent investigative information and evidence are presented clearly and accurately and the investigator's demeanor and attire are appropriate to the proceedings.

(A) Requisite Knowledge. Types of investigative findings, types of legal proceedings, professional demeanor requirements, and an understanding of due process and legal proceedings.

(B) Requisite Skills. Communication and listening skills and ability to differentiate facts from opinion and determine accepted procedures, practices, and etiquette during legal proceedings.

ProBoard Assessment Methodology Matrices for NFPA 1033, 2014 Edition

NFPA 1033 - Fire Investigator - 2014 Edition

IMPORTANT: The language from the standard on the AMMs is truncated. When completing the AMMs, the agency must refer to the NFPA standards for the complete text and a comprehensive statement of each Job Performance Requirement (JPR), Requisite Knowledge (RK), Requisite Skill (RS), and any applicable annex or explanatory information.

INSTRUCTIONS: Please review the instructions for filling out this Assessment Methodology Matrix at http://theproboard.org/AMM.htm. The submission of this form by an agency is affirmation that it is filled out in accordance with the instructions listed above.

AGENCY NAME:					DATE COMPILED:	
OBJECTIVE / JPR, RK, RS		COGNITIVE	MANIPULATIVE			FIRE INVESTIGATOR: PRINCIPLES AND PRACTICE TO NFPA 921 AND 1033, FOURTH EDITION
SECTION	ABBREVIATED TEXT	WRITTEN TEST	SKILLS STATION	PORTFOLIO	PROJECTS	
1.3.1	At least 18 years of age	Prerequisite requirements should be met according to the approved agency policy				
1.3.2	High school diploma or equivalent					Chapter 1 (p 6)
1.3.3	Background and character investigation					Chapter 1 (p 6)
4.1.2	Employ all elements of the scientific method					Chapter 2 (pp 14–17), Chapter 17 (pp 271–273)
4.2.1	Secure the fire ground					Chapter 10 (pp 172–173, 175–176), Chapter 11 (pp 188–193, 199–201), Chapter 27 (pp 408–409, 413–414)
4.2.1(A)	RK: Fire ground hazards					Chapter 10 (pp 172–173, 175–176), Chapter 11 (pp 188–193, 199–201), Chapter 27 (pp 408–409, 413–414)
4.2.1(B)	RS: Use of marking devices					Chapter 10 (pp 172–173, 175–176), Chapter 27 (pp 408–409, 413–414)
4.2.2	Conduct an exterior survey					Chapter 5 (pp 61–69, 71–72), Chapter 11 (pp 188–201)
4.2.2(A)	RK: Types of building construction					Chapter 5 (pp 61–69, 71–72), Chapter 11 (pp 188–201)
4.2.2(B)	RS: Fire ground and structural condition					Chapter 5 (pp 61–69, 71–72)
4.2.3	Conduct an interior survey					Chapter 5 (pp 61–73)
4.2.3(A)	RK: Types of building construction and interior finish					Chapter 5 (pp 61–73), Chapter 11 (pp 188–199)
4.2.3(B)	RS: Structural conditions					Chapter 5 (pp 61–73)
4.2.4	Interpret fire patterns					Chapter 4 (pp 42–55), Chapter 24 (p 352), Chapter 26 (pp 393–398)
4.2.4(A)	RK: Fire development					Chapter 4 (pp 42–55), Chapter 24 (p 352), Chapter 26 (pp 393–398)

AGENCY NAME:					DATE COMPILED:	
OBJECTIVE / JPR, RK, RS		COGNITIVE	MANIPULATIVE			FIRE INVESTIGATOR: PRINCIPLES AND PRACTICE TO NFPA 921 AND 1033, FOURTH EDITION
SECTION	ABBREVIATED TEXT	WRITTEN TEST	SKILLS STATION	PORTFOLIO	PROJECTS	
4.2.4(B)	RS: Interpret the effects of burning characteristics on different types of materials					Chapter 4 (pp 42–55), Chapter 24 (p 352), Chapter 26 (pp 393–398)
4.2.5	Interpret and analyze fire patterns					Chapter 4 (pp 42–55), Chapter 26 (pp 393–398)
4.2.5(A)	RK: Fire behavior and spread					Chapter 4 (pp 42–55), Chapter 26 (pp 393–398)
4.2.5(B)	RS: Fire patterns					Chapter 4 (pp 42–55), Chapter 26 (pp 393–398)
4.2.6	Examine and remove fire debris					Chapter 16 (pp 256–265), Chapter 17 (pp 271–273), Chapter 24 (pp 352–353)
4.2.6(A)	RK: Ignition processes					Chapter 16 (pp 256–265), Chapter 17 (pp 271–273), Chapter 24 (pp 352–353)
4.2.6(B)	RS: Search techniques					Chapter 16 (pp 256–265), Chapter 17 (pp 271–273), Chapter 24 (pp 352–353)
4.2.7	Reconstruct the area of origin					Chapter 16 (pp 256–265)
4.2.7(A)	RK: Effects of fire on different types of material					Chapter 16 (pp 256–265)
4.2.7(B)	RS: Effects of fire					Chapter 16 (pp 256–265)
4.2.8	Performance of building systems					Chapter 5 (pp 60–64), Chapter 6 (pp 78–107), Chapter 7 (pp 121–127)
4.2.8(A)	RK: Types of detection					Chapter 5 (pp 60–64), Chapter 6 (pp 78–107), Chapter 7 (pp 121–127)
4.2.8(B)	RS: Systems operation					Chapter 5 (pp 60–64), Chapter 6 (pp 78–107), Chapter 7 (pp 121–127)
4.2.9	Discriminate the effects of explosions from other types of damage					Chapter 21 (pp 304–317)
4.2.9(A)	RK: Types of explosions					Chapter 21 (pp 304–317)
4.2.9(B)	RS: Explosive effects on glass, walls					Chapter 21 (pp 304–317)
4.3.1	Diagram the scene					Chapter 14 (pp 232–235)
4.3.1(A)	RK: Symbols and legends					Chapter 14 (pp 232–235)
4.3.1(B)	RS: Sketch the scene					Chapter 14 (pp 232–235)
4.3.2	Photographically document the scene					Chapter 14 (pp 226–232)
4.3.2(A)	RK: Knowledge of high resolution camera and flash					Chapter 14 (pp 226–232)
4.3.2(B)	RS: Ability to use a high resolution camera and flash					Chapter 14 (pp 226–232)
4.3.3	Construct investigative notes					Chapter 14 (p 235)
4.3.3(A)	RK: Notes, diagrams, and photos					Chapter 14 (p 235)
4.3.3(B)	RS: Data-reduction skills, note-taking skills, and observational and correlating skills					Chapter 14 (p 235)
4.4.1	Utilize proper procedures for managing victims and fatalities					Chapter 23 (pp 336–345)

AGENCY NAME:					DATE COMPILED:	
OBJECTIVE / JPR, RK, RS		COGNITIVE	MANIPULATIVE			FIRE INVESTIGATOR: PRINCIPLES AND PRACTICE TO NFPA 921 AND 1033, FOURTH EDITION
SECTION	ABBREVIATED TEXT	WRITTEN TEST	SKILLS STATION	PORTFOLIO	PROJECTS	
4.4.1(A)	RK: Evidence associated with fire victims and fatalities and evidence preservation methods					Chapter 23 (pp 336–345)
4.4.1(B)	RS: Observational skills and the ability to apply protocols to given situations					Chapter 23 (pp 336–345)
4.4.2	Locate, document, collect, label, package, and store evidence					Chapter 15 (pp 240–241), Chapter 26 (pp 399–403), Chapter 27 (pp 413–415)
4.4.2(A)	RK: Types of evidence					Chapter 15 (pp 240–241), Chapter 26 (pp 399–403), Chapter 27 (pp 413–415)
4.4.2(B)	RS: Different types of evidence and determine evidence critical to the investigation					Chapter 15 (pp 240–241), Chapter 26 (pp 399–403), Chapter 27 (pp 413–415)
4.4.3	Select evidence for analysis					Chapter 20 (pp 290–299)
4.4.3(A)	RK: Submitting samples, types of analytical services					Chapter 20 (pp 290–299)
4.4.3(B)	RS: Evaluate the fire incident to determine forensic, engineering, or laboratory needs					Chapter 20 (pp 290–299)
4.4.4	Maintain a chain of custody					Chapter 15 (pp 244, 249), Chapter 27 (pp 413–414)
4.4.4(A)	RK: Rules of custody					Chapter 15 (pp 244, 249), Chapter 27 (pp 413–414)
4.4.4(B)	RS: Execute the chain of custody procedures					Chapter 15 (pp 244, 249), Chapter 27 (pp 413–414)
4.4.5	Dispose of evidence					Chapter 15 (p 251)
4.4.5(A)	RK: Disposal services available and common disposal procedures and problems					Chapter 15 (p 251)
4.4.5(B)	RS: Documentation skills					Chapter 15 (p 251)
4.5.1	Develop an interview plan					Chapter 12 (pp 207–208), Chapter 19 (p 285)
4.5.1(A)	RK: Persons who can provide info that furthers the fire cause determination					Chapter 12 (pp 207–208)
4.5.1(B)	RS: Planning skills, development of focused questions					Chapter 12 (pp 207–208)
4.5.2	Conduct interviews					Chapter 12 (pp 207–208)
4.5.2(A)	RK: Types of interviews					Chapter 12 (pp 207–208)
4.5.2(B)	RS: Interviewing strategies					Chapter 12 (pp 207–208)
4.5.3	Evaluate interview information					Chapter 12 (pp 207–208)
4.5.3(A)	RK: Types of interviews, report evaluation methods, and data correlating methods					Chapter 12 (pp 207–208)
4.5.3(B)	RS: Data correlating skills					Chapter 12 (pp 207–208)
4.6.1	Gather reports and records					Chapter 10 (p 176), Chapter 12 (pp 206–211)
4.6.1(A)	RK: Types of reports					Chapter 10 (p 176), Chapter 12 (pp 206–211)

AGENCY NAME:					DATE COMPILED:	
OBJECTIVE / JPR, RK, RS		COGNITIVE	MANIPULATIVE			FIRE INVESTIGATOR: PRINCIPLES AND PRACTICE TO NFPA 921 AND 1033, FOURTH EDITION
SECTION	ABBREVIATED TEXT	WRITTEN TEST	SKILLS STATION	PORTFOLIO	PROJECTS	
4.6.1(B)	RS: Proper reports and documents					Chapter 10 (p 176), Chapter 12 (pp 206–211)
4.6.2	Evaluate the investigative file					Chapter 12 (pp 206–211), Chapter 14 (pp 226–232), Chapter 20 (pp 290–299)
4.6.2(A)	RK: File assessment and/or evaluation methods					Chapter 12 (pp 206–211), Chapter 14 (pp 226–232), Chapter 20 (pp 290–299)
4.6.2(B)	RS: Information assessment and correlation skills and organizational skills					Chapter 12 (pp 206–211), Chapter 14 (pp 226–232), Chapter 20 (pp 290–299)
4.6.3	Coordinate expert resources					Chapter 13 (pp 219–220), Chapter 19 (p 285)
4.6.3(A)	RK: How to assess one's own expertise					Chapter 13 (pp 219–220), Chapter 19 (p 285)
4.6.3(B)	RS: Apply expert resources					Chapter 13 (pp 219–220), Chapter 19 (p 285)
4.6.4	Establish evidence as to motive and/or opportunity					Chapter 22 (pp 328–331)
4.6.4(A)	RK: Motives common to incendiary fire investigation					Chapter 22 (pp 328–331)
4.6.4(B)	RS: Financial analysis, records gathering and analysis					Chapter 22 (pp 328–331)
4.6.5	Formulate an opinion concerning origin, cause, or responsibility for the fire					Chapter 2 (pp 14–17), Chapter 17 (p 273), Chapter 19 (pp 285–286), Chapter 26 (pp 399–403)
4.6.5(A)	RK: Analytical methods and procedures					Chapter 2 (pp 14–17), Chapter 17 (p 273), Chapter 19 (pp 285–286), Chapter 26 (pp 399–403)
4.6.5(B)	RS: Analytical and assimilation skills					Chapter 2 (pp 14–17), Chapter 17 (p 273), Chapter 19 (pp 285–286), Chapter 26 (pp 399–403)
4.7.1	Prepare a written investigation report					Chapter 10 (pp 173–174), Chapter 14 (p 235)
4.7.1(A)	RK: Writing, typical components of a written report					Chapter 10 (pp 173–174), Chapter 14 (p 235)
4.7.1(B)	RS: Writing skills, ability to analyze information					Chapter 10 (pp 173–174), Chapter 14 (p 235)
4.7.2	Express investigative findings verbally					Chapter 10 (pp 173–175, 177–179, 181)
4.7.2(A)	RK: Investigative findings					Chapter 10 (pp 173–175, 177–179, 181)
4.7.2(B)	RS: Communication skills					Chapter 10 (pp 173–175, 177–179, 181)
4.7.3	Testify during legal proceedings					Chapter 10 (pp 173–182)
4.7.3(A)	RK: Types of investigative findings					Chapter 10 (pp 173–182)
4.7.3(B)	RS: Communication and listening skills					Chapter 10 (pp 173–182)

Fire Investigation I and II (FESHE) Correlation Guide

Fire Investigation I (FESHE) Course Outcomes	Corresponding Chapter(s)	Corresponding Page(s)
1. Identify the responsibilities of a firefighter when responding to the scene of a fire, including scene security and evidence preservation.	15, 23	240–243, 336
2. Describe the implications of constitutional amendments as they apply to fire investigations.	10	168–172, 175
3. Identify key case law decisions that have affected fire investigations.	10	170–171, 178–179
4. Define the common terms used in fire investigations.	1	9
5. Explain the basic elements of fire dynamics and how they affect cause determination.	3	24–26
6. Compare the types of building construction on fire progression.	5	65–69
7. Describe how fire progression is affected by fire protection systems and building design.	5, 6	70–73, 78–107
8. Discuss the basic principles of electricity as an ignition source.	7	127–130
9. Recognize potential health and safety hazards.	11, 25, 26, 27	188–201, 371, 398–399, 408–409, 414–415
10. Describe the process of conducting investigations using the scientific method.	2	14–18
11. Identify cause and origin and differentiate between accidental and incendiary.	16, 17, 18, 22	256–265, 271–273, 278–280, 322–329
12. Explain the procedures used for investigating vehicle fires.	25	370–387
13. Identify the characteristics of an incendiary fire and common motives of the fire setter.	22	322–331

Fire Investigation II (FESHE) Course Outcomes	Corresponding Chapter(s)	Corresponding Page(s)
1. Explain the rule of law as it pertains to arrest, search and seizure.	10	169–172
2. Interpret a fire scene.	4	42–55
3. Describe the chemistry of combustion.	3	22–23
4. Explain the nature and behavior of fire.	3	24–26, 33–35
5. Identify the combustion properties of liquid, gaseous, and solid fuels.	3	28, 30–32
6. Analyze electrical causes of fires.	7, 24, 25, 28	130–136, 350–365, 373–374, 428–429
7. List the procedures for fingerprinting and evidence collection/ preservation.	15	244–249
8. Evaluate the use of incendiary devices, explosives, and bombs.	21, 22	312–313, 324–325
9. List the procedures for fire scene documentation.	14	226–235
10. Analyze fire-related deaths and injuries.	23	336–345
11. Discuss interviewing techniques.	12	207–208
12. Explain the role of the fire investigator in courtroom demeanor and testifying.	10	173–175
13. List the sources and technology available for fire investigations.	12, 14, 20, 24	206–211, 226–227, 230, 234–235, 290–299, 352–353
14. Describe procedures for conducting background investigations.	12	208–211

Glossary

Accelerant Any fuel or oxidizer, often an ignitible liquid, used to initiate a fire or increase the rate of growth or speed the spread of fire.

Accidental fires Fires for which the proven cause does not involve an intentional human act to ignite or spread fire into an area where the fire should not be.

Accommodation space A space on a vessel designed for people to live in.

Active fire protection system Any type of fire protection system that automatically activates upon detection of a fire condition.

Addressable technology Fire alarm and detection systems in which devices and system components are assigned a data address, are capable of two-way communication, and can perform certain activities based on signal input.

Adolescent firesetters Adolescents (ages 14 to 16 years) who are often responsible for fires that occur at places other than their homes.

Adrift Loose, not on moorings or towline.

Aerial fuels All green or dead materials located in the forest canopy.

Affidavit A voluntary written statement of fact or opinion made under oath and signed by the author.

Afloat Floating in water; not sinking.

Air drop The aerial application of water, foam, or retardant mixture directly onto the fire or threatened area or along a strategic position ahead of the fire.

Air entrainment The process of air or gases being drawn into a fire, plume, or jet.

Air sampling smoke detection A method of smoke detection designed to draw air from the protected area into a detection chamber for analysis.

Alarm pressure switch A type of initiating device installed on all types of wet-based fire suppression systems; the device initiates a fire alarm signal once a water pressure threshold is met.

Alarm signal A warning signal that alerts occupants of a fire emergency.

Alloying The mixing of two or more metals in which one or more of the metals is in a liquefied state.

Alternating current (AC) The voltage varies in time with a sine wave at 60 cycles per second. A sine-wave current results if the load is resistive.

Ambient temperature Air temperature of the surrounding environment.

Ampacity The current, in amperes, that a conductor can carry continuously under the conditions of use without exceeding its temperature rating.

Amperage The unit of electric current that is equivalent to a flow of one coulomb per second; one coulomb is defined as 6.24×10^{18} electrons.

Annunciation panel A device that uses indicating lamps, alphanumeric displays, or other means to provide first responders with status, condition, and location information concerning fire alarm system components and devices.

Arc mapping The analysis of the locations where electrical arcing has caused damage and the documentation of the involved electrical circuits.

Arc survey diagrams Locations of electrical arcing are identified and plotted on a diagram of the affected area of the structure. The spatial relationship of the arc sites can create a pattern and help establish the sequence of damage. This analysis can be used on building circuits and electrical devices within a compartment to help identify or narrow the area of origin.

Arc tube A cylinder of fused silica/quartz with electrodes at either end that is used in most mercury and metal halide lamps.

Arc A high-temperature luminous electric discharge across a gap where the conductor is missing or through a medium such as charred insulation.

Arced and severed Occurs when wires are arced off on the ends and possible segemented wires result.

Arc-fault circuit interrupter (AFCI) Designed to protect against fires caused by arcing faults in home electrical wiring. The circuitry continuously monitors current flow.

Area of origin The room, building, or general area in which the point of origin is located.

Arson The crime of maliciously and intentionally, or recklessly, starting a fire or causing an explosion. Legal definitions of "arson" are defined by statutes and judicial decisions that vary among jurisdictions.

Aspect The direction the slope faces (N, E, S, W).

Assemblies Manufactured parts put together to make a completed product.

Autoignition temperature (AIT) The lowest temperature at which a combustible material ignites in air without a spark or flame.

Backflow valve Prevents gas from re-entering a container or distribution system.

Balanced pressure proportioner A proportioner that uses an atmospheric tank and pump to introduce the foam concentrate into the system with the water supply.

Balloon frame construction A construction type in which the exterior wall studs go from the foundation wall to the roofline. The floor joists are attached to the walls by the use of a ribbon board, which creates an open stud channel between floors, including the basement and attic.

Bar holes Holes driven into the surface of the ground or pavement with either weighted metal bars or drills. Gas detectors are then inserted into the holes in order to detect the location of a gas leak.

Benchmark events Events that are particularly valuable as a foundation for the timeline or may have significant relation to the cause, spread, detection, or extinguishment of a fire.

Beveling A fire pattern that indicates fire direction on wood wall studs. The bevel leans toward the direction of travel.

Bilges The lowest, interior part of a vessel's hull; the area where spilled water and oil can collect.

Blast overpressure Large quantities of gas being produced by the explosion of the material.

BLEVE Boiling liquid expanding vapor explosion.

Blow holes Visible holes in the conduit or metal panels of branch circuits. They are an indication of significant arcing and energy release inside the enclosure or conduit.

Boat Any vessel that can be placed or stored on another vessel.

Branch circuits The individual circuits that feed lighting, receptacles, and various fixed appliances.

Bulkheads The vertical separations in a vessel that form compartments and that correspond to walls in a building. The term does not refer to the vertical sections of a vessel's hull.

Burning velocity The rate of flame propagation relative to the velocity of the unburned gas ahead of it.

Calcination A fire effect realized in gypsum products, including wallboard, as a result of exposure to heat that drives off free and chemically bound water.

Carboxyhemoglobin (COHb) The carbon monoxide saturation in the blood.

Catalytic converter A device in the exhaust system that exposes exhaust gases to a catalyst metal to promote oxidation of hydrocarbon materials in the exhaust gas.

Ceiling jet A relatively thin layer of flowing hot gases that develops under a horizontal surface (e.g., ceiling) as a result of plume impingement and the flowing gas being forced to move horizontally.

Central station service A facility that receives fire alarm signals from properties that is staffed with trained, qualified, and proficient personnel who take the appropriate action to process, record, respond, and maintain monitored fire alarm systems.

Chain of custody The trail of accountability that documents the possession of evidence in an investigation.

Char Carbonaceous material that has been burned or pyrolyzed and has a blackened appearance.

Chemical explosion Explosion in which a chemical reaction is the source of the high-pressure fuel gas. The fundamental nature of the fuel is changed.

Child firesetters Children (ages 2 to 6 years) who are often responsible for fires in their homes or in the immediate area.

Circumstantial evidence Facts that usually attend other facts to be proven and are drawn by logical inference from them rather than personal knowledge.

Class A foam A mechanically generated aggregation of bubbles that have a lower density than water.

Clean burn A fire effect that appears on noncombustible surfaces after any combustible layers (such as soot, paint, and paper) have been burned away. The effect may also appear where soot was not deposited due to high surface temperatures.

Coded signal A signal that generates a predetermined visual or audible pattern to identify the location of the initiating device that is operating.

Cognitive testing The use of a person's thinking skills and judgment in order to evaluate the empirical data and challenge the conclusions of the final hypothesis.

Collapse zones Distances around a structure that may be affected by a structural collapse. The area should be identified by markers or specialized scene tape that indicates there is a potential for a structural collapse.

Combustible gas indicator An instrument that samples air and indicates whether combustible vapors are present. Some units may indicate the percentage of the lower explosive limit of the air/gas mixture.

Combustion explosion Explosion caused by the burning of combustible hydrocarbon fuels and characterized by the presence of a fuel with air as an oxidizer.

Combustion A chemical process of oxidation that occurs at a rate fast enough to produce heat and usually light in the form of either a glow or flames.

Commercial propane Derived from the refining of petroleum, a liquefied gas comprising 95 percent propane and propylene and 5 percent other gases.

Compartmentation A concept in which fire is kept confined in its room of origin, minimizing smoke movement to other areas of a building.

Complex investigation Generally includes multiple simultaneous investigations and involves a number of interested parties.

Conduction Heat transfer to another body or within a body by direct contact.

Confidential communication Those statements made under circumstances showing that the speaker intended the statements only for the ears of the person addressed.

Confirmation bias The attempt to prove a hypothesis instead of disprove a hypothesis, resulting in a failure to consider alternative hypotheses or the premature discounting of an alternative hypothesis.

Coniferous litter Primarily needles dropped from coniferous trees but may also include branches, bark, and cones.

Consent Permission from the owner or occupant to examine property.

Constitutional law Law that is based on the U.S. Constitution.

Container appurtenances The devices that are connected to the openings in tanks and other containers—such as pressure relief devices, control valves, and gauges.

Control of Hazardous Energy (Lockout/Tagout) standard The federal OSHA regulation that governs work around equipment or wiring where an unexpected energization, startup of machines, or release of stored energy could result in the injury of the investigators. This standard utilizes a lockout/tagout device that disables the electrical equipment. Specifics can be found in 29 CFR 1910.147.

Convection heater A heater that employs a fan to move air across a hot surface, heating the air and dispersing it throughout the room.

Convection Heat transfer by circulation within a medium such as a gas or a liquid.

Conventional technology The technology used in fire alarm and detection systems that provides limited communication between the fire alarm control panel and system devices and, upon activation of an alarm, provides only general information concerning the device and its location.

Crazing A complicated pattern of short cracks in glass that can be either straight or crescent shaped and can extend through the entire thickness of the glass.

Credibility The quality of being believable or trustworthy.

Cross-contamination The unintentional transfer of a substance from one fire scene or location contaminated with a residue to an evidence collection site.

Crowns Twigs, needles, or leaves of a tree or bush.

Cupping A charred surface on the fuel that is caused by the exposure of the surface to the windward side of the fire.

Dead load The weight of materials that are part of a building, such as the structural components, roof coverings, and mechanical equipment.

Decks Usually continuous, horizontal divisions running the length of a vessel and extending athwart ships. This corresponds to floors in a building on land.

Deductive reasoning The process by which conclusions are drawn by logical inference from given premises.

Defendant The entity against whom a claim is brought in a court.

Deflagration A reaction that propagates at a subsonic velocity through an unreacted medium, less than 1100 feet per second, and can be successfully vented.

Deluge sprinkler system A type of automatic fire sprinkler system equipped with open fire sprinkler heads that simultaneously discharges water from all sprinkler heads.

Demonstrative evidence Any type of tangible evidence relevant to a case, for example, diagrams and photographs.

Demonstrative evidence Photographs, maps, X-rays, visible tests, and demonstrations.

Density/area curves Graphs that establish the relationship between the required amount of water flow from a sprinkler head (density) and the area that must be covered by the water for different hazard classifications.

Deposition One type of pre-trial oral testimony made under oath.

Depth of calcination surveys Measurements of the relative depth of calcination (observable physical changes in gypsum wallboard) plotted on a detailed scene diagram to determine locations within a structure that were exposed longest to a heat source.

Depth of char surveys Measurements of the relative depth of char on identical fuels plotted on a detailed scene diagram to determine locations within a structure that were exposed longest to a heat source.

Design density The minimum predetermined amount of water that must flow from the fire sprinkler heads in the most hydraulically demanding part of the fire sprinkler system (remote area) to control or extinguish a fire.

Detonation A combustion reaction that propagates at supersonic velocities, greater than 1100 feet (340 m) per second, and cannot be vented because of its speed.

Diagrams Formal drawings that are completed after the scene investigation is completed.

Diffusion flame A flame in which the fuel and air mix or diffuse together at the region of combustion.

Direct current (DC) The voltage is a steady level that does not vary with time. The current will also be a steady level if the load is resistive.

Direct evidence Testimony of witnesses who observe acts or detect something through their five senses and surveillance equipment such as CCTV.

Documentary evidence Any type of written record or document that is relevant to the case.

Dry chemical extinguishing agents A dry powder suppression agent made from sodium bicarbonate–based, potassium-based, or ammonium phosphate chemicals that covers and smothers the burning material.

Dry pipe sprinkler system A type of automatic fire sprinkler system equipped with automatic sprinkler heads that has pressurized air or nitrogen in the pipes until a sprinkler head activates to reduce the air or nitrogen pressure, permitting the dry pipe valve clapper to open, flooding the system piping with water.

Duct smoke detectors Smoke detectors that sample the air moving through the air distribution system ductwork or plenum; upon detecting smoke, the detector sends a signal to shut down the air distribution unit, close any associated smoke dampers, or initiate smoke control system operation.

Duff The layer of decomposing organic materials lying below the litter layer of freshly fallen twigs, needles, and leaves and immediately above the mineral soil.

Electrical power The rate of doing work (using electrical energy) in an electrical circuit, such as in a hair dryer, electric motor, or lightbulb.

Electronic (or engine) control module (ECM) An electronic device that controls engine operation parameters, including fuel delivery, throttle control, and safety systems operations.

Electronic valve supervisory device A device or switch that integrates with or attaches to a water control valve that detects movement or changes in the position of the valve and then sends a signal that the normal open position has changed.

Empirical data Factual data that are based on actual measurements, observation, or direct sensory experience, rather than on theory.

Entity in control The interested party who has or represents ownership of the scene or is in effective management of the site, scene, or evidence, and is organizing, directing, or controlling the joint actions of the other interested parties.

Estimated time An approximation based on information or calculations that may or may not be relative to other events or activities.

Eutectic melting Any combination of metals with a melting point lower than that of any of the individual metals of which it consists.

Event data recorder (EDR) A device to record data before and after a crash event.

Evidence custodian Person who is responsible for managing all aspects of evidence control.

Evidence technicians Individuals who are specially trained to document, collect, preserve, and transport evidence.

Exemplar An exact duplicate of an appliance.

Exothermic reaction Reaction characterized by or formed with evolution of heat.

Expectation bias Preconceived determination or premature conclusions as to what the cause of the fire was without having examined or considered all of the relevant data.

Expert witness One who is recognized by a court as qualified by specialized knowledge, skills, experience, or education on an issue.

Explosion dynamics analysis The process of using force vectors to trace backward from the least to the most areas of damage following the general path of the explosion force vectors.

Explosion The sudden conversion of potential energy (chemical or mechanical) into kinetic energy with the production and release of gases under pressure. These high-pressure gases then do mechanical work such as moving, changing, or shattering nearby materials.

Explosives Any chemical compounds, mixtures, or devices that function by explosion.

Fact witness One who testifies about facts from their personal knowledge or experiences; also called a lay witness.

Failure analysis A logical, systematic examination of an item, component, assembly, or structure and its place and function within a system, conducted in order to identify and analyze the probability, causes, and consequences of potential and real failures.

Failure mode and effects analysis (FMEA) A technique used to identify basic sources of failure within a system and to follow the consequences of these failures in a systematic fashion.

Fault trees Logic diagrams that can be used to analyze a fire or explosion. Also known as decision trees.

Fire alarm control unit (FACU) Equipment that monitors the integrity of the fire alarm system's circuits and devices, processes manual and automatic input signals from the initiating devices, drives the notification appliances, provides an interface to control or activate other fire protection and building systems in a fire emergency, and provides the power to support all of the system devices.

Fire alarm system A group of components assembled to monitor and annunciate the status of automatic and manual initiating devices so building occupants may respond appropriately.

Fire barrier A wall, other than a fire wall, having a fire-resistance rating. Fire walls and fire barrier walls do not need to meet the same requirements as smoke barriers.

Fire breaks Natural or human-made barriers used to stop the spread or reroute the direction of the fire by separating the fuel from the fire.

Fire dynamics The detailed study of how chemistry, fire science, and the engineering disciplines of fluid mechanics and heat transfer interact to influence fire behavior.

Fire effects The observable or measurable changes in or on a material as a result of exposure to the fire.

Fire flanks The parts of a fire's perimeter that are roughly parallel to the main direction of spread.

Fire head The portion of a fire that is moving most rapidly, subject to influences of slope and other topographic features. Large fires burning in more than one drainage of fuel type can develop additional heads.

Fire heel Located at the opposite side of the fire from the head, this part of the fire is less intense and is easier to control.

Fire investigation The process of determining the origin, cause, and development of a fire or explosion.

Fire pattern analysis The process of identifying and interpreting fire patterns to determine how the patterns were created and their significance.

Fire pattern The visible or measurable physical changes or identifiable shapes formed by a fire effect or group of fire effects.

Fire patterns The visual or measurable physical effects that remain after a fire.

Fire point The lowest temperature at which a volatile combustible substance continues to burn in air after its vapors have been ignited (as when heating is continued after the flash point has been determined).

Fire scene reconstruction The process of removal of debris and replacement of contents or structural elements in their prefire positions. Reconstruction can also include recreating the physical scene during fire scene analysis investigation.

Fire spread analysis The process of identifying fire patterns relating to the movement of fire from one place to another and the sequence in which the patterns were produced to trace the fire back to an origin.

Fire storm A natural phenomenon where a fire attains such intensity that it creates and sustains its own wind system.

Fire wall A wall separating buildings or subdividing a building to prevent the spread of fire and having a fire resistance rating and structural stability.

Fireballs Momentary balls of flame during or after an explosion that may present high-intensity, short-term radiation.

Firebrands A piece of burning material normally transported by wind or convection, perhaps in an explosion.

Firing out The process of burning the fuel between a fire break and the approaching fire to extend the width of the fire barrier.

Fixed level gauge Primarily used to indicate when the filling of a tank or cylinder has reached its maximum allowable fill volume. They do not indicate liquid levels above or below their fixed lengths.

Flame detectors Radiant energy detectors that sense specific portions of the visible and invisible light spectrum.

Flame speed The local velocity of a freely propagating flame relative to a fixed point.

Flash point The lowest temperature of a liquid, as determined by specific laboratory tests, at which the liquid gives off vapors at a sufficient rate to support a momentary flame across its surface.

Flashover A transition phase in the development of a compartment fire in which surfaces exposed to thermal radiation reach ignition temperature more or less simultaneously and fire spreads rapidly throughout the space, resulting in full room involvement or total involvement of the compartment or enclosed space.

Foam concentrate A condensed form of the foam product.

Foam solution A solution that results from foam concentrate and water mixing in the correct proportions.

Freezing When leaves, needles, and small branches are heated by a passing fire front, they become very soft and pliable. Prevailing winds or drafts created by the fire bend them in the direction the fire advanced. Once cooled, they remain pointing in this direction.

Friction loss As water flows away from the source through a water system, a progressive pressure drop due to changes in the direction of the pipe, length of pipe, type of pipe, size of the pipe, types and sizes of the fittings, and other components in line with the piping.

Fuel analysis Identification of the first fuel ignited.

Fuel gas Includes natural gas, liquefied petroleum gas in the vapor phase only, liquefied petroleum gas–air mixtures, manufactured gases, and mixtures of these gases, plus gas–air mixtures within the flammable range, with the fuel gas or the flammable component of a mixture being a commercially distributed product.

Fuel items Any articles that are capable of burning.

Fuel load The total quantity of combustible contents of a building, space, or fire area, including interior finish and trim, expressed in heat units or the equivalent weight in wood.

Fuel package A collection or array of fuel items in close proximity with one another such that flames can spread throughout the array.

Fuel rail An internal passage or external tube connecting a pressurized fuel line to individual fuel injectors.

Fuel A material that will maintain combustion under specified environmental conditions.

Fuel-controlled fire A fire in which the heat release rate and growth rate are controlled by the characteristics of the fuel, such as quantity and geometry, and in which adequate air for combustion is available.

Fugitive gas Fuel gases that escape from their piping, storage, or utilization systems and serve as easily ignited fuels for fires and explosions.

Fulgurites Slender, usually tubular, bodies of glassy rock produced by electrical current striking and then fusing dry sandy soil.

Fusible plugs Thermally activated devices that open and vent the contents of a container.

Gas burner A device that allows fuel gases and air to properly mix and produce a flame.

Gas-sensing fire detection A type of detector designed to sense a specific gas, toxic gases, or a variety of gases and vapors produced when processing hydrocarbons.

Ground fuels All combustible materials below the surface litter, including duff, tree or shrub roots, punky wood, peat, and sawdust, that normally support a glowing combustion without flame.

Ground-fault circuit interrupter (GFCI) A device intended for the protection of personnel that functions to de-energize a circuit or portion thereof within an established period of time when a current to ground (possibly a human body) exceeds the values established for a Class A device.

Grounding A conducting connection, whether intentional or accidental, between an electrical circuit or equipment and earth or to some conducting body that serves in place of the earth.

Guide A document that is advisory or informative in nature and that contains only nonmandatory provisions. A guide may contain mandatory statements, such as when a guide can be used, but the document as a whole is not suitable for adoption into law.

Halocarbons (clean agents) A chemical compound made up of carbon and hydrogen, chlorine, fluorine, bromine, or iodine.

Halogenated hydrocarbons (halon) A chemical compound mixture of carbon and one or more elements from the halogen series of elements (fluorine, chlorine, bromine, or iodine).

Hard times Specific points in time that are directly or indirectly linked to a reliable clock or timing device of known accuracy.

Hatches Openings in a vessel's deck that are fitted with a watertight cover.

HAZWOPER (HAZardous Waste OPerations and Emergency Response) standard The federal OSHA regulation that governs hazardous materials waste site and response training. Specifics can be found in 29 CFR 1910.120.

Headspace The zone inside a sealed evidence can between the top of fire debris and the bottom of the lid.

Heat A form of energy characterized by vibration of molecules that is capable of initiating and supporting chemical changes and changes of state.

Heat and flame vector analysis A process to assist with fire spread analysis and origin determination in which arrows representing the investigator's assessment of the direction of heat/flame spread are placed on a detailed diagram of the fire scene. The patterns can be traced back to the fire origin.

Heat capacity The amount of heat necessary to raise the temperature of a unit mass 1 degree, under specified conditions (J/kg-K, Btu/lb-°F).

Heat detectors Initiating devices that operate after detecting a predetermined fixed temperature or sensing a specified rate of temperature change.

Heat flux The measurement of the rate of heat transfer to a surface, expressed in kilowatts/m^2, kilojoules/m^2 sec, or Btu/ft^2 sec.

Heat release rate (HRR) The rate at which heat energy is generated by burning.

Heat shadowing A pattern that results from an object blocking the travel of radiant heat from its source to a target material on which the pattern is produced.

Heat transfer models Models that allow the investigator to determine how heat was transferred from a source to a target by one or more of the common heat transfer modes: conduction, convection, or radiation.

Heat transfer The transport of heat energy from one point to another caused by a temperature difference between those points.

Heavy timber construction A construction type in which structural members (i.e., columns, beams, arches, floors, and roofs) are made of unprotected, solid, or laminated wood, with large cross-sectional areas (8 or 6 inches [200 or 150 mm] in the smallest dimension, depending on reference).

High-order damage A rapid pressure rise or high-force explosion characterized by a shattering effect on the confining structure or container and long missile distances.

High-resistance fault An unintended path for electricity that allows sufficient current for heating to occur, but not enough to reset the current protection.

High-temperature accelerants (HTA) Mixtures of fuels with Class 3 or Class 4 oxidizers and thermite mixtures.

Holds Compartments used for carrying cargo below deck in a large vessel.

Horsepower A unit of power that is used to express mechanical energy use or production. One kilowatt is equal to 1.34 horsepower.

Hot legs The two insulated conductors of a single-phase system (or three if three-phase power is used), sometimes referred to as L1, L2, or Line voltages.

Hot set Fire started in which the device is removed by the individual starting the fire.

Housing In an appliance, the outer shell that contains the working components.

Hull The outer skin of a vessel including the bottom, sides, and main deck, but not including the superstructure, masts, rigging, and other fittings.

Hybrid vehicles Vehicles that utilize a combination of internal combustion and electric motors for propulsion.

Hydraulic data nameplate A permanent and rigid sign posted on or near a fire sprinkler system riser that provides information including the hazard, design density, size of the remote area, required gallons and pressure for the water supply, and location that the system is serving.

Hydraulic design A mathematical method of determining flow and pressure at any point along the sprinkler system piping for the purpose of determining pipe size throughout the system.

Hypothesis Theory supported by the empirical data that the investigator has collected through observation and then developed into explanations for the event, which are based on the investigator's knowledge, training, experience, and expertise.

Hypoxia Condition caused by a victim breathing in a reduced-oxygen environment.

Ignition sequence The sequence of events and circumstances that allow the initial fuel and the ignition source to come together and result in a fire.

Ignition source analysis Process through which all potential ignition sources in the area of origin are identified and then considered in light of the physical properties of the first fuel ignited, fundamental scientific principles, and other available data.

Impact load A sudden added load to a structure.

Inboard Toward the interior of a vessel from the outer hull.

Incendiary devices A wide range of mechanisms used to initiate an incendiary fire.

Incendiary fire A fire that is deliberately set with the intent to cause the fire to occur in an area where it should not be.

Incident command system (ICS) The combination of facilities, equipment, personnel, procedures, and communications under a standard organizational structure to manage assigned resources effectively to accomplish stated objectives for an incident.

Inductive reasoning The process by which a person starts from a particular experience and proceeds to generalizations. The process by which hypotheses are developed based upon observable or known facts and the training, experience, knowledge, and expertise of the observer.

Inert gas A gas that could contain a mixture of helium, neon, argon, nitrogen, and small amounts of carbon dioxide and does not contaminate, react, or become flammable.

Initial scene assessment A preliminary phase of the investigation to ensure preservation of the scene; to identify safety hazards, manpower, and equipment needs; and to determine areas that require further inspection.

Initiating device A system of components and devices that provide a manual or automatic means to activate fire alarm and supervisory signals.

Intent Necessary to show proof of a crime and refers to the state of mind that exists at the time a person acts or fails to act.

Interested parties One whose legal rights or interests might be affected by the investigation of a particular incident.

Interested party Any person, entity, or organization, including their representatives, with statutory obligations or whose legal rights or interests may be affected by the investigation of a specific incident.

Interrogatory A set of questions served by one party involved in litigation to another involved party.

Interrupt current The value of current that the breaker must be able to carry and still operate.

Interrupting current rating The maximum amount of current the device is capable of interrupting.

Interstitial spaces The space between the building frame and interior walls and the exterior facade and with spaces between ceilings and the bottom face to the floor or deck above.

Intrinsically safe A protection feature for the safe operation of electronic equipment in explosive atmospheres and/or under irregular operating conditions.

Investigative file The organized collection of all of the documentary information in a fire case, including verbal, written, and visual information from the scene; the reports of investigators and other professionals; and any other investigation or research that has been conducted.

Ionization smoke detectors Smoke detectors that use a small amount of a radioactive material that electrically charges air particles to produce a measurable current in the sensing chamber; once smoke enters the sensing chamber, a current drop below a predetermined level initiates an alarm.

Isochar A line on a diagram connecting equal points of char depth.

Juvenile firesetters Juveniles (ages 7 to 13 years) who are often responsible for fires that start in their homes or in the immediate environment.

K-Factor A number assigned to represent the discharge coefficient for the orifice of a sprinkler head that is used when calculating the water flow or water pressure at a specific location in the fire sprinkler system.

Kinetic energy The energy possessed by a system or object as a result of its motion.

Laminated beams Structural elements that have the same characteristics as solid wood beams. They are composed of many wood planks that are glued or *laminated* together to form one solid beam. Designed for interior use; the effects of weathering decrease their load-bearing ability.

Lay witness One who testifies about facts from his or her personal knowledge or experiences; also called a fact witness.

Lazarettes Stowage compartments, often in the aft end of a vessel, sometimes used as a workshop.

Lightweight wood trusses Similar to other trusses in design. Individual members are fastened using nails, staples, glue, metal gusset plates, or wooden gusset plates.

Line-type detector A type of detection device that uses a tube or wires running in different directions as the sensing element to provide coverage over a wide area.

Liquefied petroleum (LP gas) (propane) Petroleum gases condensed to a liquid state with moderate pressure and normal temperatures to allow for more efficient distribution.

Live load The weight of temporary loads that need to be factored into the weight-carrying capacity of the structure, such as furniture, furnishings, equipment, machinery, snow, and rainwater.

Lividity Occurs after death and is the pooling of the blood in the lower elevations of the body caused by the effects of gravity.

Lower explosive limit (LEL) The lowest concentration of fuel in a specified oxidant in which combustion can occur; also known as the lower flammable limit (LFL).

Low-order damage A slow rate of pressure rise or low-force explosion characterized by a pushing or dislodging effect on the confining structure or container and by short missile distances.

Main disconnect Provides the master shutoff mechanism for the overall system power and provides high current level protection of downstream overcurrent protection devices and wiring.

Manual fire alarm box A type of initiating device that requires a person to pull a handle on the device to -initiate an alarm signal.

Manual initiating device A type of initiating device that requires a person to make physical contact with the device in order to operated the device.

Manufactured housing A construction technique whereby the structure is built in one or more sections. These sections are then transported to the building site and assembled on the site.

Mass arson The setting of three or more fires at the same site or location during a limited period of time.

Mechanical explosion Rupture of a vessel or container such as a cylinder, tank, or boiler, resulting in the release of pressurized gas or vapor. The pressure leading to the mechanical explosion is not due to a chemical reaction or change in chemical composition of the substances involved.

Melting A physical change caused by exposure to heat.

Meter A watt-hour meter that plugs into the meter base to measure the amount of electricity consumed at a site.

Meter base The component where the service cables come in through the weatherhead and go down the conduit to connect to an electric meter that measures the amount of electricity being used.

Mill construction An early form of heavy timber construction influenced and developed largely through insurance companies that recognized a need to reduce large fire losses in factories.

Modular home A dwelling constructed in a factory and placed on a site-built foundation, all or in part, in accordance with a standard adopted, administered, and enforced by the regulatory agency, or under reciprocal agreement with the regulatory agency, for conventional site-built dwellings.

Modus operandi (MO) The method of operation used by the offender.

Mosaic photographs Used when a sufficiently wide angle lens is not available and a panoramic view is desired. A mosaic is created by assembling a number of photographs in overlay form to give a more-than-peripheral view of an area.

Motion A request for court action regarding facts, documents, and evidence that are identified during the discovery phase of a legal proceeding. Motions may argue the admissibility of the involved facts, documents, or evidence.

Motive An inner drive or impulse that is the cause, reason, or incentive that induces or prompts a specific behavior.

Natural fire Fire caused without direct human intervention or action, such as fire resulting from lightning, earthquake, and wind.

Natural gas A naturally occurring, largely hydrocarbon gas product recovered by drilling wells into underground pockets, often in association with crude petroleum.

Negligence Failure to provide the same care that a person with similar training would provide.

Neutral plane The line where the flow of the hot gas and cooler air changes.

NFPA 1033, *Professional Qualifications for Fire Investigator* A standard that is designed to establish the minimum job performance requirements (JPRs) for service as a fire investigator.

<u>NFPA 921, *Guide for Fire and Explosion Investigations*</u> A guide that establishes guidelines and recommendations for the safe and systematic investigation or analysis of fire and explosion incidents.

<u>Noncoded signal</u> A constant visual or audible signal that remains activated until reset.

<u>Noncombustible construction</u> A construction type used primarily in commercial and industrial storage and in high-rise construction. The major structural components are noncombustible; the structure itself will not add fuel to the fire. Examples of these materials are brick, stone, steel, masonry block, cast iron, or non-reinforced concrete.

<u>Nonseated explosion</u> When the fuels in the explosion are dispersed or diffused with moderate rates of pressure rise and subsonic explosive velocities.

<u>Non-water-based fire suppression</u> A type of fire suppression system that uses agents that are gas or chemical based.

<u>Notification appliances</u> Constant visual or audible signals that remain activated until reset.

<u>Ohm's law</u> A basic law of electricity that defines the relationship among voltage, current, and resistance. If two of these three values are known, it is possible to determine the third.

<u>Ordinary construction</u> A construction type in which exterior walls are masonry or other noncombustible material and the roof, floor, and wall assemblies are wood.

<u>Outboard</u> Lying outward from the centerline of a vessel toward or beyond its sides; a type of engine that is attached to and lying aft of a vessel's stern.

<u>Overboard</u> Over the side or out of the vessel.

<u>Overcurrent</u> Any current in gross excess of the rated current of equipment or the ampacity of a conductor; it may result from an overload, short circuit, or ground fault.

<u>Overfusing</u> A dangerous condition that occurs when the circuit protection (fuse or circuit breaker) rating significantly exceeds the ampacity of the conductor, leading to a condition in which increased heat can occur in the conductors.

<u>Overload</u> Operation of equipment in excess of normal, full-load rating or of a conductor in excess of rated ampacity that, when it persists for a sufficient length of time, could cause damage or dangerous overheating.

<u>Oxidation</u> The basic chemical process associated with combustion.

<u>Oxidizing agent</u> A substance that promotes oxidation during the combustion process.

<u>Permit-Required Confined Space standard</u> The federal OSHA regulation that governs any space that an employee can bodily enter and perform assigned work, with a limited or restricted means of entrance or exit, and that is not designed for continuous employee occupancy. Specifics can be found in 29 CFR 1910.146.

<u>Personation</u> The unusual behavior by an offender, beyond that necessary to commit the crime.

<u>Phase change</u> The conversion of a material from one state of matter to another that is reversible and does not change the chemical composition of the material.

<u>Photo diagram</u> A diagram indicating the direction from which each of the photographs was taken.

<u>Photo log</u> A written chart or list of all photos taken at a fire scene, including photo numbers or other identifiers and a brief description of the item or area photographed.

<u>Photoelectric detectors</u> Smoke detectors that use a light source and receiver within the sensing chamber to initiate an alarm when the light source reflects or is obscured by smoke particles.

<u>Physical evidence</u> Any physical or tangible item that tends to prove or disprove a particular fact or issue. Also referred to as *real evidence*.

<u>Pipe schedule</u> A list of pipe sizes and the number of fire sprinkler heads that the pipe can support based on the hazard classification.

<u>Plaintiff</u> The entity who brings a claim to court.

<u>Plank-and-beam construction</u> A construction type in which a few large members replace the many small wood members used in typical wood framing; that is, large dimension beams, more widely spaced, replace the standard floor and/or roof framing of smaller dimensioned members.

<u>Platform frame construction</u> The most common construction method currently used for residential and lightweight commercial construction. In this method of construction, separate platforms or floors are developed as the structure is built.

<u>Plume</u> The column of hot gases, flames, and smoke rising above a fire; also called *convection column, thermal updraft,* or *thermal column*.

<u>Point of origin</u> The exact physical location where a heat source and a fuel come into contact with each other and a fire begins.

<u>Post-and-frame construction</u> Similar to plank-and-beam construction, the structure uses larger elements. An example is a typical barn construction where the posts provide a majority of the support and the frame provides a place for the exterior finish to be applied.

<u>Preaction sprinkler system</u> A type of automatic fire sprinkler system equipped with automatic fire sprinkler heads that interfaces with fire detection equipment and requires a fire detector to activate and an automatic fire sprinkler head to operate to flow water.

<u>Preinvestigation team meeting</u> A meeting that takes place prior to the on-scene investigation. The team leader or investigator addresses questions of jurisdictional boundaries and assigns specific responsibilities to the team members. Personnel are advised of the condition of the scene and the safety precautions required.

<u>Premixed burning</u> Burning in which the fuel and oxidizer are mixed prior to combustion, as in a laboratory Bunsen burner or a gas cooking range; propagation of the flame is governed by the interaction between flow rate, transport processes, and chemical reaction.

<u>Prescribed fire</u> A fire that resulted from intentional ignition by a person or a naturally caused fire that is allowed to continue to burn according to approved plans to achieve resource management objectives.

<u>Pressure gauge</u> A type of container appurtenance depicting the internal pressure of a tank. The gauges are connected directly to the tank or sometimes through the valve. Pressure gauges do not indicate the quantity of liquid propane in the tank.

Pressure proportioner A proportioner that redirects some of the water supply into the foam concentrate tank to either exert pressure on a collapsible bladder or push the concentrate out of the tank to the proportioner for mixing.

Pressure relief valve A valve designed to open at a specific pressure, usually around 250 psi (1700 kPa). It is generally placed in the container where it releases the vapor.

Private mode notification An alarm signal that is sent to a location within a facility so that trained individuals can interpret and implement the appropriate response procedures.

Privileged communication Those statements made by certain persons within a protected relationship such as a husband–wife, attorney–client, or priest–penitent. These communications are protected by law from forced disclosure on the witness stand at the option of the witness, spouse, client, or penitent.

Projected beam detectors Fire detectors that project a light beam to a receiver over a hazard; once the beam is obscured or scattered, it activates an alarm.

Proprietary supervising station fire alarm system A group of fire alarm systems at one location or multiple locations that are under the constant supervision and monitoring by the property owner's trained personnel.

Protocol A description of the specific procedures and methods by which a task or tasks are to be accomplished.

Public mode notification An alarm signal that propagates throughout a building to audibly or visually alert the building occupants so they can take appropriate action.

Pugilistic attitude A crouching stance with flexed arms, legs, and fingers.

Pyrolysis Process in which material is decomposed, or broken down, into simpler molecular compounds by the effects of heat alone; pyrolysis often precedes combustion.

Races A component of a rolling element bearing that contains the elements and transfers the load to the bearing; generally used in pairs (inner and outer races).

Radiant energy–sensing detectors A type of fire detector that is not dependent on smoke and heat plumes to operate; instead, the detector looks for specific portions of the visible and invisible light spectrum produced by flames, sparks, or embers.

Radiation heater A heater that transfers heat by radiation only.

Radiation Heat transfer by way of electromagnetic energy.

Rainbow effect A diffraction pattern formed when hydrocarbons float on a surface.

Real evidence Any physical or tangible item that tends to prove or disprove a particular fact or issue. Also referred to as *physical evidence*.

Recall notices A method of notifying consumers of a product defect that was identified after the product was released for consumption.

Regular current rating The level of current above which the protective device will open, such as 15 A, 20 A, or 50 A.

Regular or ordinary dry chemicals Dry chemical agents rated only for Class B and C fires.

Relative humidity The amount of moisture in a given volume of air compared with how much moisture the air could hold at that same temperature.

Relative time The chronological order of events or activities that can be identified in relation to other events or activities.

Relaxation time The amount of time for a charge to dissipate.

Remote area The minimum square footage of the most hydraulically remote pipe in a fire sprinkler system where the minimum design density must be available from all heads in that area.

Remote supervising station fire alarm system A fire alarm system in which alarm, trouble, and supervisory signals transmit to a location that is remote from the protected premises and staffed with individuals trained to take the appropriate action.

Residential fire sprinkler systems A type of automatic fire sprinkler system equipped with fast response automatic sprinkler heads specifically made for low heat release and low water pressures.

Resistance heating Occurs when current flows through a path that provides high resistance to current flow, such as a heating element (intentional) or a resistive connection (unintentional).

Resistive circuit A circuit that does not contain inductance and capacitance.

Response time index (RTI) The amount of time a head takes to activate once exposed to temperatures above the predetermined activation temperature.

Responsibility The accountability of a person or other entity for the event or sequence of events that caused the fire or explosion, spread of the fire, bodily injuries, loss of life, or property damage.

Ring flash Reduces glare and gives adequate lighting for the subject matter in close-up photography.

Root mean squared (RMS) A mathematical computation to equate the voltage level of an alternating current (AC) system to that of a more familiar direct current (DC) system.

Sabotage Intentional damage or destruction.

Safety assessment Inspection to determine whether the scene is safe to enter and the steps necessary to render the scene safe.

Schedule Used on larger projects, this details the types of equipment in lists.

Scientific method The systematic pursuit of knowledge involving the recognition and definition of a problem; the collection of data through observation and experimentation; analysis of the data; the formulation, evaluation, and testing of a hypothesis; and, when possible, selection of a final hypothesis.

Seat of the explosion A craterlike indentation created at the point of origin of an explosion.

Seismic effect The transmission of tremors through the ground as a result of the blast wave expansion causing structures to be knocked down.

Sequential pattern analysis Application of principles of fire science to the analysis of fire pattern data (including fuel packages and geometry, compartment geometry, ventilation, fire suppression operations, witness information, etc.) to determine the origin, growth, and spread of a fire.

Sequential photography Shows the relationship of a small subject to its relative position in a known area. The small subject is first photographed from a distant position, where it is shown in context with its surroundings. Additional photographs are then taken increasingly closer until the subject is the focus of the entire frame.

Serial arson Involving an offender who sets three or more fires with a cooling-off period between fires.

Serial arsonist See *serial fire-setter*.

Serial fire-setter Individual or group involved in three or more fire-sets.

Service drop The overhead service conductors from the last pole or other aerial support to and including the splices, if any, connecting to the service-entrance conductors at the structure.

Service entrance The point where the electrical service enters a building.

Service equipment The necessary equipment, usually consisting of a circuit breaker(s) or switch(s) and fuse(s) and their accessories, connected to the load end of service conductors to a building or other structure, or an otherwise designated area, and intended to constitute the main control and cutoff method of the supply.

Service lateral Wiring coming in underground.

Shore power Electrical power supplied from shore via a cord set.

Shutoff valve A valve (located in the piping system and readily accessible and operable by the consumer) used to shut off individual appliances or equipment.

Sine wave The waveform that AC voltage follows. An example is 120-V AC, which has 170-V peak, crosses 0 to –170 V, and repeats this cycle 60 times per second.

Sketches Freehand diagrams or diagrams drawn with minimal tools that are completed at the scene and can be either three- or two-dimensional.

Slope The steepness of land or a geographical feature that can greatly influence fire behavior.

Smoke alarm A single- or multi-station smoke detector with an integrated power source, sensing device, and alarm device.

Smoke barrier Continuous membrane, either vertical or horizontal, designed and constructed to restrict movement of smoke.

Smoke deposits Hot products of combustion that may adhere upon collision with a surface.

Smoke detectors Fire detectors that sense visible and invisible particles of combustion.

Snag A standing dead tree, or part of a dead tree, from which at least the smaller branches have fallen.

Soffits The horizontal undersides of the eaves or cornice.

Soft times Estimated or relative points in time.

Spalling The chipping or pitting of concrete or masonry surfaces.

Spark/ember detectors Radiant energy detectors that sense specific portions of the visible and invisible light spectrum.

Spoliation Loss, destruction, or material alteration of an object or document that is evidence or potential evidence in a legal proceeding by the one who has the responsibility for its preservation.

Spot fires Projections of flaming or burning particles (firebrands) that are found ahead of the flame front.

Spot-type detector A type of detection device that provides coverage in a specific area where the sensing element is in a fixed location.

Spree arson The setting of three or more fires at separate locations with no emotional cooling-off period between fires.

Staging Purposeful alteration of the crime scene prior to the arrival of police.

Standard A document, the main text of which contains only mandatory provisions using the word "shall" to indicate requirements and that is in a form generally suitable for mandatory reference by another standard or code or for adoption into law. Nonmandatory provisions shall be located in an appendix or annex, footnote, or fine print note and are not to be considered a part of the requirements of a standard.

Starboard The right side of a vessel when looking forward.

Statement of the danger A warning that identifies the nature and extent of the danger of a product and the gravity of the risk of injury.

Static electricity The electrical charging of materials through physical contact and separation and the various effects that result from the electrical charges formed by this process.

Steel-framed residential construction A construction type with characteristics similar to those of wood frame construction but is noncombustible. Steel framing can lose its structural capacity during extreme exposure to heat.

Step-down transformer Device that reduces the 120 V provided at the receptacle to the required voltage.

Stoichiometric The optimum ratio at which point the combustion will be most efficient.

Stoichiometric ratio The optimum ratio in a fuel air mixture at which point combustion will be most efficient (above the LEL and below the UEL).

Superstructures All vessel structures above the main deck or weather deck.

Supervisory alarm signal A type of fire alarm system alert signal that indicates when the normal ready status of other fire protection systems or devices connected to or integrated with the fire alarm panel has changed.

Surface fuels All combustible materials from the surface of the ground up to about 6 ft (2 m).

Switch loading A term used to describe a product being loaded into a tank or compartment that previously held a product of different vapor pressure and flash point.

Switch Device used to turn the appliance on or off or to change operating conditions.

System analysis An analytical approach that takes into account characteristics, behavior, and performance of a variety of elements.

Tamper switch A device that detects changes in the normal position of a fire protection system valve and causes a supervisory signal when the valve is off normal.

Temperature The degree of sensible heat of a body as measured by a thermometer or similar instrument.

Testimonial evidence Verbal testimony of a witness given under oath or affirmation and subject to cross-examination by the opposing party.

Textural signal A messaging signal that provides constant and specific information via voice, pictorial, or alphanumeric means.

Thermal conductivity (k) The measure of the amount of heat that will flow across a unit area with a temperature gradient of 1 degree per unit of length (W/m-K, Btu/hr-ft-°F).

Thermal decomposition An irreversible change in chemical composition as a result of pyrolysis.

Thermal inertia The product of thermal conductivity, density, and specific heat (or heat capacity).

Thermal inertia The properties of a material that characterize its rate of surface temperature rise when exposed to heat; related to the product of the material's thermal conductivity (κ), density (ρ), and heat capacity (c).

Thermal runaway Condition in which the heat generated exceeds the amount of heat loss within the material.

Thermistors Sensors used to measure temperature by change of their resistance.

Thermometry The study of the science, methodology, and practice of temperature measurement. The study is frequently used in fire safety or code compliance cases.

Time current curve The amount of time required for a current protection device to interrupt at a specific level of current.

Timeline A graphic or narrative representation of events related to the fire incident, arranged in chronological order.

Topside When below decks, refers to areas on or above the main deck.

Tort A civil wrong leading to a legal claim for damages.

Total burn A fire scene where a fire continued to burn until most combustibles were consumed and the fire self-extinguished because of a lack of fuel or was extinguished when the fuel load was reduced by burning and there was sufficient suppression agent application to extinguish the fire.

Traditional forensic physical evidence Includes, but is not limited to, finger and palm prints, bodily fluids such as blood and saliva, hair and fibers, footwear impressions, tool marks, soils and sand, woods and sawdust, glass, paint, metals, handwriting, questioned documents, and general types of trace evidence.

Trailers The method to spread fire to other areas by deliberately linking them together with combustible fuels or ignitible liquids.

Transformers Devices that reduce AC voltage, usually 120 or 240 V AC, to a lower voltage and can be used to isolate the appliance from its power source.

Transom The stern cross-section of a square-sterned vessel.

Triac A semiconductor switch that conducts current when the gate is turned on and blocks current when the gate is turned off.

Trial A legal proceeding that is presided over by a judge, who also makes rulings on the admissibility of evidence. Many trials utilize juries as the finder of fact.

Trolling motor A small electric motor connected to a battery and used primarily for slow-speed maneuvering of recreational boats.

Trouble alarm signal A type of fire alarm system alert signal that indicates there is a problem with the system's integrity, such as a power or component failure, device removal, communication fault or failure, ground fault, or a break in the system wiring.

Turbocharger An exhaust-driven device that compresses intake air to increase engine power.

Under way Not attached to land or mooring by tie, anchor, or grounding.

Understory vegetation The area under a forest or brush canopy that grows at the lowest height level. Plants in the understory consist of a mixture of seedlings, saplings, shrubs, grasses, and herbs.

Undetermined fire Classification of fire when the cause cannot be proven to an acceptable level of certainty.

Uninhibited chemical chain reaction One of the elements of the fire tetrahedron. This element provides for the combination and interaction of the other elements.

Vandalism Mischievous or malicious fire-setting that results in damage to property.

Vaporization A phase transition from a liquid or a solid phase to a gas phase.

Vaporizers Heaters that are used to heat and vaporize propane where larger quantities of propane are required, such as for industrial application.

Variable gauge Gauge that gives readings of the liquid contents of containers, primarily tanks or large cylinders. It gives readings at virtually any level of liquid volume.

Venting The removal of combustion products as well as process fumes (e.g., flue gases) to the outer air.

Venturi proportioner A proportioner that uses water-moving over an open orifice to create a lower pressure at the opening, which draws the foam into the water stream.

Venturi A narrowed area within a carburetor that causes air to accelerate and create a low pressure area that draws fuel into the intake manifold of the engine.

Vessels A broad grouping of every description of watercraft, other than a seaplane on the water, used or capable of being used as a means of transportation on the water.

Victimology A thorough understanding of the offender activity with the victim (or targeted property).

Video detector A type of fire detector that uses high-definition video cameras to analyze digital images for changes in the pixels due to smoke and flame generation.

Voltage sniffer A noncontact voltage monitor that outputs a beep or turns on a light when voltage is present or nearby (within about an inch or less).

Water control valve A device used to control the flow of water.

Water flow switch A type of initiating device installed on a wet pipe sprinkler system; a paddle inserted into the pipe moves to initiate an alarm signal when there is sustained water flow through the system piping.

Water mist system A fixed fire protection system that uses specialized nozzles to discharge a very fine-spray mist of water droplets that extinguish by cooling, displacing oxygen, or blocking radiant heat.

Weather history A description of atmospheric conditions over the preceding few days or several weeks.

Weatherhead The point where service entrance cables connect to the structure, which is designed to keep water out of the conduit that carries the wires.

<u>Wet chemical extinguishing agents</u> Suppression agents that mix water with potassium acetate, potassium carbonate, potassium citrate, and, in some instances, a mixture of these agents and other additives; used -primarily to suppress Class K fires.

<u>Wet pipe sprinkler system</u> A type of automatic fire sprinkler system equipped with automatic fire sprinkler heads that has water in the pipes at all times so when a sprinkler head activates, water flow is immediate.

<u>Wildfire V-shaped patterns</u> Horizontal ground surface burn patterns generated by the fire spread.

<u>Wood frame construction</u> A construction type in which exterior walls and load-bearing components are wood. This type of construction is often associated with residential construction and contemporary lightweight commercial construction.

<u>Wood I-beams</u> Constructed with small dimension or engineered lumber, as the top and bottom chord, with oriented strand board or plywood as the web of the beam.

<u>Work plan</u> An outline of the tasks to be completed as part of the investigation, including the order or timeline for completion.

Index

Figures and tables are indicated by *f* and *t* following the page number.

D

O

B01FXFGOWY - MF
By International Associatio...
1900 | Hardcover